Representations of *-Algebras, Locally Compact Groups, and Banach *-Algebraic Bundles

Volume 2
Banach *-Algebraic Bundles, Induced Representations, and the Generalized Mackey Analysis

J. M. G. Fell
Department of Mathematics
University of Pennsylvania
Philadelphia, Pennsylvania

R. S. Doran
Department of Mathematics
Texas Christian University
Fort Worth, Texas

ACADEMIC PRESS, INC.
Harcourt Brace Jovanovich, Publishers
Boston   San Diego   New York
Berkeley   London   Sydney
Tokyo   Toronto

ACADEMIC PRESS, INC.
1250 Sixth Avenue, San Diego, CA 92101

United Kingdom Edition published by
ACADEMIC PRESS INC. (LONDON) LTD.
24-28 Oval Road, London NW1 7DX

Library of Congress Cataloging-in-Publication Data

Fell, J. M. G. (James Michael Gardner), Date
  Representations of *-algebras, locally compact
groups, and Banach *-algebraic bundles.

  (Pure and applied mathematics; 125-126)
  Includes bibliographies and indexes.
  Contents: v. 1. Basic representation theory of
groups and algebras—v. 2. Banach *-algebraic
bundles, induced representations, and the generalized
Mackey analysis.
  1. Representations of algebras. 2. Banach algebras.
3. Locally compact groups. 4. Fiber bundles
(Mathematics) I. Doran, Robert S., Date
II. Title. III. Series: Pure and applied mathematics
(Academic Press); 125-126.
QA3.P8 vol. 125-126 [QA326] 510 s [512'.55] 86-30222
ISBN 0-12-252721-6 (v. 1: alk. paper)
ISBN 0-12-252722-4 (v. 2: alk. paper)

Transferred to digital printing 2005.

Contents

Chapter IX. Compact Groups   931

Chapter XII. The Generalized Mackey Analysis   1243

Topics in Volume 1

Chapter I.   Preliminaries

Chapter II.   Integration Theory and Banach Bundles

Chapter III.   Locally Compact Groups

Chapter IV.   Algebraic Representation Theory

Chapter V.   Locally Convex Representations and Banach Algebras

Chapter VI. C^*-Algebras and Their *-Representations

Chapter VII.   The Topology of the space of *-Representations

> It is unworthy of a mathematician
> to see with other people's eyes
> and to accept as true or as proven
> that for which he himself has no
> proof.
>
> —Maestlin

Introduction To Volume 2 (Chapters Eight to Twelve)

1

The main purpose of the present Volume 2 is to give a detailed presentation of the Mackey "normal subgroup analysis" for classifying the unitary representations of groups—and this not merely in the customary context of locally compact groups, but in the larger context of Banach *-algebraic bundles.

Before proceeding to sketch the background of the Mackey normal subgroup analysis, we should make special mention of Chapters IX and X, which are somewhat disjoint from the rest of the volume. They consist largely of the now classical general results on the representation theory of compact and abelian groups respectively. As regards compact groups, all the principal general results (centering around the Peter–Weyl Theorem) on their representation theory can be obtained in the context of locally convex representations rather than merely unitary ones; and this is the context in which they are presented in Chapter IX. Combining these classical results with the "Mackey analysis" of Chapter XII, we shall be able (§XII.7) to generalize much of the representation theory of compact groups (including the Frobenius Reciprocity Theorem) to a substantial class of Banach *-algebraic bundles over compact groups. As for Chapter X, the first part presents the main results on the harmonic analysis of locally compact abelian groups (following the Bourbaki approach, and leading up to the Pontryagin Duality Theorem). The last part of Chapter X is devoted to the structure of commutative saturated Banach *-algebraic bundles.

§§2 and 3 of the Introduction to Volume 1 can serve as motivation for the theory presented in Chapters X and IX respectively.

2

The main stream of this second volume develops the following sequence of topics: First, Banach *-algebraic bundles (Chapter VIII); then induced representations of Banach *-algebraic bundles (Chapter XI); and finally the Mackey normal subgroup analysis in the context of saturated Banach *-algebraic bundles (Chapter XII).

It seems best to begin our Introduction to these topics with a sketch of the vital role played by induced representations in the researches of this century on the classification of representations of groups—an enterprise whose great importance was emphasized in the Introduction to Volume 1.

The theory of induced representations of groups was born in 1898, when Frobenius [5] defined the representation T of a finite group G induced by a representation S of a subgroup H of G. While Frobenius' original definition was couched in matrix terminology, the modern formulation (equivalent to that of Frobenius) uses the terminology of linear spaces. We choose to phrase the definition in such a way that T acts by *left* translation operators: Let Y be the linear space on which S acts. Then the induced representation T acts on the linear space X of all those functions $f: G \to Y$ which satisfy

$$f(xh) = S_{h^{-1}}(f(x)) \qquad\qquad (x \in G; h \in H);$$

and the operators of T are given by left translation:

$$(T_y f)(x) = f(y^{-1}x) \qquad\qquad (f \in X; x, y \in G).$$

Why are induced representations important? It is because they enable us to give a more penetrating analysis of a group representation than is afforded by a mere direct sum decomposition of it into irreducible parts. To see this it is helpful to discuss systems of imprimitivity. Let G be a finite group and T any finite-dimensional representation of G; and suppose that the space X of T can be written as a linear space direct sum of non-zero (not necessarily T-stable) subspaces X_m:

$$X = \sum_{m \in M}^{\oplus} X_{m'} \qquad\qquad (1)$$

where the (finite) index set M is itself a left G transformation space, the action of G on X being "covariant" with the action of G on M:

$$T_x(X_m) \subset X_{xm} \qquad\qquad (x \in G; m \in M). \qquad (2)$$

Such a decomposition (1) is called a *system of imprimitivity for T over M*.

We say that T is *primitive* if the only system of imprimitivity for T is that in which M consists of a single point. Obviously one-dimensional representations are primitive; and in general G will have primitive representations of dimension greater than 1 (see Exercise 41 of Chapter IX). A direct sum decomposition of T (that is, a decomposition (1) in which the X_m are T-stable) is clearly a special kind of system of imprimitivity, in which the M is trivial. Thus a primitive representation is automatically irreducible. However the converse of this is far from true: A system of imprimitivity for an irreducible representation of G may be highly non-trivial—though it must be at least *transitive* in the sense that the index set M is acted upon transitively by G.

Now any induced representation T of G automatically carries with it a transitive system of imprimitivity. Indeed, let T be induced as above from a representation S of the subgroup H of G; and let M be the left coset space G/H. The space X of T can then be written as a direct sum in the form (1) by setting

$$X_m = \{f \in X : f \text{ vanishes outside } m\}$$

for each coset m in G/H. Since (2) clearly holds with respect to the natural G-space structure of M, we have a (transitive) system of imprimitivity for G over M.

The basic Imprimitivity Theorem for finite groups asserts just the converse of this. Let H be a subgroup of G, and M the left G-space G/H. (Thus M is essentially the most general transitive G-space.) The Imprimitivity Theorem says that every system of imprimitivity for G over M is equivalent to that obtained by inducing from an (essentially unique) representation of the subgroup H. In the context of finite groups the proof of this result is a simple and routine matter.

From the Imprimitivity Theorem one deduces easily that *every irreducible representation T of G is obtained by inducing up to G from some primitive representation S of some subgroup of G*. This gives us an important means of analyzing the structure of the irreducible representations of G. Special interest attaches to those T for which S is one-dimensional, that is, those irreducible representations of G which are induced from representations of dimension 1; such representations are called *monomial*. Blichfeldt has shown (see Lang [1], p, 478) that if G is supersolvable, then every irreducible representation of G is monomial. In the opposite direction we have Taketa's theorem (see Curtis and Reiner [1], Theorem 52.5): Every finite group whose irreducible representations are all monomial must be solvable.

As a point of interest we remark that, according to Brown [1], there exist *infinite* nilpotent discrete groups whose irreducible unitary representations are not all monomial.

The 1940's saw the birth of the general theory of unitary representations of arbitrary locally compact groups. If such a group is neither compact nor abelian, its irreducible unitary representations are in general infinite–dimensional. Several major papers of that period were devoted to classifying the irreducible unitary representations of specific non-compact non-abelian groups. We mention the work of Wigner [1] on the Poincaré group (the symmetry group of special relativity); the work of Gelfand and Naimark on the "$ax + b$" group (Gelfand and Naimark [2]) and on the classical complex semisimple Lie groups (Gelfand and Naimark [3, 5]); and the work of Bargmann [1] on $SL(2, \mathbb{R})$. In each of these investigations, the irreducible representations of the group in question were actually exhibited as being *induced* from one-dimensional representations of suitable subgroups—the inducing process used in each case being an ad hoc adaptation to the continous group context of the Frobenius construction of induced representations of finite groups.

A systematic treatment of infinite-dimensional induced representations of arbitrary locally compact groups was thus clearly called for, and was largely supplied by Mackey [2, 4, 5].

We remark incidentally that Mackey's formulation of the infinite-dimensional unitary inducing process was not quite adequate to embrace *all* the ad hoc inducing processes needed to construct unitary representations in the works of Bargmann, Gelfand and Naimark mentioned above. To catch the so-called supplementary series of irreducible representations of $SL(2, \mathbb{C})$ and other semisimple groups, one must generalize Mackey's definition by allowing the representation of the subgroup from which we induce to be non-unitary (see Mackey [10], §8). Similarly, to catch the so-called discrete series of irreducible unitary representations of $SL(2, \mathbb{R})$ and other semisimple groups, one has to require the functions on which the induced representations act to be "partially holomorphic." The resulting generalized inducing process is called holomorphic induction (see Blattner [2], Dixmier [19]). Interestingly enough, it was proved by Dixmier [19] that, if G is a connected solvable algebraic linear group, every irreducible unitary representation of G is monomial in the sense of holomorphic induction, that is, is holomorphically induced from a one-dimensional representation of some subgroup. This is an obvious analogue of the result of Blichfeldt mentioned earlier.

Indeed, it is true, not only for connected solvable algebraic linear groups but even for certain connected semisimple Lie groups, that *all* the irreducible unitary representations of the group are obtained by inducing (in a suitable generalized sense) from one-dimensional representations of some subgroup. This indicates (by comparison with Taketa's Theorem mentioned above) that induced representations are if anything even more important in the context of continuous connected groups than they are for finite groups.

3

Returning to Mackey's unitary inducing construction, we find in Mackey [2, 8] a generalization of the Imprimitivity Theorem to the context of an arbitrary separable locally compact group G. As with finite groups, this result asserts that to describe a unitary representation T of G as induced from a unitary representation of a closed subgroup H of G is the same as to specify a system of imprimitivity for T based on the transitive G-space G/H. Of course the definition of a system of imprimitivity over a continuous G-space like G/H is much more sophisticated than in the case of finite groups; and, correspondingly, the proof of the Imprimitivity Theorem, which for finite groups is a simple routine argument, becomes for continuous groups a difficult technical feat. Mackey's generalized Imprimitivity Theorem gave rise almost immediately to two very important applications. The first was Mackey's generalization of the Stone-von Neumann Theorem on the uniqueness of the operator representation of the Heisenberg commutation relations of quantum mechanics (see Mackey [1]; also our §XI.15). The second was the so-called normal subgroup analysis (sometimes called the "Mackey analysis" or the "Mackey machine") by which, given a locally compact group H and a closed normal subgroup N of H, one tries (and often succeeds!) to analyze the irreducible representations of H in terms of the irreducible representations of N and of subgroups of H/N. The reader will find a detailed introduction to this "Mackey analysis," carried out in the relatively uncluttered context of finite groups, in §XII.1.

The Mackey analysis, as applied to finite groups, was almost certainly familiar to Frobenius. For finite-dimensional representations of arbitrary groups it was worked out by Clifford [1]. For (infinite-dimensional) unitary representations of separable locally compact groups, it was systematically developed by Mackey [5, 8]. The methods that had been applied earlier by

Wigner [1] to the Poincaré group and by Gelfand and Naimark [2] to the "$ax + b$" group were simply special cases of Mackey's normal subgroup analysis.

Substantial steps forward were taken by Loomis [3] and Blattner [1, 5]. In these articles the authors define induced representations for non-separable groups, removing Mackey's standing hypothesis of separability. More importantly, they give a new proof of the Imprimitivity Theorem for locally compact groups which is completely different from Mackey's (and makes no use of separability). In a subsequent paper (Blattner [6]), Blattner also extended most of Mackey's normal subgroup analysis to the non-separable context.

In the present volume we will be following the Loomis–Blattner approach to induced representations, the Imprimitivity Theorem, and the Mackey analysis, so that assumptions of separability will play only a minimal role.

4

So far we have sketched the history of the developement of induced representations and the Mackey analysis only in the context of locally compact groups. Indeed, the theory of unitary representations of locally compact groups is the key topic whose generalizations form the subject-matter of this work. From this standpoint, the reason for the development of the *-representation theory of Banach *-algebras is that each locally compact group G gives rise to a certain Banach *-algebra (its $\mathscr{L}_1$ group algebra $\mathscr{L}_1(G)$) whose *-representation theory is essentially identical with the unitary representation theory of G (see §VIII.13). Indeed, one of the early triumphs of the theory of Banach *-algebras was the derivation, as a corollary of a corresponding theorem on general Banach *-algebras, of the important fact that every locally compact group G has enough irreducible unitary representations to distinguish its points (see §VIII.14). Accordingly, it is natural to try to derive as many facts as possible of the unitary representation theory of G by applying to $\mathscr{L}_1(G)$ corresponding theorems on general Banach *-algebras.

Evidently, however, this program cannot be entirely successful as stated, for the simple reason that some of the structure of G is lost on passing from G to $\mathscr{L}_1(G)$. (For example, two non-isomorphic finite abelian groups with the same number of elements will have isomorphic group *-algebras; see IV.6.6.) Whenever we encounter a theorem on unitary representations of groups G

which depends on that part of the structure of G which is "forgotten" by $\mathscr{L}_1(G)$, and which is therefore not a special case of a theorem on general Banach *-algebras, we are faced with a challenge: Try to equip a general Banach *-algebra with just the amount of extra structure needed so that an analogue of the given theorem about G will make sense and be true in the enriched system, and will give the original theorem as a corollary in the "group setting."

The theory of Banach *-algebraic bundles, which is the main topic of Chapter VIII, lays the groundwork for meeting this challenge with respect to the theory of induced representations and the Mackey normal subgroup analysis. Indeed, starting in the late 1960's it began to be realized that these theories do indeed have natural generalizations to contexts much wider than that of locally compact groups. The first, purely algebraic, steps in this direction were taken by Dade [1], Fell [13], and Ward [1]. According to Fell [13], the appropriate setting for this generalization is the notion of an algebraic bundle. Let us briefly sketch the simplest version of this notion (in its involutory form); for it plays a dominant role in the present work.

Let G be a finite group (with unit e). A *-algebraic bundle $\mathscr{B}$ over G is a *-algebra E together with a collection $\{B_x\}$ $(x \in G)$ of linear subspaces of E indexed by G, with the following properties:

The B_x are linearly independent and

$$\sum_{x \in G} B_x = E; \tag{3}$$

$$B_x B_y \subset B_{xy} \qquad (x, y \in G); \tag{4}$$

$$(B_x)^* = B_{x^{-1}} \qquad (x \in G). \tag{5}$$

Conditions (4) and (5) say that multiplication and involution in E are "covariant" with multiplication and inverse in G. We call B_x the *fiber over x*. Note from (4) and (5) that B_e is a *-subalgebra of E, but that the remaining B_x are not. More generally, if H is a subgroup of G, (4) and (5) imply that $B_H = \sum_{x \in H} B_x$ is a *-subalgebra of E.

Let us suppose for simplicity that E has a unit 1 (necessarily in B_e). We call $\mathscr{B}$ *homogeneous* if every fiber B_x contains an element u which is unitary in E (i.e., $u^*u = uu^* = 1$). Assume that $\mathscr{B}$ is homogeneous; let H be a subgroup of G; and let S be a *-representation of the *-subalgebra B_H, acting in some Hilbert space Y. One can then easily generalize the classical inducing process for groups to obtain an induced *-representation T of E. Indeed, let U_x stand for the set of unitary elements in B_x; thus $U = \bigcup_{x \in G} U_x$ is a group under the multiplication in E. Let X be the Hilbert space of all functions $f: U \to Y$ satisfying

$$f(ut) = S_{t^{-1}} f(u) \qquad \left(u \in U; t \in \bigcup_{x \in H} U_x\right), \qquad (6)$$

the inner product in X being given by

$$(f, g)_X = \sum_{\alpha \in G/H} (f(u_\alpha), g(u_\alpha))_Y \qquad (7)$$

where $u_\alpha \in \bigcup_{x \in \alpha} U_x$ for each α in G/H. (Notice that the particular choice of u_α is immaterial in view of (6).) Then the equation

$$(T_{au}f)(v) = S_{v^{-1}av}(f(u^{-1}v)) \qquad (a \in B_e; u, v \in U; f \in X) \qquad (8)$$

determines a *-representation T of E acting on X (cf. Fell [13], Prop. 7). We refer to T as *induced from S*. The formal similarity of this definition with the classical inducing process on groups is evident. On the basis of this definition one can prove the Imprimitivity Theorem and develop the whole Mackey normal subgroup analysis. (This is done in Fell [13], §§7–10, in the purely algebraic, non-involutive, setting.)

The most obvious example of a homogeneous *-algebraic bundle is the case when E is the group *-algebra of G, and for each x in G, B_x is the one-dimensional subspace of all elements of E which vanish except at x. The unitary elements of B_x are then just those f in B_x for which $|f(x)| = 1$. We call $\mathscr{B}$ in this case the *group bundle* of G. Inducing of representations within the group bundle, as defined in the preceding paragraph, is identical with the classical construction of induced representations of a finite group with which we began.

Somewhat more general than group bundles are group extension bundles. Let H be a finite group, N a normal subgroup, and $G = H/N$ the quotient group; and let E be the group *-algebra of H. For each coset x in G let B_x be the linear subspace $\{a \in E : a(h) = 0 \text{ for all } h \in H \setminus x\}$. It is easy to see that $\mathscr{B} = \langle E, \{B_x\}_{x \in G}\rangle$ is then a homogeneous *-algebraic bundle over G. We call $\mathscr{B}$ the *group extension bundle* of the group extension H, N.

In earlier articles (Fell [13], [14]) we showed that the Mackey analysis has a thoroughly satisfying generalization to the context of homogeneous *-algebraic bundles (or rather, their topological analogues). It turns out, however, that the entire Mackey analysis goes through for *-algebraic bundles which satisfy a property considerably weaker than homogeneity, namely the property of so-called saturation. *Our generalization of the Mackey analysis in the present work will be based entirely on the assumption of saturation*; homogeneity will hardly be mentioned.

Let us define and briefly discuss the concept of saturation in the finite group context.

A *-algebraic bundle $\mathscr{B} = \langle E, \{B_x\}_{x \in G} \rangle$ over a finite group G (with unit e) is said to be *saturated* (*in the algebraic sense*) if, for any x, y in G, the linear span of $B_x B_y$ is all of B_{xy}.

To begin with, homogeneity of $\mathscr{B}$ implies saturation; but we shall see in a moment that the converse is not true. As an example in which saturation fails, take the trivial case in which G consists of more than one element, and $B_e = E$, $B_x = \{0\}$ for $x \neq e$.

Here is a class of saturated examples whose infinite-dimensional generalization will be important in Chapter XII. Let G be a finite group (with unit e) and X a finite-dimensional Hilbert space; and let

$$ X = \sum_{y \in G}^{\oplus} X_y $$

be a Hilbert space direct sum decomposition of X indexed by the elements of G, the X_y being all non-zero (and pairwise orthogonal). Let E be the *-algebra $\mathcal{O}(X)$ of all linear operators on X; and for each $x \in G$ let

$$ B_x = \{a \in E : a(X_y) \subset X_{xy} \text{ for all } y \in G\}. $$

The reader will verify without difficulty that $\mathscr{B} = \langle E, \{B_x\} \rangle$ is a saturated *-algebraic bundle over G (the unit fiber B_e of $\mathscr{B}$ consisting of those $a \in E$ which leave each X_y stable). Furthermore, $\mathscr{B}$ is homogeneous if and only if the subspaces X_y are all of the same dimension (which certainly is not necessarily the case!). Thus saturation does not imply homogeneity.

5

This section is a digression, sketching another approach to the definition of a *-algebraic bundle over G in the case that G is a finite group.

Let G be a finite group. If S and T are two unitary representations of G, we can form the *outer tensor product* unitary representation $S \times T$ of the product group $G \times G$ by means of the formula:

$$ (S \times T)_{\langle x, y \rangle} = S_x \otimes T_y \qquad (x, y \in G). \qquad (9) $$

We can also form the *inner tensor product* unitary representation $S \otimes T$ of G itself by restricting $S \times T$ to G considered as the "diagonal" subgroup of $G \times G$:

$$ (S \otimes T)_x = S_x \otimes T_x \qquad (x \in G). \qquad (10) $$

Now let E be a *-algebra. If S and T are any two *-representations of E, we can again form an *outer tensor product* *-representation $S \times T$ of the tensor product *-algebra $E \otimes E$ as follows, in analogy with (9):

$$(S \times T)_{a \otimes b} = S_a \otimes T_b \qquad (a, b \in E). \qquad (11)$$

But in general there is no way to "restrict $S \times T$ to a diagonal *-subalgebra" so as to define an inner tensor product of *-representations of E in analogy with (10).

The possibility of defining inner tensor products of group representations but not of algebra representations is due to the fact (see §4) that a group has a richer structure than its group algebra, and that one of the capacities which G possesses but which its group *-algebra "forgets" is the capacity to form inner tensor products of *-representations.

Now, let us adjoin to $E = \mathscr{L}(G)$, as an extra piece of structure in addition to its *-algebra structure, the natural injective *-homomorphism $\delta^G : E \to E \otimes E$ given on the base elements d_x of E by

$$\delta^G(d_x) = d_x \otimes d_x \qquad (x \in G). \qquad (12)$$

(Here d_x is the element of E with value 1 at x and value 0 at all other points of G). Using δ^G we can now define an inner tensor product $S \otimes T$ of two *-representations S and T of E as follows:

$$S \otimes T = (S \times T) \circ \delta^G \qquad (13)$$

($S \times T$ being the outer tensor product defined in (11)); and this will clearly be the same as taking the inner tensor product of the unitary representations of G corresponding to S and T.

More generally, a *-algebra E together with an injective *-homomorphism $\delta : E \to E \otimes E$ satisfying appropriate postulates is called a *Hopf algebra*; and for any such Hopf algebra we can define an inner tensor product $S \otimes T$ of *-representations S and T of E by means of the relation

$$S \otimes T = (S \times T) \circ \delta. \qquad (14)$$

Thus, Hopf algebras constitute the "enrichment" of the structure of *-algebras needed in order to "algebraize" the group-theoretic concept of the inner tensor product of representations.

We note in passing that there is a large literature (see, for example, Ernest [8], Iorio [1], Kac [2, 3], Takasaki [4]) devoted to the generalization to Hopf algebras of the Tatsuuma duality theory for groups (Tatsuuma [2]), which leans heavily on the inner tensor product operation on representations. However, the theory of Hopf algebras and Tatsuuma duality is beyond the scope of the present work.

Now it turns out that the extra structure which makes a *-algebra E into a *-algebraic bundle is closely related to the extra structure which makes a *-algebra into a Hopf algebra. Indeed, let G be a finite group as before. If $\mathscr{B} = \langle E, \{B_x\}_{x \in G} \rangle$ is a *-algebraic bundle over G, one verifies that the equation

$$\delta(b) = d_x \otimes b \qquad (x \in G, b \in B_x) \qquad (15)$$

(d_x being as in (12)) defines an injective *-homomorphism $\delta: E \to \mathscr{L}(G) \otimes E$, and that δ satisfies the identity

$$(i \otimes \delta) \circ \delta = (\delta^G \otimes i) \circ \delta, \qquad (16)$$

where δ^G is as in (12), and the i on the left and right sides of (16) are the identity maps on $\mathscr{L}(G)$ and E respectively. Conversely, suppose that E is a *-algebra and that $\delta: E \to \mathscr{L}(G) \otimes E$ is an injective *-homomorphism satisfying (16). (Such a δ will be called a *comultiplication*.) Then it is easy to prove that δ is derived as in (15) from a unique *-algebraic bundle structure $\{B_x\}_{x \in G}$ for E, the fibers B_x being described in terms of δ by

$$B_x = \{b \in E: \delta(b) = d_x \otimes b\}.$$

If $\mathscr{B}$ happens to be the group bundle of G, the corresponding comultiplication is of course the δ^G of (12).

The comultiplication δ of (15) defines a generalization of the inner tensor product of representations of groups. Indeed, from a unitary representation S of G and a *-representation T of E we can form an *inner tensor product* *-representation $S \otimes T$ of E by means of the equation:

$$S \otimes T = (S \times T) \circ \delta, \qquad (17)$$

where $S \times T: \phi \otimes b \mapsto S_\phi \otimes T_b$ is the corresponding *outer* tensor product *-representation of $\mathscr{L}(G) \otimes E$. This inner tensor product *-representation is described even more simply in terms of the fibers B_x:

$$(S \otimes T)_b = S_x \otimes T_b \qquad \text{whenever } x \in G, b \in B_x. \qquad (18)$$

In view of the above relationship between the comultiplication δ and the bundle structure, it would appear that, *given a *-algebra E, the essence of a *-algebraic bundle structure for E over a finite group G lies in the possibility of forming the inner tensor product operation (17) (or (18))*. See Remark VIII.9.19.

6

What happens to the definition of a *-algebraic bundle over G if G is no longer a finite group (or even discrete), but is an arbitrary topological group?

In that case, the definition must be substantially modified. For one thing, there is no longer a natural "big" *-algebra E of which all the fibers B_x are linear subspaces. Indeed, it is best to disregard E entirely to begin with, and to confine ourselves to axiomatizing the collection of fibers $\{B_x\}_{x \in G}$. We assume then that for each x in a topological group G (with unit e) we are given a Banach space B_x, and that in fact the B_x are the fibers of a Banach bundle $\langle B, \pi \rangle$ over G (in the sense of §II.13). The multiplication and involution, which in the finite group case were "inherited" by the B_x as subspaces of the *-algebra E, must now be introduced axiomatically. We assume therefore that we are given a continuous binary operation $\cdot$ and a continuous unary operation $*$ on the bundle space B which (roughly speaking) (a) are covariant under π with the multiplication and inverse in G, and (b) satisfy all the postulates of a Banach *-algebra in so far as these make sense in a system where addition is only partially defined. Such a system $\mathscr{B} = \langle B, \pi, \cdot, * \rangle$ is called a *Banach *-algebraic bundle*. (The reader will find all the details in §§VIII.2, 3.) The unit fiber B_e is a Banach *-algebra, but the other fibers are not.

The simplest Banach *-algebraic bundle over G (as in the finite group case) is the *group bundle of G*. In the present context this is defined to have bundle space $B = G \times \mathbb{C}$, projection $\pi: \langle x, \lambda \rangle \mapsto x$, and multiplication and involution given by

$$\langle x, \lambda \rangle \cdot \langle x', \lambda' \rangle = \langle xx', \lambda\lambda' \rangle,$$

$$\langle x, \lambda \rangle^* = \langle x^{-1}, \bar{\lambda} \rangle.$$

Here of course the fibers are all one-dimensional.

There are many important special classes of Banach *-algebraic bundles $\mathscr{B}$. Most important for us is the property of saturation; we say that $\mathscr{B}$ is (*topologically*) *saturated* if for each x, y in G the linear span of $B_x B_y$ is dense in B_{xy}.

The property of homogeneity (defined for finite G in §4) can be extended to the present topological context, though the definition is technically complicated. Most of Fell [14] is devoted to working out the Mackey analysis for homogeneous Banach *-algebraic bundles. The property of homogeneity hardly appears at all in the present work.

A very useful class of Banach *-algebraic bundles are the so-called semidirect products. Given a Banach *-algebra A and a continuous action of G on A by isometric *-isomorphisms, there is a canonical way to construct from these ingredients a Banach *-algebraic bundle $\mathscr{B}$ over G whose unit fiber is A (see §VIII.4). The construction is quite analogous to that of the semidirect product of two groups. We call $\mathscr{B}$ the *semidirect product of A by G*.

As we saw in Chapter VI, C^*-algebras play a very special role in the study of *-representations of arbitrary Banach *-algebras. A similar special role among Banach *-algebraic bundles is played by the so-called C^*-*algebraic bundles* (see §VIII.16). These are Banach *-algebraic bundles $\mathscr{B}$ which satisfy the C^* norm identity ($\|b^*b\| = \|b\|^2$)—so that the unit fiber B_e is a C^*-algebra—and such that b^*b is positive in B_e for all $b \in B$. Just as for Banach *-algebras, so also any Banach *-algebraic bundle $\mathscr{B}$ over G can be "completed" to a C^*-algebraic bundle $\mathscr{C}$ over G in such a way that the *-representation theories of $\mathscr{B}$ and $\mathscr{C}$ coincide.

Finally, we should like to mention the class of saturated C^*-algebraic bundles whose fibers are all one-dimensional. These will be referred to as *cocycle bundles*. The group bundle is of course a cocycle bundle; but there are in general many cocycle bundles which are entirely different from the group bundle. We shall see in §VIII.4 that cocycle bundles are intimately related to Mackey's multipliers on groups. As we shall see especially in §XII.6, these cocycle bundles play an extremely important role in the classification of arbitrary saturated C^*-algebraic bundles.

Let us now suppose that $\mathscr{B}$ is any Banach *-algebraic bundle over G, where the group G is *locally compact*. Then by the theorem of Douady and dal Soglio-Hérault ·(Appendix C, Volume 1) $\mathscr{B}$ has enough continuous cross-sections, and we can form the $\mathscr{L}_1$ cross-sectional space $\mathscr{L}_1(\mu; \mathscr{B})$ of $\mathscr{B}$ with respect to (left) Haar measure μ on G (see §II.15). Copying the well-known definitions of convolution and involution in the $\mathscr{L}_1$ group algebra, we define convolution and involution in $\mathscr{L}_1(\mu; \mathscr{B})$ by:

$$(f * g)(x) = \int_G f(y)g(y^{-1}x)d\mu y, \tag{19}$$

$$f^*(x) = \Delta(x^{-1})(f(x^{-1}))^* \tag{20}$$

(f, $g \in \mathscr{L}_1(\mu; \mathscr{B})$, $x \in G$; notice that $f(y)g(y^{-1}x) \in B_x$ for all y). With these operations $\mathscr{L}_1(\mu; \mathscr{B})$ becomes a Banach *-algebra, called the $\mathscr{L}_1$ *cross-sectional algebra of $\mathscr{B}$*. If $\mathscr{B}$ is the group bundle of G, this is just the ordinary $\mathscr{L}_1$ group algebra of G studied in Chapter II.

Notice the special case that G is finite. In that case each fiber B_x is a linear subspace of $\mathscr{L}_1(\mu; \mathscr{B})$; and in fact the B_x are linearly independent and add up to $\mathscr{L}_1(\mu; \mathscr{B})$. Furthermore the operations (19) and (20) on $\mathscr{L}_1(\mu; \mathscr{B})$ extend the multiplication and involution on the B_x. Thus $\mathscr{L}_1(\mu; \mathscr{B})$ plays precisely the role of E in the definition in §4 of a *-algebraic bundle over a finite group. So in the case of a general locally compact group G we may think of $\mathscr{L}_1(\mu; \mathscr{B})$ as playing the role of the "big" algebra E, even though the fibers B_x now fail to be subspaces of E.

It should be mentioned here that the group extension bundles, defined in §4 for finite base group G, have a perfectly satisfactory generalization to the locally compact situation. Suppose that H is a locally compact group with closed normal subgroup N and quotient group $G = H/N$. Generalizing the construction of §4, we can set up a saturated Banach *-algebraic bundle $\mathscr{B}$ over G, called the *group extension bundle* of H, N, such that the unit fiber of $\mathscr{B}$ is exactly the $\mathscr{L}_1$ algebra of N, while the $\mathscr{L}_1$ cross-sectional algebra of $\mathscr{B}$ is isometrically *-isomorphic with the $\mathscr{L}_1$ algebra of H. (These group extension bundles are special cases of the more general "partial cross-sectional bundles" constructed in §VIII.6.)

A great deal of the past literature of functional analysis has been concerned with special cases of these $\mathscr{L}_1$ cross-sectional algebras of Banach *-algebraic bundles (even though they have not been described in these terms). For example the σ-group algebras of Mackey [8] are just the $\mathscr{L}_1$ cross-sectional algebras of cocycle bundles. We may also mention Johnson [1], Glimm [4], Effros and Hahn [1], Edwards and Lewis [1], Turumaru [1], Zeller–Meier [1, 2], and Doplicher, Kastler and Robinson [1]. Takesaki [1] developed the Mackey normal subgroup analysis for the cross-sectional algebras of semi-direct product bundles (that is, the "covariance algebras" of Doplicher, Kastler and Robinson [1]).

In addition to the works cited in the last parapraph, three concepts have arisen in the literature which are almost co-extensive with homogeneous Banach *-algebraic bundles. These are Leptin's "generalized $\mathscr{L}_1$ algebras" (see Leptin [1, 2, 3, 4, 5]), the "twisted group algebras" of Busby and Smith [1], and the "twisted covariant systems" of Philip Green [2]. Leptin [5] develops the theory of induced representations, the Imprimitivity Theorem, and a part of the Mackey analysis for his generalized $\mathscr{L}_1$ algebras; while the (later) paper of Green [2] develops the entire Mackey analysis (using the more modern methods to be followed in our Chapter XI) for his twisted covariant systems.

7

We must now introduce the concept of a *-representation of a Banach *-algebraic bundle, in preparation for a preliminary discussion of the Mackey analysis in the bundle context.

Let $\mathscr{B} = \langle B, \pi, \cdot, * \rangle$ be a Banach *-algebraic bundle over a topological group G. By a *-representation of $\mathscr{B}$, acting on a Hilbert space X, we mean a map T carrying B into the space of all continuous linear operators on X and satisfying:

(i) T is linear on each fiber B_x,
(ii) $T_b T_c = T_{bc}$ $(b, c \in B)$,
(iii) $(T_b)^* = T_{b^*}$ $(b \in B)$,
(iv) $b \mapsto T_b \xi$ is continuous on B to X for each vector ξ in X.

All the general properties of *-representations of *-algebras, such as non-degeneracy, irreducibility and so forth, can be equally well defined for *-representations of $\mathscr{B}$.

If $\mathscr{B}$ is the group bundle of G, the non-degenerate *-representations of $\mathscr{B}$ are essentially just the unitary representations of G.

Suppose G is locally compact. Then the well-known natural correspondence between unitary representations of G and non-degenerate *-representations of the $\mathscr{L}_1$ algebra of G can be generalized to a beautiful natural correspondence between the *-representations of $\mathscr{B}$ and the *-representations of the $\mathscr{L}_1$ cross-sectional algebra of $\mathscr{B}$ (see §VIII.13)—a correspondence that preserves irreducibility and other general properties of the representations.

Now the Mackey analysis in the bundle context refers to the following enterprise: *Given a Banach *-algebraic bundle $\mathscr{B}$ over a locally compact group G, we try to describe all the irreducible *-representations of $\mathscr{B}$ in terms of*

(i)   *the irreducible *-representations of the unit fiber Banach *-algebra B_e,*
(ii) *irreducible representations of subgroups of G.*

Notice that, as applied to the group extension bundles mentioned in §6, this enterprise reduces to the classical Mackey normal subgroup analysis. Indeed, let H be a locally compact group, N a closed normal subgroup, $G = H/N$ the quotient group, and $\mathscr{B}$ the group extension bundle of H, N. As we stated two paragraphs ago, the irreducible *-representations of $\mathscr{B}$ are identifiable with the irreducible *-representations of the $\mathscr{L}_1$ cross-sectional algebra of $\mathscr{B}$; the latter, as we saw in §6, is the $\mathscr{L}_1$ algebra of H; and the irreducible *-representations of the $\mathscr{L}_1$ algebra of H are identifiable with the irreducible unitary representations of H. Combining these facts, we see that the irreducible *-representations of $\mathscr{B}$ are identifiable with the irreducible unitary representations of H. Since the unit fiber B_e of $\mathscr{B}$ is the $\mathscr{L}_1$ algebra of N (see §6), the irreducible *-representations of B_e are identifiable with the irreducible unitary representations of N. So the Mackey analysis as applied to the group extension bundle $\mathscr{B}$ reduces to the following enterprise: To analyze all the irreducible unitary representations of H in terms of

(i)   the irreducible unitary representations of N, and
(ii) the irreducible representations of subgroups of G.

This is nothing but the aim of the Mackey normal subgroup analysis as developed in [8].

*Now it turns out, as we shall see in detail in Chapter XII, that the Mackey analysis can be carried through in the context of saturated Banach *-algebraic bundles over locally compact groups with just the same degree of success, and with quite similar general hypotheses, as in the classical context of locally compact group extensions.* This justifies our assertion that the natural domain of the Mackey analysis is not simply the category of locally compact group extensions, but that of saturated Banach *-algebraic bundles over locally compact groups.

Let us pursue this train of thought a little further. To begin with, because of the close connection between Banach *-algebraic bundles and their C^*-algebraic bundle completions there is really no loss of generality in restricting our attention to saturated C^*- algebraic bundles; and if we do so, the Mackey analysis admits certain technical simplifications.

Given an arbitrary C^*-algebra E and a locally compact group G, let us define a *saturated C^*-algebraic bundle structure for E over G* to mean a saturated C^*-algebraic bundle $\mathscr{B}$ over G whose $\mathscr{L}_1$ cross-sectional algebra D has a C^*-completion which is *-isomorphic with E. Since in this situation the irreducible *-representations of E, D and $\mathscr{B}$ are essentially the same objects, the Mackey analysis applied to $\mathscr{B}$ analyzes the irreducible *-representations of E. For this analysis to produce useful information, two extreme cases must be avoided—first, the case that G is the one-element group (corresponding to the choice of H itself as the normal subgroup of the group H in the classical Mackey analysis), and secondly, the case that the unit fiber C^*-algebra B_e is of dimension 1 (corresponding to the choice of the one-element normal subgroup of the group H in the classical Mackey analysis). Let us say that the saturated C^*-algebraic bundle structure $\mathscr{B}$ for E over G is *productive* if neither of these cases occur, that is, G has more than one element and B_e has dimension greater than one. *Thus the irreducible representation theory of any C^*-algebra E can be studied by means of the generalized Mackey analysis* *provided we can find some productive saturated C*-algebraic bundle structure for E.*

As an example we may mention the so-called Glimm algebras of §VI.17. In §VIII.17 we construct "Glimm bundles" which are (productive) saturated C*-algebraic bundle structures for the Glimm algebras; and in §XII.9 we apply the generalized Mackey analysis to these Glimm bundles to obtain important information (though long familiar to experts!) on the irreducible *-representations of the Glimm algebras.

A natural question (to which we do not know the answer at present) is the following: Do there exist locally compact groups H such that

(a) H has no non-trivial proper closed normal subgroup (so that the Mackey normal subgroup analysis is inapplicable to H in its classical form as developed by Mackey), and yet

(b) the C*-group algebra $C^*(H)$ of H admits some productive saturated C*-algebraic bundle structure (so that our generalized Mackey analysis can be applied to study the irreducible representation theory of $C^*(H)$ and hence of H itself)?

It should be mentioned that the Mackey analysis in the generalized context of saturated Banach *-algebraic bundles is appearing in print here in this Volume for the first time. However, the methods and techniques used to develop it here are largely similar to those used in Fell [14], where the Mackey analysis was published in the more restricted context of homogeneous Banach *-algebraic bundles. Indeed, the development of the Mackey analysis followed both here and in Fell [14] borrows heavily from Blattner's original papers [1], [3], [4], [5], [6], which develop the Mackey analysis for arbitrary (non-separable) locally compact groups.

8

As we have already asserted, the natural context for the Mackey analysis is that of saturated Banach *-algebraic bundles. In the remainder of this Introduction we should like to sketch the construction of induced representations and the Mackey analysis for finite -dimensional saturated C*-algebraic bundles over finite groups. This will serve to orient the reader toward the rather lengthy technicalities of Chapters XI and XII.

Those readers who have no previous knowledge of the Mackey normal subgroup analysis in the group context would do well at this point to peruse §XII.1, where the classical normal subgroup analysis for finite groups is presented in some detail.

Let $\mathscr{B} = \{B_x\}_{x \in G}$ be a fixed finite-dimensional saturated C^*-algebraic bundle over the finite group G.

We begin with the construction of induced representations of $\mathscr{B}$ and the Imprimitivity Theorem.

Suppose that H is a subgroup of G. Then $\mathscr{B}_H = \{B_x\}_{x \in H}$ is a saturated C^*-algebraic bundle over H (the *reduction of $\mathscr{B}$ to H*). Let S be a non-degenerate *-representation of $\mathscr{B}_H$, acting on a finite-dimensional Hilbert space X. From these ingredients we propose to construct a *-representation T of $\mathscr{B}$.

Given any coset α in G/H, put

$$L_\alpha = \sum_{x \in \alpha}^{\oplus} (B_x \otimes X);$$

and let $(\ , \ )_\alpha$ be the conjugate-bilinear form on L_α satisfying

$$(b \otimes \xi, c \otimes \eta) = (S_{c^*b}\xi, \eta)_X \tag{21}$$

whenever $\xi, \eta \in X$, $b \in B_x$, and $c \in B_y$, x and y being group elements lying in α. (Notice that under these conditions $c^*b \in B_H$, so that S_{c^*b} makes sense.) It turns out (see XI.11.10) that $(\ , \ )_\alpha$ is positive. So, factoring out the null space of $(\ , \ )_\alpha$, we obtain a Hilbert space $Y_\alpha, (\ , \ )_\alpha$. Let $b \widetilde{\otimes} \xi$ stand for the image of $b \otimes \xi$ in Y_α. One can now show that, for each α in G/H, y in G, and c in B_y, the relation

$$\tau_c(b \widetilde{\otimes} \xi) = cb \widetilde{\otimes} \xi \qquad (b \in \bigcup_{x \in \alpha} B_x; \xi \in X) \tag{22}$$

defines a linear mapping

$$\tau_c \colon Y_\alpha \longmapsto Y_{y\alpha}. \tag{23}$$

This mapping satisfies

$$\tau_{c_1 c_2} = \tau_{c_1} \tau_{c_2}, \qquad \tau_{c^*} = (\tau_c)^* \tag{24}$$

for all c_1, c_2, c in $\bigcup_{y \in G} B_y$. Now let Y be the Hilbert space direct sum $\sum_{\alpha \in G/H}^{\oplus} Y_\alpha$. In view of (24) the τ_c can be combined together to form a *-representation T of $\mathscr{B}$ on Y:

$$(T_c \phi)_\alpha = \tau_c(\phi_{x^{-1}\alpha}) \qquad (\phi \in Y; x \in G; c \in B_x; \alpha \in G/H).$$

This T is called the *-representation of $\mathscr{B}$ *induced from S*, and is denoted by Ind(S).

Let us point out how this reduces to the classical definition IX.10.4 if $\mathscr{B}$ is the group bundle. According to the classical definition we start with a unitary representation S of the subgroup H of G, acting on a Hilbert space X; we form the Hilbert space Z of all functions $f : G \to X$ satisfying

$$f(xh) = S_{h^{-1}}(f(x)) \qquad (x \in G; h \in H), \tag{25}$$

with the inner product

$$(f, g)_Z = m^{-1} \sum_{x \in G} (f(x), g(x))_x$$

(m being the order of H); and we define V to be the unitary representation of G acting by left translation on Z;

$$(V_y f)(x) = f(y^{-1}x) \qquad\qquad (f \in Z; x, y \in G).$$

This V is the classical induced representation. Suppose now that $\mathcal{B}$ is the group bundle of G. Then S can be identified with a *-representation of $\mathcal{B}_H$; and for each coset α in G/H we can form $Y_\alpha, (\, , \,)_\alpha$ as in (21) et seq. If $\xi \in X$, $x \in G$, and $h \in H$, one shows from (21) that

$$d_{xh} \widetilde{\otimes} S_{h^{-1}}\xi = d_x \widetilde{\otimes} \xi \tag{26}$$

(d_x being as in (12)). Given $f \in Z$, let $\phi = \phi(f)$ be the element of $Y = \sum_\alpha^\oplus Y_\alpha$ whose Y_α-component (for each α) is just $d_x \widetilde{\otimes} f(x)$, x being any group element in α. By (25) and (26) ϕ is well defined. The map $\Phi: f \mapsto \phi$ is an isometry of Z onto Y; in fact it sets up a unitary equivalence between V and the $T = \mathrm{Ind}(S)$ defined in the preceding paragraph. So the present definition of induced representation does generalize IX.10.4, at least in the context of finite groups.

Returning to the context of an arbitrary (finite-dimensional) saturated C^*-algebraic bundle $\mathcal{B}$ over G, we observe that, as in the group bundle case, induced representations are intimately related to systems of imprimitivity. Let M be any finite G-space (i.e., a finite set on which G acts to the left). A *system of imprimitivity for $\mathcal{B}$ over M* is a pair $T, \{P_m : m \in M\}$, where

(i) T is a non-degenerate *-representation of $\mathcal{B}$;
(ii) the P_m ($m \in M$) are pairwise orthogonal projections on $X(T)$ such that $\sum_{m \in M} P_m = 1$;
(iii) we have

$$T_b P_m = P_{xm} T_b \qquad (m \in M; x \in G; b \in B_x). \tag{27}$$

((27) amounts to saying that T_b (range(P_m)) $\subset$ range(P_{xm}) whenever $x \in G$, $b \in B_x$, $m \in M$.)

Induced representations give rise to systems of imprimitivity. Indeed, suppose that H is a subgroup of G and S is a *-representation of $\mathcal{B}_H$. Keeping the same notation as before, we construct the induced *-representation T of $\mathcal{B}$ acting on $Y = \sum_{\alpha \in G/H} Y_\alpha$. For each α in G/H let P_α be the natural projection of Y onto Y_α. From the fact that $\tau_b(Y_\alpha) \subset Y_{x\alpha}$ ($x \in G; b \in B_x; \alpha \in G/H$) one concludes that $T, \{P_\alpha : \alpha \in G/H\}$ is a system of imprimitivity for $\mathcal{B}$ over the transitive G-space $M = G/H$; we call it the *system of imprimitivity induced by S*.

The Imprimitivity Theorem asserts the converse of this: *Let H be a subgroup of G and $\mathcal{T} = \langle T, \{P_\alpha : \alpha \in G/H\} \rangle$ a system of imprimitivity for $\mathcal{B}$ over G/H. Then there exists a nondegenerate *-representation S of $\mathcal{B}_H$, unique to within unitary equivalence, such that $\mathcal{T}$ is unitarily equivalent to the system of imprimitivity induced by S.*

In this simple finite-dimensional context the proof of the Imprimitivity Theorem is a routine matter. The general Imprimitivity Theorem for bundles over arbitrary locally compact groups is proved in §XI.14; and the proof is much more difficult. It is worth mentioning that two basically different proofs of the general Imprimitivity Theorem for group bundles over locally compact groups are to be found in the literature. One is the proof given by Mackey [8]. This proof is highly measure-theoretic, and is valid for second-countable groups only. It has the advantage, however, that it gives partial information about systems of imprimitivity even over ergodic non-transitive base spaces. The other proof was initiated by Loomis [3]. It is not measure-theoretic and it is valid without any second-countability restrictions; but it gives no information about systems of imprimitivity over ergodic nontransitive spaces. In this work we have followed Loomis' approach, generalizing it of course to the context of Banach *-algebraic bundles. In fact, guided by the work of Rieffel [5], we shall in Chapter XI present the theory of induced representations and the Imprimitivity Theorem in a context much more general even than that of Banach *-algebraic bundles—namely the context of rigged modules. For a motivating sketch of Rieffel's generalization to rigged modules, see our introduction to Chapter XI.

From one standpoint, the structure of a Banach *-algebraic bundle $\mathcal{B}$ may be regarded as primarily designed to permit us to define the notion of a system of imprimitivity for $\mathcal{B}$. From this standpoint the relation between Banach *-algebras and Banach *-algebraic bundles is analogous to the relation between unitary representations of a group and systems of imprimitivity for that group.

Now let us sketch the Mackey analysis in the context of the finite-dimensional saturated C^*-algebraic bundle $\mathcal{B} = \{B_x\}_{x \in G}$ over the finite group G. We will denote by A the unit fiber C^*-algebra B_e.

The first step in the Mackey analysis is to make G act as a transformation group on the space $\hat{A}$ of all unitary equivalence classes of irreducible *-representations of A. In our present generalized situation this cannot be done by conjugation with group elements as it could in the group bundle case (see XII.1.3). Instead, we return to the construction of induced representations ((21) et seq.). In that construction suppose we take $H = \{e\}$, so that $B_H = B_e = A$, and S is a non-degenerate *-representation of A. Fix x in $G(= G/H)$. We then have a corresponding Hilbert space Y_x; and by (22) and (24) the map

$$^xS: a \mapsto \tau_a | Y_x \qquad\qquad (a \in A)$$

$$\hspace{10cm} (a \in A)$$

is a *-representation of A on Y_x. This ^xS is called the *conjugate of S under x*, or simply the *x-conjugate* of S. It turns out, because of the saturation of $\mathscr{B}$, that

$$\langle x, S \rangle \mapsto {}^xS \qquad\qquad (28)$$

is a left action of G on the collection of all unitary equivalence classes of non-degenerate *-representations of A, and that ^xS is irreducible if and only if S is. Thus, restricted to $\hat{A}$, (28) defines a left action of G on $\hat{A}$. In the group bundle case this action is just the action XII.1.3 by conjugation with group elements.

The hypothesis of saturation was actually not essential for the construction of induced representations; but it *is* essential in order to make G act on $\hat{A}$ in the manner of the preceding paragraph. it is for this reason that we always assume saturation of $\mathscr{B}$ in the development of the Mackey analysis.

The theory of the conjugation of representations for saturated bundles over arbitrary locally compact groups is developed in §XI.16.

Now let T be an arbitrary (finite-dimensional) non-degenerate *-representation of $\mathscr{B}$. To analyze the structure of T we begin by considering its restriction $T' = T|A$ to A. Being finite-dimensional, T' is the direct sum $\sum_{i \in I}^{\oplus} S^i$ of irreducible *-representations S^i of A. For each D in $\hat{A}$ let P_D be projection onto the D-space X_D of $X(T)$. (Recall from Chapter IV that X_D is the sum of those $X(S^i)$ such that S^i is of class D.) Now $\hat{A}$ was made above into a G-space; and it is not hard to check that $\langle T, \{P_D : D \in \hat{A}\} \rangle$ must be a system of imprimitivity for $\mathscr{B}$ over $\hat{A}$. (In the general case $\{P_D\}$ becomes the spectral measure of T' studied in §VII.9; and the analogue of the fact that $T, \{P_D\}$ is a system of imprimitivity is proved in XII.2.11.)

Now assume that T is irreducible. Given any orbit Θ in $\hat{A}$ under G, we set $X_\Theta = \sum_{D \in \Theta} X_D$. Condition (27) implies that X_Θ is stable not only under T' but under all of T; and so either $X_\Theta = \{0\}$ or $X_\Theta = X(T)$. Since the X_Θ (for different orbits Θ) are mutually orthogonal and add up to $X(T)$, there is *exactly one* orbit Θ_0 for which $X_{\Theta_0} = X(T)$; and $X_\Theta = \{0\}$ for all other orbits Θ. Under these conditions we say that T is *associated with* the orbit Θ_0.

This relation between elements of $\hat{\mathscr{B}}$ (the space of irreducible *-representations of $\mathscr{B}$) and orbits in $\hat{A}$ is what we call Step I of the Mackey analysis (see XII.1.28).

Let us analyze the irreducible *-representation T further. Let E be a fixed element of the orbit Θ_0 with which T is associated; and denote by H the stability subgroup $\{x \in G : {}^xE \cong E\}$ for E. The map $\gamma : xH \mapsto {}^xE$ $(x \in G)$ is a bijection of G/H onto Θ_0; in fact γ is clearly an isomorphism of G/H and Θ_0 as transitive G-spaces. So, if we set $Q_\alpha = P_{\gamma(\alpha)}$ $(\alpha \in G/H)$, then $T, \{Q_\alpha : \alpha \in G/H\}$ a system of imprimitivity for $\mathscr{B}$ over G/H. By the Imprimitivity Theorem T, $\{Q_\alpha\}$ is induced by a unique *-representation S of $\mathscr{B}_H$. In particular $T = \text{Ind}(S)$. By the same kind of argument as in XII.1.8 one shows that S is irreducible and that $S|A$ must be a direct sum of copies of E. In fact more than this is true. Let us denote by $(\mathscr{B}_H)\hat{}_E$ the family of all those S' in $(\mathscr{B}_H)\hat{}$ such that $S'|A$ is a direct sum of copies of E; and let $\hat{\mathscr{B}}_{\Theta_0}$ be the family of those T' in $\hat{\mathscr{B}}$ which are associated with the orbit Θ_0. By arguments similar to those of XII.1.9 one shows that the inducing map

$$S' \mapsto T' = \text{Ind}(S')$$

is a bijection of $(\mathscr{B}_H)\hat{}_E$ onto $\hat{\mathscr{B}}_{\Theta_0}$.

The last statement constitutes what we call Step II of the Mackey analysis (see XII.1.28). The general version of it, for bundles over arbitrary locally compact groups, is carried out in §XII.3.

The third step of the Mackey analysis consists in analyzing the set $(\mathscr{B}_H)\hat{}_E$ which figured in Step II. The procedure here is very similar to that followed in the group case in §XII.1; and we shall sketch it only very briefly. Keeping the previous notation, let us assume that E has the following special property: E can be extended to a *-representation $\tilde{E}$ of $\mathscr{B}_H$ (acting in the same space as E). Then it turns out that, given any W in $\hat{H}$ (i.e., any irreducible unitary representation W of H), the inner tensor product $W \otimes \tilde{E}$ (defined in (18)) belongs to $(\mathscr{B}_H)\hat{}_E$, and that conversely every element of $(\mathscr{B}_H)\hat{}_E$ is of the form $W \otimes \tilde{E}$ for some unique W in $\hat{H}$. Thus the (at first sight rather intractable-looking) set $(\mathscr{B}_H)\hat{}_E$ turns out to be in natural one-to-one correspondence with the structure space $\hat{H}$ of H. Hence by Step II $\hat{\mathscr{B}}_{\Theta_0}$ is also in natural one-to-one correspondence with $\hat{H}$.

In general E will not have the special property of extendability to B_H. If it does not, one can still carry through the above analysis in a modified form, obtaining a natural correspondence between $(\mathscr{B}_H)\hat{}_E$ (or $\hat{\mathscr{B}}_{\Theta_0}$) and the irreducible *projective* representations of H (see §VIII.10) of a certain projective class determined by E, the so-called Mackey obstruction of E. This analysis is carried through for general saturated bundles in §XII.4.

This then is the Mackey analysis in summary: *First, the structure space $\hat{\mathscr{B}}$ of $\mathscr{B}$ is presented as a disjoint union of subsets $\hat{\mathscr{B}}_\Theta$, one for each orbit Θ in $\hat{A}$ under the action of G. Secondly, for each orbit Θ the set $\hat{\mathscr{B}}_\Theta$ is analyzed in terms* *of the projective representation theory of the stability subgroup for any fixed element E of* Θ.

While for *finite-dimensional* saturated C^*-algebraic bundles this analysis requires no further hypotheses, in the general case certain rather broad hypotheses are necessary at different points of the development. The failure of any of these can lead to interesting phenomena, some of which are explored in the examples of the concluding section §XII.9 of this work.

It should come as no surprise that the Mackey analysis is a powerful tool for studying the structure of saturated C^*-algebraic bundles. In §XII.6, using the Mackey analysis, we will in fact completely determine the structure of an arbitrary saturated C^*-algebraic bundle $\mathscr{B}$ over a locally compact group, provided that the unit fiber C^*-algebra of $\mathscr{B}$ is of compact type. This result appears here in print for the first time.

The moving power of mathematical invention
is not reasoning but imagination.

—A. De Morgan

VIII Banach *-Algebraic Bundles and their Representations

In the Introduction to this Volume we have pointed out the importance of the notion of a Banach *-algebraic bundle as an appropriate setting for the theory of induced representations and the generalized Mackey analysis. In this chapter we develop the foundations of the theory of Banach *-algebraic bundles.

In addition to the theory of induced representations and the generalized Mackey analysis (to be studied in Chapters XI and XII), there are of course many facets of harmonic analysis on groups which have natural generalizations to Banach *-algebraic bundles. Some of these are dealt with in this chapter; see for example Theorems VIII.14.9 and VIII.16.4. Others will no doubt be topics for future research.

Here is a short summary of the contents of the sections of this chapter:

§1 deals with an important tool, the idea of a multiplier. If A is an algebra without a unit element and without annihilators, A can be canonically embedded as a two-sided ideal in a larger algebra W which in a certain sense is maximal. The elements of W are called multipliers of A; and W itself is called the multiplier algebra of A. Under certain general conditions, representations of A can be extended to representations of W. This fact will be very important when we try to recover a representation of a Banach algebraic bundle from its "integrated form" (§§12, 13).

In §2 we discuss the notion of a Banach algebraic bundle. This is a Banach bundle $\mathscr{B}$ over a topological group G, with a multiplication operation which is "covariant" with the product in G and satisfies other natural postulates. In §3 we add to $\mathscr{B}$ an involution * which is "covariant" with the operation of inverse in the base group G, obtaining the notion of a Banach *-algebraic bundle. §4 is devoted to certain basic examples of Banach *-algebraic bundles—the semidirect products and central extension bundles.

Just as from a locally compact group we construct its $\mathscr{L}_1$ group algebra, in the same way from any Banach *-algebraic bundle $\mathscr{B}$ over a locally compact group G we construct the $\mathscr{L}_1$ cross-sectional algebra $\mathscr{L}_1(\mathscr{B})$ of $\mathscr{B}$. This construction occupies §5. In a rough way, $\mathscr{L}_1(\mathscr{B})$ can be regarded as the "global" Banach *-algebra into which the Banach *-algebraic bundle structure $\mathscr{B}$ has been introduced (see §17).

§§6 and 7 contain important special constructions for passing from one Banach *-algebraic bundle to another. The transformation bundles of §7 are fundamental to the Imprimitivity Theorem of Chapter XI. Special cases of transformation bundles (or rather, of their transformation algebras) have been studied by several authors; see for example Effros and Hahn [1], Glimm [5], Zeller-Meier [1].

In §8 we begin the representation theory of Banach algebraic bundles. §8 is devoted to locally convex representations in the non-involutory context; and §9 contains elementary definitions and facts about *-representations in the involutory context (including the notion of tensor products). In §10 we show how not only unitary representations but also the so-called projective representations of groups can be included under the category of *-representations of Banach *-algebraic bundles.

In §11 we learn to pass from a locally convex representation of a Banach algebraic bundle $\mathscr{B}$ to its "integrated form." In §§12 and 13 we answer the converse question: Given a locally convex representation of the (compacted) cross-sectional algebra of $\mathscr{B}$, when is it the integrated form of some representation of $\mathscr{B}$? This is answered in §12 for the non-involutory context, and in §13 (much more powerfully) for the involutory context. This in turn is used in §14 to generalize a theorem on groups mentioned in the Introduction to this Volume: We show that a certain broad class of Banach *-algebraic bundles over a locally compact group always have enough irreducible *-representations to distinguish their non-zero points.

The most important result of §15 is the description 15.5 of a *-representation of a semidirect product bundle $A \underset{\tau}{\times} G$ as a pair $\langle S, V \rangle$, where S is a *-representation of A, and V is a unitary representation of G whose action by inner automorphisms on S follows the action τ of G on A.

Among arbitrary Banach *-algebraic bundles, the so-called C^*-algebraic bundles play the same role that C^*-algebras play among arbitrary Banach *-algebras. They are the subject of §16. At the end of §16 we explicitly construct a fairly general class of saturated C^*-algebraic bundles. The interest of this class lies in the fact (to be proved in Chapter XII, §6) that every saturated C^*-algebraic bundle whose unit fiber is of compact type must be a direct sum of C^*-algebraic bundles of the class constructed here. In §17 C^*-algebraic bundles are used to give a precise definition of what it means to start with a C^*-algebra and introduce into it the structure of a Banach *-algebraic bundle over a locally compact group G. As an example, we show in §17 how the Glimm algebras of §VI.17 can be given the structure of a semidirect product Banach *-algebraic bundle.

In §18 we return to the transformation bundles of §7, and show in detail that their *-representations are just the objects known as systems of imprimitivity. In §19 we construct certain irreducible systems of imprimitivity based on ergodic measure spaces. These special systems of imprimitivity will be important in the discussion in Chapter XII of examples of the generalized Mackey analysis.

In §20 we study the bundle analogues of positive linear functionals— namely, the so-called functionals of positive type on a Banach *-algebraic bundle. Functions of positive type on a group are special cases of these.

Finally, §21 is devoted to the regional topology of the space of *-representations of a Banach *-algebraic bundle $\mathscr{B}$. This is defined as the topology with respect to which the passage from a *-representation of $\mathscr{B}$ to its integrated form is a homeomorphism. The main results of this section are descriptions of this regional topology in terms of uniform-on-compacta convergence of associated functionals of positive type on $\mathscr{B}$. These results are immediate generalizations of well-known results about unitary representations and functions of positive type on groups.

1. Multipliers of Algebras

1.1. The theory of multipliers is an extremely helpful device in the study of extensions and representations of algebras. We introduce the notion of a multiplier in a general algebraic context.

1.2. Fix an algebra A.

Definition. A *multiplier of* A is a pair $\langle \lambda, \mu \rangle$, where λ and μ are linear mappings of A into itself satisfying (i) $a\lambda(b) = \mu(a)b$, (ii) $\lambda(ab) = \lambda(a)b$, and (iii) $\mu(ab) = a\mu(b)$ for all a, b in A.

We call λ and μ the *left* and *right action* respectively of the multiplier $\langle \lambda, \mu \rangle$. The set of all multipliers of A will be denoted by $\mathscr{W}(A)$.

One checks that $\mathscr{W}(A)$ is itself an algebra under the following operations:

$$\langle \lambda, \mu \rangle + \langle \lambda', \mu' \rangle = \langle \lambda + \lambda', \mu + \mu' \rangle;$$

$$r\langle \lambda, \mu \rangle = \langle r\lambda, r\mu \rangle \qquad\qquad (r \in \mathbb{C});$$

$$\langle \lambda, \mu \rangle \langle \lambda', \mu' \rangle = \langle \lambda \circ \lambda', \mu' \circ \mu \rangle.$$

(Notice the reversal of order $\mu' \circ \mu$ in the second member of the product multiplier.) With these operations, $\mathscr{W}(A)$ is called the *multiplier algebra* of A.

$\mathscr{W}(A)$ has the unit element $\langle i, i \rangle$, where $i: A \to A$ is the identity map.

1.3. Suppose that A is a two-sided ideal of a larger algebra B. Clearly each b in B gives rise to a multiplier $u_b = \langle \lambda_b, \mu_b \rangle$ of A defined by:

$$\lambda_b(a) = ba, \quad \mu_b(a) = ab;$$

and the map $b \mapsto u_b$ is a homomorphism of B into $\mathscr{W}(A)$. This fact was indeed the motivation for our definition of the algebra $\mathscr{W}(A)$.

Notice that $\{u_b : b \in A\}$ is a two-sided ideal of $\mathscr{W}(A)$. Indeed, one verifies without difficulty that

$$uu_b = u_{\lambda(b)}, \quad u_b u = u_{\mu(b)} \tag{1}$$

whenever $b \in A$ and $u = \langle \lambda, \mu \rangle \in \mathscr{W}(A)$.

1.4. We shall say that A *has no annihilators* if, for every $a \neq 0$ in A, there exist elements b and c of A such that $ab \neq 0$ and $ca \neq 0$. Most of the specific algebras that are important to us will have this property.

Assume that A has no annihilators. Then the relation between A and $\mathscr{W}(A)$ is much simplified.

In the first place, conditions (ii) and (iii) of 1.2 are now implied by (i), and so can be omitted from the definition of a multiplier. Indeed, assuming (i), we have $c\lambda(ab) = \mu(c)(ab) = (\mu(c)a)b = (c\lambda(a))b = c(\lambda(a)b)$ for any a, b, c in A; and by the absence of annihilators this implies $\lambda(ab) = \lambda(a)b$. Similarly $\mu(ab) = a\mu(b)$.

Secondly, if $\langle \lambda, \mu \rangle$ is a multiplier, then λ determines μ and μ determines λ. For, if λ is known, condition (i) implies that $\mu(a)b$ is known for all a and b, and hence by the absence of annihilators that $\mu(a)$ is known for all a. Similarly μ determines λ.

Thirdly, the homomorphism $a \mapsto u_a$ of A into $\mathscr{W}(A)$ (see 1.3) is now one-to-one. Thus A is isomorphic with the two-sided ideal $\{u_a : a \in A\}$ of $\mathscr{W}(A)$. If we regard this isomorphism as an identification, and write a for u_a $(a \in A)$, equations (1) take the form:

$$ua = \lambda(a), \qquad au = \mu(a) \qquad (a \in A; u = \langle \lambda, \mu \rangle \in \mathscr{W}(A)). \qquad (2)$$

1.5. Notation. If $a \in A$ and $u = \langle \lambda, \mu \rangle \in \mathscr{W}(A)$, it is convenient to write ua and au for $\lambda(a)$ and $\mu(a)$ respectively *whether or not A has no annihilators.*

Suppose that A has no annihilators. Then the associative law in $\mathscr{W}(A)$ gives

$$(ua)v = u(av) \qquad (a \in A; u, v \in \mathscr{W}(A)). \qquad (3)$$

This asserts that for any two multipliers $\langle \lambda, \mu \rangle$ and $\langle \lambda', \mu' \rangle$ of A the endomorphisms λ and μ' commute.

However, (3) fails in general if A has annihilators. Suppose for instance that A has trivial multiplication ($ab = 0$ for all a, b). Then every pair $\langle \lambda, \mu \rangle$ of linear endomorphisms of A is a multiplier, and counter-examples to (3) become trivial.

1.6. Proposition. *If A has a unit element 1, then $a \mapsto u_a$ is an isomorphism of A onto $\mathscr{W}(A)$.*

Proof. If A has a unit it obviously has no annihilators, so $a \mapsto u_a$ is one-to-one. Clearly u_1 is the unit element of $\mathscr{W}(A)$ (mentioned in 1.2); so by (1) $u = uu_1 = u_{\lambda(1)}$ whenever $u = \langle \lambda, \mu \rangle \in \mathscr{W}(A)$. Thus $a \mapsto u_a$ is onto $\mathscr{W}(A)$. ∎

Remark. Thus multiplier theory is non-trivial only for algebras which do not have a unit. Compare Example 1.8.

1.7. Proposition. *Suppose that A is Abelian and has no annihilators. Then $\lambda = \mu$ for all $\langle \lambda, \mu \rangle$ in $\mathscr{W}(A)$; and $\mathscr{W}(A)$ is Abelian.*

Proof. If $\langle \lambda, \mu \rangle \in \mathscr{W}(A)$ and $a, b \in A$, then $\lambda(a)b = \lambda(ab) = \lambda(ba) = \lambda(b)a = a\lambda(b) = \mu(a)b$; so $\lambda(a) = \mu(a)$ by the absence of annihilators. This proves the first statement. Thus two elements of $\mathscr{W}(A)$ are of the form $\langle \lambda, \lambda \rangle$ and $\langle \mu, \mu \rangle$, and by what we have just proved their product $\langle \lambda\mu, \mu\lambda \rangle$ must satisfy $\lambda\mu = \mu\lambda$. This implies that $\mathscr{W}(A)$ is Abelian. ∎

1.8. Example. Let S be any infinite set, B the commutative algebra (under pointwise addition and multiplication) of all complex functions on S, and A the ideal of B consisting of those functions which vanish except at finitely many points. Evidently A has no annihilators. The reader will verify that the homomorphism $b \mapsto u_b$ of B into $\mathscr{W}(A)$ constructed in 1.3 is in this case an isomorphism onto $\mathscr{W}(A)$. Thus $\mathscr{W}(A)$ can be identified with B.

1.9.   Proposition.   *Let S be an (algebraically) non-degenerate (algebraic) representation of A. There exists a unique (algebraic) representation T of $\mathscr{W}(A)$ acting on the same space as S and satisfying*

$$S_a = T_{u_a} \qquad\qquad (a \in A).$$

Here $a \mapsto u_a$ is the homomorphism of A into $\mathscr{W}(A)$ constructed in 1.3.

Proof.   It is clear from the non-degeneracy of S that $u_a = 0 \Rightarrow S_a = 0$. Therefore $S_a = S'_{u_a}\ (a \in A)$, where S' is a representation of the two-sided ideal $A' = \{u_a : a \in A\}$ of $\mathscr{W}(A)$ (see 1.3). By IV.3.18 S' can be extended to a representation T of $\mathscr{W}(A)$. This proves the existence of the required T. Its uniqueness is almost obvious.   ∎

The above T will be called (somewhat inaccurately) the *extension of S to* $\mathscr{W}(A)$.

Remark.   As we shall observe in 1.16, the most obvious conjectured generalization of the above proposition to locally convex representations is false.

Multipliers of Banach Algebras

1.10. Definition. Suppose now that A is a normed algebra. A multiplier $u = \langle \lambda, \mu \rangle$ of A will be called *bounded* if the linear endomorphisms λ and μ of A are bounded. The set of all bounded elements of $\mathscr{W}(A)$ will be denoted by $\mathscr{W}_b(A)$. Evidently $\mathscr{W}_b(A)$ is a subalgebra of $\mathscr{W}(A)$; it is called the *bounded multiplier algebra* of A. In fact $\mathscr{W}_b(A)$ is itself a normed algebra under the norm

$$\|u\|_0 = \max\{\|\lambda\|, \|\mu\|\} \qquad (u = \langle \lambda, \mu \rangle \in \mathscr{W}_b(A)). \qquad (4)$$

Evidently the map $a \mapsto u_a$ of 1.3 carries A into $\mathscr{W}_b(A)$ and is norm-decreasing.

If A is a Banach algebra, then $\mathscr{W}_b(A)$ is complete with respect to the norm $\| \ \|_0$, hence itself a Banach algebra. To see this it is only necessary to observe that, if $\langle \lambda_n, \mu_n \rangle \in \mathscr{W}_b(A)$ for each $n = 1, 2, \dots$ and $\lambda_n \to \lambda$ and $\mu_n \to \mu$ in the operator norm, then $\langle \lambda, \mu \rangle \in \mathscr{W}_b(A)$.

1.11.   If A is a Banach algebra with an approximate unit, then all multipliers are automatically bounded.

Proposition. *Let A be a Banach algebra with an approximate unit. Then $\mathcal{W}_b(A) = \mathcal{W}(A)$.*

Proof. Let $\{e_i\}$ be an approximate unit of A, with $\|e_i\| \le k$ for all i. Fix a multiplier $u = \langle \lambda, \mu \rangle$ of A; for each b in A define F_b to be the bounded linear endomorphism $a \mapsto \mu(b)a$ of A; and put $\mathcal{F} = \{F_b : b \in A, \|b\| \le k\}$. For fixed a in A, we have $F_b(a) = b\lambda(a)$ (by 1.2(i)), hence $\|F_b(a)\| \le k\|\lambda(a)\|$ whenever $\|b\| \le k$. By the Uniform Boundedness Principle this implies that $\mathcal{F}$ is norm-bounded; so there is a constant m such that

$$\|F_b\| \le m \qquad \text{whenever } b \in A, \|b\| \le k. \tag{5}$$

Since $\|e_i\| \le k$, we can substitute e_i for b in (5), getting

$$\|e_i \lambda(a)\| = \|F_{e_i}(a)\| \le m\|a\| \qquad \text{for all } i \text{ and } a.$$

Passing to the limit in i, we obtain $\|\lambda(a)\| \le m\|a\|$ for all a, whence λ is bounded. Similarly μ is bounded. ∎

1.12. For normed algebras we can prove the following topological version of 1.9.

Proposition. *Let A be a normed algebra with an approximate unit $\{e_i\}$; and let S be a non-degenerate locally convex representation of A, acting on a complete LCS $X(S)$, such that $\{S_a : a \in A, \|a\| \le 1\}$ is equicontinuous. Then:*

(I)   *There is a unique locally convex representation T of $\mathcal{W}_b(A)$, acting on $X(S)$, such that*

$$S_a = T_{u_a} \qquad (a \in A).$$

(II) *The family $\{T_u : u \in \mathcal{W}_b(A), \|u\|_0 \le 1\}$ is equicontinuous.*

(III) *If $\{u_j\}$ is a bounded net of elements of $\mathcal{W}_b(A)$, $u \in \mathcal{W}_b(A)$, and $u_j a \to ua$ in A for all a in A, then $T_{u_j} \to T_u$ strongly.*

Proof. Let $k = \sup\{\|e_i\| : i \text{ varying}\}$. Notice that $a \mapsto u_a$ is not only one-to-one but bicontinuous; in fact,

$$\|u_a\|_0 \ge k^{-1}\|a\|. \tag{6}$$

Let S' be the locally convex representation $u_a \mapsto S_a$ of the two-sided ideal $A' = \{u_a : a \in A\}$ of $\mathcal{W}_b(A)$. It follows from (6) and the hypothesis on S that $\{S'_u : u \in A', \|u\|_0 \le 1\}$ is equicontinuous. Therefore Theorem V.2.4 applied to S' gives all the required conclusions. ∎

*Multipliers of *-Algebras*

1.13. Now assume that A is a *-algebra. We introduce into $\mathscr{W}(A)$ the involution $u \mapsto u^*$ defined (for $u = \langle \lambda, \mu \rangle$) by:

$$u^* = \langle \lambda^*, \mu^* \rangle,$$

where

$$\lambda^*(a) = (\mu(a^*))^*, \qquad \mu^*(a) = (\lambda(a^*))^* \qquad\qquad (a \in A).$$

One verifies that $u \in \mathscr{W}(A) \Rightarrow u^* \in \mathscr{W}(A)$, and that $\mathscr{W}(A)$ becomes a *-algebra when equipped with this involution. In this context we refer to $\mathscr{W}(A)$ as the *multiplier *-algebra* of A.

If A is a (two-sided) *-ideal of a larger *-algebra B, the map $b \mapsto u_b$ of 1.3 is a *-homomorphism of B into $\mathscr{W}(A)$. Thus, as in 1.4, if A has no annihilators A can be identified with a *-ideal of $\mathscr{W}(A)$; and, if A has a unit element, A and $\mathscr{W}(A)$ coincide as *-algebras (by 1.6).

1.14. Let A be a normed *-algebra. Then $\mathscr{W}_b(A)$ is clearly a *-subalgebra of $\mathscr{W}(A)$. In fact, with the norm $\| \ \|_0$ defined in (4), $\mathscr{W}_b(A)$ is a normed *-algebra. If A is a Banach *-algebra, then by 1.10 so is $\mathscr{W}_b(A)$.

1.15. For *-representations of Banach *-algebras one obtains a topological analogue of 1.9 even without the assumption of an approximate unit which was necessary in 1.12.

Proposition. *Let A be a Banach *-algebra, and S a non-degenerate *-representation of A. Then there is a unique *-representation T of the Banach *-algebra $\mathscr{W}_b(A)$ such that*

$$S_a = T_{u_a} \qquad\qquad (a \in A).$$

Proof. We have seen in 1.14 that $\mathscr{W}_b(A)$ is a Banach *-algebra. As in 1.9, S induces a *-representation $S': u_a \mapsto S_a$ of the two-sided ideal $A' = \{u_a : a \in A\}$ of $\mathscr{W}_b(A)$. Applying VI.19.11 to S' we obtain the required T. ∎

We call this T, somewhat loosely, the *extension of S to $\mathscr{W}_b(A)$*.

Remark. Here too, as in 1.12, it is easy to show that if $\{u_j\}$ is a bounded net of elements of $\mathscr{W}_b(A)$, $u \in \mathscr{W}_b(A)$, and $u_j a \to ua$ for all a in A, then $T_{u_j} \to T_u$ strongly.

1.16. Remark. Proposition 1.15 fails if A is simply a *-algebra and $\mathscr{W}_b(A)$ is replaced by $\mathscr{W}(A)$. Indeed, consider the A and B of 1.8 as *-algebras, with complex conjugation as the involution. Let X be the Hilbert space of all complex functions f on S which are square-summable ($\sum_{s \in S} |f(s)|^2 < \infty$). The non-degenerate *-representation V of A on X defined by multiplication of functions $((V_a f)(s) = a(s)f(s))$ clearly cannot be extended to a *-representation T of B. For, if b is an unbounded function in B, T_b would have to coincide with multiplication by b; and the latter is not a bounded operator on X.

1.17. Remark. Let A be any *-algebra; and let Z be the *-algebra of *central multipliers* of A, i.e., $Z = \{u \in \mathcal{W}(A) : ua = au \text{ for all } a \text{ in } A\}$. If T is an *irreducible* *-representation of A, it follows from VII.4.8 that T can be "extended" to Z; that is, there exists a *-homomorphism $\phi : Z \to \mathbb{C}$ such that

$$T_{ua} = \phi(u)T_a \qquad \text{for all } u \text{ in } Z, a \in A.$$

Multiplier C*-Algebras

1.18. Proposition. *Suppose that A is a C*-algebra. Then $\mathcal{W}(A)$, with the norm $\| \ \|_0$ of 1.10(4), is also a C*-algebra; and the map $a \mapsto u_a$ of 1.3 is an isometric *-isomorphism of A onto a closed two-sided ideal of $\mathcal{W}(A)$.*

Proof. By 1.11 and VI.8.4 $\mathcal{W}_b(A) = \mathcal{W}(A)$; so $\mathcal{W}(A)$ is a Banach *-algebra by 1.14. In view of VI.3.2, to prove the first statement we have only to show that

$$\|u\|_0^2 \leq \|u^*u\|_0 \qquad (u \in \mathcal{W}(A)). \qquad (7)$$

To do this, we first notice that, for any bounded linear endomorphism F of A,

$$\|F\| = \sup\{\|F(a)b\| : a, b \in A, \|a\| \leq 1, \|b\| \leq 1\}$$

$$= \sup\{\|bF(a)\| : a, b \in A, \|a\| \leq 1, \|b\| \leq 1\}. \qquad (8)$$

Indeed: Clearly $\|F\|$ majorizes the two suprema in (8). Now, given $\varepsilon > 0$, choose a so that $\|a\| = 1$ and $\|F(a)\| > \|F\| - \varepsilon$; and put $b = \|F(a)\|^{-1}F(a)^*$. Then $\|b\| = 1$ and $\|F(a)b\| = \|F(a)\| > \|F\| - \varepsilon$. Thus the first supremum in (8) equals $\|F\|$. Similarly the second supremum equals $\|F\|$. So (8) is proved.

If $u = \langle \lambda, \mu \rangle \in \mathcal{W}(A)$, it follows from (8) that

$$\|\mu\| = \sup\{\|\mu(a)b\| : \|a\|, \|b\| \leq 1\}$$

$$= \{\|a\lambda(b)\| : \|a\|, \|b\| \leq 1\}$$

$$= \|\lambda\|,$$

whence

$$\|u\|_0 = \|\lambda\| = \|\mu\|. \qquad (9)$$

We now copy the proof of VI.3.10(7). Let $u = \langle \lambda, \mu \rangle \in \mathscr{W}(A)$; choose any number $0 < \gamma < 1$; and let a be an element of A with $\|a\| = 1$, $\gamma\|\lambda\| \leq \|\lambda(a)\|$. Recalling the notational convention of 1.5, we have

$$\gamma^2 \|\lambda\|^2 \leq \|\lambda(a)\|^2 = \|ua\|^2$$

$$= \|(ua)^*ua\| \qquad \text{(since } A \text{ is a } C^*\text{-algebra)}$$

$$= \|a^*(u^*ua)\|$$

$$\leq \|u^*u\|_0 \qquad \text{(since } \|a^*\| = \|a\| = 1).$$

Combining this with (9) and the arbitrariness of γ, we obtain (7).

The isometric property of $a \mapsto u_a$ follows as in the first step of the proof of VI.3.10. ∎

Definition. $\mathscr{W}(A)$ is called the *multiplier C^*-algebra* of the C^*-algebra A. We may of course think of A as identified with a closed two-sided *-ideal of $\mathscr{W}(A)$.

1.19.   Example.   Let S be a non-compact locally compact Hausdorff space, A the commutative C^*-algebra $\mathscr{C}_0(S)$, and B the commutative C^*-algebra of all bounded continuous complex functions on S (with the pointwise operations and supremum norm).

Proposition*. $\mathscr{W}(A) \cong B$. *More precisely, the map* $b \mapsto u_b$ *$(b \in B)$ is a *-isomorphism of B onto $\mathscr{W}(A)$.*

Sketch of Proof: Use 1.15 to associate to each u in $\mathscr{W}(A)$ a complex function β_u on S such that $(ua)(s) = \beta_u(s)a(s)$ for all a in A, s in S. It is easy to see that $\beta_u \in B$. Now observe that $u \mapsto \beta_u$ and $b \mapsto u_b$ are inverse to each other.

Remark.   In the above example $\hat{B}$ is the Stone-Čech compactification of $S \cong \hat{A}$. Thus the process of passing to the multiplier C^*-algebra of a C^*-algebra A can be regarded as a "non-commutative generalization" of Stone-Čech compactification.

1.20.   Proposition*.   *Let X be a Hilbert space, and A a closed *-subalgebra of $\mathscr{O}(X)$ acting non-degenerately on X. Put $B = \{b \in \mathscr{O}(X) : ba \in A$ and $ab \in A$ for all a in $A\}$. Then B is a closed *-subalgebra of $\mathscr{O}(X)$, A is a two-sided *-ideal of B, and $\mathscr{W}(A) \cong B$ (that is, the map $b \mapsto u_b$ is a *-isomorphism of B onto $\mathscr{W}(A)$).*

To show that every element of $\mathscr{W}(A)$ comes from an element of B, we use 1.15 to extend the identity representation of A to a *-representation of $\mathscr{W}(A)$ acting in X.

1.21.   Corollary*.   *If X is a Hilbert space, $\mathcal{W}(\mathcal{O}_c(X)) \cong \mathcal{O}(X)$.*

1.22.   *Remark*.   Let X, A, and B be as in 1.20. Using VI.24.2(i), one verifies easily that B is contained in the von Neumann algebra E generated by A. Corollary 1.21 provides us with an example in which $E = B$. In Example 1.19, if we take a regular Borel measure μ on S with closed support S, and identify A with the concrete C*-algebra of multiplication operators on $X = \mathcal{L}_2(\mu)$, then the von Neumann algebra E generated by A contains all functions in $\mathcal{L}_\infty(\mu)$, and so is strictly bigger than the multiplier algebra B.

1.23.   One of the most important results in the theory of the multipliers of C*-algebras is the so-called Dauns–Hofmann theorem.

Let A be a C*-algebra; and for any non-degenerate *-representation T of A let T' denote the extension of T to $\mathcal{W}(A)$ (see 1.15). Take an element u of $\mathcal{W}(A)$ which is *central*, that is, $ua = au$ for all a in A. Then T'_u is a scalar operator for each T in $\hat{A}$. Thus there is a bounded complex-valued function ϕ_u on $\hat{A}$ such that

$$T'_u = \phi_u(T)1_{X(T)} \quad \text{for all } T \text{ in } \hat{A};$$

and it follows from the continuity of the map $T \mapsto T'$ (see VII.4.2) that ϕ_u is continuous with respect to the regional topology of $\hat{A}$.

Conversely, one can prove:

Theorem* (Dauns–Hofmann).   *Let A be a C*-algebra, and f any bounded continuous complex-valued function on $\hat{A}$. Then there exists a (unique) central element u of $\mathcal{W}(A)$ such that f coincides with the ϕ_u derived as above from u.*

1.24.   Let A be a C*-algebra and u a central element of $\mathcal{W}(A)$; and keep the notation of 1.23. The following useful result extends the defining property of ϕ_u:

Proposition*.   *If T is any non-degenerate *-representation of A, and P is the spectral measure of T (see VII.9.12), then*

$$T'_u = \int_{\hat{A}} \phi_u \, dP \qquad \text{(spectral integral).}$$

The multipliers of $\mathcal{L}_1$ group algebras.

1.25. Let G be a locally compact group, and λ its left Haar measure. As usual, $\mathcal{M}_r(G)$ and $\mathcal{L}_1(\lambda)$ are the measure algebra and the $\mathcal{L}_1$ group algebra of G respectively (see III.10.15, III.11.9). In III.11.20 and III.11.22 we gave two characterizations of $\mathcal{L}_1(\lambda)$ as a subspace (in fact, a closed *-ideal) of $\mathcal{M}_r(G)$. The next theorem shows, conversely, that $\mathcal{M}_r(G)$ can be abstractly described as the multiplier algebra of $\mathcal{L}_1(\lambda)$.

Theorem (Wendel). $\mathcal{W}(\mathcal{L}_1(\lambda))$ *is isometrically *-isomorphic with* $\mathcal{M}_r(G)$. *More precisely, the map* $\mu \mapsto u_\mu$ $(\mu \in \mathcal{M}_r(G))$ *is an isometric *-isomorphism of* $\mathcal{M}_r(G)$ *onto* $\mathcal{W}(\mathcal{L}_1(\lambda))$.

Proof.   Let $\{\psi_i\}$ be an approximate unit on G (III.11.17). Since $\{\psi_i\}$ is an approximate unit of $\mathcal{L}_1(\lambda)$ (III.11.19), it follows from 1.11 that $\mathcal{W}(\mathcal{L}_1(\lambda)) = \mathcal{W}_b(\mathcal{L}_1(\lambda))$, and so is a Banach *-algebra under the norm $\| \; \|_0$ of 1.10(4).

By III.11.15 the *-homomorphism $\mu \mapsto u_\mu$ is one-to-one from $\mathcal{M}_r(G)$ to $\mathcal{W}(\mathcal{L}_1(\lambda))$. We shall complete the proof by showing that each u in $\mathcal{W}(\mathcal{L}_1(\lambda))$ is of the form u_μ, where $\mu \in \mathcal{M}_r(G)$ and $\|\mu\| = \|u\|_0$.

Take an element u of $\mathcal{W}(\mathcal{L}_1(\lambda))$. By the definition of a multiplier,

$$(uf) * g = u(f * g) \qquad (f, g \in \mathcal{L}_1(\lambda)). \quad (10)$$

Put $g_i = u\psi_i$. Since u is a bounded multiplier, the $\{g_i\}$ form a bounded net in $\mathcal{L}_1(\lambda)$, hence in $\mathcal{M}_r(G)$. Now by II.8.12 $\mathcal{M}_r(G)$ can be identified (as a Banach space) with the Banach space adjoint of $\mathscr{C}_0(G)$; and it is well known (see for example Day [1], p. 19, Lemma 3) that the unit ball of the adjoint of a Banach space is compact in the pointwise convergence topology. Therefore we can replace $\{g_i\}$ by a subnet, and assume that, for some μ in $\mathcal{M}_r(G)$,

$$\int_G g_i(x)\phi(x)d\lambda x \to \int_G \phi(x)d\mu x \quad (11)$$

for all ϕ in $\mathscr{C}_0(G)$. We claim that for this μ we have

$$uf = \mu * f \qquad \text{for all } f \text{ in } \mathcal{L}_1(\lambda). \quad (12)$$

Indeed: Since $\mathcal{L}(G)$ is dense in $\mathcal{L}_1(\lambda)$ it is enough to prove (12) for $f \in \mathcal{L}(G)$. Let $f, h \in \mathcal{L}(G)$; and define $q(y) = \int_G f(y^{-1}x)h(x)d\lambda x$ $(y \in G)$, so that $q \in \mathcal{L}(G)$. Since $\psi_i * f \to f$ in $\mathcal{L}_1(\lambda)$, we have by (10)

$$uf = \lim_i u(\psi_i * f) = \lim_i g_i * f \qquad \text{in } \mathcal{L}_1(\lambda).$$

Therefore

$$\int (uf)(x)h(x)d\lambda x = \lim_i \int (g_i * f)(x)h(x)d\lambda x$$

$$= \lim_i \int_G g_i(y)q(y)d\lambda y \qquad \text{(by Fubini's Theorem)}$$

$$= \int_G q(y)d\mu y \qquad \text{(by (11))}$$

$$= \iint f(y^{-1}x)h(x)d\lambda x\, d\mu y$$

$$= \int (\mu * f)(x)h(x)d\lambda x \qquad (13)$$

(by III.11.8 and Fubini's Theorem).

From (13) and the arbitrariness of h we conclude that (12) holds for all f in $\mathscr{L}(G)$, and hence for all f in $\mathscr{L}_1(\lambda)$.

By (12), the left actions of u and u_μ coincide on $\mathscr{L}_1(\lambda)$. So by 1.4 their right actions also coincide; and $u = u_\mu$.

It remains only to show that $\|u\|_0 = \|\mu\|$ (the $\mathscr{M}_r(G)$-norm). Since $\|\psi_i\| = 1$, we have $\|g_i\| \le \|u\|_0$ for all i; and hence $\|\mu\| \le \lim \inf_i \|g_i\| \le \|u\|_0$. On the other hand it is obvious that $\|u_\mu\|_0 \le \|\mu\|$. Therefore $\|u\|_0 = \|\mu\|$, and the proof is complete. ∎

2. Banach Algebraic Bundles

2.1. Very roughly speaking, a Banach algebraic bundle is a Banach algebra B whose multiplication is "covariant" with a given "base group" G. It is easy to make this statement meaningful if G is discrete. Let E be a Banach algebra and G a (discrete) group; and suppose that for each x in G we are given a closed linear subspace B_x of E with the following properties: (i) $\{B_x : x \in G\}$ is linearly independent, and its linear span is dense in E; (ii) $B_x B_y \subset B_{xy}$ for all x, y in G (where of course $B_x B_y = \{ab : a \in B_x, b \in B_y\}$ denotes the product in E, while xy is the product in G). Condition (ii) expresses the "covariance" of the product in E with the product in G. The specification of the B_x ($x \in G$) is what we shall mean (roughly) by a Banach algebraic bundle over G.

Putting $x = y = e$ (the unit of G) in (ii), we see that B_e is a subalgebra of E. But in general the subspaces B_x are not subalgebras of E.

As a simple example, take a finite group H with a normal subgroup N; and put $G = H/N$. Let E be the group algebra $\mathscr{L}(H)$ of H (under convolution $*$); and for each coset $x = hN$ in G let B_x be the linear subspace of E consisting of those functions b which vanish outside x. One verifies that $B_x * B_y \subset B_{xy}$ ($x, y \in G$). So conditions (i) and (ii) are satisfied; and $\mathscr{L}(H)$ has been given the structure of a Banach algebraic bundle over G.

However, if G is a non-discrete topological group, the concept of a Banach algebraic bundle $\mathscr{B}$ over G is not quite so simply related to the concept of a Banach algebra. The passage from the spaces B_x ($x \in G$) to a Banach algebra E will then consist not in taking a direct sum $\sum_x^{\oplus} B_x$, but a "direct integral" (as in the formation of the cross-sectional spaces $\mathscr{L}_p(\mu; \mathscr{B})$ in II.15.7); and the B_x are not subspaces of E any longer. To axiomatize this situation, it is better to disregard the Banach algebra E to begin with, and to concentrate on the spaces B_x. We shall assume that the spaces B_x ($x \in G$) are given to us at the outset as the fibers of a Banach bundle $\mathscr{B}$ (in the sense of II.13.4) over a topological group G. The algebraic structure will be provided by an assumed binary operation $\cdot$ on the bundle space of $\mathscr{B}$, satisfying $B_x \cdot B_y \subset B_{xy}$ for all x, y in G (see the earlier condition (ii)), and also satisfying algebraic properties which *would* hold *if* the B_x were subspaces of a Banach algebra. A structure with these ingredients is a Banach algebraic bundle. If G is locally compact, we shall then find (5.2) that the $\mathscr{L}_1$ cross-sectional space of $\mathscr{B}$ (with respect to Haar measure) becomes a Banach algebra under the natural convolution derived from the bundle product. This Banach algebra can be tentatively regarded as the non-discrete generalization of the E with which we started the discussion of the discrete case. (For a more precise discussion of this point in the involutive context, see §17.)

As in Chapters VI and VII, we are mostly concerned with structures which have an involution as well as a product. In the bundle context, these will be called Banach *-algebraic bundles. They will be defined and discussed in the sections which follow.

2.2.   For the rest of this section we fix a topological group G, with unit e.

Definition.   A *Banach algebraic bundle over* G is a Banach bundle $\mathscr{B} = \langle B, \pi \rangle$ over G (see II.13.4), together with a binary operation $\cdot$ on B satisfying:

(i) $\pi(b \cdot c) = \pi(b)\pi(c)$ for $b, c \in B$.   (Equivalently, $B_x \cdot B_y \subset B_{xy}$ for $x, y \in G$).

(ii)   For each pair of elements x, y of G, the product $\cdot$ is bilinear on $B_x \times B_y$ to B_{xy}.

(iii) The product $\cdot$ on B is associative.

(iv) $\|b \cdot c\| \le \|b\|\|c\|$ $(b, c \in B)$.

(v) The map $\cdot$ is continuous on $B \times B$ to B.

(We are adopting here the same notation as in II.13.1, II.13.4. In particular B_x is the Banach space fiber $\pi^{-1}(x)$ over x; and 0_x is the zero element of B_x.)

Usually we shall omit the product symbol $\cdot$, writing simply bc instead of $b \cdot c$.

Remark 1.   Notice that $b0_x = 0_{\pi(b)x}$, $0_x b = 0_{x\pi(b)}$ $(x \in G; b \in B)$.

Remark 2.   Condition (i) says that the multiplication operations in B and G are "covariant" under π.

Remark 3.   The fiber B_e over e is closed under $\cdot$ in virtue of (i), and so by the other postulates (ii)–(iv) is a Banach algebra. It is called the *unit fiber algebra* of $\mathscr{B}$. The other fibers B_x $(x \ne e)$ are not closed under $\cdot$

Remark 4.   Any Banach algebra can be trivially regarded as a Banach algebraic bundle over the one-element group. Thus, any definition or theorem for Banach algebraic bundles applies in particular to Banach algebras.

Remark 5.   For Banach algebras, postulate (iv) implies (v). However, this is not so for Banach algebraic bundles.

Remark 6.   If $bc = cb$ for all b, c in B, $\mathscr{B}$ is *Abelian*. This implies of course that G is Abelian.

2.3.   For the rest of this section we fix a Banach algebraic bundle $\mathscr{B} = \langle B, \pi, \cdot \rangle$ over G.

2.4.   Suppose that Γ is a fixed family of continuous cross-sections of $\mathscr{B}$ such that $\{\gamma(x): \gamma \in \Gamma\}$ is dense in B_x for every x in G. Then we claim that postulate 2.2(v) can be replaced (without changing Definition 2.2) by:

(v′) For each pair of elements β, γ of Γ, the map $\langle x, y \rangle \mapsto \beta(x)\gamma(y)$ is continuous on $G \times G$ to B.

Indeed: Evidently (v) $\Rightarrow$ (v'). To prove the converse, assume (i) − (iv) and (v'); and let $b_i \to b$ and $c_i \to c$ in B. We must show that $b_i c_i \to bc$ in B. To do this, pick $\varepsilon > 0$, and choose β, γ in such a way that

$$\|\beta(\pi(b)) - b\| < \varepsilon(4\|c\|)^{-1}$$

$$\|\gamma(\pi(c)) - c\| < \varepsilon(4\|\beta(\pi(b))\|)^{-1}. \tag{1}$$

By (v')

$$\beta(\pi(b_i))\gamma(\pi(c_i)) \to \beta(\pi(b))\gamma(\pi(c)). \tag{2}$$

By (i) and II.13.15 there is a continuous cross-section α of $\mathscr{B}$ such that

$$\alpha(\pi(bc)) = \beta(\pi(b))\gamma(\pi(c)).$$

From this and (2) we obtain:

$$\|\beta(\pi(b_i))\gamma(\pi(c_i)) - \alpha(\pi(b_i c_i))\| \to 0. \tag{3}$$

Now (1) implies that, for large i,

$$\|\beta(\pi(b_i)) - b_i\| < \varepsilon(4\|c_i\|)^{-1},$$

$$\|\gamma(\pi(c_i)) - c_i\| < \varepsilon(4\|\beta(\pi(b_i))\|)^{-1}. \tag{4}$$

So, for all large enough i,

$$\begin{aligned}
\|b_i c_i - \alpha(\pi(b_i c_i))\| &\leq \|b_i c_i - \beta(\pi(b_i))\gamma(\pi(c_i))\| \\
&\quad + \|\beta(\pi(b_i))\gamma(\pi(c_i)) - \alpha(\pi(b_i c_i))\| \\
&\leq \|b_i - \beta(\pi(b_i))\| \, \|c_i\| \\
&\quad + \|\beta(\pi(b_i))\| \, \|c_i - \gamma(\pi(c_i))\| \\
&\quad + \|\beta(\pi(b_i))\gamma(\pi(c_i)) - \alpha(\pi(b_i c_i))\| \quad \text{(by 2.2(iv))} \\
&< \tfrac{1}{4}\varepsilon + \tfrac{1}{4}\varepsilon + \tfrac{1}{2}\varepsilon = \varepsilon \quad \text{(by (3) and (4)).}
\end{aligned} \tag{5}$$

By a similar calculation,

$$\|bc - \alpha(\pi(bc))\| < \varepsilon. \tag{6}$$

By (5) and (6), the continuity of α, and the arbitrariness of ε, it follows from II.13.12 that $b_i c_i \to bc$. This establishes the claim.

   In particular cases it is sometimes much easier to check (v') than the original postulate (v).

2.5.   If H is a topological subgroup of G, the reduction of $\mathscr{B}$ to H (see II.13.3) is closed under · by 2.2(i), and so is a Banach algebraic bundle over H. We denote it simply by $\mathscr{B}_H$.

2.6.   *Example*.   Assume that G is discrete; let C be any Banach algebra; and suppose that for each x in G we are given a closed linear subspace C_x of C satisfying $C_x C_y \subset C_{xy}$ for all x, y in G. The space $B = \{\langle x, a\rangle : x \in G, a \in C_x\}$, together with the projection $\pi: \langle x, a\rangle \mapsto x$ and the product

$$\langle x, a\rangle\langle y, b\rangle = \langle xy, ab\rangle \qquad (\langle x, a\rangle, \langle y, b\rangle \in B),$$

then forms an obvious Banach algebraic bundle $\mathscr{B} = \langle B, \pi, \cdot\rangle$ over G.

This example is a slight generalization of the motivating discussion in 2.1 (the condition 2.1(i) being omitted here as unnecessary).

We shall see in 5.5 that every Banach algebraic bundle over a discrete group arises in this way.

2.7.   *Example*.   Take any Banach algebra A; and let $\mathscr{B} = A \times G$ be the trivial Banach bundle whose constant fiber is the Banach space underlying A (see II.13.6). With the multiplication

$$\langle a, x\rangle\langle b, y\rangle = \langle ab, xy\rangle \qquad (x, y \in G; a, b \in A),$$

$\mathscr{B}$ becomes a Banach algebraic bundle over G, called the *trivial Banach algebraic bundle over G with fiber algebra A*.

A very important special case arises when A is just the one-dimensional Banach algebra $\mathbb{C}$. In that case $\mathscr{B}$ is called simply the *group bundle of G*.

We mentioned in the introduction to this volume that theorems on Banach algebraic bundles are generalizations of corresponding theorems on topological groups. It is the group bundle through which the specialization of theorems from Banach algebraic bundles to topological groups takes place. Throughout the rest of this work we shall see repeatedly that a theorem about Banach algebraic bundles over a topological group G, when specialized to the group bundle of G, becomes an interesting theorem about G itself.

Saturated Banach Algebraic Bundles

2.8.   We return to a general Banach algebraic bundle $\mathscr{B}$ over G. The following property of saturation will turn out to be very important in Chapter XII.

Definition. $\mathscr{B}$ is *saturated* if, for every pair of elements x, y of G, the linear span in B_{xy} of $B_x B_y$ $(= \{bc : b \in B_x, c \in B_y\})$ is dense in B_{xy}.

Denoting linear spans by [ ], we note that $\mathscr{B}$ is saturated if and only if, for every x in G, (i) $[B_x B_{x^{-1}}]$ is dense in B_e and (ii) $[B_e B_x]$ is dense in B_x. Indeed: To see that (i) and (ii) imply saturation, notice that $[B_x B_y] \supset$

$[B_x[B_{x^{-1}}B_{xy}]] = [[B_xB_{x^{-1}}]B_{xy}]$. By (i) and the continuity of the product in $\mathscr{B}$, the last expression is dense in $[B_eB_{xy}]$; and by (ii) this is dense in B_{xy}.

2.9.   By a *unit element* of $\mathscr{B}$ we mean of course an element 1 of B such that $1b = b1 = b$ for all b in B. The unit element, if it exists, is unique and belongs to the unit fiber algebra B_e.

Suppose that $\mathscr{B}$ has a unit element 1. An element b of B has an *inverse* $c = b^{-1}$ in $\mathscr{B}$ if $cb = bc = 1$. As in the theory of groups, b^{-1}, if it exists, is unique. Clearly $\pi(b^{-1}) = (\pi(b))^{-1}$.

Suppose that $\mathscr{B}$ has a unit and satisfies the following property: For every x in G, there is an element b of B_x which has an inverse. Then, by the equivalent condition in 2.8, $\mathscr{B}$ is saturated. However, the converse is false. It is possible for $\mathscr{B}$ to have a unit and to be saturated even if there are no invertible elements outside of B_e (see 3.15).

Approximate Units

2.10.   Even if $\mathscr{B}$ has no unit, a weaker object called an approximate unit will often serve the purpose of a unit.

Definition.   An *approximate unit of* the Banach algebraic bundle $\mathscr{B}$ is a net $\{u_i\}$ of elements of the unit fiber algebra B_e such that (i) there is a constant k such that $\|u_i\| \leq k$ for all i, and (ii) $\|u_ib - b\| \to 0$ and $\|bu_i - b\| \to 0$ for all b in B.

For Banach algebras (considered as Banach algebraic bundles over $\{e\}$; see 2.2, Remark 4) this definition of course coincides with that of VI.8.1. In general, an approximate unit of $\mathscr{B}$ is an approximate unit of the reduction of $\mathscr{B}$ to any subgroup H of G (see 2.5), in particular of B_e.

If $\mathscr{B}$ has an approximate unit, part (ii) of the equivalent condition for saturation in 2.8 can of course be omitted.

2.11.   Sometimes a notion formally stronger than that of 2.10 is necessary.

Definition.   A *strong approximate unit of* $\mathscr{B}$ is an approximate unit $\{u_i\}$ such that $\|u_ib - b\| \to 0$ and $\|bu_i - b\| \to 0$ uniformly in b on each compact subset of the bundle space B of $\mathscr{B}$.

Notice that for a Banach algebra the notions of approximate unit and strong approximate unit are the same. This fact is a very special case of 2.13, and is easily verified directly. For general Banach algebraic bundles the two notions would appear to be different, though we know of no example demonstrating this.

2.12. Proposition. *Suppose that G is locally compact. Let $\{u_i\}$ be a net of elements of B_e such that $\{\|u_i\|: i \text{ varying}\}$ is bounded. Let Γ be a family of continuous cross-sections of $\mathscr{B}$ such that $\{f(x): f \in \Gamma\}$ is dense in B_x for each x in G. Then the following three conditions are equivalent:*

(i) $\{u_i\}$ *is a strong approximate unit of $\mathscr{B}$.*

(ii) *If $\{u_{i_j}\}$ ($j \in J$) is a subnet of $\{u_i\}$ and $\{b_j\}$ ($j \in J$) converges in B to b, then $u_{i_j} b_j \to b$ and $b_j u_{i_j} \to b$.*

(iii) *For every f in Γ, $\|u_i f(x) - f(x)\| \to 0$ and $\|f(x)u_i - f(x)\| \to 0$ uniformly in x on each compact subset of G.*

Proof. First we show that (ii) $\Rightarrow$ (i). Assume that (i) is false. Then there is a compact subset K of B, a positive number ε, a subnet $\{u_{i_j}\}$ of $\{u_i\}$, and a net $\{b_j\}$ of elements of K (with the same index set $\{j\}$), such that

$$\text{either} \qquad \|u_{i_j} b_j - b_j\| \geq \varepsilon \qquad \text{for all } j \tag{7}$$

$$\text{or} \qquad \|b_j u_{i_j} - b_j\| \geq \varepsilon \qquad \text{for all } j. \tag{8}$$

Passing again to a subnet, we can suppose by the compactness of K that $b_j \to b$ in K. So, if $u_{i_j} b_j \to b$, we would have $\|u_{i_j} b_j - b_j\| \to 0$, contradicting (7). Likewise, if $b_j u_{i_j} \to b$, alternative (8) fails. Hence either $u_{i_j} b_j \not\to b$ or $b_j u_{i_j} \not\to b$, showing that (ii) is false. Therefore (ii) $\Rightarrow$ (i).

Next we shall assume (iii) and prove (ii). For this it is enough to show that $u_i b_i \to b$ and $b_i u_i \to b$ when the net $\{b_i\}$ converging to b is indexed by the same directed set as $\{u_i\}$. Set $x = \pi(b)$, $k = \sup_i \|u_i\|$. Given $\varepsilon > 0$, let us choose f in Γ such that

$$\|f(x) - b\| < \tfrac{1}{2}\varepsilon k^{-1}. \tag{9}$$

This implies that

$$\|f(\pi(b_i)) - b_i\| < \tfrac{1}{2}\varepsilon k^{-1} \qquad \text{for all large } i. \tag{10}$$

Since G is locally compact, we can apply (iii) to some compact neighborhood of x, and conclude that

$$\|u_i f(\pi(b_i)) - f(\pi(b_i))\| < \tfrac{1}{2}\varepsilon \tag{11}$$

for all large enough i. Since

$$\|u_i b_i - f(\pi(b_i))\| \leq \|u_i\|\|b_i - f(\pi(b_i))\| + \|u_i f(\pi(b_i)) - f(\pi(b_i))\|,$$

we conclude from (10) and (11) that $\|u_i b_i - f(\pi(b_i))\| < \varepsilon$ for all large enough i. Combining this with (9) and II.13.12 we get $u_i b_i \to b$. Similarly $b_i u_i \to b$. Thus (iii) $\Rightarrow$ (ii).

Since the implication (i) $\Rightarrow$ (iii) is obvious, the proof is complete.   ■

2.13.   Corollary.   *Suppose that G is locally compact, and that for every x in G the linear spans of $B_e B_x$ and $B_x B_e$ are both dense in B_x. (This last condition will certainly hold if $\mathscr{B}$ is saturated.) Then any approximate unit of the Banach algebra B_e (in particular, any approximate unit of $\mathscr{B}$) is a strong approximate unit of $\mathscr{B}$.*

Proof.   Our hypotheses imply that for each x in G

$$B_e B_x B_e \text{ has dense linear span in } B_x. \tag{12}$$

Now let Γ be the linear span of the family of all continuous cross-sections of $\mathscr{B}$ of the form $x \mapsto af(x)b$ $(x \in G)$, where $a, b \in B_e$ and f is a continuous cross-section of $\mathscr{B}$. By Appendix C $\mathscr{B}$ has enough continuous cross-sections, and hence by (12) Γ satisfies the hypothesis of 2.12. Let $\{u_i\}$ be an approximate unit of B_e. The norm-boundedness of $\{u_i\}$ clearly implies that condition 2.12(iii) holds with the above Γ. So by 2.12 $\{u_i\}$ is a strong approximate unit of $\mathscr{B}$.   ■

Multipliers of Banach Algebraic Bundles

2.14.   The notion of a multiplier is often useful in connection with Banach algebraic bundles, just as it is for algebras.

Consider a mapping $\lambda: B \to B$ and an element x of G. We say that λ is *of left order x* [*of right order x*] if $\lambda(B_y) \subset B_{xy}$ $[\lambda(B_y) \subset B_{yx}]$ for all y in G. If λ is of left (or right) order x, it is called *quasi-linear* if, for each y in G, $\lambda|B_y$ is linear on B_y to B_{xy} (or B_{yx}). Also, λ is *bounded* if for some non-negative constant k we have $\|\lambda(b)\| \le k\|b\|$ for all b in B; the smallest such k is then called $\|\lambda\|$.

We now make the following definition analogous to 1.2.

Definition.   A *multiplier of $\mathscr{B}$ of order x* is a pair $\langle \lambda, \mu \rangle$, where λ and μ are continuous bounded quasi-linear maps of B into B, λ is of left order x, μ is of right order x, and the identities

$$b\lambda(c) = \mu(b)c, \qquad \lambda(bc) = \lambda(b)c, \qquad \mu(bc) = b\mu(c) \tag{13}$$

hold for all b, c in B.

We call λ and μ the *left* and *right actions* of the multiplier $\langle \lambda, \mu \rangle$.

As a matter of notation, if $u = \langle \lambda, \mu \rangle$ is a multiplier we shall usually write ub and bu instead of $\lambda(b)$ and $\mu(b)$. Thus (13) have the form of associative laws:

$$b(uc) = (bu)c, \qquad u(bc) = (ub)c, \qquad (bc)u = b(cu) \tag{14}$$

Let $\mathscr{W}_x = \mathscr{W}_x(\mathscr{B})$ denote the collection of all multipliers of $\mathscr{B}$ of order x. Then $\mathscr{W}(\mathscr{B}) = \bigcup_{x \in G} \mathscr{W}_x(\mathscr{B})$ has much of the structure of a Banach algebraic bundle over G. Indeed, each $\mathscr{W}_x$ is an obvious linear space. Setting $\|u\|_0 = \max\{\|\lambda\|, \|\mu\|\}$ whenever $u = \langle \lambda, \mu \rangle \in \mathscr{W}(\mathscr{B})$, one verifies that each $\mathscr{W}_x$ is a Banach space under $\| \ \|_0$. If $u \in \mathscr{W}_x$ and $v \in \mathscr{W}_y$ $(x, y \in G)$, the product uv defined by

$$(uv)b = u(vb), \qquad b(uv) = (bu)v \qquad\qquad (b \in B)$$

belongs to $\mathscr{W}_{xy}$; and this operation satisfies postulates 2.2(i)–(iv) on $\mathscr{W}(\mathscr{B})$. Thus $\mathscr{W}(\mathscr{B})$ lacks only a suitable topology in order to become a Banach algebraic bundle over G. In general we do not know how to find such a topology. In spite of this, it is convenient to refer to $\mathscr{W}(\mathscr{B})$ as the *multiplier bundle of $\mathscr{B}$*.

The multiplier whose left and right actions are the identity map on B is the unit element of $\mathscr{W}(\mathscr{B})$.

Each element b of B gives rise to a multiplier u_b of $\mathscr{B}$ whose order is the same as that of b:

$$u_b c = bc, \qquad cu_b = cb \qquad\qquad (c \in B).$$

The map $b \mapsto u_b$ preserves addition (on each fiber) and multiplication; and $\|u_b\|_0 \le \|b\|$.

Now assume that $\mathscr{B}$ *has no annihilators*, that is, if $0_{\pi(b)} \ne b \in B$, there exist a and c in B satisfying $ab \ne 0_{\pi(ab)}$, $bc \ne 0_{\pi(bc)}$. Then the same simplifications take place as in 1.4: The last two identities (13) can be removed from the definition of a multiplier; a multiplier is determined by its left action, and also by its right action; and the map $b \mapsto u_b$ is one-to-one. If $\mathscr{B}$ has an approximate unit it of course has no annihilators. If $\mathscr{B}$ has a unit $\mathbf{1}$, then $b \mapsto u_b$ is not only one-to-one but onto $\mathscr{W}(\mathscr{B})$; and for any w in $\mathscr{W}(\mathscr{B})$ we have $w = u_b$, where $b = w\mathbf{1} = \mathbf{1}w$. The proofs of these facts are the same as the corresponding proofs in 1.4.

2.15. Proposition. *Suppose that $\mathscr{B}$ has an approximate unit $\{u_i\}$ and that, for every x in G, there is a multiplier w of $\mathscr{B}$ of order x having an inverse (that is, a multiplier w^{-1} of $\mathscr{B}$ of order x^{-1} such that $w^{-1}w = ww^{-1} = \mathbf{1}$). Then $\mathscr{B}$ is saturated.*

Proof. Given $x \in G$, let w be an invertible multiplier of order x. For each a in B_e we have

$$a = \lim_i au_i = \lim_i [(aw)(w^{-1}u_i)],$$

where $aw \in B_x$ and $w^{-1}u_i \in B_{x^{-1}}$. So $B_x B_{x^{-1}}$ is dense in B_e. By 2.8 and the last remark of 2.10 this proves that $\mathscr{B}$ is saturated. ■

3. Banach *-Algebraic Bundles

3.1. Roughly speaking, a Banach *-algebraic bundle is a Banach algebraic bundle equipped with an involution * which (a) behaves so far as possible like the involution in a Banach *-algebra, and (b) is "covariant" with the inverse operation of the base group.

Definition. A *Banach *-algebraic bundle over* the topological group G is a Banach algebraic bundle $\mathscr{B} = \langle B, \pi, \cdot \rangle$ over G, together with a unary operation * on B (the *involution* operation) satisfying:

(i) $\pi(b^*) = (\pi(b))^{-1}$ for $b \in B$. (Equivalently, $(B_x)^* \subset B_{x^{-1}}$ for $x \in G$).
(ii) For each x in G, * restricted to B_x is conjugate-linear from B_x to $B_{x^{-1}}$.
(iii) $(bc)^* = c^* b^*$ $\qquad\qquad\qquad\qquad\qquad\qquad$ $(b, c \in B)$.
(iv) $b^{**} = b$ $\qquad\qquad\qquad\qquad\qquad\qquad\qquad$ $(b \in B)$.
(v) $\|b^*\| = \|b\|$ $\qquad\qquad\qquad\qquad\qquad\qquad$ $(b \in B)$.
(vi) $b \mapsto b^*$ is continuous on B.

Remarks. Postulate (i) says that * is covariant under π with the inverse operation on the group.

We have $(0_x)^* = 0_{x^{-1}}$ for each x in G.

The unit fiber algebra B_e is closed under * by (i), and is a Banach *-algebra by the remaining postulates. More generally, for any topological subgroup H of G, the reduction of $\mathscr{B}$ to H is a Banach *-algebraic bundle over H.

Any Banach *-algebra can trivially be regarded as a Banach *-algebraic bundle over the one-element group.

3.2. Let Γ be as in 2.4. Then postulate (vi) of Definition 3.1 can be replaced (without altering the definition) by:

(vi') For each γ in Γ, the function $x \mapsto (\gamma(x))^*$ is continuous on G *to* B.

The proof of this is similar to (though considerably simpler than) that of 2.4. We leave it to the reader.

3.3. *Definition.* Let $\mathscr{B}$ and $\mathscr{B}'$ be Banach *-algebraic bundles over topological groups G and G' respectively; and let $F: G \to G'$ be a surjective isomorphism of topological groups. By an *isomorphism of $\mathscr{B}$ and $\mathscr{B}'$ covariant* *with F* we mean an isomorphism Φ, covariant with *F*, of the Banach bundles underlying $\mathscr{B}$ and $\mathscr{B}'$ (see II.13.8) which carries the product and involution of $\mathscr{B}$ into those of $\mathscr{B}'$. If in addition Φ preserves norm, it is called an *isometric isomorphism of $\mathscr{B}$ and $\mathscr{B}'$.*

Suppose $G = G'$. Then $\mathscr{B}$ and $\mathscr{B}'$ will be said to be *isometrically isomorphic as Banach *-algebraic bundles* if there is an isometric isomorphism of $\mathscr{B}$ and $\mathscr{B}'$ covariant with the identity map on *G*.

3.4. Proposition. *If $\{u_i\}$ is a strong approximate unit of a Banach *-algebraic bundle $\mathscr{B}$ (or, more precisely, of the Banach algebraic bundle underlying $\mathscr{B}$), then $\{u_i^*\}$ and $\{u_i^*u_i\}$ are also strong approximate units of $\mathscr{B}$.*

The proof is simple and is left to the reader.

3.5. Example. Let *G*, *C*, and $\{C_x\}$ $(x \in G)$ be as in 2.6; but now assume in addition that *C* is a Banach *-algebra and that $(C_x)^* \subset C_{x^{-1}}$ $(x \in G)$. If we construct $\langle B, \pi, \cdot \rangle$ as in 2.6, and introduce the involution

$$\langle x, a \rangle^* = \langle x^{-1}, a^* \rangle \qquad\qquad (\langle x, a \rangle \in B),$$

then $\langle B, \pi, \cdot, * \rangle$ is a Banach *-algebraic bundle over the discrete group *G* (indeed, the most general such object).

Multipliers and Unitary Multipliers of Banach *-Algebraic Bundles

3.6. From here to the end of this section we fix a Banach *-algebraic bundle $\mathscr{B} = \langle B, \pi, \cdot, * \rangle$ over the topological group *G*.

3.7. In 2.14 we defined the space $\mathscr{W}_x(\mathscr{B})$ of multipliers of $\mathscr{B}$ of order *x*. Since $\mathscr{B}$ now has an involution, we can define the involution u^* of a multiplier *u* of $\mathscr{B}$ as follows:

$$u^*b = (b^*u)^*, \qquad bu^* = (ub^*)^* \qquad\qquad (b \in B)$$

(compare 1.13). If $u \in \mathscr{W}_x(\mathscr{B})$, then $u^* \in \mathscr{W}_{x^{-1}}(\mathscr{B})$; and the operation $u \mapsto u^*$ satisfies postulates (i)–(v) of 3.1 on $\mathscr{W}(\mathscr{B})$. Clearly the map $b \mapsto u_b$ of 2.14 now preserves involution. So, if $\mathscr{B}$ has no annihilators (see 2.14), each B_x can be identified with a linear subspace of $\mathscr{W}_x(\mathscr{B})$, with preservation of the linear operations, the multiplication, and the involution. If $\mathscr{B}$ has a unit, $B_x = \mathscr{W}_x(\mathscr{B})$.

3.8. It is worth observing that, if $\mathscr{B}$ has an approximate unit, $\mathscr{W}_e(\mathscr{B})$ can be identified, as a *-algebra, with $\mathscr{W}(B_e)$. The key fact here is the following:

Proposition. *Suppose that $\mathscr{B}$ has an approximate unit. If u is a multiplier of the unit fiber algebra $A = B_e$ (automatically bounded by 1.11), then there is a unique extension of u to a multiplier of $\mathscr{B}$ of order e (e being the unit of G).*

Proof. Pick an approximate unit $\{e_i\}$ of $\mathscr{B}$, with $\|e_i\| \leq p$ (for all i); and for each i set $u_i = ue_i \in A$ (so that $\|u_i\| \leq p\|u\|$). We claim that for each $b \in B$ the nets $\{u_i b\}$ and $\{bu_i\}$ are convergent.

Indeed, let $b \in B_x$ ($x \in G$). Given $\varepsilon > 0$, we use the approximate unit to choose an element a of A such that

$$\|ab - b\| < \frac{\varepsilon}{3p\|u\|}, \qquad \|ba - b\| < \frac{\varepsilon}{3p\|u\|}. \tag{1}$$

Now $u_i a = (ue_i)a = u(e_i a) \to ua$ and $au_i = a(ue_i) = (au)e_i \to au$. So we can choose an index k big enough so that, whenever $i \succ k$ and $j \succ k$, we have

$$\|u_i a - u_j a\| < \frac{\varepsilon}{3\|b\|} \quad \text{and} \quad \|au_i - au_j\| < \frac{\varepsilon}{3\|b\|},$$

whence by (1)

$$\|u_i b - u_j b\| \leq \|u_i(ab - b)\| + \|(u_i a - u_j a)b\| + \|u_j(ab - b)\|$$

$$< \frac{\varepsilon}{3} + \frac{\varepsilon}{3} + \frac{\varepsilon}{3} = \varepsilon,$$

and similarly

$$\|bu_i - bu_j\| < \varepsilon.$$

Thus $\{u_i b\}$ and $\{bu_i\}$ are Cauchy nets in B_x, hence convergent; and the claim is proved.

So we can define

$$\lambda(b) = \lim_i u_i b, \ \mu(b) = \lim_i bu_i \qquad (b \in B). \tag{}$$

Clearly $\lambda: B \to B$ and $\mu: B \to B$ are of left and right order e, quasi-linear, and bounded (with $\|\lambda\| \leq p\|u\|$, $\|\mu\| \leq p\|u\|$). If b, $c \in B$, we have $\lambda(bc) = \lim_i u_i(bc) = \lim_i (u_i b)c = \lambda(b)c$; and similarly $\mu(bc) = b\mu(c)$ and $b\lambda(c) = \mu(b)c$. To show that λ is continuous, let $b_\alpha \to b$ in B, and take $\varepsilon > 0$. Choosing $a \in A$ with $\|a\| \leq p$ so that $\|ab - b\| < \varepsilon/p\|u\|$, we have by continuity $\|ab_\alpha - b_\alpha\| < \varepsilon/p\|u\|$ α-eventually. So

$$\|\lambda(a)b - \lambda(b)\| = \|\lambda(ab - b)\| \leq \|\lambda\| \cdot \|ab - b\| < \varepsilon \tag{2}$$

and similarly

$$\|\lambda(a)b_\alpha - \lambda(b_\alpha)\| < \varepsilon \qquad \alpha\text{-eventually.} \tag{3}$$

Now by the continuity of multiplication $\lambda(a)b_\alpha \to \lambda(a)b$. From this, (2), (3), and II.13.12 we obtain $\lambda(b_\alpha) \to \lambda(b)$. So λ is continuous. Likewise μ is continuous.

We have therefore shown that $\tilde{u} = \langle \lambda, \mu \rangle$ is a multiplier of $\mathscr{B}$. Evidently $\tilde{u}$ coincides with u on A.

If $\langle \lambda', \mu' \rangle$ were any other multiplier of $\mathscr{B}$ extending u, we would have for $b \in B$

$$\lambda'(b) = \lim_i \lambda'(e_i b) = \lim_i \lambda'(e_i)b = \lim_i \lambda(e_i)b = \lim_i u_i b = \lambda(b),$$

and

$$\mu'(b) = \lim_i \mu'(b)e_i = \lim_i b\lambda'(e_i) = \lim_i bu_i = \mu(b);$$

so $\tilde{u}$ is the unique extension of u to a multiplier of $\mathscr{B}$. ∎

Remark. The preceding proof makes no use of the involution in $\mathscr{B}$, and so is valid for any Banach algebraic bundle with an approximate unit.

Corollary. *Suppose that $\mathscr{B}$ has an approximate unit. For each $w \in \mathscr{W}_e(\mathscr{B})$ (e being the unit of G) let $\rho(w)$ be the multiplier of B_e obtained by restricting w to B_e. Then $\rho: \mathscr{W}_e(\mathscr{B}) \to \mathscr{W}(B_e)$ is a *-isomorphism (onto all of $\mathscr{W}(B_e)$).*

Proof. It is obvious that ρ is a *-homomorphism. The preceding proposition states that it is one-to-one and onto $\mathscr{W}(B_e)$. ∎

3.9. Let $\mathbf{1}$ be the unit element of $\mathscr{W}(\mathscr{B})$ (see 2.14). A multiplier u of $\mathscr{B}$ will be called *unitary* if $u^*u = uu^* = \mathbf{1}$ and $\|u\|_0 \leq 1$. We denote the set of all unitary multipliers of $\mathscr{B}$ by $\mathscr{U}(\mathscr{B})$. Thus $\mathscr{U}(\mathscr{B})$ is a group under the multiplication operation of $\mathscr{W}(\mathscr{B})$, the inverse in $\mathscr{U}(\mathscr{B})$ being involution. The map $\pi_0: \mathscr{U}(\mathscr{B}) \to G$ sending each unitary multiplier into its order is a group homomorphism.

Definition. We say that $\mathscr{B}$ has *enough unitary multipliers* if for every x in G there exists a unitary multiplier of $\mathscr{B}$ of order x—that is, if the group homomorphism π_0 of the last paragraph is *onto* G.

3.10. Let u be a unitary multiplier of $\mathscr{B}$ of order x. The left action of u is norm-decreasing and its inverse is the norm-decreasing left action of u^*. Therefore the left action of u is an isometry of B onto B, and carries each fiber

B_y linearly and isometrically onto B_{xy}. Likewise the right action of u carries each fiber B_y linearly and isometrically onto B_{yx}.

It follows that, if $\mathscr{B}$ has enough unitary multipliers, the fibers of $\mathscr{B}$ are all isometrically isomorphic with each other as Banach spaces.

3.11. If $\mathscr{B}$ has an approximate unit and enough unitary multipliers, then by 2.15 it is saturated.

3.12.   *Remark.*   If $\mathscr{B}$ has a unit element 1, then by 3.7 a unitary multiplier of $\mathscr{B}$ is simply an element u of B which is "unitary" in the sense that $u^*u = uu^* = 1$ and $\|ub\| \le \|b\|$ for all b.

3.13.   *Remark.*   The property of having enough unitary multipliers is somewhat weaker than the property of homogeneity which was central in Fell [14]. The definition of homogeneity in Fell [14], §6, included a condition on the so-called strong topology of $\mathscr{U}(\mathscr{B})$. The property of homogeneity will play no role in the present work.

Examples

3.14.   Throughout the rest of this chapter, beginning with §4, we shall find many examples of Banach *-algebraic bundles having enough unitary multipliers. Here are two examples which do not have enough of them.

3.15.   As a first simple example, let C be the two-dimensional Banach *-algebra given in VI.22.8. Thus the underlying linear space of C is $\mathbb{C}^2$, and

$$\langle r, s \rangle \langle r', s' \rangle = \langle rr', ss' \rangle,$$

$$\langle r, s \rangle^* = \langle \bar{s}, \bar{r} \rangle,$$

$$\|\langle r, s \rangle\| = \max\{|r|, |s|\}.$$

Put $B_1 = \{\langle r, r \rangle : r \in \mathbb{C}\}$, $B_{-1} = \{\langle r, -r \rangle : r \in \mathbb{C}\}$. Thus B_1 and B_{-1} are subspaces of C, giving rise by the construction of 3.5 to a Banach *-algebraic bundle $\mathscr{B}$ over the (multiplicative) two-element group $G = \{1, -1\}$, with fibers B_1 and B_{-1}. $\mathscr{B}$ has a unit, but there are no unitary elements in B_{-1}, hence not enough unitary multipliers. There are of course invertible elements of B_{-1}, and $\mathscr{B}$ is saturated.

3.16. For the second example, let D be the unit disk $\{z \in \mathbb{C}: |z| \leq 1\}$, and $\mathbb{E}$ as usual its boundary $\{z \in \mathbb{C}: |z| = 1\}$. Let C be the commutative C^*-algebra $\mathscr{C}(D)$ of all continuous complex functions on D, with the pointwise operations. Let G be the two-element group $\{1, -1\}$; and define the following closed linear subspaces B_1 and B_{-1} of C:

$$B_1 = \{f \in C: f(-z) = f(z) \text{ for all } z \text{ in } \mathbb{E}\},$$

$$B_{-1} = \{f \in C: f(-z) = -f(z) \text{ for all } z \text{ in } \mathbb{E}\}.$$

Evidently B_1 and B_{-1} satisfy the conditions of 3.5, and so, by the construction of 3.5, form the fibers of a Banach *-algebraic bundle $\mathscr{B}$ over G. $\mathscr{B}$ has a unit element; and the Stone-Weierstrass Theorem (in the form A7) shows that the linear span of $B_{-1}B_{-1}$ is dense in B_1. Therefore $\mathscr{B}$ is saturated.

However B_{-1} contains no invertible elements. To see this it is enough to show that every element of B_{-1} vanishes at some point of D. Let $f \in B_{-1}$, and assume that f never vanishes on D. Then, for each z in $\mathbb{E}$ we can denote by $\alpha(z)$ the angle subtended at 0 by the image under f of the straight line path from z to $-z$. Since $f(-z) = -f(z)$, $\alpha(z) = \pi + 2\pi n_z$ for some integer n_z. By the continuity of f, α is continuous on $\mathbb{E}$. But $\alpha(z)$ must reverse its sign as z passes continuously in $\mathbb{E}$ from, say, 1 to -1. The last three facts give a contradiction. So f vanishes somewhere on D.

Thus B_{-1} contains no invertible elements, and hence no unitary elements.

Retracts of Banach *-Algebraic Bundles

3.17. Let H be another topological group, and $F: H \to G$ a continuous homomorphism. From $\mathscr{B}$ and F one can construct a new Banach *-algebraic bundle $\mathscr{C}$ over H as follows: First, let $\mathscr{C}$ be the Banach bundle retraction of $\mathscr{B}$ by F, defined as in II.13.7. Next, equip $\mathscr{C}$ with the following multiplication and involution:

$$\langle y, b \rangle \langle y', b' \rangle = \langle yy', bb' \rangle, \qquad \langle y, b \rangle^* = \langle y^{-1}, b^* \rangle$$

$(y, y' \in H; b, b' \in B; \pi(b) = F(y), \pi(b') = F(y'))$. One verifies that $\mathscr{C}$ is indeed a Banach *-algebraic bundle over H; it is called the *Banach *-algebraic bundle retraction of $\mathscr{B}$ by F*.

Evidently $\mathscr{C}$ is saturated if $\mathscr{B}$ is. If $\mathscr{B}$ has an approximate unit [strong approximate unit], so does $\mathscr{C}$.

4.   Semidirect Product Bundles and Central Extension Bundles

4.1.   In this section we will introduce two important special classes of
Banach *-algebraic bundles. This will help to make the general concept of a
Banach *-algebraic bundle more meaningful.

Semidirect Product Bundles

4.2.   Fix a topological group G with unit e, a Banach *-algebra A, and a
homomorphism τ of G into the group of isometric *-automorphisms of A. We
assume that τ is strongly continuous, that is, the map $x \mapsto \tau_x(a)$ is continuous
on G to A for each a in A. From this and the isometric character of the τ_x it
follows that $\langle a, x \rangle \mapsto \tau_x(a)$ is (jointly) continuous.

Let $\langle B, \pi \rangle$ be the trivial Banach bundle over G whose constant fiber is the
Banach space underlying A; thus $B = A \times G$, $\pi(a, x) = x$ (see II.13.6). We
now define multiplication $\cdot$ and involution $*$ in B as follows:

$$\langle a, x \rangle \cdot \langle b, y \rangle = \langle a\tau_x(b), xy \rangle, \tag{1}$$

$$\langle a, x \rangle^* = \langle \tau_{x^{-1}}(a^*), x^{-1} \rangle \tag{2}$$

$(x, y \in G; a, b \in A)$. Using the assumed properties of τ one checks easily that
$\mathscr{B} = \langle B, \pi, \cdot, * \rangle$ is a Banach *-algebraic bundle over G. For example, the
isometric character of $\tau_{x^{-1}}$ gives $\|\langle a, x \rangle^*\| = \|\tau_{x^{-1}}(a^*)\| = \|a^*\| = \|a\| = \|\langle a, x \rangle\|$. Furthermore

$$
\begin{aligned}
(\langle a, x \rangle \cdot \langle b, y \rangle)^* &= \langle a\tau_x(b), xy \rangle^* \\
&= \langle \tau_{y^{-1}x^{-1}}(a\tau_x(b))^*, y^{-1}x^{-1} \rangle \\
&= \langle \tau_{y^{-1}x^{-1}}(\tau_x(b^*)a^*), y^{-1}x^{-1} \rangle \\
&= \langle \tau_{y^{-1}}(b^*)\tau_{y^{-1}x^{-1}}(a^*), y^{-1}x^{-1} \rangle \\
&= \langle \tau_{y^{-1}}(b^*), y^{-1} \rangle \cdot \langle \tau_{x^{-1}}(a^*), x^{-1} \rangle \\
&= \langle b, y \rangle^* \cdot \langle a, x \rangle^*,
\end{aligned}
$$

proving postulate 3.1(iii). The other postulates for a Banach *-algebraic
bundle are similarly verified.

Definition.   This $\mathscr{B}$ is called the *τ-semidirect product of A and G*. It may be
denoted by $A \underset{\tau}{\times} G$.

Remark.   We shall usually write simply $\langle a, x \rangle \langle b, y \rangle$ in (1), omitting explicit
mention of $\cdot$.

Remark. The unit fiber *-algebra of this $\mathscr{B}$ is clearly identifiable with A.

Remark. Notice the formula

$$\langle a, x \rangle^* \langle a, x \rangle = \langle \tau_{x^{-1}}(a^*a), e \rangle.$$

It follows that if A is a C^*-algebra the following two properties hold in $\mathscr{B}$: (i) $\|b^*b\| = \|b\|^2$ for all $b \in B$; (ii) for any $b \in \mathscr{B}$, the element b^*b is positive in B_e ($\cong A$).

Remark. Notice the strong likeness between the above construction and that of the semidirect products of groups. Formulae (1) and (2) have a strong formal resemblance to the definitions of the product and inverse for the group semidirect product in III.4.4.

Remark. If A had been merely a Banach algebra, the same construction, omitting only the definition (2) of the involution, would have made the semidirect product $\langle B, \pi, \cdot \rangle$ into a Banach algebraic bundle.

4.3. If in 4.2 $\{u_i\}$ is an approximate unit of A, then $\{\langle u_i, e \rangle\}$ is easily seen to be a strong approximate unit of $\mathscr{B}$. In that case $\mathscr{B}$ is saturated.

Each x in G gives rise to a unitary multiplier m_x of $\mathscr{B}$, whose left and right actions are given by:

$$m_x \langle a, y \rangle = \langle \tau_x(a), xy \rangle,$$
$$\langle a, y \rangle m_x = \langle a, yx \rangle \tag{3}$$

($\langle a, y \rangle \in B$). Of course $(m_x)^{-1} = (m_x)^* = m_{x^{-1}}$. Thus the semidirect product bundles of 4.2 always have enough unitary multipliers.

4.4. Suppose in 4.2 that $\tau_x(a) = a$ for all x in G and a in A. Then formulae (1), (2) become:

$$\langle a, x \rangle \langle b, y \rangle = \langle ab, xy \rangle, \qquad \langle a, x \rangle^* = \langle a^*, x^{-1} \rangle, \tag{4}$$

and $A \underset{\tau}{\times} G$ is called the *direct product of A with G*.

The direct product of $\mathbb{C}$ with G is the involutory version of what we called in 2.7 the *group bundle of G*. This involutory version is what we shall usually mean in the future by the group bundle of G. As we pointed out in 2.7, it is through the group bundle that the unitary representation theory of groups will appear as a special case of the *-representation theory of Banach *-algebraic bundles.

4.5.   *Example.* Semidirect products of groups give rise to special semidirect product bundles.

Let G and N be two topological groups, N being locally compact; and let σ be a homomorphism of G into the group of all group automorphisms of N, such that $\langle n, x \rangle \mapsto \sigma_x(n)$ is continuous on $N \times G$ to N. Thus we have the ingredients for forming the group semidirect product $N \underset{\sigma}{\times} G$ as in III.4.4.

Let λ be a left Haar measure of N; and denote by A the $\mathscr{L}_1$ group algebra $\mathscr{L}_1(\lambda)$ of N. For each x in G let $\Gamma(\sigma_x)$ be the expansion factor of σ_x (acting on N), defined as in III.8.2. Setting

$$\tau_x(f)(n) = (\Gamma(\sigma_x))^{-1} f(\sigma_x^{-1} n) \qquad (x \in G; f \in A; n \in N), \tag{5}$$

we verify (using III.8.3, III.8.14) that τ satisfies the hypotheses of 4.2; it is a strongly continuous homomorphism of G into the group of isometric *-automorphisms of A.

We can therefore form the τ-semidirect product $A \underset{\tau}{\times} G$ as in 4.2. We shall discover in 6.6, 6.7 the intimate relation between $A \underset{\tau}{\times} G$ and the group algebra of the group semidirect product $N \underset{\sigma}{\times} G$.

4.6.   *Example.* Let G be a topological group, and M a locally compact Hausdorff left topological G-space (III.3.1). Let A be the commutative C^*-algebra $\mathscr{C}_0(M)$ (with the supremum norm). The action of G on M gives rise to an action τ of G on A:

$$\tau_x(f)(m) = f(x^{-1} m).$$

Evidently τ satisfies the hypotheses of 4.2; so we can form the τ-semidirect product bundle $A \underset{\tau}{\times} G$. This bundle is a special case of the so-called transformation bundles which we shall meet in §7.

Central Extension Bundles

4.7. Fix a Banach *-algebra A with a unit element $\mathbb{1}$ satisfying $\|\mathbb{1}\| = 1$. Let U be the multiplicative group of unitary elements of A (that is, elements u for which $\|u\| = 1$ and $u^*u = uu^* = \mathbb{1}$). We equip U with the *strong topology*, in which $u_i \to u$ if and only if $\|u_i a - ua\| \to 0$ and $\|au_i - au\| \to 0$ for all a in A. With this topology U is a topological group. Now choose a (commutative) topological subgroup N of U contained in the center of U; fix a topological group G; and let

$$N \underset{i}{\to} H \underset{j}{\to} G$$

be a central extension of N by G (see III.5.6). With these ingredients we shall now construct a Banach *-algebraic bundle $\mathscr{B} = \langle B, \pi, \cdot, * \rangle$ over G.

Note that N acts (to the left) as a topological transformation group on the Cartesian product $A \times H$ as follows:

$$u\langle a, h \rangle = \langle au^{-1}, i(u)h \rangle \qquad (u \in N; h \in H; a \in A).$$

Let B be the space of all orbits in $A \times H$ under this action of N, with the quotient topology; and let $\rho: A \times H \to B$ be the (continuous open) quotient map. One verifies that B is Hausdorff, and that the equation

$$\pi(\rho(a, h)) = j(h) \qquad (\langle a, h \rangle \in A \times H)$$

defines a continuous open surjection $\pi: B \to G$. For fixed h in H we transfer the Banach space structure of A to $\pi^{-1}(j(h)) = \{\rho(a, h): a \in A\}$ via the bijection $a \mapsto \rho(a, h)$; and this Banach space structure depends only on $j(h)$ (since $a \mapsto au^{-1}$ is an isometry on A for each u in N). When the fibers $\pi^{-1}(x)$ ($x \in G$) are thus made into Banach spaces, $\langle B, \pi \rangle$ becomes a Banach bundle over G. We now define multiplication and involution on B in analogy with the direct product formulae (4):

$$\rho(a, h) \cdot \rho(b, k) = \rho(ab, hk),$$

$$(\rho(a, h))^* = \rho(a^*, h^{-1})$$

($a, b \in A$; $h, k \in H$). Then $\mathscr{B} = \langle B, \pi, \cdot, * \rangle$ is easily seen to be a Banach *-algebraic bundle over G. It is called the *central extension bundle* constructed from these ingredients.

The unit fiber of $\mathscr{B}$ is clearly identifiable with A via the correspondence $a \mapsto \rho(a, e_H)$ (e_H being the unit of H).

$\rho(1, i(1))$ is the unit element of $\mathscr{B}$; and $\rho(1, h)$ is a unitary element of $\mathscr{B}$ for every h in H. It follows that $\mathscr{B}$ has enough unitary multipliers and is saturated.

If $N = \{1\}$, then $H \cong G$, and $\mathscr{B}$ is just the direct product bundle $A \times G$. Otherwise it is of quite a different nature from the semidirect product bundles.

Remark. Notice the formula

$$(\rho(a, h))^* \rho(a, h) = \rho(a^*a, e_H) \cong a^*a.$$

Thus if A happens to be a C^*-algebra, the following two properties hold in $\mathscr{B}$: (i) $\|b^*b\| = \|b\|^2$ for all $b \in B$; (ii) for any $b \in B$, the element b^*b is positive in $B_e (\cong A)$.

Remarks.   It was only for simplicity's sake that we assumed A to have a unit. In the absence of a unit, we would have used unitary multipliers of A instead of unitary elements, and the construction would have gone through without change.

Notice the similarity of the above construction with that of tensor bundles in differential geometry.

For a construction which generalizes both semidirect product bundles and central extension bundles, see Fell [14], §9.

4.8.   Take $A = \mathbb{C}$, and let N be the circle group $\mathbb{E}$. For any central extension $\gamma: \mathbb{E} \underset{i}{\to} H \underset{j}{\to} G$ of $\mathbb{E}$ by G, the construction of 4.7 then gives us a saturated Banach *-algebraic bundle $\mathscr{B}$ over G with the property that each fiber of $\mathscr{B}$ is one-dimensional. We shall denote this $\mathscr{B}$ by $\mathscr{B}^\gamma$.

Definition.   A Banach *-algebraic bundle over G which is isometrically isomorphic with $\mathscr{B}^\gamma$ for some central extension $\gamma: \mathbb{E} \underset{i}{\to} H \underset{j}{\to} G$ is called a *cocycle bundle* over G.

Thus a cocycle bundle $\mathscr{B}$ has a unit element and is saturated, its unit fiber is $\cong \mathbb{C}$, and all its fibers are one-dimensional. In addition it has the properties (i) and (ii) of Remark 4.7.

It will turn out in 10.10 that cocycle bundles over G are related to the so-called projective representation theory of G in just the same way that the group bundle of G is related to the ordinary unitary representation theory of G.

For another characterization of cocycle bundles see Proposition 16.2.

4.9.   The classification of the possible cocycle bundles over a given topological group G is thus the same as the classification of central extensions of $\mathbb{E}$ by G, that is, of the equivalence classes of multipliers in the sense of Mackey [8]. The importance of this problem is apparent from the role which multipliers play in the Mackey normal subgroup analysis (see Mackey [8] or our Chapter XII). Furthermore, if we knew all possible cocycle bundles over any given group, we would immediately obtain, via our Chapter XII, §6, a classification of all saturated C^*-algebraic bundles, with unit fiber of compact type, over any locally compact group.

Thus the classification of cocycle bundles appears to be a problem of some importance. Unfortunately it is beyond the scope of this work (see the Notes and Remarks at the end of the chapter for references to the literature).

5. The Cross-Sectional Algebra

5.1. Throughout this section we shall consider a fixed locally compact group G (with unit e), and a fixed Banach algebraic bundle $\mathscr{B} = \langle B, \pi, \cdot \rangle$ over G. By Appendix C $\mathscr{B}$ automatically has enough continuous cross-sections.

Let λ be a left Haar measure and Δ the modular function of G. We are going to show that the product operation on B generates a product operation on the cross-sectional space $\mathscr{L}_1(\lambda; \mathscr{B})$, making the latter a Banach algebra. This product on $\mathscr{L}_1(\lambda; \mathscr{B})$ will be called *bundle convolution*. If $\mathscr{B}$ is the group bundle, it becomes ordinary convolution; and $\mathscr{L}_1(\lambda; \mathscr{B})$ becomes just the $\mathscr{L}_1$ group algebra of G which we studied in §III.11.

5.2. We shall first introduce a multiplication in the dense subspace $\mathscr{L}(\mathscr{B})$ of $\mathscr{L}_1(\lambda; \mathscr{B})$ (see II.15.9).

Suppose $f, g \in \mathscr{L}(\mathscr{B})$. For fixed x in G, the continuity of multiplication in B (together with 2.2(i)) shows that the map $y \mapsto f(y)g(y^{-1}x)$ is continuous from G to B_x. Hence in the equation

$$(f * g)(x) = \int_G f(y)g(y^{-1}x)d\lambda y, \tag{1}$$

the right side makes sense as a B_x-valued integral (II.15.18). If x is now allowed to vary, (1) defines $f * g$ as a cross-section of $\mathscr{B}$. If the compact supports of f and g are D and E respectively, then by (1) $(f * g)(x) = 0_x$ unless $x \in DE$. So $f * g$ has compact support. Applying Proposition II.15.19 to the function $\langle x, y \rangle \mapsto f(y)g(y^{-1}x)$, we conclude that $f * g$ is continuous. Thus finally $f * g \in \mathscr{L}(\mathscr{B})$; and (1) defines a binary operation $*$ on $\mathscr{L}(\mathscr{B})$.

Clearly $f * g$ is linear in f and in g. We claim that $*$ is associative. Indeed, if $f, g, h \in \mathscr{L}(\mathscr{B})$ and $x \in G$, we have

$$((f * g) * h)(x) = \int \left[\int f(y)g(y^{-1}z)d\lambda y \right] h(z^{-1}x)d\lambda z$$

$$= \iint (f(y)g(y^{-1}z))h(z^{-1}x)d\lambda y \, d\lambda z \tag{2}$$

(by II.5.7 applied to the continuous linear map $b \mapsto bh(z^{-1}x)$ of B_z into B_x). Similarly

$$(f * (g * h))(x) = \iint f(y)(g(y^{-1}z)h(z^{-1}x))d\lambda z \, d\lambda y. \tag{3}$$

Now the right sides of (2) and (3) are equal by the associativity of the product in B and the Fubini Theorem in the form II.16.3. So (2) and (3) imply that $*$ is associative. Thus $\mathcal{L}(\mathcal{B})$ is an algebra under $*$.

Let $\| \ \|_1$ be the norm in $\mathcal{L}_1(\lambda; \mathcal{B})$. For $f, g \in \mathcal{L}(\mathcal{B})$ we have by II.5.4(2)

$$\|(f * g)(x)\| \le \int \|f(y)\| \|g(y^{-1}x)\| d\lambda y,$$

and hence

$$\|f * g\|_1 = \int \|(f * g)(x)\| d\lambda x$$

$$\le \iint \|f(y)\| \|g(y^{-1}x)\| d\lambda y \, \lambda x$$

$$= \iint \|f(y)\| \|g(x)\| d\lambda y \, d\lambda x = \|f\|_1 \|g\|_1.$$

So $\mathcal{L}(\mathcal{B}), *, \| \ \|_1$ is a normed algebra.

Since $\mathcal{L}(\mathcal{B})$ is dense in the Banach space $\mathcal{L}_1(\lambda; \mathcal{B})$, the product $*$ in $\mathcal{L}(\mathcal{B})$ can be uniquely extended to a product (also called $*$) in $\mathcal{L}_1(\lambda; \mathcal{B})$, making the latter a Banach algebra.

Definition.   This product in $\mathcal{L}_1(\lambda; \mathcal{B})$ is called the *bundle convolution*; and $\mathcal{L}_1(\lambda; \mathcal{B})$, equipped with $*$, is called the $\mathcal{L}_1$ *cross-sectional algebra of $\mathcal{B}$*. The dense subalgebra $\mathcal{L}(\mathcal{B})$ of $\mathcal{L}_1(\lambda; \mathcal{B})$ is called the *compacted cross-sectional algebra of $\mathcal{B}$*.

If $\mathcal{B}$ (and hence G) is Abelian, then $\mathcal{L}_1(\lambda; \mathcal{B})$ is commutative.

5.3.   One verifies easily that the bundle convolution $f * g$ ($f, g \in \mathcal{L}(\mathcal{B})$) is *separately* continuous in f and g with respect to the inductive limit topology of $\mathcal{L}(\mathcal{B})$ (defined in II.14.3). It is not in general *jointly* continuous in f and g, however; see Remark 1 of II.14.14.

5.4.   One feels that the bundle convolution ought to be described by formula (1) not only on $\mathcal{L}(\mathcal{B})$ but on all of $\mathcal{L}_1(\lambda; \mathcal{B})$. We shall now show that this is the case.

Proposition.   *Let f, g be in $\mathcal{L}_1(\lambda; \mathcal{B})$. Then, for λ-almost all x, (i) the function $y \mapsto f(y)g(y^{-1}x)$ belongs to $\mathcal{L}_1(\lambda; B_x)$, and (ii) the B_x-valued integral $\int_G f(y)g(y^{-1}x)d\lambda y$ is equal to $(f * g)(x)$.*

Proof. Let $f \square g$ stand for the function $\langle x, y \rangle \mapsto f(x)g(x^{-1}y)$ on $G \times G$. This is a cross-section of the Banach bundle retraction $G \times \mathscr{B}$ of $\mathscr{B}$ obtained from the projection map $\langle x, y \rangle \mapsto y$ of $G \times G$ onto G (see II.13.7). We claim that $f \square g$ is locally $(\lambda \times \lambda)$-measurable. Indeed: Choose sequences $\{f_n\}$ and $\{g_n\}$ of elements of $\mathscr{L}(\mathscr{B})$ which converge in $\mathscr{L}_1(\lambda; \mathscr{B})$ to f and g respectively. By II.3.5 we can pass to subsequences and assume that $f_n(x) \to f(x)$ and $g_n(x) \to g(x)$ in B_x for λ-almost all x in G. It follows that $f_n(x)g_n(x^{-1}y) \to f(x)g(x^{-1}y)$ for all $\langle x, y \rangle$ in $G \times G$ save for an exceptional set which by Fubini's Theorem is $(\lambda \times \lambda)$-null. Hence by II.15.4 (applied in $G \times \mathscr{B}$) $f \square g$ is locally $(\lambda \times \lambda)$-measurable. From this and the numerical Fubini Theorem we deduce that $f \square g \in \mathscr{L}_1(\lambda \times \lambda; G \times \mathscr{B})$. This and Theorem II.16.1(I) imply assertion (i) of the proposition.

To prove assertion (ii), let $\{f_n\}$ and $\{g_n\}$ be as above. We claim that

$$f_n \square g_n \to f \square g \qquad \text{in } \mathscr{L}_1(\lambda \times \lambda; G \times \mathscr{B}). \tag{4}$$

Indeed:

$$\|f \square g - f_n \square g_n\|_1 = \iint \|f(x)g(x^{-1}y) - f_n(x)g_n(x^{-1}y)\| d\lambda x \, d\lambda y$$

$$\leq \iint \|f(x)\| \|g(x^{-1}y) - g_n(x^{-1}y)\| d\lambda x \, d\lambda y$$

$$+ \iint \|f(x) - f_n(x)\| \|g_n(x^{-1}y)\| d\lambda x \, d\lambda y$$

$$= \|f\|_1 \|g - g_n\|_1 + \|g_n\|_1 \|f - f_n\|_1 \to 0 \qquad \text{as } n \to \infty.$$

So (4) holds.

In view of (4), II.16.1, and II.3.5, we can pass to a subsequence and assume that

$$\int f_n(x)g_n(x^{-1}y)d\lambda x \to \int f(x)g(x^{-1}y)d\lambda x \tag{5}$$

for λ-almost all y. On the other hand, since f_n and g_n are in $\mathscr{L}(\mathscr{B})$, the left side of (5) is $(f_n * g_n)(y)$. By the continuity of multiplication in a Banach algebra we have $f_n * g_n \to f * g$ in $\mathscr{L}_1(\lambda; \mathscr{B})$; so, again using II.3.5, we can pass to a subsequence and get

$$(f_n * g_n)(y) \to (f * g)(y) \qquad \text{for } \lambda\text{-almost } y. \tag{6}$$

Combining (5) and (6), we obtain assertion (ii) of the proposition. ∎

5.5.   Suppose that G is discrete and λ is "counting measure" ($\lambda(\{x\}) = 1$ for each x in G). Then $\mathscr{L}_1(\lambda; \mathscr{B})$ is the Banach algebra of all functions f on G such that (i) $f(x) \in B_x$ for each x in G, and (ii) $\|f\|_1 = \sum_{x \in G} \|f(x)\| < \infty$. The bundle convolution on $\mathscr{L}_1(\lambda; \mathscr{B})$ is given by:

$$(f * g)(x) = \sum_{y \in G} f(y)g(y^{-1}x).$$

Notice that in this situation, unlike the non-discrete case, each B_x can be identified with a closed linear subspace of $\mathscr{L}_1(\lambda; \mathscr{B})$; and the bundle convolution restricted to $B = \bigcup_x B_x$ is just the original product on B.

5.6.   Up to now $\mathscr{B}$ has been merely a Banach algebraic bundle, without an involution. Assume now that $\mathscr{B}$ is a Banach *-algebraic bundle. In that case $\mathscr{L}_1(\lambda; \mathscr{B})$ becomes a Banach *-algebra under the involution defined as follows:

$$f^*(x) = \Delta(x^{-1})(f(x^{-1}))^* \qquad (f \in \mathscr{L}_1(\lambda; \mathscr{B}); \, x \in G) \tag{7}$$

(We recall that Δ is the modular function of G.) To see this, notice that the f^* defined by (7) is certainly a cross-section of $\mathscr{B}$ whenever f is a cross-section of $\mathscr{B}$. By 3.1(vi) the continuity of f implies that of f^*; and hence by II.15.4 the local λ-measurability of f implies that of f^*. So, if $f \in \mathscr{L}_1(\lambda; \mathscr{B})$, $x \mapsto \|f^*(x)\|$ is locally λ-measurable and

$$\|f^*\|_1 = \int \Delta(x^{-1})\|f(x^{-1})\|d\lambda x = \|f\|_1 < \infty.$$

It follows that $f \mapsto f^*$ is an isometric map of $\mathscr{L}_1(\lambda; \mathscr{B})$ into itself. It is evidently conjugate-linear and satisfies $f^{**} = f$. Notice that $f \in \mathscr{L}(\mathscr{B}) \Rightarrow f^* \in \mathscr{L}(\mathscr{B})$. If $f, g \in \mathscr{L}(\mathscr{B})$ and $x \in G$,

$$(f * g)^*(x) = \Delta(x^{-1})\left[\int f(y)g(y^{-1}x^{-1})d\lambda y\right]^*$$

$$= \Delta(x^{-1})\int g(y^{-1}x^{-1})^* f(y)^* d\lambda y$$

$$= \int g^*(xy)f^*(y^{-1})d\lambda y = (g^* * f^*)(x).$$

(In the second of the above steps we appealed to II.5.7 for conjugate-linear maps.) Thus the identity $(f * g)^* = g^* * f^*$ holds on $\mathscr{L}(\mathscr{B})$, and hence by continuity on all of $\mathscr{L}_1(\lambda; \mathscr{B})$.

We have now shown that when $\mathcal{B}$ is a Banach *-algebraic bundle its $\mathcal{L}_1$ cross-sectional algebra, equipped with the natural involution (7), is a Banach *-algebra.

Notice that the compacted cross-sectional algebra $\mathcal{L}(\mathcal{B})$ is now a dense *-subalgebra of $\mathcal{L}_1(\lambda; \mathcal{B})$.

In the discrete case mentioned in 5.5, the involution (7) restricted to $\bigcup_x B_x$ is of course just the original involution of $\mathcal{B}$.

5.7. The form of the operations of $\mathcal{L}_1(\lambda; \mathcal{B})$ in the semidirect product context of §4 deserves special mention.

Let G, A, and τ be as in 4.2; and form the τ-semidirect product $\mathcal{B} = A \underset{\tau}{\times} G$. As usual we shall identify cross-sections of $A \underset{\tau}{\times} G$, which as a Banach bundle is trivial, with functions on G to A (so that $\mathcal{L}_1(\lambda; \mathcal{B}) = \mathcal{L}_1(\lambda; A)$). The product formulae 4.2(1) and 5.2(1) then combine to give

$$(f * g)(x) = \int_G f(y)\tau_y(g(y^{-1}x))d\lambda y \tag{8}$$

$(f, g \in \mathcal{L}_1(\lambda; A); x \in G)$; and 4.2(2) and 5.6(7) combine to give

$$f^*(x) = \Delta(x^{-1})\tau_x[(f(x^{-1}))^*] \tag{9}$$

$(f \in \mathcal{L}_1(\lambda; A); x \in G)$. The algebraic operations on the right of (8) and (9) are of course performed in A.

In the very special and important case of the group bundle (4.4), $\mathcal{L}_1(\lambda; A) = \mathcal{L}_1(\lambda)$ and every τ_x is the identity map; so (8) and (9) become

$$(f * g)(x) = \int f(y)g(y^{-1}x)d\lambda y, \qquad f^*(x) = \Delta(x^{-1})\,\mathcal{E}_1 f(x^{-1}).$$

Thus *the $\mathcal{L}_1$ cross-sectional algebra of the group bundle of G is just the $\mathcal{L}_1$ group algebra of G as defined in* III.11.9.

Multipliers

5.8. As one might expect, multipliers on a Banach algebraic bundle give rise to corresponding multipliers on the $\mathcal{L}_1$ cross-sectional algebra.

Let $\mathcal{B}$ be a Banach algebraic bundle over G (not necessarily with involution). Let u be a multiplier of $\mathcal{B}$ of order x (as in 2.14). Then u has a natural left and right action on arbitrary cross-sections of $\mathcal{B}$. Indeed, if f is any cross-section of $\mathcal{B}$, the equations

$$(uf)(y) = uf(x^{-1}y), \tag{10}$$

$$(fu)(y) = \Delta(x^{-1})f(yx^{-1})u \tag{11}$$

define cross-sections uf and fu of $\mathscr{B}$. Since the left and right actions of u on B are continuous, uf and fu will be continuous if f is. Therefore, by the boundedness of the actions of u and II.15.4, uf and fu are locally λ-measurable whenever f is. Thus, if $f \in \mathscr{L}_1(\lambda; \mathscr{B})$,

$$\|uf\|_1 = \int \|uf(x^{-1}y)\| d\lambda y \le \|u\|_0 \int \|f(x^{-1}y)\| d\lambda y$$

$$= \|u\|_0 \|f\|_1, \tag{12}$$

$$\|fu\|_1 = \Delta(x^{-1}) \int \|f(yx^{-1})u\| d\lambda y$$

$$\le \Delta(x^{-1}) \|u\|_0 \int \|f(yx^{-1})\| d\lambda y = \|u\|_0 \|f\|_1; \tag{13}$$

from which it follows that $f \mapsto uf$ and $f \mapsto fu$ are bounded linear endomorphisms of $\mathscr{L}_1(\lambda; \mathscr{B})$. It is easily verified that the three identities

$$u(f * g) = (uf) * g, \quad (f * g)u = f * (gu), \quad (fu) * g = f * (ug)$$

hold for all f, g in $\mathscr{L}(\mathscr{B})$, and hence by continuity for all f, g in $\mathscr{L}_1(\lambda; \mathscr{B})$. So the left and right actions $f \mapsto uf$ and $f \mapsto fu$ define a bounded multiplier m_u of $\mathscr{L}_1(\lambda; \mathscr{B})$.

Sometimes m_u is called the *integrated form* of u.

Evidently the map $u \mapsto m_u$ preserves the product of multipliers, and also the linear operations on each "fiber" $\mathscr{W}_x(\mathscr{B})$. By (12) and (13) $\|m_u\|_0 \le \|u\|_0$. In fact it is easy to see that

$$\|m_u\|_0 = \|u\|_0. \tag{14}$$

Indeed: Let p and q be the norms of the left actions of u and m_u respectively. We shall show that $p = q$. To do this, take $\varepsilon > 0$, and choose y in G and b in B_y such that $\|b\| = 1$ and $\|ub\| > p - \varepsilon$. Let γ be a continuous cross-section of $\mathscr{B}$ with $\gamma(y) = b$ and $\|\gamma(z)\| \le 1$ for all z. By the continuity of u there is a neighborhood U of y such that $\|u\gamma(z)\| > p - \varepsilon$ for all z in U; and we may choose a function ϕ in $\mathscr{L}_+(\mathscr{B})$ which vanishes outside U and satisfies $\int \phi d\lambda = 1$. Putting $f = \phi\gamma$, so that $f \in \mathscr{L}_1(\lambda; \mathscr{B})$, we check that $\|f\| \le 1$ and $\|uf\|_1 = \int \phi(z) \|u\gamma(z)\| d\lambda z > p - \varepsilon$. So $q > p - \varepsilon$. By the arbitrariness of ε this implies $q \ge p$. Since by (12) $q \le p$, we have $q = p$. Similarly the norms of the right actions of u and m_u are the same. Consequently (14) holds.

In particular, every element u of B_x gives rise by (10) and (11) to a multiplier m_u on $\mathscr{L}_1(\lambda; \mathscr{B})$ satisfying

$$\|m_u\|_0 \le \|u\|. \tag{15}$$

5.9. Suppose now that the $\mathscr{B}$ of 5.8 was a Banach *-algebraic bundle, so that by 5.6 $\mathscr{L}_1(\lambda; \mathscr{B})$ is a Banach *-algebra. By 1.13 and 3.7 there are involution operations on the multipliers of $\mathscr{B}$ and of $\mathscr{L}_1(\lambda; \mathscr{B})$; and one checks without difficulty that the map $u \mapsto m_u$ preserves these involutions. In particular, if u is a unitary multiplier of $\mathscr{B}$, m_u is a unitary multiplier of $\mathscr{L}_1(\lambda; \mathscr{B})$.

5.10. *Remark.* What is the form of the most general bounded multiplier on the $\mathscr{L}_1$ cross-sectional algebra of a Banach algebraic bundle $\mathscr{B}$ over G? This question is answered by Wendel's Theorem (1.23) if $\mathscr{B}$ is the group bundle of G: The multipliers of $\mathscr{L}_1(\lambda)$ are then just bounded measures on G. In the case of a general $\mathscr{B}$, for each x in G we have by 5.8 a linear isometry of $\mathscr{W}_x(\mathscr{B})$ into the Banach space of bounded multipliers of $\mathscr{L}_1(\lambda; \mathscr{B})$. In view of Wendel's Theorem one conjectures that the most general bounded multiplier of $\mathscr{L}_1(\lambda; \mathscr{B})$ is an integral, with respect to some measure on G, of a function on G whose value at each point x lies in $\mathscr{W}_x(\mathscr{B})$. The precise statement and proof of such a conjecture is an open problem at present.

Approximate Units of Cross-Sectional Algebras

5.11. Does $\mathscr{L}_1(\lambda; \mathscr{B})$ have an approximate unit? Some condition on $\mathscr{B}$ is certainly necessary to ensure this, since for example, if the multiplication in $\mathscr{B}$ is trivial ($bc = 0_{\pi(bc)}$ for all b, c), that of $\mathscr{L}_1(\lambda; \mathscr{B})$ is also trivial.

Theorem. *Assume that the Banach algebraic bundle $\mathscr{B}$ has a strong approximate unit. Then $\mathscr{L}_1(\lambda; \mathscr{B})$ has an approximate unit.*

Proof. Let $\{\psi_i\}$ ($i \in I$) be an approximate unit on G in the sense of III.11.17; and let $\{u_\alpha\}$ be a strong approximate unit of $\mathscr{B}$, with $\|u_\alpha\| \leq k$ for all α. For each α let γ_α be a continuous cross-section of $\mathscr{B}$ such that $\gamma_\alpha(e) = u_\alpha$. It is convenient, though not essential, to arrange that $\|\gamma_\alpha(x)\| \leq \|u_\alpha\|$ for all x; we can do this by the proof of II.13.15. Define $h_{\alpha, i}(x) = \psi_i(x)\gamma_\alpha(x)$ ($x \in G$). Thus $h_{\alpha, i} \in \mathscr{L}(\mathscr{B})$, and $\|h_{\alpha, i}\|_1 \leq k$.

Consider an element g of $\mathscr{L}(\mathscr{B})$ and a positive number $\varepsilon > 0$. If $x \in G$,

$$(h_{\alpha, i} * g - g)(x) = \int_G \psi_i(y)[\gamma_\alpha(y)g(y^{-1}x) - g(x)]d\lambda y. \tag{16}$$

Let K be a compact subset of G containing the compact support of g in its interior. Since $g(K)$ is compact in B and $\{u_\alpha\}$ is a strong approximate unit, there is an index α_0 such that, if $\alpha \succ \alpha_0$,

$$\|u_\alpha g(x) - g(x)\| < \varepsilon \qquad \text{for all } x \text{ in } K. \tag{17}$$

Now fix $\alpha \succ \alpha_0$. The function $\langle y, x \rangle \mapsto \|\gamma_\alpha(y)g(y^{-1}x) - g(x)\|$ is continuous on $G \times G$, and by (17) has value less than ε for $y = e$, $x \in K$. Hence there is a neighborhood U of e such that $\|\gamma_\alpha(y)g(y^{-1}x) - g(x)\| < \varepsilon$ for all x in K and y in U. By (16) this implies the existence of an index i_0 (depending of course on α) such that

$$\|(h_{\alpha,i} * g - g)(x)\| < \varepsilon \qquad \text{for all } x \in K \text{ and } i \succ i_0. \tag{18}$$

Clearly we can choose i_0 so large that $i \succ i_0$ also implies that $h_{\alpha,i} * g$ has compact support contained in K. Doing this, we deduce from (18) that

$$\|h_{\alpha,i} * g - g\|_1 \leq \varepsilon\lambda(K) \qquad \text{for } i \succ i_0. \tag{19}$$

Recalling that ε was arbitrary, that α_0 depended only on ε, and that i_0 depended on $\alpha \succ \alpha_0$, we obtain from (19):

$$\lim_\alpha \limsup_i \|h_{\alpha,i} * g - g\|_1 = 0. \tag{20}$$

If the ψ_i are chosen to be symmetric ($\psi_i(y^{-1}) = \psi_i(y)$), a similar argument shows that

$$\lim_\alpha \limsup_i \|g * h_{\alpha,i} - g\|_1 = 0. \tag{21}$$

Now (20) and (21) hold for all g in $\mathscr{L}(\mathscr{B})$. Thus, for any $\varepsilon > 0$ and any finite set $g_1, \ldots, g_n$ of elements of $\mathscr{L}(\mathscr{B})$, there are indices α, i satisfying

$$\|h_{\alpha,i} * g_r - g_r\|_1 < \varepsilon, \qquad \|g_r * h_{\alpha,i} - g_r\|_1 < \varepsilon$$

for all $r = 1, \ldots, n$. Since $\|h_{\alpha,i}\| \leq k$, this implies that from the set $\{h_{\alpha,i}\}$ we can extract an approximate unit at least for the normed algebra $\mathscr{L}(\mathscr{B})$, $\|\ \|_1$. From the fact that $\mathscr{L}(\mathscr{B})$ is dense in $\mathscr{L}_1(\lambda; \mathscr{B})$, it follows that $\mathscr{L}_1(\lambda; \mathscr{B})$ itself has an approximate unit.   ■

5.12.   Remark.   In view of (18), the above proof has actually constructed an approximate unit $\{h_\sigma\}$ of $\mathscr{L}_1(\lambda; \mathscr{B})$ with the property that, for every g in $\mathscr{L}(\mathscr{B})$, $h_\sigma * g \to g$ and $g * h_\sigma \to g$ in the inductive limit topology of $\mathscr{L}(\mathscr{B})$ (see II.14.3).

An easy modification of the argument of 5.11 shows that, in addition,

$$h_\sigma^* * g * h_\sigma \to g \qquad \text{in the inductive limit topology}$$

for all g in $\mathscr{L}(\mathscr{B})$.

5.13. **Remark.** Later on we shall find it useful to generalize Remark 5.12 as follows:

Let H be a closed subgroup of G, with left Haar measure v. If $f \in \mathscr{L}(\mathscr{B}_H)$ and $g \in \mathscr{L}(\mathscr{B})$, the equation

$$(f * g)(x) = \int_H f(h)g(h^{-1}x)dvh \qquad (x \in G)$$

defines $f * g$ as an element of $\mathscr{L}(\mathscr{B})$ (by the same argument as in 5.2). In case $H = G$ (and $v = \lambda$), this of course is the bundle convolution.

Now assume that $\{\gamma_\alpha\}$ is a net of continuous cross-sections of $\mathscr{B}_H$ with the property that $\{\gamma_\alpha(e)\}$ is a strong approximate unit of $\mathscr{B}$. Further, let $\{\psi_i\}$ be an approximate unit on H (in the sense of III.11.17). Thus $h_{\alpha,i} = \psi_i\gamma_\alpha \in \mathscr{L}(\mathscr{B}_H)$ for each i and α. Modifying the argument leading to (18), one can show that

$$\lim_\alpha \limsup_i \|h_{\alpha,i} * g - g\|_\infty = 0 \qquad (22)$$

for every g in $\mathscr{L}(\mathscr{B})$ ($\| \ \|_\infty$ as usual denoting the supremum norm). It follows from (22) that one can extract from the $\{h_{\alpha,i}\}$ a net $\{h_\sigma\}$ of elements of $\mathscr{L}(\mathscr{B}_H)$ such that

$$h_\sigma * g \to g \qquad \text{in the inductive limit topology}$$

for every g in $\mathscr{L}(\mathscr{B})$.

6. Partial Cross-Sectional Bundles

6.1. This section will present an important method of constructing one Banach algebraic bundle from another. The construction can be regarded as a generalization of the construction of the cross-sectional algebra in §5.

6.2. Throughout the section we fix a Banach *-algebraic bundle $\mathscr{B} = \langle B, \pi, \cdot, * \rangle$ over a locally compact group G with unit e and left Haar measure λ. By Appendix C $\mathscr{B}$ has enough continuous cross-sections. We shall also fix a closed normal subgroup N of G, with left Haar measure v. Our goal is to construct a certain Banach *-algebraic bundle $\mathscr{C}$ over G/N, which will have the property that the $\mathscr{L}_1$ cross-sectional algebras of $\mathscr{B}$ and of $\mathscr{C}$ are essentially the same.

As usual, the presence of an involution in $\mathscr{B}$ will not be necessary until we come to define the involution in $\mathscr{C}$.

6.3.   If $x \in G$ and $\alpha = xN$ is the corresponding coset in G/N, let v_α denote the image of v under the homeomorphism $n \mapsto xn$ of N onto α. Thus v_α is a regular Borel measure on α; and since v is left-invariant v_α depends only on α, not on the particular element x in α. We denote by $\mathscr{B}_\alpha$ the Banach bundle reduction of $\mathscr{B}$ to the closed subset α of G (see II.13.7), and by C_α the $\mathscr{L}_1$ cross-sectional space $\mathscr{L}_1(v_\alpha; \mathscr{B}_\alpha)$. If $f \in \mathscr{L}(\mathscr{B})$, let $\tilde{f}$ be the function on G/N assigning to each α in G/N the element $\tilde{f}(\alpha) = f | \alpha$ of $\mathscr{L}(\mathscr{B}_\alpha)$. Since $\mathscr{L}(\mathscr{B}_\alpha) \subset C_\alpha$, we may regard $\tilde{f}(\alpha)$ as an element of C_α.

We now claim that there is a unique Banach bundle $\mathscr{C} = \langle C, \rho \rangle$ over G/N whose fibers are the C_α ($\alpha \in G/N$), and with respect to which the $\tilde{f}$ ($f \in \mathscr{L}(\mathscr{B})$) are continuous cross-sections. Indeed, this will follow from II.13.18 if we can prove that the two hypotheses (a) and (b) of II.13.18 hold in the present case. To prove (a), we must verify the continuity of the functions $\alpha \mapsto \| \tilde{f}(\alpha) \|_1$ on G/N for each f in $\mathscr{L}(\mathscr{B})$. But, if $f \in \mathscr{L}(\mathscr{B})$ and $\alpha = xN$, $\| \tilde{f}(\alpha) \|_1 = \int_\alpha \| f(y) \| dv_\alpha y = \int_N \| f(xn) \| dvn$, and the latter expression is continuous in x by an easy uniform continuity argument (or as a special case of II.15.19). So (a) holds. To prove (b), we have only to notice from II.14.8 that $\{ \tilde{f}(\alpha) : f \in \mathscr{L}(\mathscr{B}) \}$ is all of $\mathscr{L}(\mathscr{B}_\alpha)$, and hence dense in C_α. Thus the claim is established.

6.4.   To make the Banach bundle $\mathscr{C}$ into a Banach *-algebraic bundle over G/N, we must now introduce a multiplication and an involution into $\mathscr{C}$. We first define a multiplication. Let $\phi \in \mathscr{L}(\mathscr{B}_\alpha)$ and $\psi \in \mathscr{L}(\mathscr{B}_\beta)$, where $\alpha, \beta \in G/N$; and define the *product* $\phi \psi$ as the cross-section of $\mathscr{B}_{\alpha\beta}$ given by

$$(\phi\psi)(z) = \int_\alpha \phi(x)\psi(x^{-1}z)dv_\alpha x \qquad (z \in \alpha\beta). \qquad (1)$$

Notice that if $z \in \alpha\beta$ and $x \in \alpha$ then $x^{-1}z \in \beta$, so the integrand in (1) makes sense. It is also continuous with compact support in x, with values in B_z; so the right side of (1) makes sense as a B_z-valued integral, and defines $\phi\psi$ as a cross-section of $\mathscr{B}_{\alpha\beta}$. This $\phi\psi$ clearly has compact support; and it is continuous by II.15.19. Thus $\phi\psi \in \mathscr{L}(\mathscr{B}_{\alpha\beta})$; and we have defined a bilinear map $\langle \phi, \psi \rangle \mapsto \phi\psi$ on $\mathscr{L}(\mathscr{B}_\alpha) \times \mathscr{L}(\mathscr{B}_\beta)$ to $\mathscr{L}(\mathscr{B}_{\alpha\beta})$. This map satisfies

$$\| \phi\psi \|_1 \leq \int_{\alpha\beta} \int_\alpha \| \phi(x) \| \| \psi(x^{-1}z) \| dv_\alpha x \, dv_{\alpha\beta} z$$

$$= \int_N \int_N \| \phi(rn) \| \| \psi(n^{-1}sm) \| dvn \, dvm \qquad \text{(where } r \in \alpha, s \in \beta)$$

$$= \| \phi \|_1 \| \psi \|_1;$$

so it can be extended to a bilinear map of $C_\alpha \times C_\beta$ into $C_{\alpha\beta}$ satisfying

$$\|\phi\psi\|_1 \le \|\phi\|_1 \|\psi\|_1. \tag{2}$$

Since this is the case for every α, β in G/N, we have a binary operation $\langle\phi, \psi\rangle \mapsto \phi\psi$ on the entire bundle space of $\mathscr{C}$. This operation is associative by the same argument by which we proved associativity in 5.2. Hence, to verify that $\mathscr{C}$ is a Banach algebraic bundle under this multiplication, it remains only to verify postulate 2.2(v), i.e., to show that $\langle\phi, \psi\rangle \mapsto \phi\psi$ is continuous on $C \times C$ to C.

For this, it is enough by 2.4 to take two elements f, g of $\mathscr{L}(\mathscr{B})$ and to show that $\langle\alpha, \beta\rangle \mapsto \tilde{f}(\alpha)\tilde{g}(\beta)$ is continuous on $G/N \times G/N$ to C. To do this, we shall suppose that $u_i \to u$ and $v_i \to v$ in G, and shall show that

$$\tilde{f}(u_iN)\tilde{g}(v_iN) \to \tilde{f}(uN)\tilde{g}(vN). \tag{3}$$

Defining $h(x) = \int_N f(un)g(n^{-1}u^{-1}x)dvn$ $(x \in G)$, we have (by II.15.19) $h \in \mathscr{L}(\mathscr{B})$, and also $\tilde{h}(uvN) = \tilde{f}(uN)\tilde{g}(vN)$. Hence, to prove (3), it is sufficient, by II.13.12 and the continuity of $\tilde{h}$, to show that

$$\|\tilde{f}(u_iN)\tilde{g}(v_iN) - \tilde{h}(u_iv_iN)\|_1 \underset{i}{\to} 0.$$

The proof of this is a routine calculation which we leave to the reader.

Thus $\mathscr{C}$ has become a Banach algebraic bundle over G/N.

6.5. To introduce an involution on $\mathscr{C}$, we first recall from III.8.3 that there is a (unique) continuous homomorphism Γ of G into the multiplicative group of positive reals such that

$$v(x^{-1}Wx) = \Gamma(x)v(W) \tag{4}$$

for all x in G and all Borel subsets W of N with compact closure. Now if $\alpha \in G/N$ and $\phi \in \mathscr{L}(\mathscr{B}_\alpha)$, let us define ϕ^* to be the element of $\mathscr{L}(\mathscr{B}_{\alpha^{-1}})$ given by

$$\phi^*(x) = \Gamma(x^{-1})(\phi(x^{-1}))^* \qquad (x \in \alpha^{-1}). \tag{5}$$

One verifies that $\|\phi^*\|_1 = \|\phi\|_1$. Thus $\phi \mapsto \phi^*$ extends to a conjugate-linear isometry of C_α into $C_{\alpha^{-1}}$, and by the arbitrariness of α gives rise to a unary operation $*$ on the entire bundle space C. In this context postulates 3.1(i), (ii), (iv), (v) are now evident. Postulate 3.1(iii) is verified just as in 5.6 (making use of (4)).

To prove the continuity of $*$ (postulate 3.1(vi)), it is enough by 3.2 to take an element f of $\mathscr{L}(\mathscr{B})$ and show that $\alpha \mapsto (\tilde{f}(\alpha))^*$ is continuous from G/N to C. But, setting $g(x) = \Gamma(x^{-1})(f(x^{-1}))^*$, we have $g \in \mathscr{L}(\mathscr{B})$ and $(\tilde{f}(\alpha))^* =$

$\tilde{g}(\alpha^{-1})$; and the continuity of $(\tilde{f}(\alpha))^*$ in α follows from the continuity of the cross-section $\tilde{g}$ of $\mathscr{C}$.

We have now shown that $\mathscr{C} = \langle C, \rho, \cdot, * \rangle$ is a Banach *-algebraic bundle over G/N.

Definition.   *This $\mathscr{C}$ will be called the $\mathscr{L}_1$ partial cross-sectional bundle over G/N derived from $\mathscr{B}$.*

If $n \in N$, $\Gamma(n)$ is just the modular function of N. So, comparing (1) and (5) with 5.2(1) and 5.6(7), we conclude that the unit fiber algebra of $\mathscr{C}$ is exactly the $\mathscr{L}_1$ cross-sectional algebra $\mathscr{L}_1(v; \mathscr{B}_N)$ of $\mathscr{B}_N$. In particular, if $N = G$ (and $v = \lambda$), then $\mathscr{C}$ is just the Banach *-algebra $\mathscr{L}_1(\lambda; \mathscr{B})$ (considered as a bundle over the one-element group G/N). This is what we meant in 6.1 by saying that the partial cross-sectional bundles generalize the construction of the $\mathscr{L}_1$ cross-sectional algebra.

If $N = \{e\}$, evidently $\mathscr{C} = \mathscr{B}$.

6.6.   Definition.   In the special case that $\mathscr{B}$ is the group bundle of G (see 4.4), $\mathscr{C}$ is called the *group extension bundle* (*corresponding to the group extension* $N \to G \to G/N$).

Remark.   Suppose that $\mathscr{B}$ is the group bundle of G; and assume that G is the semidirect product of N with another closed subgroup K of G. One then verifies easily that $\mathscr{C}$ is essentially the same as the $A \underset{\tau}{\times} K$ of Example 4.5 (provided that in 4.5 we replace G by K, and take $\sigma_k(n) = knk^{-1}$ for $k \in K$, $n \in N$).

6.7.   One of the most important properties of these partial cross-sectional bundles $\mathscr{C}$ is that the cross-sectional algebras of $\mathscr{B}$ and $\mathscr{C}$ are essentially the same.

We keep all the notation of 6.2–6.5.

Let μ be the left Haar measure of G/N so normalized that the integration formula III.13.17(8) holds, that is,

$$\int_G \phi(x)\,d\lambda x = \int_{G/N} d\mu(xN) \int_N \phi(xn)\,dvn \tag{6}$$

for all ϕ in $\mathscr{L}(G)$.

Now $f \mapsto \tilde{f}$, as we have seen, is a linear map of $\mathscr{L}(\mathscr{B})$ into $\mathscr{L}(\mathscr{C})$. Applying (6) to the function $\phi(x) = \| f(x) \|$, we conclude that $f \mapsto \tilde{f}$ is an isometry with respect to the norms of $\mathscr{L}_1(\lambda; \mathscr{B})$ and $\mathscr{L}_1(\mu; \mathscr{C})$. Further, $\mathscr{L}(\mathscr{B})$ is dense in

$\mathscr{L}_1(\lambda; \mathscr{B})$; and it is an easy consequence of II.15.10 that $\{\tilde{f}: f \in \mathscr{L}(\mathscr{B})\}$ is dense in $\mathscr{L}_1(\mu; \mathscr{C})$. Therefore the isometry $f \mapsto \tilde{f}$ on $\mathscr{L}(\mathscr{B})$ extends to a linear isometric bijection $\Phi: \mathscr{L}_1(\lambda; \mathscr{B}) \to \mathscr{L}_1(\mu; \mathscr{C})$.

We shall show that Φ preserves convolution. Since $\mathscr{L}(\mathscr{B})$ is dense in $\mathscr{L}_1(\lambda; \mathscr{B})$, it is enough to show that $\Phi(f * g) = \Phi(f) * \Phi(g)$, or $(f * g)^\sim = \tilde{f} * \tilde{g}$, for $f, g \in \mathscr{L}(\mathscr{B})$. Given $f, g \in \mathscr{L}(\mathscr{B})$, the convolution $\tilde{f} * \tilde{g}$ in $\mathscr{L}_1(\mu; \mathscr{C})$ was defined in 5.2(1) by:

$$(\tilde{f} * \tilde{g})(\alpha) = \int_{G/N} \tilde{f}(\beta)\tilde{g}(\beta^{-1}\alpha)d\mu\beta. \tag{7}$$

Evaluating each side of (7) at a point x belonging to the coset α, we have by (1)

$$(\tilde{f} * \tilde{g})(\alpha)(x) = \int_{G/N} \int_\beta \tilde{f}(\beta)(y)\tilde{g}(\beta^{-1}\alpha)(y^{-1}x)dv_\beta y \, d\mu\beta$$

$$= \int_{G/N} d\mu(yN) \int_N f(yn)g(n^{-1}y^{-1}x)dvn$$

$$= \int f(z)g(z^{-1}x)d\lambda z \tag{by (6)}$$

$$= (f * g)(x) = (f * g)^\sim(\alpha)(x). \tag{8}$$

It follows from (8) that $\tilde{f} * \tilde{g} = (f * g)^\sim$. So Φ preserves convolution.

A comment must be made on the validity of evaluating each side of (7) at a point x of α. Since the right side of (7) is evaluated at x *under* the integral sign, we are here proposing to apply II.5.7 to the evaluation map $\phi \mapsto \phi(x)$. At first sight this seems not legitimate, since $\phi \mapsto \phi(x)$ is not continuous with respect to the norm of C_α. However, it is not hard to check that, when f and g are in $\mathscr{L}(\mathscr{B})$, the vector-valued integral on the right of (7) exists not only with respect to the topology of C_α, but with respect to the (stronger) inductive limit topology of $\mathscr{L}(\mathscr{B}_\alpha)$, and in the latter topology the map $\phi \mapsto \phi(x)$ is continuous. Therefore II.6.3 applied to the inductive limit topology and the map $\phi \mapsto \phi(x)$ justifies the passage from (7) to (8).

Next we claim that Φ preserves *. Indeed: If Δ' denotes the modular function of G/N, we have by III.13.20

$$\Delta'(xN)\Gamma(x) = \Delta(x) \qquad (x \in G). \tag{9}$$

Hence, for $f \in \mathscr{L}(\mathscr{B})$, $x \in \alpha \in G/N$,

$$(\tilde{f})^*(\alpha)(x) = \Delta'(\alpha^{-1})[(\tilde{f}(\alpha^{-1}))^*](x) \qquad \text{(by 5.6(7))}$$

$$= \Delta'(\alpha^{-1})\Gamma(x^{-1})[\tilde{f}(\alpha^{-1})(x^{-1})]^* \qquad \text{(by (5))}$$

$$= \Delta'(\alpha^{-1})\Gamma(x^{-1})(f(x^{-1}))^*$$

$$= \Delta(x^{-1})(f(x^{-1}))^* \qquad \text{(by (9))}$$

$$= f^*(x) = (f^*)^{\tilde{}}(\alpha)(x).$$

So $(\tilde{f})^* = (f^*)^{\tilde{}}$ for $f \in \mathscr{L}(\mathscr{B})$, proving the claim.

We have now proved:

Proposition. $\mathscr{L}_1(\lambda; \mathscr{B})$ and $\mathscr{L}_1(\mu; \mathscr{C})$ are isometrically *-isomorphic under the isometry $\Phi: \mathscr{L}_1(\lambda; \mathscr{B}) \to \mathscr{L}_1(\mu; \mathscr{C})$ determined by:

$$\Phi(f) = \tilde{f} \qquad \text{for } f \in \mathscr{L}(\mathscr{B}).$$

In particular, the $\mathscr{L}_1$ cross-sectional algebra of the group extension bundle corresponding to the group extension $N \to G \to G/N$ is isometrically *-isomorphic with the $\mathscr{L}_1$ group algebra of G.

6.8.   Proposition.   *If $\mathscr{B}$ is saturated, its partial cross-sectional bundle $\mathscr{C}$ is also saturated.*

Proof.   Fix two cosets α, β in G/N, and an element x of α. If $b \in B_x$ and $\psi \in \mathscr{L}(\mathscr{B}_\beta)$, let $b\psi$ be the element $z \mapsto b\psi(x^{-1}z)$ $(z \in \alpha\beta)$ of $\mathscr{L}(\mathscr{B}_{\alpha\beta})$. Defining L as the linear span in $\mathscr{L}(\mathscr{B}_{\alpha\beta})$ of $\{b\psi : b \in B_x, \psi \in \mathscr{L}(\mathscr{B}_\beta)\}$, we see from the saturation of $\mathscr{B}$ that $\{\phi(z) : \phi \in L\}$ is dense in B_z for every z in $\alpha\beta$. Also, L is clearly closed under multiplication by continuous complex functions on $\alpha\beta$. Therefore by II.15.10 L is dense in $C_{\alpha\beta}$.

Now fix an element b of B_x. Let χ be an element of $\mathscr{L}(\mathscr{B}_\alpha)$ with $\chi(x) = b$; and let $\{\gamma_i\}$ be a net of elements of $\mathscr{L}_+(\alpha)$ such that (i) $\int \gamma_i \, dv_\alpha = 1$ (for all i), and (ii) the closed supports of the γ_i shrink down to x. Then $\phi_i = \gamma_i \chi \in \mathscr{L}(\mathscr{B}_\alpha)$; and one verifies (as in the proof of 5.11) that $\phi_i \psi \to b\psi$ in $C_{\alpha\beta}$ for every ψ in $\mathscr{L}(\mathscr{B}_\beta)$. Combining this with the preceding paragraph we conclude that the linear span of $\{\phi\psi : \phi \in \mathscr{L}(\mathscr{B}_\alpha), \psi \in \mathscr{L}(\mathscr{B}_\beta)\}$ is dense in $C_{\alpha\beta}$.   ∎

6.9. As regards multipliers, one can easily generalize 5.8 to the context of partial cross-sectional bundles. Keep the notation of 6.2–6.5.

Let u be a multiplier of $\mathcal{B}$ of order x. If $\phi \in \mathscr{L}(\mathcal{B}_\alpha)$ ($\alpha \in G/N$), define $u\phi$ and ϕu to be the elements of $\mathscr{L}(\mathcal{B}_{x\alpha})$ and $\mathscr{L}(\mathcal{B}_{\alpha x})$ respectively given by:

$$(u\phi)(y) = u\phi(x^{-1}y) \qquad\qquad (y \in x\alpha), \qquad (10)$$

$$(\phi u)(y) = \Gamma(x^{-1})\phi(yx^{-1})u \qquad (y \in \alpha x). \qquad (11)$$

One verifies without difficulty that

$$\|u\phi\|_1 \le \|u\|_0\|\phi\|_1, \qquad \|\phi u\|_1 \le \|u\|_0\|\phi\|_1;$$

so the maps $\phi \mapsto u\phi$ and $\phi \mapsto \phi u$ extend to bounded linear maps λ_u and μ_u on C_α to $C_{x\alpha}$ and $C_{\alpha x}$ respectively. It is a routine matter to check that $m_u = \langle \lambda_u, \mu_u \rangle$ is a multiplier of $\mathscr{C}$ of order xN. (To see that λ_u and μ_u are continuous, it is enough, by the argument of 3.2, to show that they carry the "standard" continuous cross-sections $\tilde{f}$ of $\mathscr{C}$, where $f \in \mathscr{L}(\mathcal{B})$, into continuous cross-sections. But this follows from the evident formulae:

$$u\tilde{f} = (uf)\tilde{\ }, \qquad \tilde{f}u = (fu)\tilde{\ }.) \qquad (12)$$

One easily checks that the map $u \mapsto m_u$ preserves the linear operations (on each fiber) as well as multiplication and involution, and is norm-decreasing. In particular, if u is unitary so is m_u. Thus we have:

Proposition. *If $\mathcal{B}$ has enough unitary multipliers, then so does $\mathscr{C}$.*

Remark. If we identify $\mathscr{L}_1(\lambda; \mathcal{B})$ and $\mathscr{L}_1(\mu; \mathscr{C})$ by the *-isomorphism Φ of 6.7, it follows from (12) that the integrated forms of a multiplier u of $\mathcal{B}$ and of the corresponding multiplier m_u of $\mathscr{C}$ are the same.

6.10. Proposition. *If $\mathcal{B}$ has a strong approximate unit, then so does $\mathscr{C}$.*

Proof. If $h \in \mathscr{L}(\mathcal{B}_N)$ and $f \in \mathscr{L}(\mathcal{B})$, let hf and fh be the elements of $\mathscr{L}(\mathcal{B})$ given by

$$(hf)(x) = \int_N h(n)f(n^{-1}x)dvn,$$

$$(fh)(x) = \int_N f(xn)h(n^{-1})dvn$$

($x \in G$). These definitions are related as follows to the multiplication in $\mathscr{C}$:

$$h\tilde{f}(\alpha) = (hf)\tilde{\ }(\alpha), \qquad \tilde{f}(\alpha)h = (fh)\tilde{\ }(\alpha) \qquad (\alpha \in G/N), \qquad (13)$$

Now, just as in 5.13, we can find a net $\{h_\sigma\}$ of elements of $\mathscr{L}(\mathscr{B}_N)$ such that

(i) $\{\|h_\sigma\|_1 : \sigma \text{ varying}\}$ is bounded, and
(ii) $h_\sigma f \to f$ and $fh_\sigma \to f$ in the inductive limit topology of $\mathscr{L}(\mathscr{B})$ for all f in $\mathscr{L}(\mathscr{B})$.

By (13), this implies in particular that, for all f in $\mathscr{L}(\mathscr{B})$,

$$h_\sigma \tilde{f}(\alpha) \to \tilde{f}(\alpha) \quad \text{and} \quad \tilde{f}(\alpha)h_\sigma \to \tilde{f}(\alpha)$$

in C_α uniformly for α in G/N. Therefore, by the equivalence of (i) and (iii) of 2.12, $\{h_\sigma\}$ is a strong approximate unit of $\mathscr{C}$.   ∎

7. Transformation Bundles

7.1. In this section we encounter another important way of forming Banach *-algebraic bundles. The full import of the bundles studied in this section will become apparent only when we develop their representation theory in §18.

7.2. For this section, we fix a Banach *-algebraic bundle $\mathscr{B} = \langle B, \pi, \cdot, * \rangle$ over a locally compact group G (with unit e), and also a locally compact Hausdorff space M on which G acts to the left as a topological transformation group. From these ingredients we are going to construct a new Banach *-algebraic bundle $\mathscr{D}$ over G.

For each x in G let $D_x = \mathscr{C}_0(M; B_x)$, the Banach space (with the supremum norm $\| \ \|_\infty$) of all continuous functions on M to B_x which vanish at infinity. We denote by E the important linear space of all continuous functions $f : G \times M \to B$ such that (i) $f(x, m) \in B_x$ for all x in G and m in M, and (ii) f has compact support, that is, $f(x, m) = 0_x$ for all $\langle x, m \rangle$ outside some compact subset of $G \times M$. If $f \in E$, let $\tilde{f}$ be the function on G assigning to each x in G the element $\tilde{f}(x) : m \mapsto f(x, m)$ of D_x. Then it is easy to see that the family $\{\tilde{f} : f \in E\}$ satisfies the hypotheses of Theorem II.13.18, and so determines a unique Banach bundle $\mathscr{D} = \langle D, \rho \rangle$ over G whose fiber over x is D_x, relative to which the cross-sections $\tilde{f}$ ($f \in E$) are all continuous.

We shall now introduce a multiplication and involution into $\mathscr{D}$ by means of the definitions:

$$(\phi\psi)(m) = \phi(m)\psi(x^{-1}m), \tag{1}$$

$$\phi^*(m) = (\phi(xm))^* \tag{2}$$

$(x, y \in G; \phi \in D_x; \psi \in D_y; m \in M)$. With these operations, we claim that $\mathscr{D}$ is a Banach *-algebraic bundle over G. Indeed: xm and $x^{-1}m$ (on the right of (1) and (2)) denote the action of G on M. Since $B_x B_y \subset B_{xy}$ and $(B_x)^* = B_{x^{-1}}$, and since $m \to \infty$ implies $x^{-1}m \to \infty$ and $xm \to \infty$ in M, it is easy to see that the $\phi\psi$ and ϕ^* defined in (1) and (2) lie in D_{xy} and $D_{x^{-1}}$ respectively. Postulates 2.2(i)–(iv) and 3.1(i)–(v) are easily checked. To prove the continuity of the product $\langle \phi, \psi \rangle \mapsto \phi\psi$, we shall make use of 2.4, taking Γ to be $\{\tilde{f}: f \in E\}$. Given $f, g, h \in E$, we first show by an easy calculation that $\langle x, y \rangle \mapsto \|\tilde{h}(xy) - \tilde{f}(x)\tilde{g}(y)\|$ is continuous on $G \times G$. Since $\langle x, y \rangle \mapsto \tilde{h}(xy)$ is continuous on $G \times G$ to D for arbitrary h in E, it follows as in the corresponding argument in 6.4 that $\langle x, y \rangle \mapsto \tilde{f}(x)\tilde{g}(y)$ is also continuous on $G \times G$ to D. So by 2.4 the product on D is continuous. Similarly we verify the continuity of the involution by means of 3.2.

Definition. This Banach *-algebraic bundle $\mathscr{D}$ over G is called the G, M *transformation bundle derived from $\mathscr{B}$.*

The unit fiber algebra of $\mathscr{D}$ is just $\mathscr{C}_0(M; B_e)$, considered as a Banach *-algebra under the pointwise operations and the supremum norm.

7.3. As an example, assume for the moment that $\mathscr{B}$ is the group bundle of G. Then $\mathscr{D}$ is called simply the G, M *transformation bundle*. The reader will verify that in this case $\mathscr{D}$ is the τ-semidirect product of $\mathscr{C}_0(M)$ and G (see 4.2), where τ_x is the natural action of the element x of G on $\mathscr{C}_0(M)$:

$$(\tau_x\phi)(m) = \phi(x^{-1}m).$$

7.4. Proposition. *If in 7.2 $\mathscr{B}$ is saturated, then $\mathscr{D}$ is also saturated.*

Proof. Fix x, y in G; and let L be the linear span in D_{xy} of $\{\phi\psi: \phi \in D_x, \psi \in D_y, \phi$ and ψ have compact support$\}$. Clearly L is closed under multiplication by elements of $\mathscr{C}(M)$. By the saturation of $\mathscr{B}$, $\{\chi(m): \chi \in L\}$ is dense in B_{xy} for each m in M. Therefore by II.14.6 L is dense in D_{xy}. ∎

7.5.   As in 6.9, multipliers of $\mathscr{B}$ give rise to multipliers of $\mathscr{D}$.

Let u be a multiplier of $\mathscr{B}$ of order x. It is easy to check that the equations

$$\left.\begin{aligned}(u\phi)(m) &= u\phi(x^{-1}m),\\[1mm](\phi u)(m) &= \phi(m)u\end{aligned}\right\} \tag{3}$$

($\phi \in D; m \in M$) define the left and right actions respectively of a multiplier m_u of $\mathscr{D}$ of the same order x. (The continuity of the left and right actions (3) on D follows, as in the analogous situation in 6.9, from formulae similar to 6.9(12) expressing the action of m_u on the "standard" continuous cross-sections $\tilde{f}$.)

The correspondence $u \mapsto m_u$ preserves multiplication, involution, the linear operations (on each fiber), and norm. In particular, if u is a unitary multiplier of $\mathscr{B}$, m_u is a unitary multiplier of $\mathscr{D}$. Thus we have:

Proposition.   *If $\mathscr{B}$ has enough unitary multipliers, so does $\mathscr{D}$.*

7.6.   Proposition.   *If in 7.2 $\mathscr{B}$ has a strong approximate unit, then so does $\mathscr{D}$.*

Proof.   Let M_0 be the one-point compactification of M (M_0 being M if M is compact). We first observe that $\langle \phi, m \rangle \mapsto \phi(m)$ is continuous on $D \times M_0$ to B. This follows by a type of argument already used several times, in which we approximate the variable ϕ by values of one of the "standard" cross-sections $\tilde{f}$ ($f \in E$).

Now let W be a compact subset of D. By the preceding paragraph, $V = \{\phi(m): \phi \in W, m \in M_0\}$ is a compact subset of B. Since $\mathscr{B}$ has a strong approximate unit, there is a constant k (independent of W) and an element u of B_e such that (i) $\|u\| \leq k$, and (ii) $\|ub - b\|$ and $\|bu - b\|$ are as small as we want for all b in V. If we now take ψ to be an element of D_e with $\|\psi\|_\infty \leq k$ and $\psi(m) = u$ for all m in a very large compact subset of M, it will follow that $\|\psi\phi - \phi\|_\infty$ and $\|\phi\psi - \phi\|_\infty$ are as small as we want for all ϕ in W.   ∎

The Transformation Algebra

7.7.   Let λ be a left Haar measure of G and Δ the modular function of G. As before, let $\mathscr{D}$ be the G, M transformation bundle derived from $\mathscr{B}$, and recall from 7.2 the definitions of E and the map $f \mapsto \tilde{f}$. The latter map is linear and one-to-one from E into $\mathscr{L}(\mathscr{D})$. It is natural to ask what the algebraic structure of $\mathscr{L}_1(\lambda; \mathscr{D})$ looks like when transferred to E via $f \mapsto \tilde{f}$.

Given two functions f, g in E, we define $f * g$ and f^* as functions on $G \times M$ to B as follows:

$$(f * g)(x, m) = \int_G f(y, m)g(y^{-1}x, y^{-1}m)d\lambda y, \qquad (4)$$

$$f^*(x, m) = \Delta(x^{-1})[f(x^{-1}, x^{-1}m)]^* \qquad (5)$$

$(x \in G, m \in M)$; and we introduce the following norm $\| \ \|$ on E:

$$\|f\| = \int_G \sup\{\|f(x, m)\| : m \in M\}d\lambda x. \qquad (6)$$

The usual arguments based on II.15.19 show that $f * g$ and f^* are in E. A uniform continuity argument shows that the integrand on the right of (6) is continuous with compact support.

Proposition. *With multiplication, involution, and norm given by (4), (5), and (6) respectively, E is a normed *-algebra. The map $f \mapsto \tilde{f}$ ($f \in E$) is an isometric *-isomorphism of E onto a dense *-subalgebra of $\mathcal{L}_1(\lambda; \mathcal{D})$.*

Proof. $\{\tilde{f} : f \in E\}$ is clearly closed under multiplication by arbitrary functions in $\mathcal{C}(G)$. For each x in G, II.14.6 applied to $\{\tilde{f}(x) : f \in E\}$ shows that the latter is dense in D_x. Therefore, by II.15.10 applied to $\{\tilde{f} : f \in E\}$, the latter is dense in $\mathcal{L}_1(\lambda; \mathcal{D})$.

The rest of the proposition is proved by verifying that

$$(f * g)\tilde{} = \tilde{f} * \tilde{g}, \qquad (f^*)\tilde{} = (f\tilde{})^*,$$

$$\|f\| = \|\tilde{f}\|_1$$

for all f, g in E. We leave this routine verification to the reader. (In verifying the first of these equalities, it will be useful to recall the comment made in 6.7 on the validity of point evaluation under an integral sign). ∎

7.8. Definition. E, with the convolution $*$ of (4) and the involution (5), is called the *compacted transformation algebra of G, M derived from $\mathcal{B}$.* The completion of E with respect to the norm (6) is called the *$\mathcal{L}_1$ transformation algebra of G, M derived from $\mathcal{B}$.* By the preceding proposition the latter is isometrically *-isomorphic with $\mathcal{L}_1(\lambda; \mathcal{D})$.

Notice that, if M is the one-element space, E and its completion become just the compacted cross-sectional algebra and the $\mathcal{L}_1$ cross-sectional algebra of $\mathcal{B}$ respectively.

7.9. Let $\mathscr{B}$ be the group bundle of G. Then the E of 7.8 and its completion are called simply the *compacted transformation algebra* and the $\mathscr{L}_1$ *transformation algebra* of G, M. In this case $E = \mathscr{L}(G \times M)$; and (4), (5), and (6) become:

$$(f * g)(x, m) = \int_G f(y, m)g(y^{-1}x, y^{-1}m)d\lambda y, \tag{7}$$

$$f^*(x, m) = \Delta(x^{-1})\overline{f(x^{-1}, x^{-1}m)}, \tag{8}$$

$$\|f\| = \int_G \sup\{|f(x, m)| : m \in M\}d\lambda x. \tag{9}$$

7.10. *Example.* Consider the special case that $\mathscr{B}$ is the group bundle and M is G itself, acted upon by itself by left multiplication. Then $E = \mathscr{L}(G \times G)$. Let $\beta : E \to E$ be the linear bijection given by $\beta(f)(x, y) = f(x^{-1}y, x^{-1})(\Delta(x^{-1}y))^{1/2}$. Under the transformation β, the operations (7), (8), and the norm (9) go into the following new operations $\cdot$, † and norm $\|\| \ \|\|$:

$$(f \cdot g)(x, y) = \int_G f(x, z)g(z, y)d\lambda z, \tag{10}$$

$$f^\dagger(x, y) = \overline{f(y, x)}, \tag{11}$$

$$\|\|f\|\| = \int_G \sup\{|f(y, yx)| : y \in G\}\Delta(x)^{-1/2}d\lambda x. \tag{12}$$

Equations (10) and (11) are quite familiar. They are the operations on integral kernels corresponding to the operations of composition and adjoint on the corresponding integral operators (see VI.15.21). In fact, if G is a *finite* group, E can be regarded as consisting of all $G \times G$ matrices, and (10) and (11) are the operations of matrix multiplication and adjoint. Thus, in this case, the G, G transformation algebra is *-isomorphic to the $n \times n$ total matrix *-algebra, where n is the order of G.

Remark. The last remark is the rudimentary motivation for expecting that transformation algebras derived from effective ergodic actions of groups will often give rise to simple C^*-algebras (or factor von Neumann algebras). Many mathematicians, from von Neumann and Murray onward, have in fact used transformation algebras to obtain most interesting examples of C^*-algebras and von Neumann algebras which are not of Type I. We shall see in §17 how the Glimm algebras of VI.17.5 can be interpreted as transformation algebras.

8. Locally Convex Representations of Banach Algebraic Bundles

8.1. We now take up the representation theory of Banach algebraic bundles—a subject which of course includes as a special case the representation theory of topological groups. Although our main interest lies in the involutory context and *-representations, we shall start with locally convex representations in the more general non-involutory context.

Throughout this section $\mathscr{B} = \langle B, \pi, \cdot \rangle$ is a Banach algebraic bundle over a (not necessarily locally compact) topological group G, with unit e.

8.2. *Definition.* Let X be an LCS. A *locally convex representation of $\mathscr{B}$ acting on X* is a mapping T of B into the space of linear endomorphisms of X such that:

(i) $T|B_x$ is linear for each x in G;
(ii) $T_b T_c = T_{bc}$ $(b, c \in B)$;
(iii) T_b is continuous on X for b in B;
(iv) for each ξ in X, the map $b \mapsto T_b \xi$ is continuous from B to X.

We call X the *space of T*, and denote it by $X(T)$. The concept of a locally convex representation includes of course the topology of the space in which it acts.

If X is a Fréchet [Banach] space, T is called a *Fréchet [Banach] representation* of $\mathscr{B}$.

Remark. If X is merely a linear space, and T satisfies the postulates 8.2(i), (ii), we refer to T as an *algebraic representation of $\mathscr{B}$ acting in X*.

8.3. The definitions of subrepresentation, quotient representation, intertwining operator, homeomorphic equivalence, Naimark-relatedness, finite direct sum, cyclic vector, non-degeneracy, irreducibility, and total irreducibility, given in §V.1, §V.3, can now be carried over from locally convex representations of algebras to locally convex representations of Banach algebraic bundles in an obvious way.

8.4. As a matter of fact it is not necessary, even formally, to repeat these definitions in the bundle context. Indeed, let us define the *discrete cross-sectional algebra of $\mathscr{B}$* to be the algebra D of all functions f on G to B such that $f(x) \in B_x$ for all x in G and $f(x) = 0_x$ for all but finitely many x; the linear operations in D are pointwise on G, and multiplication is "discrete convolution": $(fg)(x) = \sum_{y \in G} f(y)g(y^{-1}x)$. ($D$ may also be described as the compacted cross-sectional algebra of the Banach algebraic bundle obtained from

$\mathscr{B}$ by giving to G the discrete topology.) For every locally convex representation T of $\mathscr{B}$, the equation

$$T'_f = \sum_{x \in G} T_{f(x)} \tag{1}$$

defines a locally convex representation of D. We now define the concepts mentioned in 8.3 as holding for locally convex representations T of $\mathscr{B}$ if and only if they hold for the corresponding representations T' of D. Thus for example, T is defined to be *non-degenerate* [*irreducible*] *if* T' is non-degenerate [irreducible].

Notice that, if $\mathscr{B}$ has a unit $\mathbf{1}$, a locally convex representation T of $\mathscr{B}$ is non-degenerate if and only if $T_{\mathbf{1}}$ is the identity operator.

Remark. A locally convex representation S of D need not necessarily be of the form T' for some locally convex representation T of $\mathscr{B}$, since the obvious reconstruction of T need not satisfy 8.2(iv). Thus, in the above redefinition of the concepts of 8.3, it has to be observed that subrepresentations, quotient representations, and finite direct sums of locally convex representations of $\mathscr{B}$ again satisfy 8.2(iv). But this is trivial.

8.5. We shall now see how locally convex representations of the topological group G appear as special cases of Definition 8.2.

Definition. Let X be an LCS. A *locally convex representation of G acting in X* is a mapping T of G into the space of continuous linear endomorphisms of X such that (a) T_e is the identity operator, (b) $T_x T_y = T_{xy}$ for all x, y in G, and (c) for each ξ in X the map $x \mapsto T_x \xi$ is continuous from G to X.

Now let $\mathscr{B}$ be the group bundle of G (so that $B = \mathbb{C} \times G$ and $\langle r, x \rangle \langle s, y \rangle = \langle rs, xy \rangle$ for r, s in $\mathbb{C}$, x, y in G). If S is a locally convex representation of G, the equation

$$T_{\langle r, x \rangle} = r S_x \tag{2}$$

clearly defines a non-degenerate locally convex representation T of $\mathscr{B}$. Conversely, if T is a non-degenerate locally convex representation of $\mathscr{B}$, then

$$S_x = T_{\langle 1, x \rangle} \tag{3}$$

determines a locally convex representation S of G. The processes $S \mapsto T$ and $T \mapsto S$ of (2) and (3) are each other's inverses. So *the locally convex representations of G are in natural one-to-one correspondence with the non-degenerate locally convex representations of the group bundle of G.* Sometimes it is convenient to speak as if this correspondence were an identification, and thus to regard theorems on locally convex representations of groups as special cases of corresponding theorems on locally convex representations of Banach algebraic bundles.

The concepts mentioned in 8.3 now have a natural meaning for locally convex representations of groups: Each of them is defined to hold for a locally convex representation S of G if and only if it holds for the locally convex representation T of $\mathscr{B}$ given by (2). We shall not take time to write out the explicit conditions for each one.

Notice that a one-dimensional locally convex representation of G can be regarded as simply a continuous homomorphism of G into the multiplicative group of non-zero complex numbers.

Joint Continuity of Representations

8.6.   If T is a locally convex representation of $\mathscr{B}$, we notice from 8.2(iii), (iv) that $\langle b, \xi \rangle \mapsto T_b \xi$ is *separately* continuous on $B \times X(T)$ to $X(T)$.

Definition.   If $\langle b, \xi \rangle \mapsto T_b \xi$ is (jointly) continuous on $B \times X(T)$ to $X(T)$, we call T a *jointly continuous* locally convex representation of $\mathscr{B}$.

In general T is not jointly continuous. For example, let X be the space $\mathscr{C}(\mathbb{R})$ of all continuous complex functions on $\mathbb{R}$, with the topology of pointwise convergence. The translation operators T_x given by $(T_x f)(t) = f(t - x)$ ($f \in X$; $x, t \in \mathbb{R}$) then define a locally convex representation $T: x \mapsto T_x$ of the additive group $\mathbb{R}$, with $X(T) = X$. But T is not jointly continuous. Indeed, for $t > 0$ let f_t be a function in X satisfying:

$$f_t(s) = 0 \qquad \text{if } s \leq 0 \text{ or } s \geq 2t, \, f_t(t) = 1.$$

Then $f_t \to 0$ in X as $t \to 0+$; but $T_{-t} f_t \nrightarrow 0$ in X as $t \to 0+$.

8.7.   We shall show in a moment (8.8) that there are broad contexts in which joint continuity is automatic.

Lemma.   *Let T be a Fréchet representation of $\mathscr{B}$. If K is a compact subset of G and $F = \{ b \in B : \pi(b) \in K, \|b\| \leq 1 \}$, then the family $\{ T_b : b \in F \}$ is equicontinuous.*

Proof.   We abbreviate $X(T)$ to X. Observe that $\{ T_b : b \in C \}$ is equicontinuous for any compact subset C of B. Indeed, if $\xi \in X$, $\{ T_b \xi : b \in C \}$ is the continuous image of a compact set, hence compact and so bounded in X. So the observation follows from the Uniform Boundedness Principle for Fréchet spaces.

Now assume that $\{T_b : b \in F\}$ is not equicontinuous; and take a sequential base $\{U_n : n = 1, 2, \ldots\}$ of neighborhoods of 0 in X. Thus we can find an X-neighborhood W of 0 and, for each $n = 1, 2, \ldots$, elements b'_n of F and ξ'_n of $n^{-1}U_n$ satisfying

$$T_{b'_n} \xi'_n \notin W. \tag{4}$$

Putting $b_n = n^{-1}b'_n$, $\xi_n = n\xi'_n$, we have $\xi_n \in U_n$, $\|b_n\| \to 0$; and by (4)

$$T_{b_n} \xi_n \notin W. \tag{5}$$

Set $L = \{0_x : x \in K\} \cup \{b_n : n = 1, 2, \ldots\}$. We claim that L is a compact subset of B. Indeed, given any covering $\mathcal{U}$ of L by open subsets of B, the compactness of K lets us choose a finite subset $\mathcal{F}$ of $\mathcal{U}$ and a positive number ε such that $\{b \in B : \pi(b) \in K, \|b\| < \varepsilon\} \subset \bigcup \mathcal{F}$. Since $\|b_n\| \to 0$, this implies that $\bigcup \mathcal{F}$ contains all but finitely many of the b_n. Adjoining finitely many other sets of $\mathcal{U}$, we obtain a finite subcovering of L, proving the claim.

Therefore, by the first paragraph of the proof, $\{T_{b_n} : n = 1, 2, \ldots\}$ is equicontinuous. It follows that there is a positive integer m satisfying $T_{b_n}(U_m) \subset W$ for all n. Putting $n = m$ and recalling that $\xi_m \in U_m$, we see that this contradicts (5). So $\{T_b : b \in F\}$ is equicontinuous. ∎

8.8   Proposition.   *Assume that G is locally compact. Then any Fréchet representation T of $\mathcal{B}$ is jointly continuous.*

Proof.   Suppose that $b_i \to b$ in B and $\xi_i \to \xi$ in $X(T)$. For all i beyond some index i_0, the b_i are bounded in norm and lie inside $\pi^{-1}(K)$ for some compact neighborhood K of $\pi(b)$ in G. Therefore by 8.7 we may as well assume that $\{T_{b_i} : i \text{ varying}\}$ is equicontinuous.

Let W be any convex neighborhood of 0 in $X(T)$. By the equicontinuity of the $\{T_{b_i}\}$ there is a neighborhood V of 0 such that

$$\eta \in V \Rightarrow T_{b_i}\eta \in \tfrac{1}{2}W \qquad \text{for all } i.$$

Since $\xi_i \to \xi$, it follows that

$$T_{b_i}(\xi_i - \xi) \in \tfrac{1}{2}W \qquad \text{for all large enough } i. \tag{6}$$

Also $T_{bi}\xi - T_b\xi \in \tfrac{1}{2}W$ for all large i. Combining this with (6), and using the convexity of W, we have

$$T_{b_i}\xi_i - T_b\xi \in \tfrac{1}{2}W + \tfrac{1}{2}W = W$$

for all large i. Since W was an arbitrary convex neighborhood of 0, this gives $T_{b_i}\xi_i \to T_b\xi$. So T is jointly continuous. ∎

8.9 Proposition. *Suppose that $\mathscr{B}$ has an approximate unit $\{u_i\}$. If T is a non-degenerate Fréchet representation of $\mathscr{B}$, then $T_{u_i}\xi \to \xi$ for every ξ in $X(T)$.*

Proof. Let $Y = \{\xi \in X(T): T_{u_i}\xi \to \xi\}$. Clearly Y is a linear subspace of $X(T)$. Since $u_i b \to b$ for each b in B, it follows from 8.2(iv) that range$(T_b) \subset Y$ for all b in B, and hence that Y is dense in $X(T)$. Now since the $\{u_i\}$ are norm-bounded, the $\{T_{u_i}\}$ are equicontinuous by 8.7, whence it follows by a standard argument that Y is closed. So $Y = X(T)$. ∎

9. *-Representations of Banach *-Algebraic Bundles and Unitary Representations of Groups

9.1. For us the most important locally convex representations are the *-representations. In this section $\mathscr{B} = \langle B, \pi, \cdot, * \rangle$ is a Banach *-algebraic bundle over the topological group G.

Definition. A *-representation of $\mathscr{B}$ is a locally convex representation T of $\mathscr{B}$ whose space $X(T)$ is a Hilbert space, and which carries involution into the adjoint operation, that is, $T_{(b^*)} = (T_b)^*$ for all b in B.

Notice that the discrete cross-sectional algebra D of 8.4 is now a *-algebra under the involution

$$f^*(x) = (f(x^{-1}))^* \qquad\qquad (f \in D; x \in G).$$

If T is a *-representation of $\mathscr{B}$, the corresponding representation T' of D (given by 8.4(1)) is a *-representation.

9.2. Let T be a *-representation of $\mathscr{B}$. As in the case of *-representations of Banach *-algebras we have

$$\|T_b\| \le \|b\| \qquad \text{for all } b \text{ in } B. \tag{1}$$

Indeed: $T|B_e$ is a *-representation of the Banach *-algebra B_e; so by VI.3.8 (1) holds if $b \in B_e$. For arbitrary b in B we have $b^*b \in B_e$, and hence

$$\|T_b\|^2 = \|T_b^* T_b\| = \|T_{b^*b}\| \le \|b^*b\|$$

$$\le \|b^*\|\|b\| = \|b\|^2.$$

9.3. As a result of (1), condition 8.2(iv) can be weakened for *-representations.

Proposition.   *Let T be an algebraic representation (see 8.2) of $\mathscr{B}$, acting by bounded operators in a Hilbert space X, and satisfying (i) $T_{(b^*)} = (T_b)^*$ for b in B, and (ii) $b \mapsto (T_b \xi, \xi)$ is continuous on B for all ξ in some dense subset X_0 of X. Then T is a *-representation of $\mathscr{B}$.*

Proof.   It need only be shown that T satisfies 8.2(iv). To see this, notice that 8.2(iv) was not used in the proof of (1); hence (l) holds for our T. Consequently the set of ξ for which $b \mapsto (T_b \xi, \xi)$ is continuous must be closed, and so by (ii) coincides with X. This implies that, if $\xi \in X$ and $b_i \to b$ in B,

$$\| T_{b_i} \xi - T_b \xi \|^2 = (T_{b_i^* b_i} \xi, \xi) + (T_{b^* b} \xi, \xi) - (T_{b^* b_i} \xi, \xi) - (T_{b_i^* b} \xi, \xi) \underset{i}{\to} 0.$$

So 8.2(iv) is proved for T.   ∎

9.4.   Proposition.   *A *-representation T of $\mathscr{B}$ is non-degenerate if and only if the *-representation $T | B_e$ is non-degenerate.*

Proof.   The "if" part is obvious.

Assume that $T | B_e$ is degenerate. By VI.9.7 there is a non-zero vector ξ in $X(T)$ such that $T_a \xi = 0$ for all a in B_e. Hence $\| T_b \xi \|^2 = (T_{b^* b} \xi, \xi) = 0$ for any b in B (since $b^* b \in B_e$). So $0 \neq \xi \in N(T)$, and T is degenerate. This proves the "only if" part.   ∎

9.5.   Definition.   A *unitary representation of G* is a locally convex representation T of G acting in a Hilbert space $X(T)$ such that for every x in G T_x is a unitary operator on $X(T)$.

Since the unitary operators on a Hilbert space X are just those U in $\mathcal{O}(X)$ which satisfy $U^* = U^{-1}$, the reader will easily see that the unitary representations of G are just those locally convex representations of G which correspond under 8.5(2), (3) with non-degenerate *-representations of the group bundle of G (the latter being considered as a Banach *-algebraic bundle).

Since we are primarily interested in involutory contexts, unitary representations are for us by far the most important locally convex representations of a group.

9.6.   In view of the correspondence 8.4(1) between *-representations of the Banach *-algebraic bundle $\mathscr{B}$ and *-representations of its discrete cross-sectional algebra, not only the concepts of 8.3 but also the special involutory results and constructions of §VI.9 all hold for *-representations of $\mathscr{B}$. In particular, because of (1), VI.9.13(1) holds for any collection $\{T^i\}$ of *-representations of $\mathscr{B}$, and so the Hilbert direct sum $\sum_i^{\oplus} T^i$ can be formed and will again be a *-representation of $\mathscr{B}$. By VI.9.16 every non-degenerate *-representation of $\mathscr{B}$ is unitarily equivalent to a Hilbert direct sum of cyclic *-representations of $\mathscr{B}$.

By the argument of VI.9.12, if the bundle space of $\mathscr{B}$ is separable—in particular if it is second-countable—any cyclic *-representation of $\mathscr{B}$ must act in a separable space.

9.7. By VI.14.3, an irreducible *-representation of an Abelian Banach *-algebraic bundle is one-dimensional.

9.8. By VI.24.7, any irreducible *-representation of $\mathscr{B}$ is totally irreducible.

9.9. The multiplicity theory of discretely decomposable *-representations, developed in §VI.14 for *-representations of *-algebras, can be immediately transferred, via the discrete cross-sectional algebra, to *-representations of Banach *-algebraic bundles.

9.10. The above remarks all hold in particular, of course, for unitary representations.

Notice that a one-dimensional unitary representation of G can be regarded simply as a continuous homomorphism ϕ of G into the circle group $\mathbb{E}$. If $\phi(x) = 1$ for all x in G, ϕ is called the *trivial* (unitary) representation of G.

9.11. Remark. One might conjecture, in analogy with VI.22.6, that if H is any closed subgroup of G, any *-representation of $\mathscr{B}_H$ could be extended to a *-representation of $\mathscr{B}$. We shall show that this is false, even in the context of unitary representations.

Let G be the "$ax + b$" group of III.4.8(A). Thus G coincides as a topological space with $\{\langle b, a\rangle \in \mathbb{R}^2 : a > 0\}$, multiplication being given by

$$\langle b, a\rangle\langle b', a'\rangle = \langle b + ab', aa'\rangle.$$

Let N be the closed normal subgroup $\{\langle b, 1\rangle : b \in \mathbb{R}\}$ of G, and take a non-trivial one-dimensional unitary representation of N, say:

$$\phi(\langle b, 1\rangle) = e^{ib}.$$

Then we claim that there exists no unitary representation T of G such that there is a non-zero vector ξ in $X(T)$ satisfying

$$T_n\xi = \phi(n)\xi \qquad \text{for all } n \text{ in } N. \tag{2}$$

This is the assertion that ϕ has no unitary extension to G.

Assume that such a T and ξ exist. We may as well suppose that ξ is a cyclic vector for T (otherwise we replace T by its subrepresentation on the space generated by ξ), and hence by 9.6 that $X(T)$ is separable. For each real $a > 0$ set $\xi_a = T_{\langle 0,a \rangle}\xi$. Since $\langle b, 1 \rangle\langle 0, a \rangle = \langle 0, a \rangle\langle a^{-1}b, 1 \rangle$, we have by (2)

$$T_{\langle b, 1 \rangle}\xi_a = T_{\langle 0, a \rangle}T_{\langle a^{-1}b, 1 \rangle}\xi = e^{ia^{-1}b}\xi_a. \tag{3}$$

Now suppose a and a' are distinct positive numbers; and choose a real number b such that

$$e^{ia^{-1}b} \neq e^{ia'^{-1}b} \tag{4}$$

By (3)

$$\begin{aligned}
e^{ia^{-1}b}(\xi_a, \xi_{a'}) &= (T_{\langle b, 1 \rangle}\xi_a, \xi_{a'}) \\
&= (\xi_a, T_{\langle -b, 1 \rangle}\xi_{a'}) \\
&= e^{ia'^{-1}b}(\xi_a, \xi_{a'}).
\end{aligned}$$

By (4) this implies $(\xi_a, \xi_{a'}) = 0$. Thus the $\{\xi_a\}$ $(a > 0)$ form an uncountable family of pairwise orthogonal non-zero vectors in $X(T)$, contradicting the separability of $X(T)$. Therefore such T and ξ do not exist.

Tensor Products and Complex-Conjugates of Representations

9.12. The only operation which acts on arbitrary *-representations of a Banach *-algebra to give new ones is the Hilbert direct sum operation; and the same is true for general Banach *-algebraic bundles. When we come to unitary representations of groups, however, the situation is quite different: The special properties of unitary representations permit us, not only to form Hilbert direct sums, but to define two new operations on unitary representations, namely the operation of complex conjugation and the (inner) tensor product operation.

9.13. If X is a (complex) linear space, the *complex conjugate of X*, denoted by $\bar{X}$, is the linear space whose underlying set and addition coincide with those of X, but whose scalar multiplication is given by

$$(\lambda\xi)_{\bar{X}} = (\bar{\lambda}\xi)_X \qquad\qquad (\xi \in X; \lambda \in \mathbb{C}).$$

In other words, the identity map is conjugate-linear as a map from X to $\bar{X}$. (See §I.4.)

If X is a Hilbert space, the *complex conjugate $\bar{X}$ of X* is the Hilbert space whose underlying linear space is the complex conjugate of that of X, and whose inner product is given by

$$(\xi, \eta)_{\bar{X}} = (\eta, \xi)_X = \overline{(\xi, \eta)_X}.$$

The bounded linear operators on X and $\bar{X}$ are exactly the same; and so are the unitary operators.

9.14.   *Definition.* Let T be a unitary representation of the topological group G. By the *complex conjugate of* T we mean the unitary representation $\bar{T}$ of G given by:

$$X(\bar{T}) = \overline{X(T)}, \qquad \bar{T}_x = \overline{T_x} \qquad\qquad (x \in G).$$

9.15. If ϕ is a continuous homomorphism of G into the circle group, considered as a one-dimensional unitary representation of G (see 9.10), the reader should verify that the complex conjugate $\bar{\phi}$ of ϕ (in the sense of 9.14) is unitarily equivalent to the complex-conjugate homomorphism $x \mapsto \overline{\phi(x)}$ of G into $\mathbb{E}$.

9.16.   We shall now define the tensor product of a unitary representation of G and a *-representation of $\mathscr{B}$. The tensor product will be another *-representation of $\mathscr{B}$

Definition.   Let U be a unitary representation of G and S a *-representation of $\mathscr{B}$. We then define a *-representation T of $\mathscr{B}$ as follows:

$$X(T) = X(U) \otimes X(S) \qquad \text{(Hilbert tensor product)},$$

$$T_b = U_{\pi(b)} \otimes S_b \qquad\qquad (b \in B).$$

(The verification that T is a *-representation is made especially simple by 9.3.) This T is called the *(Hilbert) tensor product of U and S*, and is denoted by $U \otimes S$.

Notice that $U \otimes S$ is non-degenerate if S is.

9.17.   The case that $\mathscr{B}$ in 9.16 is the group bundle leads to the definition of the tensor product of two unitary representations of G.

Definition.   If U and S are two unitary representations of G, their *(Hilbert) tensor product $U \otimes S$* is defined by

$$X(U \otimes S) = X(U) \otimes X(S) \qquad \text{(Hilbert tensor product)},$$

$$(U \otimes S)_x = U_x \otimes S_x \qquad\qquad (x \in G).$$

This $U \otimes S$ is again a unitary representation of G.

Notice that, if ϕ and ψ are continuous homomorphisms of G into $\mathbb{E}$ (considered as one-dimensional unitary representations), their tensor product is just their pointwise product:

$$(\phi \otimes \psi)(x) = \phi(x)\psi(x) \qquad\qquad (x \in G).$$

9.18.   Direct sums, tensor products, and complex conjugates of representations obey the following rather evident laws:

$$U \otimes \sum_i{}^{\oplus} T^{(i)} \cong \sum_i{}^{\oplus}(U \otimes T^{(i)}),$$

$$\left(\sum_i{}^{\oplus} U^{(i)}\right) \otimes T \cong \sum_i{}^{\oplus}(U^{(i)} \otimes T),$$

$$(U \otimes V) \otimes T \cong U \otimes (V \otimes T),$$

$$U \otimes V \cong V \otimes U,$$

$$\bar{\bar{U}} = U,$$

$$(U \otimes V)^- \cong \bar{U} \otimes \bar{V}$$

$$\left(\sum_i{}^{\oplus} U^{(i)}\right)^- \cong \sum_i{}^{\oplus}(U^{(i)})^-$$

Here the U, $U^{(i)}$, V are unitary representations of G; and the T, $T^{(i)}$ are either all *-representations of $\mathscr{B}$ or all unitary representations of G.

The above laws are formally similar to the laws governing scalar multiplication in a complex linear space. It is suggestive to think of the tensor product operation 9.16 as giving to the space of all *-representations of $\mathscr{B}$ the structure of a "module" over the "ring" of unitary representations of G.

Remark.   The tensor product operation of 9.16–9.18 should strictly speaking be called the *inner* tensor product, to distinguish it from the *outer* tensor product of 21.25. In this work, unless the contrary is specified, tensor products of representations will always be *inner*.

9.19.   Remark.   As we indicated in §5 of the Introduction to Volume 2, the above Hilbert tensor product operation on *-representations is intimately related to the question of what it means to equip a given Banach *-algebra with a "Banach *-algebraic bundle structure over a given group G." See 17.9.

10. Projective Representations of Groups

10.1. In this section we shall fix a topological group G, with unit e, and define the notion of a projective representation of G. Projective representations are somewhat more general than unitary representations. They arise essentially in the applications of representation theory to quantum mechanics, and also in the generalized Mackey analysis of Chapter XII. It is therefore interesting to find that they too are a special case of the *-representation theory of Banach *-algebraic bundles. Whereas unitary representations of G are essentially the non-degenerate *-representations of the group bundle of G, projective representations of G are essentially non-degenerate *-representations of cocycle bundles over G.

10.2. Suppose that X is a Hilbert space; and let $\mathscr{U}(X)$ denote the multiplicative group of all unitary operators on X, equipped with the strong operator topology; thus $\mathscr{U}(X)$ is a topological group (though not locally compact unless X is finite-dimensional). Let $\mathscr{S}(X)$ be the closed central subgroup of $\mathscr{U}(X)$ consisting of the scalar operators $\xi \mapsto \lambda\xi$, where $\lambda \in \mathbb{C}, |\lambda| = 1$; and form the quotient topological group $\tilde{\mathscr{U}}(X) = \mathscr{U}(X)/\mathscr{S}(X)$.

Definition. A *projective representation of G on X* is a continuous homomorphism $T: G \to \tilde{\mathscr{U}}(X)$. We call X the *space of T* and denote it by $X(T)$.

Remark. For comparison, notice that a unitary representation of G is just a continuous homomorphism of G into $\mathscr{U}(X)$.

10.3. We make the obvious definition of the *unitary equivalence* of two projective representations T and T' of G, acting on spaces X and X' respectively: It is to mean that there is a linear isometric bijection $F: X \to X'$ such that

$$T' = \tilde{\Phi} \circ T,$$

where $\Phi: \mathscr{U}(X) \to \mathscr{U}(X')$ is the isomorphism $u \mapsto FuF^{-1}$, and $\tilde{\Phi}: \tilde{\mathscr{U}}(X) \to \tilde{\mathscr{U}}(X')$ is the derived isomorphism $u\mathscr{S}(X) \mapsto \Phi(u)\mathscr{S}(X')$.

10.4. It turns out that the theory of the projective representations of G can be reduced to the theory of the unitary representations of slightly larger groups.

To see this, let us take a central extension $\gamma: \mathbb{E} \underset{i}{\to} H \underset{j}{\to} G$ (see III.5.6) of the circle group $\mathbb{E}$ by G; and let T be any unitary representation of H with the property that

$$T_{i(\lambda)} = \lambda \math{1} \qquad\qquad (\lambda \in \mathbb{E}), \qquad (1)$$

where $\math{1}$ is the identity operator on $X = X(T)$. Let $u \mapsto \tilde{u}$ be the quotient map of $\mathscr{U}(X)$ onto $\tilde{\mathscr{U}}(X)$. By (1)

$$\tilde{T}: j(h) \mapsto (T_h)^{\sim} \qquad\qquad (2)$$

is well defined as a homomorphism of G into $\tilde{\mathscr{U}}(X)$; and the openness of j implies that $\tilde{T}$ is continuous. So $\tilde{T}$ is a projective representation of G.

A projective representation $\tilde{T}$ of G which is constructed from ingredients γ and T as above is said to be *constructible from* γ. If S is a projective representation of G which is constructible from γ, it is clearly also constructible from any central extension which is isomorphic to γ. We shall now prove the following important fact:

Proposition. *Let S be any projective representation of G (with $X(S) \neq \{0\}$). Then S is constructible from some central extension γ of $\mathbb{E}$ by G; and this γ is unique to within isomorphism.*

Proof. Let X be the space of S, and $u \mapsto \tilde{u}$ the quotient map of $\mathscr{U}(X)$ into $\tilde{\mathscr{U}}(X)$; and define $H = \{\langle u, x\rangle \in \mathscr{U}(X) \times G : S_x = \tilde{u}\}$. Since S and $u \mapsto \tilde{u}$ are homomorphisms, H is a topological subgroup of the product topological group $\mathscr{U}(X) \times G$. Evidently $i: \lambda \mapsto \langle \lambda \math{1}_X, e\rangle$ ($\lambda \in \mathbb{E}$) is a topological isomorphism of $\mathbb{E}$ into H. Since $u \mapsto \tilde{u}$ is open, the surjective homomorphism $j: \langle u, x\rangle \mapsto x$ of H onto G is also open; and obviously $\mathrm{Ker}(j) = i(\mathbb{E})$. So $\gamma: \mathbb{E} \underset{i}{\to} H \underset{j}{\to} G$ is a central extension of $\mathbb{E}$ by G. The mapping $T: \langle u, x\rangle \mapsto u$ is a unitary representation of H satisfying (1); and by the definition of H, $(T_{\langle u, x\rangle})^{\sim} = S_x = S_{j(u, x)}$. Therefore S is constructible from γ.

To prove that γ is unique to within isomorphism suppose we have another central extension $\gamma_0: \mathbb{E} \underset{i_0}{\to} H_0 \underset{j_0}{\to} G$ and a unitary representation R of H_0 acting in X and satisfying $R_{i_0(\lambda)} = \lambda \math{1}$ ($\lambda \in \mathbb{E}$) and $S_{j_0(h)} = (R_h)^{\sim}$ ($h \in H_0$). Defining

$$F(h) = \langle R_h, j_0(h)\rangle \qquad\qquad (h \in H_0),$$

we easily verify that F is an algebraic isomorphism of H_0 onto H, and is continuous. One verifies immediately that $F \circ i_0 = i$ and $j \circ F = j_0$. Therefore by Remark III.5.3 γ and γ_0 are isomorphic extensions. ∎

10.5. *Definition.* If a projective representation S of G is constructible from a central extension γ of $\mathbb{E}$ by G, we shall say that S is *of projective class* γ. By 10.4 every projective representation S is of some projective class γ, and the isomorphism class of γ is uniquely determined by S.

Obviously two unitarily equivalent projective representations are of the same projective class.

Thus the classification of all projective representations of G splits into two parts: (A) Classify the isomorphism classes of central extensions γ of $\mathbb{E}$ by G; (B) for each such γ, classify to within unitary equivalence the projective representations of projective class γ.

10.6. *Definition.* Let $\gamma: \mathbb{E} \underset{i}{\to} H \underset{j}{\to} G$ be a central extension. By a γ-*representation of* G we shall mean a unitary representation T of H satisfying (1).

If T is a γ-representation of G, let $\tilde{T}$ be the projective representation defined by (2). The following proposition is easily verified:

Proposition. *If T and V are two γ-representations of G, then $\tilde{T}$ and $\tilde{V}$ are unitarily equivalent if and only if the following condition (E) holds: There is a one-dimensional unitary representation χ of G such that T is unitarily equivalent to the unitary representation $\chi \times V: h \mapsto \chi(j(h))V_h$ of H.*

In view of this proposition, task (B) of 10.5 amounts to classifying the γ-representations of G, not with respect to unitary equivalence, but with respect to the weaker relation defined by condition (E) of the above proposition.

10.7. *Remark.* We shall see in 14.13 that, if G is locally compact, there exist γ-representations for every central extension γ of $\mathbb{E}$ by G, and hence there exist projective representations of every projective class γ.

10.8. Let β be the direct product extension of $\mathbb{E}$ by G. The reader will verify without difficulty that a projective representation S of G is of projective class β if and only if there exists a unitary representation T of G, with $X(T) = X(S)$, such that

$$S_x = (T_x)^{\sim} \qquad\qquad (x \in G).$$

A projective representation of G of projective class β will therefore be called *essentially unitary.*

In view of 10.4 and Proposition III.5.11 we have:

Proposition.   *If G is either* $\mathbb{R}$, $\mathbb{E}$, *or a finite or infinite cyclic group, then every projective representation of G is essentially unitary.*

In Chapter XII we shall find examples of projective representations which are not essentially unitary.

γ-Representations and Cocycle Representations

10.9.   When central extensions are described in terms of cocycles, as in III.5.12, the γ-representations of G become the "multiplier representations" studied by Mackey.

Since cocycles appear in this work only for clarifying the relation between our development and the work of other mathematicians, we shall in this number, as in III.5.12, content ourselves with assuming that G is discrete.

Let σ be a cocycle in $C(\mathbb{E}, G)$ (see III.5.12); and let $\gamma_\sigma : \mathbb{E} \underset{i}{\to} H_\sigma \underset{j}{\to} G$ be the central extension canonically constructed from σ as in III.5.13. If we take a γ_σ-representation T of G, and set

$$S_x = T_{\langle 1, x \rangle} \qquad\qquad (x \in G), \qquad (3)$$

then $S_e = 1$, and $S_x S_y = T_{\langle \sigma(x, y)^{-1}, xy \rangle} = T_{\langle i(\sigma(x, y)) \rangle^{-1}} \cdot S_{xy} = \sigma(x, y)^{-1} S_{xy}$, or

$$S_{xy} = \sigma(x, y) S_x S_y \qquad\qquad (x, y \in G). \qquad (4)$$

A mapping S of G into the set of unitary operators on a Hilbert space X such that $S_e = 1_X$ and (4) holds will be called a *cocycle representation of G with cocycle* σ, or simply a *σ-representation of G.* Thus, each γ_σ-representation T of G gives rise by (3) to a σ-representation S of G; and, conversely, it is easy to see that every σ-representation S of G arises via (3) from a unique γ_σ-representation T.

Combining this with 10.4, we have reduced the study of the projective representations of G to that of its cocycle representations (at least for discrete G).

γ-Representations and Banach *-Algebraic Bundles

10.10.   Let $\gamma : \mathbb{E} \underset{i}{\to} H \underset{j}{\to} G$ be a central extension of $\mathbb{E}$ by G. We have seen in 4.7 and 4.8 how to construct from γ a certain cocycle bundle $\mathscr{B} = \mathscr{B}^\gamma$ over G. The construction was as follows: $\mathbb{E}$ acts on $H \times \mathbb{C}$ according to:

$$\lambda \langle h, z \rangle = \langle h i(\lambda), \lambda^{-1} z \rangle \qquad (\lambda \in \mathbb{E}; z \in \mathbb{C}; h \in H).$$

Let $\langle h, z \rangle^{\sim}$ denote the orbit of $\langle h, z \rangle$ under this action. The bundle space B of $\mathscr{B}^\gamma$ is the quotient space of all the orbits $\langle h, z \rangle^{\sim}$; the bundle projection $\pi : B \to G$ is given by

$$\pi(\langle h, z \rangle^{\sim}) = j(h) \qquad (h \in H; z \in \mathbb{C});$$

and the multiplication and involution in $\mathscr{B}^\gamma$ are as follows:

$$\langle h, z \rangle^{\sim} \langle h', z' \rangle^{\sim} = \langle hh', zz' \rangle^{\sim},$$

$$\langle h, z \rangle^* = \langle h^{-1}, \bar{z} \rangle.$$

Each fiber of $\mathscr{B}^\gamma$ is one-dimensional; and $\langle e, 1 \rangle^{\sim}$ is the unit of $\mathscr{B}^\gamma$.

Proposition. *From each γ-representation T of G we obtain a non-degenerate *-representation T' of $\mathscr{B}^\gamma$ as follows:*

$$T'_{\langle h, z \rangle^{\sim}} = z T_h \qquad (h \in H; z \in \mathbb{C}). \qquad (5)$$

*Conversely, every non-degenerate *-representation T' of $\mathscr{B}^\gamma$ is given by (5), where T is the γ-representation of G defined by:*

$$T_h = T'_{\langle h, 1 \rangle^{\sim}} \qquad (h \in H). \qquad (6)$$

The proof is of a routine nature, and is left to the reader.

Thus the γ-representations of G are in natural one-to-one correspondence with the non-degenerate *-representations of $\mathscr{B}^\gamma$.

11. Integrable Locally Convex Representations

11.1. In this and the next section we shall study the integrated forms of locally convex representations in the non-involutory context. $\mathscr{B} = \langle B, \pi, \cdot \rangle$ will be a fixed Banach algebraic bundle over a *locally compact* group G with unit e, left Haar measure λ, and modular function Δ.

11.2. Definition. A locally convex representation T of $\mathscr{B}$ is said to be *integrable* if, for every f in $\mathscr{L}(\mathscr{B})$, there is a continuous linear operator $\tilde{T}_f$ on $X(T)$ satisfying

$$\alpha(\tilde{T}_f \xi) = \int_G \alpha(T_{f(x)} \xi) d\lambda x \qquad (1)$$

for all ξ in $X(T)$ and α in $(X(T))^*$.

Remarks.   The integrand on the right of (1) is automatically continuous with compact support.

By the Hahn-Banach Theorem there are enough α in $X(T)^*$ to distinguish points of $X(T)$; so the $\tilde{T}_f$ of (1), if it exists, is unique.

By II.6.2, (1) amounts to saying that $\tilde{T}_f = \int T_{f(x)} \, d\lambda x$, the right side being an $\mathcal{O}(X(T))$-valued integral with respect to the weak operator topology.

Remark.   As an example of a non-integrable locally convex representation we may take the locally convex representation T of $\mathbb{R}$ mentioned in 8.6 as an instance of the failure of joint continuity. Indeed: Let $\mathscr{C}(\mathbb{R})$, as in 8.6, carry the pointwise convergence topology. Define the functions f_t in $\mathscr{C}(\mathbb{R})$ as in 8.6, with the additional proviso that $f_t \geq 0$; and put $f'_t = k_t f_t$, where k_t is such a positive constant that $\int f'_t \, d\lambda = 1$ (λ being Lebesgue measure). Just as for $\{f_t\}$, we have $f'_t \to 0$ in $\mathscr{C}(\mathbb{R})$ as $t \to 0+$. Now let $0 \neq g \in \mathscr{L}(\mathbb{R})$. By (1), if T is integrable, $\tilde{T}_g f'_t$ must be the convolution $g * f'_t$. Since $\{f'_t\}$ is an approximate unit on G, we therefore have $\tilde{T}_g f'_t \to g \neq 0$ in $\mathscr{C}(\mathbb{R})$ as $t \to 0+$. Consequently $\tilde{T}_g$ cannot be continuous as an operator on $\mathscr{C}(\mathbb{R})$.

11.3.   Proposition.   *Any Fréchet representation T of $\mathscr{B}$ is integrable.*

Proof.   We abbreviate $X(T)$ to X. Let f be any element of $\mathscr{L}(\mathscr{B})$. For each ξ in X, $x \mapsto T_{f(x)}\xi$ is continuous on G to X and has compact support. So, since X is complete, $\int_G T_{f(x)}\xi \, d\lambda x$ exists as an X-valued integral (see II.6.2); call it $\tilde{T}_f \xi$. By II.5.4(2),

$$p(\tilde{T}_f \xi) \leq \int p(T_{f(x)}\xi) d\lambda x \qquad (2)$$

for any continuous seminorm p on X.

Now by 8.7 $\{T_{f(x)} : x \in G\}$ is equicontinuous. Hence, given $\varepsilon > 0$ and a continuous seminorm p on X, we can find an X-neighborhood V of 0 such that

$$\xi \in V \Rightarrow p(T_{f(x)}\xi) < \varepsilon \qquad \text{for all } x \text{ in } G.$$

From this and (2) it follows that

$$\xi \in V \Rightarrow p(\tilde{T}_f \xi) \leq \lambda(K)\varepsilon,$$

where K is the compact support of f. By the arbitrariness of p and ε this implies that $\tilde{T}_f : \xi \mapsto \tilde{T}_f \xi$ is continuous. Applying II.6.3 to the definition of $\tilde{T}_f \xi$ we obtain (1). So T is integrable.   ∎

In particular every *-representation of a Banach *-algebraic bundle over G is integrable.

11.4. Proposition. *Let T be an integrable locally convex representation of $\mathscr{B}$. Then $\tilde{T}: f \mapsto \tilde{T}_f$ ($f \in \mathscr{L}(\mathscr{B})$), where $\tilde{T}_f$ is defined by (1), is a locally convex representation of the compacted cross-sectional algebra $\mathscr{L}(\mathscr{B})$ (see 5.2).*

If $\mathscr{B}$ is a Banach *-algebraic bundle and T is a *-representation of $\mathscr{B}$, then $\tilde{T}$ is a *-representation of $\mathscr{L}(\mathscr{B})$.

Proof. $\tilde{T}_f$ is obviously linear in f. To see that it is multiplicative, fix $x \in G$, f, $g \in \mathscr{L}(\mathscr{B})$, $\xi \in X(T)$, and $\alpha \in X(T)^*$; and apply the continuous linear functional $b \mapsto \alpha(T_b \xi)$ ($b \in B_x$) to both sides of the equation 5.2(1) defining bundle convolution. By II.5.7 we get

$$\alpha(T_{(f*g)(x)} \xi) = \int \alpha(T_{f(y)g(y^{-1}x)} \xi) d\lambda y.$$

Therefore

$$\alpha(\tilde{T}_{f*g} \xi) = \int \alpha(T_{(f*g)(x)} \xi) d\lambda x$$

$$= \iint \alpha(T_{f(y)g(y^{-1}x)} \xi) d\lambda y \, d\lambda x$$

$$= \iint \alpha(T_{f(y)g(x)} \xi) d\lambda x \, d\lambda y. \qquad (3)$$

For each y in G let α_y be the linear functional $\eta \mapsto \alpha(T_{f(y)} \eta)$ belonging to $X(T)^*$. Thus (3) becomes

$$\alpha(\tilde{T}_{f*g} \xi) = \iint \alpha_y(T_{g(x)} \xi) d\lambda x \, d\lambda y$$

$$= \int \alpha_y(\tilde{T}_g \xi) d\lambda y \qquad \text{(by (1))}$$

$$= \int \alpha(T_{f(y)} \tilde{T}_g \xi) d\lambda y$$

$$= \alpha(\tilde{T}_f \tilde{T}_g \xi) \qquad \text{(by (1)).} \qquad (4)$$

By the arbitrariness of ξ and α (4) implies $\tilde{T}_{f*g} = \tilde{T}_f \tilde{T}_g$. So $\tilde{T}$ is multiplicative, and hence is a locally convex representation.

Now suppose in addition that $\mathscr{B}$ is a Banach *-algebraic bundle and T is a *-representation. If $f \in \mathscr{L}(\mathscr{B})$ and $\xi, \eta \in X(T)$,

$$(\tilde{T}_{f^*}\xi, \eta) = \int_G (T_{f^*(x)}\xi, \eta)d\lambda x \qquad \text{(by (1))}$$

$$= \int \Delta(x^{-1})(T_{(f(x^{-1}))^*}\xi, \eta)d\lambda x \qquad \text{(by 5.6(7))}$$

$$= \int \Delta(x^{-1})(\xi, T_{f(x^{-1})}\eta)dx$$

$$= \int (\xi, T_{f(x)}\eta)d\lambda x$$

$$= (\xi, \tilde{T}_f\eta).$$

It follows that $\tilde{T}_{f^*} = (\tilde{T}_f)^*$.   ∎

Definition.   This $\tilde{T}$ is called the *integrated form of T*.

Remark.   Later on we shall often fail to observe the notational distinction between T and $\tilde{T}$, writing T_f instead of $\tilde{T}_f$ when T is an integrable locally convex representation of $\mathscr{B}$ and $f \in \mathscr{L}(\mathscr{B})$.

11.5.   Let us rewrite the preceding definition for the "group case," that is, when $\mathscr{B}$ is the group bundle.

Definition.   A locally convex representation T of the locally compact group G is *integrable* if the locally convex representation T' of the group bundle, corresponding to T via 8.5(2), is integrable. By the *integrated form* $\tilde{T}$ of T we then mean of course the integrated form of T'. Thus $\tilde{T}$ is a locally convex representation of the compacted group algebra $\mathscr{L}(G)$ satisfying

$$\alpha(\tilde{T}_f\xi) = \int_G f(x)\alpha(T_x\xi)d\lambda x$$

$(f \in \mathscr{L}(G); \xi \in X(T); \alpha \in X(T)^*)$.

If T is a unitary representation of G, it is necessarily integrable, and $\tilde{T}$ is a *-representation of $\mathscr{L}(G)$.

11.6. Consider the case that $\mathscr{B}$ is a Banach *-algebraic bundle and T is a *-representation of it. If $f \in \mathscr{L}(\mathscr{B})$ and $\xi \in X(T)$,

$$\|\tilde{T}_f \xi\| \le \int \|T_{f(x)} \xi\| d\lambda x \qquad \text{(by II.5.4(2))}$$

$$\le \|\xi\| \int \|f(x)\| d\lambda x \qquad \text{(by 9.2(1))}$$

$$= \|\xi\| \|f\|_1,$$

whence

$$\|\tilde{T}_f\| \le \|f\|_1 \qquad (f \in \mathscr{L}(\mathscr{B})). \qquad (5)$$

It follows that $\tilde{T}$ can be (uniquely) extended to a *-representation of the $\mathscr{L}_1$ cross-sectional algebra $\mathscr{L}_1(\lambda; \mathscr{B})$ of $\mathscr{B}$. This extension is called the $\mathscr{L}_1$ *integrated form* of T. The integrated form as defined merely on $\mathscr{L}(\mathscr{B})$ is sometimes specified as the $\mathscr{L}(\mathscr{B})$-*integrated form*.

We leave it to the reader to show that

$$\tilde{T}_f \xi = \int_G T_{f(x)} \xi \, d\lambda x \qquad (6)$$

for any ξ in $X(T)$ and f in $\mathscr{L}_1(\lambda; \mathscr{B})$ (the right side of course being an $X(T)$-valued integral).

Remark. In the non-involutory situation, there is no reason why the integrated form of an integrable locally convex representation of $\mathscr{B}$ should be extendable to $\mathscr{L}_1(\lambda; \mathscr{B})$. Indeed, even for Banach representations of $\mathscr{B}$, 9.2(1) (which was crucial in the proof of (5)) fails to hold in general.

11.7. We shall now show that the elementary properties of representations are preserved in the passage from a locally convex representation to its integrated form. For the rest of this section T is a fixed integrable locally convex representation of $\mathscr{B}$ acting in X, and $\tilde{T}$ is its integrated form.

11.8. Let $\xi \in X$, $\alpha \in X^*$, $x \in G$, $b \in B_x$. Let f be a continuous cross-section of $\mathscr{B}$ with $f(x) = b$. Choosing a net $\{\phi_i\}$ of functions in $\mathscr{L}_+(G)$ with $\int \phi_i d\lambda = 1$ and with compact supports shrinking down to x, we have by (1)

$$\lim_i \alpha(\tilde{T}_{\phi_i f} \xi) = \alpha(T_b \xi). \qquad (7)$$

From (7) it follows that T *is uniquely determined by its integrated form.*

11.9.   Proposition.   *A closed subspace Y of X is T-stable if and only if it is*
$\tilde{T}$-*stable.*

Proof.   By the Hahn–Banach Theorem

$$Y = \{\xi \in X : \alpha(\xi) = 0 \text{ whenever } \alpha \in X^*, \alpha(Y) = \{0\}\}. \tag{8}$$

Suppose Y is T-stable; and let $\xi \in Y$, $f \in \mathcal{L}(\mathcal{B})$. Thus $T_{f(x)}\xi \in Y$ for all x; and
so

$$\alpha(\tilde{T}_f \xi) = \int \alpha(T_{f(x)}\xi) d\lambda x = 0$$

whenever $\alpha \in X^*$ and $\alpha(Y) = \{0\}$. From this and (8) it follows that $\tilde{T}_f \xi \in Y$.
Consequently Y is $\tilde{T}$-stable.

Conversely, assume Y is $\tilde{T}$-stable. Then, if $\alpha \in X^*$ and $\alpha(Y) = \{0\}$, we have
$\alpha(\tilde{T}_f \xi) = 0$ for all f in $\mathcal{L}(\mathcal{B})$ and ξ in Y, and so by (7) $\alpha(T_b\xi) = 0$ for all b in B
and ξ in Y. From this and (8) it follows that Y is T-stable.   ∎

Remark.   This proposition fails for *non-closed* subspaces Y of X. The reader
should construct a counter-example for himself.

11.10.   The proof of the following proposition is a repetition of the methods
of 11.9.

Proposition.   *The closures of the linear spans of* $\{\text{range}(T_b) : b \in B\}$ *and of*
$\{\text{range}(\tilde{T}_f) : f \in \mathcal{L}(\mathcal{B})\}$ *are the same. Furthermore, T is non-degenerate if and*
only if $\tilde{T}$ *is non-degenerate.*

11.11.   Proposition.   *A vector ξ in X is cyclic for T if and only if it is cyclic*
for $\tilde{T}$.

Proof. ξ is cyclic for T if and only if $\alpha = 0$ whenever $\alpha \in X^*$ and $\alpha(T_b\xi) = 0$
for all b in B; and similarly for $\tilde{T}$. Now use (1) and (7).   ∎

11.12.   Proposition.   *Let Y be a closed T-stable (hence* $\tilde{T}$*-stable) subspace of*
X; and let T′ and T″ be the subrepresentation and quotient representation
deduced from T, acting on Y and X/Y respectively. Let $\tilde{T}'$ *and* $\tilde{T}''$ *be the same*
deduced from $\tilde{T}$. *Then T′ and T″ are integrable, and*

$$(T')^\sim = \tilde{T}', \qquad (T'')^\sim = \tilde{T}''.$$

The proof of this is almost immediate when we recall from the Hahn–
Banach Theorem that every element of Y^* extends to an element of X^*.

11.13. Proposition. *Let $T^{(1)}, \ldots, T^{(n)}$ be integrable locally convex representations of $\mathscr{B}$. Then $\sum_{i=1}^{n\oplus} T^{(i)}$ is integrable; and its integrated form is $\sum_{i=1}^{n\oplus} (T^{(i)})^{\sim}$.*

11.14. Proposition. *$\tilde{T}$ is irreducible [totally irreducible] if and only if T is irreducible [totally irreducible].*

Proof. By V.1.9, V.1.10, 11.11, and 11.13. ∎

11.15. Proposition. *If S is another integrable locally convex representation of $\mathscr{B}$, acting on an LCS Y, then a continuous linear map $F: X \to Y$ is T, S intertwining if and only if it is $\tilde{T}$, $\tilde{S}$ intertwining.*

Proof. If we regard (the graph of) F as a closed linear subspace of $X \oplus Y$, then F is T, S intertwining if and only if it is $(T \oplus S)$-stable; and similarly for the integrated forms. So the proposition follows from 11.9 and 11.13. ∎

11.16. The same proof as in 11.15 also gives:

Proposition. *Let S be as in 11.15. Then S and T are Naimark-related (V.3.2) if and only if $\tilde{S}$ and $\tilde{T}$ are Naimark-related.*

11.17. For *-represntations, 11.13 is easily seen to be valid for infinite Hilbert direct sums.

Proposition. *Suppose $\mathscr{B}$ is a Banach *-algebraic bundle, and $\{T^{(i)}\}$ $(i \in I)$ is an indexed collection of *-representations of $\mathscr{B}$. Then*

$$\left(\sum_{i \in I}^{\oplus} T^{(i)} \right)^{\sim} = \sum_{i \in I}^{\oplus} (T^{(i)})^{\sim}.$$

The Equicontinuity Property of Integrated Forms

11.18. We begin with a lemma.

Lemma. *Let K be a closed convex subset of an LCS Y. Let μ be a measure on a σ-field of subsets of a set S, with $\mu(S) = 1$; and let $f: S \to K$ be a map such that $\xi = \int_S f(s)d\mu s$ exists as a Y-valued integral. Then $\xi \in K$.*

Proof. We may assume that Y is real. To show that $\xi \in K$ it is enough (by Eidelheit's Separation Theorem; see Day [1], p. 22, Theorem 4) to take a real number k and a continuous real linear functional α on Y satisfying $\alpha(\eta) \leq k$

for all η in K, and to show that $\alpha(\xi) \leq k$. But by II.6.3 $\alpha(\xi) = \int_S \alpha(f(s))d\mu s$. Since $\mu(S) = 1$ and $\alpha(f(s)) \leq k$ for all s, this implies $\alpha(\xi) \leq k$.   ∎

11.19.   The purpose of the following proposition is to show that hypothesis (ii) of Theorem 12.7 is a natural one.

Proposition.   *Let T be a Fréchet representation of $\mathcal{B}$. Suppose that K is a compact subset of G; and set $W = \{f \in \mathcal{L}(\mathcal{B}): f$ vanishes outside K and $\int_G \|f(x)\|d\lambda x \leq 1\}$. Then $\{\tilde{T}_f: f \in W\}$ is equicontinuous.*

Proof.   Take any closed convex neighborhood U of 0 in $X(T)$. By 8.7 there is a neighborhood V of 0 satisfying

$$T_b(V) \subset U \qquad \text{whenever } b \in \pi^{-1}(K), \|b\| \leq 1. \tag{9}$$

Let f be an element of W with $\int \|f(x)\|d\lambda x = 1$. We claim that $\tilde{T}_f(V) \subset U$.

Indeed: Let $\xi \in V$, and put $Z = \{x \in G: f(x) \neq 0\}$. Thus Z is an open set with compact closure. Define $\gamma: Z \to X(T)$ by the formula

$$\gamma(x) = (\|f(x)\|)^{-1} T_{f(x)}\xi;$$

and let $d\mu x = \|f(x)\|d\lambda x$ on Z. Thus μ is a measure on the Borel σ-field of Z satisfying $\mu(Z) = 1$; and by II.7.5 and II.6.2 $\int \gamma(x)d\mu x$ exists as an $X(T)$-valued integral and equals $\int T_{f(x)}\xi d\lambda x = \tilde{T}_f \xi$. On the other hand $\gamma(x) \in U$ for each x in Z by (9). So by 11.18 $\tilde{T}_f \xi \in U$; and the claim is proved.

Since U is convex and contains 0, it follows from this claim that

$$\tilde{T}_f(V) \subset U \qquad \text{for all } f \text{ in } W.$$

From this and the fact that the closed convex neighborhoods form a basis of neighborhoods of 0, we see that $\{\tilde{T}_f: f \in W\}$ is equicontinuous.   ∎

11.20.   Corollary.   *Let T be a Fréchet representation of $\mathcal{B}$. If $\{f_i\}$ is a net of elements of $\mathcal{L}(\mathcal{B})$ such that (i) $\|f_i\|_1 = \int \|f_i(x)\|d\lambda x \to 0$, and (ii) the f_i all vanish outside the same compact set K, then*

$$\tilde{T}_{f_i}\xi \to 0 \qquad \text{in } X(T)$$

uniformly in ξ on bounded subsets of $X(T)$.

Proof.   Let E be a bounded subset of $X(T)$, and U a neighborhood of 0 in $X(T)$. By 11.19 there is a neighborhood V of 0 such that $\tilde{T}_g(V) \subset U$ whenever $g \in \mathcal{L}(\mathcal{B})$, g vanishes outside K, and $\|g\|_1 \leq 1$. Choose $r > 0$ so that $rE \subset V$. By hypothesis (i), for all large enough i we have $\|f_i\|_1 \leq r$, hence $\|r^{-1}f_i\|_1 \leq 1$, hence (using (ii)) $\tilde{T}_{r^{-1}f_i}(V) \subset U$, and hence $\tilde{T}_{f_i}(E) = \tilde{T}_{r^{-1}f_i}(rE) \subset U$. This proves the assertion of the corollary. ∎

11.21. Suppose that T is merely an integrable locally convex representation of $\mathscr{B}$. Then the following conclusion, weaker than that of 11.20, holds: *If $\{f_i\}$ is a net of elements of $\mathscr{L}(\mathscr{B})$ such that* (i) *and* (ii) *of Corollary 11.20 hold, then $\tilde{T}_{f_i}\xi \to 0$ in $X(T)$ for every ξ in $X(T)$. In particular, for every ξ in $X(T)$, the map $f \mapsto \tilde{T}_f \xi$ is continuous on $\mathscr{L}(\mathscr{B})$ with respect to the inductive limit topology.*

Indeed, take any continuous seminorm p on $X(T)$, and any ξ in $X(T)$. We claim that $W = \{p(T_b\xi): \pi(b) \in K, \|b\| \le 1\}$ is bounded. To show this, assume that there is a net $\{b_i\}$ of elements of B such that $\pi(b_i) \in K$, $\|b_i\| \le 1$, and $p(T_{b_i}\xi) \to \infty$. Passing to a subnet and putting $c_i = p(T_{b_i}\xi)^{-1}b_i$, we may suppose that $\pi(c_i) \to x \in K$ and $\|c_i\| \to 0$, whence $c_i \to 0_x$, and $p(T_{c_i}\xi) = 1$. This contradicts the continuity of $b \mapsto p(T_b\xi)$, proving the claim. Let k be an upper bound of W. It follows then that

$$p(T_b\xi) \le k\|b\| \qquad \text{for all } b \text{ in } \pi^{-1}(K).$$

Hence for any i we have

$$p(\tilde{T}_{f_i}\xi) \le \int p(T_{f_i(x)}\xi)d\lambda x \qquad \text{(by II.5.4(2))}$$

$$\le k \int \|f_i(x)\|d\lambda x$$

$$\to 0 \qquad \text{(by hypothesis (i))}.$$

By the aribtrariness of p this shows that $\tilde{T}_{f_i}\xi \to 0$.

11.22. Let T be an integrable locally convex representation of $\mathscr{B}$. Let ϕ be a continuous cross-section of $\mathscr{B}$; and let $\{h_i\}$ be an approximate unit on G (as in III.11.17). We claim that, for every ξ in $X(T)$,

$$\lim_i \tilde{T}_{h_i\phi}\xi = T_{\phi(e)}\xi. \tag{10}$$

Indeed, choose any $\varepsilon > 0$ and any continuous seminorm p on $X(T)$. Since $x \mapsto p(T_{\phi(x)}\xi - T_{\phi(e)}\xi)$ is continuous, we may choose a G-neighborhood U of e

so small that $x \in U \Rightarrow p(T_{\phi(x)}\xi - T_{\phi(e)}\xi) < \varepsilon$. Thus, if i is so large that $\mathrm{supp}(h_i) \subset U$,

$$p(\tilde{T}_{h_i\phi}\xi - T_{\phi(e)}\xi) \le \int_G h_i(x)p(T_{\phi(x)}\xi - T_{\phi(e)}\xi)d\lambda x \quad \text{(by II.5.4(2))}$$

$$< \varepsilon.$$

By the arbitrariness of p and ε this proves (10).

11.23. As a special case of 11.22, suppose that $\mathscr{B}$ is the group bundle and that ϕ is the function identically 1 on G. Then (10) becomes the following:

Proposition. *If T is an integrable locally convex representation of G, and $\{h_i\}$ is an approximate unit on G (in the sense of III.11.17), then*

$$\lim_i \tilde{T}_{h_i}\xi = \xi$$

for all ξ in $X(T)$.

11.24. Returning to the case of a general $\mathscr{B}$, suppose that k is a positive number and T is an integrable locally convex representation of $\mathscr{B}$. Suppose also that there is a net $\{u_i\}$ of elements of B_e satisfying (i) $\|u_i\| \le k$ for all i, and (ii) $T_{u_i}\xi \to \xi$ for all ξ in $X(T)$. (By 8.9 this will be the case if $\mathscr{B}$ has an approximate unit and T is a non-degenerate Fréchet representation.) Then we claim that there exists a net $\{f_\alpha\}$ of elements of $\mathscr{L}(\mathscr{B})$ with the following three properties:

(I) The $\{\mathrm{supp}(f_\alpha)\}$ shrink down to e;
(II) $\|f_\alpha\|_1 \le k$ for all α;
(III) $\tilde{T}_{f_\alpha}\xi \to \xi$ for all ξ in $X(T)$.

Indeed, for each i let us choose a continuous cross-section ϕ_i of $\mathscr{B}$ such that $\phi_i(e) = u_i$; let $\{h_r\}$ be an approximate unit on G; and let $f_{i,r} = h_r\phi_i$. Then by 11.22 $\lim_r \tilde{T}_{f_{i,r}}\xi = T_{u_i}\xi$ for each i; and so

$$\lim_i \lim_r \tilde{T}_{f_{i,r}}\xi = \xi.$$

In view of this it is easy to extract from among the $f_{i,r}$ a net $\{f_\alpha\}$ with the required properties.

The Regular Representation of G

11.25. As we saw in Chapter VI, an important question in the study of any Banach *-algebra A is whether it is reduced, that is, whether there exist enough *-representations of A to distinguish points of A. Similar questions must be raised in the study of groups and Banach *-algebraic bundles. We should like to exhibit here one very important unitary representation of the locally compact group G, namely its regular representation R. The $\mathscr{L}_1$ integrated form of R will turn out to be faithful on $\mathscr{L}_1(\lambda)$, proving that the latter is reduced (see §14).

11.26.   Definition.   By the *left-regular representation R* of G we mean the unitary representation of G whose space is $\mathscr{L}_2(\lambda)$, and which acts by left translation:

$$(R_x f)(y) = f(x^{-1}y) \qquad (f \in \mathscr{L}_2(\lambda); x, y \in G).$$

We observe that R is indeed a unitary representation. For, by the left invariance of λ, R_x is linear and isometric on $\mathscr{L}_2(\lambda)$. Also R_e is the identity operator and $R_x R_y = R_{xy}$. Thus each R_x has an inverse and so is unitary. Finally, condition 8.5(c) holds in virtue of III.11.13.

The operation of right translation also defines a unitary representation R' on $\mathscr{L}_2(\lambda)$, provided we insert a factor involving Δ to preserve the unitariness.

Definition.   By the *right-regular representation R'* of G we mean the unitary representation of G whose space is $\mathscr{L}_2(\lambda)$, and which acts as follows:

$$(R'_x f)(y) = (\Delta(x))^{1/2} f(yx) \qquad (f \in \mathscr{L}_2(\lambda); x, y \in G).$$

As with R, one verifies that R' is indeed a unitary representation of G.

The two unitary representations R and R' are unitarily equivalent under the unitary operator Φ on $\mathscr{L}_2(\lambda)$ given by:

$$(\Phi f)(x) = \Delta(x)^{-1/2} f(x^{-1}) \qquad (f \in \mathscr{L}_2(\lambda); x \in G).$$

By the *regular representation of G* (without specification of left or right) we shall always mean the left-regular representation.

Remark.   If X is some other *LCS* consisting of complex functions on G, and if the operation of left translation defines a locally convex representation of G on X, it is usual to refer to the latter as the *left-regular representation of G on X*.

11.27. By 11.3 R and R' have integrated forms $\tilde{R}$ and $\tilde{R}'$ respectively. It follows from III.11.14(16) that

$$\tilde{R}_g f = g * f \tag{11}$$

whenever $g \in \mathscr{L}_1(\lambda)$ and $f \in \mathscr{L}_2(\lambda)$, the convolution on the right of (11) being taken in the same sense as in Proposition III.11.14.

If we assume that G is unimodular (so that λ is also left Haar measure on the reverse group $\tilde{G}$), then (11) applied to $\tilde{G}$ gives

$$\tilde{R}'_g f = f * \tilde{g} \tag{12}$$

whenever $g \in \mathscr{L}_1(\lambda)$ and $f \in \mathscr{L}_2(\lambda)$ ($\tilde{g}$ being the function $x \mapsto g(x^{-1})$).

11.28. We shall see in §14 that the integrated form of R is faithful on $\mathscr{L}_1(\lambda)$. We shall also find analogues of the regular representation for more general Banach *-algebraic bundles.

12. The Recovery of a Fréchet Representation of $\mathscr{B}$ From its Integrated Form

12.1. §11 raises a very important problem: If $\mathscr{B}$ is a Banach algebraic bundle over a locally compact group, which locally convex representations of $\mathscr{L}(\mathscr{B})$ can be obtained as the integrated forms of locally convex representations of $\mathscr{B}$? In this section we will find that, if $\mathscr{B}$ has an approximate unit, every non-degenerate Fréchet representation satisfying the equicontinuity condition of 11.19 is so obtainable. In the next section it will turn out that, in the involutory context, every *-representation of the $\mathscr{L}_1$ cross-sectional algebra of $\mathscr{B}$ is the integrated form of a *-representation of $\mathscr{B}$. The importance of this lies in the fact that it enables us, by means of the passage from bundle representations to their integrated forms, to apply the deep results of Chapters VI and VII on *-representations of Banach *-algebras to obtain corresponding deep results on *-representations of Banach *-algebraic bundles and, in particular, on unitary representations of locally compact groups.

In this section we maintain the hypotheses of 11.1.

12.2. Given a locally convex representation S of $\mathscr{L}(\mathscr{B})$, we wish to find a locally convex representation T of $\mathscr{B}$ whose integrated form is S. The technique for doing this is basically very simple: The elements of B act as multipliers on $\mathscr{L}(\mathscr{B})$ (see 5.8); so we apply a modification of 1.12 to extend S to these multipliers of $\mathscr{L}(\mathscr{B})$.

12.3. To be precise, recall that each element b of B gives rise to a multiplier of $\mathscr{B}$, and hence by 5.8 to a multiplier m_b of $\mathscr{L}_1(\lambda; \mathscr{B})$ whose left and right actions are given by:

$$(bf)(y) = bf(x^{-1}y), \tag{1}$$

$$(fb)(y) = \Delta(x^{-1})f(yx^{-1})b \tag{2}$$

($f \in \mathscr{L}_1(\lambda; \mathscr{B})$; $y \in G$; $x = \pi(b)$). It is easy to see that the left and right actions (1), (2) leave stable the subalgebra $\mathscr{L}(\mathscr{B})$ of $\mathscr{L}_1(\lambda; \mathscr{B})$; so m_b can and will be regarded as a multiplier of the compacted cross-sectional algebra $\mathscr{L}(\mathscr{B})$. As in (1) and (2) we write bf and fb instead of $m_b f$ and fm_b.

From 5.8(12), (13), we get

$$\|bf\|_1 \leq \|b\|\|f\|_1, \qquad \|fb\|_1 \leq \|b\|\|f\|_1. \tag{3}$$

12.4. For fixed f in $\mathscr{L}(\mathscr{B})$, one verifies without difficulty that the maps $b \mapsto bf$ and $b \mapsto fb$ are continuous from B to $\mathscr{L}(\mathscr{B})$ with respect to the inductive limit topology of $\mathscr{L}(\mathscr{B})$.

12.5. The bundle convolution on $\mathscr{L}(\mathscr{B})$ can be expressed in terms of the multipliers m_b. If $f, g \in \mathscr{L}(\mathscr{B})$, it follows from II.15.9 that

$$f * g = \int_G (f(x)g)d\lambda x, \tag{4}$$

the right side of (4) being an $\mathscr{L}(\mathscr{B})$-valued integral with respect to the inductive limit topology.

12.6. Let T be an integrable locally convex representation of $\mathscr{B}$. Then we claim that

$$T_b \tilde{T}_f = \tilde{T}_{bf}, \qquad \tilde{T}_f T_b = \tilde{T}_{fb} \tag{5}$$

for every b in B and f in $\mathscr{L}(\mathscr{B})$.

Indeed: If $b \in B_x$, $f \in \mathscr{L}(\mathscr{B})$, $\xi \in X(T)$, $\alpha \in X(T)^*$,

$$\alpha(\tilde{T}_{bf}\xi) = \int \alpha(T_{(bf)(y)}\xi)d\lambda y$$

$$= \int \alpha(T_b T_{f(x^{-1}y)}\xi)d\lambda y$$

$$= \int \beta(T_{f(x^{-1}y)}\xi)d\lambda y \qquad (\text{where } \beta(\eta) = \alpha(T_b\eta))$$

$$= \beta(\tilde{T}_f\xi) = \alpha(T_b\tilde{T}_f\xi).$$

By the arbitrariness of α and ξ this proves the first equality (5). The second one is proved similarly.

12.7.   Theorem.   *Suppose that $\mathscr{B}$ has a strong approximate unit. Let S be a locally convex representation of $\mathscr{L}(\mathscr{B})$ on a complete LCS X, with the following further two properties: (i) The linear span of $\{\mathrm{range}(S_f): f \in \mathscr{L}(\mathscr{B})\}$ is dense in X; (ii) for each compact subset K of G the family $\{S_f: f \in \mathscr{L}(\mathscr{B}), f \text{ vanishes outside } K, \int \|f(x)\| d\lambda x \le 1\}$ is equicontinuous. Then there exists a unique locally convex representation T of $\mathscr{B}$ on X whose integrated form is S.*

Proof.   We first prove the existence of T.

By the proof of 5.11 there is a net $\{h_i\}$ of elements of $\mathscr{L}(\mathscr{B})$ such that

(I) $\{\|h_i\|_1 : i \text{ varying}\}$ is bounded,
(II)   the h_i all vanish outside the same compact set,
(III) $h_i * f \to f$ in $\mathscr{L}_1(\lambda; \mathscr{B})$ for each f in $\mathscr{L}(\mathscr{B})$.

Let f be an element of $\mathscr{L}(\mathscr{B})$. By (II) the $h_i * f$ all vanish outside the same compact set. Hence by (III), hypothesis (ii), and the argument of 11.20,

$$S_{h_i * f}\xi \to S_f \xi \qquad\qquad (\xi \in X). \qquad (6)$$

Further, by (I), (II), and hypothesis (ii),

$$\{S_{h_i} : i \text{ varying}\} \text{ is equicontinuous.} \qquad (7)$$

Now we claim that

$$S_{h_i}\xi \to \xi \qquad \text{for all } \xi \text{ in } X. \qquad (8)$$

Indeed: Let Y be the linear space of all ξ in X for which (8) holds. By (7) Y is closed in X; and by (6) Y contains $\mathrm{range}(S_f)$ for every f in $\mathscr{L}(\mathrm{B})$. Hence by hypothesis (i) $Y = X$; and (8) is proved.

Now take any b in B. The $h_i b$ (defined by (2)) vanish outside a common compact set by (II), and by (I) and (3) $\{\|h_i b\|_1 : i \text{ varying}\}$ is bounded. Thus by hypothesis (ii)

$$\{S_{h_i b} : i \text{ varying}\} \text{ is equicontinuous.} \qquad (9)$$

It follows from (8), (9), and V.2.2 (see the remark following this proof) that there is a continuous linear operator T_b on X satisfying:

$$T_b S_f = S_{bf} \qquad\qquad (f \in \mathscr{L}(\mathscr{B})). \qquad (10)$$

By hypothesis (i), (10) determines T_b uniquely. In fact it follows from (10) that T_b is linear in b on each fiber, and that

$$T_b T_c = T_{bc} \qquad\qquad (b, c \in B). \qquad (11)$$

To prove that $T: b \mapsto T_b$ is a locally convex representation of $\mathscr{B}$, it remains to show that, for ξ in X,

$$b \mapsto T_b\xi \text{ is continuous on } B \text{ to } X. \tag{12}$$

To see this, we observe from (ii), (I), (II), and the Remark at the end of V.2.2, that

$$\{T_b : b \in \pi^{-1}(K), \|b\| \leq 1\} \text{ is equicontinuous} \tag{13}$$

for every compact subset K of G. Now let Y be the linear space of those ξ in X for which (12) holds. In view of (13) Y is closed. Further, if $b_i \to b$ in B, $f \in \mathscr{L}(\mathscr{B})$, $\eta \in X$, and $\xi = S_f\eta$, we have by 12.4 $b_i f \to bf$ in $\mathscr{L}_1(\lambda; \mathscr{B})$, and hence, from (10), (ii), and the argument of 11.20,

$$T_{b_i}\xi = S_{b_i f}\eta \to S_{bf}\eta = T_b\xi.$$

Thus Y contains the range of each S_f, and so by hypothesis (i) is dense in X. Since it is closed, it is equal to X; and (12) has been proved for all ξ. Therefore T is a locally convex representation of $\mathscr{B}$ acting in X.

All that remains now is to show that S is the integrated form of T, that is, that

$$\alpha(S_f\xi) = \int \alpha(T_{f(x)}\xi)d\lambda x \tag{14}$$

for $f \in \mathscr{L}(\mathscr{B})$, $\xi \in X$, $\alpha \in X^*$.

Fix α and f, and let Z be the linear subspace of X consisting of those ξ for which (14) holds. The left side of (14) is continuous in ξ; and by (13) so is the right side. So Z is closed. Hence by hypothesis (i) it is enough to prove (14) for $\xi = S_g\eta$ ($\eta \in X$; $g \in \mathscr{L}(\mathscr{B})$); that is, we wish to show that

$$\alpha(S_{f*g}\eta) = \int \alpha(S_{f(x)g}\eta)d\lambda x \tag{15}$$

for f, g in $\mathscr{L}(\mathscr{B})$ and η in X. But (15) results from applying the continuous linear functional $h \mapsto \alpha(S_h\eta)$ to both sides of (4) and using II.6.3.

This completes the proof of the existence of T. Its uniqueness follows from 11.8. ∎

Remark. To justify the crucial appeal to V.2.2 made in the preceding proof, we should observe that the natural map of $\mathscr{L}(\mathscr{B})$ into its multiplier algebra is one-to-one. This follows immediately from 5.11 and the fact that $\mathscr{B}$ has a strong approximate unit.

12.8.   Proposition 11.19 shows that, for Fréchet representations at least, hypothesis (ii) of 12.7 was to be expected. In fact, combining 11.3, 11.10, 11.19, and 12.7, we get:

Theorem.   *Assume that $\mathscr{B}$ has a strong approximate unit. The passage from T to its integrated form is a one-to-one correspondence between the set of all Fréchet representations T of $\mathscr{B}$ such that the linear span of $\{\text{range}(T_b): b \in B\}$ is dense in $X(T)$, and the set of all Fréchet representations S of $\mathscr{L}(\mathscr{B})$ satisfying hypotheses (i) and (ii) of 12.7.*

12.9.   We shall restate Theorem 12.8 for the "group case," that is, for the special case that $\mathscr{B}$ is the group bundle.

Theorem.   *The passage from T to its integrated form is a one-to-one correspondence between the set of all Fréchet representations of G and the set of all Fréchet representations S of $\mathscr{L}(G)$ such that (i) the linear span of $\{\text{range}(S_f): f \in \mathscr{L}(G)\}$ is dense in $X(S)$, and (ii) for each compact subset K of G, $\{S_f: f \in \mathscr{L}(G), f$ vanishes outside K, and $\int_G |f(x)|d\lambda x \leq 1\}$ is equicontinuous.*

12.10.   In the case of finite-dimensional representations one can avoid the topological intricacies of 12.7 by a direct appeal to Proposition 1.9.

Proposition.   *Let S be a non-degenerate finite-dimensional representation of $\mathscr{L}(\mathscr{B})$ which is continuous in the inductive limit topology of $\mathscr{L}(\mathscr{B})$. Then S is the integrated form of a unique locally convex representation T of $\mathscr{B}$.*

Proof.   By 1.9 S can be extended to a representation S' of $\mathscr{W}(\mathscr{L}(\mathscr{B}))$ (acting in $X(S)$). Thus by 12.3 the equation

$$T_b = S'_{m_b} \qquad\qquad (b \in B)$$

defines an algebraic representation of $\mathscr{B}$ on $X(S)$. Now for each f in $\mathscr{L}(\mathscr{B})$ the map $b \mapsto bf$ is continuous on B with respect to the inductive limit topology of $\mathscr{L}(\mathscr{B})$ (see 12.4); so it follows from the continuity of S that T is continuous on B, hence a locally convex representation of $\mathscr{B}$. The same argument as in 12.7 now shows that S is the integrated form of T. The uniqueness of T was proved in 11.8.   ∎

13.   The Recovery of a *-Representation of $\mathscr{B}$ From its Integrated Form

13.1.   In this section $\mathscr{B}$ is a Banach *-algebraic bundle over the locally compact group G with unit e, left Haar measure λ, and modular function Δ. Our aim is to show that a *-representation of $\mathscr{L}(\mathscr{B})$ satisfying a certain continuity condition must be the integrated form of a *-representation of $\mathscr{B}$. One could of course content oneself with specializing Theorem 12.8 to the involutory context. However, it turns out that in the involutory context one can obtain a stronger theorem, with much weaker hypotheses, than the one implied by Theorem 12.8; and in fact in Chapter XI we shall need this stronger version. In this section we will derive this stronger result (Theorem 13.8) independently of §12.

13.2. We begin with an involutory analogue of 12.8 which follows easily from 1.15. This is the most frequently useful involutory analogue of 12.8 though it is much weaker than what we shall prove in 13.8. Unlike 12.8, this theorem does not assume an approximate unit, nor does it confine itself to non-degenerate representations.

Theorem. *Every *-representation S of the $\mathscr{L}_1$ cross-sectional algebra of $\mathscr{B}$ is the $\mathscr{L}_1$ integrated form (see 11.6) of a unique *-representation T of $\mathscr{B}$.*

Proof. The uniqueness of T was proved in 11.8.

To prove the existence of T, it is enough to assume that S is non-degenerate. For otherwise by VI.9.14 we can write $S = S' \oplus S''$, where S' is non-degenerate and S'' is a zero representation; and, if S' is the integrated form of T' and T'' is the zero representation of $\mathscr{B}$ on $X(S'')$, S will be the integrated form of $T = T' \oplus T''$. So we shall assume that S is non-degenerate.

By 1.15 S can be "extended" to the bounded multiplier algebra of $\mathscr{L}_1(\lambda; \mathscr{B})$. In view of 5.8 (or 12.3), this means that for each b in B there is a bounded linear operator T_b on $X(S)$ *satisfying*

$$T_b S_f = S_{bf}, \qquad S_f T_b = S_{fb} \qquad (f \in \mathscr{L}_1(\lambda; \mathscr{B})), \qquad (1)$$

where bf and fb are given by 12.3(1), (2). Furthermore, 1.15 implies that $T: b \mapsto T_b$ is linear on each fiber and preserves multiplication and involution (see 5.9). To show that T is a *-representation of $\mathscr{B}$, it is enough by 8.12 to verify that $b \mapsto (T_b S_f \xi, \eta)$ is continuous on B for all ξ, η in $X(S)$ and f in $\mathscr{L}(\mathscr{B})$. But by (1) $T_b S_f \xi = S_{bf} \xi$, and the continuity of the latter in b follows from VI.3.8 and the $\mathscr{L}_1$ continuity of $b \mapsto bf$ (see 12.4). So T is a *-representation of $\mathscr{B}$.

The fact that S is the integrated form of T now follows by the same argument that was used to prove the corresponding fact in 12.7. This completes the proof. ■

Combined with 11.3, 11.9, 11.10, 11.11, 11.14, 11.15, and 11.7, this theorem asserts (i) that the passage from T to its integrated form is a one-to-one correspondence between the set of all *-representations of $\mathscr{B}$ and the set of all *-representations of $\mathscr{L}_1(\lambda; \mathscr{B})$, and (ii) that this correspondence preserves closed stable subspaces, non-degeneracy, irreducibility, intertwining operators, the commuting algebra, and Hilbert direct sums. Thus we may say that $\mathscr{B}$ and $\mathscr{L}_1(\lambda; \mathscr{B})$ have essentially identical *-representation theories.

13.3.   If $\mathscr{B}$ is the group bundle of G, we obtain (recalling the remarks on non-degeneracy in 9.5 and 11.10) the classical special case of 13.2.

Theorem.   *Every non-degenerate *-representation of the $\mathscr{L}_1$ group algebra of G is the $\mathscr{L}_1$ integrated form of a unique unitary representation of G.*

13.4.   Our next goal is to prove a stronger version of 13.2. To do this we shall need the following fact about positive functionals on $\mathscr{L}(\mathscr{B})$, which is interesting for its own sake.

The reader will recall from §§18, 19 of Chapter VI the basic facts about positive functionals, the *-representations which they generate, and Condition (R).

Proposition.   *Let p be a positive linear functional on the *-algebra $\mathscr{L}(\mathscr{B})$ which is continuous in the inductive limit topology. Then p satisfies Condition (R), and so generates a *-representation S of $\mathscr{L}(\mathscr{B})$. In fact there is a (unique) *-representation T of $\mathscr{B}$ whose integrated form is S.*

Proof.   If $b \in B$ let $f \mapsto bf$ and $f \mapsto fb$ be the left and right actions on $\mathscr{L}(\mathscr{B})$ of the multiplier m_b corresponding to b, defined as in 12.3(1), (2). Since $b \mapsto m_b$ preserves involution (5.9), we have

$$(b^*f)^* = f^*b. \qquad (2)$$

We claim that, for b in B and $f \in \mathscr{L}(\mathscr{B})$,

$$p((bf)^* * bf) \leq \|b\|^2 p(f^* * f). \qquad (3)$$

Indeed: By (2) and Schwarz's Inequality (VI.18.4(2)),

$$p((bf)^* * bf) = p(f^* * b^*bf)$$
$$\leq p(f^* * f)^{1/2} p((b^*bf)^* * (b^*bf))^{1/2}$$
$$= p(f^* * f)^{1/2} p(f^* * (b^*b)^2 f)^{1/2}.$$

Iterating this argument (with b replaced by b^*b, $(b^*b)^2$, etc.) we obtain for each positive integer n

$$p((bf)^* * bf) \leq p(f^* * f)^{1/2 + 1/4} p(f^* * (b^*b)^4 f)^{1/4}$$

$$\leq \cdots\cdots$$

$$\leq (p(f^* * f))^{1/2 + 1/4 + \cdots + 2^{-n}} (p(f^* * (b^*b)^{2^n} f))^{2^{-n}}. \qquad (4)$$

We must now fix f in $\mathscr{L}(\mathscr{B})$ and estimate the size of $p(f^* * af)$ for $a \in B_e$. Let K be the compact support of f; then $\text{supp}(f^* * af) \subset K^{-1}K$. Hence, since p is continuous in the inductive limit topology, there is a positive constant k such that

$$p(f^* * af) \leq k \| f^* * af \|_\infty \qquad\qquad (a \in B_e).$$

Now an easy calculation shows that

$$\| f^* * af \|_\infty \leq \lambda(K) \| f \|_\infty^2 \| a \| \qquad\qquad (a \in B_e).$$

Combining the last two facts we get

$$p(f^* * af) \leq k\lambda(K) \| f \|_\infty^2 \| a \| \qquad\qquad (a \in B_e). \qquad (5)$$

Putting $(b^*b)^{2^n}$ for a in (5), and substituting (5) in (4), we obtain

$$p((bf)^* * bf) \leq p(f^* * f)^{1/2 + 1/4 + \cdots + 2^{-n}} (k\lambda(K) \| f \|_\infty^2)^{2^{-n}} \| b^*b \|$$

for all $n = 1, 2, \ldots$. Passing to the limit $n \to \infty$ in this inequality, we obtain (3).

Now, as in VI.19.3, let I be the null ideal of p, construct the pre-Hilbert space $\mathscr{L}(\mathscr{B})/I$, and denote its Hilbert space completion by X. Let $\tilde{f}$ be the image in X of the element f of $\mathscr{L}(\mathscr{B})$, so that

$$(\tilde{f}, \tilde{g}) = p(g^* * f) \qquad\qquad (f, g \in \mathscr{L}(\mathscr{B})). \qquad (6)$$

The inequality (3) asserts that for each b in B the equation

$$T_b \tilde{f} = (bf)^{\tilde{}} \qquad\qquad (f \in \mathscr{L}(\mathscr{B})) \qquad (7)$$

defines a bounded linear operator T_b on X with $\| T_b \| \leq \| b \|$. Evidently $b \mapsto T_b$ is linear on each fiber, and $T_b T_c = T_{bc}$. Also $(T_b \tilde{f}, \tilde{g}) = p(g^* * bf) = p((b^*g)^* * f)$ (by (2)) $= (\tilde{f}, T_{b^*}\tilde{g})$; so $T_{b^*} = (T_b)^*$.

Thus, to show that T is a *-representation of $\mathscr{B}$, it is enough by 9.3 to show that $b \mapsto (T_b \tilde{f}, \tilde{f}) = p(f^* * bf)$ is continuous on B for each f in $\mathscr{L}(\mathscr{B})$. But this follows immediately from the following two facts: (i) $b \mapsto bf$ is continuous on B to $\mathscr{L}(\mathscr{B})$ with respect to the inductive limit topology of the latter (12.4), and (ii) $f * g$ is separately continuous in f and g on $\mathscr{L}(\mathscr{B}) \times \mathscr{L}(\mathscr{B})$ to $\mathscr{L}(\mathscr{B})$ (5.3). So we have shown that T is a *-representation of $\mathscr{B}$.

Now let S be the integrated form of T. Since S determines T (11.8), the proposition will be proved if we show that S is just the *-representation of $\mathscr{L}(\mathscr{B})$ generated by p. For this we must verify that

$$(S_h \tilde{f}, \tilde{g}) = ((h * f)\widetilde{\ }, \tilde{g}) \tag{8}$$

for all f, g, h in $\mathscr{L}(\mathscr{B})$. Now

$$(S_h \tilde{f}, \tilde{g}) = \int (T_{h(x)} \tilde{f}, \tilde{g}) d\lambda x$$

$$= \int p(g^* * h(x)f) d\lambda x. \tag{9}$$

Also, by 12.5,

$$h * f = \int (h(x)f) d\lambda x \tag{10}$$

(vector integral with respect to the inductive limit topology of $\mathscr{L}(\mathscr{B})$). Noting (by 5.3) that the linear functional $q \mapsto p(g^* * q)$ is continuous on $\mathscr{L}(\mathscr{B})$ in the inductive limit topology, and applying it to both sides of (10), we obtain by II.6.3.

$$p(g^* * h * f) = \int p(g^* * h(x)f) d\lambda x;$$

and this, together with (9), gives (8).   ■

Definition. The *-representation T of $\mathscr{B}$ constructed in the preceding proposition is said to be *generated by* the positive linear functional p on $\mathscr{L}(\mathscr{B})$.

13.5. Example III of VI.19.16 shows that the T generated by p in 13.4 need not be non-degenerate. However, we have:

Proposition. *If $\mathscr{B}$ has an approximate unit, the T of 13.4 will be non-degenerate.*

Proof. Let L_0 be the linear span of $\{af : a \in B_e, f \in \mathscr{L}(\mathscr{B})\}$. Then L_0 is closed under multiplication by complex continuous functions on G; and since $\mathscr{B}$ has an approximate unit $\{f(x) : f \in L_0\}$ is dense in B_x for every x in G. So by

II.14.6 L_0 is dense in $\mathscr{L}(\mathscr{B})$ in the inductive limit topology. Further, the continuity of p implies that the quotient map $f \mapsto \tilde{f}$ (see 13.4) is continuous from $\mathscr{L}(\mathscr{B})$ (with the inductive limit topology) to X. These facts imply that $(L_0)^{\sim}$ is dense in X. On the other hand, $(L_0)^{\sim}$ is contained in the linear span of $\{\mathrm{range}(T_a): a \in B_e\}$. So T is non-degenerate. ■

13.6. If $\mathscr{B}$ is the group bundle, then in particular $\mathscr{B}$ has a unit and 13.5 is trivial; and 13.4 becomes the following classical result on groups:

Proposition. *Let p be any positive linear functional on $\mathscr{L}(G)$ (the compacted group algebra) which is continuous in the inductive limit topology. Then p satisfies Condition (R); and there is a (unique) unitary representation T of G whose integrated form is the *-representation of $\mathscr{L}(G)$ generated by p.*

As in 13.4, T is said to be *generated by p.*

13.7. As a matter of fact, it is important to observe that the argument of 13.4 holds in somewhat greater generality. Indeed, suppose that E is any *-subalgebra of $\mathscr{L}(\mathscr{B})$ which is dense in $\mathscr{L}(\mathscr{B})$ in the inductive limit topology and such that

$$b \in B, f \in E \Rightarrow bf \in E. \tag{11}$$

Let p be a positive linear functional on E which is continuous with respect to the inductive limit topology (relativized to E). Then the same proofs as in 13.4 and 13.5 show that

(a) *p satisfies Condition (R);*
(b) *there is a unique *-representation T of $\mathscr{B}$ such that the integrated form of T, when restricted to E, is just the *-representation of E generated by p;*
(c) *if in addition $\mathscr{B}$ has an approximate unit, T is non-degenerate.*

13.8. We are now ready for our strongest answer to the question suggested in 13.1: What condition on a *-representation S of $\mathscr{L}(\mathscr{B})$ will guarantee that S is the integrated form of some *-representation of $\mathscr{B}$? By 13.2 the $\mathscr{L}_1$ continuity of S will certainly guarantee this; but in fact a much weaker condition will do. In view of 13.7 we may even work with a dense *-subalgebra of $\mathscr{L}(\mathscr{B})$ rather than $\mathscr{L}(\mathscr{B})$ itself.

Theorem.   *Let E be a *-subalgebra of $\mathscr{L}(\mathscr{B})$ which is dense in $\mathscr{L}(\mathscr{B})$ in the inductive limit topology and which satisfies* (11). *Let S be a *-representation of E and X_0 a subset of $X(S)$ with the following two properties:*

(i) *The smallest closed S-stable subspace of $X(S)$ containing X_0 is $X(S)$ itself; and*

(ii) *for every ξ in X_0, the linear functional $f \mapsto (S_f \xi, \xi)$ is continuous on E in the (relativized) inductive limit topology.*

*Then S is the restriction to E of the integrated form of a unique *-representation T of $\mathscr{B}$. In particular, S must be continuous on E in the $\mathscr{L}_1$ norm.*

Proof.   It is sufficient to assume that S is non-degenerate. Indeed, otherwise we replace S by its non-degenerate part, and X_0 by its projection onto the non-degenerate part.

Take a vector ξ in X_0; and let S^ξ be the subrepresentation of S acting on the closed S-stable subspace X_ξ of $X(S)$ generated by ξ. By VI.9.10 $\{S_f \xi : f \in E\}$ is dense in X_ξ; so the positive functional $p_\xi : f \mapsto (S_f \xi, \xi)$ on E generates S^ξ in the sense of VI.19.3. Since by hypothesis p_ξ is continuous in the inductive limit topology, it follows from 13.7 that S^ξ is the restriction of the integrated form of a *-representation of $\mathscr{B}$, and so by 11.6

$$\|S_f^\xi\| \leq \|f\|_1 \qquad\qquad (f \in E). \qquad (12)$$

Now by hypothesis (i) the linear span of $\{X_\xi : \xi \in X_0\}$ is dense in $X(S)$. So, by VI.8.14, (12) implies that

$$\|S_f\| \leq \|f\|_1 \qquad\qquad (f \in E); \qquad (13)$$

that is, S is norm-decreasing on E. Now E is dense in $\mathscr{L}(\mathscr{B})$ in the inductive limit topology, hence $\mathscr{L}_1$ dense in $\mathscr{L}_1(\lambda; \mathscr{B})$. Thus (13) permits us to extend S to a *-representation of $\mathscr{L}_1(\lambda; \mathscr{B})$. By 13.2 this extension is the $\mathscr{L}_1$ integrated form of a *-representation T of $\mathscr{B}$.

The existence of the required T is now proved. Its uniqueness is evident from 11.8 and the denseness of E in $\mathscr{L}(\mathscr{B})$.   ■

The Measure Integrated Form of a Unitary Representation

13.9.   Let T be a non-degenerate *-representation of $\mathscr{B}$, and S the (non-degenerate) $\mathscr{L}_1$ integrated form of T. By 1.15 S can be extended to a *-representation S' of the bounded multiplier algebra of $\mathscr{L}_1(\lambda; \mathscr{B})$.

Now suppose that $\mathscr{B}$ is simply the group bundle of G. In that case $\mathscr{L}_1(\lambda; \mathscr{B}) = \mathscr{L}_1(\lambda)$ and the measure algebra $\mathscr{M}_r(G)$ (see III.10.15) acts as a *-algebra of bounded multipliers on $L_1(\lambda)$. (In fact, $\mathscr{M}_r(G) \cong \mathscr{W}(\mathscr{L}_1(\lambda))$ by

Wendel's Theorem 1.23.) So S' can be regarded as a *-representation of $\mathscr{M}_r(G)$; and in this form it is called the *measure integrated form of* the unitary representation T of G.

Proposition. *The measure integrated form* S' *of the unitary representation* T *of* G *is given by*:

$$S'_\mu \xi = \int T_x \xi \, d\mu_e x \qquad (\mu \in \mathscr{M}_r(G); \, \xi \in X(T)). \qquad (14)$$

Note. μ_e is the maximal regular extension of μ; see II.8.15. The right side of (14) exists (as an $X(T)$-valued vector integral) in virtue of Remark 3 of II.8.15.

Proof. Suppose for the purposes of the proof that $S'_\mu \xi$ is *defined* by (14). Evidently $\|S'_\mu \xi\| \le \|\xi\| \, \|\mu\|$; so S'_μ is a bounded linear operator on $X(T)$. Thus it is sufficient to show that S'_μ coincides with the extension of the integrated form S of T to the multiplier algebra, that is, that

$$S'_\mu S_f = S_{\mu * f} \qquad (15)$$

for μ in $\mathscr{M}_r(G)$ and f in $\mathscr{L}_1(\lambda)$.

By III.11.14(16),

$$\mu * f = \int_G (xf) d\mu_e x \qquad (16)$$

(the right side being an $\mathscr{L}_1(\lambda)$-valued integral). Fixing ξ in $X(T)$, applying the continuous linear map $g \mapsto S_g \xi \ (g \in \mathscr{L}_1(\lambda))$ to both sides of (16), and using II.5.7, we get

$$S_{\mu * f} \xi = \int S_{xf} \xi \, d\mu_e x.$$

Now $S_{xf} = T_x S_f$; and so

$$S_{\mu * f} \xi = \int T_x S_f \xi \, d\mu_e x$$

$$= S'_\mu S_f \xi \qquad \text{(by (14))}.$$

This proves (15). ∎

14.   The Existence of Irreducible *-Representations of Banach *-Algebraic Bundles

14.1.   The preceding section will now be applied to show that, under certain general conditions, the $\mathscr{L}_1$ cross-sectional algebra of a Banach *-algebraic bundle $\mathscr{B}$ is reduced (see VI.10.2). Combined with VI.22.14, this will show that under these conditions $\mathscr{B}$ has enough irreducible *-representations to distinguish its points.

As regards the group bundle of the base group G, the integrated form of the regular representation of G turns out to be faithful on the $\mathscr{L}_1$ group algebra of G, showing that the $\mathscr{L}_1$ group algebra is reduced. Thus, to deal with more general Banach *-algebraic bundles $\mathscr{B}$, our first step will be to construct, by means of 13.4, analogues of the regular representation for $\mathscr{B}$.

14.2.   In this section we fix a Banach *-algebraic bundle $\mathscr{B}$ over a locally compact group G with unit e, left Haar measure λ, and modular function Δ. Of course, by Appendix C, $\mathscr{B}$ automatically has enough continuous cross-sections.

14.3.   Let f be any element of $\mathscr{L}_1(\lambda; \mathscr{B})$. We claim that, for any g in $\mathscr{L}(\mathscr{B})$, the cross-sections $g * f$ and $f * g$ (belonging to $\mathscr{L}_1(\lambda; \mathscr{B})$) are continuous.

Indeed: Choose a sequence $\{f_n\}$ of elements of $\mathscr{L}(\mathscr{B})$ converging to f in $\mathscr{L}_1(\lambda; \mathscr{B})$. Then one verifies easily that $g * f_n \to g * f$ and $f_n * g \to f * g$ uniformly on compact subsets of G. Since the $g * f_n$ and $f_n * g$ are continuous by 5.2, it follows that $g * f$ and $f * g$ are continuous, as we claimed.

14.4.   Lemma.   *Assume that $\mathscr{B}$ has an approximate unit and is saturated. If $f \in \mathscr{L}_1(\lambda; \mathscr{B})$ and $(h * f * g)(e) = 0$ for all $g, h \in \mathscr{L}(\mathscr{B})$, then $f = 0$.*

Remark.   Since $h * f * g$ is continuous by 14.3, the expression $(h * f * g)(e)$ makes sense.

Proof.   We first observe that if $x \in G$ and $0_x \neq b \in B_x$, then $cb \neq 0$ for some c in $B_{x^{-1}}$. Indeed: Since $\mathscr{B}$ has an approximate unit, $ab \neq 0$ for some a in B_e. Since $\mathscr{B}$ is saturated, a belongs to the closed linear span of $B_x B_{x^{-1}}$; and so $(dc)b \neq 0$ for some d in B_x and c in $B_{x^{-1}}$. This implies that $cb \neq 0$, proving the observation.

Now let $g, h \in \mathscr{L}(\mathscr{B})$. For any x in G and c in $B_{x^{-1}}$, we have

$$c[(h * f * g)(x)] = [c(h * f * g)](e)$$

$$= ((ch) * f * g)(e). \tag{1}$$

(Here c is treated as a multiplier of $\mathcal{L}_1(\lambda; \mathcal{B})$ as in 12.3(1), (2).) Since $ch \in \mathcal{L}(\mathcal{B})$, by hypothesis the right side of (1) is 0; and so $c[(h * f * g)(x)] = 0$ for all c in $B_{x^{-1}}$. By the preceding observation this implies that $(h * f * g)(x) = 0$ for all x. Thus

$$h * f * g = 0 \qquad \text{for all } g, h \text{ in } \mathcal{L}(\mathcal{B}). \tag{2}$$

Now, in view of the hypotheses and 2.13, $\mathcal{B}$ has a strong approximate unit. Hence by 5.11 $\mathcal{L}_1(\lambda; \mathcal{B})$ has an approximate unit $\{g_i\}$ consisting of elements of $\mathcal{L}(\mathcal{B})$. Thus, replacing g in (2) by g_i, we conclude from (2) that

$$h * f = 0 \qquad \text{for all } h \text{ in } \mathcal{L}(\mathcal{B}). \tag{3}$$

Replacing h in (3) by g_i we conclude from (3) that $f = 0$. ∎

14.5.   Proposition.   *Assume that $\mathcal{B}$ has enough unitary multipliers (see 3.8). If q is a continuous positive linear functional on B_e, the equation*

$$p(f) = q(f(e)) \qquad (f \in \mathcal{L}(\mathcal{B})) \tag{4}$$

defines p as a positive linear functional on $\mathcal{L}(\mathcal{B})$ which is continuous in the inductive limit topology.

Proof.   The inductive limit continuity of p follows from the continuity of q.

Suppose that $b \in B_x$ ($x \in G$). By hypothesis there is a unitary multiplier u of order x. By 3.9 we have $b = ua$ for some a in B_e, whence $b^*b = (a^*u^*)ua = a^*a$. Thus by the positivity of q,

$$q(b^*b) \geq 0 \qquad \text{for all } b \text{ in } B. \tag{5}$$

Consequently, if $f \in \mathcal{L}(\mathcal{B})$,

$$p(f^* * f) = q[(f^* * f)(e)]$$

$$= q\left[\int f^*(y)f(y^{-1})d\lambda y\right]$$

$$= q\left[\int f(y)^* f(y)d\lambda y\right]$$

$$= \int q(f(y)^* f(y))d\lambda y \qquad \text{(by II.5.7)}$$

$$\geq 0 \qquad \text{(by (5))};$$

and p is positive. ∎

14.6.   If $\mathscr{B}$, q, and p are as in 14.5, it follows from 14.5 and 13.4 that p generates a *-representation of $\mathscr{B}$ which we shall call R^q. The construction of R^q is summarized in the equations 13.4(6), (7). By 13.5 R^q is non-degenerate.

14.7.   Suppose that $\mathscr{B}$ is the group bundle of G and q is the identity map on $\mathbb{C}$, so that

$$p(f) = f(e) \qquad\qquad (f \in \mathscr{L}(\mathscr{B})). \qquad (6)$$

Then equation 13.4(6) becomes

$$(\tilde{f}, \tilde{g}) = (g^* * f)(e) = \int f(x)\overline{g(x)}d\lambda x \qquad\qquad (7)$$

$(f, g \in \mathscr{L}(G))$. Thus in this case $X(R^q)$ is the completion of $\mathscr{L}(\mathscr{B})$ with respect to the inner product (7); that is, $X(R^q) = \mathscr{L}_2(\lambda)$. Now 13.4(7) says that R_x^q operates by left translation: $(R_x^q f)(y) = f(x^{-1}y)$. Consequently R^q *is just the regular representation R of G defined in* 11.26.

   The positive linear functional (6) has already been encountered in VI.18.8 and VI.19.16. Because of its connection with the regular representation it is of great importance in functional analysis. See Chapter X.

14.8.   In view of 14.7, the *-representations R^q defined in 14.6 for the more general Banach *-algebraic bundles $\mathscr{B}$ of 14.5 may be called *generalized regular representations of $\mathscr{B}$*.

14.9.   Theorem.   *Suppose that $\mathscr{B}$ has an approximate unit and enough unitary multipliers. Then, if the unit fiber algebra B_e is reduced, $\mathscr{L}_1(\lambda; \mathscr{B})$ is reduced.*

Proof.   We shall show that the integrated forms of the generalized regular representations R^q of 14.8 separate the points of $\mathscr{L}_1(\lambda; \mathscr{B})$.

   Let f be an element of $\mathscr{L}_1(\lambda; \mathscr{B})$ and let q be a continuous positive linear functional on B_e. Denoting by S^q the integrated form of R^q, we shall show that

$$(S_f^q \tilde{g}, \tilde{h}) = q[(h^* * f * g)(e)] \qquad\qquad (8)$$

for all g, h in $\mathscr{L}(\mathscr{B})$ ($\tilde{g}$ and $\tilde{h}$ being the images of g and h in $X(R^q)$). Indeed: Fix $g, h \in \mathscr{L}(\mathscr{B})$. If $f \in \mathscr{L}(\mathscr{B})$, (8) holds in view of 13.4(8). For arbitrary f in $\mathscr{L}_1(\lambda; \mathscr{B})$, choose a sequence $\{f_n\}$ of elements of $\mathscr{L}(\mathscr{B})$ converging to f in the $\mathscr{L}_1$ norm. As in 14.3 we have $h^* * f_n * g \to h^* * f * g$ uniformly on compact subsets of G, and so

$$q[(h^* * f_n * g)(e)] \to q[(h^* * f * g)(e)]. \qquad\qquad (9)$$

On the other hand, since S^q is norm-decreasing,

$$(S_{f_n}^q \tilde{g}, \tilde{h}) \to (S_f^q \tilde{g}, \tilde{h}). \qquad\qquad (10)$$

Combining (9) and (10) with the fact that (8) holds for the f_n, we see that (8) holds for f.

Suppose now that f is an element of $\mathscr{L}_1(\lambda; \mathscr{B})$ which is not separated from 0 by the S^q; that is, $S^q_f = 0$ for all continuous positive linear functionals q on B_e. By (8) this means that $q[(h^* * f * g)(e)] = 0$ for all g, h in $\mathscr{L}(\mathscr{B})$ and all continuous positive linear functionals q on B_e. Now by assumption B_e is reduced; so there are enough such q to separate points of B_e. Hence the last equality asserts that

$$(h^* * f * g)(e) = 0 \tag{11}$$

for all g, h in $\mathscr{L}(\mathscr{B})$. Now by 3.10 the hypotheses of the theorem imply that $\mathscr{B}$ is saturated. Therefore, by 14.4, (11) gives $f = 0$.

Thus the set of all the S^q separates the points of $\mathscr{L}_1(\lambda; \mathscr{B})$. Consequently the latter is reduced. ∎

Remark. The simple example 3.14, combined with Remark VI.22.8, shows that the above theorem becomes false if the hypothesis that $\mathscr{B}$ has enough unitary multipliers is replaced by the weaker hypothesis that $\mathscr{B}$ is saturated.

14.10. Corollary. *Assume that $\mathscr{B}$ has an approximate unit and enough unitary multipliers, and that B_e is reduced. Let $\mathscr{W}$ denote the set of all irreducible *-representations of $\mathscr{B}$. Then: (I) The set of all the $\mathscr{L}_1$ integrated forms of elements of $\mathscr{W}$ separates the points of $\mathscr{L}_1(\lambda; \mathscr{B})$. (II) If b and c are two distinct elements of B which are not both zero elements (i.e., either $b \neq 0_{\pi(b)}$ or $c \neq 0_{\pi(c)}$), then there is an element T of $\mathscr{W}$ such that $T_b \neq T_c$.*

Proof.

(I) Since $\mathscr{L}_1(\lambda; \mathscr{B})$ is reduced (14.9), its irreducible *-representations S separate its points (by VI.22.14). But, by 13.2 and 11.14, such S are the integrated forms of irreducible *-representations of $\mathscr{B}$.

(II) Let b, c be as in (II); and let $\{u_i\}$ be an approximate unit of $\mathscr{B}$. By choosing an element f of $\mathscr{L}(\mathscr{B})$ which vanishes outside a small enough neighborhood of e and such that $f(e) = u_i$ for large enough i, we can ensure that $bf \neq cf$. Thus by (I) there is an irreducible *-representation T of $\mathscr{B}$ whose integrated form $\tilde{T}$ satisfies $\tilde{T}_{bf} \neq \tilde{T}_{cf}$. By 12.6(5) $\tilde{T}_{bf} = T_b \tilde{T}_f$ and $\tilde{T}_{cf} = T_c \tilde{T}_f$. The last two statements imply that $T_b \neq T_c$. ∎

14.11. Applied to the group bundle of G, Corollary 14.10 becomes a well known and important property of locally compact groups.

Corollary. *The points of G are separated by the irreducible unitary representations of G. The points of $\mathscr{L}_1(\lambda)$ are separated by the integrated forms of the irreducible unitary representations of G.*

14.12. *Remark.* If $\mathscr{B}$ is Abelian, we need not appeal to VI.22.14 in order to derive 14.10 (or 14.11) from 14.9. The theory of commutative C*-algebras (§VI.4) serves the purpose of VI.22.14 in that case.

14.13. The reader will recall from §10 that every projective representation of G comes from some γ-representation of G, where γ is a central extension of $\mathbb{E}$ by G; and that the γ-representations of G are essentially *-representations of the central extension bundle $\mathscr{B}^\gamma$ over G constructed in 4.8 from γ. Since $\mathscr{B}^\gamma$ has a unit element and enough unitary multipliers, Corollary 14.10 applied to $\mathscr{B}^\gamma$ gives:

Corollary. *For any central extension γ of $\mathbb{E}$ by the locally compact group G, there exist irreducible γ-representations of G, and hence there exist irreducible projective representations of G of projective class γ.*

Remark. Obviously a projective representation will be called *irreducible* if the corresponding γ-representation is irreducible.

Notice that in general, for given γ, the irreducible projective representations of projective class γ need not separate points of G. Indeed, if G is Abelian and γ is the direct product extension $\mathbb{E} \times G$ of $\mathbb{E}$ by G, then by 9.7 an irreducible γ-representation of G is one-dimensional, and hence the corresponding projective representation is a constant homomorphism.

15. The Extension of *-Representations to Multipliers; Application to Semidirect Products and Partial Cross-Sectional Bundles

15.1. We saw in 1.15 that a *-representation of a Banach *-algebra can be "extended" to the bounded multiplier algebra. We will begin this section by proving a similar result for Banach *-algebraic bundles.

In this section $\mathscr{B} = \langle B, \pi, \cdot, * \rangle$ is a Banach *-algebraic bundle over a (not necessarily locally compact) topological group G with unit e.

15.2. The reader will recall from 2.14 and 3.7 the definition and structure of the multiplier bundle $\mathscr{W}(\mathscr{B})$ of $\mathscr{B}$. We shall introduce a natural topology into $\mathscr{W}(\mathscr{B})$.

Definition. A net $\{u_i\}$ of elements of $\mathcal{W}(\mathcal{B})$ *converges strongly* to an element u of $\mathcal{W}(\mathcal{B})$ if $u_i b \to ub$ and $bu_i \to bu$ in B for all b in B. The topology consistent with this definition of convergence is called the *strong topology* of $\mathcal{W}(\mathcal{B})$.

Remark. In general the strong topology of $\mathcal{W}(\mathcal{B})$, taken together with the operations and norm $\| \ \|_0$ defined in 2.14, does *not* make $\mathcal{W}(\mathcal{B})$, $\{\mathcal{W}_x(\mathcal{B})\}$ into a Banach *-algebraic bundle. Indeed, as the reader will easily see from Example 1.19, the norm function $u \mapsto \|u\|_0$ fails in general to be continuous on $\mathcal{W}(\mathcal{B})$ with respect to the strong topology.

We shall refer to the set $\{u \in \mathcal{W}(\mathcal{B}): \|u\|_0 \leq 1\}$ as the *unit cylinder* $\mathcal{W}^1(\mathcal{B})$ of $\mathcal{W}(\mathcal{B})$. The following proposition has an interest in view of the preceding Remark. Its proof follows immediately from Fell [14], Proposition 5.1.

Proposition*. *Assume that G is locally compact and that $\mathcal{B}$ has a strong approximate unit. Then the multiplication in $\mathcal{W}(\mathcal{B})$ is (jointly) strongly continuous on $\mathcal{W}^1(\mathcal{B}) \times \mathcal{W}^1(\mathcal{B})$. In particular, the multiplicative group of unitary multipliers of $\mathcal{B}$ is a topological group under the strong topology.*

15.3. By a **-representation of $\mathcal{W}(\mathcal{B})$* on a Hilbert space X we shall mean a map $T: \mathcal{W}(\mathcal{B}) \to \mathcal{O}(X)$ which is linear on each fiber $\mathcal{W}_x(\mathcal{B})$, and which carries multiplication and * into product and adjoint of operators. By the same argument as in 9.2 such a T must satisfy

$$\|T_u\| \leq \|u\|_0 \qquad\qquad (u \in \mathcal{W}(\mathcal{B})). \qquad (1)$$

If $b \in B$, u_b is the multiplier of $\mathcal{B}$ corresponding to b as in 2.14.

In analogy with 1.15 we have:

Proposition. *Let T be a non-degenerate *-representation of $\mathcal{B}$. Then (i) there is a unique *-representation T' of $\mathcal{W}(\mathcal{B})$ such that $T_b = T'_{u_b}$ for all b in B, and (ii) $u \mapsto T'_u$ is continuous on the unit cylinder of $\mathcal{W}(\mathcal{B})$ with respect to the strong topology of multipliers and the strong operator topology.*

Proof. Let G_0 be the discrete group which coincides as a group with G; and let $\mathcal{B}^0$ be the unique Banach *-algebraic bundle over G_0 which coincides with $\mathcal{B}$ in all respects except for its topology (each fiber B_x being both open and closed in $\mathcal{B}^0$). We shall now consider T as a (non-degenerate) *-representation of $\mathcal{B}^0$. Let $\tilde{T}$ be its integrated form on $\mathcal{L}_1(\mathcal{B}^0)$ $(= \mathcal{L}_1(\lambda; \mathcal{B}^0))$, where λ is counting measure on G_0. Now $\mathcal{W}(\mathcal{B})$ is obviously a subset of $\mathcal{W}(\mathcal{B}^0)$, and hence (by 5.8) can be regarded as a subset of the bounded multiplier algebra of $\mathscr{L}_1(\mathscr{B}^0)$. Thus the application of 1.15 to $\tilde{T}$ gives a *-representation T' of $\mathscr{W}(\mathscr{B})$ satisfying (i) of the proposition.

It remains to prove (ii). Suppose that $u_i \to u$ strongly in $\mathscr{W}(\mathscr{B})$, where $\|u_i\|_0 \leq 1$ for all i; and denote by L the linear space of those ξ in $X(T)$ such that $T'_{u_i} \xi \to T'_u \xi$. In view of (1), $\|T'_{u_i}\| \leq 1$ for all i, and hence L is closed. If $\eta \in X(T)$ and $b \in B$, we have $T'_{u_i} T_b \eta = T_{u_i b} \eta \to T_{ub} \eta = T'_u T_b \eta$ (using the strong continuity 8.2(iv) of T); and so $T_b \eta \in L$. Since T is non-degenerate, these facts imply that $L = X(T)$. Therefore T' is continuous with respect to the strong operator topology.   ∎

Remark.   We should like to emphasize that this proposition holds even for groups G which are not locally compact.

15.4.   As in VI.9.3, the *commuting algebra* of a *-representation T of $\mathscr{B}$ or $\mathscr{W}(\mathscr{B})$ is the *-subalgebra of $\mathcal{O}(X(T))$ consisting of those D which satisfy $DT_b = T_b D$ for all b in B (or $\mathscr{W}(\mathscr{B})$).

Now let T and T' be as in Proposition 15.3, and let $\mathscr{C}$ and $\mathscr{C}'$ be the commuting algebras of T and T' respectively. Since range(T) $\subset$ range(T'), we have $\mathscr{C}' \subset \mathscr{C}$. On the other hand, if $D \in \mathscr{C}$ and $u \in \mathscr{W}(\mathscr{B})$,

$$(DT'_u)T_b = DT_{ub} = T_{ub}D$$

$$= T'_u T_b D = (T'_u D)T_b \tag{2}$$

for all b in B. Since T is non-degenerate, (2) implies that $DT'_u = T'_u D$. By the arbitrariness of D and u, this shows that $\mathscr{C} \subset \mathscr{C}'$. Thus $\mathscr{C}' = \mathscr{C}$; and the commuting algebras of T and T' are the same.

*-Representations of Semidirect Product Bundles

15.5.   As an application of 15.3 we will obtain a simplified description of the *-representations of semidirect product bundles.

Let A be a Banach *-algebra, G a topological group (with unit e), and τ a strongly continuous homomorphism of G into the group of isometric *-automorphisms of A, as in 4.2. We form the semidirect product bundle $\mathscr{B} = A \underset{\tau}{\times} G$ just as in 4.2.

15.6.   Suppose that S is a non-degenerate *-representation of A and V is a unitary representation of G with $X(V) = X(S)$, and that S and V are related by the formula:

$$V_x S_a V_x^{-1} = S_{\tau_x(a)} \qquad (x \in G; a \in A). \tag{3}$$

Then one verifies without difficulty that the equation

$$T_{\langle a, x \rangle} = S_a V_x \qquad\qquad (x \in G; a \in A) \qquad (4)$$

defines a (non-degenerate) *-representation of $\mathscr{B}$. Conversely, we have:

Proposition.   *Let T be any non-degenerate *-representation of $\mathscr{B}$. Then there is a unique non-degenerate *-representation S of A and a unique unitary representation V of G, both acting in $X(T)$, such that (3) and (4) hold.*

Proof.   We saw in 4.3 that each x in G gives rise to a unitary multiplier m_x of $\mathscr{B}$; and one observes that $x \mapsto m_x$ is continuous on G to $\mathscr{W}(\mathscr{B})$ with respect to the strong topology of the latter. Now let T' be the "extension" of T to $\mathscr{W}(\mathscr{B})$ obtained in Proposition 15.3, and put $V_x = T'_{m_x}$ ($x \in G$). By the continuity assertion of Proposition 15.3, together with the preceding sentence, V is strongly continuous and hence a unitary representation of G.

Furthermore, let

$$S_a = T_{\langle a, e \rangle} \qquad\qquad (a \in A). \qquad (5)$$

By 9.4 S is a non-degenerate *-representation of A. From the equations

$$(m_x \langle a, e \rangle) m_{x^{-1}} = \langle \tau_x(a), e \rangle,$$

$$\langle a, e \rangle m_x = \langle a, x \rangle$$

($x \in G; a \in A$), it follows that (3) and (4) hold. This proves the existence of the required S and V.

To prove the uniqueness of S and V, we notice first that (4) implies $S_a = T_{\langle a, e \rangle}$ ($a \in A$), so that T determines S. Since S is non-degenerate, (4) then determines V_x for all x.   ∎

In view of this result, non-degenerate *-representations of $\mathscr{B}$ are essentially just pairs $\langle S, V \rangle$ with the properties specified above.

15.7.   Let T be a non-degenerate *-representation of the semidirect product bundle $\mathscr{B}$ corresponding as in 15.6 with the pair S, V. It follows from (5) and 15.4 that an operator in the commuting algebra of T must lie in the commuting algebras of both S and V. Conversely, an operator which commutes with all S_a and all V_x ($x \in G; a \in A$) lies in the commuting algebra of T.

Thus, for example, T is irreducible if and only if $X(T)$ is irreducible under the combined actions of S and V.

If the group G is locally compact, 14.10 often enables us to assert the existence of many such pairs S, V:

Proposition.   *Let A, G, τ be as in 15.5, and assume that G is locally compact and that A has an approximate unit and is reduced. Then, for every $a \neq 0$ in A, there is a pair S, V such that*

(i)   *S is a non-degenerate *-representation of A with $S_a \neq 0$,*
(ii) *V is a unitary representation of G acting on $X(S)$,*
(iii) *$V_x S_a V_x^{-1} = S_{\tau_x(a)}$ $(a \in A; x \in G)$, and*
(iv) *$X(S)$ is irreducible under the combined action of S and V.*

Remark.   The hypotheses on A are automatically satisfied if A is a C^*-algebra (see VI.8.4 and VI.22.11).

Proof.   Let $\mathcal{B} = A \underset{\tau}{\times} G$. By 4.3 $\mathcal{B}$ has an approximate unit and enough unitary multipliers; and $B_e = A$ is reduced. So by 14.10 $\mathcal{B}$ has enough irreducible *-representations to distinguish points of $\mathcal{B}$. Now apply 15.6.   ■

*-Representations of Partial Cross-Sectional Bundles

15.8.   We return now to a general Banach *-algebraic bundle $\mathcal{B} = \langle B, \pi, \cdot, * \rangle$ over a *locally compact* group G (with unit e and left Haar measure λ). Let N be a given closed normal subgroup of G; and let us form the $\mathscr{L}_1$ partial cross-sectional bundle $\mathscr{C} = \langle C, \rho, \cdot, * \rangle$ over G/N derived from $\mathcal{B}$, as in 6.3–6.5. It will be important in Chapter XII to know that the "*-representation theories" of $\mathcal{B}$ and $\mathscr{C}$ are essentially the same.

15.9.   Let ν be a left Haar measure of N; and for each α in G/N let ν_α be the translated measure on α defined as in 6.3. Let μ be the left Haar measure of G/N normalized as in 6.7(6).

Proposition.   *Let S be a *-representation of $\mathcal{B}$; and for each α in G/N, ϕ in C_α, and ξ in $X(S)$ let us put*

$$T_\phi \xi = \int_\alpha S_{\phi(y)} \xi \, d\nu_\alpha y. \tag{6}$$

*The right side of (6) exists as an $X(S)$-valued integral; $T_\phi: \xi \mapsto T_\phi \xi$ is a bounded linear operator; and $T: \phi \mapsto T_\phi$ is a *-representation of $\mathscr{C}$ on $X(S)$.*
    *Furthermore, the correspondence $\Xi: S \mapsto T$ is one-to-one from the set of all *-representations of $\mathcal{B}$ onto the set of all *-representations of $\mathscr{C}$. Ξ preserves*

non-degeneracy, closed stable subspaces, intertwining operators, commuting algebras, irreducibility, and Hilbert direct sums.

Proof.   Let Ψ_1 [Ψ_2] denote the passage from a *-representation of $\mathscr{B}[\mathscr{C}]$ to its $\mathscr{L}_1$ integrated form. Let Φ be the isometric *-isomorphism of $\mathscr{L}_1(\lambda; \mathscr{B})$ onto $\mathscr{L}_1(\mu; \mathscr{C})$ established in 6.7; and let F be the corresponding map from *-representations of $\mathscr{L}_1(\lambda; \mathscr{B})$ to *-representations of $\mathscr{L}_1(\mu; \mathscr{C})$: $F(V) = V \circ \Phi^{-1}$. Denoting the composed correspondence $\Psi_2^{-1} \circ F \circ \Psi_1$ by Ξ, we see from 13.2, 11.3, 11.9, 11.10, 11.14, 11.15, and 11.17 that Ξ has the properties required in the second paragraph of the statement of the proposition. So it remains only to show that Ξ is in fact given by (6).

To see this, recall from 6.9 that an element b of B_x acts as a multiplier m_b on $\mathscr{C}$ as follows:

$$(b\phi)(y) = b\phi(x^{-1}y) \qquad\qquad (y \in x\alpha),$$

$$(\phi b)(y) = \Gamma(x^{-1})\phi(yx^{-1})b \qquad\qquad (y \in \alpha x)$$

(where $\alpha \in G/N$; $\phi \in \mathscr{L}(\mathscr{B}_\alpha)$). We claim that, if $b \in B$ and $T = \Xi(S)$, where S and T are non-degenerate, then

$$S_b = T'_{m_b} \qquad\qquad (7)$$

(T' being the extension of T to $\mathscr{W}(\mathscr{C})$ given by 15.3).

Indeed, since T and $\tilde{T} = \Psi_2(T)$ are non-degenerate, (7) will be proved if we can show that

$$S_b T_\phi \tilde{T}_w = T_{b\phi} \tilde{T}_w \qquad\qquad (8)$$

whenever $b \in B$, $\alpha \in G/N$, $\phi \in \mathscr{L}(\mathscr{B}_\alpha)$, and $w \in \mathscr{L}(\mathscr{C})$. Since $\{\tilde{f}: f \in \mathscr{L}(\mathscr{B})\}$ (see 6.3) is dense in $\mathscr{L}(\mathscr{C})$ (this was observed in 6.7), it is enough to prove (8) under the assumption that $w = \tilde{f}$ ($f \in \mathscr{L}(\mathscr{B})$). Fixing $b \in B$, $\alpha \in G/N$, $\phi \in \mathscr{L}(\mathscr{B}_\alpha)$, and $f \in \mathscr{L}(\mathscr{B})$, one verifies in a routine manner that

$$\phi\tilde{f} = \tilde{g}, \qquad \text{where } g \in \mathscr{L}(\mathscr{B}), \, g(y) = \int_\alpha \phi(z)f(z^{-1}y)dv_\alpha z, \qquad (9)$$

and

$$(b\phi)\tilde{f} = (bg)\tilde{\;}. \qquad\qquad (10)$$

If $\tilde{S} = \Psi_1(S)$, the definition of Ξ gives

$$\tilde{T}_{\tilde{g}} = \tilde{S}_g, \, \tilde{T}_{(bg)\tilde{\;}} = \tilde{S}_{bg}. \qquad\qquad (11)$$

Therefore

$$S_b T_\phi \tilde{T}_{\tilde{f}} = S_b \tilde{T}_{\phi \tilde{f}} \qquad \text{(by 12.6)}$$

$$= S_b \tilde{S}_g \qquad \text{(by (9), (11))}$$

$$= \tilde{S}_{bg} \qquad \text{(by 12.6)}$$

$$= \tilde{T}_{(b\phi)\tilde{f}} \qquad \text{(by (10), (11))}$$

$$= T_{b\phi} \tilde{T}_{\tilde{f}} \qquad \text{(by 12.6),}$$

proving (8) and therefore (7).

We shall now prove (6). It is enough to restrict our attention to the non-degenerate part of S and T. So we will assume from the beginning that S and T are non-degenerate.

Suppose that $\alpha, \beta \in G/N$, $\phi \in \mathscr{L}(\mathscr{B}_\alpha)$, $\psi \in \mathscr{L}(\mathscr{B}_\beta)$, $\eta \in X(S)$. Just as in 12.5, we can write

$$\phi\psi = \int_\alpha (\phi(y)\psi)dv_\alpha y, \qquad (12)$$

where the right side is an $\mathscr{L}(\mathscr{B}_{\alpha\beta})$-valued integral with respect to the inductive limit topology. Applying to both sides of (12) the continuous linear map $\chi \mapsto T_\chi \eta$ of $\mathscr{L}(\mathscr{B}_{\alpha\beta})$ into $X(S)$ and using II.6.3, we get

$$T_\phi(T_\psi \eta) = T_{\phi\psi} \eta$$

$$= \int_\alpha T_{\phi(y)\psi} \eta \, dv_\alpha y$$

$$= \int_\alpha S_{\phi(y)}(T_\psi \eta)dv_\alpha y \qquad \text{(by (7)).} \qquad (13)$$

This says that (6) holds for all ξ in the linear span L of $\{T_\psi \eta : \eta \in X(S), \beta \in G/N, \psi \in \mathscr{L}(\mathscr{B}_\beta)\}$. Since T is non-degenerate, L is dense in $X(S)$. Since both sides of (6) are clearly continuous in ξ, it follows that (6) holds for all ξ in $X(S)$, provided $\phi \in \mathscr{L}(\mathscr{B}_\alpha)$.

Now let ϕ be any element of $C_\alpha = \mathscr{L}_1(v_\alpha; \mathscr{B}_\alpha)$. We leave it as an exercise for the reader to verify that the right side of (6) does exist as an $X(S)$-valued integral, and is majorized in norm by $\|\phi\|_{C_\alpha}\|\xi\|$. Thus, for fixed ξ, both sides of (6) are continuous as functions of ϕ, and coincide on the dense subset $\mathscr{L}(\mathscr{B}_\alpha)$ of C_α. Therefore they coincide for all ϕ and ξ. ∎

Remark. This proposition is a generalization of 13.2, with which it coincides in the special case $N = G$.

16. *C*-Algebraic Bundles; The Bundle C*-Completion*

16.1. The analogues of C^*-algebras in the context of Banach *-algebraic bundles are the so-called C^*-algebraic bundles which we shall study in this section. Just as every Banach *-algebra A can be "completed" to a C^*-algebra A_c (see VI.10.4), so every Banach *-algebraic bundle $\mathscr{B}$ can be "completed" to a C^*-algebraic bundle $\mathscr{B}^c$. As in the context of Banach *-algebras, the bundle $\mathscr{B}^c$ can be regarded as a "photograph" of $\mathscr{B}$ using "light" which reveals the structure of its *-representation theory and nothing else.

Let G be a locally compact group with unit e and left Haar measure λ.

16.2. *Definition.* A C^*-*algebraic bundle over* G is a Banach *-algebraic bundle $\mathscr{B} = \langle B, \pi, \cdot, * \rangle$ over G such that (i) $\|b^*b\| = \|b\|^2$ for all b in B, and (ii) $b^*b \geq 0$ in B_e for every b in B.

Remark. In view of (i), the unit fiber algebra B_e of $\mathscr{B}$ is a C^*-algebra; and the positivity of b^*b postulated in (ii) is to be understood in the usual sense of VI.7.2.

Remark. Notice that postulate (ii) is independent of postulate (i). A simple example satisfying (i) but not (ii) is the $\mathscr{B}$ of 3.15.

Remark. The group bundle of G is of course a C^*-algebraic bundle. More generally, if A is a C^*-algebra, any τ-semidirect product of A with G (see 4.2) is a C^*-algebraic bundle. Likewise, if the A of 4.7 is a C^*-algebra, the central extension bundle over G constructed in 4.7 is a C^*-algebraic bundle. In particular, cocycle bundles (see 4.8) are C^*-algebraic bundles. In this connection we have:

Proposition. *Any C*-algebraic bundle whose fibers are all one-dimensional is iosmetrically *-isomorphic to a cocycle bundle.*

Proof. Let $\mathscr{B} = \langle B, \pi, \cdot, * \rangle$ be a C^*-algebraic bundle over the locally compact group G whose fibers are all one-dimensional. In particular the unit fiber B_e is isometrically *-isomorphic with $\mathbb{C}$. Let 1 be the unit of B_e.

If $x \in G$ and $0 \neq u \in B_x$, the C^*-norm identity show that $uu^*u \neq 0$. From this and the one-dimensionality of the fibers it follows (see 2.8) that $\mathscr{B}$ is saturated, and so (by 2.13) that 1 is the unit element of $\mathscr{B}$.

If $x \in G$, $u \in B_x$ and $\|u\| = 1$, then $\|u^*u\| = 1$ and $u^*u \geq 0$ in B_e, whence $u^*u = 1$; likewise $uu^* = 1$. So the unitary elements of $\mathscr{B}$ are exactly those of norm 1.

Let U be the multiplicative group of all unitary elements of $\mathscr{B}$, with the relativized topology of B. From the previous paragraph it follows that $j = \pi | U$ is a continuous open homomorphism of U onto G. So, putting $i(\lambda) = \lambda \cdot 4$ ($\lambda \in \mathbb{E}$), we have a central extension

$$\gamma: \mathbb{E} \underset{i}{\to} U \underset{j}{\to} G$$

of $\mathbb{E}$ by G. We leave to the reader the straightforward verification that $\mathscr{B}$ and $\mathscr{B}^\gamma$ are isometrically *-isomorphic.   ■

Remark.   If the $\mathscr{B}$ of 7.2 is a C^*-algebraic bundle, then the G, M transformation bundle derived from $\mathscr{B}$ (as in 7.2) is also a C^*-algebraic bundle.

16.3.   Proposition.   *A C^*-algebraic bundle $\mathscr{B}$ over G always has a strong approximate unit.*

Proof.   Since B_e is a C^*-algebra, by VI.8.4 it has an approximate unit $\{e_i\}$ satisfying $\|e_i\| \leq 1, e_i^* = e_i$. Now let $\varepsilon > 0$, and let C be any compact subset of B. Since $b \mapsto b^*b$ is continuous, $D = \{b^*b : b \in C\}$ is a norm-compact subset of B_e; and so by a remark in 2.11 there is an index i_0 such that

$$\|b^*be_i - b^*b\| < \varepsilon^2 \qquad \text{for all } i \succ i_0 \text{ and } b \text{ in } C.$$

Thus, if $i \succ i_0$ and $b \in C$, we have by 16.2(i)

$$
\begin{aligned}
\|be_i - b\|^2 &= \|(be_i - b)^*(be_i - b)\| \\
&= \|e_i b^* be_i - b^* be_i - e_i b^* b + b^* b\| \\
&\leq \|e_i\| \|b^* be_i - b^* b\| + \|b^* be_i - b^* b\| \\
&< 2\varepsilon^2;
\end{aligned}
$$

whence $\|be_i - b\| < 2\varepsilon$. Applying involution, we find that $\|e_i b^* - b^*\| < 2\varepsilon$ also. Therefore $\{e_i\}$ is a strong approximate unit of $\mathscr{B}$.   ■

16.4.   Theorem.   *Let $\mathscr{B}$ be a C^*-algebraic bundle over G. Then the $\mathscr{L}_1$ cross-sectional algebra of $\mathscr{B}$ is reduced.*

Proof.   Although $\mathscr{B}$ need not have enough unitary multipliers (in fact, it need not even be saturated), the same development that led to 14.9 also holds here.

To begin with, the conclusion of Lemma 14.4 holds here. Indeed, we have seen in 16.3 that $\mathscr{B}$ has a strong approximate unit. Apart from proving that fact, we used saturation in the proof of 14.4 only to show that $0_x \neq b \in B_x$ implies $cb \neq 0$ for some c in $B_{x^{-1}}$; but in the present context this follows from 16.2(i) on taking $c = b^*$.

Further, the conclusion of Proposition 14.5 holds in the present context, since the hypothesis of enough unitary multipliers in 14.5 served only to show that each b^*b is of the form a^*a for some a in B_e; and this follows here from 16.2(ii).

Notice that B_e, being a C^*-algebra, is reduced by VI.22.11. Thus, finally, the proof of Theorem 14.9 remains valid in the present context. ∎

16.5. Just as in 14.10 we deduce from 16.4 the following corollary:

Corollary. *Let $\mathscr{B}$ be any C^*-algebraic bundle over G; and let $\mathscr{W}$ be the set of all irreducible *-representations of $\mathscr{B}$. Then conclusions (I) and (II) of 14.10 hold.*

The Saturated Part of $\mathscr{B}$

16.6. Let $\mathscr{B}$ be any C^*-algebraic bundle over G. There is a canonical *saturated* C^*-algebraic "sub-bundle" of $\mathscr{B}$ over G, which we may call the saturated part of $\mathscr{B}$. We shall sketch its construction here:

Let A be the unit fiber C^*-algebra B_e. We denote linear spans by $[\ ]$, and closure in the bundle space B of $\mathscr{B}$ by $^-$. For each x in G, put $E_x = [B_x^* B_x]^- \subset A$; and let $E = \bigcap_{x \in G} E_x$. Then E_x and E are closed two-sided ideals of A. In particular E has an approximate unit. Thus, putting $D_x = [B_x E]^- \subset B_x$ $(x \in G)$, we have $D_e = E$.

One verifies without difficulty that

$$B_x^* E B_x \subset E \qquad (x \in G). \qquad (1)$$

Further we claim that

$$D_x = [E B_x]^- \qquad (x \in G). \qquad (2)$$

Indeed: Since $E = [EE]^-$, $E B_x \subset [EE B_x]^- \subset [B_x B_x^* E B_x]^- \subset [B_x E]^-$ (by (1)). Similarly $B_x E \subset [E B_x]^-$. So (2) holds.

Proposition*. *Let $D = \bigcup_{x \in G} D_x$; and equip D with the relativized topology of $\mathscr{B}$ and the restrictions to D of the bundle projection, norm, and linear and algebraic operations of $\mathscr{B}$. Then D is the bundle space of a saturated C^*-algebraic bundle $\mathscr{D}$ over G.*

This $\mathscr{D}$ is called the *saturated part* of $\mathscr{B}$. If $\mathscr{B}$ was already saturated, then of course $\mathscr{D} = \mathscr{B}$.

To prove the proposition, we first show that $\mathscr{D}$ is a Banach bundle. In view of II.13.18, the key to this is the existence of "enough" continuous cross-sections of $\mathscr{B}$ whose values lie in D. To establish this, we consider cross-sections of the form $x \mapsto f(x)a$, where $f \in \mathscr{C}(\mathscr{B})$ and $a \in E$.

Using (2) it is easy to check that D is closed under multiplication and $*$, and hence that $\mathscr{D}$ is a C^*-algebraic bundle. Its saturation is easily verified.

Remark.   In view of the definition of D_x and (2), each element of $\mathscr{B}$ acts by multiplication as a multiplier of $\mathscr{D}$. Hence by 15.3 every non-degenerate *-representation of $\mathscr{D}$ has a unique extension to a *-representation of $\mathscr{B}$.

The Bundle C*-Completion

16.7.   Let $\mathscr{B} = \langle B, \pi, \cdot, * \rangle$ be an arbitrary Banach *-algebraic bundle over G. We shall now show how to "complete" $\mathscr{B}$ to a C^*-algebraic bundle (see 16.1). The construction is analogous with VI.10.2-4.

For each b in B, we define

$$\|b\|_c = \sup_T \|T_b\|, \tag{3}$$

T running over all *-representations of $\mathscr{B}$. By 9.2(1) $\|b\|_c$ is finite; in fact

$$\|b\|_c \le \|b\| \qquad (b \in B). \tag{4}$$

Clearly

$$\|b^*b\|_c = \|b\|_c^2 \qquad (b \in B). \tag{5}$$

Restricted to each fiber B_x, $\| \ \|_c$ is a seminorm. If $N_x = \{b \in B_x : \|b\|_c = 0\}$, then B_x/N_x is a normed linear space (under the norm $b + N_x \mapsto \|b\|_c$), whose completion we denote by C_x, $\| \ \|_c$. Let C stand for the disjoint union of the $\{C_x\}$ ($x \in G$); let $\pi': C \to G$ be the surjection given by $\pi'^{-1}(x) = C_x$; and let $\rho: B \to C$ be the "quotient" map $b \mapsto b + N_{\pi(b)}$.

Now $b \mapsto \|b\|_c$ is continuous on B. This follows from the identity (5), the continuity of $b \mapsto b^*b$, and the continuity of $\| \ \|_c$ on B_e (see (4)). Also, $\mathscr{B}$ has enough continuous cross-sections by Appendix C. Thus the family of cross-sections of $\langle C, \pi' \rangle$ of the form

$$x \mapsto \rho(f(x)) \tag{6}$$

(where f is a continuous cross-section of $\mathscr{B}$) satisfies the hypotheses of II.13.18; and so by II.13.18 there is a unique topology on C making $\mathscr{C} = \langle C, \pi' \rangle$ a Banach bundle (with the norm $\| \ \|_c$) such that the cross-sections (6) are all continuous.

From the continuity of the cross-sections (6) and the fact that ρ is norm-decreasing by (4), we verify that $\rho: B \to C$ is continuous.

Evidently

$$\|ab\|_c \leq \|a\|_c \|b\|_c, \qquad \|b^*\|_c = \|b\|_c \tag{7}$$

($a, b \in B$). From this we see that the equations

$$\rho(a)\rho(b) = \rho(ab), \qquad (\rho(b))^* = \rho(b^*) \tag{8}$$

($a, b \in B$) determine a product $\cdot$ and involuton * on C satisfying (5) and (7) for all a, b in C. The continuity of these operations on C follows from the continuity of ρ and from 2.4 and 3.2 applied to the family of cross-sections of the form (6). Thus $\mathscr{C}$ has become a Banach *-algebraic bundle over G.

The *-representation theories of $\mathscr{B}$ and of $\mathscr{C}$ coincide in the following sense: The *-representations of $\mathscr{B}$ are precisely the maps of the form

$$T: b \mapsto T'_{\rho(b)} \qquad\qquad (b \in B), \tag{9}$$

where T' is a *-representation of $\mathscr{C}$. The correspondence $T \leftrightarrow T'$ preserves closed stable subspaces, non-degeneracy, irreducibility, intertwining operators, and Hilbert direct sums. The reader will verify this without difficulty.

Now $\mathscr{C}$ is a C*-algebraic bundle. Indeed: 16.2(i) holds by (5). To prove 16.2(ii), we recall (3) and form the Hilbert direct sum T of enough *-representations of $\mathscr{B}$ so that $\|T_b\| = \|b\|_c$ for all b in B. The *-representation T' of $\mathscr{C}$ corresponding to T by (9) thus satisfies $\|T'_b\| = \|b\|_c$ for all b in C. In particular $S = T'|C_e$ is an isomorphism of the C*-algebra C_e into the C*-algebra $\mathcal{O}(X(T))$. If $b \in C$, $S_{b^*b} = (T'_b)^*T'_b$ is a positive operator, and so by VI.7.3 is positive with respect to range(S). So b^*b is positive in C_e, and 16.2(ii) holds.

Definition.   The C*-algebraic bundle $\mathscr{C}$ which we have constructed is called the *bundle C*-completion* of $\mathscr{B}$. It will in the future be denoted by $\mathscr{B}^c$.

The analogy of this construction with VI.10.4 is transparent. As we saw above, the *-representation theories of $\mathscr{B}$ and $\mathscr{B}^c$ coincide just as they did in VI.10.5 for a Banach *-algebra and its C*-completion.

16.8.   Let $\mathscr{B}$ be as in 16.7. If $\mathscr{B}$ is saturated, then evidently so is $\mathscr{B}^c$. Furthermore we have:

Proposition.   *If $\mathscr{B}$ has enough unitary multipliers, then so does $\mathscr{B}^c$.*

Proof.   We shall stick to the notation of 16.7, and choose (as we did in 16.7) a non-degenerate *-representation T of $\mathscr{B}$ satisfying

$$\|T_b\| = \|b\|_c \qquad \text{for all } b \text{ in } B. \tag{10}$$

By 15.3 T has an extension T' to the multiplier bundle $\mathcal{W}(\mathcal{B})$. Now let u be a unitary multiplier of $\mathcal{B}$ of order x. Then T'_u is a unitary operator on $X(T)$; and hence by (10)

$$\|ub\|_c = \|T_{ub}\| = \|T'_u T_b\| = \|T_b\| = \|b\|_c$$

for all b in B. Similarly $\|bu\|_c = \|b\|_c$ for all b in B. It follows that the left and right actions of u on B extend to left and right actions λ and μ on $\mathcal{B}^c$ satisfying:

$$\lambda(\rho(b)) = \rho(ub), \qquad \mu(\rho(b)) = \rho(bu) \qquad (b \in B); \qquad (11)$$

and these will form a unitary multiplier $u' = \langle \lambda, \mu \rangle$ of $\mathcal{B}^c$ of order x provided λ and μ are continuous on B. To prove the continuity of λ, it is enough (by II.13.12) to verify that $x \mapsto \lambda(\rho(f(x)))$ is continuous for each continuous cross-section f of $\mathcal{B}$. But by (11) $\lambda(\rho(f(x))) = \rho(uf(x))$, and this is continuous in x in view of the continuity of u and ρ. So λ is continuous. Likewise μ is continuous; so u' is a unitary multiplier of $\mathcal{B}^c$ of order x. ∎

16.9. Remark. Let $\mathcal{B}$, $\mathcal{C}$ be as in 16.7. Although $B_e^c (= C_e)$ is a C*-algebra, it differs in general from the C*-completion $(B_e)_c$ of B_e. Indeed: $(B_e)_c$ is constructed from the following seminorm $\| \ \|'_c$ on B_e:

$$\|a\|'_c = \sup\{\|S_a\| : S \text{ is a *-representation of } B_e\}.$$

Clearly $\|a\|'_c \geq \|a\|_c$ for all a in B_e. So the identity map on B_e generates a norm-decreasing surjective *-homomorphism $\Phi: (B_e)_c \to B_e^c$. This Φ will be one-to-one if and only if $\| \ \|_c$ and $\| \ \|'_c$ coincide on B_e; but this, as it turns out, is not in general the case. For another condition that Φ be one-to-one see XI.11.6.

16.10. Proposition. *Suppose that $\mathcal{B}$ in 16.7 is itself a C*-algebraic bundle. Then $\|b\|_c = \|b\|$ for all b in B. Hence $\mathcal{B}$ is its own bundle C*-completion.*

Proof. Using 16.5 and forming Hilbert direct sums, we can find a *-representation T of $\mathcal{B}$ such that $T|B_e$ is faithful on the C*-algebra B_e. So by VI.8.8 T is an isometry on B_e. Hence, for any b in B,

$$\begin{aligned}
\|T_b\|^2 &= \|(T_b)^* T_b\| = \|T_{b^*b}\| \\
&= \|b^*b\| \qquad\qquad\qquad (\text{since } b^*b \in B_e) \\
&= \|b\|^2.
\end{aligned}$$

So $\|b\| = \|T_b\| \leq \|b\|_c$ for all b in B. Combining this with (4) completes the proof. ■

Remark. We have thus shown that C*-algebraic bundles over the locally compact group G are precisely the C*-completions of Banach *-algebraic bundles over G. This generalizes the analogous fact for C*-algebras.

16.11. *The case of a discrete base group.* Let $\mathscr{B} = \langle B, \{B_x\}_{x \in G} \rangle$ be a C*-algebraic bundle over a *discrete* group G; and let $\Phi: \mathscr{L}_1(\mathscr{B}) \to C^*(\mathscr{B})$ be the natural *-homomorphism of the $\mathscr{L}_1$ cross-sectional algebra $\mathscr{L}_1(\mathscr{B})$ into its C*-completion, which we always denote by $C^*(\mathscr{B})$ (see 17.2).

In virtue of 16.10, each fiber B_x, when identified with a subspace of $\mathscr{L}_1(\mathscr{B})$, is mapped by Φ *isometrically* into $C^*(\mathscr{B})$. Let $B_x' = \Phi(B_x)$; thus B_x' is a closed linear subspace of $C^*(\mathscr{B})$. We claim that *the B_x' ($x \in G$) are linearly independent in $C^*(\mathscr{B})$, and their linear span is dense in $C^*(\mathscr{B})$*.

Indeed: Clearly the B_x are linearly independent and have dense linear span in $\mathscr{L}_1(\mathscr{B})$. Since Φ is one-to-one on $\mathscr{L}_1(\mathscr{B})$ (by 16.4) and continuous and has range dense in $C^*(\mathscr{B})$, the same is true of the B_x' in $C^*(\mathscr{B})$.

Thus, regarding Φ as an *identification*, we may consider the B_x as closed linearly independent subspaces of $C^*(\mathscr{B})$, having dense linear span in $C^*(\mathscr{B})$.

16.12. Conversely, suppose E is a C*-algebra, and G is a discrete group; and suppose that for each $x \in G$ we are given a closed linear subspace B_x of E with the following properties:

(i) the B_x ($x \in G$) are linearly independent, and their linear span is dense in E;

(ii) $B_x B_y \subset B_{xy}$ ($x, y \in G$);

(iii) $(B_x)^* = B_{x^{-1}}$ ($x \in G$).

Then obviously the B_x ($x \in G$) form the fibers of a C*-algebraic bundle $\mathscr{B}$ over G. By analogy with 16.11 we might be tempted to identify E with $C^*(\mathscr{B})$; but *this identification is in general false*.

To be precise, notice that there is a natural surjective *-homomorphism $\rho: C^*(\mathscr{B}) \to E$ such that, for any *-representation S of E, $S \circ \rho$ is the *-representation of $C^*(\mathscr{B})$ corresponding to the restriction of S to $\mathscr{B}$. To say that E can be identified with $C^*(\mathscr{B})$ is the same as to say that ρ is one-to-one; but this is not in general true. (See Exercise 38 of Chapter XI.)

Direct Sums of C*-Algebraic Bundles

16.13.   For each i in an index set I let $\mathscr{B}^i = \langle B^i, \pi^i, \cdot, * \rangle$ be a C^*-algebraic bundle over the locally compact group G with unit e (the same group for all i). We construct from these a direct sum C^*-algebraic bundle $\mathscr{B} = \langle B, \pi, \cdot, * \rangle$ over G as follows.

As a Banach bundle, $\mathscr{B}$ will be just the Banach bundle C_0 direct sum $\sum_{i \in I}^{\oplus 0} \mathscr{B}^i$ constructed in II.13.20. The multiplications and involutions in the $\mathscr{B}^i$ combine in an obvious way to give a multiplication and involution in $\mathscr{B}$:

$$(a \cdot b)_i = a_i \cdot b_i, \ (a^*)_i = (a_i)^* \qquad\qquad (a, b \in B).$$

Postulates 2.2(i)–(iv) and 3.1(i)–(v) clearly hold; and the arguments of 2.4 and 3.2 (using the same family Γ of continuous cross-sections that occurred in II.13.20) show that postulates 2.2(v) and 3.1(vi) are also valid. So $\mathscr{B} = \langle B, \pi, \cdot, * \rangle$ has become a Banach C^*-algebraic bundle, in fact a C^*-algebraic bundle.

Definition.   This $\mathscr{B}$ is called the C^*-*direct sum of the* $\mathscr{B}^i$, in symbols:

$$\mathscr{B} = \sum_{i \in I}^{\oplus 0} \mathscr{B}^i$$

$\mathscr{B}$ will be saturated if and only if each $\mathscr{B}^i$ is saturated.

Notice that for each $i \in I$ the projection $p_i : B \to B_i$ (given by $p_i(b) = b_i$ for $b \in B$) gives rise to a self-adjoint idempotent multiplier $u_i = \langle p_i, p_i \rangle$ of $\mathscr{B}$ of order e; and we have: (i) $u_i u_j = 0$ for $i \neq j$; (ii) the linear span of $\bigcup_{i \in I} (u_i B_e)$ is dense in B_e; and (iii) each u_i is central, in the sense that $u_i b = b u_i$ for all $b \in B$.

Remark.   The unit fiber C^*-algebra B_e is of course just the C^* direct sum of the B_e^i, in the sense of VI.3.13.

16.14.   Conversely, suppose we are given a C^*-algebraic bundle $\mathscr{B} = \langle B, \pi, \cdot, * \rangle$ over the locally compact group G (with unit e), together with an indexed collection $\{u_i\}_{i \in I}$ of self-adjoint idempotent multipliers of $\mathscr{B}$ of order e which satisfy (i), (ii), (iii) above. From these we obtain a natural C^* direct sum decomposition as follows:

Proposition.   *For each i let $B^i = u_i B$; then $\mathscr{B}^i = \langle B^i, \pi | B^i, \cdot, * \rangle$ is a C^*-algebraic bundle over G. We have:*

$$\sum_{i \in I}^{\oplus 0} \mathscr{B}^i \cong \mathscr{B}$$

under the isometric *-isomorphism Φ which coincides with the identity on each
B^i. If $\mathscr{B}$ is saturated, each $\mathscr{B}^i$ is saturated.

Proof. Fix i. To see that $\mathscr{B}^i$ is a C*-algebraic bundle, the only slightly non-trivial step is to verify the openness of $\pi|B^i$. Let $x_\alpha \to x$ in G and $b \in B^i_x$. By the openness of π we can find a subnet $\{x'_\beta\}$ of $\{x_\alpha\}$ and elements $b'_\beta \in B$ such that $\pi(b'_\beta) = x'_\beta$ and $b'_\beta \to b$. But then $u_i b'_\beta \to u_i b = b$, and $\pi(u_i b'_\beta) = \pi(b'_\beta) = x'_\beta$. So by II.13.2 $\pi|B^i$ is open.

We have now to show that $\mathscr{B}$ is essentially the C_0 direct sum of the $\mathscr{B}^i$. Let $\tilde{\mathscr{B}} = \sum_{i \in I}^{\oplus 0} \mathscr{B}^i$.

Consider first the unit fiber C*-algebra B_e. For each finite subset F of I, the natural map

$$\sum_{i \in F}^{\oplus 0} B^i_e \to \sum_{i \in F} B^i_e \subset B_e \tag{12}$$

is a *-isomorphism of C*-algebras, hence an isometry. Thus, since the union of all the left sides of (12) (F varying) is dense in $\tilde{B}_e$, we have an isometric *-isomorphism $\Psi: \tilde{B}_e \to B_e$ which coincides with (12) for each finite subset F of I, and whose range is dense in B_e (by 16.13(ii)) and hence equal to B_e.

It follows that for each $b \in B_e$

$$\lim_{i \to \infty} \|u_i b\| = 0 \quad \text{and} \quad \|b\| = \sup_{i \in I} \|u_i b\|. \tag{13}$$

Now consider an arbitrary fiber B_x. From (13) and the C*-identity $\|b^*b\| = \|b\|^2$, it follows that (13) holds for any $b \in B_x$; and from this in turn we deduce that the equation

$$(\Phi_x(b))_i = u_i b$$

defines a linear surjective isometry $\Phi_x: B_x \to \tilde{B}_x$. Let Φ be the union of the maps Φ_x ($x \in G$), hence a surjective isometry $B \to \tilde{B}$. Since the topologies of B and $\tilde{B}$ are determined (see II.13.18) by the "dense" family Γ of cross-sections which are continuous in both, it follows that Φ is a homeomorphism. Obviously Φ preserves multiplication and involution. So Φ is an isometric *-isomorphism.

The last statement of the Proposition was already observed in 16.13.   ∎

A Class of Saturated C*-Algebraic Bundles

16.15. In paragraphs 16.15–17 we are going to construct a fairly general class of saturated C*-algebraic bundles whose unit fiber algebras are of compact type. In fact it will turn out in Chapter XII that *every* saturated

C^*-algebraic bundle whose unit fiber algebra is of compact type must be a C^*-direct sum of those to be constructed here.

The ingredients of the construction are threefold:

1) Let A be a fixed C^*-algebra of compact type (see VI.23.3); thus the structure space $\hat{A}$ is a discrete topological space (see VII.5.24). For each D in $\hat{A}$ let X_D denote the Hilbert space in which D acts.

2) Fix a locally compact group G (with unit e) which acts *transitively* (to the left) on $\hat{A}$. Further, let D_0 be some fixed element of $\hat{A}$; and denote the stability subgroup of G for D_0 by K. Since $\hat{A}$ is discrete, K is both open and closed in G.

3) Fix a central extension

$$\gamma: \mathbb{E} \underset{\sigma}{\to} L \underset{\rho}{\to} K$$

of the circle group $\mathbb{E}$ by K (see III.5.6).

From the ingredients A, G and γ we are going to construct a saturated C^*-algebraic bundle over G with unit fiber A.

Our first step is to construct (as in 4.7) the cocycle bundle $\mathscr{Q} = \langle Q, \tau, \cdot, * \rangle$ over K corresponding to γ. Thus (see 4.7) Q is the space of orbits (with the quotient topology) in $\mathbb{C} \times L$ under the following action of the circle group $\mathbb{E}$:

$$u\langle \lambda, m \rangle = \langle \lambda u^{-1}, \sigma(u)m \rangle \qquad (u \in \mathbb{E}, \lambda \in \mathbb{C}, m \in L).$$

We shall denote by $\langle \lambda, m \rangle^{\sim}$ the $\mathbb{E}$-orbit containing $\langle \lambda, m \rangle$. The norm, projection τ, multiplication and involution in $\mathscr{Q}$ are given by:

$$\| \langle \lambda, m \rangle^{\sim} \| = |\lambda|,$$

$$\tau(\langle \lambda, m \rangle^{\sim}) = \rho(m),$$

$$\langle \lambda, m \rangle^{\sim} \langle \lambda', m' \rangle^{\sim} = \langle \lambda\lambda', mm' \rangle^{\sim},$$

$$(\langle \lambda, m \rangle^{\sim})^* = \langle \bar{\lambda}, m^{-1} \rangle^{\sim}.$$

By 4.8 and 16.2 $\mathscr{Q}$ is a cocycle bundle. Notice that

$$\|pq\| = \|p\| \cdot \|q\| \qquad \text{for all } p, q \text{ in } Q \tag{14}$$

and that the unit fiber $Q_e \cong \mathbb{C}$ under the identification $\langle \lambda, \sigma(u) \rangle^{\sim} \leftrightarrow \lambda u$ ($\lambda \in \mathbb{C}$, $u \in \mathbb{E}$).

16.16. Next, for each D in $\hat{A}$ let us choose some element ξ_D of G such that $\xi_D D_0 = D$. (This is possible since G was assumed to act transitively on $\hat{A}$.) In particular set $\xi_{D_0} = e$. Thus, for any $x \in G$ and $D \in \hat{A}$ we have $x\xi_D D_0 = xD = \xi_{xD} D_0$; and hence, since K is the stability subgroup for D_0,

$$x\xi_D = \xi_{xD}\mu_{x,D} \qquad \text{for some unique } \mu_{x,D} \in K. \tag{15}$$

Proposition. *The function* $\mu: G \times \hat{A} \to K$ *defined by* (15) *has the following properties*:

(i) μ *is continuous,*

(ii) $\mu_{yx,D} = \mu_{y,xD}\mu_{x,D}$,

(iii) $\mu_{x^{-1},D} = (\mu_{x,x^{-1}D})^{-1}$,

(iv) $\mu_{e,D} = e$,

(v) $\mu_{k,D_0} = k$,

$(x, y \in G; k \in K; D \in \hat{A})$.

Proof.

(i) By (15) $\mu_{x,D} = (\xi_{xD})^{-1}x\xi_D$; and the latter is clearly continuous in x and D (since $\hat{A}$ is discrete).

(ii) $\xi_{yxD}\mu_{yx,D} = yx\xi_D$
$$= y(\xi_{xD}\mu_{x,D})$$
$$= \xi_{yxD}\mu_{y,xD}\mu_{x,D}$$
(using (15) at each step). So (ii) holds.

(iii) Applying (15) to x and $x^{-1}D$ we get

$$x\xi_{x^{-1}D} = \xi_D\mu_{x,x^{-1}D}$$

whence

$$x^{-1}\xi_D = \xi_{x^{-1}D}(\mu_{x,x^{-1}D})^{-1};$$

and by (15) this is (iii).

(iv) and (v) are obvious from (15). ∎

16.17. We shall now construct a saturated C*-algebraic bundle $\mathscr{B} = \langle B, \pi, \cdot, * \rangle$ over G.

For each $x \in G$ let B_x be the Banach space C_0 direct sum

$$B_x = \sum_{D \in \hat{A}} {}^{\oplus 0}(\mathcal{O}_c(X_D, X_{xD}) \otimes Q_{\mu_{x,D}}). \tag{16}$$

Here as usual $\mathcal{O}_c(X_D, X_{xD})$ is the Banach space of all compact operators from the Hilbert space X_D of D to the Hilbert space X_{xD} of xD. Since $Q_{\mu_{x,D}}$ is one-dimensional, the tensor product on the right of (16) is an evident Banach space under the product norm

$$\|a \otimes p\| = \|a\| \cdot \|p\| \qquad (a \in \mathcal{O}_c(X_D, X_{xD}); p \in Q_{\mu_{x,D}}). \tag{17}$$

The C_0 direct sum of Banach spaces was defined in Chapter I, §5.

Let B be the disjoint union of the B_x ($x \in G$), and $\pi: B \to G$ the projection given by $\pi^{-1}(x) = B_x$.

We first make $\langle B, \pi \rangle$ a Banach bundle. For $D, E \in \hat{A}$ let $G_{D,E}$ be the open and closed subset $\{x \in G : xD = E\}$ of G. If $\alpha : G_{D,E} \to \mathcal{O}_c(X_D, X_E)$ is norm-continuous and ϕ is a continuous cross-section of $\mathscr{Q}$, let us define $F_{D,E,\alpha,\phi}$ to be the cross-section of $\langle B, \pi \rangle$ which vanishes outside $G_{D,E}$ and has the value $\alpha(x) \otimes \phi(\mu_{x,D})$ at $x \in G_{D,E}$. The verification that the linear span Γ of the set of all $F_{D,E,\alpha,\phi}$ (with D, E, α, ϕ varying) satisfies the two hypotheses of II.13.18 is easy but messy, and we omit it. Thus by II.13.18 there is a unique topology for B making $\langle B, \pi \rangle$ a Banach bundle such that all the cross-sections $F_{D,E,\alpha,\phi}$ are continuous. We adopt this topology of B. (Observe that $\langle B, \pi \rangle$ is the C_0 direct sum over $\hat{A}$, in the sense of II.13.20, of Banach bundles whose x-fiber is $\mathcal{O}_c(X_D, X_{xD}) \otimes Q_{\mu_{x,D}}$.)

We must now introduce multiplication and involution into B. If

$$b = \sum_{D \in \hat{A}}{}^{\oplus}(b^D \otimes p^D) \in B_x, \quad c = \sum_{D \in \hat{A}}{}^{\oplus}(c^D \otimes q^D) \in B_y \tag{18}$$

$$(x, y \in G; \ b^D \in \mathcal{O}_c(X_D, X_{xD}); \ c^D \in \mathcal{O}_c(X_D, X_{yD}); \ p^D \in Q_{\mu_{x,D}}; \ q^D \in Q_{\mu_{y,D}}),$$

let us define

$$b \cdot c = \sum_{D \in \hat{A}}{}^{\oplus}(b^{yD}c^D \otimes p^{yD}q^D). \tag{19}$$

Observe that $b^{yD}c^D \in \mathcal{O}_c(X_D, X_{xyD})$ and $p^{yD}q^D \in Q_{\mu_{xy,D}}$ by Proposition 16.16 (ii). Further, by (17)

$$\|b^{yD}c^D \otimes p^{yD}q^D\| \le \|b^{yD}\| \cdot \|p^{yD}\| \cdot \|c^D\| \cdot \|q^D\| \to 0 \qquad \text{as } D \to \infty \text{ in } \hat{A} \tag{20}$$

(since $\|c^D\| \cdot \|q^D\| = \|c^D \otimes q^D\| \to 0$ and $\|b^{yD}\| \cdot \|p^{yD}\| = \|b^{yD} \otimes p^{yD}\| \to 0$ as $D \to \infty$). It follows that $b \cdot c \in B_{xy}$. Evidently this product (19) is associative and bilinear on fibers; and again by (17)

$$\|b \cdot c\| = \sup_{D \in \hat{A}} \|b^{yD}c^D \otimes p^{yD}q^D\|$$

$$\le \left\{ \sup_{D \in \hat{A}} \|b^{yD}\| \cdot \|p^{yD}\| \right\} \left\{ \sup_{D \in \hat{A}} \|c^D\| \cdot \|q^D\| \right\}$$

$$= \|b\| \cdot \|c\|. \tag{21}$$

To show that $\langle B, \pi, \cdot \rangle$ is a Banach algebraic bundle it remains only to see that $\cdot$ is continuous. By 2.4 it is enough to take two cross-sections Φ and Φ', each of the form $F_{D,E,\alpha,\phi}$ defined above, and show that $\langle x, y \rangle \mapsto \Phi(x)\Phi'(y)$ is continuous on $G \times G$ to B. The details of this are left to the reader.

As for involution, if $b \in B_x$ as in (18), we set

$$b^* = \sum_{D \in \hat{A}}{}^{\oplus}((b^{x^{-1}D})^* \otimes (p^{x^{-1}D})^*). \tag{22}$$

Since $b^{x^{-1}D} \in \mathcal{O}_c(X_{x^{-1}D}, X_D)$ and $p^{x^{-1}D} \in Q_{\mu_x, x^{-1}D}$, we have $(b^{x^{-1}D})^* \in \mathcal{O}_c(X_D, X_{x^{-1}D})$ and (by 16.16(iii)) $(p^{x^{-1}D})^* \in Q_{\mu_{x^{-1}}, D}$; and an argument similar to (20) shows that

$$\|(b^{x^{-1}D})^* \otimes (p^{x^{-1}D})^*\| \to 0 \qquad \text{as } D \to \infty \text{ in } \hat{A}.$$

So b^*, as defined in (22), belongs to $B_{x^{-1}}$. Evidently the map $*$ is conjugate-linear on fibers. An imitation of (21) shows that $\|b^*\| = \|b\|$; and the algebraic laws $b^{**} = b$ and $(bc)^* = c^*b^*$ hold in B because they hold in Q and for operators on Hilbert space. It remains only to show that $*$ is continuous on B. To see this we use 3.2 in the same way that we used 2.4 to prove the continuity of multiplication.

Thus $\mathscr{B} = \langle B, \pi, \cdot, * \rangle$ is a Banach $*$-algebraic bundle. In fact we have:

Proposition. $\mathscr{B} = \langle B, \pi, \cdot, * \rangle$, with the operations (19) and (22), is a saturated C*-algebraic bundle over G, whose unit fiber C*-algebra is just the A with which we started in 16.15.

Proof. Since $\mu_{e,D} = e$ (16.16(iv)) and $Q_e \cong \mathbb{C}$, we have

$$B_e = \sum_{D \in \hat{A}}^{\oplus 0} \mathcal{O}_c(X_D) \otimes \mathbb{C}$$

$$= \sum_{D \in \hat{A}}^{\oplus 0} \mathcal{O}_c(X_D) \cong A \qquad \text{(by VI.23.3).}$$

So the unit fiber $*$-algebra of $\mathscr{B}$ is A.

The saturation of $\mathscr{B}$ is an immediate consequence of that of $\mathscr{Q}$, together with the fact that $\mathcal{O}_c(X_2, X_3)\mathcal{O}_c(X_1, X_2)$ has dense linear span in $\mathcal{O}_c(X_1, X_3)$ whenever X_1, X_2, X_3 are three non-zero Hilbert spaces.

Finally, suppose b is as in (18). We verify that

$$b^*b = \sum_{D \in \hat{A}}^{\oplus} ((b^D)^*b^D \otimes (p^D)^*p^D). \qquad (23)$$

By (17) the argument of (21) now gives the C* norm identity

$$\|b^*b\| = \|b\|^2.$$

Since $\mathscr{Q}$ is a C*-algebraic bundle each $(p^D)^*p^D$ is a non-negative real number; and likewise each $(b^D)^*b^D$ is a positive compact operator. So (23) is a positive element of $B_e \cong A$. Hence $\mathscr{B}$ is a C*-algebraic bundle. ∎

Terminology.   For lack of a better way of referring to it, we shall speak of the above $\mathscr{B}$ as the *saturated C*-algebraic bundle canonically constructed from the ingredients A, G, D_0, K, γ.* Such a $\mathscr{B}$ will be called a *saturated C*-algebraic bundle of canonical type.*

16.18.   Remarks.

1) It is easy to check that the *-isomorphism class of the $\mathscr{B}$ of 16.17 does not depend on the particular choice of the ξ_D in 16.16. We shall see this in a much clearer light in Chapter XII, §6.

2) Consider the special case that the extension γ in 16.15 is the direct product extension. Then the $\mathscr{Q}$ of 16.15 is the direct product bundle, with $Q_k = \mathbb{C}$ for all k; and we can omit the second factors is definitions (16), (19), (22), getting:

$$B_x = \sum_{D \in \hat{A}}^{\oplus 0} \mathcal{O}_c(X_D, X_{xD}),$$

$$b \cdot c = \sum_{D \in \hat{A}}^{\oplus} b^{yD} c^D,$$

$$b^* = \sum_{D \in \hat{A}}^{\oplus} (b^{x^{-1}D})^*$$

($b = \sum_{D \in \hat{A}}^{\oplus} b^D \in B_x$; $c = \sum_{D \in \hat{A}}^{\oplus} c^D \in B_y$). Thus, in this special case, if X denotes the Hilbert space direct sum $\sum_{D \in \hat{A}}^{\oplus} X_D$, B_x *can be identified with the closed subspace of* $\mathcal{O}_c(X)$ *consisting of those b such that* $b(X_D) \subset X_{xD}$ *for all* $D \in \hat{A}$; *and multiplication and involution in* $\mathscr{B}$ *are just composition and adjoint of operators on X.*

3) Consider the special case in which A is *elementary*; that is, $\hat{A}$ consists of one element D_0 only, and $A \cong \mathcal{O}_c(X(D_0))$. Then $K = G$, $\mathscr{Q}$ is over G, and $\mathscr{B}$ reduces to the C*-algebraic bundle tensor product

$$\mathscr{B} = A \otimes \mathscr{Q}. \tag{24}$$

Here the right side of (24) is defined as a saturated C*-algebraic bundle over G in the manner one would expect: The fiber over x is $A \otimes Q_x$, which (since Q_x is one-dimensional) is a Banach space under the product norm $\|a \otimes q\| = \|a\| \cdot \|q\|$. The bundle space receives the unique topology making $A \otimes \mathscr{Q}$ a Banach bundle such that the cross-section $x \mapsto a \otimes q(x)$ is continuous for every $a \in A$ and every continuous cross-section $x \mapsto q(x)$ of $\mathscr{Q}$. The operations of multiplication and involution in $A \otimes \mathcal{Q}$ are the natural tensor product operations:

$$(a_1 \otimes q_1)(a_2 \otimes q_2) = a_1 a_2 \otimes q_1 q_2,$$

$$(a \otimes q)^* = a^* \otimes q^*.$$

17. Bundle Structures for C^*-Algebras; the Glimm Algebras

17.1. Let us go back for a moment to the motivating discussion in 2.1, by which the concept of a Banach algebraic bundle was introduced. We began there with a Banach algebra E and a discrete group G, and suggested that in that case E would become a Banach algebraic bundle when equipped with some further structure, namely a collection of closed subspaces $\{B_x\}$ indexed by G and satisfying 2.1(i), (ii). However, when we came to write down the postulates for a Banach algebraic bundle over a non-discrete group G in 2.2, we discarded the "global" Banach algebra E, and took as basic ingredients the collection of Banach spaces $\{B_x\}$ (indexed by G). For locally compact groups G a "global" Banach algebra was later (in §5) reconstructed as the $\mathcal{L}_1$ cross-sectional algebra. Nevertheless a legitimate question remains: Given any Banach algebra E and any non-discrete locally compact group G, what should it mean to equip E with the structure of a Banach algebraic bundle over G? One plausible answer would be: Such a structure consists in specifying a Banach algebraic bundle $\mathcal{B}$ over G (in the sense of 2.2) and also an isomorphism between E and the $\mathcal{L}_1$ cross-sectional algebra of $\mathcal{B}$. The trouble with this answer in general is that it depends too much on the accidental nature of the norm of the $\mathcal{L}_1$ cross-sectional algebra (and indeed in the fibers of $\mathcal{B}$ themselves). However, if we confine ourselves to C^*-algebras and C^*-algebraic bundles, this objection can be easily overcome. Such a restriction of our question is perfectly satisfactory as long as our main goal is the *-representation theory of Banach *-algebras.

With these ideas in mind we shall proceed to a precise definition of what it means to equip an abstractly given C^*-algebra with the structure of a C^*-algebraic bundle over G.

As usual G is a fixed locally compact group, with unit e and left Haar measure λ.

17.2. Definition. If $\mathcal{B}$ is a Banach *-algebraic bundle over G, the C^*-completion (VI.10.4) of the $\mathcal{L}_1$ cross-sectional algebra $\mathcal{L}_1(\lambda; \mathcal{B})$ is called the *cross-sectional C^*-algebra* of $\mathcal{B}$, and is denoted by $C^*(\mathcal{B})$.

In particular, if $\mathscr{B}$ is the group bundle of G, $C^*(\mathscr{B})$ is the C^*-completion of the $\mathscr{L}_1$ group algebra of G. It is denoted by $C^*(G)$, and is called the *group C*-algebra* of G.

17.3. Let $\mathscr{B}$ be a Banach *-algebraic bundle over G. Combining VI.10.5 with 13.2, we find that the *-representations of $\mathscr{B}$ are in natural one-to-one correspondence with the *-representations of $C^*(\mathscr{B})$, and that this correspondence preserves non-degeneracy, intertwining operators, Hilbert direct sums, and so forth. In particular, the unitary representations of G are in a similar one-to-one correspondence with the non-degenerate *-representations of $C^*(G)$.

17.4.   *Definition.*   Let A be an abstractly given C^*-algebra. By a *C*-bundle structure for A over* the locally compact group G we mean a pair $\langle \mathscr{B}, F \rangle$, where $\mathscr{B}$ is a C^*-algebraic bundle over G, and F is a *-isomorphism of $C^*(\mathscr{B})$ onto A.

If $\mathscr{B}$ is saturated, we shall say that the C^*-bundle structure $\langle \mathscr{B}, F \rangle$ is *saturated*.

If $\langle \mathscr{B}, F \rangle$ is a C^*-bundle structure for A, 17.3 shows that the *-representations of A and of $\mathscr{B}$ are essentially the same.

17.5. Two C^*-bundle structures $\langle \mathscr{B}, F \rangle$ and $\langle \mathscr{B}', F' \rangle$ for the same C^*-algebra A over the same G are *isomorphic* if there exists a (necessarily isometric) isomorphism Φ of $\mathscr{B}$ onto $\mathscr{B}'$ (see 3.3) such that $F = F' \circ \Phi_0 : C^*(\mathscr{B}) \to C^*(\mathscr{B}')$ is the *-isomorphism induced by Φ.

Obviously two isomorphic C^*-bundle structures for A should be regarded as "essentially the same".

17.6.   *If $\mathscr{B}$ is a C*-algebraic bundle over a locally compact group G, and if the cross-sectional C*-algebra $C^*(\mathscr{B})$ has a unit element, then G must be discrete.* (See Exercise 39, Chapter XI.) It follows that a C^*-algebra A which has a unit element cannot have a C^*-bundle structure over a non-discrete locally compact group.

17.7.   *Compact Abelian group actions and C*-bundle structures.*   A fertile source of C^*-bundle structures is provided by compact Abelian groups of *-automorphisms. (See Exercise 22, Chapter X.)

A C*-Bundle Structure for the Glimm Algebras

17.8. In Chapter XII we shall develop the generalized Mackey analysis of the *-representation theory of any saturated Banach *-algebraic bundle (with an approximate unit) over a locally compact group G—in particular of any saturated C^*-algebraic bundle over G. Thus, in view of 17.3, the specification of a saturated C^*-bundle structure over G for a given C^*-algebra A means that we are able to apply the Mackey analysis to the investigation of the *-representation theory of A. This makes it desirable to find interesting saturated C^*-bundle structures for as many C^*-algebras as possible.

17.9. We shall now present a C^*-bundle structure for the Glimm algebras of VI.17.5.

For each positive integer n let E_n be some (discrete) finite group of order d_n, with unit e_n. Let E be the Cartesian product $\prod_{n=1}^{\infty} E_n$; thus E is a compact group. We shall denote by G the dense subgroup of E consisting of those x in E such that $x_n = e_n$ for all but finitely many n; G will carry the discrete topology. Let A be the commutative C^*-algebra of all continuous complex functions on E (with pointwise operations and the supremum norm). We now form the τ-semidirect product $\mathscr{B} = A \underset{\tau}{\times} G$ of A and G, where τ is the action of G on A by left translation:

$$(\tau_x \phi(y)) = \phi(x^{-1}y) \qquad (x \in G; \, y \in E).$$

We have seen in 16.2 that $\mathscr{B}$ is a C^*-algebraic bundle. We shall now show that $C^*(\mathscr{B})$ is *-isomorphic with the $(d_1, d_2, \ldots)$ Glimm algebra.

For each positive integer p, let G_p be the finite subgroup of G consisting of those x for which $x_n = e_n$ for all $n > p$; thus $G_p \cong E_1 \times E_2 \times \cdots \times E_p$. Let $\pi_p : E \to G_p$ be the surjective homomorphism $x \mapsto \langle x_1, x_2, \ldots, x_p, e_{p+1}, e_{p+2}, \ldots \rangle$; and let A_p be the *-subalgebra of A consisting of those ϕ such that $\phi(x)$ depends only on $\pi_p(x)$ $(x \in E)$; thus $A_p \cong \mathscr{C}(G_p)$. Let $\mathscr{L}^p(\mathscr{B})$ be the space of those f in $\mathscr{L}(\mathscr{B})$ which vanish outside G_p and whose values all lie in A_p. Since $\tau_x(A_p) = A_p$ for $x \in G_p$, $\mathscr{L}^p(\mathscr{B})$ is a *-subalgebra of $\mathscr{L}(\mathscr{B})$. By 7.10 $\mathscr{L}^p(\mathscr{B})$ is *-isomorphic with the $G_p \times G_p$ total matrix *-algebra, and so can be regarded as a C^*-algebra. If $i_{pq} : \mathscr{L}^q(\mathscr{B}) \to \mathscr{L}^p(\mathscr{B})$ is the identity injection (for $q < p$), then $\{\mathscr{L}^p(\mathscr{B})\}, \{i_{pq}\}$ is an isometric directed system of C^*-algebras (see VI.17.2, 3); and we verify that it is isomorphic in the obvious sense with the directed system $\{A_n\}, \{F_{nm}\}$ constructed in VI.17.5. Let D be the inductive limit of the system $\{\mathscr{L}^p(\mathscr{B})\}, \{i_{pq}\}$, and $i_p : \mathscr{L}^p(\mathscr{B}) \to D$ the canonical *-homomorphism. Each i_p is an extension of the preceding ones; and D is *-isomorphic with the $(d_1, d_2, \ldots)$ Glimm algebra.

Notice that $G = \bigcup_p G_p$, and that $\bigcup_p A_p$ is dense in A (by the Stone–Weierstrass Theorem). It follows that $\mathscr{L}'(\mathscr{B}) = \bigcup_p \mathscr{L}^p(\mathscr{B})$ is dense in $\mathscr{L}_1(\mathscr{B})$ (the $\mathscr{L}_1$ cross-sectional algebra of $\mathscr{B}$). Denoting by i the union of the i_p, we deduce from formula 7.9(9) that $i: \mathscr{L}'(\mathscr{B}) \to D$ is norm-decreasing (with respect to the $\mathscr{L}_1$ norm of $\mathscr{L}'(\mathscr{B})$ and the C^*-norm of D). So i extends to a (norm-decreasing) *-homomorphism of $\mathscr{L}_1(\mathscr{B})$ into D whose range is dense in D.

If T is a *-representation of D, $S = T \circ i$ is a *-representation of $\mathscr{L}_1(\mathscr{B})$. Conversely, suppose S is a *-representation of $\mathscr{L}_1(\mathscr{B})$. For each $p = 1$, $2, \ldots, S|\mathscr{L}^p(\mathscr{B})$ is a *-representation of the C^*-algebra $\mathscr{L}^p(\mathscr{B})$ and hence continuous with respect to the C^*-norm of $\mathscr{L}^p(\mathscr{B})$. Thus $S \circ i^{-1}$ is a norm-continuous *-representation of the pre-C^*-algebra $D' = i(\mathscr{L}'(\mathscr{B}))$. Since D' is dense in D, $S \circ i^{-1}$ extends uniquely to a *-representation T of D, and we have $S = T \circ i$.

From the correspondence $S \leftrightarrow T$ set up in the preceding paragraph it follows easily that $i: \mathscr{L}_1(\mathscr{B}) \to D$ induces a *-isomorphism i_c between the C^*-completion $C^*(\mathscr{B})$ of $\mathscr{L}_1(\mathscr{B})$ and the C^*-algebra D. Since the latter is just the $(d_1, d_2, \ldots)$ Glimm algebra, we have thus introduced into the latter a saturated C^*-bundle structure $\langle \mathscr{B}, i_c \rangle$ over G.

C^*-Bundle Structures and Inner Tensor Products

17.9.   Let A be a C^*-algebra, G a locally compact group, and $\langle \mathscr{B}, F \rangle$ a C^*-bundle structure for A over G. From this we can immediately define an inner tensor product operation on representations of G and A.

Indeed: Let U be a unitary representation of G, and S a *-representation of A. Then $S' = S \circ F$ is a *-representation of $C^*(\mathscr{B})$, corresponding (see 17.3) with a *-representation S'' of $\mathscr{B}$. Now let $T'' = U \otimes S''$ be the (inner) Hilbert tensor product *-representation of $\mathscr{B}$ defined in 9.16; let T' be the corresponding *-representation of $C^*(\mathscr{B})$; and transfer T' via F^{-1} to a *-representation $T = T' \circ F^{-1}$ of A. This T is called the (*inner*) *tensor product of U and S with respect to* $\langle \mathscr{B}, F \rangle$.

As we saw in §5 of the Introduction to this Volume, in the finite-dimensional situation the bundle structure $\langle \mathscr{B}, F \rangle$ is essentially determined by this tensor product operation. It is reasonable to conjecture that an analogous statement should hold even for continuous groups G and infinite-dimensional C^*-algebras A. Indeed, one can ask two questions: (I) If $\langle \mathscr{B}, F \rangle$ and $\langle \mathscr{B}', F' \rangle$ are two C^*-bundle structures for the same A over the same G, and if the inner tensor product operations derived from the two are identical, are $\langle \mathscr{B}, F \rangle$ and $\langle \mathscr{B}', F' \rangle$ necessarily isomorphic? (II) Is there a simple set of postulates characterizing those inner tensor product operations which are derived as above from a C^*-bundle structure for A over G?

In this connection see Landstad [2] and Imai and Takai [1]. For the answer in the commutative saturated case, see X.5.17.

18. Transformation Bundles and Systems of Imprimitivity

18.1. In this section we are going to give an important alternative description of the *-representations of the transformation bundles constructed in §7. We shall find that the *-representations of transformation bundles are generalizations to the bundle context of the systems of imprimitivity which play such an important role in the "normal subgroup analysis" of group representations.

18.2. Fix a Banach *-algebraic bundle $\mathscr{B} = \langle B, \pi, \cdot, * \rangle$ (necessarily with enough continuous cross-sections) over a locally compact group G with unit e, left Haar measure λ, and modular function Δ. Let G act to the left as a topological transformation group on a locally compact Hausdorff space M; and construct the transformation bundle $\mathscr{D} = \langle D, \rho, \cdot, * \rangle$ as in 7.2. We will use the notation of 7.2 without comment.

18.3. We now observe that the elements of B, and also the bounded continuous complex functions on M, act as multipliers of $\mathscr{D}$.

Indeed, if $b \in B_x$, it was observed in 7.5 that the equations

$$(b\phi)(m) = b\phi(x^{-1}m),$$

$$(\phi b)(m) = \phi(m)b$$

($\phi \in D$; $m \in M$) define the left and right actions of a multiplier u_b of $\mathscr{D}$ of order x. The map $b \mapsto u_b$ of B into $\mathscr{W}(\mathscr{D})$ is linear on each fiber and preserves multiplication, *, and norm. It is also continuous with respect to the strong topology of multipliers (see 15.2). Indeed: To see this it is enough by II.13.12 to show that $b \mapsto b\tilde{f}(x_0)$ and $b \mapsto \tilde{f}(x_0)b$ are continuous on B to D for each fixed x_0 in G and f in E. For each b in B put $g_b(x, m) = bf(\pi(b)^{-1}x, \pi(b)^{-1}m)$, so that $g_b \in E$. Let $b_i \to b$ in B. An easy compactness argument shows that $g_{b_i} \to g_b$ uniformly on $G \times M$. Hence

$$\|b_i\tilde{f}(x_0) - (g_b)\tilde{}(\pi(b_i)x_0)\| = \|(g_{b_i})\tilde{}(\pi(b_i)x_0) - (g_b)\tilde{}(\pi(b_i)x_0)\| \to 0.$$

Since $(g_b)\tilde{}(\pi(b_i)x_0) \to (g_b)\tilde{}(\pi(b)x_0) = b\tilde{f}(x_0)$ in D, it therefore follows from II.13.12 that $b_i\tilde{f}(x_0) \to b\tilde{f}(x_0)$ in D. So $b \mapsto b\tilde{f}(x_0)$ is continuous. Similarly $b \mapsto \tilde{f}(x_0)b$ is continuous.

By 5.8 the integrated form u'_b of u_b is a multiplier of $\mathscr{L}(\mathscr{D})$. Identifying E with a *-subalgebra of $\mathscr{L}(\mathscr{D})$ via Proposition 7.7, we check that E is stable under u'_b. In fact

$$(bf)(x, m) = bf(y^{-1}x, y^{-1}m),\tag{1}$$

$$(fb)(x, m) = \Delta(y^{-1})f(xy^{-1}, m)b\tag{2}$$

($b \in B_y$; $f \in E$; $\langle x, m \rangle \in G \times M$; we write bf and fb for $u'_b f$ and fu'_b).

18.4. We shall now make the bounded elements of $\mathscr{C}(M)$ act as multipliers of $\mathscr{D}$.

Let h be a bounded continuous complex function on M. Putting

$$(h\phi)(m) = h(m)\phi(m),$$

$$(\phi h)(m) = h(x^{-1}m)\phi(m)$$

($\phi \in D_x$; $m \in M$), one checks that the maps $\phi \mapsto h\phi$ and $\phi \mapsto \phi h$ together form a multiplier v_h of $\mathscr{D}$ of order e. The correspondence $h \mapsto v_h$ is a *-homomorphism, and is norm-decreasing with respect to the supremum norm of the h.

By 5.8 the integrated form of v_h is a multiplier v'_h of $\mathscr{L}(\mathscr{D})$. The action of v'_h on E is easily seen to be given by

$$(hf)(x, m) = h(m)f(x, m), \qquad (fh)(x, m) = h(x^{-1}m)f(x, m)\tag{3}$$

($f \in E$; $\langle x, m \rangle \in G \times M$; we write hf and fh for $v'_h f$ and fv'_h).

18.5. An important relation holds between the multipliers u_b and v_h of 18.3 and 18.4, namely

$$u_b v_h = v_{yh} u_b\tag{4}$$

($b \in B_y$; h is a bounded function in $\mathscr{C}(M)$). Here yh is the translate $m \mapsto h(y^{-1}m)$ of h. The routine verification of this is left to the reader.

18.6. A similar calculation shows that if $b \in B_y$ and $h \in \mathscr{C}_0(M)$, then the product $v_h u_b$ coincides with the multiplier of $\mathscr{D}$ consisting of left and right multiplication by the following element ψ of D_y:

$$\psi(m) = h(m)b.\tag{5}$$

18.7. We shall see shortly that *-representations of $\mathscr{D}$ are intimately related to objects called systems of imprimitivity.

Definition.   A *system of imprimitivity* (*for $\mathscr{B}$ over M*) is a pair $\langle T, P \rangle$ where (i) T is a non-degenerate *-representation of $\mathscr{B}$, (ii) P is a regular $X(T)$-projection-valued Borel measure on M (see II.11.9), and (iii) we have

$$T_b P(W) = P(\pi(b)W)T_b \tag{6}$$

for all $b \in B$ and all Borel subsets W of M.

Remark 1.   If $\mathscr{B}$ is the group bundle of G, the non-degenerate *-representations of $\mathscr{B}$ are essentially just the unitary representations of G (see 9.5). In that case we can omit the reference to $\mathscr{B}$; and we obtain the more familiar notion of a *system of imprimitivity for* the group G *over* M. This is a pair $\langle T, P \rangle$, where (i) T is a unitary representation of G, (ii) P is a regular $X(T)$-projection-valued Borel measure on M, and (iii)

$$T_x P(W) = P(xW)T_x \tag{7}$$

for all x in G and all Borel subsets W of M.

Remark 2.   If X_W stands for the range of $P(W)$, (6) can be replaced by the equivalent statement:

$$T_b(X_W) \subset X_{\pi(b)W} \tag{8}$$

$(b \in B$; W is a Borel subset of M). Thus the crucial condition (6) asserts that the action of T on the subspaces X_W is "covariant" with the action of G on the space M.

18.8.   Another equivalent version of (6) is the assertion that

$$T_b P(\phi) = P(\pi(b)\phi)T_b \tag{9}$$

for all b in B and ϕ in $\mathscr{C}_0(M)$. (Here $(x\phi)(m) = \phi(x^{-1}m)$ for $x \in G$ and $\phi \in \mathscr{C}_0(M)$; and $P(\phi)$ is the spectral integral $\int \phi(m)\, dPm$ defined in II.11.7.)

To show that (6) and (9) are equivalent, we first assume (6). Given $\phi \in \mathscr{C}_0(M)$, we choose a sequence $\{W^{(r)}\}$ of finite decompositions of M into disjoint Borel sets: $M = W_1^{(r)} \cup \cdots \cup W_{n_r}^{(r)}$; and also a collection of complex numbers $\lambda_i^{(r)}$ ($r = 1, 2, \ldots$; $i = 1, \ldots, n_r$) such that

$$\phi_r = \sum_{i=1}^{n_r} \lambda_i^{(r)} \mathrm{Ch}_{W_i^{(r)}} \to \phi \qquad \text{uniformly.}$$

Then, by the definition of the spectral integral in II.11.7,

$$P(\phi_r) = \sum_{i=1}^{n_r} \lambda_i^{(r)} P(W_i^{(r)}) \to P(\phi) \qquad \text{in the norm-topology.} \tag{10}$$

Furthermore, for each x in G,

$$x\phi_r \to x\phi \qquad \text{uniformly,}$$

so that

$$P(x\phi_r) \to P(x\phi) \qquad \text{in the norm-topology.} \tag{11}$$

But by (6), if $x \in G$ and $b \in B_x$,

$$T_b P(\phi_r) = T_b \sum_{i=1}^{n_r} \lambda_i^{(r)} P(W_i^{(r)})$$

$$= \sum_{i=1}^{n_r} \lambda_i^{(r)} P(xW_i^{(r)}) T_b$$

$$= P(x\phi_r) T_b$$

for each r. Passing to the limit $r \to \infty$ and using (10) and (11), we get (9).

Conversely, assume (9). Take an element b of B with $\pi(b) = x$; and let $\mathcal{Y}$ be the family of those Borel subsets W of M such that (6) holds. Thus Y is a σ-field of Borel sets. Let W be a compact G_δ; let $\{U_n\}$ be a decreasing sequence of open sets with $\bigcap_n U_n = W$; and let $\{\phi_n\}$ be a sequence of elements of $\mathcal{L}(M)$ such that $\text{Ch}_W \leq \phi_n \leq \text{Ch}_{U_n}$ for each n. Then $P(\phi_n) \to P(W)$ weakly; and likewise $P(x\phi_n) \to P(xW)$ weakly. From this and (6) we deduce that $W \in \mathcal{Y}$; so $\mathcal{Y}$ contains all compact G_δ's. Now if U is any open subset of M, then $P(W) \to P(U)$ and $P(xW) \to P(xU)$ weakly as W runs over the upward directed set of compact G_δ's contained in U. Since the compact G_δ's are in $\mathcal{Y}$, it follows that U is in $\mathcal{Y}$. Thus $\mathcal{Y}$ contains all open sets and hence all Borel sets. So (6) holds for all b and W.

The equivalence which we have just proved can be usefully generalized by means of the following proposition.

Proposition. *Let X be a Hilbert space and P a regular X-projection-valued Borel measure on M. Let F be an algebra of bounded continuous complex functions on M such that (i) F is closed under complex conjugation, (ii) F does not vanish at any point of M, and (iii) F separates the points of M. Suppose furthermore that $x \in G$ and $Q \in \mathcal{O}(X)$, and that*

$$QP(\phi) = P(x\phi)Q \qquad \text{for all } \phi \text{ in } F.$$

Then

$$QP(W) = P(xW)Q \qquad \text{for all Borel subsets } W \text{ of } M.$$

Proof. By the proof of the equivalence of (6) and (9), it is enough to take a fixed ψ in $\mathscr{C}_0(M)$ and show that $QP(\psi) = P(x\psi)Q$.

Let K and L be compact subsets of M and ξ a vector in range $(P(L))$; and put $Q' = P(K)Q$. Given $\varepsilon > 0$, by the Stone-Weierstrass Theorem there is a function ϕ in F such that

$$|\phi(m) - \psi(m)| \leq \varepsilon \qquad \text{for all } m \text{ in } L \cup x^{-1}K.$$

This implies that $|(x\phi)(m) - (x\psi)(m)| \leq \varepsilon$ for all m in K, and hence (by II.11.8) that

$$\|P(\phi)\xi - P(\psi)\xi\| \leq \varepsilon\|\xi\|,$$

$$\|P(x\phi)Q'\xi - P(x\psi)Q'\xi\| \leq \varepsilon\|Q'\xi\| \leq \varepsilon\|Q'\|\,\|\xi\|.$$

Now since $\phi \in F$ we have

$$Q'P(\phi) = P(K)QP(\phi)$$

$$= P(K)P(x\phi)Q$$

$$= P(x\phi)Q'.$$

Therefore

$$\|Q'P(\psi)\xi - P(x\psi)Q'\xi\|$$

$$\leq \|Q'(P(\psi)\xi - P(\phi)\xi)\|$$

$$\quad + \|P(x\phi)Q'\xi - P(x\psi)Q'\xi\|$$

$$\leq 2\varepsilon\|Q\|\,\|\xi\|.$$

By the arbitrariness of ε this gives

$$Q'P(\psi)\xi = P(x\psi)Q'\xi.$$

Allowing K and L to expand indefinitely, we see from this that $QP(\psi) = P(x\psi)Q$. ∎

18.9. Concepts such as stable subspaces, Hilbert direct sums, and so forth, can now be defined for systems of imprimitivity just as they were for *-representations.

A subspace Y of $X(T)$ is *stable under* a system of imprimitivity $\langle T, P \rangle$ over M if Y is stable under both T_b and $P(W)$ for all b in B and all Borel subsets W of M. If there are no closed $\langle T, P \rangle$-stable subspaces except $\{0\}$ and $X(T)$, $\langle T, P \rangle$ is *irreducible*.

If $\mathscr{T} = \langle T, P \rangle$ and $\mathscr{T}' = \langle T', P' \rangle$ are two systems of imprimitivity over M, a bounded linear operator $F: X(T) \to X(T')$ is $\mathscr{T}, \mathscr{T}'$ *intertwining* if $FT_b =$

$T'_b F$ and $FP(W) = P'(W)F$ for all b in B and all Borel subsets W of M. If F is isometric and onto $X(T')$, $\mathcal{T}$ and $\mathcal{T}'$ are *equivalent* (*under* F).

For each i in an index set I let $\mathcal{T}^i = \langle T^i, P^i \rangle$ be a system of imprimitivity for $\mathcal{B}$ over M. Then we can form the *Hilbert direct sum* $\mathcal{T} = \langle T, P \rangle = \sum_{i \in I}^{\oplus} \mathcal{T}^i$ in the obvious way:

$$T = \sum_{i \in I}^{\oplus} T^i, \qquad P(W) = \sum_{i \in I}^{\oplus} P^i(W).$$

18.10. Our next project is to show how, from a system of imprimitivity $\langle T, P \rangle$ for $\mathcal{B}$ over M, we can construct a non-degenerate *-representation S of $\mathcal{D}$ (and hence of $\mathcal{L}_1(\lambda; \mathcal{D})$). After that we shall prove the converse—that every non-degenerate *-representation of $\mathcal{D}$ arises by this construction from a system of imprimitivity. Thus, non-degenerate *-representations of the transformation bundle $\mathcal{D}$ will be essentially the same things as systems of imprimitivity for $\mathcal{B}$ over M.

18.11. Fix a system of imprimitivity $\mathcal{T} = \langle T, P \rangle$ for $\mathcal{B}$ over M; and abbreviate $X(T)$ to X.

The corresponding representation S of $\mathcal{D}$, promised in 18.10, is going to be given by the formula:

$$S_\phi = \int_M dPm\, T_{\phi(m)}. \tag{12}$$

However the integral in (12) is of a new kind, which must be defined and discussed before (12) has meaning.

Fix an element x of G; and define $\mathcal{F}_x$ as the family of all functions $H: M \to \mathcal{O}(X)$ such that:

   (a) H is continuous with respect to the norm-topology of operators;

   (b) $\|H(m)\| \to 0$ as $m \to \infty$ in M;

   (c) $H(m)P(W) = P(xW)H(m)$ for all m in M and all Borel subsets W of M.

For $H \in \mathcal{F}_x$ we shall define an integral $\int dPm\, H(m)$, whose value will be an operator in $\mathcal{O}(X)$.

By a *partition* of M we shall mean a finite sequence

$$\sigma = \langle W_1, \ldots, W_r; m_1, \ldots, m_r \rangle, \tag{13}$$

where the $W_1, \ldots, W_r$ are pairwise disjoint non-void Borel subsets of M whose union is M, and $m_i \in W_i$ for each i. The set of all partitions σ will be considered as a directed set under the relation $\prec$ of refinement; that is, if σ is as in (13) and $\sigma' = \langle W'_1, \ldots, W'_s; m'_1, \ldots, m'_s \rangle$, then $\sigma \prec \sigma'$ means that each W'_j is contained in some W_i. Now, to the partition (13) and the function H in $\mathscr{F}_x$ we associate the operator $V_\sigma = V_\sigma(H) = \sum_{i=1}^r P(W_i)H(m_i)$. We claim that $\{V_\sigma\}$ is a Cauchy net with respect to the norm-metric of operators.

Indeed, for $\varepsilon > 0$, conditions (a) and (b) on H allow us to choose the partition σ in (13) so that on each W_i the norm-variation of H is less than ε. Now suppose $\sigma \prec \sigma' = \langle W'_1, \dots, W'_s; m'_1, \dots, m'_s \rangle$; and for each $j = 1, \dots, s$ let $\mu_j = m_i$, where i is that integer such that $W'_j \subset W_i$. Then $\|H(m'_j) - H(\mu_j)\| < \varepsilon$. Since clearly $V_\sigma = \sum_{j=1}^s P(W'_j)H(\mu_j)$, we have by condition (c) on H

$$\|(V_{\sigma'} - V_\sigma)\xi\|^2 = \sum_{j=1}^s \|P(W'_j)(H(m'_j) - H(\mu_j))\xi\|^2$$

$$= \sum_{j=1}^s \|(H(m'_j) - H(\mu_j))P(x^{-1}W'_j)\xi\|^2$$

$$\leq \varepsilon^2 \sum_{j=1}^s \|P(x^{-1}W'_j)\xi\|^2 = \varepsilon^2 \|\xi\|^2 \qquad \text{for } \xi \in X.$$

Thus $\|V_{\sigma'} - V_\sigma\| \leq \varepsilon$; and the claim is proved.

Therefore $\lim_\sigma V_\sigma(H)$ exists in the norm-topology of operators; we call it $\int_M dPm H(m)$. It is defined whenever $H \in \mathscr{F}_x$ for some x.

18.12. Here are some properties of the integral defined in 18.11.

Clearly $\int_M dPm H(m)$ is linear in H on each $\mathscr{F}_x$.

Suppose that $x \in G$ and Q is an operator in $\mathcal{O}(X)$ such that $QP(W) = P(xW)Q$ for all Borel subsets W of M. If $\psi \in \mathscr{C}_0(M)$, then $m \mapsto \psi(m)Q$ belongs to $\mathscr{F}_x$, and it is easy to see that

$$\int_M dPm\psi(m)Q = P(\psi)Q, \tag{14}$$

where $P(\psi)$ is the spectral integral $\int_M \psi(m)\, dPm$.

If $H \in \mathscr{F}_x$ and $F = \int_M dPm H(m)$, then by passing to the limit of the $V_\sigma(H)$ of 18.11 we obtain

$$FP(W) = P(xW)F \tag{15}$$

for all Borel subsets W of M.

18.13. The next proposition gives more properties of this integral.

Proposition.   *If* $x, y \in G$, $H \in \mathcal{F}_x$, *and* $K \in \mathcal{F}_y$, *then:*

(I) $$\left\| \int_M dPm H(m) \right\| \leq \sup\{\|H(m)\| : m \in M\}.$$

(II) $$\left(\int_M dPm H(m) \right)^* = \int_M dPm (H(xm))^*.$$

(III) $$\int_M dPm H(m) K(m) = \left[\int_M dPm H(m) \right]\left[\int_M dPm K(xm) \right].$$

(IV)   *If* $Q \in \mathcal{O}(X)$ *and* $QP(W) = P(xW)Q$ *for all Borel subsets* W *of* M, *then*

$$\int_M dPm Q K(m) = Q \int_M dPm K(xm).$$

Proof.   (I) If $k = \sup_m \|H(m)\|$, σ is as in (13), and $\xi \in X$, then $\|V_\sigma(H)\xi\|^2 = \sum_i \|P(W_i)H(m_i)\xi\|^2 = \sum_i \|H(m_i)P(x^{-1}W_i)\xi\|^2 \leq k^2 \sum_i \|P(x^{-1}W_i)\xi\|^2 = k^2\|\xi\|^2$; thus $\|V_\sigma(H)\| \leq k$. Passing to the limit in σ we get (I).

(II) If σ is as in (13),

$$(V_\sigma(H))^* = \left(\sum_i H(m_i)P(x^{-1}W_i) \right)^* = \sum_i P(x^{-1}W_i)(H(m_i))^* = V_{x^{-1}\sigma}(K),$$

where $K(m) = (H(xm))^*$ and $x^{-1}\sigma$ means $\langle x^{-1}W_1, \ldots, x^{-1}W_r;$ $x^{-1}m_1, \ldots, x^{-1}m_r \rangle$. Now notice that, as σ "increases indefinitely," so does $x^{-1}\sigma$. Hence $\lim_\sigma (V_\sigma(H))^* = \lim_\sigma V_{x^{-1}\sigma}(K) = \lim_\sigma V_\sigma(K)$; and this is (II).

(III) σ being as in (13), we have

$$V_\sigma(HK) = \sum_i P(W_i)H(m_i)K(m_i) = \sum_{i,j} P(W_i)P(W_j)H(m_i)K(m_j)$$

$$= \sum_{i,j} P(W_i)H(m_i)P(x^{-1}W_j)K(m_j) = V_\sigma(H)V_{x^{-1}\sigma}(L),$$

where $x^{-1}\sigma$ is as in (II) and $L(m) = K(xm)$. Passing to the limit in σ we get (III).

The proof of (IV) is just a simpler version of that of (III).   ∎

18.14.   We are now ready to justify formula 18.11(12).

If $x \in G$ and $\phi \in D_x$, the function $m \mapsto T_{\phi(m)}$ belongs to $\mathcal{F}_x$. Indeed, conditions 18.11(a), (b) follow from 9.2(1), while 18.11(c) is an immediate consequence of 18.7(6). Thus the definition of the integral in 18.11 gives meaning to the right side of (12). Let $S: \phi \mapsto S_\phi$ be the operator-valued mapping on D defined by (12).

Proposition. *S is a non-degenerate *-representation of $\mathcal{D}$.*

Proof. Clearly S is linear on each fiber. From 18.13 (and 9.2(1)) we get:

$$\|S_\phi\| \le \|\phi\|, \tag{16}$$

$$S_{\phi^*} = (S_\phi)^*, \tag{17}$$

$$S_{\phi\psi} = S_\phi S_\psi \tag{18}$$

$(\phi, \psi \in D)$.

We must now prove the strong continuity 8.2(iv) of S.

Let f belong to E, and let $\tilde{f}$ be the corresponding continuous cross-section of $\mathcal{D}$ ($\tilde{f}(x)(m) = f(x, m)$). We claim that $x \mapsto S_{\tilde{f}(x)}$ is strongly continuous on G.

Indeed, the uniform continuity of f shows that, for each $\varepsilon > 0$ and each m_0 in M, there is a neighborhood U of m_0 such that $\|f(x, m) - f(x, m_0)\| < \varepsilon$, and hence $\|T_{f(x,m)} - T_{f(x,m_0)}\| < \varepsilon$, for all m in U and x in G. Therefore, denoting by H_x the function $m \mapsto T_{f(x,m)}$, we can find a partition σ such that

$$\|V_\sigma(H_x) - S_{\tilde{f}(x)}\| < \varepsilon \qquad \text{for all } x \text{ in } G. \tag{19}$$

Now the strong continuity of T implies that $x \mapsto V_\sigma(H_x)$ is strongly continuous on G. From this and (19) we see that $x \mapsto S_{\tilde{f}(x)}$ is strongly continuous. So the claim is proved.

Now we recall that for each x $\{\tilde{f}(x): f \in E\}$ is dense in D_x (see 7.2). Suppose now that $\phi_i \to \phi$ in D, where $\pi(\phi) = x$ and $\pi(\phi_i) = x_i$. Given $\varepsilon > 0$, choose $f \in E$ so that $\|\tilde{f}(x) - \phi\| < \varepsilon$. Then $\|\tilde{f}(x_i) - \phi_i\| < \varepsilon$, and so by (16) $\|S_{\tilde{f}(x_i)} - S_{\phi_i}\| < \varepsilon$, for all large i. Combining this with the above claim, and the arbitrariness of ε, we conclude that $S_{\phi_i} \to S_\phi$ strongly.

So S is a *-representation of $\mathcal{D}$.

To show that S is non-degenerate, let η be a vector in X such that $(S_\phi \xi, \eta) = 0$ for all ϕ in D and ξ in X. Taking ϕ to be of the special form $m \mapsto h(m)b$, where $h \in \mathcal{L}(M)$ and $b \in B$, we have $S_\phi = P_h T_b$ by (14), and so

$$(P_h T_b \xi, \eta) = 0 \qquad \text{for } h \in \mathcal{L}(M), b \in B, \xi \in X. \tag{20}$$

Letting h vary so that P_h approaches the identity operator weakly, we see from (20) that $(T_b \xi, \eta) = 0$ for all b in B and ξ in X. But by the non-degeneracy of T this implies $\eta = 0$. So S is non-degenerate; and the proof is complete. ∎

Definition. The *-representation S of $\mathcal{D}$ constructed above will be called the *transformation bundle form* of $\langle T, P \rangle$.

18.15.   In the context of 18.14, suppose Y is a closed $\mathscr{T}$-stable subspace of X. Then it is clear from the definition of the integral (12) that Y is also S-stable. Let $\mathscr{T}'$ stand for the subsystem of imprimitivity acting on Y (that is, $\mathscr{T}' = \langle T', P' \rangle$, where $T'_b = T_b | Y$, $P'(W) = P(W) | Y$). Evidently the transformation bundle form of $\mathscr{T}'$ is exactly the subrepresentation acting on Y of the transformation bundle form S of $\mathscr{T}$.

Similarly for Hilbert direct sums. For each i in an index set I, let $\mathscr{T}^i$ be a system of imprimitivity for $\mathscr{B}$ over M, with transformation bundle form S^i. Then the transformation bundle form of the Hilbert direct sum $\sum_i^{\oplus} \mathscr{T}^i$ is just the Hilbert direct sum $\sum_i^{\oplus} S^i$.

We omit the verification of these simple facts.

18.16.   Keeping the notation unchanged, let us start with a system of imprimitivity $\mathscr{T} = \langle T, P \rangle$ for $\mathscr{B}$ over M acting on X, form the transformation bundle form S of $\mathscr{T}$, and denote by R the integrated form of S. Thus R is a non-degenerate *-representation of $\mathscr{L}_1(\lambda; \mathscr{D})$. Since E can be identified with a dense *-subalgebra of $\mathscr{L}_1(\lambda; \mathscr{D})$, $R | E$ is a non-degenerate *-representation of E.

Definition. $R | E$ is called the *E-integrated form* (or simply the *integrated form*) of $\mathscr{T}$. In view of (12), it is given by the formula

$$R_f \xi = \int_G \left[\int_M dPm T_{f(x,m)} \right] \xi \, d\lambda x \qquad (f \in E; \, \xi \in X). \qquad (21)$$

18.17.   Here is a simple and useful description of the integrated form R of $\mathscr{T}$. Let us denote by E_0 the linear span in E of the set of all functions f on $G \times M$ of the form

$$\langle x, m \rangle \mapsto \psi(m)\phi(x), \qquad (22)$$

where $\psi \in \mathscr{L}(M)$ and $\phi \in \mathscr{L}(\mathscr{B})$.

Proposition.   (I) *E_0 is dense in E in the inductive limit topology.* (II) *If f is the element of E_0 given by (22), then*

$$R_f = P(\psi)T_\phi \qquad (23)$$

(T_ϕ being of course the operator of the integrated form of T).

Proof.   (I) Let $\bar{E}_0$ be the inductive limit closure of E_0 in E, and denote by $\mathscr{G}$ the set of those functions g in $\mathscr{C}(G \times M)$ such that $\bar{E}_0$ is stable under multiplication by g. Evidently $\mathscr{G}$ is closed in the topology of uniform convergence on compact sets, and contains all products $\alpha \times \beta: \langle x, m \rangle \mapsto \alpha(x)\beta(m)$, where $\alpha \in \mathscr{C}(G)$, $\beta \in \mathscr{C}(M)$. Hence by the Stone-Weierstrass Theorem $\mathscr{G} = \mathscr{C}(G \times M)$. Thus statement (I) follows from the application of II.14.6 to $\bar{E}_0$.

The verification of (II), using (21), is left to the reader. ∎

18.18. Let S be the transformation bundle form of a system of imprimitivity $\mathscr{T} = \langle T, P \rangle$ for $\mathscr{B}$ over M. It is easy to reconstruct $\mathscr{T}$ from S by means of the multipliers u_b and v_h constructed in 18.3 and 18.4. Indeed, if $b \in B$, $h \in \mathscr{C}_0(M)$, and $\phi \in D$, we claim that

$$T_b S_\phi = S_{u_b \phi}, \tag{24}$$

$$P(h) S_\phi = S_{v_h \phi}. \tag{25}$$

In fact: If $x = \pi(b)$, we have by 18.13(IV)

$$T_b S_\phi = T_b \int dPm\, T_{\phi(m)} = \int dPm\, T_b\, T_{\phi(x^{-1}m)}$$

$$= \int dPm\, T_{(u_b \phi)(m)} = S_{u_b \phi},$$

proving (24). (25) is proved similarly.

In view of 15.3, (24) says that T_b is just the operator S'_{u_b} corresponding under S to the multiplier u_b of $\mathscr{D}$. Similarly, by (25) $P(h)$ is the operator S'_{v_h} corresponding to v_h under S. This immediately implies:

Proposition. *The system of imprimitivity $\mathscr{T}$ (for $\mathscr{B}$ over M) is uniquely determined by its transformation bundle form S, and hence also by its E-integrated form.*

18.19. Proposition. *Let $\mathscr{T}$ and $\mathscr{T}'$ be two systems of imprimitivity for $\mathscr{B}$ over M, with transformation bundle forms S and S' respectively, and E-integrated forms R and R' respectively. Then:*

(I) *For a closed subspace Y of the space of T, the conditions of $\mathscr{T}$-stability, S-stability, and R-stability coincide.*

(II) *For a bounded linear map F of the space of $\mathscr{T}$ into that of $\mathscr{T}'$, the conditions of being $\mathscr{T}$, $\mathscr{T}'$ intertwining, S, S' intertwining, and R, R' intertwining all coincide.*

Proof. Since E is dense in the $\mathscr{L}_1$ cross-sectional algebra of $\mathscr{D}$, the statements of equivalence between S and R follow from 11.9 and 11.15.

(I) If Y is $\mathscr{T}$-stable, it is S-stable by 18.15. Conversely, suppose Y is S-stable. Since S is non-degenerate, its subrepresentation on Y is non-degenerate by VI.9.9; that is, vectors of the form $\xi = S_\phi\eta$, where $\phi \in D$ and $\eta \in Y$, span a dense subspace of Y. But for such vectors ξ we have by (24) $T_b\xi = S_{u_b\phi}\eta \in Y$ for all b in B, and by (25) $P(h)\xi = S_{v_h\phi}\eta \in Y$ for all h in $\mathscr{C}_0(M)$. The latter is easily seen to imply that Y is stable under $P(W)$ for all Borel sets W. Therefore Y is $\mathscr{T}$-stable. Thus $\mathscr{T}$-stability and S-stability of Y are the same.

(II) This follows from Part (I) by the same argument as in 11.15.   ■

18.20. It turns out that *every* non-degenerate *-representation S of the transformation bundle $\mathscr{D}$ is the transformation bundle form of a system of imprimitivity. From this, of course, it will follow that every continuous non-degenerate *-representation R of E is the E-integrated form of a system of imprimitivity.

18.21.   Theorem. *Let S be a non-degenerate *-representation of $\mathscr{D}$. There is a unique system of imprimitivity $\mathscr{T} = \langle T, P \rangle$ for $\mathscr{B}$ over M, acting on $X(S)$, whose transformation bundle form is S.*

Proof. The uniqueness of T follows from 18.18.

To prove its existence, let us extend S to a *-representation S' of the multiplier bundle $\mathscr{W}(\mathscr{D})$ as in 15.3. Recalling the definition of the multipliers u_b and v_h (18.3 and 18.4), we set

$$T_b = S'_{u_b}, \qquad P(h) = S'_{v_h}$$

($b \in B$; $h \in \mathscr{C}_0(M)$). Now S' and $b \mapsto u_b$ both preserve all operations, and are both continuous with respect to the strong topology of multipliers (see 15.3 and 18.3). It follows that T is a *-representation of $\mathscr{B}$. Similarly P is a *-representation of $\mathscr{C}_0(M)$.

By 18.6, if $h \in \mathscr{C}_0(M)$ and $b \in B$, then

$$P(h)T_b = S_\psi, \tag{26}$$

where $\psi(m) = h(m)b$. Now it follows from II.14.6 that for each x in G, the linear span of all such ψ (constructed from h in $\mathscr{C}_0(M)$ and b in B_x) is dense in D_x. Therefore, by (26) and the non-degeneracy of S, $\{P(h)T_b : h \in \mathscr{C}_0(M), b \in B\}$ acts non-degenerately on $X(S)$. From this it follows immediately that the *-representations P and T are both non-degenerate.

Since P is non-degenerate, by II.12.8 it arises from a regular $X(S)$-projection-valued Borel measure on M, which we will also call P. By 18.5 the identity

$$T_b P(h) = P(xh)T_b$$

holds for $b \in B_x$ and $h \in \mathscr{C}_0(M)$. This implies by 18.8 that $\mathscr{T} = \langle T, P \rangle$ is a system of imprimitivity for $\mathscr{B}$ over M.

It remains only to show that the transformation bundle form of $\mathscr{T}$ is S. Let us denote by $\tilde{S}$ the transformation bundle form of $\mathscr{T}$. If $\psi(m) = h(m)b$ ($b \in B$; $h \in \mathscr{C}_0(M)$), it follows from (14) that $\tilde{S}_\psi = P(h)T_b$. Comparing this with (26), we see that $\tilde{S}_\psi = S_\psi$ for all such ψ. But we observed above that the linear span of these ψ is dense in each fiber D_x. So $\tilde{S} = S$. ∎

18.22.   As a consequence of 18.21 and 18.19 we can summarize this section as follows:

Theorem.   *The passage from a system of imprimitivity $\mathscr{T}$ for $\mathscr{B}$ over M to its transformation bundle form S, and then from S to its integrated form R, sets up a one-to-one correspondence between the following three sets of objects:*

(i)   *The set of all systems of imprimitivity for $\mathscr{B}$ over M;*

(ii)   *the set of all non-degenerate *-representations of the transformation bundle $\mathscr{D}$;*

(iii)   *the set of all non-degenerate *-representations of $\mathscr{L}_1(\lambda; \mathscr{D})$.*

This correspondence preserves closed stable subspaces, intertwining maps, and Hilbert direct sums. In particular it preserves irreducibility and unitary equivalence.

19. Ergodic Measure Transformation Spaces and Certain Irreducible Systems of Imprimitivity

19.1.   We have seen in §7 how the action of a locally compact group G on a locally compact Hausdorff space M gives rise to a transformation bundle $\mathscr{D}$; and in §18 we showed that irreducible *-representations of $\mathscr{D}$ are just irreducible systems of imprimitivity for G over M. In this section we shall see how the presence of a so-called G-ergodic measure on M generates an irreducible system of imprimitivity and hence an irreducible *-representation of $\mathscr{D}$. Examples of this construction will be needed in Chapter XII.

The transformation bundles and systems of imprimitivity considered in this section are all derived from the group bundle of G.

19.2. A triple $\langle S, \mathscr{S}, \mu \rangle$, where μ is a measure on the δ-ring $\mathscr{S}$ of subsets of the set S, is called a *measure space*.

Fix a measure space $\langle S, \mathscr{S}, \mu \rangle$ which is parabounded (i.e., μ is parabounded, see II.1.7); and let X be the Hilbert space $\mathscr{L}_2(\mu)$. For each ϕ in $\mathscr{L}_\infty(\mu)$ (see II.7.10) let α_ϕ be the operator in $\mathcal{O}(X)$ consisting of multiplication by ϕ:

$$\alpha_\phi(f) = \phi f \qquad\qquad (f \in X).$$

Evidently the map $\phi \mapsto \alpha_\phi$ is an isometric *-isomorphism of $\mathscr{L}_\infty(\mu)$ into $\mathcal{O}(X)$. Let A stand for the range of this map; that is, $A = \{\alpha_\phi : \phi \in \mathscr{L}_\infty(\mu)\}$.

Proposition. *A is its own commuting algebra. That is, any γ in $\mathcal{O}(X)$ satisfying $\gamma \alpha_\phi = \alpha_\phi \gamma$ for all ϕ in $\mathscr{L}_\infty(\mu)$ must lie in A.*

Proof. We first assume that $\mathscr{S}$ is a σ-field of subsets of S (so that $\mu(S) < \infty$). Thus the function $\mathbb{1}$ on S (satisfying $\mathbb{1}(s) \equiv 1$) is in X. Put $\psi = \gamma(\mathbb{1})$.

We claim that ψ is μ-essentially bounded, hence in $\mathscr{L}_\infty(\mu)$. Indeed: Let $k > \|\gamma\|$, and put $V = \{s \in S : |\psi(s)| \ge k\}$. We have

$$\gamma(\mathrm{Ch}_V) = \gamma(\alpha_{\mathrm{Ch}_V}(\mathbb{1})) = \alpha_{\mathrm{Ch}_V}(\gamma(\mathbb{1}))$$

$$= \mathrm{Ch}_V \psi, \qquad\qquad (1)$$

so that

$$\|\gamma(\mathrm{Ch}_V)\|_X^2 = \int_V |\psi(s)|^2 \, d\mu s \ge k^2 \mu(V)$$

$$= k^2 \|\mathrm{Ch}_V\|_X^2. \qquad\qquad (2)$$

If $\mu(V) > 0$, (2) would contradict the fact that $k > \|\gamma\|$. So $\mu(V) = 0$, and the claim is proved.

We can thus form α_ψ. If $V \in \mathscr{S}$ we have

$$\alpha_\psi(\mathrm{Ch}_V) = \mathrm{Ch}_V \psi = \gamma(\mathrm{Ch}_V) \qquad\qquad (3)$$

as in (1). Since the linear span of $\{\mathrm{Ch}_V : V \in \mathscr{S}\}$ is dense in X, it follows from (3) that $\gamma = \alpha_\psi \in A$. So we have proved the proposition in case $\mathscr{S}$ is a σ-field.

In the general case, when $\langle S, \mathscr{S}, \mu \rangle$ is merely parabounded, let $\mathscr{W}$ be the disjoint subfamily of $\mathscr{S}$ postulated in the definition of paraboundedness (II.1.7). If $X_V = \{f \in X : f \text{ vanishes outside } V\}$ for $V \in \mathscr{W}$, then $X = \sum_{V \in \mathscr{W}}^{\oplus} X_V$, and each X_V is γ-stable. So the proposition results from applying the preceding part of the proof to each X_V. ∎

19.3.   Let G be an (untopologized) group, and $\langle S, \mathscr{S}, \mu \rangle$ a measure space such that G acts as a left transformation group on S. We say that $\langle S, \mathscr{S}, \mu \rangle$ is a *measure G-transformation space* if (i) $xW \in \mathscr{S}$ whenever $x \in G$ and $W \in \mathscr{S}$, and (ii) $\mu(xW) = 0$ whenever $x \in G$, $W \in \mathscr{S}$, and $\mu(W) = 0$. Thus, in a measure transformation space $\langle S, \mathscr{S}, \mu \rangle$, each group element acts so as to preserve $\mathscr{S}$, and carries μ into a new measure $x\mu : W \mapsto \mu(x^{-1}W)$ which is measure-theoretically equivalent to μ (in the sense of II.7.7).

Definition.   The measure G-transformation space $\langle S, \mathscr{S}, \mu \rangle$ is *ergodic* (*under G*) if, whenever V, W are sets in $\mathscr{S}$ with $\mu(V) \neq 0$, $\mu(W) \neq 0$, there exists an element x of G such that $\mu(xV \cap W) \neq 0$.

Remark.   Notice that if $\{s\} \in \mathscr{S}$ and $\mu(\{s\}) > 0$ for all s in S, then ergodicity of $\langle S, \mathscr{S}, \mu \rangle$ under G says simply that G acts transitively on S. In general we can regard ergodicity as a measure-theoretic generalization of transitivity.

19.4.   We shall now obtain an important class of ergodic measure transformation spaces.

Proposition.   *Let G be a locally compact group, G_0 a dense subgroup of G, H a closed subgroup of G, and μ a non-zero G-quasi-invariant measure on the Borel δ-ring $\mathscr{S}$ of G/H (see III.14.1). Then $\langle G/H, \mathscr{S}, \mu \rangle$ is ergodic under the action of G_0.*

Proof.   Let V and W be any two sets in $\mathscr{S}$ of positive μ-measure. If λ is left Haar measure on G and $\pi : G \to G/H$ is the canonical surjection, then by III.14.8 $\pi^{-1}(V)$ and $\pi^{-1}(W)$ are not locally λ-null; so there exist subsets V_0 and W_0 of $\pi^{-1}(V)$ and $\pi^{-1}(W)$ respectively which are in the Borel δ-ring of G and are of positive λ-measure. Now by III.12.3 the function $x \mapsto \lambda(xV_0 \cap W_0)$ is continuous and not identically zero on G. Since G_0 is dense, it follows that we can choose x in G_0 to satisfy $\lambda(xV_0 \cap W_0) > 0$. Since $xV_0 \cap W_0 \subset \pi^{-1}(xV \cap W)$, this implies (again by III.14.8) that $\mu(xV \cap W) > 0$.   ∎

It follows in particular, putting $G_0 = G$, that G acts ergodically on G/H with respect to any G-quasi-invariant measure on G/H.

Likewise, putting $H = \{e\}$, we see that any dense subgroup G_0 of G acts ergodically by left multiplication on G (with respect to Haar measure on G).

19.5.   A G-space S is transitive if and only if the only G-stable subspaces of S are $\emptyset$ and S. There is a natural analogue of this fact for parabounded ergodic measure transformation spaces.

Let G be an (untopologized) group and $\langle S, \mathscr{S}, \mu \rangle$ a measure G-transformation space. We shall say that $\langle S, \mathscr{S}, \mu \rangle$ satisfies *Property (P)* if, for any locally μ-measurable subset W of S such that $xW \ominus W$ is locally μ-null for every x in G, either W is locally μ-null or $S \setminus W$ is locally μ-null.

Proposition*.   *If $\langle S, \mathscr{S}, \mu \rangle$ is ergodic it has Property (P). Conversely, if $\langle S, \mathscr{S}, \mu \rangle$ is parabounded and has Property (P), it is ergodic.*

To prove the second statement, we first show that when μ is parabounded the Boolean algebra B of locally μ-measurable sets modulo locally μ-null sets is complete. So, if $V \in \mathscr{S}$ and $\mu(V) > 0$, $\{xV : x \in G\}$ has a least upper bound W in B; and it is easy to deduce that $xW \ominus W$ is locally μ-null for all x. So by Property (P) $S \setminus W$ is locally μ-null. Thus if Z is another set in $\mathscr{S}$ with $\mu(Z) > 0$, we have $\mu(Z \cap W) > 0$ and so $\mu(Z \cap xV) > 0$ for some x.

Remark.   As an illustration, let R_0 be the additive group of rational numbers. Since R_0 is dense in $\mathbb{R}$, 19.4 and the above proposition show that there cannot exist a subset V of $\mathbb{R}$ with the following properties:

(i) V is a union of cosets in $\mathbb{R}/R_0$;
(ii) V is Lebesgue-measurable, and neither V nor $\mathbb{R} \setminus V$ is Lebesgue-null.

This conclusion is surprising at first sight, considering the vast number of subsets V of $\mathbb{R}$ satisfying property (i).

19.6.   We now return to representation theory. Let G be a discrete group with unit e, and M a locally compact Hausdorff topological G-transformation space. Consider a regular Borel measure μ on M (defined of course on the compacted Borel δ-ring $\mathscr{S}(M)$ of M) such that $\langle M, \mathscr{S}(M), \mu \rangle$ is a measure G-transformation space. From these ingredients we can construct a unitary representation U of G on $\mathscr{L}_2(\mu)$. Indeed, let x be an element of G. Since μ and $x^{-1}\mu$ are measure-theoretically equivalent, by the Radon-Nikodym Theorem II.7.8 and II.8.7 there is a positive-valued locally μ-summable function $m \mapsto \gamma(x, m)$ on M such that $x^{-1}\mu = \gamma(x, m)\, d\mu m$, or

$$d\mu(xm) = \gamma(x, m)\, d\mu m. \tag{4}$$

Evidently $\gamma(e, m) \equiv 1$. If $x, y \in G$, we have by (4) and II.7.6

$$\gamma(xy, m)d\mu m = (xy)^{-1}\mu = y^{-1}(x^{-1}\mu)$$
$$= \gamma(x, ym)d(y^{-1}\mu)m$$
$$= \gamma(x, ym)\gamma(y, m)d\mu m,$$

whence

$$\gamma(xy, m) = \gamma(x, ym)\gamma(y, m) \qquad \text{for locally } \mu\text{-almost all } m. \qquad (5)$$

In particular, putting $y = x^{-1}$ in (5), we get

$$\gamma(x^{-1}, m)\gamma(x, x^{-1}m) = 1 \qquad \text{for locally } \mu\text{-almost all } m. \qquad (6)$$

Now, if $x \in G$ and $f \in \mathscr{L}_2(\mu)$, set

$$(U_x f)(m) = (\gamma(x, x^{-1}m))^{-1/2} f(x^{-1}m) \qquad (m \in M). \qquad (7)$$

Then

$$\|U_x f\|_2^2 = \int |(U_x f)(m)|^2 d\mu m$$

$$= \int \gamma(x, x^{-1}m)^{-1} |f(x^{-1}m)|^2 d\mu m$$

$$= \int \gamma(x, m)^{-1} |f(m)|^2 d\mu(xm)$$

$$= \int |f(m)|^2 d\mu m \qquad \text{(by (4) and II.7.5)}$$

$$= \|f\|_2^2;$$

and so $U_x : f \mapsto U_x f$ is a linear isometric operator on $\mathscr{L}_2(\mu)$. Given $x, y \in G$ and $f \in \mathscr{L}_2(\mu)$, we have for locally μ-almost all m

$$(U_x U_y f)(m) = \gamma(x, x^{-1}m)^{-1/2} (U_y f)(x^{-1}m)$$

$$= \gamma(x, x^{-1}m)^{-1/2} \gamma(y, y^{-1}x^{-1}m)^{-1/2} f(y^{-1}x^{-1}m)$$

$$= \gamma(xy, y^{-1}x^{-1}m)^{-1/2} f(y^{-1}x^{-1}m) \qquad \text{(by (5))}$$

$$= (U_{xy} f)(m),$$

and hence

$$U_x U_y = U_{xy}. \qquad (8)$$

Since U_e is the identity operator, (8) shows that U_x has an inverse $U_{x^{-1}}$, and so is a unitary operator. Thus by (8) $U : x \mapsto U_x$ is a unitary representation of the discrete group G on $\mathscr{L}_2(\mu)$.

In addition to U, μ gives rise to a canonical regular $\mathscr{L}_2(\mu)$-projection-valued measure P on M:

$$P(W)f = \text{Ch}_W f$$

$(f \in \mathcal{L}_2(\mu)$; W a Borel subset of M). Using (6), one verifies the important relation:

$$U_x P(W) = P(xW)U_x$$

$(x \in G$; W a Borel subset of M). Thus $\langle U, P \rangle$ is a system of imprimitivity for G over M (see 18.7); let us denote it by $\mathcal{T}^\mu$.

19.7. Proposition. *Assume that the measure G-transformation space* $\langle M, \mathcal{S}(M), \mu \rangle$ *of 19.6 is ergodic. Then* $\mathcal{T}^\mu$ *is irreducible.*

Proof. Let p be projection onto a closed $\mathcal{T}^\mu$-stable subspace of $\mathcal{L}_2(\mu)$. Thus in particular p commutes with all $P(W)$. It follows that p commutes with multiplication by all functions in $\mathcal{L}_\infty(\mu)$, and so by 19.2 that $p = P(V)$, where V is some locally μ-measurable set (which, by the paraboundedness of μ, may be taken to be a Borel set). Also p commutes with all U_x. Since $p = P(V)$ this implies by (7) that $xV \setminus V$ and $x^{-1}V \setminus V$ are locally μ-null for all x in G. Since $x(x^{-1}V \setminus V)$ is therefore also locally μ-null, we find that $xV \ominus V$ is locally μ-null for all x. Since μ is ergodic, the first (and very easy) part of Proposition 19.5 tells us that either V or $M \setminus V$ is locally μ-null, and hence that either $p = 0$ or $p = 1$. ∎

We have thus constructed an interesting class of irreducible systems of imprimitivity for the discrete group G.

If G is not discrete, exactly the same development goes through except that the U defined by (7) must be shown to be strongly continuous before we can assert that it is a unitary representation of G. In general, the argument proving this is fairly intricate, and we shall omit it since it will not be needed. There is one special case in which no such intricacy arises, namely, when μ is actually invariant under G, i.e., $x\mu = \mu$ for all x in G. The reader should check the strong continuity of U in this case.

Thus, if the group G is locally compact, and if μ is both G-ergodic and G-invariant, the above construction gives us an irreducible system of imprimitivity $\mathcal{T}^\mu$ for G over M, and therefore by 18.22 an irreducible *-representation of the G, M transformation bundle.

20. Functionals of Positive Type on Banach *-Algebraic Bundles

20.1. In the present section we will investigate the objects on Banach *-algebraic bundles which correspond to positive linear functionals on *-algebras.

We fix a Banach *-algebraic bundle $\mathscr{B} = \langle B, \pi, \cdot, * \rangle$ over a topological group G with unit e.

By a *linear functional* on $\mathscr{B}$ we shall mean a function $p: B \to \mathbb{C}$ whose restriction to B_x is linear for each x in G. If in addition p is continuous on B, then by II.13.11 it is norm-continuous on each fiber B_x; and we can form the norms $\|p|B_x\|$. We shall say that the continuous linear functional p is *bounded* if

$$\sup\{\|p|B_x\| : x \in G\} < \infty; \tag{1}$$

and in that case the supremum in (1) is denoted by $\|p\|$.

20.2. Definition. By a *functional of positive type on $\mathscr{B}$* we mean a continuous linear functional p on $\mathscr{B}$ satisfying the inequality

$$\sum_{i,j=1}^{n} p(b_j^* b_i) \geq 0 \tag{2}$$

for any finite sequence $b_1, \ldots, b_n$ of elements of B.

Remark. In particular, $p|B_e$ is a continuous positive linear functional on B_e.

Remark. If $G = \{e\}$, so that $B = B_e$, then the left side of (2) is $p((\sum_{i=1}^{n} b_i)^*(\sum_{i=1}^{n} b_i))$; and a functional of positive type becomes just a continuous positive linear functional on B_e.

20.3. In this number we suppose that $\mathscr{B}$ is the group bundle of G. If p is a functional of positive type on $\mathscr{B}$, the equation $q(x) = p(1, x)$ $(x \in G)$ defines $q: G \to \mathbb{C}$ as a continuous function satisfying

$$\sum_{i,j=1}^{n} \lambda_i \bar{\lambda}_j q(x_j^{-1} x_i) \geq 0 \tag{3}$$

for all $n = 1, 2, \ldots$, all $x_1, \ldots, x_n$ in G, and all $\lambda_1, \ldots, \lambda_n$ in $\mathbb{C}$. Conversely, if $q: G \to \mathbb{C}$ is a continuous function satisfying (3) for all $n, x_1, \ldots, x_n, \lambda_1, \ldots, \lambda_n$, the function $p: B \to \mathbb{C}$ given by $p(\lambda, x) = \lambda q(x)$ $(x \in G; \lambda \in \mathbb{C})$ is a functional of positive type on $\mathscr{B}$.

Definition. A *function of positive type on G* is a continuous function $q: G \to \mathbb{C}$ satisfying (3) for all positive integers n, all $x_1, \ldots, x_n$ in G, and all $\lambda_1, \ldots, \lambda_n$ in $\mathbb{C}$. (In particular $q(e) \geq 0$.)

Using the above correspondence $p \leftrightarrow q$, we can essentially identify functions of positive type on G with functionals of positive type on the group bundle of G.

20.4. We return now to the general $\mathscr{B}$. Let $D = D(\mathscr{B})$ be the discrete cross-sectional *-algebra of $\mathscr{B}$ defined in 8.4 and 9.1. If p is a functional of positive type on $\mathscr{B}$, the definition

$$p'(f) = \sum_{x \in G} p(f(x)) \qquad\qquad (f \in D) \qquad (4)$$

makes p' a linear functional on D, and the condition (2) asserts that p' is positive on D. Applying VI.18.4(1), (2) to p' we obtain in particular

$$p(b^*c) = \overline{p(c^*b)}, \qquad\qquad\qquad (5)$$

$$|p(b^*c)|^2 \le p(b^*b)p(c^*c) \qquad\qquad\qquad (6)$$

$(b, c \in B)$.

20.5. In general (for example, if $\mathscr{B}$ has trivial multiplication) a functional of positive type on $\mathscr{B}$ need not be bounded. However we have:

Proposition. *Assume that $\mathscr{B}$ has an approximate unit $\{u_i\}$. Then every functional p of positive type on $\mathscr{B}$ is bounded, and also satisfies $p(b^*) = \overline{p(b)}$ ($b \in B$).*

In particular, a function q of positive type on G is bounded and satisfies $q(x^{-1}) = \overline{q(x)}$ ($x \in G$).

Proof. Suppose that $\|u_i\| \le k$ for all i, and let $r = \|p|B_e\|$. Putting u_i for b in (6), we have $|p(u_i^*c)|^2 \le r^2k^2\|c^*c\| \le r^2k^2\|c\|^2$. Letting i go to infinity in this inequality (and noting that $\{u_i^*\}$ is also an approximate unit) we get $|p(c)| \le rk\|c\|$ ($c \in B$), or

$$\|p\| \le rk = k\|p|B_e\|. \qquad\qquad\qquad (7)$$

Similarly, putting u_i for b in (5) and passing to the limit in i, we obtain $p(c^*) = \overline{p(c)}$. ∎

Remark. If $\|u_i\| \le 1$ in the above proposition, then by (7), $\|p\| = \|p|B_e\|$. In particular, for a function q of positive type on G, we have $|q(x)| \le q(e)$ ($x \in G$).

20.6. The most important functionals of positive type are those which arise from *-representations. If T is a *-representation of $\mathscr{B}$ and $\xi \in X(T)$, the map $p : B \to \mathbb{C}$ given by

$$p(b) = (T_b\xi, \xi) \qquad\qquad (b \in B) \qquad (8)$$

is evidently a bounded functional of positive type. Conversely we have:

Proposition.   *Assume that $\mathscr{B}$ has an approximate unit. Then every functional p of positive type on $\mathscr{B}$ is of the form (8) for some cyclic *-representation T of $\mathscr{B}$ and some cyclic vector ξ for T.*

Proof.   Notice that the discrete cross-sectional algebra $D(\mathscr{B})$ is a normed *-algebra under the norm

$$\|f\|_1 = \sum_{x \in G} \|f(x)\|.$$

Let p' be the positive extension of p to $D(\mathscr{B})$, as in (4). By 20.5 p is bounded; so, for $f \in D(\mathscr{B})$,

$$|p'(f)| \le \sum_{x \in G} |p(f(x))| \le \|p\| \sum_{x \in G} \|f(x)\| = \|p\|\|f\|_1;$$

that is, p' is continuous with respect to the norm of $D(\mathscr{B})$. So p' extends to a continuous positive linear functional on the Banach *-algebra completion $D_1(\mathscr{B})$ of $D(\mathscr{B})$. Since $\mathscr{B}$ has an approximate unit, so does $D_1(\mathscr{B})$. Therefore by VI.19.9 there is a cyclic *-representation T' of $D_1(\mathscr{B})$, with cyclic vector ξ, satisfying

$$p'(f) = (T'_f\xi, \xi) \qquad (f \in D_1(\mathscr{B})). \qquad (9)$$

Now define T as the restriction of T' to B (identified with a subset of $D(\mathscr{B})$). To show that T is a *-representation of $\mathscr{B}$, it is sufficient by 9.3 to show that $b \mapsto (T_b T'_f\xi, T'_f\xi)$ is continuous on B for all f in $D(\mathscr{B})$. But $(T_b T'_f\xi, T'_f\xi) = p'(f^*bf) = \sum_{x,y \in G} p((f(x))^*bf(y))$, which is continuous in b by virtue of the continuity of p on B. So T is a *-representation. Since ξ is a cyclic vector for T', it is also cyclic for T (see 8.4). Finally, from (9) follows

$$p(b) = (T_b\xi, \xi) \qquad (b \in B),$$

completing the proof.   ∎

20.7.   A *cyclic pair* for $\mathscr{B}$ is a pair $\langle T, \xi \rangle$, where ξ is a cyclic vector for the cyclic *-representation T of $\mathscr{B}$. Two cyclic pairs $\langle T, \xi \rangle$ and $\langle T', \xi' \rangle$ for $\mathscr{B}$ are *unitarily equivalent* if there is a unitary equivalence for T and T' carrying ξ into ξ'. Applying VI.19.8 to the corresponding cyclic pairs for $D(\mathscr{B})$, we see that the two cyclic pairs $\langle T, \xi \rangle$ and $\langle T', \xi' \rangle$ for $\mathscr{B}$ are unitarily equivalent if and only if

$$(T_b\xi, \xi) = (T'_b\xi', \xi') \qquad \text{for all } b \text{ in } B. \qquad (10)$$

Thus, in analogy with VI.19.9, we get:

Theorem.   *Assume that $\mathscr{B}$ has an approximate unit. Then there is a natural one-to-one correspondence between the set of all functionals p of positive type on $\mathscr{B}$ and the set of all unitary equivalence classes of cyclic pairs $\langle T, \xi \rangle$ for $\mathscr{B}$. The correspondence (in the direction $\langle T, \xi \rangle \mapsto p$) is given by (8).*

In particular, the equation

$$q(x) = (T_x \xi, \xi) \qquad\qquad (x \in G) \qquad (11)$$

sets up a one-to-one correspondence between the set of all functions q of positive type on G and the set of all unitary equivalence classes of cyclic pairs $\langle T, \xi \rangle$ for G (that is, ξ is a cyclic vector for the cyclic unitary representation T of G).

20.8.   Proposition.   *Assume that $\mathscr{B}$ has an approximate unit. If $0 \le c \in \mathbb{R}$, p and p' are functionals of positive type on $\mathscr{B}$, and ϕ is a function of positive type on G, then cp, $p + p'$, and ϕp are functionals of positive type on $\mathscr{B}$. (Here $(\phi p)(b) = \phi(\pi(b))p(b)$ for $b \in B$.)*

Proof.   The first two are evident. To prove that ϕp is of positive type, we use 20.6 to find a unitary representation U of G, a *-representation T of $\mathscr{B}$, and vectors ξ, η in $X(U)$ and $X(T)$ respectively, such that

$$\phi(x) = (U_x \xi, \xi), \quad p(b) = (T_b \eta, \eta) \qquad\qquad (x \in G; b \in B).$$

Now form the tensor product *-representation $U \otimes T$ of $\mathscr{B}$ as in 9.16. For $b \in B$ we have $((U \otimes T)_b(\xi \otimes \eta), \xi \otimes \eta) = (U_{\pi(b)}\xi, \xi)(T_b \eta, \eta) = \phi(\pi(b))p(b)$. So ϕp is of positive type.   ∎

In particular, sums and products of functions of positive type on G are of positive type on G.

Functionals of Positive Type and Positive Functionals on $\mathscr{L}(\mathscr{B})$.

20.9.   *From here to the end of the section G is assumed to be locally compact.* Let λ be a left Haar measure on G. We can then form the compacted cross-sectional algebra $\mathscr{L}(\mathscr{B})$ of $\mathscr{B}$.

20.10.   For each continuous linear functional p on $\mathscr{B}$, the equation

$$\alpha_p(f) = \int_G p(f(x))d\lambda x \qquad\qquad (f \in \mathscr{L}(\mathscr{B})) \qquad (12)$$

defines a linear functional α_p on $\mathscr{L}(\mathscr{B})$ which is continuous in the inductive limit topology.

Proposition. *Assume that $\mathcal{B}$ has an approximate unit; and let p be a continuous linear functional on $\mathcal{B}$. The following three conditions are then equivalent:*

(i) *p is of positive type on $\mathcal{B}$.*
(ii) *α_p is a positive linear functional on $\mathcal{L}(\mathcal{B})$.*
(iii) *p is of the form $b \mapsto (T_b\xi, \xi)$ $(b \in B)$, where T is some cyclic *-representation of $\mathcal{B}$ with cyclic vector ξ.*

Proof. The equivalence of (i) and (iii) was proved in 20.6 (even without local compactness). The new feature of the present proposition is (ii).

Assume (iii). If $f \in \mathcal{L}(\mathcal{B})$, we have

$$\alpha_p(f^* * f) = \int p((f^* * f)(x))d\lambda x$$

$$= \int (T_{(f^* * f)(x)}\xi, \xi)d\lambda x$$

$$= (T_{f^* * f}\xi, \xi)$$

$$= (T_f\xi, T_f\xi) \geq 0$$

(denoting also by T the integrated form of T). So (iii) $\Rightarrow$ (ii).

It remains to show that (ii) $\Rightarrow$ (i). Assume (ii), and let $b_1, \ldots, b_n$ be a finite sequence of elements of B, with $x_i = \pi(b_i)$. Choose a continuous cross-section f_i of $\mathcal{B}$ through b_i, and an approximate unit $\{\phi_\nu\}$ on G; and set

$$g_\nu(y) = \sum_{i=1}^n \phi_\nu(x_i^{-1}y)f_i(y) \qquad (y \in G).$$

Thus $g_\nu \in \mathcal{L}(\mathcal{B})$; so by (ii)

$$0 \leq \alpha_p(g_\nu^* * g_\nu) = \int_G p((g_\nu^* * g_\nu)(y))d\lambda y$$

$$= \sum_{i,j=1}^n \int_G \int_G \phi_\nu(x_i^{-1}z)\phi_\nu(x_j^{-1}zy)p[(f_i(z))^*f_j(zy)]d\lambda z \, d\lambda y$$

$$= \sum_{i,j=1}^n \int_G \int_G \phi_\nu(z)\phi_\nu(y)p[f_i(x_iz)^*f_j(x_jy)]d\lambda z \, d\lambda y.$$

Since $\{\phi_\nu\}$ is an approximate unit, and since $p[f_i(x_iz)^*f_j(x_jy)]$ is continuous in z and y, the ν-limit of the right side of the last equality is $\sum_{i,j=1}^n p[f_i(x_i)^*f_j(x_j)]$; thus the latter is non-negative. On the other hand

$f_i(x_i) = b_i$; so we have shown that $\sum_{i,j=1}^{n} p(b_i^* b_j) \geq 0$. Hence (ii) $\Rightarrow$ (i), and the proof is complete.   ■

20.11.   The set of all positive linear functionals on $\mathscr{L}(\mathscr{B})$ of the form (12) is very easily characterized.

Proposition.   *Assume that $\mathscr{B}$ has a strong approximate unit; and let β be a positive linear functional on $\mathscr{L}(\mathscr{B})$ which is continuous in the inductive limit topology. Then the following three conditions are equivalent:*

(i) $\beta = \alpha_p$ *for some functional p of positive type on $\mathscr{B}$.*
(ii) β *is continuous in the $\mathscr{L}_1(\lambda; \mathscr{B})$ norm.*
(iii) β *is extendable (in the sense of VI.18.6).*

Proof.   Assume that $\beta = \alpha_p$ for some functional p of positive type. Since p is bounded by 20.5,

$$|\beta(f)| = \left| \int p(f(x)) d\lambda x \right|$$

$$\leq \int |p(f(x))| d\lambda x$$

$$\leq \|p\| \int \|f(x)\| d\lambda x$$

$$= \|p\| \|f\|_1$$

for all f in $\mathscr{L}(\mathscr{B})$. Thus (i) $\Rightarrow$ (ii).

Assume (ii); and let β' be the continuous positive extension of β to $\mathscr{L}_1(\lambda; \mathscr{B})$. Since $\mathscr{B}$ has a strong approximate unit, $\mathscr{L}_1(\lambda; \mathscr{B})$ has an approximate unit by 5.11; and so by VI.18.11 β' is extendable. It follows that $\beta = \beta'|\mathscr{L}(\mathscr{B})$ is extendable; and we have shown that (ii) $\Rightarrow$ (iii).

We shall now assume (iii) and prove (i). By 13.4 β satisfies condition (R), and the *-representation S of $\mathscr{L}(\mathscr{B})$ generated by β is the integrated form of a *-representation T of $\mathscr{B}$. Since in addition β is extendable, by VI.19.6 there is a vector ξ in $X(T)$ such that

$$\beta(f) = (S_f \xi, \xi) \qquad\qquad (f \in \mathscr{L}(\mathscr{B})). \qquad (13)$$

Put $p(b) = (T_b \xi, \xi)$ $(b \in B)$. Thus p is a functional of positive type on $\mathscr{B}$; and, for $f \in \mathscr{L}(\mathscr{B})$,

$$\alpha_p(f) = \int p(f(x)) d\lambda x$$

$$= \int (T_{f(x)} \xi, \xi) d\lambda x$$

$$= (S_f \xi, \xi)$$

$$= \beta(f) \qquad\qquad\qquad \text{(by (13)).}$$

So $\beta = \alpha_p$; and (i) is proved. Consequently (iii) $\Rightarrow$ (i); and the proof of the proposition is complete. ■

20.12.   If $\mathscr{B}$ has a strong approximate unit, it follows from the equivalence of (i) and (ii) in 20.11 that the map $p \mapsto$ (extension of α_p to $\mathscr{L}_1(\lambda; \mathscr{B})$) is a one-to-one correspondence between the set of all functionals p of positive type on $\mathscr{B}$ and the set of all (continuous) positive linear functionals on $\mathscr{L}_1(\lambda; \mathscr{B})$.

20.13.   Let us say that a functional p of positive type on $\mathscr{B}$ is *indecomposable* if the relation $p = p_1 + p_2$, where p_1 and p_2 are functionals of positive type on $\mathscr{B}$, implies that $p_i = \lambda_i p$ $(i = 1, 2; 0 \le \lambda_i \in \mathbb{R})$. Applying VI.20.4 to the continuous positive linear functionals on $\mathscr{L}_1(\lambda; \mathscr{B})$, and using 20.11, we obtain:

Theorem.   *Assume that $\mathscr{B}$ has a strong approximate unit. If $\langle T, \xi \rangle$ is a cyclic pair for $\mathscr{B}$, and p is the corresponding functional (8) of positive type, then T is irreducible if and only if p is indecomposable.*

21. The Regional Topology of *-Representations of Banach *-Algebraic Bundles

21.1.   As usual, $\mathscr{B}$ will be a fixed Banach *-algebraic bundle (automatically having enough continuous cross-sections) over a locally compact group G with unit e and left Haar measure λ. We have seen in 13.2 that the passage from a *-representation T of $\mathscr{B}$ to its $\mathscr{L}_1$ integrated form $\tilde{T}$ is a one-to-one correspondence between the set $\mathscr{T}$ of all *-representations of $\mathscr{B}$ and the set $\mathscr{S}$ of all *-representations of $\mathscr{L}_1(\lambda; \mathscr{B})$.

21.2.   *Definition*.   Let $\mathcal{T}$ and $\mathcal{S}$ be as in 21.1. By the *regional topology* of $\mathcal{T}$ we mean the topology which makes the bijection $T \mapsto \tilde{T}$ a homeomorphism with respect to the regional topology of $\mathcal{S}$ as defined in VII.1.3.

Likewise, a *-representation T of $\mathcal{B}$ is *weakly contained* in a family W of *-representations of $\mathcal{B}$ if $\tilde{T}$ is weakly contained in $\tilde{W} = \{\tilde{S} : S \in W\}$ in the sense of VII.1.21.

***Remark*.**   By VII.1.16, the definition of the regional topology and weak containment of *-representations of $\mathcal{B}$ would not have been altered had we used the integrated forms as defined on $\mathcal{L}(\mathcal{B})$ instead of on $\mathcal{L}_1(\lambda; \mathcal{B})$.

***Definition*.**   In analogy with VII.3.1, we denote by $\hat{\mathcal{B}}$ the space of all unitary equivalence classes of irreducible *-representations of $\mathcal{B}$, equipped with the (relativized) regional topology. This topological space $\hat{\mathcal{B}}$ is called the *structure space* of $\mathcal{B}$.

We shall have much to say about $\hat{\mathcal{B}}$ in Chapter XII.

21.3.   For each non-degenerate *-representation S of $\mathcal{L}_1(\lambda; \mathcal{B})$, let S' be its extension to the bounded multiplier algebra $\mathcal{W}(\mathcal{L}_1(\lambda; \mathcal{B}))$ (see 15.3). By VII.4.2 the map $S \mapsto S'$ is a homeomorphism with respect to the regional topologies. Thus, regarding the discrete cross-sectional *-algebra $D(\mathcal{B})$ (defined in 8.4 and 9.1) as a *-subalgebra of $\mathcal{W}(\mathcal{L}_1(\lambda; \mathcal{B}))$, and using the first statement of VII.1.17, we find:

***Proposition*.**   *The map sending each T in $\mathcal{T}$ into the *-representation T^D of $D(\mathcal{B})$ corresponding to it by 8.4(1) is continuous with respect to the regional topologies.*

***Remark*.**   We shall see in X.1.15 that the map $T \mapsto T^D$ is *not* in general a homeomorphism.

***Remark*.**   Let G_0 be the discrete group whose group structure coincides with that of G; and let $\mathcal{B}^0$ be the Banach *-algebraic bundle over G_0 coinciding except for its topology with $\mathcal{B}$. For each *-representation T of $\mathcal{B}$ let T^0 be the *-representation of $\mathcal{B}^0$ which, as a mapping, coincides with T. The preceding proposition and remark together say that $T \mapsto T^0$ is continuous with respect to the regional topologies of the spaces of *-representations of $\mathcal{B}$ and $\mathcal{B}^0$ respectively, but is not in general a homeomorphism.

21.4.   The above definitions apply of course to the unitary representations of G, considered as non-degenerate *-representations of the group bundle of G. We need not rewrite all the definitions explicitly for this case.

Definition. The space of all unitary equivalence classes of irreducible unitary representations of the locally compact group G, equipped with the (relativized) regional topology, is called the *structure space* of G, and is denoted by $\hat{G}$.

21.5. The reader will verify without difficulty that the correspondence $T \leftrightarrow T'$ of 16.7(9), between the *-representations of $\mathcal{B}$ and of $\mathcal{B}^c$, is a homeomorphism with respect to the regional topologies of the spaces of *-representations of $\mathcal{B}$ and of $\mathcal{B}^c$.

21.6. Let N be a closed normal subgroup of G, and $\mathcal{C}$ the $\mathcal{L}_1$ partial cross-sectional bundle over G/N derived from $\mathcal{B}$ as in 6.5. Then the correspondence $S \mapsto T$ of 15.9(6) between *-representations of $\mathcal{B}$ and of $\mathcal{C}$ is a homeomorphism with respect to the regional topologies. This follows immediately from the equivalent definition of this correspondence, given at the beginning of the proof of 15.9, in terms of the integrated forms of the *-representations.

The Regional Topology and Functionals of Positive Type

21.7. Our next goal is to describe the regional topology of *-representations of $\mathcal{B}$ in terms of the functionals of positive type which are associated with the *-representations. It will turn out that the regional convergence of *-representations is intimately related to the uniform convergence on compact sets of the corresponding functionals of positive type.

21.8. If T is a *-representation of $\mathcal{B}$ and $\xi \in X(T)$, let $p_{T,\xi}$ be the functional of positive type on $\mathcal{B}$ given by:

$$p_{T,\xi}(b) = (T_b\xi, \xi) \qquad\qquad (b \in B). \qquad (1)$$

Definition. We shall denote $\{p_{T,\xi} : T$ is a non-degenerate *-representation of $\mathcal{B}, \xi \in X(T), \|\xi\| = 1\}$ by $\mathscr{P}(\mathcal{B})$.

Evidently $\|p\| \leq 1$ for every p in $\mathscr{P}(\mathcal{B})$.

Let γ be the natural *-homomorphism of $\mathcal{L}_1(\lambda; \mathcal{B})$ into its C^*-completion $C^*(\mathcal{B})$ (see 17.2). Each p in $\mathscr{P}(\mathcal{B})$ gives rise to positive linear functionals α_p and α_p^c on $\mathcal{L}_1(\lambda; \mathcal{B})$ and $C^*(\mathcal{B})$ respectively:

$$\alpha_p(f) = \int_G p(f(x))d\lambda x \qquad (f \in \mathcal{L}_1(\lambda; \mathcal{B})), \qquad (2)$$

$$\alpha_p = \alpha_p^c \circ \gamma. \qquad (3)$$

Since $\|p\| \leq 1$, we have $\|\alpha_p\| \leq 1$ for $p \in \mathscr{P}(\mathscr{B})$. Furthermore, by VI.19.12,

$$p \in \mathscr{P}(\mathscr{B}) \Rightarrow \|\alpha_p^c\| = 1. \tag{4}$$

21.9. Let us consider $\mathscr{P}(\mathscr{B})$ as equipped with the topology of uniform convergence on compact subsets of B. We claim that, with this topology, $p_i \to p$ in $\mathscr{P}(\mathscr{B})$ if and only if

$$p_i(f(x)) \to p(f(x)) \qquad \text{uniformly on compact subsets of } G \tag{5}$$

for every f in $\mathscr{C}(\mathscr{B})$.

Indeed: If C is a compact subset of G and $f \in \mathscr{C}(\mathscr{B})$, $f(C)$ is compact in B, so (5) is certainly necessary. Conversely, assume (5); and let D be any compact subset of B. Thus $C = \pi(D)$ is compact in G. Given $\varepsilon > 0$, the compactness of D permits us to choose finitely many continuous cross-sections $f_1, \dots, f_n$ such that every element b of D satisfies

$$\|b - f_j(\pi(b))\| < \varepsilon \tag{6}$$

for some $j = 1, \dots, n$. Now by (5) we have for all sufficiently large i

$$|p_i(f_j(x)) - p(f_j(x))| < \varepsilon \qquad \text{for all } x \text{ in } C \text{ and all } j = 1, \dots, n. \tag{7}$$

Combining (6) and (7) with the fact that $\|p_i\| \leq 1$ and $\|p\| \leq 1$ we get for all sufficiently large i

$$|p_i(b) - p(b)| < 3\varepsilon \qquad \text{for all } b \text{ in } D.$$

This and the arbitrariness of D and ε show that $p_i \to p$ in $\mathscr{P}(\mathscr{B})$; and the claim is proved.

21.10. Before proceeding to the next proposition we need two lemmas.

Lemma. *If X is a Hilbert space, $\varepsilon > 0$, ξ is a unit vector in X, $a \in \mathcal{O}(X)$, $\|a\| \leq 1$, and $|(a\xi, \xi) - 1| < \varepsilon$, then*

$$\|a\xi - \xi\|^2 < \varepsilon^2 + 2\varepsilon.$$

Proof. We can write $a\xi$ in the form $\lambda\xi + \eta$, where $\lambda \in \mathbb{C}$ and $\eta \perp \xi$. Then $\lambda = (a\xi, \xi)$ and $|\lambda - 1| < \varepsilon$. Since $|\lambda|^2 + \|\eta\|^2 = \|a\xi\|^2 \leq 1$, we have

$$\|\eta\|^2 \leq 1 - |\lambda|^2 = (1 - |\lambda|)(1 + |\lambda|) < 2\varepsilon.$$

So $\|a\xi - \xi\|^2 = |\lambda - 1|^2 + \|\eta\|^2 < \varepsilon^2 + 2\varepsilon.$ ∎

21.11. Lemma. *Let X be a Banach space, and $\{\alpha_i\}$ a norm-bounded net of elements of X^* such that $\alpha_i \to 0$ pointwise on X. Then $\alpha_i(\xi) \to 0$ uniformly for ξ in any norm-compact subset of X.*

Proof. Suppose that $\|\alpha_i\| \le k$ for all i. Let $\varepsilon > 0$, and let Y be a norm-compact subset of X. Choose finitely many elements $\xi_1, \dots, \xi_n$ of Y such that for every η in Y

$$\|\eta - \xi_j\| < \tfrac{1}{2}\varepsilon k^{-1} \qquad \text{for some } j; \tag{8}$$

and then choose i_0 so large that

$$i \succ i_0 \Rightarrow |\alpha_i(\xi_j)| < \tfrac{1}{2}\varepsilon \qquad \text{for } j = 1, \dots, n. \tag{9}$$

Then, for any η in Y, if j is chosen so that (8) holds, we have by (9) for all $i \succ i_0$

$$
\begin{aligned}
|\alpha_i(\eta)| &\le |\alpha_i(\eta - \xi_j)| + |\alpha_i(\xi_j)| \\
&\le k\|\eta - \xi_j\| + |\alpha_i(\xi_j)| \\
&< k\tfrac{1}{2}\varepsilon k^{-1} + \tfrac{1}{2}\varepsilon = \varepsilon. \quad\blacksquare
\end{aligned}
$$

21.12.   Proposition.   *Keep the notation of 21.8. The map $p \mapsto \alpha_p$ is a homeomorphism on $\mathscr{P}(\mathscr{B})$ with respect to the topology of pointwise convergence of linear functionals on $\mathscr{L}_1(\lambda; \mathscr{B})$ (or, equivalently, on $\mathscr{L}(\mathscr{B})$). Likewise, the map $p \mapsto \alpha_p^c$ is a homeomorphism on $\mathscr{P}(\mathscr{B})$ with respect to the topology of pointwise convergence of linear functionals on $C^*(\mathscr{B})$.*

Proof.   It is enough to prove the continuity of three maps: (I) the map $p \mapsto \alpha_p$; (II) the map $\alpha_p \mapsto \alpha_p^c$; (III) the map $\alpha_p^c \mapsto p$. (In all of these p runs over $\mathscr{P}(\mathscr{B})$.)

(I) Let $p_i \to p$ in $\mathscr{P}(\mathscr{B})$; then evidently $\alpha_{p_i}(f) \to \alpha_p(f)$ for every f in $\mathscr{L}(\mathscr{B})$. Since $\{\|\alpha_{p_i}\|\}$ is bounded and $\mathscr{L}(\mathscr{B})$ is dense in $\mathscr{L}_1(\lambda; \mathscr{B})$, it follows that $\alpha_{p_i} \to \alpha_p$ pointwise on $\mathscr{L}_1(\lambda; \mathscr{B})$. So map (I) is continuous on $\mathscr{P}(\mathscr{B})$.

(II) Let $\alpha_{p_i} \to \alpha_p$ pointwise on $\mathscr{L}_1(\lambda; \mathscr{B})$ (where $p_i, p \in \mathscr{P}(\mathscr{B})$). By (3) $\alpha_{p_i}^c \to \alpha_p^c$ pointwise on range (γ). Since range (γ) is dense in $C^*(\mathscr{B})$, it follows from (4) that $\alpha_{p_i}^c \to \alpha_p^c$ pointwise on $C^*(\mathscr{B})$. So map (II) is continuous.

(III) To prove the continuity of (III), we have to observe that each b in B acts as a natural bounded multiplier on $C^*(\mathscr{B})$. Indeed: b acts as a multiplier on $\mathscr{L}_1(\lambda; \mathscr{B})$ by 5.8; and by the argument of 16.8 (applied with $G = \{e\}$), the equations

$$b\gamma(f) = \gamma(bf), \qquad \gamma(f)b = \gamma(fb) \tag{10}$$

$(f \in \mathscr{L}_1(\lambda; \mathscr{B}))$ define a multiplier m_b of $C^*(\mathscr{B})$ with

$$\|m_b\| \le \|b\|. \tag{11}$$

Furthermore, the map $b \mapsto m_b$ is continuous on B with respect to the strong multiplier topology (see 15.2). Indeed: If $\phi \in \mathscr{L}(\mathscr{B})$, then by 12.4 the maps

$b \mapsto b\phi$ and $b \mapsto \phi b$ are continuous with respect to the $\mathscr{L}_1(\lambda; \mathscr{B})$ norm and hence also with respect to the $C^*(\mathscr{B})$ norm. The strong continuity of $b \mapsto m_b$ now follows from this fact together with (11) and the denseness of $\gamma(\mathscr{L}(\mathscr{B}))$ in $C^*(\mathscr{B})$.

Now let $\{p_i\}$ and p be elements of $\mathscr{P}(\mathscr{B})$ given by

$$p(b) = (T_b\xi, \xi), \qquad p_i(b) = (T_b^i\xi^i, \xi^i) \qquad (b \in B), \qquad (12)$$

where $\{T^i\}$, T are non-degenerate *-representations of $\mathscr{B}$ and ξ^i and ξ are unit vectors in $X(T^i)$ and $X(T)$ respectively. We assume that $\alpha_{p_i}^c \to \alpha_p^c$ pointwise on $C^*(\mathscr{B})$. To prove that $p_i \to p$ in $\mathscr{P}(\mathscr{B})$, we shall take a compact subset D of B and show that

$$p_i(b) \to p(b) \qquad \text{uniformly for } b \in D. \qquad (13)$$

Clearly we may as well assume $\|b\| \leq 1$ for $b \in D$.

Now by VI.8.4 $C^*(\mathscr{B})$ has a self-adjoint approximate unit with norm no greater than 1. Hence, given $\varepsilon > 0$, and picking $\delta > 0$ so that $\delta^2 + 2\delta < \varepsilon^2$, we can find an element ϕ of $C^*(\mathscr{B})$ satisfying

$$\|\phi\|_c \leq 1, \quad \phi^* = \phi, \qquad (14)$$

$$|(T_\phi\xi, \xi) - 1| < \delta \qquad \text{(see 8.9).} \qquad (15)$$

From (12), (15), and the assumption that $\alpha_{p_i}^c \to \alpha_p^c$ pointwise, we conclude that

$$|(T_\phi^i\xi^i, \xi^i) - 1| < \delta \qquad \text{for all large enough } i. \qquad (16)$$

Since $\|\xi\| = \|\xi^i\| = 1$, $\|T_\phi\| \leq 1$, and $\|T_\phi^i\| \leq 1$ (by (14)), it follows from (15), (16), Lemma 21.10, and the definition of δ, that

$$\|T_\phi\xi - \xi\| < \varepsilon, \|T_\phi^i\xi^i - \xi^i\| < \varepsilon \qquad \text{for all large } i. \qquad (17)$$

Now by the continuity of $b \mapsto \phi b$ mentioned above, the set $\{\phi b : b \in D\}$ is norm-compact in $C^*(\mathscr{B})$. From this and 21.11 applied to the net $\{\alpha_{p_i}^c - \alpha_p^c\}$, we find:

$$(T_{\phi b}^i\xi^i, \xi^i) \to (T_{\phi b}\xi, \xi) \qquad \text{uniformly for } b \in D.$$

Since $\phi^* = \phi$ this implies that for all large enough i

$$|(T_b^i\xi^i, T_\phi^i\xi^i) - (T_b\xi, T_\phi\xi)| < \varepsilon \qquad \text{for all } b \text{ in } D. \qquad (18)$$

Combining (12), (17), and (18) (and recalling that $\|b\| \leq 1$ for $b \in D$) we find that for all large i

$$|p_i(b) - p(b)| < 3\varepsilon \qquad \text{for all } b \text{ in } D.$$

Thus (13) has been proved. Therefore $p_i \to p$; and the mapping (III) is continuous.

The continuity of the maps (I), (II), (III) establishes the proposition. ∎

21.13. By means of Proposition 21.12, many of the results of Chapter VII, when transferred to the context of Banach *-algebraic bundles, can be stated in terms of uniform-on-compacta convergence of functionals of positive type.

To begin with, we shall reformulate the original definition 21.2 of the regional topology of the space of *-representations of $\mathscr{B}$ in the light of 21.12.

Let $\mathscr{T}^0$ be the family of all non-degenerate *-representations of $\mathscr{B}$. Let T be a fixed *-representation in $\mathscr{T}^0$, and let Z be a fixed family of vectors in $X(T)$ which generates $X(T)$ (i.e., the smallest closed T-stable subspace containing Z is $X(T)$ itself). If $\varepsilon > 0$, D is a compact subset of B, and $\xi_1, \ldots, \xi_n$ is any finite sequence of vectors in $X(T)$, let $U''(T; \varepsilon; \{\xi_i\}; D)$ be the set of all T' in $\mathscr{T}^0$ such that there exist vectors $\xi'_1, \ldots, \xi'_n$ in $X(T')$ satisfying

(i) $|(\xi'_i, \xi'_j) - (\xi_i, \xi_j)| < \varepsilon$ for all $i, j = 1, \ldots, n$, and
(ii) $|T'_b \xi'_i, \xi'_j) - (T_b \xi_i, \xi_j)| < \varepsilon$ for all $i, j = 1, \ldots, n$ and all b in D.

Theorem.   *The set $\mathscr{U}''$ of all $U''(T; \varepsilon; \{\xi_i\}; D)$, where ε runs over all positive numbers, D over all compact subsets of B, and $\{\xi_i\}$ over all finite sequences of vectors in Z, is a basis of neighborhoods of T in the regional topology of $\mathscr{T}^0$.*

Proof.   For each S in $\mathscr{T}^0$ let $\tilde{S}$ be the integrated form of S on $\mathscr{L}(\mathscr{B})$. It is clear from VII.1.4 that every regional neighborhood of $\tilde{S}$ in $\mathscr{T}^0$ contains the image under $S \mapsto \tilde{S}$ of some set in $\mathscr{U}''$. So it is enough to show that every set in $\mathscr{U}''$ is in fact a regional neighborhood of T. That is, we fix $U'' = U''(T; \varepsilon, \{\xi_i\}; D) \in \mathscr{U}''$, and also a net $\{T^\nu\}$ converging to T in the regional topology of $\mathscr{T}^0$; and we shall show that some subnet of $\{T^\nu\}$ lies entirely in U''.

By VII.1.8 (applied to the integrated forms) we can replace $\{T^\nu\}$ by a subnet and find vectors $\{\xi^\nu_i\}$ $(i = 1, \ldots, n)$ in $X(T^\nu)$ such that: (i) $(\xi^\nu_i, \xi^\nu_j) \xrightarrow{\nu} (\xi_i, \xi_j)$ for all $i, j = 1, \ldots, n$, and (ii) $(\tilde{T}^\nu_f \xi^\nu_i, \xi^\nu_j) \xrightarrow{\nu} (\tilde{T}_f \xi_i, \xi_j)$ for all $i, j = 1, \ldots, n$ and f in $\mathscr{L}_1(\lambda; \mathscr{B})$. It follows from (i) and (ii) that if $c_1, \ldots, c_n$ are any fixed complex numbers, and if we put $\eta^\nu = \sum_{i=1}^n c_i \xi^\nu_i$, $\eta = \sum_{i=1}^n c_i \xi_i$, then $\|\eta^\nu\| \to \|\eta\|$ and $(\tilde{T}^\nu_f \eta^\nu, \eta^\nu) \to (\tilde{T}_f \eta, \eta)$ for all f in $\mathscr{L}_1(\lambda; \mathscr{B})$. From this and 21.12 (trivially extended to cover the case that $\|\eta^\nu\| \to \|\eta\|$ instead of $\|\eta^\nu\| = \|\eta\| = 1$) we deduce that

$$(T^\nu_b \eta^\nu, \eta^\nu) \to (T_b \eta, \eta) \tag{19}$$

uniformly in b on compact subsets of B. In particular, for fixed i, j, (19) holds when $\eta = \xi_i \pm \xi_j$ and $\eta^\nu = \xi^\nu_i \pm \xi^\nu_j$, and also when $\eta = \xi_i \pm i\xi_j$ and $\eta^\nu =$

$\xi_i^\nu \pm i\xi_j^\nu$. Hence the polarization identity applied to (19) shows that for each $i, j = 1, \ldots, n$

$$(T_b^\nu \xi_i^\nu, \xi_j^\nu) \to (T_b \xi_i, \xi_j) \qquad \text{uniformly in } b \text{ on compact sets.} \qquad (20)$$

Now from (i) and (20) it follows that T^ν lies in U'' for all large ν. This completes the proof.   ■

21.14.   A functional p of positive type on $\mathscr{B}$ will be said to be *associated with* a *-representation T of $\mathscr{B}$ if $p = p_{T,\xi}$ (see (1)) for some ξ in $X(T)$.

   If T is cyclic, 21.13 takes the following form:

Corollary.   *Let T be a cyclic *-representation of $\mathscr{B}$, with cyclic vector ξ; and let $\mathscr{V}$ be a family of non-degenerate *-representations of $\mathscr{B}$. Then T belongs to the regional closure of $\mathscr{V}$ if and only if there is a net $\{q_i\}$ of elements of $\mathscr{P}(\mathscr{B})$, each associated with some *-representation in $\mathscr{V}$, such that*

$$q_i(b) \to p_{T,\xi}(b)$$

uniformly in b on each compact subset of B.

21.15.   A one-dimensional *-representation of $\mathscr{B}$ will be regarded as a continuous map $\phi : B \to \mathbb{C}$ which is linear on each fiber and satisfies $\phi(bc) = \phi(b)\phi(c)$, $\phi(b^*) = \overline{\phi(b)}$ ($b, c \in B$). Thus ϕ itself is the unique element of $\mathscr{P}(\mathscr{B})$ associated with ϕ. Therefore we have the following corollary of 21.14:

Corollary.   *The regional topology, relativized to the space of all non-zero one-dimensional *-representations of $\mathscr{B}$, coincides with the topology of uniform convergence on compact subsets of B.*

21.16.   Applied to the group bundle of G, Corollary 21.15 becomes:

Corollary.   *The regional topology of the space of all one-dimensional unitary representations of G is just the topology of uniform convergence on compact subsets of G.*

21.17.   It would be useful to characterize the elements $p_{T,\xi}$ of the set $\mathscr{P}(\mathscr{B})$ (which plays such an important role in 21.12) in terms of their behavior on B, rather than in terms of T and ξ. We are able to do this provided $\mathscr{B}$ has an approximate unit $\{u_i\}$ with $\|u_i\| \leq 1$.

Proposition. *Assume that $\mathscr{B}$ has an approximate unit $\{u_i\}$ satisfying $\|u_i\| \leq 1$ for all i. Then a functional p of positive type on $\mathscr{B}$ belongs to $\mathscr{P}(\mathscr{B})$ if and only if $\|p\| = 1$, or, equivalently, if and only if $\|p|B_e\| = 1$.*

Proof. Let p be a non-zero functional of positive type on $\mathscr{B}$. By 20.5 p is bounded; and by Remark 20.5

$$\|p\| = \|p|B_e\|. \tag{21}$$

By 20.6 $p = p_{T,\xi}$ (see 21.8), where T is cyclic with cyclic vector ξ. Thus

$$p \in \mathscr{P}(\mathscr{B}) \Leftrightarrow \|\xi\| = 1. \tag{22}$$

By 8.9 $T_{u_i} \to \mathbb{1}$ strongly. So

$$\|\xi\|^2 = \lim_i (T_{u_i^* u_i}\xi, \xi) = \lim_i p(u_i^* u_i). \tag{23}$$

Now, since $\|u_i\| \leq 1$,

$$|p(u_i^* u_i)| \leq \|p\|. \tag{24}$$

On the other hand, for any i and any b in B, 20.4(6) gives

$$|p(u_i^* b)|^2 \leq p(u_i^* u_i)p(b^*b). \tag{25}$$

Suppose that $\lim \inf_i p(u_i^* u_i) \leq m < \|p\|$. Then (25) and the fact that $p(u_i^* b) \to p(b)$ imply that

$$|p(b)|^2 \leq mp(b^*b) \leq m\|p\|\|b\|^2 \qquad \text{for all } b.$$

Since $m\|p\| < \|p\|^2$, this contradicts the definition of $\|p\|$. So $\lim \inf_i p(u_i^* u_i) \geq \|p\|$, which combines with (24) to give

$$\lim_i p(u_i^* u_i) = \|p\|.$$

By (23) this implies $\|p\| = \|\xi\|^2$; and this together with (21) and (22) completes the proof. ∎

21.18. If $\mathscr{B}$ is the group bundle, we shall denote $\mathscr{P}(\mathscr{B})$ by $\mathscr{P}(G)$. Thus, by 20.3 and 21.17, $\mathscr{P}(G)$ consists of those functions p of positive type on G such that $p(e) = 1$.

*Continuity of Operations on *-Representations*

21.19.   We conclude this section by pointing out that the basic operations on *-representations of $\mathscr{B}$ are continuous.

It follows from VII.1.7 and 11.17 that the operation of taking Hilbert direct sums of *-representations of $\mathscr{B}$ is regionally continuous. The precise formulation of this fact is essentially the same as VII.1.7; and we shall not repeat it.

21.20. Proposition.   *Let H be a closed subgroup of G, and $\mathscr{B}_H$ the reduction of $\mathscr{B}$ to H. For any non-degenerate *-representation T of $\mathscr{B}$, the restriction $T|\mathscr{B}_H$ of T to $\mathscr{B}_H$ is a non-degenerate *-representation of $\mathscr{B}_H$; and the map $T \mapsto T|\mathscr{B}_H$ is continuous with respect to the regional topologies of the spaces of *-representations of $\mathscr{B}$ and of $\mathscr{B}_H$.*

Proof.   The non-degeneracy of $T|\mathscr{B}_H$ follows from 9.4. The continuity of the map $T \mapsto T|\mathscr{B}_H$ is an immediate consequence of 21.13.   ∎

In particular, the operation of restricting a unitary representation of G to H is continuous with respect to the regional topologies of the spaces of unitary representations of G and of H.

21.21.   One can generalize 21.20 as follows: Let H be any other locally compact group, and $F: H \to G$ a continuous homomorphism; and construct the Banach *-algebraic bundle retraction $\mathscr{C} = \langle C, \rho, \cdot, * \rangle$ of $\mathscr{B}$ by F, as in 3.16. The map $\Phi: C \to B$ sending $\langle y, b \rangle$ into b ($\langle y, b \rangle \in C$) is continuous, linear on each fiber, and preserves multiplication and *. Thus, if T is a non-degenerate *-representation of $\mathscr{B}$, $T' = T \circ \Phi$ is a *-representation of $\mathscr{C}$ and is non-degenerate by 9.4. We say that T' is *lifted from* T. It is easy to see from 21.13 that the map $T \mapsto T'$ is continuous with respect to the regional topologies of the spaces of non-degenerate *-representations of $\mathscr{B}$ and of $\mathscr{C}$.

21.22. Proposition.   *Keep the notation of 21.21. If in addition F is an open homomorphism onto G, then the map $T \mapsto T'$ is a homeomorphism with respect to the regional topologies.*

Proof.   The continuity of $T \mapsto T'$ was observed in 21.21. To prove the continuity of its inverse, it is sufficient, in view of 21.13, to verify the following claim: For each compact subset D of B, there is a compact subset E of C such that $D \subset \Phi(E)$. To prove this, we first obtain from III.2.5 a compact subset W

of H such that $F(W) \supset \pi(D)$. Then $E = (W \times D) \cap C$ has the property required in the claim. ∎

In particular, let N be a closed normal subgroup of G; and for each unitary representation T of G/N, let T' be the lifted unitary representation $x \mapsto T_{xN}$ of G. It follows from the above proposition that the map $T \mapsto T'$ is a homeomorphism with respect to the regional topologies of the spaces of unitary representations of G and of G/N. The range of the map $T \mapsto T'$ is just the set of those unitary representations of G whose kernels contain N.

21.23. It is an almost trivial matter to verify from 21.13 that the operation $T \mapsto \bar{T}$ of complex conjugation of unitary representations of G (see 9.14) is continuous with respect to the regional topology.

21.24. We shall now establish the continuity of the Hilbert tensor product operation $\otimes$ studied in 9.16.

Let $\mathcal{U}$ be the space of all unitary representations of G; and as before let $\mathcal{T}^0$ be the space of all non-degenerate *-representations of $\mathcal{B}$.

Proposition. *The Hilbert tensor product operation* $\gamma : \langle U, T \rangle \mapsto U \otimes T$ *is continuous on* $\mathcal{U} \times \mathcal{T}^0$ *to* $\mathcal{T}^0$ *with respect to the regional topologies of* $\mathcal{U}$ *and* $\mathcal{T}^0$.

Proof. Suppose that $U^\nu \to U$ in $\mathcal{U}$ and $T^\nu \to T$ in $\mathcal{T}^0$, and that W is an element of some basis of neighborhoods of $U \otimes T$. It is enough to show that some subnet of $\{U^\nu \otimes T^\nu\}$ lies entirely in W. Thus, by 21.13 it is enough to take n-termed sequences $\xi_1, \ldots, \xi_n$ and $\eta_1, \ldots, \eta_n$ of elements of $X(U)$ and $X(T)$ respectively, and show that, on passing to a subnet of $\{U^\nu \otimes T^\nu\}$, we can find elements $\xi_1^\nu, \ldots, \xi_n^\nu$ of $X(U^\nu)$ and $\eta_1^\nu, \ldots, \eta_n^\nu$ of $X(T^\nu)$ such that for all i, j

$$(\xi_i^\nu \otimes \eta_i^\nu, \xi_j^\nu \otimes \eta_j^\nu) \underset{\nu}{\to} (\xi_i \otimes \eta_i, \xi_j \otimes \eta_j) \tag{26}$$

and

$$((U^\nu \otimes T^\nu)_b(\xi_i^\nu \otimes \eta_i^\nu), \xi_j^\nu \otimes \eta_j^\nu) \underset{\nu}{\to} ((U \otimes T)_b(\xi_i \otimes \eta_i), \xi_j \otimes \eta_j) \tag{27}$$

uniformly in b on compact subsets of B.

To obtain such vectors ξ_i^v and η_i^v, we apply 21.13 to U and T separately. Thus we can pass twice to a subnet and find elements $\xi_1^v, \ldots, \xi_n^v$ of $X(U^v)$ and $\eta_1^v, \ldots, \eta_n^v$ of $X(T^v)$ such that for all i, j

$$(\xi_i^v, \xi_j^v) \to (\xi_i, \xi_j), \tag{28}$$

$$(U_x^v \xi_i^v, \xi_j^v) \to (U_x \xi_i, \xi_j) \qquad \text{uniformly on compact sets,} \tag{29}$$

$$(\eta_i^v, \eta_j^v) \to (\eta_i, \eta_j), \tag{30}$$

$$(T_b^v \eta_i^v, \eta_j^v) \to (T_b \eta_i, \eta_j) \qquad \text{uniformly on compact sets.} \tag{31}$$

Now multiplying (28) and (30), we get (26). Similarly, multiplying (29) and (31), we get (27). ∎

21.25. We will close by defining a modification of the Hilbert tensor product of 9.16.

Let H be another locally compact topological group; and let $F: H \times G \to G$ be projection onto G (that is, $F(y, x) = x$). Thus we can form the retraction $\mathscr{C}$ of $\mathscr{B}$ by F, and this will be a Banach *-algebraic bundle over $H \times G$. As in 21.21, each *-representation T of $\mathscr{B}$ lifts to a *-representation T' of $\mathscr{C}$. Likewise, each unitary representation U of H lifts to a unitary representation $U': \langle y, x \rangle \mapsto U_y$ of $H \times G$. So we can now form the Hilbert tensor product $S = U' \otimes T'$; this will be a *-representation of $\mathscr{C}$.

Definition. This S is called the *outer (Hilbert) tensor product* of U and T, and will be denoted by $U \underset{\circ}{\otimes} T$.

If $\mathscr{B}$ is the group bundle of G, the outer tensor product operation carries the pair $\langle U, T \rangle$, where U and T are unitary representations of H and G respectively, into a unitary representation of $H \times G$:

$$(U \underset{\circ}{\otimes} T)_{\langle y, x \rangle} = U_y \otimes T_x \qquad (x \in G; \, y \in H).$$

Proposition. *The outer tensor product operation $\langle U, T \rangle \mapsto U \underset{\circ}{\otimes} T$ is continuous with respect to the appropriate regional topologies.*

Proof. The lifting maps $T \mapsto T'$ and $U \mapsto U'$ are continuous (in fact homeomorphisms) by 21.22. Therefore $\langle U, T \rangle \mapsto U \underset{\circ}{\otimes} T = U' \otimes T'$ is continuous by 21.24. ∎

21.26. *Remark.* The continuity results of 21.19–21.25 have immediate corollaries regarding weak containment. For brevity we will leave it to the reader to formulate these.

22. Exercises for Chapter VIII

1. Verify that the set $\mathcal{W}(A)$ of all multipliers of the algebra A with the operations defined in 1.2 is an algebra.

2. Let A be an algebra, and $\lambda: A \to A$ a linear map satisfying $\lambda(ab) = \lambda(a)b$ for all a, $b \in A$. Does there necessarily exist another linear map $\mu: A \to A$ making $\langle \lambda, \mu \rangle$ a multiplier of A (see 1.2)?

3. Show in Example 1.8 that the homomorphism $b \mapsto u_b$ of B into $\mathcal{W}(A)$ is an isomorphism.

4. Verify that the representation T of the multiplier algebra $\mathcal{W}(A)$ in 1.9 is unique.

5. Give an example of an algebra A with no annihilators, and a Banach representation T of A, which *cannot* be extended to a Banach representation of the multiplier algebra $\mathcal{W}(A)$ acting in the same Banach space $X(T)$.

6. Let T be an algebraically non-degenerate (algebraic) representation of the algebra A on the linear space X, and T' the extension of T to a representation of $\mathcal{W}(A)$ on X (see 1.9). Let Y be a T-stable subspace of X.

   (a) Give an example showing that Y need not be stable under T'.

   (b) Show that, if in addition the subrepresentation of T acting on Y is (algebraically) non-degenerate, then Y is stable under T'.

[*Note*: In connection with this exercise, see Exercise 1 of Chapter V.]

7. Give a detailed proof of Proposition 1.19.

8. Prove Proposition 1.20 and its Corollary 1.21. Also show in Remark 1.22 that B is contained in the von Neumann algebra E generated by A.

9. Give a direct proof that in the case of a Banach algebra the notions of approximate unit and strong approximate unit coincide (see 2.11).

10. Give a proof of Proposition 3.4.

11. Verify that the Banach *-algebraic bundle retraction $\mathcal{C}$ as defined in 3.17 is indeed a Banach *-algebraic bundle over H.

12. Complete the verification of the remaining postulates of a Banach *-algebraic bundle for the semidirect product bundle defined in 4.2.

13. Show that, as stated in 4.3, if $\{u_i\}$ is an approximate unit of A, then $\{\langle u_i, e \rangle\}$ is a strong approximate unit of $\mathscr{B}$.

14. Verify each of the following in the construction of the central extension bundle in 4.7.

   (a) The multiplicative group U of unitary elements of the Banach *-algebra A with the strong topology is a topological group.

   (b) The space B of all orbits in $A \times H$ under the given action, with the quotient topology, is Hausdorff, and the map $\pi: B \to G$ as defined is an open surjection.

(c) $\mathscr{B} = \langle B, \pi, \cdot, * \rangle$, with the given operations, is a Banach *-algebraic bundle over G.

15. Verify properties (i) and (ii) in 4.8 for the cocycle bundle $\mathscr{B}^\gamma$.

16. Show that the bundle convolution $f * g$ $(f, g \in \mathscr{L}(\mathscr{B}))$ is separately continuous in f and g with respect to the inductive limit topology of $\mathscr{L}(\mathscr{B})$ (see 5.3).

17. Show in 5.9 that the map $u \mapsto m_u$ preserves involutions.

18. Prove, as stated in 5.12, that $h_\sigma^* * g * h_\sigma \to g$ in the inductive limit topology for all g in $\mathscr{L}(\mathscr{B})$.

19. Prove equation (22) in 5.13.

20. Show that $\| \tilde{f}(u_i N) \tilde{g}(v_i N) - \tilde{h}(u_i v_i N) \|_1 \underset{i}{\to} 0$ as asserted in 6.4.

21. Verify the statement made in Remark 6.6 concerning the group extension bundle $\mathscr{C}$.

22. Prove the assertions made in 6.9 concerning multipliers of partial cross-sectional bundles.

23. Verify the details of the construction of the transformation bundle described in 7.2.

24. Verify Remark 7.3.

25. Prove the assertions in 7.5.

26. Complete the proof of Proposition 7.7.

27. The continuity requirement in the definition of a group representation can often be substantially weakened. For example, we have the following result:

Let G be a locally compact group, with unit e and left Haar measure λ. Let X be a separable Hilbert space; and let T be a mapping of G into the set of all unitary operators on X such that:

(i) $T_{xy} = T_x T_y$ $(x, y \in G)$; $T_e = 1$;

(ii) for each $\xi, \eta \in X$, the function $x \mapsto (T_x \xi, \eta)$ is locally λ-measurable.

Show that T is a unitary representation of G.

[*Hint*: Show that T can be "integrated" to give a *-representation T' of the $\mathscr{L}_1$ group algebra. The separability of X enables us to show that T' is non-degenerate. Thus T' is the integrated form of a unitary representation S of G. Now T and S must coincide on a locally λ-measurable subgroup H whose complement is locally λ-null; but such an H must equal G.]

28. Let G be a σ-compact locally compact group. Show that, if T is a cyclic unitary representation of G, then $X(T)$ is separable.

29. Let G and H be two topological groups, and S and T be unitary representations of G and H respectively. Then $S \times T : \langle x, y \rangle \mapsto S_x \otimes T_y$ is a unitary representation of the Cartesian product group $G \times H$.

Prove that $S \times T$ is irreducible if and only if both S and T are irreducible.

Formulate and prove a corresponding statement about Banach *-algebraic bundles.

30. Establish the elementary properties of direct sums, tensor products, and complex conjugates of representations in 9.18.

31. Let σ be a norm-function on the locally compact group G; and let T be a Banach representation of $\mathscr{L}(G)$ satisfying

$$\|T_f\| \le p\|f\|_\sigma \qquad \text{for all } f \in \mathscr{L}(G).$$

where p is some constant independent of f. Show that T is the integrated form of a Banach representation S of G satisfying

$$\|S_x\| \le q\sigma(x) \qquad \text{for all } x \in G$$

(q being some constant independent of x).

(See Exercise 65 of Chapter III for the definition of a norm function and of $\| \ \|_\sigma$.)

32. Let G be a locally compact group with unit e, and X a locally convex space. Let T be a function assigning to each x in G a continuous linear operator on X such that

(i) $T_x T_y = T_{xy}$ $(x, y \in G)$; $T_e = 1$;
(ii) $x \mapsto \alpha(T_x \xi)$ is continuous on G to $\mathbb{C}$ for each $\xi \in X$ and $\alpha \in X^*$;
(iii) for each compact subset K of G the family of operators $\{T_x : x \in K\}$ is equicontinuous on X.

Show that T is then a locally convex representation of G (i.e., $x \mapsto T_x \xi$ is continuous on G to X for every $\xi \in X$).

33. Show that the group C^*-algebra $C^*(G)$ of a non-discrete locally compact group G has no unit element.

34. Prove Proposition 10.6.

35. Prove the result in 10.8 concerning projective representations.

36. Give a proof of Proposition 10.10.

37. Verify equation (6) in 11.6.

38. Show that Proposition 11.9 fails if the subspace Y is not closed by constructing an appropriate counter-example.

39. Write out the details of the proof of Proposition 11.10.

40. Establish Propositions 11.12, 11.13, and 11.17.

41. Show that the left and right actions described in 12.3 by (1) and (2) leave stable the subalgebra $\mathscr{L}(\mathscr{B})$ of $\mathscr{L}_1(\lambda; \mathscr{B})$.

42. Verify the statements make in 12.4 about continuity of the maps $b \mapsto bf$ and $b \mapsto fb$.

43. Show that $T_{\langle a, x \rangle} = S_a V_x$ defines a non-degenerate *-representation of $\mathscr{B}$ as described in 15.6.

44. Verify equations (9) and (10) in the proof of Proposition 15.9. Show also in this proof that the right hand side of (6) exists as an $X(S)$-valued integral, and is majorized in norm by $\|\phi\|_{C_a}\|\xi\|$.

45. Give a detailed proof of Proposition 16.6.

46. Let $\mathscr{B}$ be a C^*-algebraic bundle and $\mathscr{D}$ its saturated part (see 16.6). Show that $C^*(\mathscr{D})$ can be indentified with a closed two-sided ideal of $C^*(\mathscr{B})$. (In particular, then, the structure space of $C^*(\mathscr{D})$ is an open subset of the structure space of $C^*(\mathscr{B})$.)

47. Verify the relations given in (3) of 18.4; (4) of 18.5; and (5) of 18.6.

48. Verify that equation (14) holds in 18.12.

49. Prove part (II) of Proposition 18.17.

50. Show, as stated following the proof of 19.7, that if μ is invariant under G then the U defined by (7) is strongly continuous, hence a unitary representation.

51. Let G be a locally compact group, and $\phi: G \to \mathbb{E}$ a continuous function. Show that ϕ is of positive type on G (see 20.3) if and only if ϕ is a homomorphism of G into $\mathbb{E}$.

52. Let ϕ be a function of positive type on a locally compact group G.

(I) Show that ϕ is both left- and right-uniformly continuous; that is, for each $\varepsilon > 0$, there is a neighborhood U of e such that

$$|\phi(xy) - \phi(x)| < \varepsilon \quad \text{and} \quad |\phi(yx) - \phi(x)| < \varepsilon$$

whenever $x \in G$ and $y \in U$.

(II) If $\phi \in \mathcal{L}_p(\lambda)$ ($1 \le p < \infty$, λ is left Haar measure on G), show that $\phi(x) \to 0$ as $x \to \infty$ in G.

53. If G is a locally compact group with left Haar measure λ and modular function Δ, show that $f * (f^* \cdot \Delta)$ is a function of positive type on G whenever $f \in \mathcal{L}(G)$.

54. Verify 21.5.

55. Use 21.13 to prove 21.23.

56. In this Exercise we present a construction of Banach *-algebraic bundles generalizing both the semidirect product bundles and the central extension bundles of §VIII.4. It is taken from Fell [14], §9.

The ingredients for this construction are G, A, N, γ, τ, as follows:

(I) G is a topological group.

(II) A is a Banach *-algebra with no annihilators (see 1.4). As usual $\mathcal{W}_b(A)$ will be the Banach *-algebra of bounded multipliers of A (see 1.10, 1.14). Let U be the multiplicative group of all unitary elements of $\mathcal{W}_b(A)$. (An element u of $\mathcal{W}_b(A)$ will be called *unitary* if $u^*u = uu^* = 1$ and $\|u\|_0 \le 1$, where $\| \ \|_0$ is the norm in $\mathcal{W}_b(A)$.) We equip U with the *strong topology* (in which $u_i \to u$ if and only if $\lim_i \|u_i a - au\| = \lim_i \|au_i - au\| = 0$ for all $a \in A$). Show that U then becomes a topological group.

(III) N is a closed subgroup of U.

(IV) $\gamma: N \xrightarrow{i} H \xrightarrow{j} G$ is an extension of N by G (see III.5.1).

(V) τ is a homomorphism of H into the group of all isometric *-automorphisms of A satisfying: (i) For each $a \in A$, the map $h \mapsto \tau_h(a)$ is continuous on H to A; (ii) if $u \in N$ and $a \in A$, then $\tau_u(a) = uau^{-1}$ (multiplication in $\mathcal{W}_b(A)$); (iii) if $h \in H$ and $u \in N$, then $\tau_h'(u) = huh^{-1}$ (multiplication in H; here τ_h' is the unique extension of τ_h to an isometric *-automorphism of $\mathcal{W}_b(A)$).

To construct a Banach *-algebraic bundle from these ingredients, we proceed as follows (in analogy with 4.7):

In the Cartesian product space $A \times H$ define the equivalence relation $\sim$ as follows: $\langle a_1, h_1 \rangle \sim \langle a_2, h_2 \rangle$ if and only if there exists an element u in N such that $a_2 = a_1 u$, $h_2 = u^{-1} h_1$. Let B be the space of $\sim$ —equivalence classes (with the quotient topology), and denote by $\langle a, h \rangle^\sim$ the equivalence class containing $\langle a, h \rangle$.

Prove that B is a Hausdorff space, and that the map $\pi: B \to G$ given by $\pi(\langle a, h \rangle^{\sim}) = j(h)$ is a continuous open surjection.

If $x \in G$ and $h \in j^{-1}(x)$, then $a \mapsto \langle a, h \rangle^{\sim}$ is a bijection of A onto $B_x = \pi^{-1}(x)$. Let us transfer to B_x the Banach space structure of A via this bijection.

Show that this Banach space structure of B_x is independent of the choice of h, and that with this Banach space structure on each fiber $\langle B, \pi \rangle$ becomes a Banach bundle.

We now define multiplication and $*$ on B as follows:

$$\langle a, h \rangle^{\sim} \cdot \langle b, k \rangle^{\sim} = \langle a\tau_h(b), hk \rangle^{\sim},$$

$$(\langle a, h \rangle^{\sim})^* = \langle \tau_{h^{-1}}(a^*), h^{-1} \rangle^{\sim}.$$

Show that these definitions are legitimate, and that they make $\mathscr{B} = \langle B, \pi, \cdot, * \rangle$ into a Banach *-algebraic bundle.

Notice that if $N = \{1\}$ $\mathscr{B}$ is just the τ-semidirect product bundle (4.2); while if A has a unit and τ is trivial (so that by (ii) and (iii) N is central in both A and H) then $\mathscr{B}$ is the central extension bundle constructed from G, A, N, γ (see 4.7).

Notes and Remarks

The Introduction to this second Volume (see especially Sections 4, 6, and 7) contains an account of the origins and history of Banach *-algebraic bundles, together with the principal contributors; we refer the reader to it for information. The theory as presented in this chapter is largely due to Fell [14, 15, 17]; for this reason Banach *-algebraic bundles are often referred to in the literature as Fell bundles.

For the results on multipliers (also called centralizers) of algebras in §1 see Busby [2] and B. E. Johnson [1]. The theory of multipliers has a long history; we refer the reader to Hewitt and Ross [2, pp. 412–415] for an excellent account. The reader may also wish to consult the book of Larsen [1].

Theorem 1.23 was proved in 1968 by Dauns and Hofmann [1]. For a short simple proof of the theorem see Elliott and Olsen [1]. (The book by Dupré and Gillett [1] also contains a nice self-contained proof.) Theorem 1.25 was established in 1952 by Wendel [1].

For information on the classification of cocycle bundles (see 4.9) over a given group see, for example, Backhouse [2], Backhouse and Bradley [3], Baggett [6], Baggett and Kleppner [1], Baggett, Mitchell, and Ramsay [1], L. Brown [2], Carey and Moran [2], Cattaneo [3, 4] Hannabuss [1], Holzherr [1, 2], Kleppner [3, 5, 6], Mackey [16], and Sund [6, 7].

Both the semidirect product bundles and the central extension bundles of §4 of this chapter are special cases of the homogeneous Banach *-algebraic bundles dealt with by Fell [14].

It is worth noticing that §9 of Fell [14] presents an explicit construction for obtaining the most general homogeneous Banach *-algebraic bundle, and that in the C^*-algebra context this construction is essentially identical with the definition of a twisted covariant system in the paper of Philip Green [2]. (This construction is given in Exercise 56 of the present chapter.) Thus Green's twisted covariant systems are essentially the same (disregarding conditions of second countability, etc.) as C^*-algebraic bundles which are homogeneous in the sense of Fell [14].

The transformation algebras defined in 7.7 were first considered by Glimm [5]. Later they were studied by Effros and Hahn [1] and several others. The $\mathscr{L}_1$ cross-sectional algebras (see 5.2) are discussed in Fell [14, §9], [17, §13], and generalize the "covariance algebras" of Doplicher, Kastler, and Robinson [1]. (Covariance algebras are essentially just cross-sectional algebras of semidirect product bundles, and have applications in physics.) The transformation bundles introduced in §7 are discussed in Fell [17, §29].

For a survey of the theory of projective representations of finite groups see Farmer [1]. Projective representations of the inhomogeneous Lorentz group were considered in 1939 by Wigner [1] and multipliers for projective representations of Lie groups were systematically studied by Bargmann [2] in 1954. The theory of projective unitary representations of second countable locally compact groups was developed in the important 1958 paper of Mackey [8].

Most of the remaining results in this chapter are developed in Fell [17]. The C^*-algebraic bundle structure for the Glimm algebras described in §17 is due to Takesaki [4].

The Notes and Remarks to Chapter II of Volume I of this work contain additional remarks on Banach bundles. References which deal with various aspects of bundle theory (not necessarily Banach *-algebraic bundles) are the following: Busby [3, 5], Dade [1], Dauns and Hofmann [1, 2], Dixmier and Douady [1], Douady and dal Soglio-Hérault [1], Dupré [1, 2, 3, 4], Dupré and Gillette [1], Evans [1], Fabec [2], Fell [3, 11, 13, 14, 15, 16, 17], Gelbaum [1, 2, 3], Gelbaum and Kyriazis [1], Gierz [1], Green [1, 2, 3, 4, 5], Hofmann [2, 3, 4], Hofmann and Liukkonen [1], Kitchen and Robbins [1, 2, 3], R. Lee [1, 2], Leinert [1], Leptin [1, 2, 3, 4, 5], Mayer [1, 2], Raeburn and Williams [1], Rieffel [5, 9, 12, 13], Rousseau [10], Schochetman [9, 10], Seda [4, 5 6], Sen [1], Tomiyama [1], Tomiyama and Takesaki [1], and Varela [1, 2].

IX   Compact Groups

Chapter VIII was devoted to general Banach algebraic bundles over locally compact groups. In the present chapter we restrict our attention to a very special case of this, namely the group bundle of a compact group. That is, we shall study locally convex representations of a compact group. In Chapter XII we shall obtain certain extensions of the results of this chapter to Banach *-algebraic bundles over compact groups.

The most obvious compact groups are the finite groups (that is, compact discrete groups). At the other topological extreme are the compact connected groups. We met several examples of these in III.1.20—namely, the series $U(n)$ $(n \geq 1)$, $SU(n)$ $(n \geq 1)$, and $SO(n)$ $(n \geq 2)$. These are all Lie groups (see 3.14 of this chapter). Surprisingly enough, it turns out that, unlike finite groups, the compact connected Lie groups can be completely classified. Indeed, this classification is one of the outstanding achievements of modern mathematics. Of course we shall not attempt to carry it out here. The reader who is interested in this may consult Pontryagin [6] or Price [2].

As examples of compact groups which are totally disconnected but not finite, we may cite the additive groups I_p of p-adic integers encountered in III.6.13.

The material of this chapter is almost entirely classical. In §2 we prove the fundamental theorem on the representation theory of a compact group G, namely the Peter–Weyl Theorem (2.11). In §3 we derive another form of it (3.9), which asserts that the space $\mathscr{I}(G)$ of all irreducible finite-dimensional representations of G distinguishes the points of G. §4 is devoted to the important orthogonality relations for the matrix elements associated with $\mathscr{I}(G)$. These essentially determine the structure of the group algebra of G. In §5 the results of §4 are specialized to the characters of elements of $\mathscr{I}(G)$. §6 contains a useful theorem on subsets of $\mathscr{I}(G)$ which is analogous to the Stone–Weierstrass Theorem. In §§11 and 12 this theorem is applied to classify the irreducible representations of two very important non-commutative compact groups, namely the special 2×2 unitary group $SU(2)$ and the special 3×3 orthogonal group $SO(3)$.

§§7, 8, 9 develop the elementary theory of arbitrary integrable locally convex representations of a compact group G. The three main results of this development are as follows:

(I)   Every irreducible integrable locally convex representation of G (in particular every irreducible unitary representation of G) is finite-dimensional.

(II)   Every integrable locally convex representation of G is "topologically completely reducible"; in particular, every unitary representation of G is discretely decomposable.

(III)   The multiplicity of each element T of $\mathscr{I}(G)$ in the regular representation of G is equal to the dimension of T.

In the process of obtaining these results we show that, for any integrable locally convex representation T of G, the harmonic analysis of the vectors in $X(T)$ converges with respect to a "generalized Fejér summation process," analogous to the Fejér summation process of classical Fourier analysis.

In §10, after a motivating discussion, we define induced unitary representations of compact groups, and prove the famous Frobenius Reciprocity Theorem. This section is partly intended to serve as an orienting introduction to the much more general theory of induced representations given in Chapter XI. No completely satisfactory generalization of the Frobenius Reciprocity Theorem beyond the sphere of compact groups has yet been found; but the search for partial generalizations has played an important role in the modern investigation of induced representations.

1. Preliminary Remarks on Function Spaces

1.1. *Throughout this entire chapter we shall fix a compact group G, with unit e. By III.8.8 G is unimodular. Its Haar measure will be denoted by λ. We assume that λ is normalized by the requirement that $\lambda(G) = 1$.*

1.2. Let δ_x be the measure consisting of unit mass at the point x of G.

If $x \in G$ and f is a complex function on G, it will be convenient to write f_x and f^x for the left and right translates of f:

$$f_x(y) = f(xy), \qquad f^x(y) = f(yx) \qquad\qquad (y \in G).$$

Notice that $f_x = \delta_{x^{-1}} * f$ and $f^x = f * \delta_{x^{-1}}$; so that by the associative law for convolution we have

$$(f * \mu)_x = f_x * \mu, \qquad (\mu * f)^x = \mu * (f^x) \tag{1}$$

whenever $f \in \mathscr{L}_1(\lambda)$, $\mu \in \mathscr{M}_r(G)$, and $x \in G$.

1.3. Let $1 \le p < \infty$. By Hölder's Inequality II.2.6 (applied with $g \equiv 1$), we have $\mathscr{L}_p(\lambda) \subset \mathscr{L}_1(\lambda)$, and

$$\|f\|_1 \le \|f\|_p \tag{2}$$

whenever $f \in \mathscr{L}_p(\lambda)$.

We also note that, if $f \in \mathscr{L}_1(\lambda)$ and $g \in \mathscr{L}_p(\lambda)$, then $f * g$ and $g * f$ are in $\mathscr{L}_p(\lambda)$ and

$$\|f * g\|_p \le \|f\|_1 \|g\|_p, \tag{3}$$

$$\|g * f\|_p \le \|f\|_1 \|g\|_p. \tag{4}$$

Indeed: By III.11.14

$$f * g = \int f(x)g_{x^{-1}} \, d\lambda x \tag{5}$$

($\mathscr{L}_p(\lambda)$-valued integral); and (3) follows from (5) and II.5.4(2). The same argument applied to the reverse group of G gives (4).

From (2) and (3) we deduce that not only $\mathscr{L}_1(\lambda)$ but every $\mathscr{L}_p(\lambda)$ $(1 \le p < \infty)$ is a Banach *-algebra under the usual convolution and involution of functions. This fact of course is peculiar to compact groups, and fails entirely in the locally compact context.

One easily verifies (without appealing to III.11.14) that (2), (3), and (4) also hold when $p = \infty$. Likewise $\mathscr{L}(G)$ $(= \mathscr{C}(G))$ is a Banach *-algebra under convolution, involution, and the supremum norm.

1.4. The reader will verify without trouble that, if $f \in \mathscr{L}_1(\lambda)$ and $g, h \in \mathscr{L}_2(\lambda)$, then

$$(f * g, h) = (g, f^* * h), \tag{6}$$

$$(g * f, h) = (g, h * f^*), \tag{7}$$

where (,) denotes the inner product of $\mathscr{L}_2(\lambda)$. (Indeed, by III.11.23, (6) holds for any locally compact group. So, applying (6) to the reverse group of G, we conclude that (7) is true for any unimodular locally compact group.)

2.   The Peter–Weyl Theorem

2.1.   *Definition*.   Let $\mathscr{R} = \mathscr{R}(G)$ be the set of all those functions f in $\mathscr{L}(G)$ such that the linear span of $\{f_x : x \in G\}$ (the set of all left translates of f) is finite-dimensional.

The elements of $\mathscr{R}$ are called *representative functions*.

2.2.   One verifies without difficulty the following fact.

Proposition.   *$\mathscr{R}$ is a linear subspace of $\mathscr{L}(G)$. Further, $\mathscr{R}$ is closed under pointwise multiplication and under complex conjugation, and contains the constant functions.*

2.3.   *Remark*.   The Peter–Weyl Theorem will assert that there are plenty of functions in $\mathscr{R}$, in fact that $\mathscr{R}$ is dense in $\mathscr{L}(G)$. In view of 2.2 and the Stone–Weierstrass Theorem (Appendix A), this would be proved if it could be shown that $\mathscr{R}$ separates points of G. This observation, however, does not seem to help much in proving the theorem.

2.4.   It follows from 1.2(1) that $\mathscr{R}$ is a right ideal of $\mathscr{M}_r(G)$.

2.5.   Proposition.   *A function f in $\mathscr{L}(G)$ belongs to $\mathscr{R}$ if and only if there exist a positive integer n and elements $g_1, \ldots, g_n, h_1, \ldots, h_n$ of $\mathscr{L}(G)$ satisfying*

$$f(xy) = \sum_{i=1}^{n} g_i(x) h_i(y) \tag{1}$$

for all x, y in G.

Proof.   The "if" part is evident.

Conversely, suppose that $0 \neq f \in \mathscr{R}$; and let $f_{x_1}, \ldots, f_{x_n}$ be a maximal linearly independent set of left translates of f. Thus, for each x in G there are unique numbers $g_1(x), \ldots, g_n(x)$ such that

$$f_x = \sum_{i=1}^{n} g_i(x) f_{x_i}. \tag{2}$$

The continuity of $x \mapsto f_x$, together with the uniqueness of the topology of a finite-dimensional LCS, implies that the functions g_i defined by (2) are continuous on G. If we put $h_i = f_{x_i}$, (2) becomes (1).   ∎

2.6. Corollary. *If $f \in \mathcal{L}(G)$, the following three conditions are equivalent:*

(i) $f \in \mathcal{R}$;

(ii) *the linear span of the set $\{f^x : x \in G\}$ of right translates of f is finite-dimensional;*

(iii) *the linear span of the set $\{(f_x)^y : x, y \in G\}$ of two-sided translates of f is finite-dimensional.*

Proof. Proposition 2.5 applied to G and its reverse group shows that (i) $\Leftrightarrow$ (ii). Obviously (iii) $\Rightarrow$ (i). To prove that (i) $\Rightarrow$ (iii), let us assume (i), and choose a maximal linearly independent family $\{f_{x_1}, \ldots, f_{x_n}\}$ of left translates of f. Now each f_{x_i} is in $\mathcal{R}$; and so, since (i) $\Rightarrow$ (ii), the linear span L_i of $\{(f_{x_i})^y : y \in G\}$ is finite-dimensional for each i. The space $\sum_{i=1}^{n} L_i$ is therefore finite-dimensional, and clearly contains all $(f_x)^y$.   ∎

2.7. Corollary. *$\mathcal{R}$ is closed under involution $*$, and is a two-sided ideal of the measure algebra $\mathcal{M}_r(G)$.*

Proof. Let $f \in \mathcal{R}$. By 2.5 f can be represented as in (1). Thus

$$f^*(xy) = \overline{f(y^{-1}x^{-1})} = \sum_{i=1}^{n} \overline{g_i(y^{-1})h_i(x^{-1})}$$

$$= \sum_{i=1}^{n} h_i^*(x)g_i^*(y),$$

from which it follows by 2.5 that $f^* \in \mathcal{R}$. So $\mathcal{R}$ is closed under involution. Since $\mathcal{R}$ is a right ideal of $\mathcal{M}_r(G)$ (2.4) and closed under $*$, it is also a left ideal.   ∎

2.8. If $h \in \mathcal{L}(G)$ let us define γ_h to be the bounded linear operator $f \mapsto h * f$ on $\mathcal{L}_2(\lambda)$. We have

$$\gamma_{h_1}\gamma_{h_2} = \gamma_{h_1 * h_2}, \qquad (\gamma_h)^* = \gamma_{h^*}$$

(see 1.4). In view of III.12.2, the range of γ_h ($h \in \mathcal{L}(G)$) is contained in $\mathcal{L}(G)$.

Notice that $\gamma ; h \mapsto \gamma_h$ is one-to-one on $\mathcal{L}(G)$. For, if $h \in \mathcal{L}(G)$, then $\gamma_h(h^*)$ is continuous (by the preceding paragraph) and $\gamma_h(h^*)(e) = \int |h(x)|^2 \, d\lambda x$; so $\gamma_h = 0 \Rightarrow h = 0$.

2.9.   Proposition.   *For each h in $\mathscr{L}(G)$, γ_h is a compact operator (in fact a Hilbert–Schmidt operator) on $\mathscr{L}_2(\lambda)$.*

Proof.   The formula

$$(\gamma_h f)(x) = \int h(xy^{-1})f(y)d\lambda y$$

shows that γ_h is an integral operator with integral kernel $\langle x, y \rangle \mapsto h(xy^{-1})$ (see VI.15.21). Thus γ_h is a Hilbert–Schmidt operator (VI.15.21) and hence (by VI.15.19) compact.   ■

2.10.   The crucial argument for the Peter–Weyl Theorem is contained in the following proof.

Proposition.   *$\mathscr{R}$ is dense in $\mathscr{L}_2(\lambda)$.*

Proof.   It suffices to contradict the assumption that

$$0 \neq f \in \mathscr{L}_2(\lambda), \qquad f \perp \mathscr{R}. \tag{3}$$

Assume (3). Then for any g in $\mathscr{R}$ and x in G,

$$(f * g)(x) = \int f(y)g(y^{-1}x)d\lambda y$$

$$= \int f(y)\overline{g^*(x^{-1}y)}d\lambda y$$

$$= (f, (g^*)_{x^{-1}});$$

and the right side of this is 0 since $(g^*)_{x^{-1}} \in \mathscr{R}$ (by (3) and 2.7). Therefore

$$f * g = 0 \qquad \text{for all } g \text{ in } \mathscr{R}. \tag{4}$$

Now $f^* * f \in \mathscr{L}(G)$ (by III.12.2); hence we can form $A = \gamma_{f^* * f}$. Since $f \neq 0$ we have $(f^* * f)(e) = \int |f(x)|^2 \, d\lambda x \neq 0$, and so by 2.8 $A \neq 0$. Also $A^* = A$ since $(f^* * f)^* = f^* * f$ (see 2.8). Thus A is a non-zero Hermitian operator; and is compact by 2.9. Therefore, by the spectral theorem VI.15.10, there exists a non-zero real number λ such that

$$W = \{h \in \mathscr{L}_2(\lambda); Ah = \lambda h\}$$

is of finite positive dimension.

Now, if $h \in W$, $h = A(\lambda^{-1}h) \in \text{range}(A)$. So by 2.8

$$W \subset \mathscr{L}(G).$$

Also, if $x \in G$ and $h \in W$, it follows from 1.2(1) that $A(h^x) = (Ah)^x = \lambda h^x$; so $h^x \in W$. Thus W is a finite-dimensional subspace of $\mathcal{L}(G)$ and is stable under all right translations. This implies by 2.6 that $W \subset \mathcal{R}$. Consequently, by (4),

$$f * h = 0 \qquad \text{for all } h \text{ in } W. \tag{5}$$

On the other hand, choosing a non-zero vector h in W, we have $0 \neq \lambda h = Ah = f^* * (f * h)$, so that $f * h \neq 0$, contradicting (5). We have now contradicted the assumption (3). ∎

2.11. Theorem (Peter–Weyl). $\mathcal{R}$ *is dense in* $\mathcal{L}(G)$ *(with respect to the supremum norm).*

Proof. Let $f \in \mathcal{L}(G)$, $\varepsilon > 0$. By III.11.19(I) we can find a function g in $\mathcal{L}(G)$ satisfying

$$\|g * f - f\|_\infty < \tfrac{1}{2}\varepsilon. \tag{6}$$

By 2.10 there is an element h of $\mathcal{R}$ such that

$$\|h - f\|_2 < \varepsilon(2\|g\|_2)^{-1}. \tag{7}$$

By Schwarz's Inequality and (7)

$$\|g * (h - f)\|_\infty \leq \|g\|_2 \|h - f\|_2 < \tfrac{1}{2}\varepsilon.$$

From this and (6) we get

$$\|f - (g * h)\|_\infty < \varepsilon. \tag{8}$$

But $g * h \in \mathcal{R}$ in view of 2.7. So f has been approximated arbitrarily closely (with respect to $\| \ \|_\infty$) by elements of $\mathcal{R}$; and the proof is complete. ∎

3. Finite-dimensional Representations of G

3.1. In this chapter the phrase "finite-dimensional representation T of G" will always imply the continuity of $T: x \mapsto T_x$, so that T is a locally convex representation of G in the sense of VIII.8.5.

It should be clear that the Peter–Weyl Theorem is going to assert the existence of "many" finite-dimensional representations of G.

3.2. Our first observation is that every finite-dimensional representation of G is "essentially" unitary.

Proposition. *Every finite-dimensional representation T of G is (algebraically) equivalent to a unitary representation of G, and hence is completely reducible.*

Proof.   Choose an arbitrary inner product (,) in $X(T)$ making the latter a finite-dimensional Hilbert space; and set

$$(\xi, \eta)_0 = \int_G (T_x\xi, T_x\eta)d\lambda x \qquad\qquad (\xi, \eta \in X(T)).$$

It is easy to verify that (,)$_0$ is again an inner product making $X(T)$ a Hilbert space, and that

$$(T_x\xi, T_x\eta)_0 = (\xi, \eta)_0 \qquad\qquad (x \in G; \xi, \eta \in X(T)).$$

Thus T is unitary with respect to (,)$_0$.

We have shown that T can be made unitary by the introduction of a suitable inner product; and this is the same as the first part of the proposition. The complete reducibility of T now follows from VI.14.6.   ∎

Remark.   If T is one-dimensional, the fact that it is "essentially unitary" simply means that $|T_x| = 1$ for all x. This of course follows directly from the fact that the image of G under a continuous homomorphism $T: G \to \mathbb{C}_*$ is compact and so contained in $\mathbb{E}$.

3.3.   Remark.   Here is a different proof of the complete reducibility of a finite-dimensional representation T of the compact group G.

Let Y be a T-stable subspace of $X = X(T)$. It is enough to show that there is a T-stable subspace Z of $X(T)$ complementary to Y. To do this, choose any idempotent linear operator p on X with range Y; and if $\xi \in X$ set

$$p_0(\xi) = \int_G T_x(p(T_x^{-1}\xi))d\lambda x.$$

Thus p_0 is a linear operator on X, range(p_0) $\subset Y$, and $p_0(\xi) = \xi$ if $\xi \in Y$. Further,

$$T_y(p_0(\xi)) = \int_G T_{yx}(p(T_x^{-1}\xi))d\lambda x$$

$$= \int T_x(p(T_x^{-1}T_y\xi))d\lambda x = p_0(T_y\xi).$$

So p_0 is idempotent with range Y, and commutes with all T_y. It follows that $Z = \text{Ker}(p_0)$ is a T-stable subspace complementary to Y.

This proof was given by Hurwitz [1] in 1897 for the special case of the orthogonal group.

3.4. Remark. Proposition 3.2 should be compared with Maschke's Theorem for finite groups (IV.6.1). The essential step in the proof of Maschke's Theorem was a *summation* over the group elements. Both in the proof in 3.2 and in the alternative proof given in 3.3, the essential step is *integration* over G with respect to the finite Haar measure.

Recall that Maschke's Theorem asserted the complete reducibility of arbitrary (not just finite-dimensional) representations of a finite group. Analogously, we shall show in 8.5 that an arbitrary integrable locally convex representation T of the compact group G is at least "topologically completely reducible"—that is, the sum of its irreducible subspaces is dense in $X(T)$.

Matrix Elements of Representations

3.5. Definition. Let T be a representation of G of finite dimensional r, and $\xi_1, \ldots, \xi_r$ a fixed basis of $X(T)$. For each x in G let $(T_{ij}(x))_{i,j=1,\ldots,r}$ be the matrix of T_x with respect to this basis. The functions $T_{ij}: x \mapsto T_{ij}(x)$ $(i, j = 1, \ldots, r)$ belong to $\mathscr{L}(G)$ and are called the *matrix elements of T with respect to* $\xi_1, \ldots, \xi_r$.

Notice that the linear span of $\{T_{ij}: i, j = 1, \ldots, r\}$ is independent of the particular choice of the basis $\xi_1, \ldots, \xi_r$. We denote this linear span by $\mathscr{E}(T)$.

3.6. Proposition. *If S and T are two finite-dimensional representations of G, then $\mathscr{E}(S \oplus T) = \mathscr{E}(S) + \mathscr{E}(T)$, and $\mathscr{E}(S \otimes T)$ is the linear span of $\{\sigma\tau: \sigma \in \mathscr{E}(S), \tau \in \mathscr{E}(T)\}$.*

This is easily verified.

3.7. From any finite-dimensional representation T of G we can construct another finite-dimensional representation T^* of G, the representation *adjoint to T*, as follows:

$$X(T^*) = (X(T))^*,$$

$$(T^*)_x(\alpha) = \alpha \circ T_x^{-1} \qquad (x \in G; \alpha \in (X(T))^*).$$

The reader will check that $T^{**} = T$; and that T is irreducible if and only if T^* is.

If the finite-dimensional representation T is unitary, the inner product of $X(T)$ enables us to identify $X(T)^*$ with $\overline{X(T)}$ (see VIII.9.13). Under this identification $(T^*)_x$ becomes the same mapping as T_x. Therefore, if T is unitary, T^* and $\overline{T}$ (see VIII.9.13) are the same.

Let T be a finite-dimensional representation of G; let $\xi_1, \ldots, \xi_r$ be a basis of $X(T)$; and form the dual basis $\alpha_1, \ldots, \alpha_r$ of $X(T)^*: \alpha_i(\xi_j) = \delta_{ij}$ $(i, j = 1, \ldots, r)$.

Then the matrix $(T_{ij}^*(x))_{i,j=1,...,r}$ of $(T^*)_x$ with respect to $\alpha_1,...,\alpha_r$ is connected with the matrix $(T_{km}(x))_{k,m=1,...,r}$ of T_x with respect to $\xi_1,...,\xi_r$ by the relation

$$T_{ij}^*(x) = T_{ji}(x^{-1}) \qquad (x \in G; i,j = 1,...,r). \qquad (1)$$

One verifies without difficulty that (1) holds and that it implies the following consequence:

Proposition. *For any finite-dimensional representation T of G we have $\mathscr{E}(T^*) = \{\tilde{\sigma}; \sigma \in \mathscr{E}(T)\}$ (where $\tilde{\sigma}(x) = \sigma(x^{-1})$).*

Remark. The definitions and results 3.5–3.7 are valid of course without the restriction that G is compact.

3.8. Proposition. *$\mathscr{R}$ (defined in 2.1) coincides with the linear span of the set of all matrix elements of irreducible finite-dimensional representations of G.*

Proof. Let $\mathscr{R}'$ be the linear span of the set of all matrix elements of irreducible finite-dimensional representations of G.

Suppose $f \in \mathscr{R}$; and choose $x_1,...,x_r$ in G so that $x_1 = e$ and so that the left translates $f_{x_1},...,f_{x_r}$ form a basis of the finite-dimensional space X spanned by $\{f_x : x \in G\}$. As in 2.5(2) we have:

$$f(xy) = \sum_{i=1}^{r} g_i(x)f_{x_i}(y) \qquad (x, y \in G), \qquad (2)$$

where $g_i \in \mathscr{L}(G)$. Denoting by R the (finite-dimensional) subrepresentation acting on X of the regular representation of G (see VIII.11.26), we get from (2)

$$R_u(f_{x_j}) = f_{x_j u^{-1}} = \sum_{i=1}^{r} g_i(x_j u^{-1})f_{x_i}. \qquad (3)$$

So the matrix element R_{ij} of R with respect to the basis $\{f_{x_i}\}$ is given by

$$R_{ij}(u) = g_i(x_j u^{-1}) \qquad (u \in G). \qquad (4)$$

From (2) and (4) we get

$$f(u) = f_{x_1 u}(x_1) = \sum_{i=1}^{r} g_i(x_i u)f_{x_i}(x_1)$$

$$= \sum_{i=1}^{r} f(x_i)R_{i1}(u^{-1}) \qquad (u \in G). \qquad (5)$$

Since R is completely reducible (3.2), R_{i1} is in $\mathscr{R}'$ (see 3.6), and hence by Proposition 3.7 $u \mapsto R_{i1}(u^{-1})$ is in $\mathscr{R}'$. So by (5) $f \in \mathscr{R}'$. By the arbitrariness of f we have shown that $\mathscr{R} \subset \mathscr{R}'$.

Conversely, let T be a finite-dimensional representation of G with matrix elements $\{T_{ij}\}$ (with respect to some fixed basis). Since

$$T_{ij}(yx) = \sum_k T_{ik}(y)T_{kj}(x),$$

it follows that any left translate of a matrix element of T is a linear combination of matrix elements of T. This shows that $\mathscr{R}' \subset \mathscr{R}$.

Combining the last two paragraphs we obtain $\mathscr{R} = \mathscr{R}'$, and the proof is complete. ■

3.9. The Peter–Weyl Theorem combined with 3.8 shows that the set of all matrix elements of irreducible finite-dimensional representations of G distinguishes the points of G. We therefore have:

Theorem. *For any element x of G different from e, there is an irreducible finite-dimensional representation T of G such that $x \notin \mathrm{Ker}(T)$ (i.e., $T_x \neq 1_{X(T)}$).*

3.10. *Remark.* If we take 3.8 and the Stone–Weierstrass Theorem as known, then, as we remarked in 2.3, the Peter–Weyl Theorem is an immediate consequence of Theorem 3.9. For this reason Theorem 3.9 is essentially just another form of the Peter–Weyl Theorem.

3.11. *Definition.* We shall denote by $\mathscr{J}(G)$ the family of all (algebraic) equivalence classes of irreducible finite-dimensional representations of G.

In view of 3.2 every class τ in $\mathscr{J}(G)$ contains some unitary representation T; and by VI.13.16 (or IV.7.10) this T is unique to within unitary equivalence. Therefore $\mathscr{J}(G)$ can be identified with the family of all *unitary* equivalence classes of irreducible finite-dimensional *unitary* representations of G.

Remark. When no ambiguity can arise, we shall for the sake of brevity deliberately confuse elements τ of $\mathscr{J}(G)$ with representations belonging to the class τ.

Remark. We shall see in 8.6 that every irreducible unitary representation of the compact group G is finite-dimensional. Therefore $\mathscr{J}(G)$ in fact coincides with the set $\hat{G}$ of all unitary equivalence classes of irreducible unitary representations of G (see VIII.21.4).

3.12. *Corollary.* G *is Abelian if and only if every element of $\mathscr{J}(G)$ is one-dimensional.*

Proof. Suppose that every element of $\mathscr{I}(G)$ is one-dimensional. It follows that $xyx^{-1}y^{-1} \in \mathrm{Ker}(\phi)$ for every x, y in G and every ϕ in $\mathscr{I}(G)$, and hence by 3.9 that $xyx^{-1}y^{-1} = e$ for all x, y in G. So G is Abelian. This proves the "if" part. The "only if" part follows from VI.14.3 (or IV.4.11). ∎

3.13. Corollary. *Let U be a neighborhood of e in G. There exists a finite-dimensional representation T of G such that $\mathrm{Ker}(T) \subset U$.*

Proof. We may assume that U is open. By 3.9, for each x in $G \setminus U$ there is a finite-dimensional representation T^x of G such that $x \notin \mathrm{Ker}(T^x)$. Since $\mathrm{Ker}(T^x)$ is closed and $G \setminus U$ is compact, this implies the existence of finitely many points $x^{(1)}, \ldots, x^{(r)}$ of $G \setminus U$ such that $G \setminus U$ does not intersect $\bigcap_{i=1}^{r} \mathrm{Ker}(T^{x^{(i)}})$, that is, $\mathrm{Ker}(\sum_{i=1}^{\oplus r} T^{x^{(i)}}) \subset U$. ∎

Compact Lie Groups

3.14. Corollary 3.13 has an important bearing on the characterization of compact *Lie* Groups.

Let us say that a topological group H (with unit u) *has no small subgroups* if there exists an H-neighborhood U of u such that the only subgroup of H which is contained in U is the one-element subgroup $\{u\}$.

Definition. A *Lie group* is a locally compact group which has no small subgroups.

Notice that a closed subgroup of a Lie group is a Lie group.

Proposition.* *The following three conditions on the compact group G are equivalent:*

(i) *G is a Lie group;*
(ii) *G has a faithful finite-dimensional representation;*
(iii) *there is a positive integer n such that G is isomorphic, as a topological group, with some compact subgroup of $GL(n, \mathbb{C})$.*

To prove this, one would first verify that $GL(n, \mathbb{C})$ is a Lie group in the sense of the preceding definition. From this and the sentence preceding the proposition it would follow that (iii) $\Rightarrow$ (i). The implication (i) $\Rightarrow$ (ii) follows from 3.13; and (ii) $\Rightarrow$ (iii) immediately.

Remark. This proposition is a very special case of the profound Gleason–Yamabe theory of Hilbert's Fifth Problem, according to which the above definition of a Lie group is equivalent to the usual differential-geometric definition.

4. Orthogonality Relations

4.1. In this section we are going to derive the important orthogonality relations between the matrix elements of the irreducible finite-dimensional representations of G. For finite groups these orthogonality relations are a special case of IV.5.26, which in turn is derived from the structure of finite-dimensional semisimple algebras. For infinite G, however, another approach is needed; and we shall follow a beautiful method discovered by I. Schur, based on integration over the group. The orthogonality relations, once obtained, will give the complete structure of the convolution algebra $\mathscr{R}$ of representative functions (defined in 2.1).

4.2. Theorem (Orthogonality relations). *Let S and T be two irreducible finite-dimensional representations of G, of dimension s and t respectively. Let $(S_{kr}(x))_{k,r=1,\ldots,s}$ and $(T_{ij}(x))_{i,j=1,\ldots,t}$ be the matrices of S_x and T_x with respect to fixed bases of $X(S)$ and $X(T)$ respectively. Then:*

(I) *If S and T are inequivalent,*

$$\int_G T_{ij}(x)S_{kr}(x^{-1})d\lambda x = 0 \qquad \text{for all } i,j = 1,\ldots,t$$

$$\text{and } k, r = 1,\ldots, s. \qquad (1)$$

(II) *We have:*

$$\int_G S_{ij}(x)S_{kr}(x^{-1})d\lambda x = s^{-1}\delta_{jk}\delta_{ir} \qquad \text{for all } i, j, k, r = 1,\ldots, s. \qquad (2)$$

Proof. Let $A: X(S) \to X(T)$ be an arbitrary linear map; and set

$$B = \int_G T_x A S_{x^{-1}} \, d\lambda x. \qquad (3)$$

Thus $B: X(S) \to X(T)$ is a linear map; and if $y \in G$, the left-invariance of λ gives:

$$T_y B = \int T_{yx} A S_{x^{-1}} \, d\lambda x = \int T_x A S_{x^{-1}y} \, d\lambda x = B S_y.$$

So, no matter what A we started with, B intertwines S and T.

Now assume that S and T are inequivalent. Then by Schur's Lemma (IV.1.9) $B = 0$. In particular, suppose we started with $A = E^{jk}$ ($j = 1, \ldots, t$; $k = 1, \ldots, s$), where the matrix of E^{jk} (with respect to the given bases of $X(S)$ and $X(T)$) has 1 in the j, k place and 0 elsewhere. We then have for all i and r:

$$0 = B_{ir} = \int (T_x E^{jk} S_{x^{-1}})_{ir}\, d\lambda x$$

$$= \sum_{p, q} \int T_{ip}(x) E^{jk}_{pq} S_{qr}(x^{-1}) d\lambda x$$

$$= \int T_{ij}(x) S_{kr}(x^{-1}) d\lambda x,$$

proving (I).

Next take $T = S$ in (3). Since B is S, S intertwining, IV.4.6 and IV.4.10 show that

$$B = \lambda \mathbf{1} \tag{4}$$

where λ is a complex number depending on A. Taking the trace of both sides of (4), we find $\lambda = s^{-1} \operatorname{Tr}(B)$. From (3) we have $\operatorname{Tr}(B) = \int \operatorname{Tr}(S_x A S_{x^{-1}}) d\lambda x = \operatorname{Tr}(A)$. So $\lambda = s^{-1} \operatorname{Tr}(A)$, and (4) becomes

$$B = s^{-1} \operatorname{Tr}(A) \mathbf{1}. \tag{5}$$

Taking $A = E^{jk}$ as before, we find from (3) and (5) that for all i, j, k, r

$$s^{-1} \delta_{jk} \delta_{ir} = B_{ir} = \sum_{p, q} \int S_{ip}(x) E^{jk}_{pq} S_{qr}(x^{-1}) d\lambda x$$

$$= \int S_{ij}(x) S_{kr}(x^{-1}) d\lambda x;$$

and this is (II).   ■

4.3. Suppose in 4.2 that S and T are unitary, and that the bases in $X(S)$ and $X(T)$ are taken to be orthonormal. Then of course the matrices $(S_{kr}(x))$ and $(T_{ij}(x))$ are unitary; and we have

$$S_{kr}(x^{-1}) = \overline{S_{rk}(x)}. \tag{6}$$

Thus in this case the orthogonality relations become:

(I') $$\int T_{ij}(x)\overline{S_{rk}(x)}d\lambda x = 0 \text{ for all } i, j, k, r \text{ if } S \ncong T;$$ (7)

(II') $$\int S_{ij}(x)\overline{S_{rk}(x)}d\lambda x = (\dim(S))^{-1}\delta_{ir}\delta_{jk} \text{ for all } i, j, k, r.$$ (8)

It follows from these relations that the matrix elements of inequivalent irreducible finite-dimensional representations of G are orthogonal in $\mathscr{L}_2(\lambda)$, and that different matrix elements (with respect to the same orthonormal basis) of a given unitary irreducible finite-dimensional representation of G are orthogonal in $\mathscr{L}_2(\lambda)$. Notice also from (8) that the $\mathscr{L}_2(\lambda)$-norm of S_{ij} is $(\dim(S))^{-1/2}$. Thus, combining (7) and (8) with 3.8 and the Peter–Weyl Theorem 2.11, we obtain the following:

Theorem.   *From each class σ in $\mathscr{J}(G)$ choose a unitary representation S^σ, and an orthonormal basis of $X(S^\sigma)$; and let $(\sigma_{ij}(x))_{i, j=1,\ldots,\dim(\sigma)}$ be the matrix of S_x^σ with respect to this basis. Then the set of all $(\dim(\sigma))^{1/2}\sigma_{ij}$ $(\sigma \in \mathscr{J}(G);$ $i, j = 1, \ldots, \dim(\sigma))$ is an orthonormal basis of $\mathscr{L}_2(\lambda)$.*

4.4   From (1) and (2) we obtain the structure of $\mathscr{R}$ under convolution.

Theorem.   *Keep the notation of 4.2. Then:*

(I)   *If $S \ncong T$ and $i, j = 1, \ldots, t$ and $k, r = 1, \ldots, s$,*

$$T_{ij} * S_{kr} = 0.$$ (8)

(II)   *If $i, j, k, r = 1, \ldots, s$,*

$$S_{ij} * S_{kr} = s^{-1}\delta_{jk}S_{ir}.$$ (9)

Proof.   We have

$$(T_{ij} * S_{kr})(x) = \int T_{ij}(y)S_{kr}(y^{-1}x)d\lambda y$$

$$= \sum_{p=1}^{s} S_{pr}(x) \int T_{ij}(y)S_{kp}(y^{-1})d\lambda y.$$ (10)

If $T \ncong S$, the right side of (10) is 0 by 4.2(I), proving (I). Replacing T by S in (10), we have by 4.2(II)

$$(S_{ij} * S_{kr})(x) = \sum_p S_{pr}(x)s^{-1}\delta_{ip}\delta_{jk}$$

$$= s^{-1}\delta_{jk}S_{ir}(x);$$

and (II) is proved.   ∎

4.5.   Recall from 3.5 that $\mathscr{E}(S)$ is the linear span of the set of matrix elements of S. Keeping the notation of 4.2, let us put $S'_{ij} = \dim(S)S_{ij}$. Then relation (9) becomes

$$S'_{ij} * S'_{kr} = \delta_{jk}S'_{ir}. \tag{11}$$

This says that $\mathscr{E}(S)$ is closed under convolution, and that in fact $\mathscr{E}(S)$, as an algebra under convolution, is isomorphic with the $s \times s$ total matrix algebra (the S'_{ij} forming a canonical basis). If the S_{ij} are *unitary* matrix elements as in 4.3, equation (6) says that $S_{ji} = S^*_{ij}$, or $S'_{ji} = (S'_{ij})^*$. Thus we have the first part of the following theorem.

Theorem.   *For each S in $\mathscr{J}(G)$, $\mathscr{E}(S)$ is a *-ideal of $\mathscr{R}$ and is *-isomorphic with the $\dim(S) \times \dim(S)$ total matrix *-algebra. The family $\{\mathscr{E}(S): S \in \mathscr{J}(G)\}$ is linearly independent, in fact pairwise orthogonal in $\mathscr{L}_2(\lambda)$; and its algebraic sum is $\mathscr{R}$. If S and T are inequivalent elements of $\mathscr{J}(G)$, $\mathscr{E}(S) * \mathscr{E}(T) = \{0\}$.*

The second and last parts of the theorem were observed in 4.3, 3.8, and (8).

Thus $\mathscr{R}$, as a *-algebra, is the direct sum of the total matrix *-algebras $\mathscr{E}(S)$ $(S \in \mathscr{J}(G))$.

4.6.   Let A be one of the Banach *-algebras $\mathscr{L}(G)$ or $\mathscr{L}_p(\lambda)$ $(1 \leq p < \infty$; see §1). Since $\mathscr{R}$ is dense in A by the Peter–Weyl Theorem, it follows from 4.5 that each $\mathscr{E}(S)$ $(S \in \mathscr{J}(G))$ is a *-ideal of A. In fact $\mathscr{E}(S)$ is also a *-ideal of the measure algebra $\mathscr{M}_r(G)$ of G; the reader will easily verify this.

The case that $A = \mathscr{L}_2(\lambda)$ is of particular interest. In that case A is the Hilbert direct sum of the $\mathscr{E}(S)$ $(S \in \mathscr{J}(G))$ not only as a Hilbert space but also as a Banach *-algebra. In §9 we shall return to a closer examination of $\mathscr{L}_2(\lambda)$.

4.7.   We close this section with further easy consequences of the orthogonality relations.

Proposition.   *The regional topology of $\mathscr{J}(G)$ (see VIII.21.4) is the discrete topology.*

Proof. Fix $S \in \mathcal{J}(G)$. Since $\mathcal{E}(T)$ and $\mathcal{E}(S)$ are orthogonal in $\mathscr{L}_2(\lambda)$ whenever $S \ncong T \in \mathcal{J}(G)$, a non-zero matrix element of S cannot be the uniform limit of matrix elements of representations in $\mathcal{J}(G)$ different from S. Therefore by VIII.21.14 $\{S\}$ is regionally open in $\mathcal{J}(G)$. ∎

In view of the result to be proved in 8.6, this says that *the structure space of a compact group (as defined in VIII.21.4) is discrete.*

4.8. Proposition. *$\mathcal{J}(G)$ is finite [countable] if and only if G is finite [second-countable].*

Proof. By the Peter–Weyl Theorem G is finite if and only if $\mathscr{R}$ is finite-dimensional. By 4.5 this happens if and only if $\mathcal{J}(G)$ is finite. Thus $\mathcal{J}(G)$ is finite if and only if G is.

If G is second-countable, $\mathscr{L}_2(\lambda)$ is separable by II.15.11; so $\mathcal{J}(G)$ is countable by the orthogonality of the $\mathcal{E}(S)$ (4.3). Conversely, suppose that $\mathcal{J}(G)$ is countable. Then by 4.5 $\mathscr{R}$ is of countable Hamel dimension, and so is separable with respect to the supremum norm. Therefore by the Peter–Weyl Theorem $\mathscr{L}(G)$ is separable with respect to the supremum norm. It is well known that this implies that G is second-countable. This completes the proof. ∎

5. Characters and Central Functions

5.1. As in Chapter IV, the *character* χ_T of a finite-dimensional representation T of G is defined by:

$$\chi_T(x) = \mathrm{Tr}(T_x) \qquad (x \in G).$$

Thus χ_T is an element of $\mathscr{L}(G)$ depending only on the equivalence class of T. In particular, if T is irreducible, χ_T depends only on the class τ in $\mathcal{J}(G)$ to which T belongs, and is often called χ_τ.

5.2. Since by 3.2 each finite-dimensional representation T of G is equivalent to a unitary representation, we have $\chi_T(x^{-1}) = \overline{\chi_T(x)}$ for all x in G, that is,

$$(\chi_T)^* = \chi_T. \qquad (1)$$

If T^* is the representation adjoint to T (see 3.7), then by (1) and 3.7(1)

$$\chi_{T^*}(x) = \chi_T(x^{-1}) = \overline{\chi_T(x)} \qquad (x \in G). \qquad (2)$$

Notice also that, if S and T are any finite-dimensional representations of G,

$$\chi_{S \otimes T}(x) = \chi_S(x)\chi_T(x) \qquad (x \in G). \qquad (3)$$

5.3.   The next two results should be compared with IV.6.14.

Proposition.   *If $\sigma, \tau \in \mathcal{J}(G)$, then:*

(I) $\displaystyle\int \chi_\sigma(x)\overline{\chi_\tau(x)}d\lambda x = 0$ *if* $\sigma \neq \tau$;

(II) $\displaystyle\int |\chi_\sigma(x)|^2\, d\lambda x = 1.$

Proof.   Apply 4.3(I′), (II′), remembering that $\chi_S = \sum_i S_{ii}$.   ■

Thus $\{\chi_\sigma : \sigma \in \mathcal{J}(G)\}$ is an orthonormal set in $\mathcal{L}_2(\lambda)$.

5.4.   Corollary.   *Let T and U be finite-dimensional representations of G; and, for $\sigma \in \mathcal{J}(G)$, let μ_σ and ν_σ be the multiplicities of σ in T and U respectively (see VIII.9.9). Then:*

$$\int_G \chi_T(x)\overline{\chi_U(x)}d\lambda x = \sum_{\sigma \in \mathcal{J}(G)} \mu_\sigma \nu_\sigma.$$

In particular, T is irreducible if and only if

$$\int_G |\chi_T(x)|^2\, d\lambda x = 1.$$

Proof.   Notice that $\chi_T = \sum_\sigma \mu_\sigma \chi_\sigma$, $\chi_U = \sum_\sigma \nu_\sigma \chi_\sigma$; and apply 5.3.   ■

5.5.   Corollary.   *If $\sigma \in \mathcal{J}(G)$ and T is a finite-dimensional representation of G, the multiplicity of σ in T is*

$$\int_G \chi_T(x)\overline{\chi_\sigma(x)}d\lambda x.$$

Remark.   This corollary implies that the equivalence class of a finite-dimensional representation T of G is determined by χ_T—a fact already known from IV.4.16.

5.6.   Combining Corollary 5.5 with (3), we can express the multiplicity m of ρ in $\sigma \otimes \tau$ for any elements ρ, σ, τ of $\mathcal{J}(G)$ as follows:

$$m = \int \chi_\sigma(x)\chi_\tau(x)\overline{\chi_\rho(x)}d\lambda x. \tag{4}$$

Formula (4) combined with 5.2(2) implies immediately:

Proposition.   *Let ρ, σ, τ be any elements of $\mathcal{J}(G)$. Then the multiplicity of ρ^* in $\sigma \otimes \tau$ is equal to the multiplicity of τ^* in $\rho \otimes \sigma$, and to the multiplicity of σ^* in $\rho \otimes \tau$.*

In particular, if ρ is the trivial representation, ρ occurs in $\sigma \otimes \tau$ if and only if $\tau = \sigma^*$, and in that case with multiplicity 1.

5.7.   A finite-dimensional representation T of G is called *self-adjoint* if $T^* \cong T$. From 5.2(2) we have:

Corollary.   *A finite-dimensional representation T of G is self-adjoint if and only if χ_T is real-valued.*

5.8.   Proposition.   *If $\sigma \in \mathcal{J}(G)$, $(\dim(\sigma))\chi_\sigma$ is the unit element of the total matrix algebra $\mathcal{E}(\sigma)$ (see 4.5).*

Proof.   By 4.5(11) the unit element of $\mathcal{E}(\sigma)$ is $\sum_i S'_{ii} = \dim(\sigma) \sum_i S_{ii} = \dim(\sigma)\chi_\sigma$.   ∎

Notation.   In the future we will denote the unit element $\dim(\sigma)\chi_\sigma$ of $\mathcal{E}(\sigma)$ by u_σ.

Central Functions

5.9.   A function f in $\mathcal{L}(G)$ is *central* if it is constant on conjugacy classes, that is, if $f(xyx^{-1}) = f(y)$ (or, equivalently, $f(xy) = f(yx)$) for all x, y in G. The space $\mathcal{Z}(G)$ of all central functions is a closed linear subspace of $\mathcal{L}(G)$.

In view of the formulae for convolution (see III.11.5), a central function f satisfies $\mu * f = f * \mu$ for all μ in $\mathcal{M}_r(G)$, and so belongs to the center of $\mathcal{M}_r(G)$.

Clearly every character χ_T of a finite-dimensional representation T of G is central.

5.10.   If $f \in \mathcal{L}(G)$, let

$$(\Gamma f)(x) = \int_G f(yxy^{-1})d\lambda y \qquad\qquad (x \in G). \qquad (5)$$

Thus the operator Γ averages f over conjugacy classes; and we verify that for each f in $\mathcal{L}(G)$:

  (i) Γf is a central function in $\mathcal{L}(G)$;
  (ii) $\Gamma f = f$ if and only if f is central;
  (iii) $\|\Gamma f\|_\infty \le \|f\|_\infty$.
  (iv) If $\sigma \in \mathcal{J}(G)$ and $f \in \mathcal{E}(\sigma)$, then Γf is a scalar multiple of χ_σ.

(To prove (iv), let S belong to the class σ and let $\{S_{ij}\}$ be as in 4.2. Then

$$(\Gamma(S_{ij}))(x) = \int S_{ij}(yxy^{-1})d\lambda y = \sum_{k,r} S_{kr}(x) \int S_{ik}(y)S_{rj}(y^{-1})d\lambda y$$

$$= \dim(\sigma)^{-1}\delta_{ij}\chi_\sigma(x)$$

(by 4.2(II)).)

5.11.   Theorem.   *The $\| \ \|_\infty$-closure of the linear span of $\{\chi_\sigma : \sigma \in \mathcal{J}(G)\}$ is equal to $\mathscr{Z}(G)$.*

Proof.   We have only to show that every central function f in $\mathscr{L}(G)$ can be approximated in the $\| \ \|_\infty$ norm by linear combinations of the χ_σ ($\sigma \in \mathcal{J}(G)$).

Take $\varepsilon > 0$. By the Peter–Weyl Theorem there is a representative function g such that $\| f - g \|_\infty < \varepsilon$. From this, applying Γ and using 5.10(ii), (iii), we find that $\| f - \Gamma g \|_\infty < \varepsilon$. By 4.5 and 5.10(iv) Γg is a linear combination of the χ_σ ($\sigma \in \mathcal{J}(G)$). This proves the theorem.   ∎

5.12.   The following fact is often useful.

Proposition.   *If $\mu \in \mathcal{M}_r(G)$ and $\mu * \chi_\sigma = 0$ for all σ in $\mathcal{J}(G)$, then $\mu = 0$.*

Proof.   By 5.8 u_σ is the unit element of $\mathscr{E}(\sigma)$. So, by the hypothesis, $\mu * \mathscr{E}(\sigma) = (\mu * u_\sigma) * \mathscr{E}(\sigma) = 0$ for every σ in $\mathcal{J}(G)$. Since $\sum_\sigma \mathscr{E}(\sigma) = \mathscr{R}$, we obtain $\mu * \mathscr{R} = 0$. Since $\mathscr{R}$ is $\| \ \|_\infty$-dense in $\mathscr{L}(G)$, this implies that

$$\mu * \mathscr{L}(G) = 0. \tag{6}$$

Now, for any f in $\mathscr{L}(G)$ we observe that $\int f d\mu = (\mu * \tilde{f})(e)$ (where $\tilde{f}(x) = f(x^{-1})$). By (6) this gives $\int f d\mu = 0$ for all f in $\mathscr{L}(G)$, whence $\mu = 0$.   ∎

5.13.   Corollary.   *Let B be any subalgebra of $\mathcal{M}_r(G)$ containing $\mathscr{R}$. Then the minimal non-zero two-sided ideals of B are exactly those of the form $\mathscr{E}(\sigma)$ ($\sigma \in \mathcal{J}(G)$).*

The proof follows easily from 5.12, and is left to the reader.

5.14.   Proposition.*   *Equipped with convolution, $*$, and the supremum norm, $\mathscr{Z}(G)$ is a commutative Banach $*$-algebra (the center of $\mathscr{L}(G)$). Each σ in $\mathcal{J}(G)$ gives rise to a complex homomorphism λ_σ of $\mathscr{Z}(G)$:*

$$S_f = \lambda_\sigma(f)\mathbb{1}_{X(S)} \qquad (S \in \sigma \in \mathcal{J}(G); f \in \mathscr{Z}(G)).$$

The map $\sigma \mapsto \lambda_\sigma$ is a homeomorphism of the discrete space $\mathscr{J}(G)$ onto the Gelfand space of $\mathscr{Z}(G)$.

6. The Stone–Weierstrass Theorem for Representations

6.1. The following theorem is essentially the Stone–Weierstrass Theorem as applied to the representations of compact groups. It is often useful for determining $\mathscr{J}(G)$.

Theorem. *Suppose that Γ is a non-void subset of $\mathscr{J}(G)$ with the following three properties*:

  (i) Γ *distinguishes points of G, i.e.,* $\bigcap\{\mathrm{Ker}(\sigma) : \sigma \in \Gamma\} = \{e\}$;
  (ii) *if $\sigma \in \Gamma$ then $\sigma^* \in \Gamma$*;
  (iii) *if $\sigma, \tau \in \Gamma$ and ρ is an element of $\mathscr{J}(G)$ which occurs in $\sigma \otimes \tau$, then $\rho \in \Gamma$.*

Then $\Gamma = \mathscr{J}(G)$.

Proof. Since $\Gamma \neq \emptyset$, we can choose an element σ of Γ. By 5.6 the trivial representation occurs in $\sigma \otimes \sigma^*$, and so by hypotheses (ii) and (iii) belongs to Γ.

Now let $\mathscr{R}_0$ be the linear span of the set of all matrix elements of representations belonging to Γ. By the preceding paragraph $\mathscr{R}_0$ contains the constant functions. By hypothesis (i) $\mathscr{R}_0$ separates points of G. By hypothesis (ii) (and the fact that if S is unitary $S^* \cong \bar{S}$; see 3.7) $\mathscr{R}_0$ is closed under complex conjugation. By hypothesis (iii) $\mathscr{R}_0$ is closed under multiplication. Hence the Stone–Weierstrass Theorem (A8) asserts that $\mathscr{R}_0$ is uniformly dense in $\mathscr{L}(G)$.

Suppose some τ belongs to $\mathscr{J}(G) \setminus \Gamma$. By 4.3 χ_τ is then orthogonal (in $\mathscr{L}_2(\lambda)$) to all of $\mathscr{R}_0$. Since $\mathscr{R}_0$ is dense in $\mathscr{L}(G)$, this implies that χ_τ is orthogonal to all of $\mathscr{L}(G)$, in particular to itself. This contradicts the fact that $\chi_\tau \neq 0$. So there is no such τ; and $\Gamma = \mathscr{J}(G)$. ∎

6.2. Suppose that G is Abelian. Then the elements of $\mathscr{J}(G)$ are one-dimensional (3.12); that is, they are continuous homomorphisms $\chi : G \to \mathbb{E}$. In this case tensor products are pointwise products, and χ^* is the complex conjugate of χ. Thus 6.1 becomes:

Corollary. *Let G be Abelian; and let Γ be a non-void subset of $\mathscr{J}(G)$ which separates points of G and is closed under the operations of pointwise product and complex conjugation. Then $\Gamma = \mathscr{J}(G)$.*

6.3. As a simple example of 6.2 consider the n-dimensional torus $\mathbb{E}^n$. Any n-termed sequence of integers $\langle r_1, \ldots, r_n \rangle$ determines an element $\chi_{r_1, \ldots, r_n}$ of $\mathscr{I}(\mathbb{E}^n)$:

$$\chi_{r_1, \ldots, r_n}(u_1, \ldots, u_n) = u_1^{r_1} \cdots u_n^{r_n}.$$

The family of all $\chi_{r_1, \ldots, r_n}$ evidently satisfies the hypotheses of 6.2; and so we have:

Corollary. *Every element of* $\mathscr{I}(\mathbb{E}^n)$ *is of the form*

$$\langle u_1, \ldots, u_n \rangle \mapsto u_1^{r_1} u_2^{r_2} \cdots u_n^{r_n}$$

for some n-termed sequence of integers $\langle r_1, \ldots, r_n \rangle$.

6.4. Before mentioning other corollaries of 6.1, we should like to develop an interesting and non-trivial special consequence (6.5) of 6.3. It is based on the following general result:

Proposition. *Suppose G is Abelian, and let* $x \in G$. *The following two conditions are equivalent*:

  (i) $\{x^n : n = 1, 2, 3, \ldots\}$ *is dense in* G;
  (ii) *the only element* χ *of* $\mathscr{I}(G)$ *satisfying* $\chi(x) = 1$ *is the trivial representation of* G.

Proof. Evidently (i) $\Rightarrow$ (ii). To prove the converse, assume that (i) fails, and let H be the closure of $\{x^n : n = 1, 2, \ldots\}$. Thus H is closed under multiplication, and so by III.1.17 is a closed subgroup of G. Since (i) fails, $H \neq G$; so G/H is a compact Abelian group with more than one element. By the Peter–Weyl Theorem 3.9 $\mathscr{I}(G/H)$ has a non-trivial element, which when composed with the quotient homomorphism $G \to G/H$ gives a nontrivial element χ of $\mathscr{I}(G)$ satisfying $\chi(x) = 1$. Thus (ii) fails; and we have shown that (ii) $\Rightarrow$ (i). ∎

6.5. Proposition. *Let* $n = 1, 2, \ldots$; *and let F be the natural open homomorphism* $\langle t_1, \ldots, t_n \rangle \mapsto \langle e^{2\pi i t_1}, \ldots, e^{2\pi i t_n} \rangle$ *of* $\mathbb{R}^n$ *onto* $\mathbb{E}^n$.

  (I) *Let* $t = \langle t_1, \ldots, t_n \rangle \in \mathbb{R}^n$, *where* $t_1, \ldots, t_n, 1$ *are linearly independent over the rational field. Then* $\{F(nt) : n = 1, 2, \ldots\}$ *is dense in* $\mathbb{E}^n$.
  (II) *Let* $t = \langle t_1, \ldots, t_n \rangle \in \mathbb{R}^n$, *where* $t_1, \ldots, t_n$ *are linearly independent over the rational field. Then* $\{F(rt) : r \in \mathbb{R}\}$ *is dense in* $\mathbb{E}^n$.

Proof. (I) Let χ be an element of $\mathcal{J}(\mathbb{E}^n)$ satisfying $\chi(F(t)) = 1$. To prove (I) it is enough by 6.4 to show that χ is trivial. Now by 6.3 $\chi \circ F$ is of the form

$$\chi(F(s)) = e^{2\pi i(r_1 s_1 + \cdots + r_n s_n)} \qquad (s \in \mathbb{R}^n),$$

where $r_1, \ldots, r_n$ are integers. The assumption that $\chi(F(t)) = 1$ therefore implies:

$$r_1 t_1 + \cdots + r_n t_n \text{ is an integer.}$$

Since $t_1, \ldots, t_n, 1$ are independent over the rationals this is impossible unless $r_1 = \cdots = r_n = 0$. So χ is trivial, and (I) is proved.

(II) Since $t_1, \ldots, t_n$ are independent over the rationals, we can find another real number w such that $t_1, \ldots, t_n, w$ are independent over the rationals. Then $w^{-1} t_1, \ldots, w^{-1} t_n, 1$ are independent over the rationals, whence by (I) $\{F(nw^{-1}t): n = 1, 2, \ldots\}$ is dense in $\mathbb{E}^n$. This implies (II). ∎

6.6. As another application of 6.1 we mention the following:

Proposition. *Let $\mathcal{N}$ be a downward directed set of closed normal subgroups of G such that $\bigcap \mathcal{N} = \{e\}$. Then every T in $\mathcal{J}(G)$ satisfies $N \subset \mathrm{Ker}(T)$ for some N in $\mathcal{N}$.*

Proof. Let Γ be the set of those T in $\mathcal{J}(G)$ which satisfy $N \subset \mathrm{Ker}(T)$ for some N in $\mathcal{N}$. The reader will easily verify that this Γ satisfies the hypotheses of 6.1, and so coincides with $\mathcal{J}(G)$. ∎

6.7. Let $G_1, \ldots, G_n$ be compact groups. If $S^i \in \mathcal{J}(G_i)$ for each $i = 1, \ldots, n$, we can form the *outer tensor product* $S = S^1 \underset{o}{\otimes} \cdots \underset{o}{\otimes} S^n$, which will be a representation of $G = G_1 \times \cdots \times G_n$:

$$X(S) = X(S^1) \otimes \cdots \otimes X(S^n),$$

$$S_{\langle x_1, \ldots, x_n \rangle} = S^1_{x_1} \otimes \cdots \otimes S^n_{x_n}.$$

(Compare VIII.21.25.) Now we claim that S is irreducible. To see this, we may assume that the S^i are unitary, and hence that S is unitary. So by VI.14.1 it is enough to take an operator A in the commuting algebra of S and show that it is a scalar operator. Now A must commute with all $S^1_{x_1} \otimes 1 \otimes \cdots \otimes 1$ $(x_1 \in G_1)$; and so by VI.14.5 $A = 1 \otimes A'$, where A' acts on $X(S^2) \otimes \cdots \otimes X(S^n)$ and belongs to the commuting algebra of $S^2 \underset{o}{\otimes} \cdots \underset{o}{\otimes} S^n$. Repeating this argument we conclude eventually that $A = 1 \otimes 1 \otimes \cdots \otimes B$, where B is in the commuting algebra of S^n, and so by VI.14.1 is a scalar operator. So A is a scalar operator, and the claim is proved.

Remark 1.   The above argument has shown that *if* $H_1, \ldots, H_n$ *are any topological groups and* $T^1, \ldots, T^n$ *are arbitrary (not necessarily finite-dimensional) irreducible unitary representations of* $H_1, \ldots, H_n$ *respectively, then the outer tensor product* $T^1 \underset{o}{\otimes} \cdots \underset{o}{\otimes} T^n$ *is an irreducible unitary representation of* $H_1 \times \cdots \times H_n$.

Proposition.   *Every element of* $\mathcal{J}(G_1 \times \cdots \times G_n)$ *is of the form* $S^1 \underset{o}{\otimes} \cdots \underset{o}{\otimes} S^n$, *where* $S^i \in \mathcal{J}(G_i)$ *for each i.*

Proof.   By 6.1 it is only necessary to verify that $\{S^1 \underset{o}{\otimes} \cdots \underset{o}{\otimes} S^n : S^i \in \mathcal{J}(G_i)$ for each $i\}$ satisfies the hypotheses of Theorem 6.1. We leave this verification to the reader.   ■

Remark 2.   If G_1 and G_2 are arbitrary locally compact groups, it need not be true that every irreducible unitary representation of $G_1 \times G_2$ is unitarily equivalent to an outer tensor product $S^1 \underset{o}{\otimes} S^2$, where S^i is an irreducible unitary representation of G_i. Counterexamples illustrating this involve non-Type I factors, and will not be given in this work.

Remark 3.   Combining the above proposition with 6.6, one easily deduces a corresponding result on *infinite* Cartesian products of compact groups. Details are left to the reader.

6.8. Proposition.*   *Let K be a closed subgroup of G. Every element S of* $\mathcal{J}(K)$ *occurs in* $T|K$ *for some T in* $\mathcal{J}(G)$.

This can be proved by applying 6.1 to the set of those S in $\mathcal{J}(K)$ which occur in $T|K$ for some T in $\mathcal{J}(G)$.

This proposition will be proved by a different method in 10.12.

7.   σ-Subspaces and Multiplicity for Locally Convex Representations of G

7.1.   Up till now we have dealt exclusively with finite-dimensional representations of G. In this and the next two sections we will derive the main facts about arbitrary (in general infinite-dimensional) integrable locally convex representations of G.

If S is an irreducible finite-dimensional representation of G, then by 3.7 so is its adjoint S^*; and the class σ of S determines the class of S^*, which we naturally call σ^*.

We recall from 5.8 the definition of u_σ.

Definition. For each σ in $\mathscr{J}(G)$, put

$$w_\sigma = u_{\sigma^*}.$$

Thus

$$w_\sigma = \dim(\sigma)\chi_{\sigma^*}, \tag{1}$$

or, by 5.2(2),

$$w_\sigma(x) = \dim(\sigma)\overline{\chi_\sigma(x)} \qquad (x \in G). \tag{2}$$

By 5.8 w_σ is the unit element of the total matrix *-algebra $\mathscr{E}(\sigma^*)$. It is thus a self-adjoint idempotent belonging to the center of $\mathscr{M}_r(G)$, and (by 4.5)

$$w_\sigma * w_\tau = 0 \qquad \text{if } \sigma \neq \tau. \tag{3}$$

7.2. Definition. Let T be any locally convex representation of G. As in §IV.2, if $\sigma \in \mathscr{J}(G)$ we denote by $X_\sigma(T)$ the linear span of the set of all those finite-dimensional T-stable subspaces Y of $X(T)$ such that the subrepresentation of T acting on Y is irreducible and of class σ; $X_\sigma(T)$ is called the σ-subspace of $X(T)$; it is evidently T-stable. A vector ξ in $X(T)$ is of type σ (relative to T) if $\xi \in X_\sigma(T)$.

The picking out of the subspaces $X_\sigma(T)$ is called the *harmonic analysis* of T.

7.3. Suppose now that T is an *integrable* locally convex representation of G. (We shall agree not to distinguish notationally between T and its integrated form.) The $X_\sigma(T)$ can then be elegantly characterized in terms of the w_σ of 7.1.

By 7.1 the T_{w_σ} $(\sigma \in \mathscr{J}(G))$ are continuous idempotent operators on $X(T)$ satisfying

$$T_{w_\sigma} T_{w_\tau} = 0 \qquad \text{if } \sigma \neq \tau. \tag{4}$$

If T is a unitary representation, the T_{w_σ} are pairwise orthogonal projections on the Hilbert space $X(T)$.

7.4. Proposition. *Let T be an integrable locally convex representation of G. For each σ in $\mathscr{J}(G)$,*

$$X_\sigma(T) = \text{range}(T_{w_\sigma}). \tag{5}$$

Proof. Let S be a representation of class $\sigma \in \mathscr{J}(G)$. Put $\dim(S) = r$; and let $\{S_{ij}\}$ $(i, j = 1, \ldots, r)$ be the matrix elements of S with respect to some basis of $X(S)$. Then the functions $S_{ij}^*(x) = S_{ji}(x^{-1})$ are the matrix elements of S^* with respect to the dual basis (see 3.7(1)). The integrated form of S sends the function S_{ij}^* into the operator whose k, m matrix element is

$$(S_{S_{ij}^*})_{km} = \int S_{km}(x)S_{ij}^*(x)d\lambda x$$

$$= \int S_{km}(x)S_{ji}(x^{-1})d\lambda x$$

$$= r^{-1}\delta_{ik}\delta_{jm} \qquad \text{(by 4.2).} \qquad (6)$$

Putting $w_{ij} = rS_{ij}^*$, we get from (6)

$$S_{w_{ij}} = E^{ij}, \qquad (7)$$

where E^{ij} is the operator on $X(S)$ whose matrix with respect to the given basis has i, j entry 1 and all other entries 0. In particular, since $w_\sigma = \sum_i w_{ii}$, (7) implies

$$S_{w_\sigma} = \text{identity operator.} \qquad (8)$$

Thus, if Y is a stable subspace of $X(T)$ such that the subrepresentation of T on Y is of class σ, (8) says that T_{w_σ} is the identity operator on Y, whence $Y \subset \text{range}(T_{w_\sigma})$. Since $X_\sigma(T)$ is the sum of all Y of this form, it follows that

$$X_\sigma(T) \subset \text{range}(T_{w_\sigma}). \qquad (9)$$

To prove the converse, recall from 4.5 that the $\{w_{ij}\}$ form a canonical basis of the total matrix algebra $\mathscr{E} = \mathscr{E}(\sigma^*)$. Let $N = \text{range}(T_{w_\sigma})$. Since w_σ belongs to the center of $\mathscr{M}_r(G)$ (see 5.9), N is stable under the integrated form of T. Let $V_f = T_f|N$ $(f \in \mathscr{E})$. Thus V is a non-degenerate representation of $\mathscr{E}$ on N. Since $\mathscr{E}$ is a total matrix algebra, IV.5.7 asserts that V is completely reducible; and we can write N as an algebraic direct sum:

$$N = \sum_\alpha^\oplus N_\alpha \qquad (10)$$

of r-dimensional V-stable subspaces N_α, such that V restricted to each N_α is equivalent to the unique irreducible representation of the total matrix algebra $\mathscr{E}$. Thus, fixing α, we can choose a basis $\xi_1, \ldots, \xi_r$ of N_α such that

$$T_{w_{ij}}\xi_k = V_{w_{ij}}\xi_k = \delta_{jk}\xi_i. \qquad (11)$$

It follows that for any x in G and any $k = 1, \ldots, r$,

$$T_x\xi_k = T_x T_{w_{kk}}\xi_k = T_{\delta_x * w_{kk}}\xi_k. \qquad (12)$$

But

$$(\delta_x * w_{kk})(y) = w_{kk}(x^{-1}y) = rS^*_{kk}(x^{-1}y) = rS_{kk}(y^{-1}x) = r\sum_{i=1}^{r} S_{ki}(y^{-1})S_{ik}(x)$$

$$= \sum_{i=1}^{r} S_{ik}(x)w_{ik}(y).$$

Consequently

$$\delta_x * w_{kk} = \sum_{i=1}^{r} S_{ik}(x)w_{ik}. \qquad (13)$$

Combining (11), (12), and (13), we have:

$$T_x\xi_k = \sum_{i=1}^{r} S_{ik}(x)T_{w_{ik}}\xi_k = \sum_{i=1}^{r} S_{ik}(x)\xi_i.$$

From this it follows that each N_α is stable under T, and that T restricted to N_α is equivalent to S. This and (10) imply that

$$\text{range}(T_{w_\sigma}) \subset X_\sigma(T). \qquad (14)$$

Now (9) and (14) complete the proof of (5). ■

Remark. This proposition should be compared with IV.6.12.

7.5. Corollary. *For any integrable locally convex representation T of G and any σ in $\mathcal{I}(G)$, $X_\sigma(T)$ is closed.*

7.6. Corollary. *For any integrable locally convex representation T of G and any σ in $\mathcal{I}(G)$, the multiplicity of σ in T (in the sense of IV.2.21) is equal to $(\dim(\sigma))^{-1}$ times the (Hamel) dimension of* $\text{range}(T_{w_\sigma})$.

7.7 Remark. Let w_{ij} $(i, j = 1, \ldots, \dim(\sigma))$ be defined as in the proof of 7.4. Since the w_{ij} form a canonical basis of $\mathscr{E}(\sigma^*)$, it follows from 7.5 that, *for each i, the multiplicity of σ in T is equal to the Hamel dimension of* $\text{range}(T_{w_{ii}})$.

7.8. It is sometimes convenient *not* to distinguish between different infinite multiplicities. With this in view we make the following definition:

Definition. If T and σ are as in 7.6, let $v(T;\sigma)$ denote the algebraic multiplicity of σ in T if this multiplicity is finite, and otherwise ∞.

One advantage of this definition is that it leads to the following result:

Proposition. *Let T be an integrable locally convex representation of G; and let $\{Y^\alpha\}$ be a linearly independent indexed collection of closed T-stable linear subspaces of $X(T)$ whose linear span is dense in $X(T)$. Then, if T^α is the subrepresentation of T acting on Y^α, and if $\sigma \in \mathcal{J}(G)$, we have:*

$$v(T;\sigma) = \sum_\alpha v(T^\alpha;\sigma).$$

The proof of this follows easily from 7.7, and is left to the reader.

7.9.   Remark.   Let $\mathcal{J}$ be the set of all minimal non-zero two-sided ideals of $\mathcal{L}(G)$. There are two natural one-to-one correspondences between $\mathcal{J}(G)$ and $\mathcal{J}$. The first of these is the correspondence

$$\sigma \mapsto \mathcal{E}(\sigma) \tag{15}$$

(see 5.13). To obtain the second one, let us start with an element σ of $\mathcal{J}(G)$. Since $\sum \mathcal{J}$ is dense in $\mathcal{L}(G)$, there must be some J in $\mathcal{J}$ for which $J \not\subset \mathrm{Ker}(\sigma)$. In view of the fact that $J * K = \{0\}$ whenever $J \neq K \in \mathcal{J}$, this J is unique; let us denote it by J_σ. The mapping

$$\sigma \mapsto J_\sigma \tag{16}$$

is the second correspondence referred to above.

The two correspondences (15) and (16) are not in general the same. They differ by the adjoint operation on $\mathcal{J}(G)$; that is, we have:

$$J_\sigma = \mathcal{E}(\sigma^*) \qquad\qquad (\sigma \in \mathcal{J}(G)). \tag{17}$$

To see this it is enough to observe from 7.4(8) that the unit element w_σ of $\mathcal{E}(\sigma^*)$ is carried by σ into the identity operator.

Here is a useful description of the mapping inverse to (16): Let J be an element of $\mathcal{J}$. Thus J is a total matrix algebra, and so has a unit element w and a unique irreducible representation D. The map

$$\mu \mapsto D_{\mu * w} \qquad\qquad (\mu \in \mathcal{M}_r(G)) \tag{18}$$

is then the measure integrated form of the element σ of $\mathcal{J}(G)$ which satisfies $J_\sigma = J$. We leave the verification of this to the reader.

8. Weak Féjer Summation

8.1. Let T be an integrable locally convex representation of G, and ξ a vector in $X(T)$. If $\sigma \in \mathscr{J}(G)$, we shall refer to the vector $\xi_\sigma = T_{w_\sigma}\xi$ (see 7.1) as the σ-*component of* ξ (*with respect to* T). By 7.4 ξ_σ is of type σ relative to T.

Now we should like to be able to assert that ξ is the sum of its components:

$$\xi = \sum_{\sigma \in \mathscr{J}(G)} \xi_\sigma. \tag{1}$$

This is true if G is finite; indeed, in that case $\mathscr{J}(G)$ is finite, and $\sum_{\sigma \in \mathscr{J}(G)} w_\sigma$ is the unit element of $\mathscr{L}(G)$. We shall see in 8.9 that, if T is unitary, (1) holds in the sense of unconditional convergence. In general, however, (1) need not hold in the sense of unconditional convergence (see Remark 8.12). Nevertheless there is always a "weakened summation method" for the series $\sum_\sigma \xi_\sigma$ which makes (1) true. One might indeed expect this from the theory of ordinary Fourier series. If f is a continuous complex periodic function on $\mathbb{R}$, the ordinary Fourier series for f need not converge uniformly to f. However, if we sum the Fourier series by the more "liberal" method of Fejér, the summation does converge uniformly to f (see 8.11).

8.2. To obtain the generalization of Fejér summation to arbitrary integrable locally convex representations of G, we begin with a lemma:

Lemma. *G has an approximate unit* $\{h_\nu\}$ *consisting of central functions.*

Proof. We must show that, if U is a fixed neighborhood of e, there is a non-negative central function h on G, vanishing outside U, such that $\int h \, d\lambda = 1$.

For each y in G there is an open neighborhood W_y of y and a neighborhood V_y of e such that $W_y V_y W_y^{-1} \subset U$. Using the compactness of G to cover G by finitely many of the W_y, and denoting by V the intersection of the corresponding V_y, we obtain a neighborhood V of e such that

$$xVx^{-1} \subset U \qquad \text{for all } x \text{ in } G. \tag{2}$$

Let g be any function in $\mathscr{L}_+(G)$ vanishing outside V such that $\int g \, d\lambda = 1$. An easy verification shows that $h = \Gamma g$ (see 5.10) has the required properties. ∎

Definition. The $\{h_\nu\}$ of the above lemma is called a *central approximate unit on G*.

8.3.   Theorem (Weak Fejér Summation).   *There exists a net $\{\gamma^\alpha\}$ with the following properties:*

(i)   *For each index α, γ^α is a complex-valued function on $\mathcal{I}(G)$, vanishing except at finitely many points of $\mathcal{I}(G)$.*

(ii) *For any integrable locally convex representation T of G and any ξ in $X(T)$,*

$$\sum_{\sigma \in \mathcal{I}(G)} \gamma^\alpha(\sigma) T_{w_\sigma} \xi \xrightarrow[\alpha]{} \xi \tag{3}$$

in the topology of $X(T)$.

Proof.   Let $\{h_\nu\}$ $(\nu \in N)$ be a central approximate unit on G (see 8.2). For each $\varepsilon > 0$ and each ν in N, Theorem 5.11 permits us to choose complex coefficients $\gamma^{\nu, \varepsilon}(\sigma)$ $(\sigma \in \mathcal{I}(G))$, equal to 0 for all but finitely many σ, such that

$$\left\| h_\nu - \sum_{\sigma \in \mathcal{I}(G)} \gamma^{\nu, \varepsilon}(\sigma) w_\sigma \right\|_\infty < \varepsilon. \tag{4}$$

Now let A be the directed set of all pairs $\langle \nu, \phi \rangle$, where $\nu \in N$ and ϕ is a function on N to the set of positive reals, with the directing relation given by:

$$\langle \nu, \phi \rangle \prec \langle \nu', \phi' \rangle \Leftrightarrow \nu \prec \nu' \quad \text{and} \quad \phi(\mu) \geq \phi'(\mu) \qquad \text{for all } \mu \text{ in } N.$$

Let $\{\gamma^\alpha\}$ $(\alpha \in A)$ be the net given by

$$\gamma^{\langle \nu, \phi \rangle} = \gamma^{\nu, \phi(\nu)}.$$

We claim that $\{\gamma^\alpha\}$ has the properties required by the theorem.

Property (i) is immediate. To prove (ii), take an integrable locally convex representation T of G, a vector ξ in $X(T)$, an $X(T)$-neighborhood U of 0, and a neighborhood V of 0 such that $V + V \subset U$. By VIII.11.23, there is an index ν_0 in N such that

$$\nu \succ \nu_0 \Rightarrow T_{h_\nu} \xi - \xi \in V. \tag{5}$$

Further, by (4) and VIII.11.21, for each index ν we can find a positive number $\phi_0(\nu)$ such that

$$0 < \varepsilon \leq \phi_0(\nu) \Rightarrow \left(\sum_{\sigma \in \mathcal{I}(G)} \gamma^{\nu, \varepsilon}(\sigma) T_{w_\sigma} \xi - T_{h_\nu} \xi \right) \in V. \tag{6}$$

It follows from (5) and (6) that, if $\alpha = \langle v, \phi \rangle \succ \langle v_0, \phi_0 \rangle = \alpha_0$, we have:

$$\sum_{\sigma \in \mathscr{I}(G)} \gamma^\alpha(\sigma) T_{w_\sigma} \xi - \xi$$

$$= \left(\sum_{\sigma \in \mathscr{I}(G)} \gamma^{v, \phi(v)}(\sigma) T_{w_\sigma} \xi - T_{h_v} \xi \right) + (T_{h_v} \xi - \xi) \in V + V \subset U.$$

By the arbitrariness of U, this proves (3).   ■

8.4. Theorem 8.3 implies some very useful corollaries.

Corollary. *Let T be an integrable locally convex representation of G, and ξ a vector in $X(T)$. Then ξ belongs to the closure of the linear span of the set $\{ T_{w_\sigma} \xi : \sigma \in \mathscr{I}(G) \}$ of its components. In particular, if $T_{w_\sigma} \xi = 0$ for all σ in $\mathscr{I}(G)$, then $\xi = 0$.*

8.5. Corollary. *If T is an integrable locally convex representation of G, the linear span of $\{ X_\sigma(T) : \sigma \in \mathscr{I}(G) \}$ is dense in $X(T)$.*

Thus, any such T is "topologically completely reducible" in the sense that the sum of its (finite-dimensional) stable irreducible subspaces is dense in $X(T)$.

8.6. Theorem. *Every irreducible integrable locally convex representation T of G is finite-dimensional.*

Proof. By 8.4 $T_{w_\sigma} \neq 0$ for some σ in $\mathscr{I}(G)$. From this and 7.4 it follows that $X(T)$ has a non-zero finite-dimensional (hence closed) T-stable subspace Y (on which T acts equivalently to σ). By the irreducibility of T this implies that $X(T) = Y$. So T is finite-dimensional.   ■

8.7. From 8.6 and 3.11 we obtain:

Corollary. *For the compact group G, the structure space $\hat{G}$ (defined in VIII.21.4) coincides (as a set) with the $\mathscr{I}(G)$ of this chapter.*

Notation. In view of this, *we may in the future write $\hat{G}$ instead of $\mathscr{I}(G)$.*

8.8.   Remark.   Here are two problems, related to 8.6 and 8.5, to which we do not know the answer:

  (I)   Can the hypothesis that T is integrable be omitted from 8.6? We strongly suspect that the answer is "no."
  (II) Is it true that any integrable locally convex representation of G is completely reducible in the purely algebraic sense? Again the answer is presumably "no."

8.9.   Theorem.   *Suppose that T is a unitary representation of G. Then the $X_\sigma(T)$ ($\sigma \in \mathscr{I}(G)$) are pairwise orthogonal closed subspaces of $X(T)$ whose linear span is dense in $X(T)$. For each σ, T_{w_σ} is projection onto $X_\sigma(T)$; and, for each ξ in $X(T)$, equation (1) holds in the sense of unconditional convergence.*

Proof.   By 7.3, 7.4, and 8.5.   ■

8.10.   Corollary.   *Every unitary representation of G is discretely decomposable (see VIII.9.9).*

Classical Fourier Series and Theorem 8.3

8.11.   The classical special case of Theorem 8.3, that of the regular representation of the circle group $\mathbb{E}$ on $\mathscr{L}(\mathbb{E})$, was worked out by Fejér. In this number, using Fejér's methods, we will obtain a simple specific sequence having the properties of the $\{\gamma^\alpha\}$ of 8.3 for the circle group.

By 6.3 the elements of $\mathscr{I}(\mathbb{E})$ are indexed by the set $\mathbb{Z}$ of all integers, the element of $\mathscr{I}(\mathbb{E})$ corresponding to n being $\chi_n : u \mapsto u^n$. If T is any integrable locally convex representation of $\mathbb{E}$, and $\xi \in X(T)$, by 8.1 the χ_n-component of ξ with respect to T is

$$\xi_n = T_{\chi_{-n}}\xi. \tag{7}$$

For each positive integer q and each integer n, define

$$\gamma_n^{(q)} = \begin{cases} 0 & \text{if } |n| \geq q, \\ 1 - q^{-1}|n| & \text{if } |n| < q. \end{cases}$$

Proposition.   *Let T be an integrable locally convex representation of $\mathbb{E}$. For any ξ in $X(T)$, we have*

$$\xi = \lim_{q \to \infty} \sum_{n=-\infty}^{\infty} \gamma_n^{(q)} \xi_n \tag{8}$$

(where ξ_n is the χ_n-component of ξ with respect to T).

Remark. Fix ξ in $X(T)$, and let η_p $(p = 0, 1, 2, \ldots)$ be the partial sum $\sum_{n=-p}^{p} \zeta_n$. One verifies that, for each $q = 1, 2, \ldots$,

$$\sum_{n=-\infty}^{\infty} \gamma_n^{(q)} \zeta_n = q^{-1} \sum_{p=0}^{q-1} \eta_p;$$

so (8) asserts that

$$\xi = \lim_{q \to \infty} \left(q^{-1} \sum_{p=0}^{q-1} \eta_p \right).$$

This is the more familiar form of Fejér's classical result.

Proof of Proposition. From the formulae

$$
\left.
\begin{aligned}
\sum_{n=-\infty}^{\infty} \gamma_n^{(q)} e^{-in\theta} &= q^{-1} \sum_{p=0}^{q-1} \sum_{n=-p}^{p} e^{-in\theta}, \\[6pt]
\sum_{n=-p}^{p} e^{-in\theta} &= \left(\sin \frac{1}{2}\theta \right)^{-1} \sin\left(p + \frac{1}{2} \right)\theta, \\[6pt]
\sum_{p=0}^{q-1} \sin\left(p + \frac{1}{2} \right)\theta &= \left(\sin \frac{\theta}{2} \right)^{-1} \sin^2\left(\frac{q\theta}{2} \right),
\end{aligned}
\right\}
\tag{9}
$$

one obtains for each positive integer q:

$$\sum_{n=-\infty}^{\infty} \gamma_n^{(q)} e^{-in\theta} = \left(q \sin^2 \frac{\theta}{2} \right)^{-1} \sin^2\left(\frac{q\theta}{2} \right). \tag{10}$$

Let us denote either side of (10) by $\rho_q(e^{i\theta})$. We notice from the right side of (10) that ρ_q is non-negative on $\mathbb{E}$. From (9) it follows easily that

$$\int_{\mathbb{E}} \rho_q \, d\lambda = 1 \tag{11}$$

(λ as usual being normalized Haar measure on $\mathbb{E}$). Furthermore, by (7),

$$\sum_{n=-\infty}^{\infty} \gamma_n^{(q)} \zeta_n = T_{\rho_q}\xi. \tag{12}$$

Now we see from the right side of (10) that, for large q, ρ_q is "peaked" around 1 and very small elsewhere. More precisely, in view of (11), we can write $\rho_{\bar{q}} = \sigma_q + \tau_q$, where $\sigma_q, \tau_q \in \mathscr{L}_+(\mathbb{E})$ and (i) $\sigma_q \to 0$ uniformly on E, (ii) $\int \tau_q \, d\lambda \to 1$, and (iii) the compact support of τ_q shrinks down to 1 as $q \to \infty$. Thus $T_{\sigma_q}\xi \to 0$ by VIII.11.21; and $T_{\tau_q}\xi \to \xi$ by VIII.11.23. So $T_{\rho_q}\xi \to \xi$, which by (12) implies (8). ∎

8.12. *Remark.*   The T to which the preceding proposition was first applied by Fejér was the representation of $\mathbb{E}$ on $\mathscr{L}(\mathbb{E})$ (equipped with the supremum norm) by left translation. It should be noticed that in this case the partial sums $\eta_p = \sum_{n=-p}^{p} \xi_n$ do not in general converge to ξ (in $\mathscr{L}(\mathbb{E})$). There are well-known examples of functions ξ in $\mathscr{L}(\mathbb{E})$ whose Fourier series are divergent at certain points of $\mathbb{E}$ (see for instance Titchmarsh [1], §13.4).

9.   The Structure of the Regular Representation and the Plancherel Formula

9.1.   Let X denote the Hilbert space $\mathscr{L}_2(\lambda)$, and R and R' the left-regular and right-regular representations respectively of G on X (see VIII.11.26). Let $\mathscr{E}(\sigma)$ be as in 3.5; and let w_σ be the unit element of $\mathscr{E}(\sigma^*)$ (as in 7.1).

The following theorem gives the harmonic analysis of R and R'.

Theorem.   *Suppose that $\sigma \in \mathscr{J}(G)$. Then*:

(I)   *The σ-subspace of X with respect to R is $\mathscr{E}(\sigma^*)$.*

(II)   *The σ-subspace of X with respect to R' is $\mathscr{E}(\sigma)$.*

(III)   *If $f \in X$, $w_\sigma * f$ is the σ-component of f with respect to R, and also the σ^*-component of f with respect to R'.*

(IV)   *The multiplicity of σ in both R and R' is equal to the dimension of σ.*

Proof.   By VIII.11.27(12)

$$R_{w_\sigma} f = w_\sigma * f. \tag{1}$$

Applying (1) to the reverse group of G, and using 5.2(2) and the centrality of the w_σ, we get

$$R'_{w_\sigma} f = f * w_{\sigma^*} = w_{\sigma^*} * f. \tag{2}$$

Equations (1) and (2) assert (III) (see 8.1). Since w_σ is the unit element of the two-sided ideal $\mathscr{E}(\sigma^*)$, (1), (2), and 7.4 imply (I) and (II). Since $\mathscr{E}(\sigma)$ and $\mathscr{E}(\sigma^*)$ are total matrix algebras of dimension $(\dim(\sigma))^2$, (IV) follows from 7.6.   ∎

Remark.   For finite groups this theorem is of course a special case of IV.5.20.

9.2. Let S be an irreducible representation of G, of class σ in $\mathscr{J}(G)$. As usual, S also denotes the integrated form of S. If $\mu \in \mathscr{M}_r(G)$, we have for $x \in G$

$$(w_\sigma * \mu)(x) = \int w_\sigma(xy^{-1})d\mu y$$

$$= \dim(\sigma) \int \overline{\chi_\sigma(xy^{-1})}d\mu y$$

$$= \dim(\sigma) \int \chi_\sigma(yx^{-1})d\mu y \qquad \text{(see 5.2(2))}$$

$$= \dim(\sigma) \int \operatorname{Tr}(S_y S_{x^{-1}})d\mu y$$

$$= \dim(\sigma) \operatorname{Tr}(S_\mu S_{x^{-1}}). \qquad (3)$$

In particular, if we replace μ by a function f in X, we obtain from (3) and 9.1 the following interesting formula for the σ-component f_σ of f with respect to the left-regular representation:

$$f_\sigma(x) = \dim(\sigma) \operatorname{Tr}(S_f S_{x^{-1}}) \qquad (x \in G). \qquad (4)$$

In view of the relation $f_\sigma = w_\sigma * f$, we have, for $f, g \in X$,

$$(f * g)_\sigma = f_\sigma * g_\sigma, \qquad (f^*)_\sigma = (f_\sigma)^*. \qquad (5)$$

Remark. In the special case that σ is one-dimensional, (4) becomes

$$f_\sigma(x) = \sigma(f)\overline{\sigma(x)}, \qquad \text{where } \sigma(f) = \int f(x)\sigma(x)d\lambda x. \qquad (6)$$

9.3. If $f \in X$, then by 8.9, the f_σ ($\sigma \in \mathscr{J}(G)$) are orthogonal in X; and

$$f = \sum_{\sigma \in \mathscr{J}(G)} f_\sigma \quad \text{(unconditional convergence)}. \qquad (7)$$

In this context the harmonic analysis (7) is referred to as the *Fourier analysis* of f.

9.4. Remark. Everything said in 9.1, 9.2 continues to hold if X is replaced by any one of the Banach spaces $\mathscr{L}_p(\lambda)$ ($1 \le p < \infty$) or $\mathscr{L}(G)$, and R and R' by the left-regular and right-regular representations of G on X. As we have pointed out in §8, however, in that case (7) must be replaced by the weaker statement (see 8.3):

$$f = \lim_\alpha \sum_{\sigma \in \mathscr{J}(G)} \gamma^\alpha(\sigma)f_\sigma. \qquad (8)$$

9.5.   We have seen in 9.1 that $\mathscr{E}(\sigma)$ is the σ^*-subspace of X with respect to R and the σ-subspace of X with respect to R'. It is easy to pick out subspaces of $\mathscr{E}(\sigma)$ on which R and R' act *irreducibly*. This will provide an alternative proof of 9.1.

Let S be a representation of G of class σ in $\mathscr{J}(G)$, and put $n = \dim(\sigma)$. Let $\{S_{ij}(x)\}$ $(i, j = 1, \ldots, n)$ be the matrix of S_x with respect to some fixed basis. For each $i = 1, \ldots, n$ let J_i be the linear span of $\{S_{ji} : j = 1, \ldots, n\}$, and J_i' the linear span of $\{S_{ij} : j = 1, \ldots, n\}$. Thus J_i and J_i' are a minimal left ideal and a minimal right ideal respectively of $\mathscr{E}(\sigma)$. By 4.4

$$J_i = \mathscr{L}(G) * S_{ii}, \qquad J_i' = S_{ii} * \mathscr{L}(G). \tag{9}$$

If $x \in G$, we have for each $i, j = 1, \ldots, n$ and y in G

$$(R_x(S_{ij}))(y) = S_{ij}(x^{-1}y)$$

$$= \sum_{k=1}^{n} S_{ik}(x^{-1})S_{kj}(y)$$

$$= \sum_{k=1}^{n} S_{ki}^*(x)S_{kj}(y) \qquad \text{(see 3.7(1)),}$$

or

$$R_x(S_{ij}) = \sum_{k=1}^{n} S_{ki}^*(x)S_{kj}. \tag{10}$$

This shows that for each $j = 1, \ldots, n$ the subrepresentation of R acting on J_j is of class σ^*. Thus

$$\mathscr{E}(\sigma) = \sum_{j=1}^{n} {}^{\oplus} J_j$$

is a direct sum decomposition of $\mathscr{E}(\sigma)$ into subspaces which are irreducible of class σ^* under R.

Similarly

$$(R_x'(S_{ij}))(y) = S_{ij}(yx) = \sum_{k=1}^{n} S_{ik}(y)S_{kj}(x),$$

whence

$$R_x'(S_{ij}) = \sum_{k=1}^{n} S_{kj}(x)S_{ik}. \tag{11}$$

Thus for each i the subrepresentation of R' acting on J_i' is of class σ; and

$$\mathscr{E}(\sigma) = \sum_{i=1}^{n} {}^{\oplus} J_i'$$

is a direct sum decomposition of $\mathscr{E}(\sigma)$ into subspaces which are irreducible of class σ under R'.

9.6. Theorem (Plancherel). *For each σ in $\mathscr{I}(G)$ let us choose a unitary representation $S^{(\sigma)}$ of G of class σ. Then, if $f \in \mathscr{L}_2(\lambda)$,*

$$\|f\|_2^2 = \sum_{\sigma \in \mathscr{I}(G)} \dim(\sigma) \mathrm{Tr}((S_f^{(\sigma)})^* S_f^{(\sigma)}). \tag{12}$$

Proof. Let f_σ be as in 9.2. One checks that $\|g\|_2^2 = (g^* * g)(e)$ for each g in $\mathscr{L}_2(\lambda)$. Consequently, if $\sigma \in \mathscr{I}(G)$,

$$\|f_\sigma\|_2^2 = (f_\sigma^* * f_\sigma)(e)$$

$$= (f^* * f)_\sigma(e) \qquad \text{(by (5))}$$

$$= \dim(\sigma)\mathrm{Tr}(S_{f^* * f}^{(\sigma)}) \qquad \text{(by (4))}$$

$$= \dim(\sigma)\mathrm{Tr}((S_f^{(\sigma)})^* S_f^{(\sigma)}). \qquad \text{(13)}$$

Now it follows from (7) that

$$\|f\|_2^2 = \sum_{\sigma \in \mathscr{I}(G)} \|f_\sigma\|_2^2.$$

Combining this with (13), we obtain (12). ∎

Equation (12) is known as the *Plancherel formula* for G.

9.7. If G is Abelian, so that the elements of $\mathscr{I}(G)$ are all one-dimensional, the Plancherel formula becomes

$$\|f\|_2^2 = \sum_{\chi \in \mathscr{I}(G)} |\chi(f)|^2 \qquad (f \in \mathscr{L}_2(\lambda)), \qquad (14)$$

where $\chi(f) = \int f(x)\chi(x)d\lambda x$.

10. Induced Representations and the Frobenius Reciprocity Theorem

10.1. In this section we define induced representations of compact groups and prove the Frobenius Reciprocity Theorem. The discussion of induced representations given here, in the context of compact groups, is intended to serve as a motivating introduction to the much more general discussion of induced representations in the context of arbitrary Banach *-algebraic bundles over locally compact groups, given in Chapter XI. Our reason for inserting this introductory discussion here rather than elsewhere is that the Frobenius Reciprocity Theorem—one of the most important theorems on induced representations—does not have a perfectly satisfactory generalization much beyond the compact context. Other theorems on induced representations which are valid in non-compact situations are postponed until Chapter XI.

10.2.   Let us fix a closed subgroup H of the compact group G. The inducing operation is going to be a construction by which one passes from a representation V of H to a representation S of the bigger group G. We will say that S is *induced* from V. The following discussion is designed to motivate this construction.

Let us denote G/H by M, and the special coset H in M by m_0. Thus M is a compact Hausdorff space on which G acts transitively; in fact it is essentially the most general transitive G-space (see III.3.10). Now let X be the linear space of complex functions on M. The action of G on M generates a corresponding representation T of G on X:

$$(T_x f)(m) = f(x^{-1}m) \qquad (f \in X; x \in G; m \in M). \qquad (1)$$

Notice that X can also be naturally exhibited as a space of functions on G. Indeed, let $\pi: G \to M$ be the quotient surjection $x \mapsto xm_0$ of G onto M. Then $X_0 = \{f \circ \pi : f \in X\}$ is the linear space of all complex functions g on G satisfying

$$g(xh) = g(x) \qquad\qquad (x \in G; h \in H); \qquad (2)$$

and X and X_0 are linearly isomorphic under $\Phi: f \to f \circ \pi$. Under Φ, the operators T_x go into the operators T_x^0 of left translation on the group:

$$(T_x^0 g)(y) = g(x^{-1}y) \qquad (g \in X_0; x, y \in G). \qquad (3)$$

So far we have said nothing very new. But now let us look at functions on M of a more general kind suggested by physics and differential geometry. Suppose for concreteness that G is the orthogonal group $O(3)$ acting in $\mathbb{R}^3$, and that M is the 2-sphere $\{u \in \mathbb{R}^3 : u_1^2 + u_2^2 + u_3^2 = 1\}$. Certainly G acts transitively on M. Now for each point u of M we can form the two-dimensional real linear space $N(u)$ of all vectors at u which are tangent to M; $N(u)$ is called the *tangent space to M at u*. The $N(u)$ vary of course as u varies; and though they are linearly isomorphic to each other, there is no one *natural* isomorphism connecting $N(u)$ with $N(v)$ for $u \neq v$. A function ϕ on M whose value at each point u of M lies in $N(u)$ is called a *vector field on M*; and the set Y of all vector fields is a real linear space under pointwise addition. Observe that each element x of G, along with its action on M, has a corresponding natural action, which we shall call σ_x, on the space $N = \bigcup \{N(u) : u \in M\}$ of all tangent vectors. Without writing down this action explicitly, we note that

$$\sigma_x(N(u)) = N(xu) \qquad (x \in G; u \in M), \qquad (4)$$

and that $\sigma_x | N(u): N(u) \to N(xu)$ is a linear isomorphism. The action $x \mapsto \sigma_x$ of G on N now generates a natural action $S: x \mapsto S_x$ of G on the space Y of all vector fields:

$$(S_x \phi)(u) = \sigma_x(\phi(x^{-1}u)) \qquad (\phi \in Y; x \in G; u \in M) \qquad (5)$$

This S is of course a representation of G (over the real field) on Y.

Our next goal is to exhibit (5) in a different form, analogous to the passage from (1) to (2), (3). To do this, we consider the fixed point $u^0 = \langle 0, 0, 1 \rangle$ of M (the "north pole"); and denote by H the stability subgroup $\{h \in G : hu^0 = u^0\}$ for u^0. If $h \in H$, then by (4) $\sigma_h(N(u^0)) = N(u^0)$; and the equation $V_h = \sigma_h | N(u^0)$ $(h \in H)$ defines V as a two-dimensional real representation of H on $N(u_0)$. Now, given $\phi \in Y$, set

$$\gamma(x) = \sigma_x^{-1}(\phi(xu^0)) \qquad (x \in G). \qquad (6)$$

By (4) γ maps G into $N(u^0)$. If $x \in G$ and $h \in H$, $\gamma(xh) = \sigma_{xh}^{-1}(\phi(xhu^0)) = V_{h^{-1}}\sigma_x^{-1}(\phi(xu^0)) = V_{h^{-1}}(\gamma(x))$; or

$$\gamma(xh) = V_{h^{-1}}(\gamma(x)) \qquad (x \in G; h \in H). \qquad (7)$$

Conversely, suppose that $\gamma: G \to N(u^0)$ is a function satisfying (7); then the equation

$$\phi(xu^0) = \sigma_x(\gamma(x)) \qquad (x \in G) \qquad (8)$$

defines a vector field ϕ on M, since by (7) the right side of (8) is unaltered on replacing x by xh $(h \in H)$ and so depends only on xu^0. One checks that the correspondences $\phi \mapsto \gamma$ of (6) and $\gamma \mapsto \phi$ of (8) are inverse to each other, and hence that they define a linear isomorphism of Y onto the linear space Y_0 of all those functions $\gamma: G \to N(u^0)$ which satisfy (7). One also verifies that, under this correspondence, S_x goes into the operation S_x^0 of left translation on Y_0:

$$(S_x^0 \gamma)(y) = \gamma(x^{-1}y) \qquad (\gamma \in Y_0; x, y \in G). \qquad (9)$$

Thus S^0, which is equivalent to S, acts by left translation on the space of all those functions $\gamma: G \to N(u^0)$ which satisfy (7).

Examining the above description of S^0 we uncover a surprising fact: In order to describe S to within equivalence, it was not necessary to know all about σ; it was only necessary to know the action V of the stability subgroup H on the tangent space $N(u^0)$ at the single point u^0. This suggests an important formal generalization of the above train of thought. Notice that H is simply the orthogonal group in 2-space. Now in tensor analysis we learn to recognize vectors as simply one of a large class of quantities (vectors, pseudovectors, tensors of rank p, etc.) each of which is characterized by a certain (real) representation of the orthogonal group of the space with which the quantity is associated. Suppose that in the definition (7), (9) of S^0 we were to replace V (acting in $N(u^0)$) by another real representation W (acting in a different space $X(W)$) of H. For example W might be the representation of H defining tensors of rank p. With this replacement, the representation S^0 of G obtained from (7) and (9) might be expected to be the action of G on the space of all "rank p tensor fields" on M; and this is indeed the case.

As a very simple example, suppose that W is taken to be the trivial representation of H on the one-dimensional space, which characterizes *scalars*. Then (7) becomes (2); and the present construction gives us the representation T by "scalar fields," i.e., numerical functions on M, with which the discussion started.

This discussion may help to suggest the importance of the construction which we will now obtain by placing (7) and (9) in their most general natural context. Since our main interest is in unitary representations, the context of the above discussion will be modified in two respects: First, the representations of G and H will be unitary (in particular, over the complex field rather than the real field). Second, in order to achieve this, we must replace Y_0 by a space of *square-integrable* functions on G. We then obtain the following general definition of induced unitary representations of compact groups.

10.3.   As before, G is any fixed compact group. Throughout this section H is a fixed closed subgroup of G, with normalized Haar measure ν.

10.4.   Suppose that V is a unitary representation of the subgroup H. Since $X(V)$ is a Hilbert space, the space $\mathscr{L}_2(\lambda; X(V))$ of all functions $f: G \to X(V)$ which are square-integrable with respect to λ is a Hilbert space (see II.4.7). By the argument of VIII.11.26 one shows that the operators of left translation define a unitary representation T of G on $\mathscr{L}_2(\lambda; X(V))$:

$$(T_x f)(y) = f(x^{-1}y) \qquad (f \in \mathscr{L}_2(\lambda; X(V)); \; x, y \in G). \qquad (10)$$

Let $\mathscr{X}_0$ be the linear space of those continuous functions $f: G \to X(V)$ which satisfy

$$f(xh) = V_{h^{-1}}(f(x)) \tag{11}$$

for all x in G and h in H; and let $\mathscr{X}$ denote the closure of $\mathscr{X}_0$ in $\mathscr{L}_2(\lambda; X(V))$. Thus $\mathscr{X}$, with the inner product of $\mathscr{L}_2(\lambda; X(V))$, is a Hilbert space in its own right. Now $\mathscr{X}_0$ is stable under T. For, if $f \in \mathscr{X}_0$, $x, y \in G$, and $h \in H$, then $(T_x f)(yh) = f(x^{-1}yh) = V_{h^{-1}}(f(x^{-1}y)) = V_{h^{-1}}((T_x f)(y))$; and so $T_x f \in \mathscr{X}_0$. Consequently $\mathscr{X}$ is also T-stable.

Definition.   The subrepresentation S of T acting on $\mathscr{X}$ is called the (*unitary*) *representation of G induced by V*, and is denoted by $\mathrm{Ind}_{H \uparrow G}(V)$, or simply by $\mathrm{Ind}(V)$ (if no confusion can arise concerning the groups involved).

Remark 1.   Equations (11) and (10) are just (7) and (9) respectively in the present context.

Remark 2.   If $H = \{e\}$ and V is the trivial representation of $\{e\}$, then $\mathrm{Ind}(V)$ is just the left-regular representation of G.

Remark 3.   If $H = G$, the reader will easily check that $\mathrm{Ind}(V) \cong V$.

Remark 4.   Evidently the unitary equivalence class σ of $\mathrm{Ind}_{H \uparrow G}(V)$ depends only on the unitary equivalence class ρ of V. Therefore we often write

$$\sigma = \underset{H \uparrow G}{\mathrm{Ind}}\,(\rho).$$

10.5. Proposition*.   *$\mathscr{X}$ coincides with the set of those g in $\mathscr{L}_2(\lambda; X(V))$ such that, for some f in $\mathscr{L}_2(\lambda; X(V))$ coinciding with g except on a λ-null set, (11) holds for all x in G and h in H.*

Remark.   The above proposition shows that the use of the space $\mathscr{X}_0$ as an intermediary in defining $\mathscr{X}$ in 10.4 was not really necessary. Our purpose in proceeding as we did was to make it clear with a minimum amount of argument that the set of continuous functions in $\mathscr{X}$ is dense in $\mathscr{X}$.

10.6.   Notice that the natural injection of $\mathscr{M}_r(H)$ into $\mathscr{M}_r(G)$ (see II.8.5) is a one-to-one *-homomorphism of the measure algebras. We shall in fact identify $\mathscr{M}_r(H)$ with a *-subalgebra of $\mathscr{M}_r(G)$ via the natural injection. Since $\mathscr{L}(H)$ is already identified with a *-ideal of $\mathscr{M}_r(H)$, $\mathscr{L}(H)$ thus becomes a

*-subalgebra of $\mathcal{M}_r(G)$. From the definition of convolution in $\mathcal{M}_r(G)$ (see III.11.5), we verify that, if $f \in \mathscr{L}(G)$ and $\phi \in \mathscr{L}(H)$, then $\phi * f$ and $f * \phi$ are in $\mathscr{L}(G)$ and, for $x \in G$,

$$(\phi * f)(x) = \int_H \phi(h) f(h^{-1}x) dvh, \tag{12}$$

$$(f * \phi)(x) = \int_H f(xh^{-1}) \phi(h) dvh. \tag{13}$$

10.7.   The Frobenius Reciprocity Theorem is concerned with the harmonic analysis of $\mathrm{Ind}(V)$ when V is an irreducible unitary representation of H. The main step in proving it will be to express $\mathrm{Ind}(V)$, in that case, as a subrepresentation of the regular representation of G.

Let V be a fixed irreducible unitary representation of H, of dimension n. Let $(V_{ij}(h))$ $(i, j = 1, \ldots, n)$ be the matrix of V_h with respect to some fixed orthonormal basis of $X(V)$; and put $v_{ij} = nV_{ij}$ $(i, j = 1, \ldots, n)$. By 4.5 the $\{v_{ij}\}$ form a canonical basis of the total matrix *-algebra $\mathscr{E}(V) \subset \mathscr{L}(H)$.

Let $\tilde{\ }$ denote the inversion of a function on H: $\tilde{\phi}(h) = \phi(h^{-1})$.

Since v_{11} is a self-adjoint idempotent element in $\mathscr{L}(H)$, the same is true of $\tilde{v}_{11}$; and so $f \mapsto f * \tilde{v}_{11}$ is a projection operator on $\mathscr{L}_2(\lambda)$.

Proposition.   *The closed subspace $Z = \mathscr{L}_2(\lambda) * \tilde{v}_{11}$ of $\mathscr{L}_2(\lambda)$ is stable under the left-regular representation R of G; and the subrepresentation of R acting on Z is unitarily equivalent to $\mathrm{Ind}_{H \uparrow G}(V)$.*

Proof.   Let J be the right ideal of $\mathscr{L}(H)$ spanned by $v_{11}, v_{12}, \ldots, v_{1n}$; and let W be the right-regular representation of H. By 9.5 the subrepresentation of W acting on J is equivalent to V. We shall in fact identify $X(V)$ with J, and V with the action of W on J. Thus the space $\mathscr{X}_0$ (which appeared in the definition of $\mathrm{Ind}(V)$ in 10.4) now consists of all continuous functions $\phi : G \to J$ satisfying

$$\phi(xh) = W_{h^{-1}}(\phi(x)) \qquad (x \in G; h \in H). \tag{14}$$

We now define a linear map $F : \mathscr{X}_0 \to \mathscr{L}(G)$ by means of the formula:

$$F_\phi(x) = \phi(x)(e) \qquad (\phi \in \mathscr{X}_0; x \in G).$$

Let $\phi \in \mathscr{X}_0$. If $x \in G$ and $h \in H$, we have

$$F_\phi(xh) = \phi(xh)(e) = [W_{h^{-1}}(\phi(x))](e)$$

$$= \phi(x)(h^{-1}). \tag{15}$$

Hence

$$\|F_\phi\|_2^2 = \int |F_\phi(x)|^2 \, d\lambda x$$

$$= \iint |F_\phi(xh)|^2 \, dvh \, d\lambda x$$

$$= \iint |\phi(x)(h^{-1})|^2 \, dvh \, d\lambda x \qquad \text{(by (15))}$$

$$= \int \|\phi(x)\|_2^2 \, d\lambda x$$

$$= \|\phi\|_2^2;$$

and $F: \mathscr{X}_0 \to \mathscr{L}(G)$ is a linear isometry with respect to the $\mathscr{L}_2$ norms. Thus F extends to a linear isometry (also called F) of $\mathscr{X} = X(\mathrm{Ind}(V))$ into $\mathscr{L}_2(\lambda)$.

Putting $S = \mathrm{Ind}(V)$, we have for $\phi \in \mathscr{X}_0$ and $x, y \in G$:

$$(FS_y\phi)(x) = [(S_y\phi)(x)](e) = \phi(y^{-1}x)(e)$$

$$= F_\phi(y^{-1}x) = (R_y F_\phi)(x).$$

This shows that $FS_y = R_y F$; so F intertwines S and R.

It remains only to show that

$$F(\mathscr{X}) = Z. \tag{16}$$

If $\phi \in \mathscr{X}_0$ and $x \in G$, we have

$$(F_\phi * \tilde{v}_{11})(x) = \int F_\phi(xh^{-1})\tilde{v}_{11}(h)dvh \qquad \text{(by (13))}$$

$$= \int \phi(x)(h)v_{11}(h^{-1})dvh \qquad \text{(by (15))}$$

$$= (v_{11} * \phi(x))(e)$$

$$= \phi(x)(e) \qquad \text{(since } \phi(x) \in J)$$

$$= F_\phi(x).$$

Therefore $F_\phi * \tilde{v}_{11} = F_\phi$, whence $\phi \in \mathscr{X}_0 \Rightarrow F_\phi \in Z$. Since $\mathscr{X}_0$ is dense in $\mathscr{X}$, this gives

$$F(\mathscr{X}) \subset Z. \tag{17}$$

Conversely, suppose that $f \in \mathscr{L}(G) * \tilde{v}_{11}$; that is, $f \in \mathscr{L}(G)$ and $f * \tilde{v}_{11} = f$. Put $\phi(x)(h) = f(xh^{-1})$ $(x \in G; h \in H)$; so that $\phi: G \to \mathscr{L}(H)$. For each $x \in G$ and h in H,

$$(v_{11} * \phi(x))(h) = \int v_{11}(k) f(xh^{-1}k) dvk$$

$$= (f * \tilde{v}_{11})(xh^{-1})$$

$$= f(xh^{-1})$$

$$= \phi(x)(h)$$

Consequently $v_{11} * \phi(x) = \phi(x)$ for all x, showing that ϕ maps G into J. Clearly $\phi: G \to J$ is continuous. If $x \in G$ and $h, k \in H$,

$$\phi(xh)(k) = f(xhk^{-1}) = \phi(x)(kh^{-1})$$

$$= [W_{h^{-1}}(\phi(x))](k),$$

whence $\phi(xh) = W_{h^{-1}}(\phi(x))$. Combining these facts we see that $\phi \in \mathscr{X}_0$. Evidently $F_\phi = f$. So we have shown that $f \in F(\mathscr{X})$. Now f runs over $\mathscr{L}(G) * \tilde{v}_{11}$, which is dense in Z. Therefore

$$Z \subset F(\mathscr{X}).$$

Combining this with (17) we get (16). ■

10.8. Frobenius Reciprocity Theorem. *Let H be a closed subgroup of the compact group G; and let $\rho \in \mathscr{J}(H)$, $\tau \in \mathscr{J}(G)$. Then the multiplicity of τ in $\mathrm{Ind}_{H \uparrow G}(\rho)$ is equal to the multiplicity of ρ in $\tau | H$.*

Remark. By 8.10 $\mathrm{Ind}_{H \uparrow G}(\rho)$ is discretely decomposable, and the multiplicity theory of §VI.14 can be applied to it (see VIII.9.9). It is clear that, as regards finite multiplicities (which are all that appear in the present theorem), there is no difference between the involutory multiplicity theory of §VI.14 and the purely algebraic multiplicity theory of §IV.2.

Proof. Choose a unitary representation V of H of class ρ; and let the v_{ij} be as in 10.7. Similarly, let T be a unitary representation of G of class τ; and put $t_{ij} = \dim(\tau) T_{ij}$, where $(T_{ij}(x))$ is the matrix of T_x with respect to some fixed orthonormal basis of $X(T)$. By 4.5 the $\tilde{t}_{ji}$ form a canonical basis of the total matrix *-algebra $\mathscr{E}(\tau^*)$.

Let R and R' be the left-regular and right-regular representation of G respectively. Since $R_{\tilde{t}_{11}} f = \tilde{t}_{11} * f$ (see VIII.11.27(11)), the combination of 7.7 and 10.7 shows that the multiplicity of τ in $\mathrm{Ind}(\rho)$ is

$$\dim(\tilde{t}_{11} * \mathscr{L}_2(\lambda) * \tilde{v}_{11}). \tag{18}$$

On the other hand, applying 9.5 to τ^*, we conclude that the subrepresentation of R' acting in $\tilde{t}_{11} * \mathscr{L}_2(\lambda)$ is of class τ^*. Since $(R'|H)_{v_{11}} f = f * \tilde{v}_{11}$ (compare VIII.11.27(12)), it follows from 7.7 and the preceding sentence that the multiplicity of ρ^* in $\tau^*|H$ is the quantity (18). Since the multiplicity of ρ^* in $\tau^*|H$ is the same as the multiplicity of ρ in $\tau|H$, we have shown that the latter quantity is equal to (18), that is, to the multiplicity of τ in $\mathrm{Ind}(\rho)$. ∎

10.9. Here is a very simple illustration of the Frobenius Reciprocity Theorem. If $\tau \in \mathscr{I}(G)$, the multiplicity of the trivial representation of $\{e\}$ in $\tau|\{e\}$ is just $\dim(\tau)$. Hence, applying the Frobenius Reciprocity Theorem to the regular representation R of G (and remembering Remark 2 of 10.4), we conclude that the multiplicity of τ in R is equal to $\dim(\tau)$–a fact already known from 9.1.

10.10. Let μ be the normalized G-invariant measure on G/H (given of course by $\mu(A) = \lambda(\pi^{-1}(A))$, where $\pi: G \to G/H$ is the canonical quotient map and A is a Borel subset of G/H). Let S be the natural unitary representation of G on $\mathscr{L}_2(\mu)$:

$$(S_x f)(yH) = f(x^{-1}yH) \qquad (f \in \mathscr{L}_2(\mu); \ x, y \in G).$$

The reader will verify that S is unitarily equivalent to $\mathrm{Ind}_{H \uparrow G}(\rho)$, where ρ is the trivial representation of H. Hence the Frobenius Reciprocity Theorem gives the following information about the harmonic analysis of S, generalizing 10.9:

For each τ in $\mathscr{I}(G)$, the multiplicity of τ in S is equal to the multiplicity in $\tau|H$ of the trivial representation of H.

10.11. As an example of 10.10, we consider the so-called *two-sided regular representation* W of $G \times G$, acting by both left and right translation on $\mathscr{L}_2(\lambda)$:

$$X(W) = \mathscr{L}_2(\lambda),$$

$$(W_{\langle x, y \rangle} f)(z) = f(x^{-1}zy) \qquad (f \in \mathscr{L}_2(\lambda); \ x, y, z \in G).$$

Notice that, if D denotes the diagonal subgroup $\{\langle x, x \rangle : x \in G\}$ of $G \times G$, the homogeneous space $(G \times G)/D$ is homeomorphic with G under the homeomorphism $\Phi: \langle r, s \rangle D \mapsto rs^{-1}$, and that under Φ the natural action of $\langle x, y \rangle$

on $(G \times G)/D$ goes into the permutation $z \mapsto xzy^{-1}$ of G. It follows from this, as in 10.10, that W is unitarily equivalent to $\mathrm{Ind}_{D \uparrow (G \times G)} (\rho)$, where ρ is the trivial representation of D.

Now by 6.7 $\mathscr{J}(G \times G)$ consists of all $S \underset{\circ}{\otimes} T$, where S and T run over $\mathscr{J}(G)$. If we identify D with G (via the isomorphism $x \mapsto \langle x, x \rangle$), $(S \underset{\circ}{\otimes} T)|D$ becomes the (inner) tensor product $S \otimes T$. By 5.6 the trivial representation of G occurs in $S \otimes T$ if and only if $T \cong S^*$, and then with multiplicity 1. So the application of the Frobenius Reciprocity Theorem shows that

$$W \cong \sum_{\tau \in \mathscr{J}(G)}^{\oplus} (\tau \underset{\circ}{\otimes} \tau^*).$$

Notice that, in view of 9.1, the $(\tau \underset{\circ}{\otimes} \tau^*)$-subspace of $\mathscr{L}_2(\lambda)$ (with respect to W) is $\mathscr{E}(\tau^*)$.

10.12. As another example of the use of the Frobenius Reciprocity Theorem, we point out an alternative proof of 6.8. If $V \in \mathscr{J}(H)$, then by 8.10 some T in $\mathscr{J}(G)$ occurs in $\mathrm{Ind}_{H \uparrow G} (V)$. By the Frobenius Reciprocity Theorem, this implies that V occurs in $T|H$.

An Algebraic Generalization of Induced Representations
and Frobenius Reciprocity for Finite Groups

10.13. For finite groups at least, the concept of induced representations and the Frobenius Reciprocity Theorem emerge as a special case of a much more general purely algebraic theory, of which we shall here sketch the beginnings. The reader should bear in mind the algebraic construction to be sketched in this number when he comes to study the somewhat different generalization of the inducing construction to be presented in Chapter XI.

Let A be any algebra with unit 1, B any subalgebra of A, and S an (algebraic) representation of B. Putting X for $X(S)$, we form the tensor product $A \otimes X$, and denote by J the linear span in $A \otimes X(S)$ of the set of all elements of the form

$$a \otimes S_b \xi - ab \otimes \xi \qquad (a \in A; b \in B; \xi \in X).$$

The quotient $(A \otimes X)/J$ is called the *B-tensor product* of A and X, and is denoted by $A \otimes_B X$. If $a \in A$ and $\xi \in X$, the image of $a \otimes \xi$ in $A \otimes_B X$ is written $a \otimes_B \xi$.

Now for each c in A the linear endomorphism $a \otimes \xi \mapsto ca \otimes \xi$ of $A \otimes X$ leaves J stable. So the equation

$$T_c(a \otimes_B \xi) = ca \otimes_B \xi \quad (c, a \in A; \xi \in X) \qquad (19)$$

defines T as an algebraic representation of A on $X(T) = A \otimes_B X$. We shall refer to T as (algebraically) induced by S, and write

$$T = \operatorname*{Ind}_{B \uparrow A} (S).$$

Notice that $T_{\downarrow}$ is the identity endomorphism of $X(T)$.

This simple construction contains a built-in Frobenius Reciprocity relation. Indeed, let us take any (algebraic) representation V of A whatsoever satisfying

$$V_{\downarrow} = \text{identity endomorphism of } X(V). \tag{20}$$

We claim that there is a natural linear isomorphism between the space $\mathscr{I}(S, V|B)$ of all linear $S, V|B$ intertwining maps and the space $\mathscr{I}(T, V)$ of all linear T, V intertwining maps. To see this, we first suppose that $\phi \in \mathscr{I}(S, V|B)$. Then, in view of the intertwining property of ϕ, the linear map $f_0 : A \otimes X \to X(V)$ given by $f_0(a \otimes \xi) = V_a \phi(\xi)$ vanishes on J; and hence the equation

$$f(a \otimes_B \xi) = V_a \phi(\xi) \qquad (a \in A; \xi \in X) \tag{21}$$

defines a linear map $f : X(T) \to X(V)$. Evidently $f \in \mathscr{I}(T, V)$. Conversely, let $f \in \mathscr{I}(T, V)$; and define $\phi : X \to X(V)$ as follows:

$$\phi(\xi) = f(1 \otimes_B \xi) \qquad (\xi \in X). \tag{22}$$

Then a simple calculation shows that $\phi \in \mathscr{I}(S, V|B)$. The two constructions $\phi \mapsto f$ of (21) and $f \mapsto \phi$ of (22) are easily seen to be each other's inverses. Therefore they set up a linear isomorphism of $\mathscr{I}(S, V|B)$ and $\mathscr{I}(T, V)$. In particular,

$$\dim \mathscr{I}(S, V|B) = \dim \mathscr{I}(T, V). \tag{23}$$

Equation (23) is a Frobenius Reciprocity relation. To see why, let us assume that S and V are both finite-dimensional and irreducible; and suppose it is somehow known that $V|B$ and T are completely reducible. Then it follows from IV.2.16 and IV.4.10 that $\dim \mathscr{I}(S, V|B)$ is equal to the multiplicity $m(S; V|B)$ of S in $V|B$, and $\dim \mathscr{I}(T, V)$ is equal to the multiplicity $m(V; T)$ of V in T. Thus (23) takes the form of the Frobenius assertion 10.8:

$$m(S; V|B) = m\left(V; \operatorname*{Ind}_{B \uparrow A} (S) \right) \tag{24}$$

10.14. We shall now observe that the definition of induced representations given in 10.4 is equivalent, in the case of finite groups, to that of 10.13. Thus, for finite groups we shall have given an alternative proof of the Frobenius Reciprocity Theorem.

Let G be a finite group, H a subgroup of G, and Γ a fixed transversal for G/H. Let A and B be the group algebras (under convolution $*$) of G and H respectively (B being considered as a subalgebra of A). For each x in Γ let δ_x be the element $y \mapsto \delta_{xy}$ of A. Then it is easy to verify that $\{\delta_x : x \in \Gamma\}$ is a "basis" of A with respect to B, in the sense that every element f of A can be written in one and only one way in the form

$$f = \sum_{x \in \Gamma} (\delta_x * b_x), \qquad \text{where } b_x \in B. \tag{25}$$

Now let S be an algebraic representation of H, acting in the linear space X; and denote also by S the corresponding representation of B. Denote by $\mathscr{X}$ the linear space of all functions $\phi : G \to X$ satisfying

$$\phi(xh) = S_{h^{-1}}(\phi(x)) \qquad (x \in G; h \in H); \tag{26}$$

and let $F : A \otimes X \to \mathscr{X}$ be the linear map given by

$$[F(f \otimes \xi)](x) = \sum_{h \in H} f(xh) S_h \xi \tag{27}$$

($f \in A; \xi \in X; x \in G$). If the f in (27) is of the form (25), one verifies that, for $x \in \Gamma$,

$$[F(f \otimes \xi)](x) = S_{b_x} \xi. \tag{28}$$

Since an element of $\mathscr{X}$ is determined by its values on Γ, one deduces easily from (28) that

$$\text{range}(F) = \mathscr{X}, \tag{29}$$

$$J \subset \text{Ker}(F) \tag{30}$$

(J being defined as in 10.13). Furthermore, if $\zeta = \sum_{i=1}^{n} (f_i \otimes \xi_i) \in A \otimes X$, where $f_i = \sum_{x \in \Gamma} (\delta_x * b_x^i)$ ($b_x^i \in B$), then

$$\zeta - \sum_{x \in \Gamma} \left[\delta_x \otimes \left(\sum_{i=1}^{n} S_{b_x^i} \xi_i \right) \right] \in J,$$

so that

$$\sum_{i=1}^{n} S_{b_x^i} \xi_i = 0 \quad \text{for each } x \text{ in } \Gamma \Rightarrow \zeta \in J. \tag{31}$$

Combining (28) and (31) we see that $F(\zeta) = 0 \Rightarrow \zeta \in J$, or $\mathrm{Ker}(F) \subset J$. This and (30) imply

$$J = \mathrm{Ker}(F). \tag{32}$$

From (29) and (32) we obtain a linear isomorphism $\tilde{F}$ of $A \otimes_B X$ onto $\mathscr{X}$:

$$[\tilde{F}(f \otimes_B \zeta)](x) = \sum_{h \in H} f(xh) S_h \zeta \tag{33}$$

($f \in A$; $\xi \in X$; $x \in G$).

Now $\mathscr{X}$ is the space of the induced representation $\mathrm{Ind}_{H \uparrow G}(S)$ defined in 10.4; and one checks from (33) that $\tilde{F}$ intertwines the algebraically induced representation $\mathrm{Ind}_{B \uparrow A}(S)$ (defined in 10.13) and $\mathrm{Ind}_{H \uparrow G}(S)$. So $\mathrm{Ind}_{B \uparrow A}(S)$ and $\mathrm{Ind}_{H \uparrow G}(S)$ are equivalent, as we claimed.

Remark. Suppose in 10.13 that A and B are *-algebras and that S is a *-representation. Can one then give to $\mathrm{Ind}_{B \uparrow A}(S)$, as defined in 10.13, the structure of a *-representation? There appears to be no obvious way to do this without invoking some extra ingredient of structure. For example, in the finite group situation of this number, let $(\ ,\ )_0$ be the inner product on $A \otimes_B X$ corresponding under $\tilde{F}$ with the natural inner product on $\mathscr{X}$:

$$(\zeta, \zeta')_0 = \sum_{x \in \Gamma} (\tilde{F}(\zeta)(x), \tilde{F}(\zeta')(x))_X$$

($\zeta, \zeta' \in A \otimes_B X$). Thus $\mathrm{Ind}_{B \uparrow A}(S)$ is a unitary representation of G with respect to $(\ ,\ )_0$. It turns out that $(\ ,\ )_0$ can be elegantly described as follows in terms of the restriction mapping $\rho : f \mapsto f|H$ of A into B:

$$(\tilde{F}(f \otimes_B \zeta), \tilde{F}(g \otimes_B \eta))_0 = (S_{\rho(g^* * f)} \zeta, \eta) \tag{34}$$

($f, g \in A$; $\xi, \eta \in X$). This observation is proved in XI.4.18(20), and constitutes the point of departure for the "involutive" generalization of the inducing process to be presented in Chapter XI.

11. The Representations of $SU(2)$

11.1. In this section we are going to apply the general theory of this chapter to one of the most important of all non-commutative compact groups, namely $SU(2)$, the multiplicative group of all unitary 2×2 complex matrices of determinant 1.

The compactness of $SU(2)$ was proved in III.1.20. The importance of $SU(2)$ in physics arises from the fact that it is the "simply connected covering group" of $SO(3)$, the proper orthogonal group of Euclidean 3-space. The relationship between $SU(2)$ and $SO(3)$ will be treated in §12.

11.2. We shall obtain all the irreducible representations of $SU(2)$. It will turn out that there is exactly one such irreducible representation of each positive dimension. The procedure for showing this will consist of the following steps:

  (A) Exhibit a certain family of irreducible representations of $SU(2)$.
  (B) Calculate their characters, and thus show that the family obtained in (A) satisfies the hypotheses of Theorem 6.1.
  (C) Invoke Theorem 6.1 to show that we have obtained *all* the irreducible representations of $SU(2)$.

11.3. We easily verify that $SU(2)$ consists of all matrices of the form

$$a = \begin{pmatrix} \alpha & \beta \\ -\bar{\beta} & \bar{\alpha} \end{pmatrix}, \tag{1}$$

where α and β are complex numbers satisfying

$$|\alpha|^2 + |\beta|^2 = 1. \tag{2}$$

As usual e stands for the unit matrix $\begin{pmatrix} 1 & 0 \\ 0 & 1 \end{pmatrix}$

Observe that, according to (2), $SU(2)$ is homeomorphic with the 3-sphere. Thus $SU(2)$ is connected (in fact simply connected).

11.4. Let L be the polynomial algebra (over the complex field) in two commuting indeterminates u and v. For each non-negative integer n let L_n be the linear subspace of L consisting of the homogeneous elements of order n. Thus $\dim(L_n) = n + 1$; L_0 is spanned by the unit element 1 of L; and for each positive integer n L_n has the basis $\{u^{n-r}v^r : r = 0, 1, \ldots, n\}$.

Suppose now that a is the element (1) of $SU(2)$. Let D_a be the homomorphism of the algebra L into itself which sends

$$1 \quad \text{into} \quad 1,$$

$$u \quad \text{into} \quad \alpha u - \bar{\beta}v,$$

$$v \quad \text{into} \quad \beta u + \bar{\alpha}v.$$

Evidently $D_{ab} = D_a D_b$ $(a, b \in SU(2))$, and D_e is the identity. So D is a representation of $SU(2)$ on L. It is clear that L_n is D-stable for each non-negative integer n. Let the subrepresentation of D acting on L_n be denoted by $D^{(n)}$. Thus $D^{(0)}$ is the trivial one-dimensional representation of $SU(2)$; and $D^{(1)}$ is the two-dimensional identity representation of $SU(2)$. In general, $D^{(n)}$ is of dimension $n + 1$. Of course $D^{(n)}$ is continuous on $SU(2)$.

It is important to notice the structure of $D^{(n)}|E$, where E is the diagonal subgroup of $SU(2)$ consisting of all the matrices (1) with $\beta = 0$. If $a = \begin{pmatrix} \alpha & 0 \\ 0 & \bar{\alpha} \end{pmatrix} \in E$, we clearly have

$$D_a^{(n)}(u^{n-r}v^r) = \alpha^{n-2r}(u^{n-r}v^r) \qquad (r = 0, 1, \ldots, n). \qquad (3)$$

Consequently, $D^{(n)}|E$ contains each one-dimensional representation

$$\begin{pmatrix} \alpha & 0 \\ 0 & \bar{\alpha} \end{pmatrix} \mapsto \alpha^{n-2r}$$

of E, where $r = 0, 1, \ldots, n$, exactly once—and no others.

If $\alpha = -1$, we find from (3) that

$$D_{\begin{pmatrix} -1 & 0 \\ 0 & -1 \end{pmatrix}}^{(n)} = (-1)^n 1_n \qquad (4)$$

(1_n being the identity operator on L_n).

11.5. We know from 3.2 that $D^{(n)}$ is unitary with respect to some inner product. The following proposition tells us what this inner product is.

Proposition. $D^{(n)}$ is unitary with respect to the following inner product on L_n:

$$\left(\sum_{r=0}^{n} c_r u^{n-r} v^r, \sum_{r=0}^{n} d_r u^{n-r} v^r \right) = \sum_{r=0}^{n} r!(n-r)! c_r \bar{d}_r \qquad (5)$$

$(c_r, d_r \in \mathbb{C})$.

Proof. The following proof involves very little calculation.

Let $Q = L_1 \otimes \cdots \otimes L_1$ (n times) and $F = D^{(1)} \otimes \cdots \otimes D^{(1)}$ (n times). Let σ be the linear operation of *symmetrization* on Q: $\sigma(x_1 \otimes \cdots \otimes x_n) = (n!)^{-1} \sum_\pi x_{\pi(1)} \otimes \cdots \otimes x_{\pi(n)}$ (π running over all permutations of $\{1, \ldots, n\}$). Thus σ is an idempotent operator on Q, whose range Q_σ is the subspace of *symmetric* elements of Q. Also let $\rho: Q \to L_n$ be the linear map given by $\rho(x_1 \otimes \cdots \otimes x_n) = x_1 x_2 \cdots x_n$ ($x_i \in L_1$). Notice that $\rho \circ \sigma = \rho$, σ commutes with all F_a, and ρ intertwines F and $D^{(n)}$. It follows that Q_σ is F-stable, and $\rho(Q_\sigma) = \rho(Q) = L_n$. Since $\dim(L_1) = 2$, one checks that $\dim(Q_\sigma) = n + 1$; thus ρ is a linear isomorphism of Q_σ onto L_n, and so sets up an equivalence between $D^{(n)}$ and the subrepresentation $F^{(\sigma)}$ of F acting on Q_σ.

Now F, hence $F^{(\sigma)}$, is unitary under the inner product

$$((x_1 \otimes \cdots \otimes x_n), (y_1 \otimes \cdots \otimes y_n)) = \prod_{i=1}^{n} (x_i, y_i) \tag{6}$$

$((x_i, y_i)$ being the obvious inner product in L_1). Hence by the preceding paragraph $D^{(n)}$ is unitary with respect to the inner product $(\ ,\ )_0$ into which (6) is mapped by $\rho | Q_\sigma$. Let

$$w_r = \underbrace{u \otimes \cdots \otimes u}_{n-r \text{ times}} \otimes \underbrace{v \otimes \cdots \otimes v}_{r \text{ times}}.$$

We have $u^{n-r}v^r = \rho(w_r) = \rho(\sigma(w_r))$; and so, if $r, s = 0, 1, \ldots, n$,

$$(u^{n-r}v^r, u^{n-s}v^s)_0 = (\sigma(w_r), \sigma(w_s)) = \begin{cases} 0 & \text{if } r \neq s, \\ (n!)^{-2}(r!(n-r)!)^2 v & \text{if } r = s, \end{cases}$$

where v is the number of ways of dividing a set of n objects into two subsets with r and $n - r$ objects respectively; that is, $v = n!(r!(n-r)!)^{-1}$. Therefore:

$$(u^{n-r}v^r, u^{n-r}v^r)_0 = (n!)^{-1}r!(n-r)!,$$

and $(u^{n-r}v^r, u^{n-s}v^s)_0 = 0$ if $r \neq s$. This proves the proposition. ∎

Remark. In the preceding proof we have shown that $D^{(n)}$ is equivalent to the subrepresentation of $D^{(1)} \otimes \cdots \otimes D^{(1)}$ (n times) acting on the subspace Q_σ of symmetric elements of $L_1 \otimes \cdots \otimes L_1$.

11.6. Proposition. $D^{(n)}$ is irreducible for $n = 0, 1, 2, \ldots$

Proof.   Our proof will essentially use the Lie algebra of $SU(2)$, though we shall make no explicit mention of the Lie algebra.

Let ε be a small real number, and put

$$a = \begin{pmatrix} (1 - \varepsilon^2)^{1/2} & \varepsilon \\ -\varepsilon & (1 - \varepsilon^2)^{1/2} \end{pmatrix} \in SU(2).$$

Then, if $w_r = u^{n-r}v^r$, we have

$$D_a^{(n)}(w_r) = ((1 - \varepsilon^2)^{1/2}u - \varepsilon v)^{n-r}(\varepsilon u + (1 - \varepsilon^2)^{1/2}v)^r$$

$$= w_r + \varepsilon[rw_{r-1} - (n - r)w_{r+1}] + \text{higher powers of } \varepsilon,$$

whence

$$\lim_{\varepsilon \to 0} \varepsilon^{-1}(D_a^{(n)}w_r - w_r) = rw_{r-1} - (n - r)w_{r+1}.$$

It follows that any subspace M of L_n which is stable under $D^{(n)}$ must be stable under the operator A on L_n defined by

$$Aw_r = rw_{r-1} - (n-r)w_{r+1} \qquad\qquad (r = 0, 1, \ldots, n).$$

In the same way, starting from elements

$$b = \begin{pmatrix} (1-\varepsilon^2)^{1/2} & i\varepsilon \\ i\varepsilon & (1-\varepsilon^2)^{1/2} \end{pmatrix} \in SU(2),$$

we can show that any $D^{(n)}$-stable subspace M of L_n must be stable under the operator B defined by

$$Bw_r = rw_{r-1} + (n-r)w_{r+1} \qquad\qquad (r = 0, 1, \ldots, n).$$

Put $J = A + B$, $K = B - A$. Then

$$Jw_r = 2rw_{r-1}, \qquad Kw_r = 2(n-r)w_{r+1}; \qquad\qquad (7)$$

and any $D^{(n)}$-stable subspace must be stable under J and K.

Let M be a $D^{(n)}$-stable subspace of L_n, and suppose $0 \neq \xi \in M$. Writing $\xi = \sum_{r=r_0}^{n} c_r w_r$, where $0 \leq r_0 \leq n$ and $c_{r_0} \neq 0$, we see from (7) that $K^{n-r_0}\xi$ is a non-zero multiple of w_n, and hence that $w_n \in M$. Applying to w_n the various powers of J, and using (7), we conclude that all the w_r are in M, and hence that $M = L_n$. So $D^{(n)}$ is irreducible. ∎

11.7. Thus we have constructed one irreducible unitary representation $D^{(n)}$ of $SU(2)$ for each non-negative integer n. Since $\dim(D^{(n)}) = n + 1$, the $D^{(n)}$ are pairwise inequivalent.

Let us compute the characters $\chi^{(n)} = \chi_{D^{(n)}}$. Since every element a of $SU(2)$ is conjugate in $SU(2)$ to a diagonal element of $SU(2)$, it is sufficient to compute $\chi^{(n)}\begin{pmatrix} e^{i\theta} & 0 \\ 0 & e^{-i\theta} \end{pmatrix}$. It follows from (3) that

$$\chi^{(n)}\begin{pmatrix} e^{i\theta} & 0 \\ 0 & e^{-i\theta} \end{pmatrix} = \sum_{r=0}^{n} e^{i\theta(n-2r)}$$

$$= \frac{\sin(n+1)\theta}{\sin\theta}. \qquad\qquad (8)$$

Remark. The right side of (8) is real; so by 5.7 $D^{(n)}$ is self-adjoint.

11.8.   Our next goal is to decompose $D^{(n)} \otimes D^{(m)}$ into irreducible parts.

We shall fix non-negative integers n and m, and assume that $m \leq n$. Let ψ stand for $\chi_{D^{(n)} \otimes D^{(m)}}$. By (8)

$$\psi \begin{pmatrix} e^{i\theta} & 0 \\ 0 & e^{-i\theta} \end{pmatrix} = (\chi^{(n)} \chi^{(m)}) \begin{pmatrix} e^{i\theta} & 0 \\ 0 & e^{-i\theta} \end{pmatrix}$$

$$= \sum_{r=0}^{n} \sum_{s=0}^{m} e^{i\theta(n+m-2(r+s))}. \tag{9}$$

We now replace r and s by new variables p and q defined as follows:

If $r + s \leq m$,   then $p = r, q = s$;

If $r + s > m$,   then $p = m - s, q = r + s - (m - s)$.

The following figure illustrates these variables for the case $m = 3$, $n = 4$. The rectilinear curves represent constant p and variable q:

In general p and q determine r and s, and we have $r + s = p + q$. The variable p runs through $0, 1, \ldots, m$; and, for given p, q runs through $0, 1, \ldots, n + m - 2p$. Thus (9) becomes:

$$\psi \begin{pmatrix} e^{i\theta} & 0 \\ 0 & e^{-i\theta} \end{pmatrix} = \sum_{p=0}^{m} \sum_{q=0}^{n+m-2p} e^{i\theta(n+m-2p-2q)}$$

$$= \sum_{p=0}^{m} \chi^{(n+m-2p)} \begin{pmatrix} e^{i\theta} & 0 \\ 0 & e^{-i\theta} \end{pmatrix} \qquad \text{(by (8)).} \tag{10}$$

Since (as we pointed out above) every element of $SU(2)$ is conjugate to a diagonal matrix, any central function on $SU(2)$ is determined by its values on the diagonal matrices in $SU(2)$; and so (10) implies

$$\psi = \sum_{p=0}^{m} \chi^{(n+m-2p)} \tag{11}$$

Since finite-dimensional representations are uniquely determined by their characters (5.5), (11) implies:

Proposition. *If m, n are non-negative integers and $m \leq n$, then*

$$D^{(n)} \otimes D^{(m)} \cong \sum_{p=0}^{m} \oplus D^{(n+m-2p)}. \tag{12}$$

Remark 1. Equation (12) is not a complete description of the harmonic analysis of $D^{(n)} \otimes D^{(m)}$. What is still missing is the description of how the $D^{(n+m-2p)}$-subspaces ($p = 0, 1, \ldots, m$) are situated within the tensor product space $X(D^{(n)}) \otimes X(D^{(m)})$. This information is given by the Clebsch-Gordan formulae (see for example Gelfand, Minlos, and Shapiro [1], p. 142).

Remark 2. We notice from (12) that $D^{(n)} \otimes D^{(m)}$ is never irreducible when n and m are both positive.

11.9. We come now to the main result of this section.

Theorem. *Every irreducible unitary representation of $SU(2)$ is unitarily equivalent to $D^{(n)}$ for some $n = 0, 1, 2, \ldots$. That is to say, $\mathscr{I}(SU(2)) = \{D^{(n)} : n = 0, 1, 2, \ldots\}$.*

Proof. $D^{(1)}$ has kernel $\{e\}$. Thus, in view of (12) and Remark 11.7, $\{D^{(n)} : n = 0, 1, \ldots\}$ satisfies the hypotheses of Theorem 6.1, and so exhausts $\mathscr{I}(SU(2))$. ■

12. The Representations of $SO(3)$

12.1. We have already encountered the compact group $SO(3)$ in III.1.20. We shall begin this section by showing that

$$SO(3) \cong SU(2)/Z, \tag{1}$$

where Z is the two-element central subgroup $\left\{ \begin{pmatrix} 1 & 0 \\ 0 & 1 \end{pmatrix}, \begin{pmatrix} -1 & 0 \\ 0 & -1 \end{pmatrix} \right\}$ of $SU(2)$. The classification of $\mathscr{I}(SO(3))$ will follow immediately from (1) and 11.9.

12.2. Let F be the following one-to-one real-linear map of $\mathbb{R}^3$ into the space of 2×2 complex matrices:

$$F(x_1, x_2, x_3) = \begin{pmatrix} x_3 & x_1 - ix_2 \\ x_1 + ix_2 & -x_3 \end{pmatrix} \tag{2}$$

Thus the range of F, which we denote by H, consists of all 2×2 Hermitian matrices of trace 0. If $a \in SU(2)$, it is easy to see that

$$h \mapsto aha^{-1} \qquad (h \in H) \qquad (3)$$

is a linear transformation of H onto itself, and so induces via F a linear transformation τ_a of $\mathbb{R}^3$:

$$\tau_a(x_1, x_2, x_3) = F^{-1}(aF(x_1, x_2, x_3)a^{-1}).$$

Since $\det(F(x_1, x_2, x_3)) = -x_1^2 - x_2^2 - x_3^2$, τ_a must preserve $x_1^2 + x_2^2 + x_3^2$, and so must lie in $O(3)$. In fact, $\tau: a \mapsto \tau_a$ is a continuous homorphism of $SU(2)$ into $O(3)$. Since $SU(2)$ is connected (see 11.3), and since $\det(\rho) = \pm 1$ for $\rho \in O(3)$, we must have $\det(\tau_a) = 1$ for $a \in SU(2)$; that is, τ is a continuous homomorphism of $SU(2)$ into $SO(3)$.

Easy calculations show that

$$\tau \begin{pmatrix} e^{-i\theta} & 0 \\ 0 & e^{i\theta} \end{pmatrix} = \begin{pmatrix} \cos 2\theta & -\sin 2\theta \\ \sin 2\theta & \cos 2\theta \\ 0 & 0 \end{pmatrix} \begin{pmatrix} 0 \\ 0 \\ 1 \end{pmatrix}, \qquad (4)$$

$$\tau \begin{pmatrix} \cos \theta & -\sin \theta \\ \sin \theta & \cos \theta \end{pmatrix} = \begin{pmatrix} \cos 2\theta & 0 & \sin 2\theta \\ 0 & 1 & 0 \\ -\sin 2\theta & 0 & \cos 2\theta \end{pmatrix}, \qquad (5)$$

$$\tau \begin{pmatrix} \cos \theta & -i\sin \theta \\ -i\sin \theta & \cos \theta \end{pmatrix} = \begin{pmatrix} 1 & 0 & 0 \\ 0 & \cos 2\theta & -\sin 2\theta \\ 0 & \sin 2\theta & \cos 2\theta \end{pmatrix} \qquad (6)$$

From (4), (5), and (6) it follows that the range of τ contains all rotations about the x-, y-, and z-axes. Consequently the range of τ is all of $SO(3)$.

Definition. The τ thus constructed will be called the *canonical homomorphism* of $SU(2)$ onto $SO(3)$.

Notice that the only non-unit element of $SU(2)$ commuting with all the elements of H is $\begin{pmatrix} -1 & 0 \\ 0 & -1 \end{pmatrix}$. Hence the kernel of τ is the center $Z = \left\{ \begin{pmatrix} 1 & 0 \\ 0 & 1 \end{pmatrix}, \begin{pmatrix} -1 & 0 \\ 0 & -1 \end{pmatrix} \right\}$ of $SU(2)$; and we have:

Proposition. τ induces an isomorphism of the compact groups $SU(2)/Z$ and $SO(3)$.

12.3. We are now in a position to determine all the irreducible representations of $SO(3)$. Let $\tau: SU(2) \to SO(3)$ be the canonical homomorphism; and let $D^{(n)}$ be as in 11.4.

Proposition *For each non-negative even integer n, the equation*

$$\Delta^{(n)}_{\tau(a)} = D^{(n)}_a \qquad\qquad (a \in SU(2)) \qquad (7)$$

defines an irreducible representation $\Delta^{(n)}$ of $SO(3)$. Further, every element of $\mathcal{I}(SO(3))$ is equivalent to $\Delta^{(n)}$ for some nonnegative even integer n.

Proof. If n is even, then $\mathrm{Ker}(\tau) = Z \subset \mathrm{Ker}(D^{(n)})$ by 11.4(4); so $D^{(n)}$ factors into the form $\Delta^{(n)} \circ \tau$, where $\Delta^{(n)}$ is an irreducible representation of $SO(3)$. Conversely, if E is any element of $\mathcal{I}(SO(3))$, then $E \circ \tau$ is an irreducible representation of $SU(2)$, and so by 11.9 is of the form $D^{(n)}$. Since $Z \subset \mathrm{Ker}(E \circ \tau) = \mathrm{Ker}(D^{(n)})$, n must be even by 11.4(4); and $E = \Delta^{(n)}$. ∎

12.4. Remark. Since $\dim(\Delta^{(n)}) = \dim(D^{(n)}) = n + 1$, it follows from 12.3 that $SO(3)$ *has exactly one irreducible representation of each positive odd dimension, and no irreducible representations of even dimension.*

The Representation of $SO(3)$ by Functions on the Sphere

12.5. Let M be the 2-sphere $\{t \in \mathbb{R}^3 : t_1^2 + t_2^2 + t_3^2 = 1\}$. The natural action of $SO(3)$ on $\mathbb{R}^3$, when restricted to M, gives a transitive action of $SO(3)$ on M. This action in turn generates a unitary representation T of $SO(3)$ on the space $\mathcal{L}_2(\mu)$ of all square-integrable complex functions on M (μ being the usual $SO(3)$-invariant measure of area on M):

$$(T_a f)(m) = f(a^{-1}m) \qquad (f \in \mathcal{L}_2(\mu); a \in SO(3); m \in M).$$

The representation T is of great importance in applied mathematics. Its harmonic analysis is a special case of 10.10.

Proposition. *Every element of $\mathcal{I}(SO(3))$ occurs in T with multiplicity exactly 1. That is,*

$$T \cong \sum_{m=0}^{\infty} {}^{\oplus} \Delta^{(2m)}. \qquad (8)$$

Proof. Let u be the north pole $\langle 0, 0, 1 \rangle$ of M. The stability subgroup of $SO(3)$ for u consists of all rotations around the z-axis.

Composing the action of $SO(3)$ on M with the homomorphism τ of 12.2, we obtain a transitive action of $SU(2)$ on M. By (4) and the preceding paragraph, the stability subgroup of $SU(2)$ for u is precisely the diagonal subgroup E of

$SU(2)$. Denoting by T' the natural representation of $SU(2)$ on $\mathscr{L}_2(\mu)$ (so that T' is induced from the trivial represenation of E), we see from 10.10 that the multiplicity of each $D^{(n)}$ in T' is equal to the multiplicity of the trivial representation ρ of E in $D^{(n)}|E$. But the latter can be read off from 11.4(3): The multiplicity of ρ in $D^{(n)}|E$ is 1 if n is even and 0 if n is odd. Therefore

$$T' \cong \sum_{m=0}^{\infty} {}^{\oplus} D^{(2m)}. \tag{9}$$

Now evidently $T' = T \circ \tau$. Also $D^{(2m)} = \Delta^{(2m)} \circ \tau$ (see (7)). So, since τ is surjective, (9) becomes (8). ∎

Remark. The functions belonging to the $\Delta^{(2m)}$-subspace of $\mathscr{L}_2(\mu)$ (with respect to T) are called *spherical harmonics of order m*. Their exact description is very important in classical analysis. The reader will find it in almost any text on advanced analysis; see for example Whittaker and Watson [1], Chapter XV.

13. Exercises for Chapter IX

1. Let G be an infinite compact group with normalized Haar measure λ. We have seen in 1.3 that $\mathscr{L}_p(\lambda)$ is a Banach *-algebra (under the usual convolution and involution of functions) for every $1 \le p < \infty$. Show that $\mathscr{L}_p(\lambda)$ does *not* have an approximate unit if $p > 1$. [Hint: An approximate unit $\{u_i\}$ of $\mathscr{L}_p(\lambda)$ $(p > 1)$ must also be an approximate unit of $\mathscr{L}_1(\lambda)$. It follows that for any open neighborhood V of e we have $\int_V |u_i(x)| d\lambda x > \frac{1}{2}$ i-eventually. Hence, if G is not discrete, we have $\|u_i\|_p \underset{i}{\to} \infty$ by Hölder's inequality.]

2. Prove proposition 2.2.

3. Let G be a compact group. Show that any dense two-sided ideal of the $\mathscr{L}_1$ group algebra of G must contain $\mathscr{R}(G)$ (see 2.1).

4. Prove Proposition 3.6.

5. Verify that the representation T^* which is adjoint to a finite-dimensional representation T, as defined in 3.7, satisfies $T^{**} = T$; show also that T is irreducible if and only if T^* is.

6. Let G_0 be any group (considered as carrying the discrete topology), and T a finite-dimensional representation of G_0 which is uniformly bounded (i.e., there is a number k such that $\|T_x\| \le k$ for all $x \in G_0$). Show that T is algebraically equivalent to some unitary representation of G_0.

7. Prove that (1) holds in 3.7, and establish the Proposition in that section.

8. Let I be a directed set, with directing relation $\prec$. For each $i \in I$ let a compact group G_i be given; and, for each pair i, j of elements of I with $i \prec j$, let a continuous surjective homomorphism $\phi_{ij} : G_j \to G_i$ be given satisfying:

(i) ϕ_{ii} is the identity map on G_i for all $i \in I$;

(ii) $\phi_{ij} \circ \phi_{jk} = \phi_{ik}$ whenever $i \prec j \prec k$.

Show that there exists a compact group G, and for each $i \in I$ a continuous surjective homomorphism $\psi_i : G \to G_i$, such that $\phi_{ij} \circ \psi_j = \psi_i$ for all $i \prec j$, and such that the following "universal" property holds:

Given a topological group K and for each $i \in I$ a continuous homomorphism $f_i : K \to G_i$ satisfying $\phi_{ij} \circ f_j = f_i$ whenever $i \prec j$, there exists a unique continuous homomorphism $f : K \to G$ such that $f_i = \psi_i \circ f$ for all $i \in I$.

Formulate and prove the fact that the objects G, $\{\psi_i\}$ are "essentially" uniquely determined by the above universal property.

The above G is called the *inverse limit* of the *inverse directed system* $\{G_i\}$, $\{\phi_{ij}\}$.

9. Let p be a positive prime number. For each positive integer n let G_n be the cyclic group of order p^n (with generating element a_n); and if $n \le m$ let $\phi_{nm} : G_m \to G_n$ be the homomorphism sending a_m into a_n. Obviously $\{G_n\}$, $\{\phi_{nm}\}$ (with the usual ordering of the integers) is an inverse directed system of compact groups. Show that the inverse limit of this system is the additive group of p-adic integers.

10. Show that every compact group is an inverse limit (in the sense of Exercise 8) of compact Lie groups (see 3.14).

11. Give a detailed proof of Proposition 3.14.

12. Show in 4.6 that $\mathscr{E}(S)$ is a *-ideal of the measure algebra $\mathscr{M}_r(G)$ of the compact group G.

13. Show that if $\mathscr{L}(G)$ is separable with respect to the supremum norm, then G is second countable (see the proof of 4.8).

14. Prove Proposition 5.14.

15. Fill in the details of the proof of Proposition 6.7.

16. Formulate and establish a result which corresponds to 6.7 for infinite Cartesian products of compact groups (see Remark 3 of 6.7).

17. Let T be an irreducible unitary representation of the compact group G. By the representation $\bar{T}$ *complex-conjugate* to T we mean the representation consisting of the same operators (i.e., $(\bar{T})_x = T_x$ for $x \in G$) but acting in the space $\overline{X(T)}$ complex-conjugate to $X(T)$. Clearly $\bar{T}$ is also irreducible.

(a) Show that $\bar{T} \cong T$ if and only if there exists a non-zero bilinear form β on $X(T) \times X(T)$ which is T-invariant (i.e., $\beta(T_x \xi, T_x \eta) = \beta(\xi, \eta)$ for all $x \in G$ and $\xi, \eta \in X(T)$).

(b) If $\bar{T} \cong T$, show that the β of (a) is unique to within a non-zero multiplicative constant, and is *either* symmetric *or* antisymmetric.

(c) Show that, if λ is normalized Haar measure on G,

$$\int_G \chi_T(x^2) d\lambda x = \begin{cases} 0, & \text{if } T \ncong \bar{T}, \\ 1, & \text{if } T \cong \bar{T} \text{ and the } \beta \text{ of (a) is symmetric,} \\ -1, & \text{if } T \cong \bar{T} \text{ and } \beta \text{ in (a) is antisymmetric.} \end{cases}$$

[Hint: Let B be the space of all bilinear complex functionals on $X(T) \times X(T)$, and B^s and B^a the subspace of symmetric and antisymmetric elements of B

respectively. Thus B is the space of the representation Q of G given by $(Q_x \beta)(\xi, \eta) = \beta(T_{x^{-1}}\xi, T_{x^{-1}}\eta)$; and B^s and B^a are stable under Q. Let ψ_s and ψ_a be the characters of the subrepresentations of Q on B^s and B^a respectively. If $x \in G$ and T_x is diagonal in some basis with diagonal elements $\lambda_1, \ldots, \lambda_n$ ($n = \dim(T)$), we compute that $\psi_a(x) = \sum_{i<j} \bar{\lambda}_i \bar{\lambda}_j$, $\psi_s(x) = \sum_{i \leq j} \bar{\lambda}_i \bar{\lambda}_j$; so that $\psi_s(x) - \psi_a(x) = \sum_{i=1}^{n} \bar{\lambda}_i^2 = \chi_T(x^{-2})$, whence $\int \chi_T(x^2) d\lambda x = \int \psi_s(x) d\lambda x - \int \psi_a(x) d\lambda x$. Now use Cor. 5.5.]

(d) Parts (b) and (c) above classify all irreducible representations T of compact groups into three mutually exclusive types, according to the three permissible values of $\int \chi_T(x^2) d\lambda x$. Find the types of each of the irreducible representations of the groups whose structure spaces we have studied, namely S_3, S_4, A_4, S_5, A_5, $SU(2)$, $U(2)$, $O(2)$, $SO(3)$, $O(3)$. (See Exercise 26 of Chapter IV; Exercises 36–40 below; and §§11, 12 of this Chapter).

18. Show that, *if G is finite*, hypothesis (ii) of Theorem 6.1 can be omitted; but that it cannot be omitted if G is infinite.

19. Show that Theorem 6.1 remains valid if hypothesis (iii) is replaced by: (iii′) If x and y are any two elements of G which are not conjugate in G, there is an element σ of Γ such that $\chi_\sigma(x) \neq \chi_\sigma(y)$.

20. Let G be a compact group, with normalized Haar measure λ. Let R^t be the *two-sided* regular representation of $G \times G$ acting on $\mathscr{L}_2(\lambda)$, given by

$$(R^t_{\langle x, y \rangle} f)(u) = f(x^{-1}uy) \qquad (f \in \mathscr{L}_2(\lambda); \; x, y \in G).$$

Let $\sigma, \tau \in \mathscr{I}(G)$, so that $\sigma \otimes \tau$ is the typical member of $\mathscr{I}(G \times G)$ (see 6.7). Show that the multiplicity of $\sigma \underset{\circ}{\otimes} \tau$ in R^t is 1 if $\tau \cong \bar{\sigma}$ (see Exercise 17 above), and 0 if $\tau \ncong \bar{\sigma}$.

21. Prove Proposition 7.8.

22. Prove that the map $\mu \mapsto D_{\mu * w}$ $(\mu \in \mathscr{M}_r(G))$ described in 7.9(18) is the measure integrated form of the element σ of $\mathscr{I}(G)$ which satisfies $J_\sigma = J$.

23. Let T be any integrable locally convex representation of the compact group G, and ξ any vector in $X(T)$. Show that the following two conditions are equivalent:
  (i) The linear span of $\{T_x \xi : x \in G\}$ is finite-dimensional.
  (ii) There is a central idempotent ψ in $\mathscr{L}(G)$ such that $T_\psi \xi = \xi$.

24. Let G be a compact group and T an integrable locally convex representation of G. Suppose $\mathscr{I}(G) = P \cup Q$, where $P \cap Q = \emptyset$; and define Y and Z to be the closed linear spans in $X(T)$ of $\bigcup\{X_\sigma(T) : \sigma \in P\}$ and $\bigcup\{X_\sigma(T) : \sigma \in Q\}$ respectively.
  (i) Show that $Y \cap Z = \{0\}$ and $Y + Z$ is dense in $X(T)$.
  (ii) Show by an example that $Y + Z$ need not equal $X(T)$, even if T is a Banach representation.

[Hint for part (ii): Let G be the circle group and T the regular representation of G acting on $\mathscr{L}(G)$. For each integer n let $\chi_n(u) = u^n$ $(u \in G)$; and take Y and Z to be the closed linear spans in $\mathscr{L}(G)$ of $\{\chi_n : n \geq 0\}$ and $\{\chi_n : n < 0\}$ respectively. We claim that $Y + Z \neq \mathscr{L}(G)$. To see this, assume that $Y + Z = \mathscr{L}(G)$; and let P_Y and P_Z be the resulting projections of $\mathscr{L}(G)$ onto Y and Z respectively. By the Closed Graph Theorem P_Y and P_Z are continuous. Now for each $0 < r < 1$ consider the function

$g_r : u \mapsto \sum_{n=0}^{\infty} r^n u^n$ in Y and $h_r : u \mapsto \sum_{n=1}^{\infty} r^n u^{-n}$ in Z, and put $f_r = g_r + h_r$; and show that

$$f_r * \phi \to \phi \quad \text{in} \quad \mathscr{L}(G) \quad \text{as} \quad r \to 1- \tag{*}$$

for every ϕ in $\mathscr{L}(G)$. Since $g_r * \phi \in Y$ and $h_r * \phi \in Z$, the continuity of P_Y together with (*) shows that

$$g_r * \phi \to P_Y \phi \quad \text{in} \quad \mathscr{L}(G) \tag{**}$$

for all ϕ in $\mathscr{L}(G)$. But $(g_r * \phi)(1) = \int_G \phi(\bar{u})/(1 - ru)du$; and (**) is contradicted by exhibiting a function ϕ in $\mathscr{L}(G)$ such that $(g_r * \phi)(1) \to \infty$ as $r \to 1-$.]

25. Let G be a compact group; and let $\{\gamma^\alpha\}$ be the net of Theorem 8.3. Show that:

(i) $\lim_\alpha \gamma^\alpha(\sigma) = 1$ for each $\sigma \in \hat{G}$;

(ii) The convergence in 8.3(3) is uniform in ξ on norm-compact subsets of $X(T)$.

Show also that it is possible to choose the net $\{\gamma^\alpha\}$ so that the $\gamma^\alpha(\sigma)$ are all real and non-negative.

26. Let G be a compact group with Haar measure λ, and f a continuous complex function on G which is of the form $g * h$ where $g, h \in \mathscr{L}_2(\lambda)$. Show that in this case

$$f(x) = \sum_{\sigma \in \hat{G}} f_\sigma(x) \tag{see 9.3(7)}$$

absolutely uniformly on G.

27. Let G be a compact group, and M a subset of $\mathscr{L}(G)$ which is separable in the $\| \ \|_\infty$-norm. Show that G has a closed normal subgroup K such that

(i) G/K satisfies the second axiom of countability,

(ii) every f in M is constant on all K-cosets.

28. Let $I, \{G_i\}, \{\phi_{ij}\}, G, \{\psi_i\}$ be as in Exercise 8 above. (Thus G is an inverse limit of an inverse directed system of compact groups.) Characterize $\mathscr{J}(G)$ by showing that every element T of $\mathscr{J}(G)$ is of the form $S * \psi_i$, where $i \in I$ and $S \in \mathscr{J}(G_i)$.

29. Show that a compact group G is a Lie group if and only if there exists a finite subset Γ of $\mathscr{J}(G)$ such that every σ in $\mathscr{J}(G)$ occurs in $\tau_1 \otimes \tau_2 \otimes \cdots \otimes \tau_r$ for some positive integer r and some $\tau_1, \tau_2, \ldots, \tau_r$ in Γ (repetitions being allowed of course among the τ_i). (Thus, G is a Lie group if and only if $\mathscr{J}(G)$ is "finitely generated" under tensor products.)

30. Let G be a finite group, K a subgroup of G, S a finite-dimensional representation of K (over the complex field), and $T = \text{Ind}(S)$ the representation of G induced from S.

Prove that the character of the induced representation is given by:

$$\chi_T(x) = \frac{\gamma(x)}{m} \sum_{\substack{k \in K \\ k \sim x}} \chi_S(k) \tag{$x \in G$}.$$

Here m is the order of K; for each $x \in G$, $\gamma(x)$ is the number of distinct elements y of G such that $xy = yx$; and the symbol $x \sim y$ means that the elements x and y are conjugate in G.

31. (A symmetric form of the Frobenius Reciprocity Theorem.) Let G_1 and G_2 be any two closed subgroups of the compact group G; and let σ and τ be elements of $\mathscr{J}(G_1)$ and $\mathscr{J}(G_2)$ respectively. Show that the multiplicity of τ in $\mathrm{Ind}_{G_1 \uparrow G}(\sigma)|G_2$ is equal to the multiplicity of σ in $\mathrm{Ind}_{G_2 \uparrow G}(\tau)|G_1$.

32. Let K be the diagonal subgroup $\{\langle x, x\rangle : x \in G\}$ of $G \times G$ (G being a compact group); and let I be the trivial one-dimensional representation of K. Show that $\mathrm{Ind}_{K \uparrow (G \times G)}(I)$ is precisely the two-sided regular representation of $G \times G$ defined in Exercise 20 above.

Thus the Frobenius Reciprocity Theorem provides a new approach to Exercise 20.

33. If S and S' are unitary representations of the closed subgroup K of the compact group G, show that

$$\mathrm{Ind}(S \oplus S') \cong \mathrm{Ind}(S) \oplus \mathrm{Ind}(S').$$

34. Let G be a compact group, K a closed subgroup of G, S a unitary representation of K, and $x \in G$. Let $K' = xKx^{-1}$; and let S' be the unitary representation of K' given by $S'_k = S_{x^{-1}kx}$ ($k \in K'$). Show that

$$\underset{K' \uparrow G}{\mathrm{Ind}} (S') \cong \underset{K \uparrow G}{\mathrm{Ind}} (S).$$

35. (The "Stages" Theorem.) Let G be a compact group, and K and L closed subgroups of G with $K \subset L$. If S is a unitary representation of K, show that

$$\underset{K \uparrow G}{\mathrm{Ind}} (S) \cong \underset{L \uparrow G}{\mathrm{Ind}} \left(\underset{K \uparrow L}{\mathrm{Ind}} (S) \right).$$

36. In this Exercise we will obtain the character table of the symmetric group $G = S_5$ on five objects $\{a, b, c, d, e\}$. We will consider S_4 (the symmetry group on four objects) as the subgroup of G consisting of those permutations which leave a fixed. The conjugacy classes $\xi_1, \xi_2, \ldots, \xi_7$ are 7 in number, and correspond to the possible partitions of 5 as a sum of positive integers:

$$
\begin{aligned}
\{e\} = \xi_1: \quad & 5 = 1 + 1 + 1 + 1 + 1 \\
\xi_2: \quad & 5 = 1 + 1 + 1 + 2 \\
\xi_3: \quad & 5 = 1 + 2 + 2 \\
\xi_4: \quad & 5 = 1 + 1 + 3 \\
\xi_5: \quad & 5 = 2 + 3 \\
\xi_6: \quad & 5 = 1 + 4 \\
\xi_7: \quad & 5 = 5.
\end{aligned}
$$

So there are seven elements of $\hat{G}$.

(i) Using Exercise 30 above, and keeping the notation of Exercise 26(C) of Chapter IV for elements of $(S_4)\hat{\ }$, verify the following table of characters of the Ind(S), $S \in (S_4)\hat{\ }$:

	ζ_1	ζ_2	ζ_3	ζ_4	ζ_5	ζ_6	ζ_7
Ind(I)	5	3	1	2	0	1	0
Ind(A)	5	-3	1	2	0	-1	0
Ind(T)	10	0	2	-2	0	0	0
Ind(U)	15	3	-1	0	0	-1	0
Ind(V)	15	-3	-1	0	0	1	0

(ii) Using the above character table and 5.4, show that Ind(I) [resp. Ind(A), Ind(T), Ind(U), Ind(V)] is a direct sum of 2 [resp. 2, 2, 3, 3] inequivalent irreducible representations.

(iii) Let I' be the one-dimensional trivial representation of G, and A' the one-dimensional representation sending even and odd permutations into $+1$ and -1 respectively. Show from 5.5 that Ind(I) and Ind(A) contain I' and A' respectively just once. Thus, by (ii) above, subtracting I' and A' from the characters of Ind(I) and Ind(A) we have obtained the characters of the first *four* elements I', A', Q_1, Q_2 of $\hat{G}$ (see the table below).

(iv) To find the remaining three elements of $\hat{G}$, apply 5.4 to the characters of Ind(T), Ind(U) and Ind(V), to show that:

Ind(T) does not contain I', A', Q_1 or Q_2;
Ind(U) contains Q_1 once, but does not contain I', A', or Q_2;
Ind(V) contains Q_2 once, but does not contain I', A', or Q_1.

From this and (ii) it follows that each of the known quantities

$$\chi_{\text{Ind}(T)}, \qquad \chi_{\text{Ind}(U)} - \chi_{Q_1}, \qquad \chi_{\text{Ind}(V)} - \chi_{Q_2}$$

is a sum of the characters of *two* out of the three as yet unknown elements Q_3, Q_4, Q_5 of $\hat{G}$. Show that these three quantities are all different, so that we may write

$$\chi_{\text{Ind}(T)} = \chi_{Q_3} + \chi_{Q_4}$$

$$\chi_{\text{Ind}(U)} - \chi_{Q_1} = \chi_{Q_3} + \chi_{Q_5}$$

$$\chi_{\text{Ind}(V)} - \chi_{Q_2} = \chi_{Q_4} + \chi_{Q_5}.$$

Solving these equations for χ_{Q_3}, χ_{Q_4}, χ_{Q_5}, we obtain the final character table of $G = S_5$:

	ξ_1	ξ_2	ξ_3	ξ_4	ξ_5	ξ_6	ξ_7
I'	1	1	1	1	1	1	1
A'	1	-1	1	1	-1	-1	1
Q_1	4	2	0	1	-1	0	-1
Q_2	4	-2	0	1	1	0	-1
Q_3	5	1	1	-1	1	-1	0
Q_4	5	-1	1	-1	-1	1	0
Q_5	6	0	-2	0	0	0	1

37. In this Exercise we are going to determine the character table of A_5, the 60-element group of all *even* permutations of 5 objects. This is a group of special interest, being the smallest non-commutative simple group.

(i) A_5 consists of the even conjugacy classes ξ_1, ξ_3, ξ_4, ξ_7 of S_5. Show that, of these, ξ_1, ξ_3 and ξ_4 are conjugacy classes with respect to A_5, while ξ_7 splits into two A_5-conjugacy classes which we call ξ_7^e and ξ_7^o: ξ_7^e [resp. ξ_7^o] consists of all those cyclic permutations

$$a' \to b' \to c' \to d' \to e' \to a'$$

of $\{a, b, c, d, e\}$ such that $\langle a', b', c', d', e' \rangle$ is an even [resp. odd] permutation of $\langle a, b, c, d, e \rangle$.

   Thus A_5 has five conjugacy classes, and hence there are five elements of $(A_5)\hat{\,}$.

(ii) Let I'' be the trivial one-dimensional representation of A_5; and, for $i = 1, \ldots, 5$, let P_i be the restriction of Q_i to A_5 ($Q_1, \ldots, Q_5$ being as in Exercise 36). Using 5.4 and the character table of S_5, show that:

   $P_1 \cong P_2$, $P_3 \cong P_4$;
   P_1 and P_3 are irreducible (and inequivalent);
   $P_5 \cong P_5' \oplus P_5''$, where P_5' and P_5'' are two irreducible representations
       of A_5, inequivalent to each other and to I'', P_1, P_3.

   Thus $(A_5)\hat{\,} = \{I'', P_1, P_3, P_5', P_5''\}$; and we have only to determine the characters of P_5' and P_5''.

(iii) Since the sums of the squares of the dimensions of the elements of $(A_5)\hat{\,}$ must be 60, show that

$$\dim(P_5') = \dim(P_5'') = 3.$$

Using the orthogonality relations 5.3 for the characters, verify finally the following character table for A_5:

	ζ_1	ζ_3	ζ_4	ζ_7^e	ζ_7^o
I''	1	1	1	1	1
P_1	4	0	1	-1	-1
P_3	5	1	-1	0	0
P_5'	3	-1	0	$\dfrac{1+\sqrt{5}}{2}$	$\dfrac{1-\sqrt{5}}{2}$
P_5''	3	-1	0	$\dfrac{1-\sqrt{5}}{2}$	$\dfrac{1+\sqrt{5}}{2}$

38. Classify the structure space of $O(2)$.
39. Classify the structure space of $U(2)$. (See §11.)
40. Classify the structure space of $O(3)$. (See §12.)
41. Let G be a finite group. A unitary representation T of G is called *primitive* if it is irreducible and there is no proper subgroup H of G having a unitary representation S such that

$$T \cong \underset{H \uparrow G}{\mathrm{Ind}}\ (S).$$

(By the Imprimitivity Theorem of XII.1.7 this is equivalent to the definition given in the Introduction to Volume II.)

In Exercise 37 above we determined the five irreducible representations I'', P_1, P_3, P_5', P_5'' of the 60-element alternating group A_5. Prove that I'', P_1, P_5' and P_5'' are primitive, but that P_3 is not primitive.

Notes and Remarks

The fundamental paper which laid the foundations for the representation theory of compact groups was published in 1927 by Peter and Weyl [1]. This paper, which treats compact Lie groups, contains the important Peter–Weyl Theorem 2.11, the orthogonality relations 4.2, a discussion of the characters of unitary representations, and several applications of these results. Somewhat earlier the orthogonality relations were given for finite groups in a 1905 paper by Schur [2]. In 1924 Schur [3] observed that one could apply ideas of Hurwitz [1] on integration over manifolds to define integration of continuous functions on compact Lie groups. Using this as a substitute for summing over the group he was able to prove the orthogonality relations 4.2 for compact Lie groups (the proof of 4.2 which we have given is essentially Schur's original argument).

For arbitrary compact groups the Peter–Weyl Theorem and the orthogonality relations appear in the 1940 book of Weil [1]. It should be pointed out however that Haar [1] observed in 1933 that, using his invariant measure, the Peter–Weyl Theorem and the theory of unitary representations of compact Lie groups could be extended to arbitrary second countable compact Lie groups automatically.

The Stone–Weierstrass formulation of the Peter–Weyl Theorem given in §6 is due to Stone [5]. Further information and historical notes on the Peter–Weyl Theorem can be found, for example, in Gaal [1], Gross [1], Hawkins [3], Hewett and Ross [2], Mackey [10, 15, 21, 22, 23], Naimark [8], Robert [2], Weiss [1], Želobenko [3].

We remark that the orthogonality relations were extended to square-integrable unitary representations of $SL(2, \mathbb{R})$ by Bargmann [1]; and to square-integrable representations of arbitrary locally compact groups by Godement [2, 3]. A systematic study of characters and their generalizations was undertaken by Godement [8, 10, 11].

The results in §§7 and 8 on integrable locally convex representations of compact groups generalize corresponding well known results on Banach (and unitary) representations (see, for example, G. Warner [1, pp. 218–252]). The Plancherel formula 9.6 for compact groups may be found in Dixmier [24, p. 322] and Naimark [8, p. 445]. Various generalizations of the formula are given in Andler [1], Dixmier [24], Kajiwara [1], Kleppner and Lipsman [1], Mauceri and Ricardello [1], Segal [7], and Tatsuuma [3].

For finite groups the notion of induced representation was introduced in 1898 by Frobenius [5]; this paper also contains the Frobenius Reciprocity Theorem for finite groups. The definition of induced representation and the Frobenius Reciprocity Theorem were extended from finite to compact groups by Weil [1]. Although the Reciprocity Theorem does not hold in the form stated in 10.8 for non-compact groups, there are many variants and analogues which have been proposed in the literature. For example, see Armacost [1], Felix, Henrichs, and Skudlarek [1], Fell [8], Gaal [1, pp. 409–411], Henrichs [3], Howe [1], Kirillov [6, pp. 207–208], Kunze [4], Lange [1],

Lange, Ramsay, and Rota [1], Mackey [6, 10, 15], Mautner [6], C. Moore [1], C. Moore and Repka [1], Moscovici [2], Nielsen [1, 4], Penney [1], Rosenberg [4], Shin'ya [3, 4], Szmidt [1], and Wawrzyńczyk [1, 2].

The classification of the irreducible representations of $SU(2)$ and $SO(3)$ can be found in Hewitt and Ross [2, pp. 125–141], Sugiura [1, pp. 43–93], and Želobenko [3].

> The perfection of mathematical beauty is such
> ... that whatsoever is most beautiful and regular
> is also found to be most useful and excellent.
>
> —D'Arcy W. Thompson

X Abelian Groups and Commutative Banach *-Algebraic Bundles

Since the development of the calculus it would be hard to find another discovery which has so profoundly influenced both pure and applied mathematics as did the discovery of Fourier series. Indeed, a decisive step in the history of the subject-matter of this work was the realization (in the 1930's) that Fourier series and integrals are simply special cases of the harmonic analysis of unitary representations of an arbitrary locally compact Abelian group G. To obtain the classical Fourier series one takes G to be the circle group; to obtain the Fourier integral, one takes $G = \mathbb{R}$.

The fundamental building-blocks in the analysis of the unitary representations of G are of course the irreducible ones; and when G is Abelian these are one-dimensional. The one-dimensional unitary representations of a locally compact Abelian group G (that is, the continuous homomorphisms of G into $\mathbb{E}$) are called the characters of G. Under pointwise multiplication these form a locally compact Abelian group $\hat{G}$ called the character group or dual group of G. The object of fundamental interest in the theory of Fourier series and integrals (as generalized to arbitrary locally compact Abelian groups) is the map $f \mapsto \hat{f}$ which carries a (summable) function f on G into the function $\hat{f}$ on $\hat{G}$ defined by

$$\hat{f}(\chi) = \int_G f(x)\chi(x)d\lambda x \qquad (\chi \in \hat{G}),$$

where λ is Haar measure on G. We call $\hat{f}$ the Fourier transform of f. Since $\hat{G}$ turns out to be the Gelfand space of the $\mathscr{L}_1$ group algebra of G, the Fourier transform of f is just a special case of the Gelfand transform as defined in V.7.7. This shows that the modern theory of normed algebras is a lineal descendant of Fourier's classic solution of the heat conduction equation.

Of course, since $\hat{G}$, like G, is a locally compact Abelian group (and not just a locally compact Hausdorff space), the harmonic analysis of functions on G has much more structure than that of an arbitrary commutative Banach *-algebra.

Throughout this chapter G is a fixed locally compact Abelian group. In §1 we introduce the character group $\hat{G}$ of G; and in §2 we set down the group versions of the Stone, Bochner, and Plancherel Theorems, whose more abstract versions were proved in Chapter VI. Thus these two sections are largely specializations to the group context of more general concepts and results.

In §3 we take up that part of the harmonic analysis of G which depends on the structure of G and $\hat{G}$ as Abelian groups. The principal results of §3 are three—the Fourier Inversion Theorem (3.8), the regularity of the $\mathscr{L}_1$ group algebra of G (3.9), and above all the Pontryagin Duality Theorem (3.11). The first and third of these show that the Fourier transform operation is highly symmetric: Iterating the construction of $\hat{G}$ from G brings us back to the original G; and iterating the Fourier transform (with a slight variation) brings us back to the original function on G.

The regularity of the $\mathscr{L}_1$ group algebra leads to the important subject of Tauberian theorems and "spectral synthesis"—topics which we do not have space to include here.

In §§4 and 5 we investigate the structure of a saturated commutative Banach *-algebraic bundle $\mathscr{B}$ over G. We know from Chapter VIII that such an object has more structure than a mere commutative Banach *-algebra, and on the other hand is more general than the group G. It is shown in §4 that the bundle structure of $\mathscr{B}$ causes the structure space $\hat{\mathscr{B}}$ to become a locally compact principal $\hat{G}$-bundle over the structure space $\hat{A}$ of its unit fiber algebra A (see Definition 4.14). In the special case that $\mathscr{B}$ is the group bundle of G, this principal $\hat{G}$-bundle is just $\hat{G}$ itself (acting on itself by left multiplication, and considered as a bundle over a one-element base space). Conversely, we show in §5 that every locally compact principal $\hat{G}$-bundle arises in this way from some such $\mathscr{B}$. In fact, if we restrict ourselves to saturated commutative C^*-algebraic bundles, the correspondence $\mathscr{B} \mapsto \hat{\mathscr{B}}$ is one-to-one; and we discover that saturated commutative C^*-algebraic bundles over G are essentially the same category of objects as locally compact principal $\hat{G}$-bundles. This duality generalizes on the one hand the Pontryagin duality for locally compact Abelian groups, and on the other hand the duality (proved in VI.4.3) between commutative C*-algebras and locally compact Hausdorff spaces.

1. The Character Group

1.1. *Throughout this chapter G is a fixed locally compact Abelian group, with unit e and Haar measure λ.* Of course G is unimodular. For brevity we shall denote the $\mathcal{L}_1$ group algebra $\mathcal{L}_1(\lambda)$ of G by A. Thus A is a commutative Banach *-algebra.

1.2. By the proof of VIII.14.9 the integrated form of the regular representation of G is faithful on A, and so A is reduced. By VI.10.8 this implies the important conclusion:

Theorem.   *A is semisimple.*

1.3. Theorem.   *A is symmetric.*

Proof.   Let ϕ be a complex homomorphism of A. By V.7.3 ϕ is continuous on A. Thus by VIII.12.10 ϕ is the integrated form of a one-dimensional representation χ of G.

We claim that χ (considered as a continuous homomorphism of G into $\mathbb{C}_*$) is bounded. For, otherwise, for each positive integer n there would be a point x_n in G at which $|\chi(x_n)| > n$; and hence there would be a function f_n in $\mathcal{L}_+(G)$ (peaked around x_n) satisfying

$$\|f_n\|_1 = \int f_n \, d\lambda = 1 \text{ and } |\phi(f_n)| = \left| \int \chi f_n \, d\lambda \right| > n.$$

This would contradict the boundedness of ϕ on A, proving the claim.

Since χ is bounded, we must have $|\chi(x)| = 1$ for all x in G; for, if $|\chi(x)| \neq 1$ for some x, then $\{\chi(x^n) : n \in \mathbb{Z}\}$ would be an unbounded set of numbers contained in range(χ). So

$$\chi(x^{-1}) = \chi(x)^{-1} = \overline{\chi(x)} \qquad \text{for all } x;$$

and this implies that for all f in A

$$\phi(f^*) = \int \chi f^* \, d\lambda = \int \chi(x)\overline{f(x^{-1})} d\lambda$$

$$= \int \overline{\chi(x^{-1})f(x^{-1})} d\lambda = \left(\int \chi f \, d\lambda\right)^-$$

$$= \overline{\phi(f)}. \qquad \blacksquare$$

1.4.   Definition.   *A character* of G is a continuous homomorphism of G into the circle group $\mathbb{E}$.

Thus the characters of G are essentially just the one-dimensional unitary representations of G.

By the proof of 1.3, all the elements of the Gelfand space $A^\dagger$ of A are of the form $f \mapsto \int \chi f \, d\lambda$, where χ is a character of G. So, applying to 1.2 the argument of VIII.14.10(II), we obtain:

Theorem.   *The set of all characters of G separates the points of G.*

Remark.   At first glance one might have expected that the proof of this theorem would have required a detailed analysis of the structure of locally compact Abelian groups. One is quite surprised to find it emerging from nothing more special than the existence of Haar measure (which enables us to construct the regular representation) and the general theory of commutative C^*-algebras.

Remark.   The above theorem is of course a very special case of VIII.14.10, once we recall the first sentence of 1.5.

1.5.   Since G is Abelian, the elements of the structure space $\hat{A}$ ($\cong \hat{G}$; see VIII.21.4) are all one-dimensional (VIII.9.7). So $\hat{G}$ is just the set of all characters of G, and by 1.3 coincides as a set with $A^\dagger$. In fact by VII.1.13 $\hat{G}$ and $A^\dagger$ coincide as topological spaces. Thus V.7.5 gives:

Theorem. $\hat{G}$ *is a locally compact Hausdorff space.*

Remark.   The local compactness of $\hat{G}$ is of course a very special case of VII.6.9.

1.6.   As an immediate special case of VIII.21.12 we have the following direct description of the topology of $\hat{G}$ in terms of the characters of G.

Theorem. *The regional topology of $\hat{G}$ coincides with the topology of uniform convergence on compact subsets of G.*

Remark. The full intricacy of the proof of VIII.21.12 is not required in order to obtain merely the above theorem. To prove the latter one can proceed more simply as follows:

The regional topology is, as we know, the topology of pointwise convergence on A; and this is clearly contained in the topology of uniform convergence on compact subsets of G. To prove the converse, we start with a net $\{\chi_i\}$ of elements of $\hat{G}$ converging regionally to χ in $\hat{G}$, and we take a compact subset K of G. Choose an element f of A such that $\chi(f)\,(=\int \chi f\, d\lambda) \neq 0$. Denoting by xf the left translate $y \mapsto f(x^{-1}y)$ of f, we check that, for all i and all x in G,

$$\chi_i(xf) = \chi_i(x)\chi_i(f), \qquad \chi(xf) = \chi(x)\chi(f). \tag{1}$$

Now $x \mapsto xf$ is continuous from G to A (by III.11.13); so $\{xf : x \in K\}$ is compact in A. Also the characters of G, as elements of A^*, are of norm 1. Therefore by Lemma VIII.21.11

$$\chi_i(xf) \to \chi(xf) \qquad \text{uniformly for } x \text{ in } K. \tag{2}$$

Combining (1) and (2) with the fact that $\chi_i(f) \to \chi(f) \neq 0$, we see that $\chi_i(x) \to \chi(x)$ uniformly on K.

1.7. Since $\hat{G}$ and $A^{\dagger}$ essentially coincide as topological spaces, the Gelfand transform $\hat{f}$ of a function f in A can be considered as defined on $\hat{G}$ by the formula:

$$\hat{f}(\chi) = \int_G f(x)\chi(x)d\lambda x \qquad (\chi \in \hat{G}) \tag{3}$$

Definition. In this context the function $\hat{f}$ defined by (3) is usually called the *Fourier transform* of f.

We recall that, if $f, g \in A$,

$$(f * g)^{\hat{}} = \hat{f}\hat{g}, \tag{4}$$

$$(f^*)^{\hat{}} = \bar{\hat{f}}, \tag{5}$$

$$\|\hat{f}\|_\infty \leq \|f\|_1. \tag{6}$$

Applying the Stone–Weierstrass Theorem as in the proof of VI.10.7, we conclude:

Proposition. $\{\hat{f} : f \in \mathcal{L}(G)\}$ *is dense in* $\mathcal{C}_0(\hat{G})$.

Remark. It is worth observing that in general, if $f \in \mathscr{L}(G)$, $\hat{f}$ will not be in $\mathscr{L}(\hat{G})$; that is, the Fourier transform of a continuous function with compact support need not have compact support.

Consider for example the additive group $\mathbb{R}$ of the reals, and let λ be Lebesgue measure. We shall see in 1.11 that $\hat{\mathbb{R}}$ can be identified with $\mathbb{R}$. When this is done, the Fourier transform $\hat{f}$ of an element f of $\mathscr{L}_1(\lambda)$ is given by

$$\hat{f}(s) = \int_{-\infty}^{\infty} f(t)e^{ist}\, d\lambda t \qquad\qquad (s \in \mathbb{R}).$$

Now let f be in $\mathscr{L}(\mathbb{R})$. Then this integral exists for all *complex s*, and in fact defines an extension of $\hat{f}$ to an entire function on the complex plane. Hence, if $\hat{f}$ vanishes identically on any non-void open subinterval of $\mathbb{R}$, it must vanish everywhere, and so f must be 0. In particular, if $0 \neq f \in \mathscr{L}(\mathbb{R})$, then $\hat{f}$ cannot have compact support.

The Group Structure of $\hat{G}$

1.8. Now $\hat{G}$ has a natural group structure. Indeed, the pointwise product of two characters of G is again a character of G; and the pointwise product operation makes $\hat{G}$ into an Abelian group, the inverse of a character χ being the complex-conjugate character $\bar{\chi}$. The operations of pointwise product and complex conjugation are clearly continuous with respect to the topology of uniform convergence on compact subsets of G; so $\hat{G}$ is an Abelian topological group with respect to this topology. By 1.6 this is the regional topology of $\hat{G}$; and by 1.5 the latter is locally compact and Hausdorff. Thus $\hat{G}$ becomes a locally compact Abelian group.

Definition. Considered as a locally compact Abelian group under the regional topology and the pointwise product operation, $\hat{G}$ is called the *dual group* or the *character group* of G.

Remark. If χ and ϕ are two characters of G, their pointwise product $\chi\phi$ is just the tensor product $\chi \otimes \phi$ (see VIII.9.17), and the complex-conjugate $\bar{\chi}$ is just the complex-conjugate in the sense of VIII.9.15. Thus the continuity of the operations of pointwise product and complex conjugation with respect to the regional topology is a special case of VIII.21.24.

1.9. Proposition. *If* $G_1, G_2, \ldots, G_n$ *are finitely many locally compact Abelian groups and G is their Cartesian product, then $\hat{G}$ is isomorphic with the Cartesian product* $(G_1)\hat{}\, \times \cdots \times (G_n)\hat{}$ *under the natural isomorphism which sends the* *n-tuple* $\langle \chi_1, \ldots, \chi_n \rangle$ $(\chi_i \in (G_i)\hat{\ })$ *into the character* $\langle x_1, \ldots, x_n \rangle \mapsto \prod_{i=1}^{n} \chi_i(x_i)$ *of G.*

The reader can easily verify this for himself.

1.10.   *Example.*   Let G be a finite cyclic group of order n, with generator u. Then each nth root w of unity gives us a character $\chi_w: u^r \mapsto w^r$ of G; and the correspondence $w \mapsto \chi_w$ is an isomorphism between $\hat{G}$ and the cyclic group (of order n) of all nth roots of unity. Thus in this case G and $\hat{G}$ are isomorphic.

It is well known that every finite Abelian group is the Cartesian product of cyclic groups. Therefore, by 1.9 and the preceding paragraph, $\hat{G} \cong G$ for every finite Abelian group.

1.11.   *Example.*   The character group of $\mathbb{R}$.

As usual $\mathbb{R}$ denotes the additive group of all real numbers. For each real number s, the equation $\chi_s(t) = e^{ist}$ $(t \in \mathbb{R})$ defines a character χ_s of $\mathbb{R}$. We claim that every character of $\mathbb{R}$ is of this form. Indeed, let ψ be any character of $\mathbb{R}$. Since ψ is continuous, there is a positive number w such that

$$\psi([-w, w]) \subset \left\{ e^{ir}: -\frac{\pi}{4} < r < \frac{\pi}{4} \right\}. \tag{7}$$

So there is a unique real number s satisfying

$$\psi(w) = e^{isw}, \qquad -\frac{\pi}{4} < sw < \frac{\pi}{4}. \tag{8}$$

Now suppose we have shown that

$$\psi(2^{-n}w) = e^{is2^{-n}w} \tag{9}$$

for some positive integer n. Then, since ψ is a homomorphism, $\psi(2^{-n-1}w)$ must be $\pm e^{is2^{-n-1}w}$. By (7) $\psi(2^{-n-1}w) \neq -e^{is2^{-n-1}w}$, since the latter number does not belong to the right side of (7). So (9) holds when n is replaced by $n + 1$. In view of (8) we have proved by induction that (9) holds for all positive integers n. Hence the closed subgroup $\{t: \psi(t) = e^{ist}\}$ of $\mathbb{R}$ contains $2^{-n}w$ for all positive integers n, and so is dense in $\mathbb{R}$ and hence equal to $\mathbb{R}$. This shows that $\psi = \chi_s$.

We have shown that the algebraic homomorphism $s \mapsto \chi_s$ of $\mathbb{R}$ into $\hat{\mathbb{R}}$ is actually onto $\hat{\mathbb{R}}$. This homomorphism is clearly one-to-one and continuous (with respect to the regional topology of $\hat{\mathbb{R}}$; see 1.6). So by III.3.11 *it is a topological isomorphism of $\mathbb{R}$ and $\hat{\mathbb{R}}$.*

1.12.   *Example*.   The character group of $\mathbb{R}^n$ ($n = 1, 2, \ldots$).

It follows from 1.11 and 1.9 that every character of $\mathbb{R}^n$ is of the form

$$\chi_{s_1, \ldots, s_n} \colon \langle t_1, \ldots, t_n \rangle \mapsto e^{i(s_1 t_1 + \cdots + s_n t_n)},$$

where $s_i \in \mathbb{R}$. In fact, the map $\langle s_1, \ldots, s_n \rangle \mapsto \chi_{s_1, \ldots, s_n}$ is a topological isomorphism of $\mathbb{R}^n$ and $(\mathbb{R}^n)\hat{\ }$.

1.13.   *Remark*.   By 1.10 and 1.12, G and $\hat{G}$ are isomorphic as locally compact Abelian groups whenever G is either finite or $\mathbb{R}^n$ (for some positive integer n). We shall see in 3.16 that the same is true when G is the additive group of any non-discrete locally compact field. In general, however, it is far from true that $\hat{G} \cong G$. This is brought out by the next result.

1.14.   Proposition.   *If G is discrete, $\hat{G}$ is compact. If G is compact, $\hat{G}$ is discrete.*

Proof.   The second of these statements was proved in IX.4.7.

To prove the first, assume that G is discrete. Then the topology of $\hat{G}$ is just the topology of pointwise convergence on G. With this topology $\hat{G}$ is a closed subset of the space of all functions from G to $\mathbb{E}$, and the latter is compact by Tihonov's Theorem. So $\hat{G}$ is compact.   ∎

1.15.   *Remark*.   Suppose that G is infinite and compact; and let G_d be the discrete group coinciding as a group with G. Evidently the characters of G are also characters of G_d; so $\hat{G}$, as a group, can be identified with a subgroup (in fact a dense subgroup, as we shall see in 3.14) of $(G_d)\hat{\ }$. Now by 1.14 the regional topology of $\hat{G}$ is discrete. But we claim that the topology of $\hat{G}$ relativized from the regional topology of $(G_d)\hat{\ }$ (that is, the topology of pointwise convergence on G) is *not* discrete. Indeed, suppose that the latter topology were discrete. Then by III.1.16 $\hat{G}$ would be a closed subgroup of $(G_d)\hat{\ }$. But $(G_d)\hat{\ }$ is compact by 1.14; so $\hat{G}$ would be a compact subgroup of $(G_d)\hat{\ }$. Being discrete and compact, $\hat{G}$ would be finite; and so by IX.4.8 G would be finite, contrary to the original hypothesis. Thus we have proved the claim.

This claim provides counter-examples to substantiate the Remarks VIII.21.3.

2. The Theorems of Stone, Bochner, and Plancherel

Stone's Theorem

2.1. Stone's Theorem VI.10.10 has the following important restatement in the context of locally compact Abelian groups:

Theorem (Stone). *If X is a Hilbert space and P is a regular X-projection-valued Borel measure on $\hat{G}$, then the equation*

$$T_x = \int_{\hat{G}} \chi(x)dP\chi \qquad\qquad (x \in G) \qquad (1)$$

defines a unitary representation T of G on X. Conversely, every unitary representation T of G on a Hilbert space X determines a unique regular X-projection-valued Borel measure P on $\hat{G}$ such that (1) holds.

Remark. The right side of (1) is of course a spectral integral (II.11.7).

Proof. If $x \in G$, let $\hat{x}$ denote the function $\chi \mapsto \chi(x)$ on $\hat{G}$. It follows from 1.6 that, if $x_i \to x$ in G and $\chi_i \to \chi$ in $\hat{G}$, then $\chi_i(x_i) \to \chi(x)$. From this we see that, if $x_i \to x$ in G, then $(x_i)\hat{} \to \hat{x}$ uniformly on compact subsets of $\hat{G}$. An easy argument based on this fact shows that the map $x \mapsto T_x$ defined by (1) is strongly continuous. Since by II.11.8 T is multiplicative and satisfies $T_{x^{-1}} = (T_x)^*$, it is thus a unitary representation of G; and the first statement of the theorem is verified.

To prove the second statement, let T be a unitary representation of G on X. Applying VI.10.10 to the integrated form S of T, we deduce the existence of a regular X-projection-valued Borel measure P on $\hat{G}$ such that

$$S_f = \int \hat{f}(\chi)dP\chi \qquad\qquad (f \in A). \qquad (2)$$

Now let T' be the unitary representation obtained from P by means of (1):

$$T'_x = \int \chi(x)dP\chi \qquad\qquad (x \in G). \qquad (3)$$

If $f \in A$, $x \in G$, and $g(y) = f(x^{-1}y)$ $(y \in G)$, we verify that $\hat{g}(\chi) = \chi(x)\hat{f}(\chi)$; so by (2), (3), and II.11.8,

$$T'_x S_f = \int \chi(x)\hat{f}(\chi)dP\chi$$

$$= \int \hat{g}(\chi)dP\chi = S_g$$

$$= T_x S_f.$$

Since S is non-degenerate, this implies that $T'_x = T_x$, whence $T' = T$. So (3) becomes (1); and the existence of P has been shown.

To prove the uniqueness of P, we suppose that Q is any other regular X-projection-valued Borel measure on $\hat{G}$ satisfying

$$T_x = \int \chi(x)dQ\chi \qquad\qquad (x \in G). \qquad (4)$$

Fix $\xi, \eta \in X$; and let $(P\xi, \eta)$ and $(Q\xi, \eta)$ stand for the measures

$$W \mapsto (P(W)\xi, \eta) \quad \text{and} \quad W \mapsto (Q(W)\xi, \eta);$$

these are maximal regular extensions (see II.8.15) of bounded regular complex Borel measures on $\hat{G}$. If $f \in A$ and S again denotes the integrated form of T, we have by (4), II.11.8(VII), and the Fubini Theorem II.9.16,

$$(S_f\xi, \eta) = \int f(x)d\lambda x \int \chi(x)d(Q\xi, \eta)\chi$$

$$= \int d(Q\xi, \eta)\chi \int f(x)\chi(x)d\lambda x$$

$$= \int \hat{f}(\chi)d(Q\xi, \eta)\chi. \qquad (5)$$

Similarly, replacing Q by P, we also have

$$(S_f\xi, \eta) = \int \hat{f}(\chi)d(P\xi, \eta)\chi. \qquad (6)$$

Since $(P\xi, \eta)$ and $(Q\xi, \eta)$ are bounded, and since $\{\hat{f} : f \in A\}$ is dense in $\mathscr{C}_0(\hat{G})$ by 1.7, it follows from (5) and (6) that $(P\xi, \eta) = (Q\xi, \eta)$. This holds for all ξ and η, whence it follows that $P = Q$. This proves the uniqueness of the P satisfying (1), and completes the proof of the theorem. ∎

Definition. As in VI.10.10, the projection-valued measure P satisfying (1) is called the *spectral measure* of T.

Remark. The set $\mathcal{N}$ of all null sets of the spectral measure P of T is an important "unitary invariant" of the unitary representation T. If ρ is a regular Borel measure on $\hat{G}$ and if $\mathcal{N} \cap \mathcal{S}(\hat{G}) = \{W \in \mathcal{S}(\hat{G}): \rho(W) = 0\}$, we shall say that T is *of the same measure class as* ρ. (Here $\mathcal{S}(\hat{G})$, as usual, is the compacted Borel δ-ring of $\hat{G}$.)

2.2. An important special case of Stone's Theorem is that of the additive group $\mathbb{R}$ of the reals. We have seen in 1.11 that $\mathbb{R}$ and $\hat{\mathbb{R}}$ are isomorphic under the map that sends s into $t \mapsto e^{ist}$. Thus in this case, if X is a Hilbert space, 2.1 asserts a one-to-one correspondence between unitary representations T of $\mathbb{R}$ on X and X-projection-valued Borel measures P on $\mathbb{R}$, the correspondence being given by

$$T_t = \int_{\mathbb{R}} e^{ist}\, dPs. \tag{7}$$

(Note that by II.11.10 the condition that P is regular can be omitted.)

Now by VI.12.1 (and VI.12.18) the X-projection-valued Borel measures P on $\mathbb{R}$ are in one-to-one correspondence with the (possibly unbounded) self-adjoint operators E on X, the correspondence being

$$E = \int s\, dPs \tag{8}$$

(where the integral on the right is defined as in II.11.15). Combining (7) and (8), and using the definition VI.12.6 of functions of unbounded normal operators, we obtain:

Theorem. *Let X be a Hilbert space. There is a natural one-to-one correspondence between the set of all unitary representations T of $\mathbb{R}$ acting on X and the set of all (possibly unbounded) self-adjoint operators E on X. The correspondence is given by*

$$T_t = e^{itE} \qquad\qquad (t \in \mathbb{R}). \tag{9}$$

Definition. In the context of this theorem we say that E is the *self-adjoint generator* of T.

2.3.   Proposition*.   *Let X be a Hilbert space; and let T and E be related as in Theorem 2.2. A vector ξ in X belongs to* domain(E) *if and only if* $\lim_{t \to 0} t^{-1}(T_t\xi - \xi)$ *exists (in the norm topology of X); and in that case*

$$iE\xi = \lim_{t \to 0} t^{-1}(T_t\xi - \xi).$$

2.4.   Proposition*.   *Let T and E be related as in Theorem 2.2. Then E is a bounded operator if and only if $T: t \mapsto T_t$ is continuous on $\mathbb{R}$ with respect to the norm-topology of operators.*

2.5.   *Remark*.   Unitary representations of $\mathbb{R}$ are of great importance in physics. Indeed, take a physical system, and suppose that at any instant its "state" is described by a point in some "state space" X (which to begin with we think of as merely a set). The principle of causality for the system can be formulated as follows: For each real t there is a well-defined permutation T_t of X such that, if ξ is the state of the system at time r, then $T_t\xi$ is the state of the system at time $r + t$. This implies that $t \mapsto T_t$ must be a homomorphism of $\mathbb{R}$ into the group of permutations of X. We now assume that the system is a quantum-mechanical one. Then, roughly speaking, the state space will be a complex Hilbert space, and the T_t will have to preserve the linear structure and the inner product, that is, they will be unitary. If we add the reasonable requirement that $T: t \mapsto T_t$ be strongly continuous, we find that the development of the system with changing time is given by a unitary representation T of $\mathbb{R}$ on the Hilbert state space of the system. Let E be the self-adjoint generator of T. The reader who has a little acquaintance with quantum theory will remember that the (possibly unbounded) self-adjoint operators on the state space are just the *physical observables* of the system. Thus E, the self-adjoint generator of T, is a physical observable, presumably an important one. It is in fact the fundamental quantity known as *energy*!

The above discussion was inaccurate in one respect. Strictly speaking, the "states" of a quantum-mechanical system are not the vectors ξ of a Hilbert space X, but the "rays" $\mathbb{C}\xi$ ($\xi \in X$). As a consequence of this correction, it can be shown that the development of the system with changing time should strictly be given, not by a unitary representation, but by a *projective* representation T^0 of $\mathbb{R}$ on X (see VIII.10.2). This correction turns out to be of minor importance, however; for we have seen in VIII.10.8 that every projective representation of $\mathbb{R}$ is essentially unitary. Thus T^0 still arises from a unitary representation T of $\mathbb{R}$, and hence (as in 2.2) from a self-adjoint operator E. It should be observed that E is determined by T^0 only to within an additive constant (since the addition of a constant k to E changes T to

$t \mapsto e^{ikt}T_t$, and so leaves T^0 unchanged). This corresponds to the physical fact that energy is only meaningful to within an additive constant.

The Bochner Theorem

2.6. We have seen in 1.3 that the $\mathscr{L}_1$ group algebra A of G is symmetric. Furthermore, by VIII.20.11 the (continuous) positive linear functionals p on A are just those which arise from functions ϕ of positive type on G:

$$p(f) = \int_G \phi(x)f(x)d\lambda x \qquad (f \in A). \qquad (10)$$

Hence the Generalized Bochner Theorem VI.21.2 gives rise in the group context to the following more classical result:

Bochner Theorem. *Every bounded regular (non-negative) Borel measure v on $\hat{G}$ gives rise to a function ϕ of positive type on G as follows:*

$$\phi(x) = \int_{\hat{G}} \chi(x)dv_e\chi \qquad (x \in G) \qquad (11)$$

(v_e being the maximal regular extension of v; see II.8.15). Conversely, for every function ϕ of positive type on G there is a unique bounded regular Borel measure v on $\hat{G}$ such that (11) holds.

We shall leave the details of the proof to the reader. It can be obtained either directly from VI.21.2 by an argument in the spirit of 2.1, or from 2.1 in much the same way that VI.21.2 was obtained from VI.10.10.

We observe that, if ϕ is a function of positive type on G and p is the functional on A obtained from ϕ by (10), then p is related to the v of the preceding theorem as one would expect:

$$p(f) = \int_{\hat{G}} \hat{f}(\chi)dv\chi \qquad (f \in A). \qquad (12)$$

The Plancherel Theorem

2.7. Our next task is to specialize Theorem VI.21.4 to the group context. As before, we take A to be the $\mathscr{L}_1$ group algebra of G; and we denote by B the dense *-subalgebra $\mathscr{L}(G)$ of A.

Definition. A regular complex Borel measure μ on G is said to be *of positive type* if the corresponding linear functional $f \mapsto \int f \, d\mu$ on the *-algebra B is positive.

Fix a regular complex Borel measure μ of positive type on G; and put $p(f) = \int f \, d\mu \, (f \in B)$. We claim that the positive functional p satisfies the two conditions (i), (ii) of VI.21.4. Indeed: p is continuous in the inductive limit topology of B, and so by VIII.13.6 generates a *-representation of B which extends to a *-representation T of A. If $\sim$ denotes the canonical quotient map of B into $X(T)$, this implies that $p(b^*a^*ab) = (T_a\tilde{b}, \; T_a\tilde{b}) \leq \|T_a\|^2 \|\tilde{b}\|^2 \leq \|a\|_1^2 p(b^*b) \, (a, b \in B)$, proving VI.21.4(i). Condition VI.21.4(ii) is an easy consequence of the existence of approximate units on G; we leave its proof to the reader.

Thus, Theorem VI.21.4 tells us that p has a Gelfand transform, that is:

Theorem.   *There is a unique regular (non-negative) Borel measure $\hat{\mu}$ on $\hat{G}$ with the property that, for every pair f, g of elements of B, $\hat{f}\hat{g}$ is $\hat{\mu}$-summable and*

$$\int_G (f * g)(x)d\mu x = \int_{\hat{G}} \hat{f}(\chi)\hat{g}(\chi)d\hat{\mu}\chi. \tag{13}$$

Remark.   Of course $\hat{\mu}$ need not be bounded.

Definition.   In the present group context, we refer to $\hat{\mu}$ as the *Fourier transform* of μ.

Remark.   In general it is false that $\hat{f}$ is $\hat{\mu}$-summable for every f in B. Therefore (13) cannot be strengthened to the assertion that $\int f(x)d\mu x = \int \hat{f}(\chi)d\hat{\mu}\chi \, (f \in B)$. As an example, let $G = \mathbb{E}$, and let μ be the unit mass at 1, so that (by 2.10) $\hat{\mu}$ is counting measure on $\hat{\mathbb{E}} \cong \mathbb{Z}$. In Remark IX.8.11 we mentioned the existence of a function f in $\mathcal{L}(\mathbb{E})$ whose Fourier series does not converge pointwise to f. For this f it follows easily that $\sum_{n \in \mathbb{Z}} |\hat{f}(n)| = \infty$, that is, $\hat{f} \notin \mathcal{L}_1(\hat{\mu})$.

2.8.   Suppose that ϕ is a function of positive type on G and that $d\mu x = \phi(x)d\lambda x$. By VIII.20.10 μ is a regular complex Borel measure of positive type. Equation (12) then shows that the Fourier transform of μ is exactly the ν of Theorem 2.6, and so is *bounded*. Conversely, by Theorem 2.6 every bounded regular Borel measure ν on $\hat{G}$ is the Fourier transform of $\phi(x)d\lambda x$, where ϕ is the function of positive type on G given by (11).

Remark.   In analogy with the preceding paragraph, one might at first sight conjecture that every unbounded regular Borel measure on $\hat{G}$ is the Fourier transform of some regular complex Borel measure of positive type on G. This however is far from true. To obtain a counter-example, we recall from

Remark 1.7 that, if $f \in \mathcal{L}(G)$, the Fourier transform $\hat{f}$ of f need not have compact support in $\hat{G}$. Choose G, and a function f in $\mathcal{L}(G)$, so that $\hat{f}$ does not have compact support. Then we can find an (unbounded) regular Borel measure ν on $\hat{G}$ such that $\int |\hat{f}(\chi)|^2 \, d\nu\chi = \infty$; and by Theorem 2.7 ν cannot be the Fourier transform of any complex Borel measure of positive type on G.

This leads us to refine the conjecture as follows: Let ν be any regular Borel measure on $\hat{G}$ such that $|\hat{f}|^2$ is ν-summable for every f in $\mathcal{L}(G)$. Then is ν necessarily the Fourier transform of some regular complex Borel measure on G of positive type? We do not know.

2.9.   We shall now translate VI.21.6 into the group context.

Let μ be a regular complex Borel measure of positive type on G, and $\hat{\mu}$ its Fourier transform. By VIII.13.6 μ generates a unitary representation T of G. Let $\rho: B \to X(T)$ denote the natural quotient map (so that $(\rho(f), \rho(g))_{X(T)} = \int (g^* * f) d\mu$). Furthermore, let S be the unitary representation of G on $\mathcal{L}_2(\hat{\mu})$ given by:

$$(S_x g)(\chi) = \chi(x)g(\chi) \qquad (g \in \mathcal{L}_2(\hat{\mu}); \, x \in G; \, \chi \in \hat{G}). \qquad (14)$$

We leave it to the reader to check that (14) does indeed define a unitary representation of G. Notice from (13) that

$$\hat{f} \in \mathcal{L}_2(\hat{\mu}) \qquad \text{for all } f \text{ in } \mathcal{L}(G). \qquad (15)$$

Theorem.   *With the above notation, the equation*

$$\Phi(\rho(f)) = \hat{f} \qquad (f \in \mathcal{L}(G)) \qquad (16)$$

determines a linear isometry Φ of $X(T)$ onto $\mathcal{L}_2(\hat{\mu})$ which intertwines T and S.

We leave to the reader the minor technical problems of deducing this theorem from VI.21.6.

2.10.   We come now to an extremely important special case. Suppose that μ is the unit mass at the unit element e of G:

$$\int f \, d\mu = f(e) \qquad (f \in B).$$

We have already observed in VI.18.8 that μ is then of positive type. So μ has a Fourier transform $\hat{\mu}$ which is a regular Borel measure on the locally compact Abelian group $\hat{G}$. What is this $\hat{\mu}$? The answer is as follows:

Theorem. $\hat{\mu}$ *is a Haar measure on $\hat{G}$.*

Proof. If $f, g \in \mathscr{L}(G)$, we have $\mu(g^* * f) = (g^* * f)(e) = \int f(x)\overline{g(x)}d\lambda x$. So (13), together with 1.7(5), asserts that

$$\int_G f(x)\overline{g(x)}d\lambda x = \int_{\hat{G}} \hat{f}(\phi)\overline{\hat{g}(\phi)}d\hat{\mu}\phi \qquad (17)$$

for $f, g \in \mathscr{L}(G)$.

Now fix an element χ of $\hat{G}$, and let ν be the χ-translate of $\hat{\mu}$:

$$\nu(W) = \hat{\mu}(\chi^{-1}W).$$

Obviously $\hat{f}(\chi\phi) = (f\chi)\hat{}(\phi)$ ($f \in \mathscr{L}(G)$). So, for any two elements f, g of $\mathscr{L}(G)$ we have:

$$\int \hat{f}(\phi)\overline{\hat{g}(\phi)}d\nu\phi = \int \hat{f}(\chi\phi)\overline{\hat{g}(\chi\phi)}d\hat{\mu}\phi$$

$$= \int (f\chi)\hat{}(\phi)\overline{(g\chi)\hat{}(\phi)}d\hat{\mu}\phi$$

$$= \int (f\chi)(x)\overline{(g\chi)(x)}d\lambda x \qquad \text{(by (17))}$$

$$= \int f(x)\overline{g(x)}d\lambda x = \mu(g^* * f).$$

From this and the uniqueness statement in Theorem 2.7 we conclude that $\nu = \hat{\mu}$. Thus $\hat{\mu}$ is invariant under translation by arbitrary elements of $\hat{G}$. It must therefore be a Haar measure. ∎

2.11. We have observed in VIII.14.7 that the unitary representation of G generated by the unit mass at the unit element is just the regular representation. Combining this with 2.9 and 2.10, we get the extremely important Plancherel Theorem for a locally compact Abelian group G. As usual λ is a fixed Haar measure on G.

Plancherel Theorem. *With a proper normalization of the Haar measure ν on $\hat{G}$, the following statements hold:*

(I) *If $f \in \mathscr{L}(G)$, then the Fourier transform $\hat{f}$ of f, given by $\hat{f}(\chi) = \int f(x)\chi(x)d\lambda x$ ($\chi \in \hat{G}$), belongs to $\mathscr{L}_2(\nu)$; and*

$$\int_G |f(x)|^2 \, d\lambda x = \int_{\hat{G}} |\hat{f}(\chi)|^2 \, d\nu\chi. \qquad (18)$$

(II) *The map $f \mapsto \hat{f}$ ($f \in \mathscr{L}(G)$) extends to a (unique) linear isometry F of $\mathscr{L}_2(\lambda)$ onto $\mathscr{L}_2(\nu)$. This F is called the Fourier transform map on $\mathscr{L}_2(\lambda)$.*

(III) *The left-regular representation of G is unitarily equivalent, under the Fourier transform map, with the unitary representation S of G on $\mathscr{L}_2(v)$ given by*

$$(S_x f)(\chi) = \chi(x) f(\chi) \qquad (x \in G; f \in \mathscr{L}_2(v); \chi \in \hat{G}).$$

Thus we have an illuminating description of the regular representation of a locally compact Abelian group. In view of (III), *the regular representation of G has the same measure class as Haar measure on $\hat{G}$* (see Remark 2.1).

Remark. In the next section we will extend (I) of the above theorem to arbitrary functions in $\mathscr{L}_1(\lambda) \cap \mathscr{L}_2(\lambda)$.

Cognateness

2.12. By the Plancherel Theorem 2.11, each normalization λ of Haar measure on G gives rise to a unique normalization v of Haar measure on the character group $\hat{G}$ such that (18) holds. If λ and v are thus related, we say that they are *cognate*, or that v is the Haar measure on $\hat{G}$ *cognate* to λ.

We close this section with a few elementary facts about cognateness.

2.13. If λ and v are cognate and k is any positive constant, one verifies that $k\lambda$ and $k^{-1}v$ are also cognate.

2.14. Let $G_1, \ldots, G_n$ be locally compact Abelian groups, with Haar measures $\lambda_1, \ldots, \lambda_n$ respectively; and let v_i be the Haar measure on $(G_i)\hat{\ }$ which is cognate to λ_i. Then the product Haar measure $v_1 \times \cdots \times v_n$ of $(G_1)\hat{\ } \times \cdots \times (G_n)\hat{\ }$ (see III.9.3) is cognate to the product Haar measure $\lambda_1 \times \cdots \times \lambda_n$ of $G_1 \times \cdots \times G_n$.

2.15. Suppose that G is compact, so that by 1.14 $\hat{G}$ is discrete. Then it follows from IX.9.7 that the Haar measure λ of G which is normalized to satisfy $\lambda(G) = 1$ is cognate with the "counting" measure on $\hat{G}$ (which assigns measure 1 to each one-element set).

2.16. We shall now answer the question of cognateness on $\mathbb{R}$. Let λ be ordinary Lebesgue measure on $\mathbb{R}$. Let Γ be the isomorphism of $\mathbb{R}$ and $\hat{\mathbb{R}}$ which sends s into the character $t \mapsto e^{ist}$ (see 1.11); and let λ' be the image of λ under Γ. Then we claim that λ and $(2\pi)^{-1}\lambda'$ are cognate.

Indeed: Let us regard Γ as an identification. Then the Fourier transform $\hat{f}$ of an element f of $\mathscr{L}_1(\mathbb{R})$ is again defined on $\mathbb{R}$:

$$\hat{f}(s) = \int_{-\infty}^{\infty} f(t) e^{ist} \, d\lambda t.$$

Consider the particular function $g(t) = e^{-(1/2)t^2}$ in $\mathscr{L}_1(\mathbb{R})$. We have

$$\hat{g}(s) = \int_{-\infty}^{\infty} e^{-(1/2)t^2 + ist}\, d\lambda t$$

$$= e^{-(1/2)s^2} \int_{-\infty}^{\infty} e^{-(1/2)(t - is)^2}\, d\lambda t. \tag{19}$$

A simple argument using contour integrals shows that $\int_{-\infty}^{\infty} e^{-(1/2)(t - is)^2}\, d\lambda t = \int_{-\infty}^{\infty} e^{-(1/2)t^2}\, d\lambda t$ for any real s; and it is a well-known fact of analysis that the latter integral is just $(2\pi)^{1/2}$. So by (19)

$$\hat{g}(s) = (2\pi)^{1/2} g(s) \qquad\qquad (s \in \mathbb{R}). \tag{20}$$

Now it will be proved in 3.2 that the Fourier transform map F of 2.11(II) coincides with the map $f \mapsto \hat{f}$ not only on $\mathscr{L}(\mathbb{R})$ but on all of $\mathscr{L}_1(\lambda) \cap \mathscr{L}_2(\lambda)$; for the present discussion we shall assume this. Consequently, if k is such a positive constant that λ and $k\lambda'$ are cognate, applying 2.11(II) to g we obtain

$$\int_{-\infty}^{\infty} g(t)^2\, d\lambda t = k \int_{-\infty}^{\infty} \hat{g}(s)^2\, d\lambda s$$

$$= 2\pi k \int_{-\infty}^{\infty} g(s)^2\, d\lambda s \qquad\qquad \text{(by (20)).}$$

It follows that $2\pi k = 1$, or $k = (2\pi)^{-1}$; and the claim is proved.

3.   The Fourier Inversion Theorem, Regularity and Pontryagin Duality

3.1.   Let v be the Haar measure on $\hat{G}$ which is cognate to λ (2.12), and let F be the Fourier transform map as defined in 2.11, which sends $\mathscr{L}_2(\lambda)$ isometrically onto $\mathscr{L}_2(v)$.

3.2.   We shall begin by answering an obvious question: Do $f \mapsto \hat{f}$ and F coincide on all of their common domain? The answer is "yes."

Proposition.   *If $g \in \mathscr{L}_1(\lambda) \cap \mathscr{L}_2(\lambda)$, then $F(g) = \hat{g}$.*

Proof.   By definition this is true for $g \in \mathscr{L}(G)$. Now let g be an arbitrary function in $\mathscr{L}_1(\lambda) \cap \mathscr{L}_2(\lambda)$. We claim that there exists a sequence $\{g_n\}$ of elements of $\mathscr{L}(G)$ which approaches g in both the $\mathscr{L}_1$ and the $\mathscr{L}_2$ norms. To see this, we first observe that, for each n, there is a bounded measurable complex function h_n on G with compact support such that $\|h_n - g\|_1 < n^{-1}$

and $\|h_n - g\|_2 < n^{-1}$. Fixing n, let us take a compact subset C of G whose interior contains the compact support of h_n. Then there is an element g_n of $\mathscr{L}(G)$ with compact support contained in C such that $\|g_n - h_n\|_2 < k$, where $k = \min\{n^{-1}, (\lambda(C))^{-1/2}n^{-1}\}$. This implies by Schwarz's Inequality that $\|g_n - h_n\|_1 < n^{-1}$ as well as $\|g_n - h_n\|_2 < n^{-1}$. Combining this with the definition of h_n, we obtain both $\|g_n - g\|_1 < 2n^{-1}$ and $\|g_n - g\|_2 < 2n^{-1}$; and the claim is proved.

Since $g_n \to g$ in $\mathscr{L}_2(\lambda)$, we have $F(g_n) \to F(g)$ in $\mathscr{L}_2(\nu)$. So, passing to a subsequence, we can assume (by II.3.5) that

$$F(g_n)(\chi) \to F(g)(\chi) \qquad \text{for } \nu \text{ almost all } \chi. \tag{1}$$

On the other hand, since $g_n \in \mathscr{L}(G)$ and $g_n \to g$ in $\mathscr{L}_1(\lambda)$, it follows that $F(g_n) = \hat{g}_n \to \hat{g}$ pointwise on $\hat{G}$. Combining this with (1) we see that $F(g)$ and $\hat{g}$ coincide ν almost everywhere, that is, as elements of $\mathscr{L}_2(\nu)$. ∎

Remark. In view of this, we may sometimes write $F(g)$ instead of $\hat{g}$ for arbitrary members g of $\mathscr{L}_1(\lambda)$.

3.3. Corollary *If $g \in \mathscr{L}_1(\lambda) \cap \mathscr{L}_2(\lambda)$, then $\hat{g} \in \mathscr{L}_2(\nu)$.*

3.4. *Notation*. It will be convenient to denote by $\tilde{}$ the image of a function on G (or $\hat{G}$) under the inverse map: $\tilde{g}(x) = g(x^{-1})$. We shall also write

$$\tilde{F}(g) = F(\tilde{g}) \qquad (g \in \mathscr{L}_1(\lambda) \text{ or } \mathscr{L}_2(\lambda)). \tag{2}$$

Thus we have the formulae (for $g, g_i \in \mathscr{L}_1(\lambda)$):

$$\tilde{F}(g)(\chi) = \int \overline{\chi(x)} g(x) d\lambda x, \tag{3}$$

$$\tilde{F}(g_1 * g_2) = \tilde{F}(g_1)\tilde{F}(g_2), \tag{4}$$

$$\tilde{F}(g^*) = \overline{\tilde{F}(g)}. \tag{5}$$

If f is a function in $\mathscr{L}_1(\lambda)$ and $x \in G$, put $f_x(y) = f(xy)$. Then, for $x \in G$, $\chi \in \hat{G}$,

$$(f * g)_x = f_x * g, \tag{6}$$

$$(f_x)\hat{}(\chi) = \overline{\chi(x)}\hat{f}(\chi). \tag{7}$$

Notice also that, if $\chi, \chi_0 \in \hat{G}$,

$$(\chi_0 f)\hat{}(\chi) = \hat{f}(\chi_0 \chi). \tag{8}$$

3.5. *Definition*. Let us denote by $\mathscr{S}$ the linear span of $\{f * g : f, g \in \mathscr{L}_1(\lambda) \cap \mathscr{L}_2(\lambda)\}$. We notice (see III.11.13) that the functions in $\mathscr{S}$ are continuous functions in $\mathscr{L}_1(\lambda)$.

Using approximate units, we easily verify that $\mathscr{S}$ is dense in both $\mathscr{L}_1(\lambda)$ and $\mathscr{L}_2(\lambda)$.

3.6. Inversion Theorem (Preliminary form). *If $f \in \mathscr{S}$, then $\hat{f} \in \mathscr{L}_1(v)$, and*

$$f(x) = \int_{\hat{G}} \overline{\chi(x)} \hat{f}(\chi) dv\chi \qquad \text{for all } x \text{ in } G. \qquad (9)$$

Proof. Suppose that $f, g \in \mathscr{L}_1(\lambda) \cap \mathscr{L}_2(\lambda)$. Then $(f * g)\hat{} = \hat{f}\hat{g}$, and $\hat{f}$ and $\hat{g}$ are in $\mathscr{L}_2(v)$ by 3.3. So $(f * g)\hat{} \in \mathscr{L}_1(v)$. This proves the first statement of the theorem.

Notice   that $(f * g^*)(e) = \int f(y)\overline{g(y)}d\lambda y = (f, g)_2 = (Ff, Fg)_2 = (\hat{f}, \hat{g})_2$ (by 3.2). Hence for any x in G we get

$$(f * g^*)(x) = (f_x * g^*)(e) \qquad \qquad \text{(by (6))}$$

$$= ((f_x)\hat{}, \hat{g})_2$$

$$= \int \overline{\chi(x)} \hat{f}(\chi)\overline{\hat{g}(\chi)}dv\chi \qquad \text{(by (7))}$$

$$= \int \overline{\chi(x)}(f * g^*)\hat{}(\chi)dv\chi.$$

By the arbitrariness of f and g this proves (9).   ∎

3.7. Our next aim is to enlarge the class of functions for which (9) holds.

Proposition. *Let $\phi \in \mathscr{L}_1(v) \cap \mathscr{L}_2(v)$; and define*

$$f(x) = \int \overline{\chi(x)}\phi(\chi)dv\chi \qquad (x \in G). \qquad (10)$$

Then

  (I) *f is bounded and continuous on G and belongs to $\mathscr{L}_2(\lambda)$, and*
  (II) *$F(f) = \phi$.*

Proof. The boundedness and continuity of f are an easy exercise. To prove that $f \in \mathscr{L}_2(\lambda)$, take any function $g \in \mathscr{L}_1(\lambda) \cap \mathscr{L}_2(\lambda)$. Then by Fubini's Theorem and (3),

$$\int f(x)g(x)d\lambda x = \iint g(x)\overline{\chi(x)}\phi(\chi)dv\chi \, d\lambda x$$

$$= \int \phi(\chi)(\tilde{F}g)(\chi)dv\chi; \qquad (11)$$

and so

$$\left| \int f(x)g(x)d\lambda x \right| \le \|\phi\|_2 \|\tilde{F}g\|_2 = \|\phi\|_2 \|g\|_2. \tag{12}$$

Now, if $f \notin \mathscr{L}_2(\lambda)$, we can choose g so that $f(x)g(x) \ge 0$, $|g(x)| \le |f(x)|$ for all x, and $\|g\|_2 > \|\phi\|_2$. Then (12) implies that $\|g\|_2^2 \le \|\phi\|_2 \|g\|_2$, or $\|g\|_2 \le \|\phi\|_2$, a contradiction. So $f \in \mathscr{L}_2(\lambda)$. This proves (I).

Consequently, for any g in $\mathscr{L}_1(\lambda) \cap \mathscr{L}_2(\lambda)$, we have by 2.11

$$\int f(x)g(x)d\lambda x = \int (Ff)(\chi)\overline{(F\bar{g})(\chi)}dv\chi$$

$$= \int (Ff)(\chi)(\tilde{F}g)(\chi)dv\chi \qquad (\text{since } \overline{F\bar{g}} = \tilde{F}g).$$

Comparing this with (11) and noting that the $\tilde{F}g$ run over a dense subspace of $\mathscr{L}_2(v)$, we conclude that $\phi = F(f)$. So the proposition is proved.   ∎

3.8   Inversion Theorem.   *Suppose that $f \in \mathscr{L}_2(\lambda)$ and $F(f) \in \mathscr{L}_1(v)$. Then:*

(I)   *f is λ-almost everywhere equal to a continuous function f' on G, and*
(II)   *$f'(x) = \int_{\hat{G}} \overline{\chi(x)}(Ff)(\chi)dv\chi$ for all x in G.*

Proof.   Apply 3.7 to $\phi = F(f) \in \mathscr{L}_1(v) \cap \mathscr{L}_2(v)$. If we set $f_0(x) = \int \overline{\chi(x)}(Ff)(\chi)dv\chi$ $(x \in G)$, then by 3.7 f_0 is continuous and in $\mathscr{L}_2(\lambda)$ and $F(f_0) = \phi = F(f)$. Since F is one-to-one, it follows that $f = f_0$. This proves the theorem.   ∎

3.9.   Our next step is to show that $\mathscr{L}_1(\lambda)$ has the following important property which is usually referred to as regularity.

Theorem.   *If $\chi \in \hat{G}$ and U is an open neighborhood of χ in $\hat{G}$, there exists a function f in $\mathscr{L}_1(\lambda)$ such that $\hat{f}$ vanishes outside U and $\hat{f}(\chi) \ne 0$.*

Proof.   If $\chi_0 \in \hat{G}$, $f \in \mathscr{L}_1(\lambda)$, $\hat{f}(\chi) \ne 0$, and $\hat{f} \equiv 0$ outside U, then by (8) the Fourier transform of $\chi_0 f$ vanishes outside $\chi_0^{-1}U$ but not at $\chi_0^{-1}\chi$. So it is sufficient to assume from the beginning that $\chi = \textup{+} = $ the unit of $\hat{G}$. Take an open symmetric neighborhood V of $\textup{+}$ in $\hat{G}$ such that $V^2 \subset U$. Let ϕ_1 be a function in $\mathscr{L}_+(\hat{G})$, non-zero at $\textup{+}$ and vanishing outside V. Then $\phi_2 = \phi_1 * \phi_1$ vanishes outside U and $\phi_2(\textup{+}) > 0$. It is enough to show that $\phi_2 = F(f)$ for some f in $\mathscr{L}_1(\lambda)$.

Let $f_i(x) = \int \overline{\chi(x)}\phi_i(\chi)dv\chi$ $(x \in G; i = 1, 2)$. By 3.7 $f_i \in \mathscr{L}_2(\lambda)$ and $F(f_i) = \phi_i$. By (4) (applied on $\hat{G}$ instead of G) we have $f_2 = f_1^2$. Since $f_1 \in \mathscr{L}_2(\lambda)$ this implies that $f_2 \in \mathscr{L}_1(\lambda)$. Since $F(f_2) = \phi_2$, the proof is complete.   ∎

3.10. Notice that every point x in G gives rise to an element η_x of the character group $\hat{\hat{G}}$ of $\hat{G}$, namely,

$$\eta_x(\chi) = \chi(x) \qquad\qquad (\chi \in \hat{G}). \qquad (13)$$

Clearly the map $\eta: x \mapsto \eta_x$ is a group homomorphism of G into $\hat{\hat{G}}$. We know from 1.4 that it is one-to-one. Also, we observed in the proof of 2.1 that η is continuous.

3.11. Pontryagin Duality Theorem. *The η of 3.10 is an isomorphism and homeomorphism onto $\hat{\hat{G}}$.*

Proof. We shall first show that η is a homeomorphism onto its range $\eta(G)$. Since it is one-to-one and continuous (3.10), and since η^{-1} is a homomorphism, we need only show that η^{-1} is continuous at the unit $\bar{e}$ of $\hat{\hat{G}}$. That is, we need only fix a neighborhood U of e, and find a neighborhood W of $\bar{e}$ such that $\eta^{-1}(W) \subset U$.

Let V be a compact symmetric G-neighborhood of e with $V^2 \subset U$. Let f be a non-zero function in $\mathcal{L}_+(G)$ with compact support contained in V; and put $g = f^* * f$. Then g has compact support contained in U, $g(e) > 0$, and $\hat{g} \in \mathcal{L}_1(\nu)$ by 3.6. Since the topology of $\hat{\hat{G}}$ is that of pointwise convergence on $\mathcal{L}_1(\nu)$, there is a neighborhood W of $\bar{e}$ in $\hat{\hat{G}}$ such that

$$\alpha \in W \Rightarrow |\alpha(\hat{g}) - \bar{e}(\hat{g})| < \tfrac{1}{2}g(e). \qquad (14)$$

Now let $x \in \eta^{-1}(W)$, that is, $\eta_x \in W$. Since $\eta_x(\hat{g}) = \int \hat{g}(\chi)\chi(x)d\nu\chi = \int \hat{g}(\chi)\overline{\chi(x)}d\nu\chi = g(x)$ by 3.6, and similarly $\bar{e}(\hat{g}) = g(e)$, we see from (14) that $|g(x) - g(e)| < \tfrac{1}{2}g(e)$. In particular $g(x) \neq 0$; and this implies that x is in the support of g, hence in U. So we have shown that $\eta^{-1}(W) \subset U$. Therefore η is a homeomorphism of G onto $\eta(G)$.

It follows that $\eta(G)$ with the relativized topology of $\hat{\hat{G}}$ is a locally compact subgroup of $\hat{\hat{G}}$. By III.1.16 $\eta(G)$ is therefore *closed* in $\hat{\hat{G}}$.

It remains only to show that η is onto $\hat{\hat{G}}$. Suppose it is not. Since $\eta(G)$ is closed, Theorem 3.9 (applied to $\hat{G}$) gives us a non-zero element ϕ of $\mathcal{L}_1(\nu)$ such that $\hat{\phi}$ vanishes on $\eta(G)$, that is,

$$\int \phi(\chi)\chi(x)d\nu\chi = 0 \qquad \text{for all } x \text{ in } G. \qquad (15)$$

Thus for every f in $\mathcal{L}_1(\lambda)$, (15) and Fubini's Theorem give

$$\int \phi(\chi)\hat{f}(\chi)d\nu\chi = \iint \phi(\chi)\chi(x)f(x)d\lambda x \, d\nu\chi = 0. \qquad (16)$$

On the other hand, by 1.7 $\{\hat{f}: f \in \mathscr{L}_1(\lambda)\}$ is dense in $\mathscr{C}_0(\hat{G})$. Therefore (16) implies that $\int \phi(\chi)\tau(\chi)dv\chi = 0$ for all τ in $\mathscr{C}_0(\hat{G})$; and this in turn implies that $\phi = 0$, a contradiction. So $\eta(G) = \hat{\hat{G}}$; and the theorem is completely proved. ■

Definition. The isomorphism $\eta: G \to \hat{\hat{G}}$ of the above theorem is called the *canonical isomorphism* of G and $\hat{\hat{G}}$.

3.12. Let us denote by $\mathscr{A}$ the family of all isomorphism classes of locally compact Abelian groups. The Pontryagin Duality Theorem asserts that the map $G \mapsto \hat{G}$, considered as a mapping of $\mathscr{A}$ into $\mathscr{A}$, is a permutation of $\mathscr{A}$ of order 2.

If $\mathscr{B}$ is a subfamily of $\mathscr{A}$, the subfamily $\{\hat{G}: G \in \mathscr{B}\}$ is said to be *dual to $\mathscr{B}$*. If $\mathscr{C}$ is dual to $\mathscr{B}$, then $\mathscr{B}$ is dual to $\mathscr{C}$. It is interesting to ask what are the duals of certain natural subfamilies of $\mathscr{A}$. Clearly the class of all finite Abelian groups is self-dual. By V.7.6 the class of all second-countable locally compact Abelian groups is self-dual.

By 1.14 the class of all compact Abelian groups and the class of all discrete Abelian groups are each other's duals. This says that the map $G \mapsto \hat{G}$ is a one-to-one correspondence between the class of all compact Abelian groups and the class of all discrete Abelian groups.

Many other examples of dual subclasses of $\mathscr{A}$ will be found in Hewitt and Ross [1], §24.

3.13. As an addendum to the Pontryagin Duality Theorem, we should observe that, when G and $\hat{\hat{G}}$ are identified via the canonical isomorphism η, the relation of cognateness (2.12) is symmetric, and the inverse of the Fourier transform F is just $\tilde{F}$ (applied on $\hat{G}$). More precisely, we have:

Proposition. *Let λ and v be Haar measures on G and $\hat{G}$ respectively which are cognate, and let λ' be the η-image of λ in $\hat{\hat{G}}$. Then*

(I) *v and λ' are cognate.*
(II) *If $F: \mathscr{L}_2(\lambda) \to \mathscr{L}_2(v)$ and $F': \mathscr{L}_2(v) \to \mathscr{L}_2(\lambda')$ are the Fourier transform maps, and $\tilde{F}'$ is defined as in 3.4(2), then*

$$f = (\tilde{F}'F(f)) \circ \eta \qquad \text{for all } f \text{ in } \mathscr{L}_2(\lambda). \qquad (17)$$

Proof. Theorem 3.6 asserts that (17) holds for all f in the space $\mathscr{S}$ of 3.5. Since $\mathscr{S}$ is dense in $\mathscr{L}_2(\lambda)$, and since $f \mapsto (\tilde{F}'F(f)) \circ \eta$ is clearly continuous on $\mathscr{L}_2(\lambda)$ to itself, (17) must hold for all f in $\mathscr{L}_2(\lambda)$.

Since F is an isometry, it follows from (17) that $\tilde{F}'$ is an isometry from $\mathscr{L}_2(\mu)$ to $\mathscr{L}_2(\lambda')$. Hence μ and λ' are cognate.   ∎

Applications of Pontryagin Duality

3.14.   We shall conclude this section with some interesting applications of Pontryagin duality.

Proposition.   *Let Γ be a subgroup of $\hat{G}$ which separates the points of G. Then Γ is dense in $\hat{G}$.*

Proof.   Suppose that the closure $\bar{\Gamma}$ of Γ is not equal to $\hat{G}$. Then by 1.4 $\hat{G}/\bar{\Gamma}$ has a non-trivial character; that is, there is a non-constant character α of $\hat{G}$ such that $\alpha(\chi) = 1$ for all χ in Γ. By 3.11 $\alpha = \eta_x$ for some x in G with $x \neq e$. Thus $\chi(x) = 1$ for all χ in Γ; and this contradicts the hypothesis that Γ separates the points of G.   ∎

3.15.   The next result deals with the extension of characters.

Theorem.   *Let H be a closed subgroup of the locally compact Abelian group G. Then every character ϕ of H can be extended to a character of G.*

Proof.   Let $K = \{\chi \in \hat{G}: \chi(h) = 1 \text{ for all } h \text{ in } H\}$. Applying 1.4 to the quotient group G/H, we see that

$$H = \{x \in G: \chi(x) = 1 \quad \text{for all } \chi \text{ in } K\}. \tag{18}$$

We shall now define a continuous surjective homomorphism

$$\alpha: (\hat{G}/K)^{\hat{}} \to H.$$

Let $\eta: G \to \hat{\hat{G}}$ be the canonical isomorphism; and let $\gamma \in (\hat{G}/K)^{\hat{}}$. The composition γ' of γ with the quotient homomorphism $\hat{G} \to \hat{G}/K$ belongs to $\hat{\hat{G}}$, and so $x = \eta^{-1}(\gamma')$ is in G. In fact, if $\chi \in K$, $\chi(x) = \eta_x(\chi) = \gamma'(\chi) = 1$; so by (18) $x \in H$. Putting $x = \alpha(\gamma)$, we obtain a continuous homomorphism $\alpha: (\hat{G}/K)^{\hat{}} \to H$. To see that it is surjective, observe that each x in H gives rise to a character γ_x of $\hat{G}/K$ by means of the formula $\gamma_x(\chi K) = \chi(x)$; and obviously $\alpha(\gamma_x) = x$.

Now let ϕ be a character of H. Thus $\phi' = \phi \circ \alpha \in (\hat{G}/K)^{\hat{\hat{}}}$. By Pontryagin duality applied to $\hat{G}/K$, there is an element χ of $\hat{G}$ such that χK goes into ϕ' under the canonical map, that is, $\phi'(\gamma) = \gamma(\chi K)$ or

$$\phi(\alpha(\gamma)) = \chi(\alpha(\gamma)) \quad \text{for all } \gamma \text{ in } (\hat{G}/K)^{\hat{\hat{}}}.$$

Since α is onto H, this implies that ϕ and χ coincide on H, proving the proposition. ∎

Remark. In 4.8 we will prove a generalization of this extension theorem.

The Character Groups of the Additive Groups of Locally Compact Fields

3.16. Let F be any non-discrete locally compact division ring (see §III.6). As another application of Pontryagin duality we shall show that the character group of the additive group F^+ of F is isomorphic to F^+ (a fact already well known to us for $\mathbb{R}$ and $\mathbb{C}$). We shall write simply $\hat{F}$ instead of $(F^+)\hat{\ }$.

Fix an element ψ of $\hat{F}$ which is non-trivial, i.e., $\psi(x) \neq 1$ for some $x \in F$. For each u in F we can define an element ψ_u of $\hat{F}$ by means of the formula

$$\psi_u(x) = \psi(ux) \qquad (x \in F). \qquad (19)$$

Evidently $u \mapsto \psi_u$ is a homomorphism of F^+ into $\hat{F}$. If $\psi_u = \psi_v$, then $\psi((u - v)x) = 1$ for all x in F, implying by the non-triviality of ψ that $u - v = 0$; so $u \mapsto \psi_u$ is one-to-one. It is also easily seen to be continuous.

Theorem. *The map $u \mapsto \psi_u$ ($u \in F$) is a homeomorphism and isomorphism of F^+ onto $\hat{F}$.*

Proof. It remains to show that $\{\psi_u : u \in F\} = \hat{F}$, and that the inverse of $u \mapsto \psi_u$ is continuous. We shall prove the latter first.

Suppose that $\{u_i\}$, u are in F and that $\psi_{u_i} \to \psi_u$ in $\hat{F}$. We wish to show that $u_i \to u$. If this were not the case, then by the local compactness of F we could pass to a subnet and assume that either $u_i \to v \neq u$ or $u_i \to \infty$ in F. By the continuity and one-to-oneness of $u \mapsto \psi_u$, the first alternative would imply $u = v$, a contradiction. So we assume that $u_i \to \infty$. Choose w so that $\psi(w) \neq 1$. By III.6.9 $u_i^{-1}w \to 0$. Therefore, by the uniform convergence of ψ_{u_i} to ψ_u on compact sets we get

$$1 \neq \psi(w) = \psi_{u_i}(u_i^{-1}w) \to \psi_u(0) = 1,$$

a contradiction. Thus $u_i \to u$; and we have shown that $u \mapsto \psi_u$ is a homeomorphism onto its range.

Let $L = \{\psi_u : u \in F\}$. Then by the preceding paragraph L is a locally compact subgroup of $\hat{F}$, and hence by III.1.16 is closed in $\hat{F}$. On the other hand L clearly separates the points of F, and so by 3.14 is dense in $\hat{F}$. So $L = \hat{F}$; and the proof is complete. ∎

Remark. This theorem is also valid for finite fields F. Indeed, the above proof used the non-discreteness of F only to prove the continuity of $\psi_u \mapsto u$; and this is trivial if F is finite.

The theorem fails, however, if F is infinite and discrete. For in that case $\hat{F}$ is infinite and compact and so cannot be homeomorphic with F.

In spite of this one can prove the following interesting weaker statement in case F is a countably infinite and discrete field. Notice that (19) defines an action $\langle u, \psi \rangle \mapsto \psi_u$ of the multiplicative group $F_* = F \setminus \{0\}$ on $\hat{F}$. If F is non-discrete, the above theorem asserts that this action is transitive on $\hat{F} \setminus \{1\}$. If F is a countably infinite discrete field, it can be shown that this action, though not transitive, is ergodic (in the sense of VIII.19.3) with respect to the Haar measure on $\hat{F}$.

4. Commutative Banach *-Algebraic Bundles

4.1. Throughout this section G is a fixed locally compact Abelian group, with unit e and Haar measure λ; and $\mathscr{B} = \langle B, \pi, \cdot, * \rangle$ is a fixed commutative Banach *-algebraic bundle over G. We shall denote the unit fiber algebra B_e by A. The following two further assumptions on $\mathscr{B}$ will remain in force throughout this section:

(I)　If $b \in B$ and $\tau \in \hat{A}$, then $\tau(b^*b) \geq 0$.

(II)　$\mathscr{B}$ is saturated.

Notice that assumption (I) holds automatically if $\mathscr{B}$ either is a C^*-algebraic bundle or has enough unitary multipliers; for, in either of these cases, for every b in B there is an element a of A such that $b^*b = a^*a$. (This was pointed out in the proof of VIII.14.5 for the case that $\mathscr{B}$ has enough unitary multipliers.) Thus, by VIII.2.15, if $\mathscr{B}$ has enough unitary multipliers and also an approximate unit, both conditions (I) and (II) are satisfied.

4.2. The following useful observation follows from the saturation of $\mathscr{B}$: If ϕ is a one-dimensional locally convex representation of $\mathscr{B}$ which is not identicially zero, then $\phi|B_x$ cannot be 0 for any x in G. Indeed, if $y \in G$ and $\phi|B_y = 0$, then $\phi(bc) = \phi(b)\phi(c) = 0$ for every b in B_y and c in B. Since $\mathscr{B}$ is saturated, $\{bc : b \in B_y, c \in B\}$ has dense linear span in every fiber of $\mathscr{B}$; and so ϕ, being continuous, must vanish on all of B, a contradiction.

4.3. We have the following generalization of 1.3.

Proposition. *If $\mathscr{B}$ is a (commutative saturated) C^*-algebraic bundle, then $\mathscr{L}_1(\lambda; \mathscr{B})$ is symmetric.*

Proof. Let ψ be an (automatically continuous) complex homomorphism of $\mathscr{L}_1(\lambda; \mathscr{B})$. By VIII.12.10 ψ is the integrated form of a non-zero one-dimensional representation ϕ of $\mathscr{B}$. Since ψ is continuous in the $\mathscr{L}_1$ norm, an easy argument (like that used in 1.3) shows that ϕ is bounded, that is, there is a constant k satisfying

$$|\phi(b)| \le k\|b\| \qquad\qquad (b \in B). \qquad (1)$$

If for some b we have $|\phi(b)| > \|b\|$, then, choosing $r > 1$ to satisfy $|\phi(b)| > r\|b\|$, we get $|\phi(b^n)| = |\phi(b)|^n > r^n\|b\|^n \ge r^n\|b^n\|$ for all $n = 1, 2, \ldots$; and this contradicts (1) since $r^n > k$ for large n. So

$$|\phi(b)| \le \|b\| \qquad\qquad (b \in B). \qquad (2)$$

Now we claim that there exists a homomorphism σ of G into the multiplicative group of positive reals such that

$$\phi(b^*) = \sigma(\pi(b))\overline{\phi(b)} \qquad\qquad (b \in B). \qquad (3)$$

Indeed: Fix $x \in G$. Notice that $\phi(a^*) = \overline{\phi(a)}$ for all a in $A = B_e$ (since A is a C^*-algebra and so symmetric). Therefore

$$\phi(b^*c) = \overline{\phi(c^*b)} \qquad \text{for all } b, c \text{ in } B_x.$$

Hence

$$(\overline{\phi(b)})^{-1}\phi(b^*) = \phi(c)^{-1}\overline{\phi(c^*)} \qquad\qquad (4)$$

whenever $b, c \in B_x$ and $\phi(b) \ne 0 \ne \phi(c)$. From (4) we deduce (putting $b = c$) that $(\overline{\phi(b)})^{-1}\phi(b^*)$ is real, and also that it is independent of b as long as $b \in B_x$ and $\phi(b) \ne 0$. Denoting either side of (4) by $\sigma(x)$ we obtain

$$\phi(b^*) = \sigma(x)\overline{\phi(b)} \qquad\qquad (5)$$

whenever $b \in B_x$ and $\phi(b) \ne 0$. Now by 4.2 $\{b \in B_x : \phi(b) \ne 0\}$ is dense in B_x. So by continuity (5) holds for all b in B_x. Evidently $\sigma(e) = 1$ and σ is multiplicative on G; so in particular $\sigma(x) \ne 0$. If $b \in B_x$ and $\phi(b) \ne 0$ we thus have by (5) and the positivity of b^*b:

$$0 \le \phi(b^*b) = \phi(b^*)\phi(b) = \sigma(x)|\phi(b)|^2 \ne 0,$$

whence $\sigma(x) > 0$. This completes the verification of the claim.

We shall now show that $\sigma(x) \equiv 1$. If this is not the case, there must be some x in G such that $\sigma(x) < 1$. Take such an x. We shall denote the element $\phi|A$ of $\hat{A}$ by τ. Since the linear span of $\{b^*b : b \in B_x\}$ is dense in A (by saturation), we can pick an element b of B_x such that $\tau(b^*b) = 1$. Put $g = (b^*b)\hat{\ } ($ $\hat{\ }$ denoting the Gelfand transform on A), so that $g \geq 0$ and $g(\tau) = 1$. Define f and h as real functions on $\hat{A}$ as follows:

$$f(\tau') = \begin{cases} g(\tau')^{-1} \min\{g(\tau'), 1\} & \text{if } g(\tau') \neq 0, \\ 1 & \text{if } g(\tau') = 0, \end{cases}$$

$$h = f^{1/2} - 1.$$

Thus $h \in \mathcal{L}(\hat{A})$ and is real. Let a be the element of A such that $\hat{a} = h$, and put $c = b + ba$. Thus $a^* = a$, $c \in B_x$, and

$$c^*c = b^*b + 2ab^*b + a^2b^*b,$$

whence

$$(c^*c)^{\hat{}} = (1 + h)^2 g = fg = \min\{g, 1\}. \tag{6}$$

This implies

$$\|c\|^2 = \|c^*c\| = \|(c^*c)^{\hat{}}\| \leq 1, \tag{7}$$

and also, since $g(\tau) = 1$,

$$\phi(c^*c) = (c^*c)^{\hat{}}(\tau) = \min\{g(\tau), 1\} = 1. \tag{8}$$

Since $\sigma(x) < 1$, it follows from (5) and (8) that

$$1 = \phi(c^*c) = \phi(c^*)\phi(c) = \sigma(x)|\phi(c)|^2 < |\phi(c)|^2 \tag{9}$$

But (7) and (9) together contradict (2). So we have shown that $\sigma(x) \equiv 1$, or (by (5))

$$\phi(b^*) = \overline{\phi(b)} \qquad\qquad (b \in B). \tag{10}$$

An easy integration shows that (10) implies

$$\psi(f^*) = \overline{\psi(f)} \qquad\qquad (f \in \mathcal{L}_1(\lambda; \mathcal{B})).$$

So $\mathcal{L}_1(\lambda; \mathcal{B})$ is symmetric.   ■

Remark.   The above proposition becomes false if $\mathcal{B}$ merely satisfies the hypotheses of 4.1 (even if we assume in addition that A is symmetric). Here is an example. Let $G = \mathbb{Z}$; and let $\mathcal{B}$ coincide with the group bundle of $\mathbb{Z}$ in all respects except that the norm of an element $\langle u, n \rangle$ of B is $e^{|n|}|u|$ (instead of $|u|$ as in the case of the group bundle). The relations $e^{|n+m|} \leq e^{|n|}e^{|m|}$, $e^{|-n|} = e^{|n|}$ assure us that $\mathcal{B}$ is indeed a Banach *-algebraic bundle satisfying the conditions of 4.1. If λ is counting measure on $\mathbb{Z}$, $\mathcal{L}_1(\lambda; \mathcal{B})$ can be identified with the Banach *-algebra (under ordinary convolution and involution) of all complex functions f on $\mathbb{Z}$ satisfying $\sum_{n=-\infty}^{\infty} e^{|n|}|f(n)| < \infty$. Now it is clear that $\psi: f \mapsto \sum_{n=-\infty}^{\infty} e^n f(n)$ is a complex homomorphism of $\mathscr{L}_1(\lambda; \mathscr{B})$ which fails to satisfy the identity $\psi(f^*) = \overline{\psi(f)}$.

Certain Generalized Regular Representations

4.4.   We shall be very interested in the generalized regular representations T^τ of $\mathscr{B}$ (see VIII.14.8) which arise from elements τ of $\hat{A}$.

Let τ be an element of $\hat{A}$. If $f \in \mathscr{L}(\mathscr{B})$ we have

$$\tau[(f^* * f)(e)] = \int \tau(f(x)^* f(x)) d\lambda x \geq 0 \qquad \text{(by Condition 4.1(I)).} \qquad (11)$$

Hence the equation

$$p^\tau(f) = \tau(f(e)) \qquad\qquad (f \in \mathscr{L}(\mathscr{B}))$$

defines a positive linear functional p^τ on $\mathscr{L}(\mathscr{B})$ which is continuous in the inductive limit topology. By VIII.13.4 p^τ generates a *-representation T^τ of $\mathscr{B}$. We denote by q^τ the natural quotient map of $\mathscr{L}(\mathscr{B})$ into $X(T^\tau)$, so that

$$(q^\tau(f), q^\tau(g)) = p^\tau(g^* * f) \qquad (f, g \in \mathscr{L}(\mathscr{B})). \qquad (12)$$

Notice that $X(T^\tau) \neq \{0\}$. Indeed, since $\tau \neq 0$, we can choose f in $\mathscr{L}(\mathscr{B})$ to satisfy $\tau(f(e)) \neq 0$, whence $\tau(f(e)^* f(e)) = |\tau(f(e))|^2 \neq 0$. From this, (11), and (12), we deduce that $q^\tau(f) \neq 0$.

Proposition.   *If $a \in A$ then*

$$T_a^\tau = \tau(a)\mathbf{1} \qquad (13)$$

(**1** *being the identity operator on $X(T^\tau)$). In particular, T^τ is non-degenerate.*

Proof.   We shall abbreviate $q^\tau(f)$ to $\tilde{f}$.

Notice that

$$p^\tau(af) = \tau(a)p^\tau(f) \qquad (a \in A; f \in \mathscr{L}(\mathscr{B})). \qquad (14)$$

If $a \in A$ and $f, g \in \mathscr{L}(\mathscr{B})$ we have $T_a^\tau \tilde{f} = (af)^{\sim}$; and so

$$
\begin{aligned}
(T_a^\tau \tilde{f}, \tilde{g}) &= ((af)^{\sim}, \tilde{g}) \\
&= p^\tau(g^* * af) && \text{(by (12))} \\
&= \tau(a)p^\tau(g^* * f) && \text{(by (14))} \\
&= \tau(a)(\tilde{f}, \tilde{g}) && \text{(by (12)).}
\end{aligned}
$$

By the denseness of $\{\tilde{f}: f \in \mathscr{L}(\mathscr{B})\}$ in $X(T^\tau)$, this implies (13).   ∎

4.5. Proposition. *The map* $\tau \mapsto T^{\tau}$ *is continuous on* $\hat{A}$ *with respect to the regional topology of* *-*representations of* $\mathcal{B}$.

Proof. Suppose that $\tau^{\alpha} \to \tau$ in $\hat{A}$. For brevity we write p^{α}, p for $p^{\tau^{\alpha}}$, p^{τ}; T^{α}, T for $T^{\tau^{\alpha}}$, T^{τ}; q^{α}, q for $q^{\tau^{\alpha}}$, q^{τ}.

Recalling VII.1.3, let $U = U(T; \varepsilon; \{\xi_i\}; F)$, where $\varepsilon > 0$, F is a finite subset of $\mathcal{L}(\mathcal{B})$, and $\xi_i = q(f_i)$ for $i = 1, \ldots, p$ ($f_i \in \mathcal{L}(\mathcal{B})$). By VII.1.16 such U form a basis of regional neighborhoods of T. So the proposition will be proved if we show that $T^{\alpha} \in U$ for all large enough α.

For each α and each $i = 1, \ldots, p$ put $\xi_i^{\alpha} = q^{\alpha}(f_i)$. Then $\xi_i^{\alpha} \in X(T^{\alpha})$; and

$$(\xi_i^{\alpha}, \xi_j^{\alpha}) = p^{\alpha}(f_j^* * f_i) = \tau^{\alpha}[(f_j^* * f_i)(e)]$$

$$\xrightarrow[\alpha]{} \tau[(f_j^* * f_i)(e)] = p(f_j^* * f_i) = (\xi_i, \xi_j). \tag{15}$$

Similarly, if $g \in \mathcal{L}(\mathcal{B})$,

$$(T_g^{\alpha} \xi_i^{\alpha}, \xi_j^{\alpha}) = \tau^{\alpha}[(f_j^* * g * f_i)(e)]$$

$$\xrightarrow[\alpha]{} \tau[(f_j^* * g * f_i)(e)] = (T_g \xi_i, \xi_j). \tag{16}$$

It follows from (15) and (16) that $T^{\alpha} \in U$ for all large enough α. ∎

$\hat{\mathcal{B}}$ is a Bundle Over $\hat{A}$ and a $\hat{G}$-space

4.6. The structure space of a commutative Banach *-algebra is, as we know, a locally compact Hausdorff space. In particular, the structure space $\hat{\mathcal{B}}$ of $\mathcal{B}$ is a locally compact Hausdorff space (essentially coinciding with $(\mathcal{L}_1(\lambda; \mathcal{B}))\hat{\ }$). It consists of all non-zero one-dimensional *-representations of $\mathcal{B}$, that is, of continuous maps $\psi : B \to \mathbb{C}$ which are not identically zero, are linear on each fiber, preserve multiplication, and satisfy $\psi(b^*) = \overline{\psi(b)}$ ($b \in B$); and the topology of $\hat{\mathcal{B}}$ is that of uniform convergence on compact subsets of B (see VIII.21.12).

Now $\hat{\mathcal{B}}$ has more structure than merely its topology. For one thing, it turns out to be a bundle over $\hat{A}$ under the restriction map $\rho : \psi \mapsto \psi | A$. Also, it is a $\hat{G}$-space under the tensor product action of $\hat{G}$: $\langle \chi, \psi \rangle \mapsto \chi \otimes \psi$ ($\chi \in \hat{G}$; $\psi \in \hat{\mathcal{B}}$). We shall show that these two structures are intimately related; indeed, the fibers $\rho^{-1}(\tau)$ ($\tau \in \hat{A}$) are exactly the orbits in $\hat{\mathcal{B}}$ under the action of $\hat{G}$.

4.7. **Definition.** Let $\rho : \mathscr{\hat{B}} \to \hat{A}$ be the restriction map: $\rho(\psi) = \psi | A$.

Notice from 4.2 that an element ψ of $\mathscr{\hat{B}}$ does not vanish on A, and hence that $\rho(\mathscr{\hat{B}}) \subset \hat{A}$. From the description of $\mathscr{\hat{B}}$ recalled in 4.6 it is obvious that ρ is continuous. We are going to show that it is surjective and open.

4.8. **Proposition.** $\rho(\mathscr{\hat{B}}) = \hat{A}$.

Proof. Take any element τ of $\hat{A}$. Since T^τ is a non-zero representation (see 4.4), it weakly contains some element ψ of $\mathscr{\hat{B}}$ (see VII.5.16). This implies by VIII.21.20 that $T^\tau | A$ weakly contains the element $\rho(\psi)$ of $\hat{A}$. Now by Proposition 4.4 $T^\tau | A$ is a direct sum of copies of τ; and therefore by VII.1.21 τ weakly contains $T^\tau | A$. The last two sentences imply that τ weakly contains $\rho(\psi)$. By VII.5.10 this means that $\tau = \rho(\psi)$; and the proposition is proved. ■

4.9. Before proving that ρ is open we shall look at the natural action of $\hat{G}$ on $\mathscr{\hat{B}}$.

If $\chi \in \hat{G}$ and $\psi \in \mathscr{\hat{B}}$, let us set

$$(\chi\psi)(b) = \chi(\pi(b))\psi(b) \qquad\qquad (b \in B). \qquad (17)$$

Clearly $\chi\psi \in \mathscr{\hat{B}}$; in fact $\chi\psi$ is the tensor product $\chi \otimes \psi$ defined in VIII.9.16. If 1 is the unit character of G, we have

$$1\psi = \psi, \qquad \chi_1(\chi_2\psi) = (\chi_1\chi_2)\psi \qquad (\chi_i \in \hat{G}; \psi \in \mathscr{\hat{B}}).$$

Since the topologies of $\hat{G}$ and $\mathscr{\hat{B}}$ are those of uniform convergence on compact sets, the map $\langle \chi, \psi \rangle \mapsto \chi\psi$ is continuous on $\hat{G} \times \mathscr{\hat{B}}$ to $\mathscr{\hat{B}}$, and so defines $\mathscr{\hat{B}}$ as a topological $\hat{G}$-transformation space.

In speaking of $\mathscr{\hat{B}}$ as a $\hat{G}$-transformation space, it is always this action which we have in mind.

We observe that the action of $\hat{G}$ on $\mathscr{\hat{B}}$ is *effective*, that is,

$$\chi \in \hat{G}, \quad \psi \in \mathscr{\hat{B}}, \quad \chi\psi = \psi \Rightarrow \chi = 1. \qquad (18)$$

Indeed, assume that $\psi \in \mathscr{\hat{B}}$ and $1 \neq \chi \in \hat{G}$; and let x be an element of G such that $\chi(x) \neq 1$. By 4.2 there exists an element b of B_x such that $\psi(b) \neq 0$. We then have $(\chi\psi)(b) = \chi(x)\psi(b) \neq \psi(b)$; so $\chi\psi \neq \psi$, and (18) is proved.

4.10. **Proposition.** *Two elements ϕ and ψ of $\mathscr{\hat{B}}$ satisfy $\rho(\phi) = \rho(\psi)$ if and only if they belong to the same $\hat{G}$-orbit.*

Proof. If $\psi = \chi\phi$ ($\chi \in \hat{G}$), then $\psi(a) = \chi(e)\phi(a) = \phi(a)$ for all a in A; and so $\rho(\psi) = \rho(\phi)$.

Conversely, assume that $\rho(\psi) = \rho(\phi) = \tau \in \hat{A}$. We must find an element χ of $\hat{G}$ such that $\psi = \chi\phi$. Fix an element x of G. By 4.2 $\phi(c) \neq 0$ for some c in $B_{x^{-1}}$. So, if $b, b' \in B_x$ (so that $cb, cb' \in A$), we have:

$$\phi(c)\psi(b)\phi(b') = \phi(cb')\psi(b) = \tau(cb')\psi(b)$$

$$= \psi(cb')\psi(b) = \psi(cb'b)$$

$$= \psi(cb)\psi(b') = \tau(cb)\psi(b')$$

$$= \phi(cb)\psi(b') = \phi(c)\phi(b)\psi(b'),$$

whence, since $\phi(c) \neq 0$,

$$\psi(b)\phi(b') = \phi(b)\psi(b') \qquad \text{for all } b, b' \text{ in } B_x. \tag{19}$$

By 4.2 we can choose b' in B_x to satisfy $\phi(b') \neq 0$. Then (19) gives

$$\psi(b) = \chi(x)\phi(b) \qquad \text{for all } b \text{ in } B_x, \tag{20}$$

where we have put $\chi(x) = (\phi(b'))^{-1}\psi(b')$. By 4.2 the number $\chi(x)$ satisfying (20) is non-zero and unique (i.e., independent of its construction), and satisfies $\chi(xy) = \chi(x)\chi(y)$, $\chi(e) = 1$. If $b \in B_x$ then $b^* \in B_{x^{-1}}$, and we have by (20) $\chi(x^{-1})\phi(b^*) = \psi(b^*) = \overline{\psi(b)} = \overline{\chi(x)\phi(b)} = \overline{\chi(x)}\phi(b^*)$; so $\chi(x^{-1}) = \overline{\chi(x)}$. Thus χ is a homomorphism of G into $\mathbb{E}$. Furthermore it follows from (20) and the continuity of ϕ and ψ that χ is continuous. Consequently χ is a character of G, and (20) asserts that $\psi = \chi\phi$. ∎

4.11. Let us define $\theta = \{\langle \phi, \psi \rangle \in \hat{\mathscr{B}} \times \hat{\mathscr{B}} : \rho(\phi) = \rho(\psi)\}$. Thus θ is a closed subset of $\hat{\mathscr{B}} \times \hat{\mathscr{B}}$. By 4.9 and 4.10, for each $\langle \phi, \psi \rangle$ in θ there is a unique character χ in $\hat{G}$ such that $\psi = \chi\phi$; denote this χ by $\gamma(\phi, \psi)$.

Proposition. *The mapping $\gamma : \theta \to \hat{G}$ is continuous.*

Proof. Suppose that $\phi_i \to \phi$ and $\psi_i \to \psi$ in $\hat{\mathscr{B}}$, where $\psi_i = \chi_i\phi_i$, $\psi = \chi\phi$ ($\chi_i, \chi \in \hat{G}$). It must be shown that $\chi_i \to \chi$ uniformly on compact subsets of G. To do this, it is enough to show that each point x_0 of G has a neighborhood on which $\chi_i \to \chi$ uniformly.

Fix a point x_0 of G. By 4.2 we may choose a continuous cross-section f of $\mathscr{B}$ such that $\phi(f(x_0)) \neq 0$. Since ϕ is continuous we can choose a compact neighborhood V of x_0 such that

$$\phi(f(x)) \neq 0 \qquad \text{for } x \in V. \tag{21}$$

Now by the definition of χ_i and χ

$$\psi(f(x)) = \chi(x)\phi(f(x)),$$
$$\psi_i(f(x)) = \chi_i(x)\phi_i(f(x)) \tag{22}$$

for all x and all i. Also, since $f(V)$ is compact in B,

$$\phi_i(f(x)) \to \phi(f(x))$$

and
$$\psi_i(f(x)) \to \psi(f(x)) \tag{23}$$

uniformly for x in V. From (21), (22), and (23) we deduce that $\chi_i \to \chi$ uniformly on V.   ∎

4.12.   Corollary.   *For each fixed ψ in $\hat{\mathscr{B}}$, the map $\chi \mapsto \chi\psi$ is a homeomorphism of $\hat{G}$ into $\hat{\mathscr{B}}$.*

4.13.   Proposition.   *ρ is open from $\hat{\mathscr{B}}$ onto $\hat{A}$.*

Proof.   Let U be an open subset of $\hat{\mathscr{B}}$, and put $W = \rho(U)$. We must show that W is open in $\hat{A}$. Notice by 4.10 that $\rho^{-1}(W) = \bigcup\{\chi U : \chi \in \hat{G}\}$; and the latter is open since each χU is open. So $\rho^{-1}(W)$ is open.

Suppose W is not open. Then there is an element τ of W belonging to the closure $(\hat{A} \setminus W)^-$ of $\hat{A} \setminus W$. According to the proof of 4.8, T^τ weakly contains some element ψ of $\hat{\mathscr{B}}$, and $\rho(\psi) = \tau$.

Put $\mathscr{V} = \{T^\sigma : \sigma \in \hat{A} \setminus W\}$. Since $\tau \in (\hat{A} \setminus W)^-$, 4.5 implies that T^τ is weakly contained in $\mathscr{V}$. This and the preceding paragraph show that $\mathscr{V}$ weakly contains ψ.

We recall from VII.5.16 that each T^σ is equivalent to its spectrum, which we shall denote by Z_σ. Also, the proof of 4.8 has shown that

$$Z_\sigma \subset \rho^{-1}(\sigma) \qquad\qquad (\sigma \in \hat{A}). \tag{24}$$

From (24) and the preceding paragraph it follows that $\rho^{-1}(\hat{A} \setminus W)$ weakly contains ψ. By VII.5.10 this implies that ψ lies in the closure of $\rho^{-1}(\hat{A} \setminus W)$. Since we have seen that $\rho^{-1}(W)$ is open, $\rho^{-1}(\hat{A} \setminus W)$ must be closed. So

$$\psi \in \rho^{-1}(\hat{A} \setminus W). \tag{25}$$

On the other hand, $\tau \in W$ and $\psi \in Z_\tau$; so by (24) $\psi \in \rho^{-1}(W)$. This contradicts (25), showing that W is open.   ∎

4.14.   The results obtained so far in this section are summarized in the statement that *$\hat{\mathscr{B}}$ has the structure of a principal $\hat{G}$-bundle over $\hat{A}$* in the sense of the following general definition:

Definition. Let X be any Hausdorff space and H any topological group. By a *principal H-bundle over* X we mean a Hausdorff space Y, together with a continuous open surjection $\pi_0 \colon Y \to X$ and a left action $\langle h, y \rangle \mapsto hy$ of H on Y making Y a left topological H-space, such that the following properties hold:

(i)   The orbits of Y under H are exactly the fibers $\pi_0^{-1}(x)$ $(x \in X)$; that is, for any two points y, y' of Y we have

$$\pi_0(y') = \pi_0(y) \Leftrightarrow y' \in Hy.$$

(ii) H acts effectively on Y; that is,

$$y \in Y, \quad h \in H, \quad hy = y \Rightarrow h = \text{unit of } H.$$

(iii) If $y_i \to y$ in Y, $\{h_i\}$ is a net of elements of H indexed by the same directed set as $\{y_i\}$, and $h_i y_i \to z$ in Y, then $h_i \to h$ in H, where $z = hy$.

Remark. Condition (iii) is of course the general assertion of Proposition 4.11. It implies in particular the following condition (iv): For each fixed y in Y, the map $h \mapsto hy$ is a homeomorphism of H onto the orbit of y. It is not too difficult to construct examples showing that conditions (i), (ii), (iii), taken together, are strictly stronger than (i), (ii), (iv) (even when $H = \mathbb{Z}$).

Remark. If the group H is finite, condition (iii) can of course be omitted.

Example. If H is a closed subgroup of a topological group L, then L, when equipped with the quotient projection $L \to L/H$ and the action of H on L by left multiplication, is a principal H-bundle over L/H.

The Harmonic Analysis of T^τ

4.15. We shall conclude this section with a description of the harmonic analysis of the generalized regular representations T^τ. This description generalizes the classical Plancherel Theorem for Abelian groups. Like the Plancherel Theorem, it will be deduced from Theorem VI.21.4.

4.16. Fix an element τ of $\hat{A}$. By 4.4 the linear functional $p^\tau \colon f \mapsto \tau(f(e))$ is positive on the dense *-subalgebra $\mathscr{L}(\mathscr{B})$ of $\mathscr{L}_1(\lambda; \mathscr{B})$, and generates the *-representation T^τ of $\mathscr{B}$. If $f, g \in \mathscr{L}(\mathscr{B})$ we thus have:

$$p^\tau(g^* * f^* * f * g) = ((f * g)^{\tilde{}}, (f * g)^{\tilde{}})$$

$$= \| T_f^\tau \tilde{g} \|^2 \leq \| T_f^\tau \|^2 \| \tilde{g} \|^2$$

$$\leq \| f \|_1^2 p^\tau(g^* * g)$$

(where $\tilde{g}$ and $(f * g)\tilde{\,}$ denote the canonical images in $X(T^\tau)$). This shows that p^τ satisfies condition (i) of VI.21.4. Since we already know from Proposition 4.4 that T^τ is non-degenerate, Remark VI.21.5 shows that we can apply Theorem VI.21.4 to p^τ even without verifying condition (ii) of VI.21.4. Hence by Theorem VI.21.4 p^τ has a Gelfand transform, which will be denoted by ζ_τ. This is the unique regular Borel measure on $\hat{\mathscr{B}}$ such that, for every pair f, g of elements of $\mathscr{L}(\mathscr{B})$, $\hat{f}\hat{g}$ is ζ_τ-summable and

$$p^\tau(f * g) = \int \hat{f}\hat{g} \, d\zeta_\tau. \tag{26}$$

(Here $\hat{f}$ is the Gelfand transform $\psi \mapsto \int \psi(f(x))d\lambda x$ of f on $\hat{\mathscr{B}}$.)

4.17. Our next theorem will describe what ζ_τ is. Recall from 4.10 that $\rho^{-1}(\tau)$ is a $\hat{G}$-orbit in $\hat{\mathscr{B}}$, and so by 4.12 is homeomorphic with $\hat{G}$ under the map $\chi \mapsto \chi\psi$, where ψ is any fixed element of $\rho^{-1}(\tau)$. If ν is a Haar measure on $\hat{G}$, let ν_τ denote the injection onto $\hat{\mathscr{B}}$ (see II.8.5) of the image of ν under the homeomorphism $\chi \mapsto \chi\psi$, where $\psi \in \rho^{-1}(\tau)$. Since ν is invariant, ν_τ is clearly independent of the particular choice of ψ. We shall call ν_τ *the natural image of ν on $\rho^{-1}(\tau)$*.

Theorem. *The Gelfand transform ζ_τ of p^τ is the natural image on $\rho^{-1}(\tau)$ of some Haar measure on $\hat{G}$.*

Remark. This of course describes ζ_τ up to a multiplicative constant.

Proof. By VII.5.17 the spectrum Z_τ of T^τ is the closed support of its spectral measure. On the other hand, by Remark VI.21.7 the closed support of the spectral measure of T^τ is the same as the closed support of ζ_τ. Combining these facts with (24), we conclude that the closed support of ζ_τ is contained in $\rho^{-1}(\tau)$, and hence can be identified with a regular Borel measure on $\rho^{-1}(\tau)$.

Since $\rho^{-1}(\tau)$ is isomorphic as a $\hat{G}$-space with $\hat{G}$ itself, all that remains to be proved is that ζ_τ is invariant under the action of $\hat{G}$ on $\rho^{-1}(\tau)$.

To see this, fix χ in $\hat{G}$; and let ζ' be the χ-translate of ζ_τ. Let $\mathscr{S}$ be the set of all $f * g$ ($f, g \in \mathscr{L}(\mathscr{B})$). By the definition of ζ_τ, ζ' is that unique regular Borel measure on $\rho^{-1}(\tau)$ such that

$$p^\tau(f) = \int (\chi\hat{f})(\psi)d\zeta'\psi \tag{27}$$

for all f in $\mathscr{S}$ (where $(\chi\hat{f})(\psi) = \hat{f}(\chi^{-1}\psi)$). Now it is easy to check that $\chi\hat{f} = (\chi^{-1}f)^{\widehat{}}$ (where $(\chi^{-1}f)(x) = \chi^{-1}(x)f(x)$), and that if $f \in \mathscr{S}$ then $\chi^{-1}f \in \mathscr{S}$. Hence, replacing $f * g$ by $\chi^{-1}f$ in (26), we get

$$\int (\chi\hat{f})(\psi)d\zeta_\tau\psi = \int (\chi^{-1}f)^{\widehat{}}\,d\zeta_\tau$$

$$= p^\tau(\chi^{-1}f) = p^\tau(f) \tag{28}$$

for all f in $\mathscr{S}$. Comparing (27) and (28), we deduce that $\zeta' = \zeta_\tau$. So ζ_τ is $\hat{G}$-invariant, and the theorem is proved. ∎

4.18.   Our next result describes the harmonic analysis of T^τ.

Theorem.   *Let τ be an element of $\hat{A}$; and let v be the natural image on $\rho^{-1}(\tau)$ of a Haar measure on $\hat{G}$ (see 4.17). Then T^τ is unitarily equivalent to the *-representation W of $\mathscr{B}$ defined as follows:*

$$X(W) = \mathscr{L}_2(v),$$

$$(W_b f)(\psi) = \psi(b)f(\psi) \qquad (b \in B; \psi \in \hat{\mathscr{B}}; f \in \mathscr{L}_2(v)).$$

This theorem is deduced from VI.21.6 and 4.17 in the same way that Theorem 2.11(III) was deduced in the group case. We leave the details to the reader.

Application to Characters of Subgroups

4.19.   Theorems 4.17 and 4.18 have an interesting and important application to characters of subgroups of Abelian groups.

Let H be any locally compact Abelian group, N a closed subgroup of H, and G the quotient group H/N. We shall now suppose that $\mathscr{B}$ is the group extension bundle corresponding to the group extension $N \to H \to G$ (see VIII.6.6). Thus $\mathscr{B}$ is a commutative Banach *-algebraic bundle over G. By VIII.6.9 and VIII.6.10 $\mathscr{B}$ has an approximate unit and enough unitary multipliers, and so satisfies the hypotheses of 4.1. By VIII.6.7 the unitary representations of H and the non-degenerate *-representations of $\mathscr{B}$ are essentially identical. In particular $\hat{H} \cong \hat{\mathscr{B}}$. The unit fiber algebra A of $\mathscr{B}$ is the $\mathscr{L}_1$ group algebra of N; so $\hat{N} \cong \hat{A}$. In this context the results of this section showed the following: Every character of N can be extended to a character of H; and the restriction map $\rho: \hat{H} \to \hat{N}$ (carrying ψ into $\psi|N$) is open.

Thus 4.8 is a generalization of 3.15. Notice that the proof of 4.8 as given in this section was independent of 3.15.

There is a natural injection $i: \hat{G} \to \hat{H}$, sending χ into the character $h \mapsto \chi(hN)$ of H. Evidently i is a homeomorphism, and its range is the kernel of the restriction homomorphism $\rho: \hat{H} \to \hat{N}$. Thus, in view of the surjectivity and openness of ρ,

$$\hat{G} \underset{i}{\to} \hat{H} \underset{\rho}{\to} \hat{N} \tag{29}$$

is a group extension in the sense of III.5.1. We call (29) the *dual* of the original extension

$$N \to H \to G. \tag{30}$$

It is easy to see that, when $\hat{\hat{N}}$, $\hat{\hat{H}}$, and $\hat{\hat{G}}$ are identified with N, H, and G respectively by means of the canonical isomorphisms set up in 3.11, the dual extension of the dual extension (29) is just the original extension (30).

4.20.   Keeping the context of 4.19, let us take a character τ of N. As in 4.19, the *-representation T^τ of $\mathscr{B}$ (defined in 4.4) can be regarded as a unitary representation of H. In view of Theorem 4.18, the harmonic analysis of T^τ is described as follows: Choose an element ψ of $\hat{H}$ belonging to the $i(\hat{G})$-coset $\rho^{-1}(\tau)$ (see (29)). Let v be a Haar measure on $\hat{G}$; and let ζ be the image of v under the homeomorphism $\sigma \mapsto \psi i(\sigma)$ (so that ζ is a regular Borel measure on $\rho^{-1}(\tau)$).

Proposition.   *T^τ is unitarily equivalent to the unitary representation W of H acting on $\mathscr{L}_2(\zeta)$ which is given by the following formula:*

$$(W_h f)(\psi) = \psi(h) f(\psi) \qquad (f \in \mathscr{L}_2(\zeta); h \in H; \psi \in \rho^{-1}(\tau)).$$

Remark.   If N is the one-element subgroup of H, this proposition of course becomes the harmonic analysis of the regular representation of H, already known from 2.11.

5. Commutative C*-Algebraic Bundles

5.1.   As before, G is a fixed locally compact Abelian group, with unit e and Haar measure λ.

We have seen in 4.14 that every commutative Banach *-algebraic bundle satisfying the conditions of 4.1 gives rise to a locally compact principal $\hat{G}$-bundle. The question naturally arises whether *every* locally compact principal $\hat{G}$-bundle Y arises in this way from a commutative Banach *-algebraic bundle $\mathscr{B}$ over G. The answer is "yes." In fact, the purpose of this section is to show that, if we restrict ourselves to commutative saturated C^*-algebraic bundles $\mathscr{B}$, the correspondence $\mathscr{B} \mapsto Y$ is not only onto but one-to-one. Thus commutative saturated C^*-algebraic bundles over G are in a sense the same objects as locally compact principal $\hat{G}$-bundles.

Since this section is not crucial for the following chapters, the proofs of the main results will be omitted or only sketched.

5.2.   Let $\mathscr{B}$ be a fixed saturated commutative C^*-algebraic bundle over G. By 4.1 the results of §4 are all valid for $\mathscr{B}$. We denote its unit fiber C^*-algebra by A, and keep in force the rest of the notation of §4. By 4.3 $\mathscr{L}_1(\lambda; \mathscr{B})$ is symmetric.

5.3.   In analogy with V.7.7 we make the following definition:

Definition.   For each b in B, the complex-valued function $\hat{b}: \psi \mapsto \psi(b)$ on $\hat{\mathscr{B}}$ is called the *Gelfand transform* of b; and $b \mapsto \hat{b}$ is called the *Gelfand transform map*.

Notice that $b \mapsto \hat{b}$ is linear on each fiber, preserves multiplication, and satisfies $(b^*)\hat{} = \overline{\hat{b}}$.

5.4.   Proposition.   *If $b \in B$, $\hat{b}$ is bounded on $\hat{\mathscr{B}}$; in fact*

$$\|\hat{b}\|_\infty = \|b\|. \tag{1}$$

Proof.   Since A is a C^*-algebra and since by 4.8 every element of $\hat{A}$ extends to an element of $\hat{\mathscr{B}}$, (1) holds when $b \in A$. Hence, for any b in B,

$$\|b\|^2 = \|b^*b\| = \|(b^*b)\hat{}\|_\infty = \|\hat{b}\|_\infty^2. \qquad \blacksquare$$

5.5.   Since the topology of $\hat{\mathscr{B}}$ is that of uniform convergence on compact sets (4.6), the Gelfand transforms $\hat{b}$ $(b \in B)$ are continuous on $\hat{\mathscr{B}}$.

If $b \in B_x$ $(x \in G)$, then by the definition of the action of $\hat{G}$ on $\hat{\mathscr{B}}$ (see 4.9) we have

$$\hat{b}(\chi\psi) = \hat{x}(\chi)\hat{b}(\psi) \qquad (\chi \in G; \psi \in \hat{\mathscr{B}}) \tag{2}$$

(where $\hat{x}: \chi \mapsto \chi(x)$ is the character of $\hat{G}$ corresponding to x).

In view of (2), $|\hat{b}(\psi)|$ depends only on the $\hat{G}$-orbit to which ψ belongs, that is, on $\rho(\psi)$. Since $\hat{b}$ is continuous, this implies that each element b of B gives rise to a continuous function $\beta_b: \rho(\psi) \mapsto |\hat{b}(\psi)|$ on $\hat{A}$ to the non-negative reals.

Proposition.   *If $b \in B$, β_b vanishes at infinity on $\hat{A}$ and so belongs to $\mathscr{C}_0(\hat{A})$.*

Proof. If $a \in A$, β_a is the absolute value of the Gelfand transform of a on $\hat{A}$, and so belongs to $\mathscr{C}_0(\hat{A})$. Further, for any b in B, β_b is bounded by 5.4. So $\beta_{ab} = \beta_a \beta_b \in \mathscr{C}_0(\hat{A})$ whenever $a \in A$ and $b \in B$. But by the saturation of $\mathscr{B}$, $\{ab : a \in A, b \in B\}$ has dense linear span in each fiber. Combining these facts with 5.4, we find that, for all b in B, β_b is the limit (in the supremum norm) of a sequence of functions in $\mathscr{C}_0(\hat{A})$, and so itself belongs to $\mathscr{C}_0(\hat{A})$. ∎

Remark. This proposition says that the Gelfand transforms $\hat{b}$ of elements of B "do their best to vanish at infinity" consistently with relation (2).

5.6. Definition. For each x in G let E_x be the linear space of all continuous complex functions f on $\mathscr{B}$ such that

(i)   the relation $f(\chi\psi) = \hat{x}(\chi)f(\psi)$ holds for all χ in $\hat{G}$ and $\psi \in \hat{\mathscr{B}}$,
(ii) the real-valued function $\rho(\psi) \mapsto |f(\psi)|$ $(\psi \in \hat{\mathscr{B}})$, which is well-defined on $\hat{A}$ in virtue of (i), vanishes at ∞ on $\hat{A}$.

Functions in E_x are necessarily bounded. In fact, it is easy to see that E_x is a Banach space under the supremum norm $\|f\|_\infty = \sup\{|f(\psi)| : \psi \in \hat{\mathscr{B}}\}$.

Proposition. *If $x \in G$, the Gelfand transform map $b \mapsto \hat{b}$ carries B_x linearly and isometrically onto E_x.*

Sketch of proof. We have seen in 5.4 and 5.5 that the Gelfand transform map carries B_x linearly and isometrically onto a subspace E'_x of E_x. By VI.4.3 $E'_e = E_e$. Hence, since $b \mapsto \hat{b}$ is multiplicative (5.3), E'_x is closed under multiplication by arbitrary functions in E_e. Also, by 4.2, there is no point of $\hat{\mathscr{B}}$ at which the functions in E'_x all vanish. Therefore the same argument as was used to prove II.14.1 and II.14.6 shows here that E'_x is dense in E_x and hence equal to E_x. ∎

Remark. Given $x \in G$ and $\tau \in \hat{A}$, let F^x_τ be the one-dimensional space of complex functions g on $\rho^{-1}(\tau)$ satisfying $g(\chi\psi) = \hat{x}(\chi)g(\psi)$ $(\chi \in \hat{G}; \psi \in \rho^{-1}(\tau))$. It is possible and very natural to construct a Banach bundle $\mathscr{F}^x$ over $\hat{A}$ whose fiber over τ is F^x_τ, and whose topology is such that the continuous cross-sections of $\mathscr{F}^x$ are identified with the continuous complex functions f on $\hat{\mathscr{B}}$ satisfying

$$f(\chi\psi) = \hat{x}(\chi)f(\psi) \qquad (\chi \in \hat{G}; \psi \in \hat{\mathscr{B}}). \qquad (3)$$

When this is done, we have $E_x = \mathscr{C}_0(\mathscr{F}^x)$; and Proposition II.14.6 is directly applicable to the proof of the above proposition.

5.7. Remark. Suppose for the moment that $\mathscr{B}$ is not a C^*-algebraic bundle but merely satisfies the hypotheses of 4.1. In that case the developments of 5.3 to 5.6 are valid when weakened as follows: The Gelfand transform map $b \mapsto \hat{b}$ is norm-decreasing (instead of an isometry); and the image of B_x under the Gelfand transform map is dense in E_x (rather than equal to E_x).

5.8. We shall now express the topology of B in terms of the Gelfand transforms of the elements of B.

Proposition*. *Let $\{b_i\}$ be a net of elements of B and b an element of B. Then $b_i \to b$ in B if and only if the following conditions* (i), (ii), (iii) *hold. Also, $b_i \to b$ in B if and only if conditions* (i), (ii), (iv) *hold:*

(i) $\pi(b_i) \to \pi(b)$ *in G.*
(ii) $\hat{b}_i \to \hat{b}$ *uniformly on compact subsets of $\hat{\mathscr{B}}$.*
(iii) $|\hat{b}_i| \to |\hat{b}|$ *uniformly on $\hat{\mathscr{B}}$.*
(iv) *For each $\varepsilon > 0$, there is a compact subset K of $\hat{A}$ and an index i_0 such that*

$$i \succ i_0, \quad \psi \in \hat{\mathscr{B}}, \quad \rho(\psi) \notin K \Rightarrow |\hat{b}_i(\psi)| < \varepsilon.$$

The Duality Between Saturated Commutative C^*-algebraic Bundles and Principal Bundles

5.9. In paragraphs 5.3 to 5.8 we have seen how the whole structure of the saturated commutative C^*-algebraic bundle $\mathscr{B}$ over G can be recovered from the structure of the principal $\hat{G}$-bundle $\hat{\mathscr{B}}$ over $\hat{A}$. This suggests the right approach for answering the converse question raised in 5.1: Given any locally compact Hausdorff principal $\hat{G}$-bundle Y over a locally compact Hausdorff space X, does it come from some saturated commutative C^*-algebraic bundle over G? The answer will be "yes".

5.10. Let M and T be locally compact Hausdorff spaces; and let $\langle T, \rho, \alpha \rangle$ be a principal $\hat{G}$-bundle over M ($\rho: T \to M$ being the bundle projection, and α the action of $\hat{G}$ on T; as usual we abbreviate $\alpha(\chi, t)$ to χt).

For each x in G let us define B_x to be the space of all those continuous complex functions f on T such that (i) the relation

$$f(\chi t) = \hat{x}(\chi) f(t) \qquad (\chi \in \hat{G}; t \in T) \qquad (4)$$

holds, and (ii) the real-valued function $\rho(t) \mapsto |f(t)|$ ($t \in T$), which is well-defined on M in virtue of (i), vanishes at ∞ on M.

This definition of course is essentially the same as 5.6; and as in 5.6 B_x is a Banach space under the supremum norm $\| \ \|_\infty$.

5.11. Let B be the disjoint union $\bigcup_{x \in G} B_x$, and $\pi: B \to G$ the surjection given by $\pi^{-1}(x) = B_x$ ($x \in G$). Motivated by 5.8, we topologize B as follows: A net $\{f_i\}$ of elements of B will converge to f in B if and only if

(i) $\pi(f_i) \to \pi(f)$ in G,
(ii) $f_i \to f$ uniformly on compact subsets of T,
(iii) $|f_i| \to |f|$ uniformly on T.

We also introduce into B the operations $\cdot$ of pointwise multiplication and $^-$ of pointwise complex conjugation, and observe that

$$B_x \cdot B_y \subset B_{xy}, \qquad (B_x)^- = B_{x^{-1}} \qquad\qquad (x, y \in G).$$

Proposition*. *The topological space B, together with the fibers B_x ($x \in G$) and the above operations $\cdot$ and $^-$, forms a saturated commutative C*-algebraic bundle $\mathscr{B}$ over G.*

The proof of this proposition is of quite a routine nature, and we omit it.

5.12. Notice that the unit fiber C*-algebra of $\mathscr{B}$ is just $\{g \circ \rho : g \in \mathscr{C}_0(M)\}$, and so is *-isomorphic with $\mathscr{C}_0(M)$. Hence $(B_e)^\wedge$ can and will be identified with M. Thus the theory of §4 gives to $\hat{\mathscr{B}}$ the structure of a principal $\hat{G}$-bundle over M. The next proposition will show that this structure is just the same as the principal $\hat{G}$-bundle $\langle T, \rho, \alpha \rangle$ with which we started in 5.10.

Proposition*. *For each t in T let $\psi_t : f \mapsto f(t)$ be the element of $\hat{\mathscr{B}}$ consisting of evaluation at t. Then $t \mapsto \psi_t$ is a homeomorphism of T onto $\hat{\mathscr{B}}$.*

The crucial part of this proposition is the fact that $t \mapsto \psi_t$ is onto $\hat{\mathscr{B}}$. We shall sketch the proof of this. Let ψ be an element of $\hat{\mathscr{B}}$. Then $\psi | B_e$ is identified with an element m of M; and one shows that, for any f in B, $\psi(f)$ depends only on $f | \rho^{-1}(m)$. Fix a point t_0 of $\rho^{-1}(m)$. The preceding sentence shows that, if $f \in B$, $\psi(f)$ depends only on $\pi(f)$ and $f(t_0)$; in fact, there will be a function $\chi_0 : G \to \mathbb{C}$ such that

$$\psi(f) = f(t_0)\chi_0(\pi(f)) \qquad\qquad (f \in B). \qquad (5)$$

An easy verification shows that χ_0 is multiplicative and satisfies $\chi_0(x^{-1}) = \overline{\chi_0(x)}$; and it can also be shown that χ_0 is continuous. So $\chi_0 \in \hat{G}$; and (4) and (5) imply that

$$\psi(f) = f(\chi_0 t_0) \qquad \text{for all } f \text{ in } B.$$

This shows that $\psi = \psi_{\chi_0 t_0}$, hence that the map $t \mapsto \psi_t$ is onto $\hat{\mathcal{B}}$.

5.13. With the notation of 5.12, one verifies that $\psi_t | B_e$ is just the element of $(B_e)^{\hat{}}$ which was identified with the element $\rho(t)$ of M. Also, equation (4) shows that

$$\chi\psi_t = \psi_{\chi t} \qquad\qquad (\chi \in \hat{G}; t \in T).$$

From these two facts, together with Proposition 5.12, we see that the construction of 5.10–5.11 for passing from the principal $\hat{G}$-bundle T, ρ, α to the saturated commutative C^*-algebraic bundle $\mathcal{B}$ over G, followed by the construction of §4 for passing from $\mathcal{B}$ to the principal $\hat{G}$-bundle $\hat{\mathcal{B}}$, brings us back essentially to the T, ρ, α with which we started. Similarly, if we start from a saturated commutative C^*-algebraic bundle $\mathcal{B}$ over G, form its associated principal $\hat{G}$-bundle $\hat{\mathcal{B}}$, and then perform on $\hat{\mathcal{B}}$ the construction of 5.10–5.11, it follows from 5.6, 5.8 that we essentially return to the original $\mathcal{B}$. Thus we arrive at the following duality theorem which generalizes the Pontryagin duality for locally compact Abelian groups:

Theorem*. *Let G be a locally compact Abelian group, and M a locally compact Hausdorff space. The map $\mathcal{B} \mapsto \hat{\mathcal{B}}$ sets up a one-to-one correspondence between the following two families of objects:*

(i) *The family of all isomorphism classes of saturated commutative C^*-algebraic bundles $\mathcal{B}$ over G having $\mathscr{C}_0(M)$ as unit fiber algebra,*

(ii) *The family of all isomorphism classes of locally compact principal $\hat{G}$-bundles over M.*

5.14. Let T be any locally compact Hausdorff space. It follows from 5.13 (and the definition VIII.17.4) that *the possible ways of introducing a saturated C^*-algebraic bundle structure over G into the C^*-algebra $\mathscr{C}_0(T)$ are in one-to-one correspondence with the principal $\hat{G}$-bundles which have T as their underlying topological space.*

5.15. Remark. In particular, suppose that T is a locally compact Hausdorff space with the property that every homeomorphism of T onto itself has at least one fixed point. Then no group with more than one element can act effectively on T; and so it follows from 5.14 that $\mathscr{C}_0(T)$ admits no saturated C*-algebraic bundle structure over a group with more than one element.

The unit ball $\{t \in \mathbb{R}^n : t_1^2 + \cdots + t_n^2 \leq 1\}$ in n-space has this property (by the Brouwer Fixed Point Theorem). So does the compact space $\{0\} \cup \{n^{-1} : n = 1, 2, \ldots\}$, since any homeomorphism of this space onto itself must leave 0 fixed.

5.16. Remark. Let M, T, ρ, α be as in 5.10; and let us construct from these ingredients the saturated commutative C*-algebraic bundle $\mathscr{B}$ as in 5.10, 5.11. If $x \in G$, an easy application of VIII.15.3 shows that a unitary multiplier of $\mathscr{B}$ of order x is essentially just a continuous complex function f on T such that (4) holds and $|f(t)| = 1$ for all t in T. Thus, $\mathscr{B}$ *will have enough unitary multipliers if and only if such a function f exists for every x in G.*

As an example, suppose that T is the 2-sphere $\{t \in \mathbb{R}^3 : t_1^2 + t_2^2 + t_3^3 = 1\}$, that G (and hence also $\hat{G}$) is a two-element group, and that the non-unit element of $\hat{G}$ sends t into $-t$ (for $t \in T$). The argument of VIII.3.15 shows that any continuous complex function f on T satisfying $f(-t) = -f(t)$ (for all t in T) must vanish somewhere on T. Hence in this case $\mathscr{B}$ is a saturated C*-algebraic bundle over the two-element group which has no unitary multipliers of non-unit order.

The reader will verify that this example is identical with the example of VIII.3.15.

5.17. Remark. The Duality Theorem 5.13 is the solution, in the saturated commutative case, of the following important general duality problem. Let G be any locally compact group (Abelian or not) and C any C*-algebra. If we equip C with a C*-bundle structure over G (in the sense of VIII.17.4), this automatically generates a corresponding tensor product operation $\langle U, T \rangle \mapsto U \otimes T$ (see VIII.9.16), sending a pair $\langle U, T \rangle$, where U is a unitary representation of G and T is a *-representation of C, into the tensor product *-representation $U \otimes T$ of C. Conversely, one conjectures that the tensor product operation $\langle U, T \rangle \mapsto U \otimes T$ uniquely determines the C*-bundle structure (up to isomorphism); and it is natural to ask for an abstract characterization of those operations $\langle U, T \rangle \mapsto U \otimes T$ which arise as tensor products in this way from some C*-bundle structure for C over G. If G and C are commutative, one can confine one's attention to one-dimensional representations, and Theorem 5.13 provides a complete answer (at least for saturated C*-bundle structures). But in the non-commutative situation the complete answer is not yet known; see the discussion in VIII.17.9.

6.   Exercises for Chapter X

1. Let χ be a character of the circle group $\mathbb{E}$. Show that if χ is injective, then $\chi(x) = x$ for all $x \in \mathbb{E}$ or $\chi(x) = \bar{x}$ for all $x \in \mathbb{E}$.

2. Let G be a locally compact Abelian group; and let U denote the neighborhood $\{z \in \mathbb{C} : z = e^{2\pi i \alpha}, |\alpha| < \frac{1}{3}\}$ of 1 in $\mathbb{E}$. Show that if $\chi \in \hat{G}$ and $\chi(x) \in U$ for all $x \in G$, then $\chi \equiv 1$.

3. Let G be a connected locally compact Abelian group. Show that each $\chi \in \hat{G}$, $\chi \neq 1$, is surjective.

4. Let G be a locally compact Abelian group; and let $\chi \in \hat{G}$. Show that if G is totally disconnected, then $\mathrm{Ker}(\chi)$ is open in G.

5. We have seen in 1.14 that Pontryagin duality (see 3.11) sets up a one-to-one correspondence between isomorphism classes of compact Abelian groups and of discrete Abelian groups. It is interesting to observe how the possible properties of these groups correspond to each other.

Let G be a compact Abelian group. Show that:
   (a) G is connected if and only if no element of $\hat{G}$ except the unit element is of finite order.
   (b) G is totally disconnected if and only if $\hat{G}$ is a torsion group, i.e., every element of $\hat{G}$ is of finite order.
   (c) G is a Lie group if and only if $\hat{G}$ is finitely generated.

6. Prove Proposition 1.9.

7. For each i in an index set I let G_i be a compact Abelian group, and let G be the (compact Abelian) Cartesian product $\prod_{i \in I} G_i$. Then

$$\hat{G} = \sum_{i \in I}{}^{\oplus} (G_i)\hat{}$$

(consisting of all those γ in $\prod_{i \in I} (G_i)\hat{}$ such that γ_i is the unit of $(G_i)\hat{}$ for all but finitely many i).

8. Let H be any discrete Abelian torsion group. It is a known fact that

$$H = \sum_p{}^{\oplus} H_p,$$

where p runs over the primes > 1 and $H_p = \{h \in H : \text{the order of } h \text{ in } H \text{ is a power of } p\}$. We call H_p the *p-component* of H.

Show that any totally disconnected compact Abelian group G has a canonical Cartesian product decomposition

$$G = \prod_p G_p \qquad (p \text{ running over all primes} > 1),$$

where each G_p is a compact Abelian group satisfying $(G_p)\hat{} = (\hat{G})_p$.

9. Let H be the additive group of the rationals modulo the integers (considered as a discrete group); and let G be the compact Abelian group satisfying $\hat{G} \cong H$. Note that

H is a torsion group, and that for each prime p, H_p is the additive group of all rationals of the form mp^{-r} (r, m integers) modulo the integers.

(a) Show that, for each prime p, $(I_p)\hat{\ } = H_p$, where I_p is the compact Abelian group of all p-adic integers (under addition).

(b) Show that $G = \prod_p I_p$ (p prime).

(c) Let a be the element of G corresponding to the character $r + \mathbb{Z} \mapsto e^{2\pi i r}$ of H. Show that $n \mapsto a^n$ maps $\mathbb{Z}$ onto a *dense* subgroup of G; and that the projection of a onto each factor I_p (see (b)) is just the multiplicative unit of I_p.

10. Let G be a locally compact Abelian group and H a closed subgroup of G. By the *annihilator of H* in $\hat{G}$ we mean the closed subgroup $H^\perp = \{\chi \in \hat{G} : \chi(x) = 1 \text{ for all } x \in H\}$ in $\hat{G}$. Show that:

(a) $(G/H)\hat{\ } \cong H^\perp$.

(b) $\hat{H} \cong \hat{G}/H^\perp$.

(Here $\cong$ means the two sides are naturally isomorphic as topological groups.)

11. Prove Proposition 2.4.

12. Establish Bochner's Theorem 2.6.

13. Let G be a locally compact Abelian group. Let $\hat{G}_d$ be the group $\hat{G}$ with the discrete topology, and $\bar{G}$ the dual of $\hat{G}_d$. By 1.14 $\bar{G}$ is a compact Abelian group which is called the *Bohr compactification* of G. Let β be the mapping of G into $\bar{G}$ defined by $\beta_x(\chi) = \chi(x)$ ($x \in G$, $\chi \in \hat{G}$). Show that β is a continuous isomorphism of G onto a dense subgroup $\beta(G)$ of $\bar{G}$. [*Note*: In general $\beta(G)$ is not a locally compact subset of $\bar{G}$ and β is not a homeomorphism.]

14. Let G be a locally compact Abelian group. Prove that *every* homomorphism $\psi : G \to \mathbb{E}$ is the pointwise limit of characters on G; that is, for every $\varepsilon > 0$ and every finite subset $\{x_1, x_2, \ldots, x_n\}$ of G, there is $\chi \in \hat{G}$ such that

$$|\chi(x_i) - \psi(x_i)| < \varepsilon \qquad\qquad (i = 1, \ldots, n).$$

15. Let A be the $\mathscr{L}_1$ group algebra of $\mathbb{R}$, and B the dense *-subalgebra of A consisting of all twice continuously differentiable complex functions with compact support on $\mathbb{R}$. Let us define the linear functional p on B as follows:

$$p(f) = -f''(0) \qquad\qquad (f \in B).$$

Show that p is a positive functional on B satisfying the conditions (i), (ii) of VI.21.4. What measure on $\hat{\mathbb{R}} \cong \mathbb{R}$ is the Gelfand transform of p (see 2.7)?

16. Let G be a locally compact Abelian group, with Haar measure λ. Let σ be a fixed norm-function on G (see Exercise 65 of Chapter III). We have seen in Exercise 65 of Chapter III that $\mathscr{L}(G)$ is a normed algebra under the norm

$$\|f\|_\sigma = \int_G |f(x)| \sigma(x) d\lambda x \qquad\qquad (f \in \mathscr{L}(G)).$$

The completion of $\mathscr{L}(G)$ with respect to $\|\ \|_\sigma$ is therefore a Banach algebra, called the σ-group algebra of G and denoted by $\mathscr{A}_\sigma(G)$. We shall determine the Gelfand space $(\mathscr{A}_\sigma(G))^\dagger$ of $\mathscr{A}_\sigma(G)$.

Let $G_\sigma^\dagger$ denote the family of all continuous homomorphisms χ of G into the multiplicative group of non-zero complex numbers, such that

$$|\chi(x)| \le \sigma(x) \qquad \text{for all } x \in G.$$

(a) Show that if $\chi \in G_\sigma^\dagger$, the equation

$$\chi'(f) = \int_G f(x)\chi(x)d\lambda x \qquad\qquad (f \in \mathscr{L}(G))$$

defines a complex homomorphism χ' of $\mathscr{L}(G)$ which is continuous with respect to $\| \ \|_\sigma$ and so extends to a complex homomorphism (also called χ') of $\mathscr{A}_\sigma(G)$.

(b) Prove that the map $\chi \mapsto \chi'$ ($\chi \in G_\sigma^\dagger$) is a surjective homeomorphism of $G_\sigma^\dagger$ onto $(\mathscr{A}_\sigma(G))^\dagger$, with respect to the Gelfand topology of $(\mathscr{A}_\sigma(G))^\dagger$ and the topology of uniform convergence on compact sets for $G_\sigma^\dagger$. (Note: The special case $\sigma \equiv 1$ has of course already been dealt with in this Chapter.)

17. Let T be a unitary representation of $\mathbb{R}$ (the additive reals) on a Hilbert space X. Let E be projection of X onto $\{\xi \in X : T_t\xi = \xi \text{ for all } t \in \mathbb{R}\}$ (the T-invariant subspace). Show that, for each a and each $\xi \in X$,

$$\lim_{s \to \infty} \frac{1}{s} \int_a^{a+s} T_t\xi \, dt = E\xi.$$

(This is the von Neumann Ergodic Theorem.)

18. Prove Proposition 5.8.

19. Prove Proposition 5.11.

20. Prove Theorem 5.13.

21. We begin this Exercise with a definition: Let A be a Banach *-algebra and G a topological group. We shall say that τ is a *continuous action of G by *-automorphisms* on A if τ is a homomorphism of G into the multiplicative group of all *-automorphisms of A such that, for each $a \in A$, the map $x \mapsto \tau_x(a)$ ($x \in G$) is continuous on G to A.

Suppose now that $\mathscr{B}$ is a C^*-algebraic bundle over a locally compact Abelian group G. Prove that the equation

$$\tau_\chi(a)(x) = \chi(x)a(x) \qquad (\chi \in \hat{G}, a \in \mathscr{L}(\mathscr{B}), x \in G) \qquad (*)$$

defines a continuous action τ of the character group $\hat{G}$ by *-automorphisms on $C^*(\mathscr{B})$.

Remark. This Exercise suggests a converse conjecture, namely: If G is a locally compact Abelian group, A is a C^*-algebra, and τ is a continuous action of $\hat{G}$ by *-automorphisms on A, then there is a C^*-bundle structure for A over G such that τ is given by (*). But this conjecture is *false*. Indeed, suppose that A has a unit and that τ is a (continuous) action of the discrete Abelian group $\mathbb{Z} \cong \hat{\mathbb{E}}$ on A. If the conjecture were true, this action would come from a C^*-bundle structure for A over the non-discrete group $\mathbb{E}$, contradicting Exercise 39, Chapter XI. However, the conjecture *is* true if $\hat{G}$ is compact (so that G is discrete); see Exercise 22.

22. Let A be a C^*-algebra, G a compact Abelian group, and τ a continuous action of G by *-automorphisms on A. For each $\chi \in \hat{G}$, let B_χ be the (closed) χ-subspace of A

relative to τ (treating τ as a Banach representation of G on the Banach space A; see IX.7.2). Show that: (a) $\mathscr{B} = \{B_\chi\}_{\chi \in \hat{G}}$ is a C^*-algebraic bundle over $\hat{G}$; (b) A can be identified with $C^*(\mathscr{B})$ in the sense of VIII.16.12 (so that $\mathscr{B}$ becomes a C^*-bundle structure for A over $\hat{G}$).

[Hint: (a) is easy. To prove (b) show that the map $\rho: C^*(\mathscr{B}) \to A$ of VIII.16.12 is one-to-one. Notice that G has a continuous action τ' by *-automorphisms on $C^*(\mathscr{B})$ (as in Exercise 21). For each χ in $\hat{G}$ let π_χ and π'_χ be the standard projections of A and $C^*(\mathscr{B})$ respectively onto their χ-subspaces B_χ (with respect to τ and τ'). Verify that $\rho \circ \pi'_\chi = \pi_\chi \circ \rho$. It follows that $\mathrm{Ker}(\rho)$ is stable under the π'_χ. On the other hand $\mathrm{Ker}(\rho) \cap B_\chi = \{0\}$ by VIII.16.11; so $\rho(a) = 0$ implies $\pi'_\chi(a) = 0$ for all χ. From this and IX.8.3 we obtain $\mathrm{Ker}(\rho) = \{0\}$.]

23. It would be of great interest to have a simple necessary and sufficient condition for the C^*-bundle structure of the preceding Exercise to be *saturated*. We know of no such condition unless A is of compact type and G acts by *inner* automorphisms on A. The present Exercise is devoted to this case.

Let A be a C^*-algebra, and $\mathscr{W}(A)$ its multiplier C^*-algebra (see VIII.1.18). A unitary element u of $\mathscr{W}(A)$ is called a *unitary multiplier* of A. If u is a unitary multiplier of A, then $a \mapsto uau^*$ ($a \in A$) is a *-automorphism of A; such a *-automorphism is called an *inner* *-automorphism of A.

Now let G_0 be a compact Abelian group, and σ a homomorphism of G_0 into the multiplicative group of all unitary multipliers of A, which is continuous in the sense that for each $a \in A$, the maps $x \mapsto \sigma(x)a$ and $x \mapsto a\sigma(x)$ are continuous on G_0 to A. For each $x \in G_0$ let $\tau'(x)$ be the (inner) *-automorphism $a \mapsto \sigma(x)a\sigma(x)^*$ of A. Show that τ' is a continuous action of G_0 by *-automorphisms on A. In fact, let C be the closed normal subgroup of G_0 consisting of those x such that $\sigma(x)$ is a *central* multiplier (i.e., $\sigma(x)a = a\sigma(x)$ for all $a \in A$); and put $G = G_0/C$. Then the equation $\tau_{xC}(a) = \tau'_x(a)$ defines τ as a continuous action of G by *-automorphisms on A. Denoting by B_χ the χ-subspace of A with respect to τ (for each $\chi \in \hat{G}$), we see from Exercise 22 that $\mathscr{B} = \{B_\chi\}_{\chi \in \hat{G}}$ is a C^*-bundle structure for A over the discrete Abelian group $\hat{G}$.

To proceed further, we shall now assume in addition that A is *of compact type* (see VI.23.3).

As usual $\hat{A}$ denotes the structure space of A; let us in fact identify $\hat{A}$ with a family of concrete *-representations of A containing exactly one element from each class in $\hat{A}$.

Take any $T \in \hat{A}$. For each unitary multiplier u of A, there is (by VIII.1.15) a unique unitary operator T'_u on $X(T)$ such that

$$T_{ua} = T'_u T_a, \quad T_{au} = T_a T'_u \qquad (a \in A).$$

Show that $T''; x \mapsto T'_{\sigma(x)}$ ($x \in G_0$) is a unitary representation of G_0 satisfying

$$T_{\tau'_x(a)} = T''_x T_a T''^{-1}_x \qquad (x \in G_0, a \in A).$$

For each $\chi \in (G_0)\hat{\ }$ and each $T \in \hat{A}$, let $X_\chi(T)$ be the χ-subspace of $X(T)$ with respect to T''.

Since G is a quotient group of G_0, we can (and will) identify $\hat{G}$ with a subgroup of $(G_0)\hat{}$.

Now prove the following:

(i)   For each $\chi \in \hat{G}$, B_χ consists precisely of those $a \in A$ such that, for each $T \in \hat{A}$ and $\psi \in (G_0)\hat{}$,

$$T_a(X_\psi(T)) \subset X_{\psi\chi}(T).$$

(ii) The unit fiber algebra B_1 of $\mathscr{B}$ is commutative if and only if, for each $T \in \hat{A}$ and $\psi \in (G_0)\hat{}$, the multiplicity of ψ in T'' is at most 1.

(iii) For each $T \in \hat{A}$, the family $\{\psi \in (G_0)\hat{}: X_\psi(T) \neq \{0\}\}$ is contained in some coset $\psi\hat{G}$ (belonging to $(G_0)\hat{}/\hat{G}$).

(iv) Let $\mathscr{D}$ be the saturated part of $\mathscr{B}$ (see VIII.16.6). Thus $C^*(\mathscr{D})$ is a closed two-sided ideal of $C^*(\mathscr{B}) \cong A$ (see Exercise 46, Chapter VIII)). Show that $C^*(\mathscr{D})$ consists of those elements a of A such that $T_a = 0$ for all those T in $\hat{A}$ such that $\{\psi \in (G_0)\hat{}: X_\psi(T) \neq \{0\}\}$ is *properly* contained in some coset belonging to $(G_0)\hat{}/\hat{G}$.

(v)   In particular, $\mathscr{B}$ is saturated if and only if, for every $T \in \hat{A}$, the family $\{\psi \in (G_0)\hat{}: X_\psi(T) \neq \{0\}\}$ is a coset belonging to $(G_0)\hat{}/\hat{G}$; that is to say, if $T \in \hat{A}$, $\psi \in (G_0)\hat{}$, $\chi \in \hat{G}$, $X_\psi(T) \neq \{0\}$, then $X_{\psi\chi}(T) \neq \{0\}$.

As Corollaries of (iv) and (v) we obtain:

(vi) If the (finite) positive integer n is the smallest of the dimensions of the elements of $\hat{A}$, then $\mathscr{B}$ cannot be saturated unless the order of G is $\leq n$.

(vii) If G is infinite and all elements of $\hat{A}$ are finite-dimensional (for example, if $A = C^*(H)$ for some compact group H), then the saturated part of $\mathscr{B}$ is zero (i.e., $D_\chi = \{0\}$ for all $\chi \in \hat{G}$).

24. Suppose H is a compact group and G_0 is a closed Abelian subgroup of H; and let A be the group C^*-algebra $C^*(H)$. Each element x of G_0 generates a unitary multiplier $\sigma(x)$ of A (via left and right multiplication), and this σ is a homomorphism. Also, by Chapter IX, A is of compact type (in fact, all its irreducible *-representations are finite-dimensional). So we are in the situation of Exercise 23; and (in the notation of Exercise 23) $\mathscr{B}$ is never saturated unless $G = G_0/C$ is the one-element group.

Consider the following two special cases of this situation:

(i) $H = SU(2)$, G_0 is the diagonal subgroup of H. Show that the unit fiber algebra of $\mathscr{B}$ is then commutative (see Exercise 23 and §IX.11) and that the saturated part of $\mathscr{B}$ is zero.

(ii) Let H be the alternating group A_5 (i.e., the simple group of all even permutations of the five objects 1, 2, 3, 4, 5). Let G_0 be the five-element cyclic subgroup of H generated by the cyclic permutation $1 \to 2 \to 3 \to 4 \to 5 \to 1$. Show that the unit fiber algebra of $\mathscr{B}$ is then commutative. If $\mathscr{D}$ is the saturated part of $\mathscr{B}$, show that $C^*(\mathscr{D})$ is the two-sided ideal of A consisting of those a such that $T_a = 0$ for all irreducible unitary representations T of H *except* the unique 5-dimensional one (see Exercise 37, Chap. IX).

Notes and Remarks

The development of harmonic analysis on locally compact Abelian groups as we have presented it in §§1–3 of the present chapter closely follows the treatment given in Bourbaki [12, Chapter II]. This theory is also treated in various ways and at various levels of generality in Dieudonné [1], Helson [2], Hewitt and Ross [1, 2], Loomis [2], Naimark [8], Rudin [1], Pontryagin [6], and Weil [1]. Of these references, the two volume work of Hewitt and Ross contains the most complete and comprehensive coverage, and also has excellent historical notes on nearly every aspect of the theory.

For the convenience of the reader we mention a few of the historical highlights of the main results.

The original version of Stone's Theorem (2.1) was proved in 1932 for one parameter unitary groups of the real line (see Stone [1, 2]). Stone's Theorem for $\mathbb{R}$ was also proved by von Neumann [9]. The theorem, essentially presented as in 2.1, for locally compact Abelian groups is due, independently and simultaneously, to Ambrose [1] and Godement [1]. A similar but not identical result was proved slightly earlier by Naimark [1]. The role of Stone's Theorem in theoretical physics is discussed, for example, in Emch [1], Bratteli and Robinson [1], and Mackey [12, 21].

The Bochner Theorem 2.6 was proved for the case $G = \mathbb{R}$ by Bochner [1] in 1932; earlier work by Carathéodory, Toeplitz, Herglotz, and others played an important role in the evolution of the theorem and set the stage for Bochner's result (see Hewitt and Ross [2, p. 235]). For general locally compact Abelian groups Bochner's Theorem was proved independently and nearly simultaneously by Weil [1, pp. 120–122], Raikov [1], and Povzner [1].

The Plancherel Theorem 2.11 was proved for the case $G = \mathbb{R}$ in 1910 by Plancherel [1]. A short elegant proof of Plancherel's Theorem for $\mathbb{R}$ was given in 1927 by F. Riesz [4]; and another was given by Wiener [2] in 1933 based on special properties of the Hermite functions. For general locally compact Abelian groups Plancherel's Theorem is due to Weil [1, pp. 111–118]. Various proofs of the theorem have been given by several workers in the field, among them M. G. Krein, Cartan and Godement, M. H. Stone, and Povzner (see Hewitt and Ross [2, pp. 251–252]). A few additional remarks concerning the Plancherel Theorem are given in the Notes and Remarks to Chapter VI in Volume 1.

The Pontryagin Duality Theorem 3.11 was proved in 1934 by L. S. Pontryagin [4] for second countable locally compact Abelian groups. The next year van Kampen [1] showed that Pontryagin's result could be extended to arbitrary locally compact Abelian groups. The proofs of Pontryagin and van Kampen depend on the structure theory of locally compact Abelian groups and are quite different from our proof of 3.11 which follows, at least in spirit, the proof of H. Cartan and Godement [1]. For a proof which follows van Kampen's original argument see Hewitt and Ross [1, pp. 379–380].

We wish to emphasize that the above mentioned paper of H. Cartan and Godement [1], which developed much of the theory of Abelian harmonic analysis from the point of view of the Fourier transform, has played a very important and influential role in modern treatments of the theory.

Further references which deal specifically with the historical origins and development of harmonic analysis on locally compact Abelian groups are Dieudonné [3, pp. 194–207], Hawkins [2, 4], and Mackey [22, 23]. For the classical theory of Fourier and trigonometrical series, which inspired much of the abstract theory, the reader should consult the two volume treatise of Zygmund [1].

The results in §§4 and 5 involving the structure of saturated commutative Banach *-algebraic bundles, and the simultaneous generalization of the Pontryagin Duality Theorem and the duality between commutative C^*-algebras and locally compact spaces are due to the first named author and are published here for the first time. However, it should be mentioned that a weaker version of the results in §5 can be found in H. A. Smith [2].

XI   Induced Representations and the Imprimitivity Theorem

We have already encountered induced representations of compact groups in §IX.10. This extremely important construction was generalized by Mackey to arbitrary second-countable locally compact groups G. Along with many other results, Mackey proved the fundamental Imprimitivity Theorem, which characterizes induced representations as being those which are accompanied by a system of imprimitivity (see VIII.18.7) over a *transitive* G-space. This result and the Frobenius Reciprocity Theorem (IX.10.8) are perhaps the two most striking results in the entire theory of induced representations.

In accordance with the philosophy of Chapter VIII one naturally asks whether the construction and theory of induced representations has a natural generalization to Banach *-algebraic bundles, of which the classical theory of Mackey is simply the specialization to the group case. If $\mathscr{B}$ is a Banach *-algebraic bundle over a locally compact group G, H is a closed subgroup of G, and S is a *-representation of $\mathscr{B}_H$, is there a natural way to construct an "induced" *-representation Ind(S) of $\mathscr{B}$ which reduces to Mackey's construction in the group case? It turns out that Mackey's construction for groups has a natural formulation in terms of Hilbert bundles, and that this Hilbert bundle approach can be easily generalized to a satisfactory definition of Ind(S) in the bundle context—provided S has a special property called $\mathscr{B}$-positivity. In the group case, or more generally if $\mathscr{B}$ has enough unitary multipliers, S is automatically $\mathscr{B}$-positive (8.14), so that Ind(S) can always be formed.

§8 of this chapter studies the property of $\mathscr{B}$-positivity; §9 is devoted to the construction of Ind(S) when S is a $\mathscr{B}$-positive *-representation of $\mathscr{B}_H$; and in §10 we show how Ind(S) reduces to Mackey's induced representation in the group case. §11 develops certain further properties of $\mathscr{B}$-positivity. In §§12 and 13 several elementary properties of induced representations of Banach *-algebraic bundles are obtained, prominent among which is the theorem on "inducing in stages" (12.14). In §14 we deduce the bundle generalization of the Mackey Imprimitivity Theorem (14.17), which is undoubtedly the high point of the present chapter. It is particularly elegant in the special case that the Banach *-algebraic bundle is saturated (see 14.18). §15 is devoted to a bundle version of one of the most beautiful applications of the Imprimitivity Theorem for groups—the generalization by Mackey of the theorem of Stone and von Neumann on the uniqueness of the position and momentum operators of quantum mechanics.

In §16 we discuss a generalization to saturated Banach *-algebraic bundles of the simple notion of conjugation of group representations. This generalization will be of vital importance in Chapter XII. It is worth observing here one feature of this generalized conjugation which may possibly be of far-reaching significance. In the group case, or more generally in the case of Banach *-algebraic bundles $\mathscr{B}$ having enough unitary multipliers, conjugation of representations follows in the wake of the conjugation of bundle elements by unitary multipliers (see 16.16). But if $\mathscr{B}$ does not have unitary multipliers, the conjugation of representations, though still possible, is not described in terms of the conjugation of bundle elements. It seems plausible in fact that, given an arbitrary action Γ of the group G on the "representation theory" of a Banach *-algebra A (and such an action might be quite unobtainable from automorphisms of A itself!), there is a saturated Banach *-algebraic bundle $\mathscr{B}$ over G, with unit fiber algebra A, such that Γ is realized as conjugation in $\mathscr{B}$. *Indeed, the possibility of describing Γ in terms of $\mathscr{B}$ may turn out in the long run to be the greatest justification for the concept of a saturated Banach *-algebraic bundle.* See 16.30.

We shall see in XII.6.3 that, in the special case that A is a C^*-algebra *of compact type* (so that $\hat{A}$ is discrete), there is indeed a saturated C^*-algebraic bundle $\mathscr{B}$ over G whose action on $\hat{A}$ by conjugation coincides with any preassigned action of G on $\hat{A}$. In fact, in §XII.6 we shall manage to describe explicitly the structure of all the possible $\mathscr{B}$ which achieve this.

The entire exposition outlined above is based on the techniques of Loomis and Blattner for avoiding any assumptions of second countability. In fact we make no assumptions of second countability anywhere in this chapter.

We have outlined the contents of the second half of this chapter. The first half owes its existence to the desire, and the possibility, of placing the entire theory of induced representations, along with the Imprimitivity Theorem, in a setting far more general and abstract even than that of Banach *-algebraic bundles. To show how this is done, we would first remind the reader of the observation made in IX.10.13, 14. It was pointed out there that, for *finite* groups, the classical definition of induced representations is a special case of a general purely algebraic construction by which, given a subalgebra B of an algebra A (with unit), one passes from an algebraic representation S of B to an algebraic representation T of A. It was also pointed out in IX.10.13 that this construction is ideally suited to the verification of a generalized Frobenius Reciprocity. One naturally seeks to generalize this observation to unitary representations of arbitrary locally compact groups. Is Mackey's construction of unitary induced representations of arbitrary locally compact groups a special case of a more abstract and more general construction for passing from *-representations of a *-subalgebra B to *-representations of a bigger *-algebra A?

Two obstacles arise when we try to adapt the tensor product construction of IX.10.13 to the context of locally compact groups. First, if H is a closed (non-open) subgroup of a locally compact group G, the group algebra of H is not a *-subalgebra of the group algebra of G. Possibly this obstacle is not essential, since the measure algebra of H *is* a *-subalgebra of the measure algebra of G. But a second obstacle turns out to be more serious: Even if the A and B of IX.10.13 are *-algebras and S is a *-representation of B, there seems to be no natural way of making $A \otimes_B X(S)$ into a pre-Hilbert space so that T, acting on its completion, could become a *-representation of A.

It seems to have been M. Rieffel who first saw the way out of this difficulty (see Rieffel [5]). Taking his cue from IX.10.14(34), he decided to consider not merely a *-algebra A and a *-subalgebra B of A, but also a linear map $p: A \rightarrow B$ with the properties

(i) $p(a^*) = (p(a))^*$ $(a \in A)$,

(ii) $p(ab) = p(a)b$ $(a \in A; b \in B)$.

Such a p is called an A, B *conditional expectation*. Suppose now that S is a *-representation of B which is p-positive in the sense that

$$S_{p(a^*a)} \geq 0 \qquad \text{for all } a \text{ in } A. \tag{1}$$

One can then introduce into $A \otimes X(S)$ the conjugate-bilinear form $(\ ,\ )_0$ satisfying

$$(a_1 \otimes \xi_1, a_2 \otimes \xi_2)_0 = (S_{p(a_2^*a_1)}\xi_1, \xi_2)_{X(S)} \tag{2}$$

$(a_i \in A; \xi_i \in X(S))$; and it follows from (1) that $(\; , \;)_0$ is positive. So $A \otimes X(S)$ can be completed to a Hilbert space $\mathscr{X}$ with respect to $(\; , \;)_0$ (after factoring out the null space of the latter); and under rather general conditions the equation

$$T_c(a \widetilde{\otimes} \xi) = ca \widetilde{\otimes} \xi \qquad (c, a \in A; \xi \in X(S)) \qquad (3)$$

will then define a *-representation T of A on $\mathscr{X}$. (Here $a \widetilde{\otimes} \xi$ stands for the natural image of $a \otimes \xi$ in $\mathscr{X}$.) We refer to T as *induced from S via p*.

We saw in IX.10.14 (especially IX.10.14(34)) that, if A and B are the group algebras of the finite group G and of its subgroup H respectively, and if p is the A, B conditional expectation which sends each f in A into its restriction to H, then (2) and (3) exactly duplicate the definition of $T = \text{Ind}_{H \uparrow G}(S)$ as a *unitary* (not merely an algebraic) representation.

Actually, the construction described above is only a special case of Rieffel's inducing process. Suppose that B is not necessarily a *-subalgebra of A, but that it acts on A to the right so that A is a right B-module; and let $p: A \to B$ be a linear map such that the above properties (i) and (ii) hold. We again call p an A, B conditional expectation. With these ingredients, the same construction (2), (3) enables us to pass from a *-representation S of B which is p-positive (in the sense of (1)) to a *-representation T of A, which we refer to as induced by S via p. This construction is now broad enough to embrace the theory of unitary induced representations of arbitrary locally compact groups. Indeed, let H be a closed subgroup of the locally compact group G; and let A and B be the compacted group algebras of G and H respectively. Assuming that G and H are unimodular, we make B act on A to the right by convolution, and define $p: A \to B$ as restriction: $p(f) = f | H$ ($f \in A$). Then p is an A, B conditional expectation; and, if S is any unitary representation of H, the *-representation of A induced via p by the integrated form of S turns out to be just the integrated form of Mackey's induced representation $\text{Ind}_{H \uparrow G}(S)$. If G and H are not unimodular, the same construction goes through when modular functions are appropriately inserted. Indeed, if $\mathscr{B}$ is an arbitrary Banach *-algebraic bundle over G, it is shown in §§8, 9 that essentially the same construction leads to the bundle generalization of Mackey's inducing construction which we mentioned earlier; and $\mathscr{B}$-positivity and p-positivity then coincide.

Even abstract conditional expectations do not form the most general natural context for Rieffel's abstract inducing process. It will be observed from (2) that what is important in this process is not p itself but the conjugate-bilinear map $\langle a_1, a_2 \rangle \mapsto p(a_2^* a_1)$ of $A \times A$ into B. Thus, given two *-algebras A and B, we introduce in §4 the more general notion of a right B-rigged left

A-module. This is a linear space L which is both a left A-module and a right B-module, together with a "rigging" map $[\ ,\ ]: L \times L \to B$ satisfying the following properties:

$[r, s]$ is linear in s and conjugate-linear in r,

$[r, s]^* = [s, r]$,

$[r, sb] = [r, s]b$,

$[ar, s] = [r, a^*s]$

$(r, s \in L; a \in A; b \in B)$. If p is an A, B conditional expectation, then A itself, together with the "rigging" $\langle a_1, a_2 \rangle \mapsto p(a_1^* a_2)$, becomes a right B-rigged left A-module. The abstract inducing process described by (2), (3) can now be generalized to any right B-rigged left A-module $L, [\ ,\ ]$: We introduce into $L \otimes X(S)$ the conjugate-bilinear form $(\ ,\ )_0$ by means of

$$(r_1 \otimes \xi_1, r_2 \otimes \xi_2)_0 = (S_{[r_2, r_1]}\xi_1, \xi_2)$$

$(r_i \in L; \xi_i \in X(S))$. Let S be $[\ ,\ ]$-positive in the sense that $S_{[r,r]} \geq 0$ for all r in L. Then $(\ ,\ )_0$ is positive, and we can complete $L \otimes X(S)$ to the Hilbert space $\mathscr{X}$ with respect to $(\ ,\ )_0$. We now define the "induced" *-representation T of A on $\mathscr{X}$ (if possible) by

$$T_c(r \overset{\sim}{\otimes} \xi) = cr \overset{\sim}{\otimes} \xi$$

$(c \in A; r \in L; \xi \in X(S))$. This is the so-called Rieffel inducing process, defined and discussed in §§4, 5 of the present chapter. Formulated in these more general terms, it embraces not only the Mackey induced representations of groups but also various operations on representations which up till now have not been thought of as inducing processes (for example the operation of conjugation of representations in Banach *-algebraic bundles; see §16).

It should be remarked that, in passing from the purely algebraic tensor product construction of IX.10.13 to the involutory Rieffel construction of induced representations, one loses the built-in Frobenius Reciprocity which was so noteworthy in IX.10.13. To see this, take any non-compact locally compact Abelian group G; and let τ be the trivial one-dimensional representation of the one-element subgroup H of G. Then $\text{Ind}_{H \uparrow G}(\tau)$ is just the regular representation R of G (see 10.12), which (as one verifies from X.2.11) has no one-dimensional stable subspaces. Let χ be any character of G. If the Frobenius Reciprocity relation of IX.10.13 continued to hold in the context of unitary induced representations, the existence of the trivial non-zero $\tau, \chi|H$ intertwining operator would result in the existence of a non-zero R, χ

intertwining operator; and this would imply the false assertion that $X(R)$ has a one-dimensional stable subspace.

But the loss of Frobenius Reciprocity is compensated by the fact that the Imprimitivity Theorem has a very elegant and general formulation in terms of the Rieffel inducing process. Indeed, suppose that A and B are two *-algebras, and that L is both a right B-rigged left A-module with respect to a rigging $[\ , \ ]$, and also a left A-rigged right B-module (the same concept with left and right interchanged) with respect to another rigging $[\ , \ ]'$. If in addition the associative law

$$[r, s]'t = r[s, t] \qquad\qquad (r, s, t \in L)$$

holds, we have what is called an A, B imprimitivity bimodule. If now S is a $[\ , \ ]$-positive *-representation of B, we can hope to induce it via $L, [\ , \ ]$ to a *-representation T of A; and similarly, if T is a $[\ , \ ]'$-positive *-representation of A, we can hope to induce it via $L, [\ , \ ]'$ to a *-representation S of B. The Imprimitivity Theorem in its abstract form now makes the following valuable assertion: Suppose in addition that the linear spans of the ranges of $[\ , \ ]$ and $[\ , \ ]'$ are B and A respectively. Then the above constructions $S \mapsto T$ and $T \mapsto S$ are each other's inverses, and set up a very natural one-to-one correspondence between the set of all equivalence classes of non-degenerate $[\ , \ ]$-positive *-representations of B on the one hand, and the set of all equivalence classes of non-degenerate $[\ , \ ]'$-positive *-representations of A on the other. This theorem is the backbone of §§6 and 7.

To see how this abstract Imprimitivity Theorem is related to the Mackey Imprimitivity Theorem for groups, we recall what it is that the latter asserts. Given a locally compact group G and a closed subgroup H of G, it asserts a one-to-one correspondence between equivalence classes of unitary representations of H and equivalence classes of systems of imprimitivity for G over G/H. Now by VIII.18.22, the latter objects are essentially *-representations of the transformation algebra E for G and G/H. So the Imprimitivity Theorem for groups can be regarded as asserting that two *-algebras— namely the group algebra B of H and the transformation algebra E—have isomorphic *-representation theories. According to the abstract Imprimitivity Theorem this will be proved if we can set up an E, B imprimitivity bimodule of the right sort. The construction of such an E, B imprimitivity bimodule, not only in the group case but also in the context of Banach *-algebraic bundles, is our main occupation in §14.

We have still not reached the level of greatest generality in our discussion of the abstract Imprimitivity Theorem. In its most general form it is not a statement about *-algebras at all, but about so-called operator inner products. Given a linear space L and a Hilbert space X, by an operator inner product in L, acting in X, we mean a map $V: L \times L \to \mathcal{O}(X)$ which is linear in its first variable and conjugate-linear in its second, and such that the conjugate-bilinear form $(\ ,\ )_0$ on $L \otimes X$ given by

$$(r \otimes \xi, s \otimes \eta)_0 = (V_{r,s}\xi, \eta)_X$$

$(r, s \in L; \xi, \eta \in X)$ is positive. Such a V of course gives rise to a new Hilbert space $\mathcal{X}$, the completion of $L \otimes X$ with respect to $(\ ,\ )_0$. It also gives rise in a canonical manner to a new operator inner product $W: \bar{L} \times \bar{L} \to \mathcal{O}(\mathcal{X})$, in the complex-conjugate space $\bar{L}$ and acting on $\mathcal{X}$. We say that W is deduced from V. The germ of the abstract Imprimitivity Theorem now lies in the following very general assertion: If the operator inner product V is non-degenerate, and if W is deduced from V and V' is deduced from W, then V' and V are unitarily equivalent; thus the map $V \mapsto W$ becomes a duality. This statement is the climax of §§1 and 2. The abstract Imprimitivity Theorem of §6 is an easy consequence of this.

We have now summarized the main contents of this chapter (proceeding largely in reverse order for purposes of motivation). In the final §17 we have collected together a few facts about imprimitivity bimodules and the inducing process in the non-involutory context. As one might expect, relatively little can be said about the non-involutory context at the present time.

1. Operator Inner Products

1.1 Let L be a (complex) linear space and X a Hilbert space.

Definition. An *operator inner product on L, acting in X*, is a map $V: L \times L \to \mathcal{O}(X)$ such that

(i) $V_{s,t}$ is linear in s and conjugate-linear in t $(s, t \in L)$;
(ii) we have

$$\sum_{i,j=1}^{n} (V_{t_i, t_j}\xi_i, \xi_j) \geq 0 \qquad (1)$$

for any positive integer n, any $t_1, \ldots, t_n$ in L, and any $\xi_1, \ldots, \xi_n$ in X.

We refer to X as the *space of V*, and denote it by $X(V)$.
Condition (1) is called the *complete positivity condition*.
Taking $n = 1$ in (1) we obtain $(V_{t,t}\xi, \xi) \geq 0$ $(t \in L; \xi \in X)$, or

$$V_{t,t} \geq 0 \qquad \text{for all } t \text{ in } L. \qquad (2)$$

This implies in particular that

$$V_{t,s} = (V_{s,t})^* \qquad (s, t \in L). \qquad (3)$$

Indeed, condition (i) gives $V_{s+t,s+t} = V_{s,s} + V_{t,t} + V_{s,t} + V_{t,s}$. Since each $V_{r,r}$ is Hermitian by (2), this shows that $V_{s,t} + V_{t,s}$ is Hermitian, that is, $(V_{s,t})^* + (V_{t,s})^* = V_{s,t} + V_{t,s}$. Replacing t by it in this equation, using (i) and cancelling i, and then adding the resulting two equations, we obtain (3).

1.2. We shall see in 3.10 that in general (1) is a stronger condition than (2). In some special situations, however, (1) can be replaced by (2) without changing Definition 1.1. Suppose for example that $X = \mathbb{C}$ and that $V: L \times L \to \mathbb{C}$ satisfies 2.1(i) and (2). Then the ξ_i of (1) are just complex numbers, and

$$\sum_{i,j} (V_{t_i,t_j}\xi_i, \xi_j) = \sum_{i,j} \xi_i \bar{\xi}_j V_{t_i,t_j}$$

$$= V_{\sum_i \xi_i t_i, \sum_j \xi_j t_j} \geq 0.$$

Thus V is an operator inner product on L.

This shows that a numerical inner product making L a pre-Hilbert space is an example of an operator inner product.

1.3. In view of (3) the range $\{V_{s,t} : s, t \in L\}$ of an operator inner product V is a self-adjoint collection of operators, and we can carry over to it the elementary definitions and results of *-representation theory as contained in §VI.9. Such applications of §VI.9 in the context of operator inner products will often be made without explicit mention.

Definition. An operator inner product V is *non-degenerate* if its range acts non-degenerately on $X(V)$, that is, if the linear span of $\{V_{s,t}\xi : s, t \in L, \xi \in X(V)\}$ is dense in $X(V)$—or equivalently, if $\xi = 0$ whenever $V_{s,t}\xi = 0$ for all s, t.

Let V be an arbitrary operator inner product. If Y is a closed subspace of $X(V)$ which is V-stable in the nautral sense, then (as in VI.9.4) $Y^\perp$ is also V-stable. In particular, let Y be the closed linear span of $\{V_{s,t}\xi : s, t \in L, \xi \in X(V)\}$. Then $V' : \langle s, t \rangle \mapsto V_{s,t} | Y$ is non-degenerate and $V_{s,t} | Y^\perp = 0$ for all s, t. We call V' the *non-degenerate part of V*.

1.4. Let V and V' be two operator inner products on L.

Definition. By a V, V' *intertwining map* we mean a bounded linear map $F: X(V) \to X(V')$ such that

$$F \circ V_{s,t} = V'_{s,t} \circ F \qquad (s, t \in L).$$

If there is a V, V' intertwining map which is onto $X(V')$ and is an isometry, then V and V' are *unitarily equivalent*—in symbols, $V \cong V'$.

The space of all V, V intertwining maps is a closed *-subalgebra of $\mathcal{O}(X(V))$, called as usual the *commuting algebra of* V, and often denoted by $\mathscr{I}(V)$. A closed subspace Y of $X(V)$ is V-stable if and only if the projection onto Y belongs to $\mathscr{I}(V)$. If there are no closed V-stable subspaces except $\{0\}$ and $X(V)$, V is *irreducible*. This happens if and only if $\mathscr{I}(V)$ consists of the scalar operators only (see VI.14.1).

1.5. The following important proposition generalizes the Schwarz Inequality to operator inner products.

Proposition. *Let V be an operator inner product on L. If $t_1, \ldots, t_n$, $s \in L$ and $\xi_1, \ldots, \xi_n \in X(V)$, then:*

$$\left\| \sum_{i=1}^{n} V_{t_i, s} \xi_i \right\|^2 \leq \|V_{s,s}\| \sum_{i,j=1}^{n} (V_{t_i, t_j} \xi_i, \xi_j). \tag{4}$$

Proof. Let η be any unit vector in $X(V)$; let λ be a real number; and let μ be a complex number such that $|\mu| = 1$ and

$$\bar{\mu} \sum_{i=1}^{n} (V_{t_i, s} \xi_i, \eta) = - \left| \sum_{i=1}^{n} (V_{t_i, s} \xi_i, \eta) \right|. \tag{5}$$

In the complete positivity condition (1) we shall replace n by $n + 1$; $t_1, \ldots, t_n$ by $\lambda t_1, \ldots, \lambda t_n, \mu s$; and $\xi_1, \ldots, \xi_n$ by $\xi_1, \ldots, \xi_n, \eta$. Putting $k = \sum_{i,j=1}^{n} (V_{t_i, t_j} \xi_i, \xi_j)$ and recalling (3), we then obtain from (1) and (5):

$$0 \leq \lambda^2 k + |\mu|^2 (V_{s,s} \eta, \eta) + 2 \operatorname{Re} \left[\lambda \bar{\mu} \sum_{i=1}^{n} (V_{t_i, s} \xi_i, \eta) \right]$$

$$= \lambda^2 k - 2\lambda \left| \sum_{i=1}^{n} (V_{t_i, s} \xi_i, \eta) \right| + (V_{s,s} \eta, \eta).$$

Since this holds for all real λ,

$$\left| \sum_{i=1}^{n} (V_{t_i, s} \xi_i, \eta) \right|^2 \leq k(V_{s,s} \eta, \eta).$$

Letting η run over all unit vectors in this relation, we obtain (4). ∎

1.6.   Taking $n = 1$ in (4) and letting ξ_1 run over all unit vectors we obtain as a special case of (4):

$$\|V_{s,t}\|^2 \le \|V_{s,s}\|\,\|V_{t,t}\| \qquad\qquad (s, t \in L). \qquad (6)$$

This directly generalizes the numerical Schwarz Inequality.

Remark.   If V is an operator inner product on L, the inequality (6) implies that the function $\gamma: t \mapsto \|V_{t,t}\|^{1/2}$ satisfies the triangle inequality and so is a seminorm on L. In particular, if $V_{t,t} = 0 \Rightarrow t = 0$, L becomes a normed linear space under γ.

Deduced Hilbert Spaces

1.7.   From each operator inner product V there arises naturally a new Hilbert space, to be called the Hilbert space deduced from V. It is on these deduced Hilbert spaces that the abstractly induced *-representations, which form the main topic of this chapter, will act. They will therefore be of great importance.

1.8.   Fix an operator inner product V on the linear space L, acting on a Hilbert space X. As usual the inner product of X is denoted by $(\ ,\ )$. Now the expression $(V_{s,t}\xi, \eta)$ is linear in s and ξ and conjugate-linear in t and η; so the equation

$$(s \otimes \xi, t \otimes \eta)_0 = (V_{s,t}\xi, \eta) \qquad\qquad (7)$$

$(s, t \in L; \xi, \eta \in X)$ determines a conjugate-bilinear form $(\ ,\ )_0$ on $M = L \otimes X$. Furthermore, the complete positivity condition (1) says that $(\ ,\ )_0$ is positive:

$$(\zeta, \zeta)_0 \ge 0 \qquad \text{for all } \zeta \text{ in } M.$$

Thus, if $N = \{\zeta \in M : (\zeta, \zeta)_0 = 0\}$, the quotient $\mathscr{X}' = M/N$ becomes a pre-Hilbert space under the lifted inner product (also denoted by $(\ ,\ )_0$):

$$(\zeta + N, \zeta' + N)_0 = (\zeta, \zeta')_0.$$

Completing $\mathscr{X}'$ with respect to $(\ ,\ )_0$, we obtain a Hilbert space $\mathscr{X}$.

Definition.   This $\mathscr{X}$ is called the *Hilbert space deduced from* V. It is denoted by $\mathscr{X}(L, V)$, or $\mathscr{X}(V)$ for short.

The norm in $\mathscr{X}$ is generally denoted by $\|\ \ \|_0$. We shall refer to the image $(s \otimes \xi) + N$ of $s \otimes \xi$ in $\mathscr{X}$ as $s \,\widetilde{\otimes}\, \xi$. Thus (7) becomes

$$(s \,\widetilde{\otimes}\, \xi, t \,\widetilde{\otimes}\, \eta)_0 = (V_{s,t}\xi, \eta) \qquad\qquad (8)$$

$(s, t \in L; \xi, \eta \in X)$. Notice the inequality

$$\|s \widetilde{\otimes} \xi\|_0^2 \leq \|V_{s,s}\| \|\xi\|^2 \qquad (s \in L; \xi \in X). \qquad (9)$$

This shows that, for fixed s in L, the map $\xi \mapsto s \widetilde{\otimes} \xi$ of X into $\mathscr{X}$ is continuous. In particular, if Y is a dense subset of X, then $\{s \widetilde{\otimes} \xi : s \in L, \xi \in Y\}$ has dense linear span in $\mathscr{X}$.

Remark. The complete positivity of V is clearly *equivalent* to the positivity of the conjugate-bilinear form $(\ ,\ )_0$ defined by (7). This remark will often be useful in what follows.

1.9. Example. Suppose in 1.8 that $X = \mathbb{C}$, so that V itself is just a positive conjugate-bilinear form on L (see 1.2). Then $M \cong L$ and $(\ ,\ )_0 \cong V$. In this case, therefore, $\mathscr{X}(V)$ is just the Hilbert space completion of the quotient of L modulo the null space of V.

1.10. Example. Let X be a non-zero Hilbert space (with inner product $(\ ,\ )$); and if $\xi, \eta \in X$, let $V_{\xi,\eta}$ be the operator on X of rank 1 (or 0) given by

$$V_{\xi,\eta}(\zeta) = (\zeta, \xi)\eta \qquad (\zeta \in X). \qquad (10)$$

If $\xi_1, \ldots, \xi_n, \eta_1, \ldots, \eta_n$ are in X, we have

$$\sum_{i,j} (V_{\eta_i, \eta_j} \xi_i, \xi_j) = \left| \sum_i (\xi_i, \eta_i) \right|^2 \geq 0. \qquad (11)$$

Notice that $V_{\xi,\eta}$ is conjugate-linear in ξ and linear in η. So, if ξ and η are considered as belonging to the complex-conjugate space $\bar{X}$ (see VIII.9.13) rather than to X itself, $V_{\xi,\eta}$ becomes linear in ξ and conjugate-linear in η. Consequently, in view of (11), V is an operator inner product on $\bar{X}$, acting in X.

As regards the Hilbert space $\mathscr{X}(V)$ deduced from V, equation (11) shows that $\mathscr{X}(V)$ *is one-dimensional*. In fact, identifying $\mathscr{X}(V)$ with $\mathbb{C}$, we see from (11) that

$$\eta \widetilde{\otimes} \xi = (\xi, \eta) \qquad (\xi \in X; \eta \in \bar{X}).$$

Remark. It might at first sight seem more appropriate to interchange ξ and η in (10); for then $V_{\xi,\eta}$ would be linear in ξ and conjugate-linear in η, and it would not be necessary to introduce the complex-conjugate space $\bar{X}$. The trouble is that, if we do that, V is no longer completely positive. Indeed, instead of (11) we would then have

$$\sum_{i,j=1}^{n} (V_{\eta_i, \eta_j} \xi_i, \xi_j) = \sum_{i,j} (\xi_i, \eta_j)(\eta_i, \xi_j) \qquad (12)$$

Take $n = 2$. Let ρ and σ be orthogonal unit vectors in X, and put $\xi_1 = \eta_2 = \rho$, $\xi_2 = -\eta_1 = \sigma$. Then by (12)

$$\sum_{i,j=1}^{2} (V_{\eta_i,\eta_j}\xi_i, \xi_j) = \sum_{i,j=1}^{2} (\xi_i, \eta_j)(\eta_i, \xi_j) = -2.$$

So V is not completely positive when we interchange its variables.

1.11. The following important result shows how certain operators on $X(V)$—namely those in the commuting algebra of V—give rise to operators on $\mathcal{X}(V)$.

Proposition. *Let V and V' be two operator inner products on L; and put $X = X(V)$, $X' = X(V')$, $\mathcal{X} = \mathcal{X}(V)$, $\mathcal{X}' = \mathcal{X}(V')$. If $F: X \to X'$ is a V, V' intertwining map, the equation*

$$\tilde{F}(s \widetilde{\otimes} \xi) = s \widetilde{\otimes} F(\xi) \qquad (s \in L; \xi \in X) \qquad (13)$$

determines a unique bounded linear map $\tilde{F}: \mathcal{X} \to \mathcal{X}'$.

Note. $s \widetilde{\otimes} \xi$ and $s \widetilde{\otimes} F(\xi)$ are the elements of $\mathcal{X}$ and $\mathcal{X}'$ respectively defined in 1.8.

Proof. Let I be the identity map on L; and let $(\ ,\ )_0$ and $(\ ,\ )_0'$ be the inner products of $\mathcal{X}$ and $\mathcal{X}'$. We have for $r, s \in L$, $\xi \in X$ and $\eta \in X'$,

$$(r \otimes F\xi, s \otimes \eta)_0' = (V'_{r,s}F\xi, \eta)$$

$$= (FV_{r,s}\xi, \eta) \qquad \qquad \text{(since F is intertwining)}$$

$$= (V_{r,s}\xi, F^*\eta) = (r \otimes \xi, s \otimes F^*\eta)_0.$$

From this it follows that

$$((I \otimes F)\zeta, \zeta')_0' = (\zeta, (I \otimes F^*)\zeta')_0 \qquad (14)$$

for all ζ in $L \otimes X$ and ζ' in $L \otimes X'$.

To prove the proposition it is enough to show that

$$((I \otimes F)\zeta, (I \otimes F)\zeta)_0' \le \|F\|^2 (\zeta, \zeta)_0 \qquad (15)$$

for all ζ in $L \otimes X$.

Now $\|F\|^2 - F^*F$ is a positive operator on X, and so has a positive square root S (see VI.7.15) which belongs to the C^*-algebra of V, V intertwining operators. By the positivity of $(\ ,\ )_0$, if $\zeta \in L \otimes X$

$$((I \otimes S)\zeta, (I \otimes S)\zeta)_0 \ge 0.$$

Hence

$$0 \leq ((I \otimes S^2)\zeta, \zeta)_0 \qquad \text{(by (14) applied to } S)$$

$$= ((I \otimes (\|F\|^2 - F^*F))\zeta, \zeta)_0$$

$$= \|F\|^2(\zeta, \zeta)_0 - ((I \otimes F^*F)\zeta, \zeta)_0$$

$$= \|F\|^2(\zeta, \zeta)_0 - ((I \otimes F)\zeta, (I \otimes F)\zeta)_0 \qquad \text{(by (14))}.$$

But this is (15). ∎

The inequality (15) asserts that

$$\|\tilde{F}\| \leq \|F\|. \tag{16}$$

Notice that the correspondence $F \mapsto \tilde{F}$ is linear and preserves compositions and adjoints, and sends 1_X into $1_{\mathscr{X}}$. To prove that it preserves adjoints, we observe that

$$(\tilde{F}(s \tilde{\otimes} \xi), t \tilde{\otimes} \eta)_0 = (s \tilde{\otimes} F\xi, t \tilde{\otimes} \eta)_0$$

$$= (V'_{s,t}F\xi, \eta) = (FV_{s,t}\xi, \eta)$$

$$= (V_{s,t}\xi, F^*\eta) = (s \tilde{\otimes} \xi, t \tilde{\otimes} F^*\eta)_0$$

$$= (s \tilde{\otimes} \xi, (F^*)^{\tilde{}}(t \tilde{\otimes} \eta))_0.$$

Therefore $(\tilde{F})^* = (F^*)^{\tilde{}}$; and adjoints are preserved.

We recall that $\mathscr{I}(V)$ is a norm-closed *-subalgebra of $\mathscr{O}(X)$. By the preceding paragraph $F \mapsto \tilde{F}$ is a *-homomorphism of $\mathscr{I}(V)$ into $\mathscr{O}(\mathscr{X}(V))$. From this and the general result VI.3.7 we obtain of course an independent proof of (16) (once the boundedness of $\tilde{F}$ is known).

1.12. Suppose now that V is a non-degenerate operator inner product on L. Then we claim that $F \mapsto \tilde{F}$ is one-to-one on $\mathscr{I}(V)$. Indeed, let $F \in \mathscr{I}(V)$, $\tilde{F} = 0$. Then (as in the proof of the preservation of adjoints) we have $0 = (\tilde{F}(s \otimes \xi), t \otimes \eta)_0 = (V_{s,t}\xi, F^*\eta)$ for all s, t in L and ξ, η in X. Fixing η and letting s, t, ξ vary, we conclude from the non-degeneracy of V that $F^*\eta = 0$. Since this holds for all η we have $F^* = 0$, or $F = 0$.

Thus, if V is non-degenerate, $F \mapsto \tilde{F}$ is a *-isomorphism of $\mathscr{I}(V)$ into $\mathscr{O}(\mathscr{X}(V))$; and so by VI.8.8

$$\|\tilde{F}\| = \|F\| \qquad \text{for all } F \text{ in } \mathscr{I}(V). \tag{17}$$

1.13. *Remark.* The construction $F \mapsto \tilde{F}$ is of a categorical nature: If the operator inner products on a fixed L are regarded as the objects of a category $\mathscr{C}$ whose morphisms are the intertwining maps, then the two mappings

$X(V) \mapsto \mathscr{X}(V)$ and $F \mapsto \tilde{F}$ together form a functor from $\mathscr{C}$ to the category of Hilbert spaces.

Direct Sums and Tensor Products of Operator Inner Products

1.14. For each j in an index set J let V^j be an operator inner product on the fixed linear space L, acting in $X_j = X(V^j)$. We shall assume that, for each pair s, t of elements of L, the set $\{\|V^j_{s,t}\| : j \in J\}$ is bounded, so that we can form the direct sum operator $V_{s,t} = \sum_{j \in J}^{\oplus} V^j_{s,t}$ acting in the Hilbert direct sum $X = \sum_{j \in J}^{\oplus} X_j$. Then we claim that $V: \langle s, t \rangle \mapsto V_{s,t}$ is again an operator inner product. Indeed, if $t_1, \ldots, t_n \in L$ and $\xi_1, \ldots, \xi_n \in X$ (with $\xi_i = \sum_j^{\oplus} \xi_i^j$, $\xi_i^j \in X_j$), then

$$\sum_{i,k=1}^{n} (V_{t_i, t_k} \xi_i, \xi_k) = \sum_{j \in J} \sum_{i,k=1}^{n} (V^j_{t_i, t_k} \xi_i^j, \xi_k^j)$$

$$\geq 0:$$

and the claim is now evident.

Definition. V is called the *direct sum of the V^j*, and is denoted by $\sum_{j \in J}^{\oplus} V^j$.

1.15.   The construction of deduced Hilbert spaces preserves direct sums.

Proposition.   *Let $\{V^j\}$ $(j \in J)$ be an indexed collection of operator inner products on L whose Hilbert direct sum $V = \sum_{j \in J}^{\oplus} V^j$ can be formed. Then $\mathscr{X}(V)$ is essentially the same as the Hilbert direct sum $\sum_{j \in J}^{\oplus} \mathscr{X}(V^j)$.*

The precise meaning of "essentially," as well as the proof of the proposition, are left to the reader.

We have seen in 1.3 that any operator inner product V is the direct sum of a non-degenerate operator inner product V' and a zero operator inner product V''. Evidently $\mathscr{X}(V'') = \{0\}$. So by the preceding proposition $\mathscr{X}(V) \cong \mathscr{X}(V')$. That is, *the deduced Hilbert space is unchanged on replacing an operator inner product by its non-degenerate part.*

1.16.   We shall next show that the tensor product of two operator inner products is again an operator inner product.

To be precise, let V and W be operator inner products on linear spaces L and M respectively. Put $X = X(V)$, $Y = X(W)$; and let Z be the Hilbert tensor product $X \otimes Y$. Let P be the algebraic tensor product $L \otimes M$.

Proposition. *The equation*

$$U_{s_1 \otimes t_1, s_2 \otimes t_2} = V_{s_1, s_2} \otimes W_{t_1, t_2} \tag{18}$$

$(s_i \in L; t_i \in M)$ *defines an operator inner product U on P, acting in Z.*

Proof. Let us first assume that $X = Y = \mathbb{C}$, so that V and W are positive conjugate-bilinear forms on L and M respectively. Let N_1 and N_2 be the null spaces of V and W respectively. Then $\mathscr{X}(V)$ and $\mathscr{X}(W)$ are the Hilbert space completions of L/N_1 and M/N_2 respectively (see 1.9), and the inner product of $\mathscr{X}(V) \otimes \mathscr{X}(W)$, when composed with the quotient map $P \to \mathscr{X}(V) \otimes \mathscr{X}(W)$, is just the conjugate-bilinear form U defined by (18). So the latter is positive. In fact, this argument shows that

$$\mathscr{X}(U) \cong \mathscr{X}(V) \otimes \mathscr{X}(W) \tag{19}$$

(Hilbert tensor product).

We now discard the assumption that $X = Y = \mathbb{C}$. Equation (18) clearly defines a conjugate-bilinear map $U: P \times P \to \mathcal{O}(Z)$; and the condition that U be completely positive is equivalent to the condition that the conjugate-bilinear form U^0 on $P \otimes Z$ given by

$$U^0(c \otimes \zeta, c' \otimes \zeta') = (U_{c,c'}\zeta, \zeta') \tag{20}$$

$(c, c' \in P; \zeta, \zeta' \in Z)$ be positive (see Remark 1.8). Now let V^0 and W^0 be the positive conjugate-bilinear forms on $L \otimes X$ and $M \otimes Y$ respectively derived from V and W respectively as in (7):

$$V^0(s \otimes \xi, s' \otimes \xi') = (V_{s,s'}\xi, \xi'), \tag{21}$$

$$W^0(t \otimes \eta, t' \otimes \eta') = (W_{t,t'}\eta, \eta') \tag{22}$$

$(s, s' \in L; \xi, \xi' \in X; t, t' \in M; \eta, \eta' \in Y)$. Combining (20), (21), and (22), we see that the complex-valued forms U^0, V^0, and W^0 are related to each other just as U, V, and W are related in (18). Hence by the preceding paragraph U^0 is positive; and so U is completely positive, proving the proposition. ∎

Definition. This operator inner product U on $L \otimes M$, acting on the Hilbert tensor product $X(V) \otimes X(W)$, is called the *tensor product of V and W*, and is denoted by $V \otimes W$.

Keeping the notation of the preceding proof, we notice that $\mathscr{X}(U) = \mathscr{X}(U^0)$, $\mathscr{X}(V) = \mathscr{X}(V^0)$, $\mathscr{X}(W) = \mathscr{X}(W^0)$, and (by (19)) $\mathscr{X}(U^0) \cong \mathscr{X}(V^0) \otimes \mathscr{X}(W^0)$. Hence

$$\mathscr{X}(V \otimes W) \cong \mathscr{X}(V) \otimes \mathscr{X}(W) \quad \text{(Hilbert tensor product).} \tag{23}$$

That is, the process of forming deduced Hilbert spaces preserves tensor products.

2.   Duality for Operator Inner Products

2.1.   The abstract Imprimitivity Theorem 6.15 will actually be obtained as a corollary of an even more general duality theorem on operator inner products, which we present in this section.

2.2.   Fix a linear space L and an operator inner product V on L. We shall write X and Y for $X(V)$ and the deduced Hilbert space $\mathscr{X}(V)$ respectively, and shall denote by $(\ ,\ )_0$ the inner product in Y. As in 1.8, $s \widetilde{\otimes} \xi$ is the image of $s \otimes \xi$ in Y.

By the generalized Schwarz Inequality 1.5(4) each element s of L gives rise to a unique bounded linear map $U_s: Y \to X$ such that

$$U_s(t \widetilde{\otimes} \xi) = V_{t,s}\xi \qquad\qquad (t \in L; \xi \in X) \qquad (1)$$

U_s is conjugate-linear in s (since $V_{t,s}$ is), and by 1.5(4)

$$\|U_s\|^2 \le \|V_{s,s}\|. \qquad\qquad (2)$$

If we fix another element t of L, and compose U_s with the bounded linear map $\xi \mapsto t \widetilde{\otimes} \xi$ of X into Y (see 1.8(9)), we obtain a bounded linear map $W_{s,t}: Y \to Y$;

$$W_{s,t}(\zeta) = t \widetilde{\otimes} U_s(\zeta) \qquad\qquad (\zeta \in Y), \qquad (3)$$

that is,

$$W_{s,t}(r \widetilde{\otimes} \xi) = t \widetilde{\otimes} V_{r,s}\xi \qquad\qquad (r \in L; \xi \in X). \qquad (4)$$

Combining (2) and 1.8(9) we get for $s, t \in L$:

$$\|W_{s,t}\|^2 \le \|V_{s,s}\|\,\|V_{t,t}\|. \qquad\qquad (5)$$

Notice that $W_{s,t}$ is linear in t and conjugate-linear in s. We also observe that

$$(W_{s,t})^* = W_{t,s} \qquad\qquad (s, t \in L). \qquad (6)$$

Indeed: If $s, t, p, q \in L$ and $\xi, \eta \in X$,

$$(W_{s,t}(p \widetilde{\otimes} \xi), q \widetilde{\otimes} \eta)_0 = (t \widetilde{\otimes} V_{p,s}\xi, q \widetilde{\otimes} \eta)_0$$

$$= (V_{t,q}V_{p,s}\xi, \eta) \qquad \text{(by 1.8(8))}$$

$$= (V_{p,s}\xi, V_{q,t}\eta) \qquad \text{(by 1.1(3))}$$

$$= (p \widetilde{\otimes} \xi, s \widetilde{\otimes} V_{q,t}\eta)_0 \qquad \text{(by 1.8(8))}$$

$$= (p \widetilde{\otimes} \xi, W_{t,s}(q \widetilde{\otimes} \eta))_0 \qquad \text{(by (4))}.$$

This proves (6).

2.3. By (6), $\{W_{s,t}: s, t \in L\}$ is a self-adjoint collection of operators. We claim that this collection acts non-degenerately on Y (see 1.3).

To see this, notice first (by the last part of 1.15) that V may be replaced by its non-degenerate part without altering Y or the $W_{s,t}$. So we may assume from the beginning that V is non-degenerate. Then by (4), for each fixed t, the linear span of $\{\text{range}(W_{s,t}): s \in L\}$ consists of the $t \widetilde{\otimes} \eta$ where η runs over a dense subset of X. From this and the last statement of 1.8 we conclude that the linear span of $\{\text{range}(W_{s,t}): s, t \in L\}$ is dense in Y.

2.4. *Example.* Suppose that L itself is a Hilbert space and that V is the inner product $(\, , \,)$ of L, so that $X = \mathbb{C}$ and (by 1.9) $Y = L$. Then (4) becomes:

$$W_{\xi,\eta}(\zeta) = (\zeta, \xi)\eta \qquad (\xi, \eta, \zeta \in L).$$

So in this example W coincides with the V of Example 1.10.

2.5. *Remark.* By 1.10 the W of Example 2.4 is an operator inner product on the complex-conjugate space $\bar{L}$. In a moment we are going to generalize this fact and show that the W constructed in 2.2 is always an operator inner product on $\bar{L}$. Thus the remark at the end of 1.10 clearly motivates the choice of $W_{s,t}$ rather than $W_{t,s}$ for the left side of (3).

2.6. The crux of the abstract Imprimitivity Theorem lies in the following proposition:

Proposition.

(I) *The W constructed in 2.2 is an operator inner product on $\bar{L}$. We shall speak of it as deduced from the operator inner product V on L.*

(II) *In view of (I), replacing L and V by $\bar{L}$ and W, we can form the operator inner product $\tilde{V}$ on $\bar{\bar{L}} = L$ deduced from W. Then, if V is non-degenerate, we have $\tilde{V} \cong V$.*

Proof. If $s, s', t, t' \in L$ and $\xi, \xi' \in X$,

$$(W_{s,s'}(t \tilde{\otimes} \xi), t' \tilde{\otimes} \xi')_0 = (s' \tilde{\otimes} V_{t,s}\xi, t' \tilde{\otimes} \xi')_0 \qquad \text{(by (4))}$$

$$= (V_{s',t'} V_{t,s}\xi, \xi')$$

$$= (V_{t,s}\xi, V_{t',s'}\xi') \qquad \text{(by 1.1(3)).} \qquad (7)$$

Now to prove (I) it is sufficient (by the boundedness of the $W_{s,s'}$) to show that $\sum_{i,j} (W_{t_i,t_j}\zeta_i, \zeta_j)_0 \geq 0$ when the ζ_i run over a dense subset of Y, for example over the linear combinations of the $s \tilde{\otimes} \xi$. Thus it is sufficient to prove that

$$\sum_{i,j=1}^{n} (W_{t_i,t_j}(s_i \tilde{\otimes} \xi_i), s_j \tilde{\otimes} \xi_j)_0 \geq 0 \qquad (8)$$

whenever $n = 1, 2, \ldots$, the s_i and t_i are in L, and the ξ_i are in X. But by (7) the left side of (8) is

$$\sum_{i,j=1}^{n} (V_{s_i,t_i}\xi_i, V_{s_j,t_j}\xi_j) = \left\| \sum_{i=1}^{n} V_{s_i,t_i}\xi_i \right\|^2 \geq 0.$$

So (I) has been proved.

By (I) we may form the Hilbert space $\tilde{X} = \mathscr{X}(W)$ deduced from W. This is obtained by factoring and completing the space $\bar{L} \otimes Y$ with respect to the positive conjugate-bilinear form $(\ ,\ )_{00}$ on $\bar{L} \otimes Y$ given by:

$$(s \otimes \zeta, t \otimes \zeta')_{00} = (W_{s,t}\zeta, \zeta')_0 \qquad (9)$$

$(s, t \in \bar{L}; \zeta, \zeta' \in Y)$. As usual the image of $s \otimes \zeta$ in $\tilde{X}$ is called $s \tilde{\otimes} \zeta$. Combining (7) and (9) we get

$$(s \tilde{\otimes} (t \tilde{\otimes} \xi), s' \tilde{\otimes} (t' \tilde{\otimes} \xi'))_{00} = (V_{t,s}\xi, V_{t',s'}\xi') \qquad (10)$$

$(s, s', t, t' \in L; \xi, \xi' \in X)$. In view of the denseness in $\tilde{X}$ of the linear span of $\{s \tilde{\otimes} (t \tilde{\otimes} \xi): s, t \in L, \xi \in X\}$ (see 1.8), (10) implies that the equation

$$F(s \tilde{\otimes} (t \tilde{\otimes} \xi)) = V_{t,s}\xi \qquad (s, t \in L; \xi \in X) \qquad (11)$$

defines a linear isometry F of $\tilde{X}$ into X. Since V is non-degenerate, F is onto X.

Thus, to prove (II), it is sufficient to show that F intertwines $\tilde{V}$ and V. But, if $r, s, p, q \in L$ and $\xi \in X$, the equation (4) applied to $\tilde{V}$ and W gives:

$$F\tilde{V}_{p,q}(r \tilde{\otimes} (s \tilde{\otimes} \xi)) = F(q \tilde{\otimes} W_{r,p}(s \tilde{\otimes} \xi))$$

$$= F(q \tilde{\otimes} (p \tilde{\otimes} V_{s,r}\xi))$$

$$= V_{p,q}V_{s,r}\xi \qquad \text{(by (11))}$$

$$= V_{p,q}F(r \tilde{\otimes} (s \tilde{\otimes} \xi)).$$

Since the $r \widetilde{\otimes} (s \widetilde{\otimes} \xi)$ span a dense subspace of $\tilde{X}$, this shows that $F\tilde{V}_{p,q} = V_{p,q}F$. So F is $\tilde{V}, V$ intertwining; and (II) is proved. ∎

2.7. The preceding proposition gives rise immediately to the following duality theorem. Let L be a fixed linear space.

Duality Theorem for Operator Inner Products. *The map sending each operator inner product V on L into the operator inner product W on $\bar{L}$ deduced from V sets up a bijection $\Phi: \mathscr{V} \to \mathscr{W}$ between the set $\mathscr{V}$ of all unitary equivalence classes of non-degenerate operator inner products on L, and the set $\mathscr{W}$ of all unitary equivalence classes of non-degenerate operator inner products on $\bar{L}$. The inverse Φ^{-1} of Φ is the map which sends an element of $\mathscr{W}$ into the element of $\mathscr{V}$ deduced from it.*

2.8 *Example.* As we saw in 2.4, if L itself is a Hilbert space, the operator inner product on $\bar{L}$ corresponding under Φ with the (numerical) inner product $(\, , \,)$ of L is the V defined in Example 1.10.

Thus the correspondence Φ is very far from preserving the dimensionality of the spaces in which the operator inner products act.

2.9. If V and W correspond under the Φ of 2.7, we have

$$\|V_{t,t}\| = \|W_{t,t}\| \qquad (t \in L). \qquad (12)$$

Indeed, putting $s = t$ in (5) we obtain $\|W_{t,t}\| \le \|V_{t,t}\|$. Since by 2.6 or 2.7 V and W play symmetric roles, the reverse inequality also holds. Thus we get (12).

Remark. In the example of 2.8, if s and t are orthogonal non-zero vectors in L, then $(s, t) = 0$ while $V_{s,t} \ne 0$. So the equality $\|V_{s,t}\| = \|W_{s,t}\|$ fails in general if $t \ne s$.

2.10. The correspondence Φ of Theorem 2.7 preserves direct sums. More precisely, let $\{V^{\alpha}\}$ $(\alpha \in I)$ be an indexed collection of operator inner products on L whose direct sum $V = \sum_{\alpha \in I}^{\oplus} V^{\alpha}$ can be formed (see 1.14). If W^{α} is the operator inner product on $\bar{L}$ deduced from V^{α}, then $W = \sum_{\alpha \in I}^{\oplus} W^{\alpha}$ can be formed; and W is the operator inner product on $\bar{L}$ deduced from V.

The verification of this follows easily from 1.15, and is left to the reader.

2.11.   Similarly, Φ preserves tensor products. Suppose that L, M, P, V, W, U are as in 1.16; and let V', W', U' be the operator inner products on $\bar{L}, \bar{M}, \bar{P}$ deduced from V, W, U respectively. Then it follows without difficulty from 1.16(23) that

$$U' \cong V' \otimes W'.$$

Isomorphism of the Commuting Algebras

2.12.   One of the most important facts about the correspondence Φ of 2.7 is that two operator inner products which correspond under Φ have isomorphic commuting algebras.

Let V be an operator inner product on L, and let W be the operator inner product on $\bar{L}$ deduced from V. We shall keep the notation of 2.2; in particular, X and Y will stand for $X(V)$ and $X(W)$ respectively.

By 1.11 every operator A in the commuting algebra $\mathscr{I}(V)$ of V gives rise to a unique bounded linear operator $\tilde{A}$ on Y such that

$$\tilde{A}(t \widetilde{\otimes} \xi) = t \widetilde{\otimes} A\xi \qquad\qquad (t \in L; \xi \in X).$$

Notice that $\tilde{A}$ belongs to the commuting algebra $\mathscr{I}(W)$ of W. Indeed, if $r, s, t \in L$ and $\xi \in X$,

$$\begin{aligned}
\tilde{A}W_{r,s}(t \widetilde{\otimes} \xi) &= \tilde{A}(s \widetilde{\otimes} V_{t,r}\xi) \\
&= s \widetilde{\otimes} AV_{t,r}\xi = s \widetilde{\otimes} V_{t,r}A\xi \\
&= W_{r,s}(t \widetilde{\otimes} A\xi) = W_{r,s}\tilde{A}(t \widetilde{\otimes} \xi);
\end{aligned}$$

and this shows that $\tilde{A}W_{r,s} = W_{r,s}\tilde{A}$.

Proposition.   *Assume that V is non-degenerate. Then $A \mapsto \tilde{A}$ is an isometric *-isomorphism of $\mathscr{I}(V)$ onto $\mathscr{I}(W)$.*

Proof.   We have seen in 1.11 and 1.12 that $A \mapsto \tilde{A}$ is an isometric *-isomorphism. By the observation preceding this proposition it is into $\mathscr{I}(W)$. It remains only to show that it is *onto* $\mathscr{I}(W)$.

Let $\tilde{V}$ be the operator inner product on L deduced from W. We shall keep the notation of the proof of 2.6. If $B \in \mathscr{I}(W)$, we shall also denote by $\tilde{B}$ the operator in $\mathscr{I}(\tilde{V})$ constructed from B:

$$\tilde{B}(t \widetilde{\otimes} \zeta) = t \widetilde{\otimes} B\zeta \qquad\qquad (t \in \bar{L}; \zeta \in Y).$$

Because of the symmetric roles of V and W, it will be enough to show that $B \mapsto \tilde{B}$ maps $\mathscr{I}(W)$ onto $\mathscr{I}(\tilde{V})$.

Now according to the proof of 2.6, $\tilde{V}$ and V are unitarily equivalent under the F defined in (11). Thus the most general element of $\mathscr{I}(\tilde{V})$ is of the form $F^{-1}AF$, where $A \in \mathscr{I}(V)$. Hence, to prove that $B \mapsto \tilde{B}$ maps $\mathscr{I}(W)$ onto $\mathscr{I}(\tilde{V})$, it will suffice to show that

$$\tilde{A} = F^{-1}AF \qquad \text{for all } A \text{ in } \mathscr{I}(V). \tag{13}$$

To see this, let $A \in \mathscr{I}(V)$, s, $t \in L$, $\xi \in X$. Then

$$\begin{aligned}
F\tilde{A}(s \widetilde{\otimes} (t \widetilde{\otimes} \xi)) &= F(s \widetilde{\otimes} \tilde{A}(t \widetilde{\otimes} \xi)) \\
&= F(s \widetilde{\otimes} (t \widetilde{\otimes} A\xi)) \\
&= V_{t,s}A\xi &&\text{(by (11))} \\
&= AV_{t,s}\xi \\
&= AF(s \widetilde{\otimes} (t \widetilde{\otimes} \xi)).
\end{aligned}$$

Since the $s \widetilde{\otimes} (t \widetilde{\otimes} \xi)$ span a dense subspace of $\tilde{X}$, this implies that $F\tilde{A} = AF$, which is (13). ∎

Compact Operator Inner Products

2.13. Definition.   An operator inner product V on L is called *compact* if $V_{s,t}$ is a compact operator for all s, t in L.

2.14.   The V defined in Example 1.10 is certainly compact, and by 2.8 corresponds under the duality 2.7 with the numerical inner product. This is a special case of a more general phenomenon, namely that the operator inner product dual to a compact operator inner product is itself compact. To see this we first prove the following slightly stronger result:

Proposition.   *Let V be an operator inner product on a linear space L, and let W be the operator inner product on $\bar{L}$ deduced from V. If $t \in L$ and $V_{t,t}$ is a compact operator, then $V_{s,t}$ and $W_{s,t}$ are compact for all s in L.*

Proof.   We recall from VI.15.8 that a bounded linear operator F on a Hilbert space Z is compact if and only if $\|F\zeta_\alpha\| \to 0$ whenever $\{\zeta_\alpha\}$ is a norm-bounded net of vectors in Z such that $\zeta_\alpha \to 0$ weakly.

Let us keep the notation of 2.2. We shall first show that $W_{s,t}$ is compact for each s in L.

Let $\{\zeta_\alpha\}$ be a norm-bounded net of vectors in $Y = X(W)$ approaching 0 weakly. By (3)

$$\| W_{s,t}\zeta_\alpha \|^2 = (t \stackrel{\sim}{\otimes} U_s\zeta_\alpha, t \stackrel{\sim}{\otimes} U_s\zeta_\alpha)_0$$

$$= (V_{t,t}U_s\zeta_\alpha, U_s\zeta_\alpha). \tag{14}$$

Since U_s is continuous, the net $\{U_s\zeta_\alpha\}$ is norm-bounded and converges weakly to 0. So the assumed compactness of $V_{t,t}$ implies that $\| V_{t,t}U_s\zeta_\alpha \| \underset{\alpha}{\to} 0$. From this, (14), and the norm-boundedness of $\{U_s\zeta_\alpha\}$, it follows that $\| W_{s,t}\zeta_\alpha \| \underset{\alpha}{\to} 0$. This and the first sentence of the proof establish the compactness of $W_{s,t}$.

In particular $W_{t,t}$ is compact. Applying to this the result already proved, with V and W reversed, and using the duality 2.7, we conclude that $V_{s,t}$ is compact for all s. ∎

2.15.   As a corollary of this we get immediately:

Theorem.   *If V and W correspond under the duality Φ of 2.7, then W is compact if and only if V is compact.*

3.   *V*-Bounded Endomorphisms and Modules; Completely Positive Maps

3.1.   Let V be an operator inner product on a linear space L; and let us form the deduced Hilbert space $\mathscr{X}(V)$ as in 1.8. We shall use the rest of the notation of 1.8 without explicit mention.

Definition.   Let γ be a linear endomorphism of L. If there exists a bounded linear operator g on $\mathscr{X}(V)$ satisfying

$$g(t \stackrel{\sim}{\otimes} \xi) = \gamma(t) \stackrel{\sim}{\otimes} \xi \qquad (t \in L; \xi \in X(V)), \tag{1}$$

then γ will be said to be *V-bounded.*

The g of (1), if it exists, is uniquely determined by γ, and is called the *expansion of γ to $\mathscr{X}(V)$.*

The V-boundedness of γ is clearly equivalent to the existence of a constant number $k \geq 0$ such that

$$\sum_{i,j=1}^{n} (V_{\gamma(t_i),\,\gamma(t_j)}\xi_i, \xi_j) \leq k^2 \sum_{i,j=1}^{n} (V_{t_i,t_j}\xi_i, \xi_j) \tag{2}$$

for all $n = 1, 2, \ldots$ and all $t_1, \ldots, t_n$ in L and $\xi_1, \ldots, \xi_n$ in $X(V)$.

Remark. If L is a pre-Hilbert space and V is the inner product of L, a linear endomorphism of L is of course V-bounded in the above sense if and only if it is bounded in the usual sense with respect to the norm derived from V.

3.2. Let γ be a linear endomorphism of L as in 3.1; and suppose that there exists another linear endomorphism β of L which is V-adjoint to γ in the sense that

$$V_{\gamma(s),t} = V_{s,\beta(t)} \qquad (3)$$

for all s, t in L. Then we claim that γ can be "expanded" at least as far as the linear span $\mathcal{X}'$ of $\{t \widetilde{\otimes} \xi : t \in L, \xi \in X(V)\}$ (see 1.8). Indeed: Suppose that $\sum_{i=1}^{n} t_i \widetilde{\otimes} \xi_i = 0$ $(t_i \in L; \xi_i \in X(V))$. Then for all s in L and η in $X(V)$,

$$\left(\sum_{i=1}^{n} \gamma(t_i) \widetilde{\otimes} \xi_i, s \widetilde{\otimes} \eta \right)_0 = \sum_{i=1}^{n} (V_{\gamma(t_i),s} \xi_i, \eta)$$

$$= \sum_{i=1}^{n} (V_{t_i, \beta(s)} \xi_i, \eta) \qquad \text{(by (3))}$$

$$= \left(\sum_{i=1}^{n} t_i \widetilde{\otimes} \xi_i, \beta(s) \widetilde{\otimes} \eta \right)_0 = 0. \qquad (4)$$

By the arbitrariness of b and η this implies that $\sum_{i=1}^{n} \gamma(t_i) \widetilde{\otimes} \xi_i = 0$. So we have shown that

$$\sum_{i=1}^{n} t_i \widetilde{\otimes} \xi_i = 0 \Rightarrow \sum_{i=1}^{n} \gamma(t_i) \widetilde{\otimes} \xi_i = 0;$$

and from this it follows that the equation (1) defines a linear endomorphism g of $\mathcal{X}'$.

If it happens that $\mathcal{X}(V)$ is finite-dimensional (in particular, if $X(V)$ and L are finite-dimensional), then $\mathcal{X}' = \mathcal{X}(V)$; and we have shown that in this case the possession of a V-adjoint implies V-boundedness. If $\mathcal{X}(V)$ is not finite-dimensional, however, the example of Remark 3.1 shows that neither of the two properties, V-boundedness and possession of a V-adjoint, implies the other.

Hermitian A-modules

3.3. We now introduce a fixed *-algebra A, and assume that the L of 3.1 is a left A-module. (This means that we are given an algebraic representation τ of the algebra underlying A, acting on L. But in using "module language" instead of "representation language," we omit explicit reference to τ, writing simply at instead of $\tau_a(t)$.)

Definition. We say that L is a *Hermitian A-module (with respect to V)* if

$$V_{as,t} = V_{s,a^*t} \tag{5}$$

for all a in A and s, t in L.

Assume that L is a Hermitian A-module with respect to V. According to (5), the endomorphism $t \mapsto at$ of L has a V-adjoint (see (3)), namely $t \mapsto a^*t$. Thus by 3.2 the action of A on L gives rise to an algebraic representation T' of A on the dense subspace $\mathscr{X}'$ of $\mathscr{X}(V)$ (see 3.2):

$$T'_a(t \overset{\sim}{\otimes} \xi) = at \overset{\sim}{\otimes} \xi \qquad (t \in L; \xi \in X(V)); \tag{6}$$

and it follows from the calculation (4) that

$$(T'_a\zeta, \zeta') = (\zeta, T'_{a^*}\zeta') \qquad (\zeta, \zeta' \in \mathscr{X}'; a \in A). \tag{7}$$

That is, T' is a pre-*-representation of A on $\mathscr{X}'$.

3.4. Definition. We say that *L as an A-module is V-bounded* if, for every a in A, the endomorphism $t \mapsto at$ of L is V-bounded in the sense of 3.1.

Assume that L is a Hermitian A-module and is V-bounded. Then, for each a in A, the operator T'_a of (6) extends uniquely to a bounded linear operator T_a on $\mathscr{X}(V)$. Since T' is a pre-*-representation, $T : a \mapsto T_a$ is a *-representation of A on $\mathscr{X}(V)$.

Definition. This T is called the **-representation of A deduced from* the A-module L and V. It is defined whenever V is an operator inner product on L and L is an A-module which is Hermitian and bounded with respect to V.

3.5. Let L be a Hermitian A-module with respect to V. Under certain circumstances L is automatically V-bounded. For example, we have seen in 3.2 that this is the case if $\mathscr{X}(V)$ is finite-dimensional. More importantly, applying Proposition VI.18.16 to the pre-*-representation T' of 3.3, we obtain:

Proposition. *If the A of 3.3 is a Banach *-algebra, then L (as an A-module) is automatically V-bounded, and so gives rise to a deduced *-representation of A.*

3.6. Let L be an A-module which is Hermitian and bounded with respect to the operator inner product V on L; and let T be the corresponding deduced *-representation of A on $\mathscr{X}(V)$. We have seen in 1.11 that each element F

of the commuting algebra of V gives rise to a bounded linear operator $\tilde{F}$ on $\mathscr{X}(V)$:

$$\tilde{F}(t \tilde{\otimes} \xi) = t \tilde{\otimes} F\xi \qquad\qquad (t \in L; \xi \in X(V)).$$

Notice that $\tilde{F}$ lies in the commuting algebra of T. Indeed: If $a \in A$, $t \in L$, and $\xi \in X(V)$,

$$\tilde{F}T_a(t \tilde{\otimes} \xi) = \tilde{F}(at \tilde{\otimes} \xi) = at \tilde{\otimes} F\xi$$
$$= T_a(t \tilde{\otimes} F\xi) = T_a\tilde{F}(t \tilde{\otimes} \xi).$$

So $\tilde{F}T_a = T_a\tilde{F}$ for all a. Thus $F \mapsto \tilde{F}$ is a *-homomorphism of $\mathscr{I}(V)$ into the commuting algebra $\mathscr{I}(T)$ of T, and is one-to-one if V is non-degenerate (by 1.12). It follows in particular that, if V is non-degenerate, T cannot be irreducible unless V is irreducible.

Of course, the *-homomorphism $F \mapsto \tilde{F}$ will not in general map $\mathscr{I}(V)$ *onto* $\mathscr{I}(T)$; and the irreducibility of V does not imply the irreducibility of T.

*Completely Positive Functions on a *-algebra*

3.7.　The context of 3.3 is often realized in the following special situation. Let A be a fixed *-algebra, X a Hilbert space, and $P: A \to \mathcal{O}(X)$ a linear map.

Definition.　We say that P is *completely positive* if the map $\langle a, b \rangle \mapsto P(b^*a)$ $(a, b \in A)$ is an operator inner product on A, that is, if

$$\sum_{i,j=1}^{n} (P(a_j^*a_i)\xi_i, \xi_j) \geq 0 \qquad\qquad (8)$$

for all positive integers n, all $a_1, \ldots, a_n$ in A, and all $\xi_1, \ldots, \xi_n$ in X.

3.8.　As we saw in 1.1, (8) implies that

$$P(a^*a) \geq 0 \qquad\qquad (a \in A) \qquad (9)$$

and

$$P(b^*a) = (P(a^*b))^* \qquad\qquad (a, b \in A). \qquad (10)$$

If $X = \mathbb{C}$, so that P is simply a linear functional on A, then (by 1.2) (9) implies complete positivity. Thus, if X is one-dimensional, the completely positive linear maps $P: A \to \mathcal{O}(X)$ are simply the positive linear functionals on A. Much of what we shall say from here to the end of this section will in fact be a generalization of the theory of positive linear functionals (§§VI.18, 19) to the case that X is of dimension more than 1.

3.9.   We shall start by expressing the condition (8) of complete positivity in a different form.

For each positive integer n let M_n be the $n \times n$ total matrix *-algebra. If A is a *-algebra and n is a positive integer, A_n will denote the tensor product *-algebra $A \otimes M_n$. Thus the elements of A_n can be regarded as $n \times n$ matrices with entries in A, the multiplication and involution being given by the usual matrix rules:

$$(ab)_{ij} = \sum_{k=1}^{n} a_{ik} b_{kj},$$

$$(a^*)_{ij} = (a_{ji})^*.$$

If X is a Hilbert space, then $(\mathcal{O}(X))_n$ can be identified with $\mathcal{O}(nX)$, where $nX = X \oplus \cdots \oplus X$ (n times).

Let A be a *-algebra, X a Hilbert space, and $P \colon A \to \mathcal{O}(X)$ a linear map. For each positive integer n we construct the linear map $P_n \colon A_n \to (\mathcal{O}(X))_n \cong \mathcal{O}(nX)$ in the natural way:

$$P_n(a \otimes m) = P(a) \otimes m \qquad\qquad (a \in A; m \in M_n).$$

Thus, if $\alpha \in A_n$,

$$(P_n(\alpha))_{ij} = P(\alpha_{ij}) \qquad (i, j = 1, \ldots, n). \qquad (11)$$

Proposition.   *P is completely positive if and only if, for each $n = 1, 2, \ldots$,*

$$P_n(\alpha^*\alpha) \geq 0 \qquad \text{for all } \alpha \text{ in } A_n. \qquad (12)$$

Note.   (12) means of course that $P_n(\alpha^*\alpha)$ is a positive operator on nX.

Thus (9) is sufficient for complete positivity *provided it is assumed not merely for P but for all the P_n.*

Proof.   Assume that (12) holds for all n; and take a positive integer n, elements $a_1, \ldots, a_n$ of A, and vectors $\xi_1, \ldots, \xi_n$ of X. We shall show that (8) holds, and hence that P is completely positive. Consider the matrix α in A_n given by: $\alpha_{ij} = \delta_{i1} a_j$ ($i, j = 1, \ldots, n$). Then $(\alpha^*\alpha)_{ij} = a_i^* a_j$. Denoting by ξ the vector $\langle \xi_1, \ldots, \xi_n \rangle$ of nX, we have, since by (12) $P_n(\alpha^*\alpha)$ is positive,

$$0 \leq (P_n(\alpha^*\alpha)\xi, \xi) = \sum_{i,j} ((P_n(\alpha^*\alpha))_{ij}\xi_j, \xi_i)$$

$$= \sum_{i,j} (P(a_i^* a_j)\xi_j, \xi_i) \qquad\qquad \text{(by (11))}.$$

So (8) holds, and P is completely positive.

Conversely, let us assume that P is completely positive. We shall take a positive integer n and an element α of A_n, and show that (12) holds. For each $i = 1, \ldots, n$ let α^i be the element of A_n given by: $(\alpha^i)_{jk} = \delta_{ij}\alpha_{jk}$. Then $\alpha = \sum_i \alpha^i$; and since $(\alpha^i)^*\alpha^j = 0$ for $i \neq j$ we get

$$\alpha^*\alpha = \sum_i (\alpha^i)^*\alpha^i. \tag{13}$$

Thus, if $\xi = \langle \xi_1, \ldots, \xi_n \rangle \in nX$, we have

$$(P_n(\alpha^*\alpha)\xi, \xi) = \sum_{j,k} ((P_n(\alpha^*\alpha))_{jk}\xi_k, \xi_j)$$

$$= \sum_i \left[\sum_{j,k} (P((\alpha_{ij})^*\alpha_{ik})\xi_k, \xi_j) \right] \quad \text{(by (11) and (13)).} \tag{14}$$

Now the complete positivity of P asserts that for each i the quantity in square brackets on the right of (14) is non-negative. So by (14) $(P_n(\alpha^*\alpha)\xi, \xi) \geq 0$ for all ξ in nX, proving (12). ∎

3.10. Example. The following example shows that (9) is a strictly weaker condition than complete positivity.

Let n be a fixed integer greater than 1; and put $A = M_n$. We shall denote by e^{ij} the matrix in A having 1 in the i, j place and 0 elsewhere. Let $P: A \to A$ be the linear map given by

$$(P(e^{ij}))_{rs} = (r - j)(s - i)$$

$(i, j, r, s = 1, \ldots, n)$. If $a \in A$ and $\xi \in \mathbb{C}^n$,

$$(P(a)\xi, \xi) = \sum_{i,j} a_{ij}(P(e^{ij})\xi, \xi)$$

$$= \sum_{i,j,r,s} a_{ij}(r - j)(s - i)\xi_s\bar{\xi}_r \tag{15}$$

$$= \sum_{i,j} a_{ij}y_j\bar{y}_i,$$

where $y_i = \sum_{s=1}^n (s - i)\bar{\xi}_s$. If a is positive in $A = M_n$, then $\sum_{i,j} a_{ij}y_j\bar{y}_i \geq 0$, so that by (15) $P(a)$ is positive. Thus P satisfies (9).

Keeping the notation of 3.9, we have $A_n = M_n \otimes M_n$; and the elements of A_n are $n \times n$ matrices with entries in M_n. If P were completely positive then by Proposition 3.9 we would have $P_n(\alpha) \geq 0$ for all positive α in A_n. We shall now show that this is not true.

Consider the element α of A_n given by $\alpha_{ij} = e^{ij}$ $(i, j = 1, \ldots, n)$. If $\xi = \langle \xi^{(1)}, \ldots, \xi^{(n)} \rangle \in n\mathbb{C}^n$ (where $\xi^{(i)} = \langle \xi_1^{(i)}, \ldots, \xi_n^{(i)} \rangle \in \mathbb{C}^n$), we have

$$(\alpha\xi, \xi) = \sum_{i,j} (\alpha_{ij}\xi^{(j)}, \xi^{(i)}) = \sum_{i,j} (e^{ij}\xi^{(j)}, \xi^{(i)})$$

$$= \sum_{i,j} \xi_j^{(j)}\overline{\xi_i^{(i)}} \geq 0.$$

So α is a positive element of A_n. But we claim that $P_n(\alpha)$ is not positive. Indeed: If $\xi = \langle \xi^{(1)}, \ldots, \xi^{(n)} \rangle$, where $\xi^{(i)} \in \mathbb{C}^n$, we have

$$(P_n(\alpha)\xi, \xi) = \sum_{i,j} (P(e^{ij})\xi^{(j)}, \xi^{(i)})$$

$$= \sum_{i,j,r,s} (r - j)(s - i)\xi_s^{(j)}\overline{\xi_r^{(i)}}.$$

In particular, if we put $\xi_r^{(i)} = \delta_{ir}$, we get (since $n > 1$)

$$(P_n(\alpha)\xi, \xi) = \sum_{i,j} (i - j)(j - i) < 0.$$

So $P_n(\alpha)$ is not positive. Since α is positive, this implies, as we have seen, that P is not completely positive.

Thus condition (9) is strictly weaker than complete positivity.

The above example is due to Stinespring [1].

3.11.   The following result generalizes VI.18.12.

Proposition.   *If A is a Banach *-algebra with unit 1, and $P: A \to \mathcal{O}(X)$ is a completely positive linear map (as in 3.7), then P is norm-continuous; in fact*

$$\|P(a)\| \leq \|P(1)\| \|a\| \qquad\qquad (a \in A). \qquad (16)$$

Proof.   Applying 1.6(6) with $V_{a,b} = P(b^*a)$, and taking b to be 1, we get

$$\|P(a)\|^2 \leq \|P(1)\| \|P(a^*a)\|. \qquad\qquad (17)$$

By (9) $P(a^*a) \geq 0$. Hence, for each unit vector ξ in X, the functional $a \mapsto (P(a)\xi, \xi)$ is positive on A, and so by VI.18.12

$$|(P(a)\xi, \xi)| \leq |(P(1)\xi, \xi)| \|a\|$$

$$\leq \|P(1)\| \|a\| \qquad\qquad (a \in A). \qquad (18)$$

Replacing a by a^*a in (18) and recalling from VI.11.8 that $\|P(a^*a)\| = \sup\{(P(a^*a)\xi, \xi): \|\xi\| = 1\}$, we get

$$\|P(a^*a)\| \leq \|P(1)\| \|a^*a\| \qquad\qquad (a \in A). \qquad (19)$$

From (17) and (19) we obtain

$$\|P(a)\|^2 \leq \|P(1)\|^2 \|a^*a\| \leq \|P(1)\|^2 \|a\|^2;$$

and this completes the proof. ∎

*-Representations Generated by a Completely Positive Map

3.12. Let A be a *-algebra, X a Hilbert space, and $P: A \to \mathcal{O}(X)$ a completely positive map. Thus $V: \langle a, b \rangle \mapsto P(b^*a)$ is an operator inner product. Now A is of course a left A-module under left multiplication; and the simple calculation

$$V_{ab,c} = P(c^*ab) = P((a^*c)^*b) = V_{b, a^*c}$$

$(a, b, c \in A)$ shows that this left A-module is Hermitian with respect to V.

Definitions. If A, as a left A-module, is bounded with respect to V in the sense of 3.4, we shall say that P satisfies *Condition (R)*. If this is so, the *-representation of A deduced from A (as a left A-module) and V will be said to be *generated by P*.

Remark. If X is one-dimensional, so that P is simply a positive functional on A (see 3.7), these definitions coincide with the definitions of Condition (R) and the generated *-representation given in VI.19.3.

3.13. Generalizing VI.19.5, we have by Proposition 3.5:

Proposition. *If A is a Banach *-algebra and X is a Hilbert space, every completely positive linear map $P: A \to \mathcal{O}(X)$ satisfies Condition (R) and so generates a *-representation of A.*

3.14. Our next result generalizes VI.19.6, at least for Banach *-algebras with unit.

Theorem. *Let A be a Banach *-algebra with unit 1, X a Hilbert space, and $P: A \to \mathcal{O}(X)$ a linear map. Then the following two conditions are equivalent:*

(i) *P is completely positive.*
(ii) *There exists a *-representation T of A and a bounded linear map $F: X \to X(T)$ such that*

$$P(a) = F^* T_a F \qquad \text{for all } a \text{ in } A. \tag{20}$$

Proof.   Assume (ii). Then for any elements $a_1, \ldots, a_n$ of A and $\xi_1, \ldots, \xi_n$ of X we have by (20)

$$\sum_{i,j=1}^{n} (P(a_j^* a_i)\xi_i, \xi_j) = \sum_{i,j=1}^{n} (T_{a_i}F\xi_i, T_{a_j}F\xi_j) \geq 0.$$

Therefore (i) holds.

Conversely, assume (i). By 3.13 P generates a *-representation T of A, acting in the Hilbert space $\mathcal{X}(V)$ deduced from $V: \langle a, b \rangle \mapsto P(b^*a)$. As in 1.8(9), the equation

$$F(\xi) = 1 \widetilde{\otimes} \xi \qquad\qquad (\xi \in X)$$

defines a bounded linear map $F: X \to X(V)$; and if $\xi, \eta \in X$ and $(\ ,\ )_0$ is the inner product on $\mathcal{X}(V)$,

$$(F^* T_a F\xi, \eta) = (T_a F\xi, F\eta)_0$$
$$= (a \widetilde{\otimes} \xi, 1 \widetilde{\otimes} \eta)_0$$
$$= (P(a)\xi, \eta).$$

So (20) holds, and (ii) is proved.   ∎

Remark.   Following the pattern of §VI.19, one could now look for generalizations of the above result to *-algebras without unit. One can also assert the essential uniqueness of the pair F, T provided it is "cyclic." We shall leave these questions as exercises for the reader, since they will not be important for the rest of this volume.

4.   Rigged Modules and Abstractly Induced Representations

4.1.   In this section we will describe the abstract Rieffel inducing process (sketched in the Introduction to this Chapter) by which one passes from a *-representation S of a *-algebra B to a *-representation T of another *-algebra A. The essential ingredient of this process is the notion of a B-rigged A-module.

4.2.   Definition.   A *left B-rigged space* is a left B-module L, together with a map $[\ ,\ ]: L \times L \to B$ such that:

(i) $[s, t]$ is linear in s and conjugate-linear in t $(s, t \in L)$;

(ii) $[t, s] = [s, t]^*$ $(s, t \in L)$;

(iii) $[bs, t] = b[s, t]$ for all s, t in L and b in B.

Note. The "bs" in condition (iii) is of course the left action of B on L, and "$b[s, t]$" is the product in B.

Conditions (ii) and (iii) imply that $[s, t]b = (b^*[t, s])^* = [b^*t, s]^* = [s, b^*t]$; or

$$[s, t]b = [s, b^*t] \qquad (s, t \in L; b \in B). \qquad (1)$$

From (1) and (ii), (iii) it follows that the linear span of $\{[r, s] : r, s \in L\}$ is a *-ideal of B.

4.3. Similarly we define a right B-rigged space.

Definition. A *right B-rigged space* is a right B-module L, together with a map $[\ , \ ] : L \times L \to B$ such that

(i) $[s, t]$ is linear in t and conjugate-linear in s $(s, t \in L)$;
(ii) $[t, s] = [s, t]^*$ $(s, t \in L)$;
(iii) $[s, tb] = [s, t]b$ for all s, t in L and b in B.

Notice that in a right B-rigged space $[s, t]$ is linear in the second variable instead of the first. Analogously with (1), we observe that

$$b[s, t] = [sb^*, t] \qquad (s, t \in L; b \in B). \qquad (2)$$

Just as in 4.2, the linear span of $\{[r, s] : r, s \in L\}$ is a *-ideal of B.

The map $[\ , \ ]$ in 4.2 and 4.3 is called the B-*rigging* of the space L.

4.4. It is important to notice that a left B-rigged space becomes a right B-rigged space when L is replaced by its complex-conjugate $\bar{L}$, and vice versa.

Indeed, suppose that L is a left B-module. Then the equation

$$s : b = b^*s \qquad (s \in \bar{L}; b \in B) \qquad (3)$$

defines a right B-module structure : for $\bar{L}$. (The replacement of L by $\bar{L}$ is necessary in order to make $s : b$, as defined in (3), linear in b.) We shall call $\bar{L}_,$: the *right B-module complex-conjugate to L*. Similarly, if L is a right B-module, the equation

$$b : s = sb^* \qquad (s \in \bar{L}; b \in B) \qquad (4)$$

defines a left B-module structure : for $\bar{L}$; and $\bar{L}_,$: is called the *left B-module complex-conjugate to L*.

Now let $L, [\ , \ ]$ be a left B-rigged space. Condition 4.2(i) becomes 4.3(i) when $[\ , \ ]$ is considered as defined on $\bar{L} \times \bar{L}$ instead of on $L \times L$; and by (1)

$$[s, t : b] = [s, t]b$$

$(s, t \in L; b \in B)$. Thus 4.3(iii) is satisfied by [,] and the right B-module $\bar{L}$ complex-conjugate to L. So $\bar{L}$, [,] is a right B-rigged space; we call it the *right B-rigged space complex-conjugate to L, [,]*.

Similarly, suppose that L, [,] is a right B-rigged space. If $\bar{L}$ denotes the left B-module complex-conjugate to L, then $\bar{L}$, [,] is a left B-rigged space; we call it the *left B-rigged space complex-conjugate to L, [,]*.

4.5.   Rigged spaces are closely related to operator inner products.

Let L, [,] be a right B-rigged space, and S any *-representation of B; and let us put

$$V_{s,t} = S_{[t,s]} \qquad\qquad (s, t \in L). \qquad (5)$$

Then the map $V: L \times L \to \mathcal{O}(X(S))$ is linear in the first variable and conjugate-linear in the second.

Proposition.   *V is an operator inner product on L if and only if*

$$S_{[t,t]} \geq 0 \quad \text{for all } t \text{ in } L. \qquad (6)$$

Proof.   The condition (6) is obviously necessary. We shall assume that (6) holds, and shall prove that V is an operator inner product. Since by (5) V is linear in s and conjugate-linear in t, it is only necessary to show that V is completely positive (see 1.1(1)).

By VI.9.14 and VI.9.16 S can be written as a Hilbert direct sum $\sum_i^{\oplus} S^i$, where each S^i is either cyclic or a zero representation. By 1.14 it is enough to show that $V^i: \langle s, t \rangle \mapsto S_{[t,s]}^i$ is an operator inner product for each i. This is evident if S^i is a zero representation; so it is enough to prove it when S^i is cyclic. That is, we may as well assume from the very beginning that S is cyclic, with cyclic vector ξ.

In proving 1.1(1), it is enough to have the ξ_i run over the dense subset $\{S_b \xi : b \in B\}$ of $X(S)$. If $t_1, \ldots, t_n$ are elements of L and $\xi_i = S_{b_i} \xi$ $(i = 1, \ldots, n; b_i \in B)$, we have

$$\sum_{i,j=1}^{n} (V_{t_i, t_j} \xi_i, \xi_j) = \sum_{i,j=1}^{n} (S_{[t_j, t_i]} S_{b_i} \xi, S_{b_j} \xi)$$

$$= \sum_{i,j=1}^{n} (S_{b_j^* [t_j, t_i] b_i} \xi, \xi)$$

$$= \sum_{i,j=1}^{n} (S_{[t_j b_j, t_i b_i]} \xi, \xi) \qquad \text{(by 4.3(iii) and (2))}$$

$$= (S_{[r,r]} \xi, \xi) \geq 0 \qquad \text{(by (6))},$$

where $r = \sum_{i=1}^{n} t_i b_i$. So 1.1(1) holds for V; and V is an operator inner product. ∎

Remark. This proof is just an elaboration of the simple argument of 1.2.

Definition. A *-representation S of B which satisfies (6) (and therefore gives rise by (5) to an operator inner product V on L) will be said to be *positive with respect to the right B-rigged space L, [,]*.

Let S be a *-representation of B which is positive with respect to L, [,] and V the corresponding operator inner product (5) on L; and let Y denote the Hilbert space deduced from V. Denoting as usual by $r \widetilde{\otimes} \xi$ the image of $r \otimes \xi$ in Y, we observe that

$$rb \widetilde{\otimes} \xi = r \widetilde{\otimes} S_b \xi \qquad (7)$$

for all r in L, b in B, and ξ in $X(S)$.

Indeed: Fixing r, b, and ξ, we have for any s in L and η in $X(S)$

$$(rb \widetilde{\otimes} \xi - r \widetilde{\otimes} S_b \xi, s \widetilde{\otimes} \eta)_Y = (S_{[s, rb]} \xi, \eta) - (S_{[s, r]} S_b \xi, \eta) = 0$$

in virtue of 4.3(iii). Since the $s \widetilde{\otimes} \eta$ span a dense subspace of Y, this implies (7).

4.6. Suppose that L, [,] is a *left* B-rigged space. Let S be a *-representation of B; and put

$$W_{s, t} = S_{[t, s]} \qquad (s, t \in L). \qquad (8)$$

Applying 4.5 to the complex-conjugate right B-rigged space $\bar{L}$, [,] (see 4.4), we obtain:

Proposition. *W is an operator inner product on $\bar{L}$ if and only if*

$$S_{[t, t]} \geq 0 \qquad \textit{for all } t \textit{ in } L. \qquad (9)$$

Definition. If (9) holds we shall say that S is *positive with respect to the left B-rigged space L, [,]*.

Rigged Modules

4.7. Let us now consider *two* fixed *-algebras A and B.

Definition. A *right B-rigged left A-module* is a linear space L which is both a left A-module and a right B-module, together with a map [,]$: L \times L \to B$ making L, [,] a right B-rigged space and satisfying the following "Hermitian" condition:

$$[as, t] = [s, a^*t] \qquad (s, t \in L; a \in A). \qquad (10)$$

Similarly, a *left B-rigged right A-module* is a linear space L which is both a left B-module and a right A-module, together with a map $[\ , \ ]: L \times L \to B$ making $L, [\ , \ ]$ a left B-rigged space and satisfying

$$[sa, t] = [s, ta^*] \qquad (s, t \in L; a \in A). \qquad (11)$$

Remark. We do *not* assume that the actions of A and B on L commute, though this will be the case in the most interesting applications.

4.8. As in 4.4 we observe that complex conjugation of L inverts left and right. Indeed, let $L, [\ , \ ]$ be a right B-rigged left A-module. Then $\bar{L}$, considered as a left B-module and a right A-module according to 4.4, becomes together with $[\ , \ ]$ a left B-rigged right A-module, which we refer to as the *complex conjugate of* $L, [\ , \ ]$. Similarly, if $L, [\ , \ ]$ is a left B-rigged right A-module, $\bar{L}$ (considered as a right B-module and left A-module according to 4.4) becomes together with $[\ , \ ]$ a right B-rigged left A-module, called the *complex conjugate of* $L, [\ , \ ]$.

4.9. Let $\mathscr{L} = \langle L, [\ , \ ] \rangle$ be a right B-rigged left A-module; and let S be any *-representation of B. If S is positive with respect to $\mathscr{L}$ (i.e., $S_{[t,t]} \geq 0$ for all t in L), by 4.5 we can use (5) to define an operator inner product V on L; and by (10) the left A-module structure of L is Hermitian with respect to V (see 3.3). We can then ask whether L as a left A-module is V-bounded (in the sense of 3.4). If it is, we can form the *-representation T of A deduced as in 3.4 from L (as a left A-module) and V.

Definition. We shall say that the *-representation S of B is *inducible to A (via $\mathscr{L}$)* if

(i) S is positive with respect to $\mathscr{L}$,

(ii) L as a left A-module is bounded with respect to the V of (5).

If this is so, the above T is called the *-*representation of A induced from S (via $\mathscr{L}$)* and is denoted by $\mathrm{Ind}^{\mathscr{L}}_{B \uparrow A}(S)$ or simply $\mathrm{Ind}_{B \uparrow A}(S)$ (if the context makes it clear which rigged module $\mathscr{L}$ is intended).

This is the fundamental abstract *Rieffel inducing process*, whose ramifications will occupy us for the rest of this volume.

In view of its importance, it will be well to review the construction of $T = \mathrm{Ind}_{B \uparrow A}(S)$. Keeping the notation of this number and assuming that S is inducible to A, we introduce into $L \otimes X(S)$ the positive conjugate-bilinear form $(\ , \ )_0$ given by

$$(s \otimes \xi, t \otimes \eta)_0 = (S_{[t, s]}\xi, \eta). \qquad (12)$$

Factoring $L \otimes X(S)$ by the null-space of $(\, , \,)_0$ and completing with respect to $(\, , \,)_0$, we obtain the Hilbert space $\mathscr{X}$. Let the image of $s \otimes \xi$ in $\mathscr{X}$ ($s \in L$; $\xi \in X(S)$) be denoted by $s \widetilde{\otimes} \xi$. Then the operators of the *-representation T of A are given by

$$T_a(s \widetilde{\otimes} \xi) = as \widetilde{\otimes} \xi \qquad (a \in A; s \in L; \xi \in X(S)). \tag{13}$$

4.10. Interchanging left and right, suppose that $L, [\, , \,]$ is a left B-rigged right A-module, and that S is a *-representation of B.

Definition. We shall say that S is *inducible to A via $L, [\, , \,]$* if it is so via the complex-conjugate right B-rigged left A-module $\bar{L}, [\, , \,]$. In that case the *-representation T of A induced by S via $\bar{L}, [\, , \,]$ will also be said to be *induced from S via* the original $L, [\, , \,]$, and will as before be denoted by $\mathrm{Ind}_{B \uparrow A}(S)$.

4.11. Let $\mathscr{L} = \langle L, [\, , \,] \rangle$ be a right B-rigged left A-module and S a *-representation of B. If L and $X(S)$ are both finite-dimensional, it follows from an observation in 3.4 that for S to be inducible to A (via $\mathscr{L}$) it is sufficient that S be positive with respect to $\mathscr{L}$.

A simple example of a situation where S is not positive (and hence not inducible) will be given in 8.6, Remark 3.

4.12. From Proposition 3.5 we obtain the following important fact:

Proposition. *Assume that A is a Banach *-algebra. If $\mathscr{L}$ is a right B-rigged left A-module, and S is a *-representation of B which is positive with respect to $\mathscr{L}$, then S is inducible to A (via $\mathscr{L}$).*

Conditional Expectations

4.13. An important special situation in which rigged modules (and hence induced representations) arise is provided by what we shall call conditional expectations.

As before let A and B be two *-algebras. We shall now assume that the underlying linear space L of A has the structure of a right B-module. This will be the case automatically, for example, if B is a *-subalgebra of A or more generally of the multiplier *-algebra of A (see VIII.1.13). Thus L is now a right B-module and a left A-module (under left multiplication in A). Under these conditions we make the following definition:

Definition.   An A, B *conditional expectation* is a linear mapping $p: A \to B$ such that

   (i) $p(a^*) = (p(a))^*$ for $a \in A$,
   (ii) $p(ab) = p(a)b$ for $a \in A, b \in B$ (*ab* referring of course to the right B-module structure of L).

4.14.   Let $p: A \to B$ be an A, B conditional expectation. Defining

$$[a_1, a_2] = p(a_1^* a_2) \qquad\qquad (a_1, a_2 \in A), \qquad (14)$$

we see from conditions (i) and (ii) on p that $[\ ,\ ]$ satisfies conditions (i), (ii), and (iii) of 4.3; so $A, [\ ,\ ]$ is a right B-rigged space. In addition the condition (10) is satisfied by the left action of A on itself:

$$[ca_1, a_2] = p((ca_1)^* a_2) = p(a_1^* c^* a_2)$$
$$= [a_1, c^* a_2] \qquad\qquad (a_1, a_2, c \in A).$$

Thus $A, [\ ,\ ]$ is a right B-rigged left A-module.

Definition.   A *-representation S of B is said to be *positive with respect to p* [*inducible to A via p*] if it is positive with respect to $A, [\ ,\ ]$ [inducible to A via $A, [\ ,\ ]$].

If S is inducible to A via p we can form the *induced* *-representation $\mathrm{Ind}_{B \uparrow A}(S)$ of A as in 4.9. It will sometimes be denoted by $\mathrm{Ind}_{B \uparrow A}^p(S)$.

If the *-representation S of B is positive with respect to p, it follows from Proposition 4.5 that $P: a \mapsto S_{p(a)}$ $(a \in A)$ is a completely positive map in the sense of 3.6.

4.15.   By 4.12 we obtain:

Proposition.   *In the context of* 4.14 *assume that A is a Banach *-algebra. Then every *-representation of B which is positive with respect to p is inducible to A via p.*

4.16.   Remark.   The term "conditional expectation" stems from a concept of classical probability theory. Let M be a finite set, μ a function from M to the positive reals, and $\sim$ an equivalence relation on M; and denote by A the *-algebra (under pointwise multiplication and complex conjugation) of all complex functions on M. The equation

$$(p(f))(m) = \left(\sum_{n \sim m} \mu(n) \right)^{-1} \left(\sum_{n \sim m} f(n)\mu(n) \right) \qquad (15)$$

($f \in A$; $m \in M$) defines an "averaging" map p of A onto the *-subalgebra B of A consisting of all g such that $m \sim n \Rightarrow g(m) = g(n)$. This p is an idempotent map preserving the operation of complex conjugation and carrying non-negative functions into non-negative functions. Though not a homomorphism, it has the following partial homomorphism property:

$$p(fg) = p(f)g \qquad \text{whenever } f \in A, g \in B. \tag{16}$$

Such maps p are called *conditional expectations*, since, if M, μ is interpreted as a probability space, p carries the random variable f into another random variable g expressing the expectation of $f(m)$ under the condition that m lies in some given $\sim$-equivalence class. It is because of the formal identity of (16) with condition (ii) of Definition 4.13 that we have used the term "conditional expectation" in 4.13, even though the latter context has no probabilistic associations.

Example; Induced Representations of Finite Groups

4.17. We shall conclude this section by showing how one recovers the Frobenius inducing process for finite groups as a special case of the abstract Rieffel inducing process just defined. Later in this chapter (starting with §9) we will see how the Rieffel inducing process also embraces a natural generalization of the Frobenius inducing process not only to arbitrarily locally compact groups (the Mackey construction) but even to Banach *-algebraic bundles over locally compact groups.

In fact, Rieffel induction also embraces a variety of other processes by which new representations are constructed from old, and which do not look at all like the classical inducing process. See for example 7.12 (on hereditary *-subalgebras) and §16 (on conjugation in Banach *-algebraic bundles).

4.18. The construction of induced representations of compact groups was encountered in IX.10.4. We shall now point out that, at least for finite groups, this construction is a special case of the abstract inducing process of this section. This will help to orient the reader toward §§9, 10, where it will be shown that the same holds for the classical construction of induced representations of arbitrary locally compact groups.

The reader should also compare the present number with IX.10.13.

Let G be a finite group and H a subgroup of G; and denote by A and B the group *-algebras $\mathscr{L}(G)$ and $\mathscr{L}(H)$ of G and H respectively (under convolution $*$ and involution $*$). Let S be a finite-dimensional unitary representation of H, and T the unitary representation of G induced from S as in IX.10.4. Thus the space $X(T)$ of T consists of all functions $\phi: G \to X(S)$ which satisfy

$$\phi(xh) = S_{h^{-1}}(\phi(x)) \qquad (x \in G; h \in H); \qquad (17)$$

and T acts on $X(T)$ by left translation:

$$(T_y \phi)(x) = \phi(y^{-1}x) \qquad (x, y \in G; \phi \in X(T)).$$

If Γ is a transversal for G/H (i.e., a subset of G intersecting each coset xH in just one point), then in view of (17) T will be unitary under the inner product on $X(T)$ given by

$$(\phi, \psi) = \sum_{x \in \Gamma} (\phi(x), \psi(x)); \qquad (18)$$

and in fact the inner product (18) is independent of which transversal Γ we take.

Now let $F: A \otimes X(S) \to X(T)$ be the linear map given by

$$F(f \otimes \xi)(x) = \sum_{h \in H} f(xh) S_h \xi \qquad (19)$$

$(f \in A; \xi \in X(S); x \in G)$. To begin with, we observe that, if $k \in H$,

$$F(f \otimes \xi)(xk) = \sum_{h \in H} f(xkh) S_h \xi = \sum_{h \in H} f(xh) S_{k^{-1}h} \xi$$

$$= S_{k^{-1}}(F(f \otimes \xi)(x));$$

so range(F) is indeed contained in $X(T)$. In fact we claim that range(F) = $X(T)$. Indeed, let ϕ be any element of $X(T)$; and set $\gamma = \sum_{x \in \Gamma} \delta_x \otimes \phi(x)$ $\in A \otimes X(S)$.   Then,   if $y \in G$, $F(\gamma)(y) = \sum_{x \in \Gamma} F(\delta_x \otimes \phi(x))(y) =$ $\sum_{x \in \Gamma} \sum_{h \in H} \delta_x(yh) S_h(\phi(x)) = \sum_{x \in \Gamma} \sum_{h \in H} \delta_{xh^{-1}}(y)\phi(xh^{-1})$ (by (17)). Now as x runs over Γ and h over H, xh^{-1} runs over G; so the last equation becomes $F(\gamma)(y) = \sum_{x \in G} \delta_x(y)\phi(x) = \phi(y)$. Thus $\phi = F(\gamma) \in$ range(F); and this proves the claim. (These same facts were observed in IX.10.14.)

Now let $\rho: A \to B$ be the restriction map: $\rho(f) = f | H$. Then $\rho(f^*) = (\rho(f))^*$ $(f \in A)$; and if B is identified with the *-subalgebra $\{g \in A: g(x) = 0$ for all $x \in G \setminus H\}$ of A, we have

$$\rho(f * g) = \rho(f) * g \qquad (f \in A; g \in B).$$

So ρ is an A, B conditional expectation. As in (14) we set

$$[f, g] = \rho(f^* * g) \qquad (f, g \in A).$$

Then, if $f, g \in A$ and $\xi, \eta \in X(S)$,

$$(S_{[g, f]}\xi, \eta) = \sum_{h \in H} (g^* * f)(h)(S_h\xi, \eta)$$

$$= \sum_{h \in H} \sum_{x \in G} \overline{g(x)} f(xh)(S_h\xi, \eta)$$

$$= \sum_{x \in \Gamma} \sum_{h, k \in H} \overline{g(xk)} f(xkh)(S_h\xi, \eta)$$

$$= \sum_{x \in \Gamma} \sum_{h, k \in H} \overline{g(xk)} f(xh)(S_{k^{-1}h}\xi, \eta)$$

$$= \sum_{x \in \Gamma} \sum_{h, k \in H} \overline{g(xk)} f(xh)(S_h\xi, S_k\eta)$$

$$= \sum_{x \in \Gamma} (F(f \otimes \xi)(x), F(g \otimes \eta)(x))$$

$$= (F(f \otimes \xi), F(g \otimes \eta))_{X(T)}. \qquad (20)$$

From this it follows that $(S_{[f, f]}\xi, \xi) \geq 0$ for all f and ξ; so S is positive with respect to the conditional expectation ρ (in the sense of 4.14). Thus S (or rather its integrated form on B, which we do not distinguish from S) is inducible to A via ρ (see 4.11). Let W denote $\text{Ind}^\rho_{B \uparrow A}(S)$; and as usual let $f \widetilde{\otimes} \xi$ be the image of $f \otimes \xi$ in $X(W)$. Since F is onto $X(T)$, (20) implies that $X(W)$ and $X(T)$ are isometrically isomorphic under the bijection $\tilde{F}: X(W) \to X(T)$ given by:

$$\tilde{F}(f \widetilde{\otimes} \xi) = F(f \otimes \xi) \qquad (f \in A; \xi \in X(S)). \qquad (21)$$

One verifies that

$$F((g * f) \otimes \xi) = T_g(F(f \otimes \xi)) \qquad (f, g \in A; \xi \in X(S)). \qquad (22)$$

Hence by (21) $\tilde{F}$ intertwines W and T, showing that W is unitarily equivalent to (the integrated form of) T.

We conclude that the representation of $A (= \mathcal{L}(G))$ abstractly induced from S via the conditional expectation $\rho: A \to B$ is the same, up to unitary equivalence, as the integrated form of the representation $\text{Ind}_{H \uparrow G}(S)$ of G "concretely" induced from S by means of the construction IX.10.4.

5. General Properties of Abstractly Induced Representations

Direct Sums

5.1. Let A and B be two *-algebras; and let $\mathscr{L}$ be a right B-rigged left A-module.

Proposition. *Suppose that $\{S^i\}$ $(i \in I)$ is an indexed set of *-representations of B whose Hilbert direct sum $S = \sum_{i \in I}^{\oplus} S^i$ exists and is inducible to A via $\mathscr{L}$. Then each S^i is inducible to A via $\mathscr{L}$; and*

$$\operatorname*{Ind}_{B \uparrow A} (S) \cong \sum_{i \in I}^{\oplus} \operatorname*{Ind}_{B \uparrow A} (S^i).$$

This follows immediately from 1.15.

Thus the inducing operation preserves direct sums whenever it is defined.

If S is any inducible *-representation of B, notice that $\operatorname{Ind}_{B \uparrow A}(S)$ is unaffected when S is replaced by its non-degenerate part.

Tensor Products

5.2. Let A, B, C, D be four *-algebras; let $L, [\ ,\ ]$ be a right B-rigged left A-module; and let $M, [\ ,\ ]'$ be a right D-rigged left C-module. Then $L \otimes M$ is a left $(A \otimes C)$-module and a right $(B \otimes D)$-module:

$$(a \otimes c)(r \otimes u) = ar \otimes cu, \qquad (r \otimes u)(b \otimes d) = rb \otimes ud$$

$(a \in A; b \in B; c \in C; d \in D; r \in L; u \in M)$; and the equation

$$[r \otimes u, s \otimes v]'' = [r, s] \otimes [u, v]'$$

$(r, s \in L; u, v \in M)$ defines $L \otimes M, [\ ,\ ]''$ as a right $(B \otimes D)$-rigged left $(A \otimes C)$-module.

Proposition. *Let S be a *-representation of B which is inducible to A via $L, [\ ,\ ]$, and W a *-representation of D which is inducible to C via $M, [\ ,\ ]'$. Then the (outer) tensor product *-representation $S \otimes W$ of $B \otimes D$ is inducible to $A \otimes C$ via $L \otimes M, [\ ,\ ]''$; and*

$$\operatorname{Ind}(S \otimes W) \cong \operatorname{Ind}(S) \otimes \operatorname{Ind}(W).$$

This follows easily from 1.16.

Thus the inducing operation, when it can be performed, preserves (outer) tensor products.

Continuity of the Inducing Operation

5.3. Let A and B be two *-algebras, and $\mathscr{L} = \langle L, [\ , \ ] \rangle$ a right B-rigged left A-module. We shall denote by $\mathscr{S}$ the set of all those *-representations of B which are inducible to A via $L, [\ , \ ]$. We recall from VII.1.3 the definition of the regional topology of the space of all *-representations of B or of A.

Proposition. *The inducing map* $\Phi: S \mapsto \mathrm{Ind}_{B \uparrow A}^{\mathscr{L}}(S)$ *is continuous on* $\mathscr{S}$ *with respect to the regional topologies.*

Proof. Take a *-representation S in $\mathscr{S}$, with space X; and let $T = \mathrm{Ind}_{B \uparrow A}(S)$, $Y = X(T)$. Let $Z = \{r \widetilde{\otimes} \xi : r \in L, \xi \in X\}$ (where as usual $r \widetilde{\otimes} \xi$ is the image of $r \otimes \xi$ in Y). Thus Z has dense linear span in Y; and so by VII.1.4 the set of all

$$U^0 = U(T; \varepsilon; \zeta_1, \ldots, \zeta_n; F) \tag{1}$$

(see VII.1.3), where $\varepsilon > 0$, $\zeta_1, \ldots, \zeta_n$ are in Z, and F is a finite subset of A, is a basis of regional neighborhoods of T. Thus, to prove the continuity of Φ, we need only fix a set U^0 of the form (1) with $\zeta_i = r_i \widetilde{\otimes} \xi_i$ $(r_i \in L; \xi_i \in X)$, and exhibit a regional neighborhood U of S such that $\Phi(U \cap \mathscr{S}) \subset U^0$.

Set

$$U = U(S; \varepsilon; \xi_1, \ldots, \xi_n; G), \tag{2}$$

where G is the (finite) subset of B consisting of all $[r_j, r_i]$ and all $[r_j, ar_i]$ $(i, j = 1, \ldots, n; a \in F)$. This is a regional neighborhood of S; and we claim that $\Phi(U \cap \mathscr{S}) \subset U^0$. Indeed: Let $S' \in U \cap \mathscr{S}$; and put $X' = X(S')$, $T' = \mathrm{Ind}_{B \uparrow A}(S')$, $Y' = X(T')$. Thus by (2) there exist $\xi'_1, \ldots, \xi'_n$ in X' such that

$$|(S'_{[r_j, r_i]} \xi'_i, \xi'_j) - (S_{[r_j, r_i]} \xi_i, \xi_j)| < \varepsilon$$

and

$$|(S'_{[r_j, ar_i]} \xi'_i, \xi'_j) - (S_{[r_j, ar_i]} \xi_i, \xi_j)| < \varepsilon$$

for all i, j and all a in F. If we set $\zeta'_i = r_i \widetilde{\otimes} \xi'_i \in Y'$, these inequalities become

$$|(\zeta'_i, \zeta'_j)_{Y'} - (\zeta_i, \zeta_j)_Y| < \varepsilon$$

and

$$|(T'_a \zeta'_i, \zeta'_j)_{Y'} - (T_a \zeta_i, \zeta_j)_Y| < \varepsilon$$

for all i, j and all a in F. This says that $T' \in U^0$. So the claim is proved and the proposition established. ∎

Inducing in Stages

5.4. Our next proposition will assert that, under suitable conditions, the result of iterating the abstract inducing process is the same as the result of a single abstract inducing process. For induced representations of finite groups this proposition takes the form of Theorem 5.10.

5.5. We begin with a lemma on operator inner products.

Let K be a linear space; and denote the linear space of all linear endomorphisms of K by $\mathrm{End}(K)$. Let V be an operator inner product on K, acting in the Hilbert space X. Let L be another linear space; and consider a map $F: L \times L \to \mathrm{End}(K)$ which is linear in its first variable and conjugate-linear in its second, and such that $F(r, s)$ is V-bounded (see 3.1) for all r, s in L. Thus, if $r, s \in L$, we can form the bounded linear operator $\Phi(r, s)$ on the Hilbert space $\mathscr{X}(V)$ deduced from V:

$$\Phi(r, s)(k \mathbin{\widetilde{\otimes}} \xi) = (F(r, s)k) \mathbin{\widetilde{\otimes}} \xi \qquad (r, s \in L; k \in K; \xi \in X).$$

Further, put $P = L \otimes K$. Since the expression $V(F(r, s)k, m)$ $(r, s \in L; k, m \in K)$ is linear in r and k and conjugate-linear in s and m, the equation

$$\Gamma(r \otimes k, s \otimes m) = V(F(r, s)k, m)$$

defines a map $\Gamma: P \times P \to \mathcal{O}(X)$ which is linear in the first variable and conjugate-linear in the second.

Lemma. *Φ is an operator inner product on L if and only if Γ is an operator inner product on P. If this is the case, the two deduced Hilbert spaces $\mathscr{X}(P, \Gamma)$ and $\mathscr{X}(L, \Phi)$ are isomorphic under the linear isometric bijection $\beta: X(P, \Gamma) \to \mathscr{X}(L, \Phi)$ given by*

$$\beta((r \otimes k) \mathbin{\widetilde{\otimes}} \xi) = r \mathbin{\widetilde{\otimes}} (k \mathbin{\widetilde{\otimes}} \xi) \qquad (r \in L; k \in K; \xi \in X).$$

Proof. Let us define conjugate-bilinear forms $(\ ,\ )_0$ and $(\ ,\ )_{00}$ on $P \otimes X$ and $L \otimes \mathscr{X}(V)$ respectively as follows:

$$((r \otimes k) \otimes \xi, (s \otimes m) \otimes \eta)_0 = (\Gamma(r \otimes k, s \otimes m)\xi, \eta)_X,$$

$$(r \otimes \zeta, s \otimes \zeta')_{00} = (\Phi(r, s)\zeta, \zeta')_{\mathscr{X}(V)}.$$

Then, if $r, s \in L$, $k, m \in K$, $\xi, \eta \in X$,

$$((r \otimes k) \otimes \xi, (s \otimes m) \otimes \eta)_0 = (V(F(r, s)k, m)\xi, \eta)_X$$

$$= ((F(r, s)k) \widetilde{\otimes} \xi, m \widetilde{\otimes} \eta)_{\mathscr{X}(V)}$$

$$= (\Phi(r, s)(k \widetilde{\otimes} \xi), m \widetilde{\otimes} \eta)_{\mathscr{X}(V)}$$

$$= (r \otimes (k \widetilde{\otimes} \xi), s \otimes (m \widetilde{\otimes} \eta))_{00}. \qquad (3)$$

Now Γ is an operator inner product if and only if $(\ ,\ )_0$ is positive, that is, if and only if

$$\sum_{i, j=1}^{n} ((r_i \otimes k_i) \otimes \xi_i, (r_j \otimes k_j) \otimes \xi_j)_0 \geq 0 \qquad (4)$$

whenever $r_i \in L$, $k_i \in K$, $\xi_i \in X$. By (3), (4) is equivalent to the condition that

$$\sum_{i, j=1}^{n} (r_i \otimes (k_i \widetilde{\otimes} \xi_i), r_j \otimes (k_j \widetilde{\otimes} \xi_j))_{00} \geq 0. \qquad (5)$$

Since $K \widetilde{\otimes} X$ is dense in $\mathscr{X}(V)$, (5) is the same as the condition that $\sum_{i,j=1}^{n} (r_i \otimes \zeta_i, r_j \otimes \zeta_j)_{00} \geq 0$ whenever $r_i \in L$, $\zeta_i \in \mathscr{X}(V)$. That is, (5) holds if and only if $(\ ,\ )_{00}$ is positive, or, equivalently, Φ is an operator inner product. Thus the first statement of the proposition is proved. Equation (3) now implies the second statement.   ∎

5.6.   The preceding lemma will now be applied to the context of rigged modules.

Let A and B be two *-algebras, and $\mathscr{L} = \langle L, [\ ,\ ] \rangle$ a right B-rigged left A-module. Further, let V be an operator inner product on a linear space K; and let K be also a bounded Hermitian left B-module with respect to V (see 3.3, 3.4). Thus, by 3.4, we can form the *-representation S of B deduced from K and V. Putting $P = L \otimes K$, we can also form the map $\Gamma : P \times P \to \mathcal{O}(X)$ given by

$$\Gamma(r \otimes k, s \otimes m) = V([s, r]k, m) \qquad (r, s \in L; k, m \in K),$$

which is linear in the first variable and conjugate-linear in the second. We then have:

Proposition

(I)   *S is positive with respect to $\mathscr{L}$ if and only if Γ is an operator inner product on P.*

(II)   *S is inducible to A via $\mathscr{L}$ if and only if Γ is an operator inner product and P, as a left A-module, is Γ-bounded.*

(III) *If the conditions of* (II) *hold,* $\mathrm{Ind}_{B \uparrow A}^{\mathscr{G}} (S)$ *is unitarily equivalent to the*
  -representation of A *deduced from* P *and* Γ.

Note. The A-module structure of L automatically provides an A-module
structure for P:

$$a(r \otimes k) = ar \otimes k \qquad\qquad (a \in A; r \in L; k \in K).$$

By 4.7(9) P, as an A-module, is Hermitian with respect to Γ.

Proof. Putting $F(r, s)(k) = [s, r]k$ $(s, r \in L; k \in K)$, we see that the hypotheses of 5.5 hold. So statement (I) is an immediate consequence of Lemma 5.5.

To prove (II), suppose that S is inducible to A. This means that for each a
in A there exists a bounded linear operator T_a on $X(\mathrm{Ind}(S))$ such that
$T_a(r \widetilde{\otimes} \zeta) = ar \widetilde{\otimes} \zeta$ $(r \in L; \zeta \in X(S))$. Passing to $\mathscr{X}(P, \Gamma)$ via the isometry β of
Lemma 5.5, we conclude that there exists a bounded linear operator T_a' on
$\mathscr{X}(P, \Gamma)$ satisfying

$$T_a'((r \otimes k) \widetilde{\otimes} \xi) = (ar \otimes k) \widetilde{\otimes} \xi \qquad (r \in L; k \in K; \xi \in X).$$

So P, as a left A-module, is Γ-bounded. Since the argument works equally
well in the opposite direction, statement (II) is proved. The unitary equivalence required for statement (III) is just the β of Lemma 5.5. ∎

5.7. Proposition 5.6 is the abstract version of the theorem on "inducing in
stages" for group representations. To obtain this and other useful applications of 5.6, a few preliminary remarks are in order.

5.8. Let A be a *-algebra; let L and K be left A-modules; and let $\phi: L \to K$
be a surjective A-module homomorphism. Suppose further than X is a
Hilbert space and that the map $V: K \times K \to \mathcal{O}(X)$ is linear in the first variable
and conjugate-linear in the second; and define $V'(r, s) = V(\phi(r), \phi(s))$
$(r, s \in L)$. The reader will easily verify the following result:

Proposition

(I) V' *is an operator inner product on* L *if and only if* V *is an operator*
     inner product on K.

(II) *If the conditions of* (I) *hold,* L *(as an* A-*module) is Hermitian and*
     bounded with respect to V' *if and only if* K *(as an* A-*module) is*
     Hermitian and bounded with respect to V.

(III) *If the conditions of* (II) *hold, the* *-representations* T *and* T' *of* A
     deduced from K, V *and* L, V' *respectively are unitarily equivalent*
     under the obvious correspondence $r \widetilde{\otimes} \xi \leftrightarrow \phi(r) \widetilde{\otimes} \xi$ $(r \in L; \xi \in X)$.

5.9.   Returning to the context of 5.6, let us assume in addition that the actions of A and B on L commute:

$$a(rb) = (ar)b \qquad (a \in A; r \in L; b \in B). \qquad (6)$$

Let $P_B = L \otimes_B K$ denote the B-tensor product of L and K, that is, the quotient of $L \otimes K$ by the linear span Q of $\{rb \otimes k - r \otimes bk : r \in L, k \in K, b \in B\}$. Since by (6) Q is A-stable, P_B, like P, is a left A-module (under the quotient module structure).

Notice that $\Gamma(\sigma, \tau)$, as defined in 5.6, vanishes if either σ or τ lies in Q. Indeed, if $r, s \in L$, $k, m \in K$, and $b \in B$,

$$\Gamma(rb \otimes k - r \otimes bk, s \otimes m) = V([s, rb]k - [s, r]bk, m) = 0$$
$$\text{(by 4.3(iii));} \qquad (7)$$

and

$$\Gamma(r \otimes k, sb \otimes m - s \otimes bm) = V([sb, r]k, m) - V([s, r]k, bm)$$
$$= V([sb, r]k - b^*[s, r]k, m)$$
$$\text{(since } K \text{ is } V\text{-Hermitian)}$$
$$= 0 \qquad \text{(by 4.3(2)).} \qquad (8)$$

Let $r \otimes_B k$ denote the image of $r \otimes k$ in P_B ($r \in L$; $k \in K$). It follows from (7) and (8) that the equation

$$\Gamma^0(r \otimes_B k, s \otimes_B m) = \Gamma(r \otimes k, s \otimes m)$$
$$= V([s, r]k, m) \quad (r, s \in L; k, m \in K) \qquad (9)$$

defines a map $\Gamma^0 : P_B \times P_B \to \mathcal{O}(X)$ which is linear in the first variable and conjugate-linear in the second. Applying 5.8 to 5.6 we conclude:

Proposition 5.6 holds when P is replaced by P_B and Γ by Γ^0.

5.10.   *Example: Inducing in Stages in Finite Groups.*   Let G be a finite group, and H and J two subgroups of G with $J \subset H$. Take A and B to be the group algebras $\mathscr{L}(G)$ and $\mathscr{L}(H)$ of G and H respectively; and let L and K be the linear spaces underlying A and B respectively. Identifying B with a *-subalgebra of A, we see that L is a left A-module and a right B-module, while K is a left B-module (under the convolution multiplication in A and B). Further (6) holds; and

$$P_B = L \otimes_B K \cong L \qquad (10)$$

under the identification $a \otimes_B b \cong a * b$ ($a \in A; b \in B$).

Now let R be a finite-dimensional unitary representation of J. As in 4.18 let $\rho : A \to B$ be the restriction map; and let ρ_1 be the restriction map of B onto

$\mathscr{L}(J)$. According to 4.18, the map $V:\langle\phi,\psi\rangle\mapsto R_{\rho_1(\psi^**\phi)}$ $(\phi,\psi\in K)$ is an operator inner product on K; and the unitary representation S of H deduced from K, V is just $\text{Ind}_{J\uparrow H}(R)$. Also, if $[f, g] = \rho(f^* * g)$ $(f, g\in L)$, then by 4.18 $L, [\, , \,]$ is a right B-rigged left A-module; and the unitary representation of G abstractly induced from S via $L, [\, , \,]$ is $\text{Ind}_{H\uparrow G}(S)$. On the other hand, if $f, g\in L$, equation (9) (together with the identification (10)) gives:

$$\Gamma^0(f, g) = \Gamma^0(f \otimes_B 1, g \otimes_B 1)$$
$$= V(\rho(g^* * f), 1) = R_{(\rho_1 \circ \rho)(g^* * f)} \qquad (11)$$

(where 1 is the unit of A and B). Since $\rho_1 \circ \rho$ is the map consisting of restriction from G to J, (11) and 4.18 say that the unitary representation of G deduced from L, Γ^0 is equivalent to $\text{Ind}_{J\uparrow G}(R)$. Combining these facts with 5.6 (and 5.9), we obtain the "stages" theorem for induced representations of finite groups.

Theorem. *Let G be a finite group, and let H and J be two subgroups of G with $J \subset H$. If R is any finite-dimensional unitary representation of J,*

$$\underset{J\uparrow G}{\text{Ind}}\,(R) \cong \underset{H\uparrow G}{\text{Ind}} \left(\underset{J\uparrow H}{\text{Ind}}\,(R) \right). \qquad (12)$$

Remark. The simplest proof of this theorem is a direct verification of (12) starting from the definition IX.10.4 of the "concrete" inducing process. The reader should certainly work out this direct verification for himself (see Chap. IX, Exercise 35). The more intricate proof given above, exhibiting the "stages" theorem as a special case of 5.6, will, we hope, help to clarify the meaning of 5.6 and to orient the reader toward the more general application of 5.6 to Banach *-algebraic bundles over locally compact groups (see 12.15).

5.11. The next observation is useful for the application of 5.6 to infinite-dimensional topological situations.

Let A be a *-algebra; and let L be a left A-module which is also an LCS such that the endomorphism $r\mapsto ar$ of L is continuous for each a in A. Also, let X be a Hilbert space; let $V:L \times L \to \mathscr{O}(X)$ be linear in the first variable and conjugate-linear in the second; and assume that V is (jointly) continuous on $L \times L$ with respect to the topology of L and the weak operator topology of $\mathscr{O}(X)$. Furthermore, let L_0 be a dense A-submodule of L; and put $V^0 = V|(L_0 \times L_0)$. The following relationships (similar in form to 5.8) then hold.

Proposition

(I)   *V is an operator inner product on L if and only if V^0 is an operator inner product on L_0.*

(II)   *If the conditions of (I) hold, L (as an A-module) is bounded and Hermitian with respect to V if and only if L_0 is bounded and Hermitian with respect to V^0.*

(III)   *If the conditions of (II) hold, the *-representations of A deduced from L, V and from L_0, V^0 are unitarily equivalent.*

Proof.   (I) is evident from the continuity of V. Assuming the conditions of (I), observe that for each fixed ξ the map $r \mapsto r \widetilde{\otimes} \xi$ of L into $\mathscr{X}(V)$ is continuous (again by the continuity of V). From this and the denseness of L_0 it follows that $\mathscr{X}(V) = \mathscr{X}(V_0)$. So, if L_0 is V^0-bounded, each a in A gives rise to a bounded linear operator T_a on $\mathscr{X}(V)$ satisfying $T_a(r \widetilde{\otimes} \xi) = ar \widetilde{\otimes} \xi$ for all r in L_0 and ξ in X. This and the continuity of $r \mapsto ar$ imply that $T_a(r \widetilde{\otimes} \xi) = ar \widetilde{\otimes} \xi$ for all r in L and ξ in X. Hence L is V-bounded; and (II) is essentially proved. (III) follows from the identification of $\mathscr{X}(V)$ and $\mathscr{X}(V^0)$ already observed.   ■

6.   Imprimitivity Bimodules and the Abstract Imprimitivity Theorem

6.1.   Let A and B be two *-algebras. We have seen in §4 that the natural context for passing from *-representations of B to *-representations of A, by means of the abstract inducing process, is a B-rigged A-module. Suppose we have a right B-rigged left A-module L, $[\ ,\ ]_B$, and also a left A-rigged right B-module L, $[\ ,\ ]_A$ (having the same underlying left A-module and right B-module L). Then we can induce via L, $[\ ,\ ]_B$ from *-representations of B to *-representations of A, and also via L, $[\ ,\ ]_A$ (see 4.10) from *-representations of A to *-representations of B. It turns out that, under certain further simple conditions on $[\ ,\ ]_B$ and $[\ ,\ ]_A$, we can use the Duality Theorem 2.7 for operator inner products to show that these two inducing operations are inverse to each other, and so provide an isomorphism between the *-representation theories of A and B. This is the so-called abstract Imprimitivity Theorem. The conditions which must be assumed for L, $[\ ,\ ]_A$, $[\ ,\ ]_B$ will amount to saying that the latter is a strict A, B imprimitivity biomodule.

6.2.   *Definition.* Let A and B be two *-algebras. An A, B *imprimitivity bimodule* is a system $L, [\ ,\ ]_A, [\ ,\ ]_B$, where

(i) L is both a left A-module and a right B-module,
(ii) $L, [\ ,\ ]_B$ is a right B-rigged left A-module,
(iii) $L, [\ ,\ ]_A$ is a left A-rigged right B-module,
(vi) the associative relation

$$[r, s]_A t = r[s, t]_B \qquad (r, s, t \in L] \qquad (1)$$

holds between the riggings.

If in addition the linear span of $\{[r, s]_B : r, s \in L\}$ coincides with B, and the linear span of $\{[r, s]_A : r, s \in L\}$ coincides with A, then $L, [\ ,\ ]_A, [\ ,\ ]_B$ is called a *strict A, B imprimitivity bimodule*.

Remark 1.   As we remarked in 4.2 and 4.3, even without strictness the linear spans of range($[\ ,\ ]_A$) and range($[\ ,\ ]_B$) are *-ideals of A and B respectively.

Remark 2.   If $L, [\ ,\ ]_A, [\ ,\ ]_B$ is a strict A, B imprimitivity bimodule, the actions of A and B on L commute, that is,

$$(ar)b = a(rb) \qquad (a \in A; b \in B; r \in L). \qquad (2)$$

Indeed, by the strictness property we need only prove (2) for the case that $a = [s, t]_A$ $(s, t \in L)$. But then

$$
\begin{aligned}
(ar)b &= ([s, t]_A r)b \\
&= (s[t, r]_B)b && \text{(by (1))} \\
&= s([t, r]_B b) \\
&= s[t, rb]_B && \text{(by 4.3(iii))} \\
&= a(rb) && \text{(by (1))}.
\end{aligned}
$$

6.3.   The definition of an imprimitivity bimodule has of course a high degree of symmetry between left and right. Indeed, let $L, [\ ,\ ]_A, [\ ,\ ]_B$ be an A, B imprimitivity bimodule. As in 4.4, we shall regard $\bar{L}$ as carrying the right A-module and left B-module structures complex-conjugate to those of L. Then by 4.8 $\bar{L}, [\ ,\ ]_A$ is a right A-rigged left B-module and $\bar{L}, [\ ,\ ]_B$ is a left

B-rigged right A-module. In addition, denoting the module structures of $\bar{L}$ by : (as in 4.4), we have for $r, s, t \in L$:

$$[r, s]_B : t = t([r, s]_B^*)$$

$$= t[s, r]_B = [t, s]_A r$$

$$= r : ([t, s]_A^*) = r : [s, t]_A.$$

Therefore relation (1) holds for the module structures :, and we conclude that $\bar{L}, [\ ,\ ]_B, [\ ,\ ]_A$ is a B, A imprimitivity bimodule.

Definition. $\bar{L}, [\ ,\ ]_B, [\ ,\ ]_A$ is called the *complex conjugate* of the imprimitivity bimodule $L, [\ ,\ ]_A, [\ ,\ ]_B$.

6.4. *Example; Banach *-Algebraic Bundles.* Let $\mathcal{B}$ be a Banach *-algebraic bundle over the topological group G. Let A be the unit fiber algebra of $\mathcal{B}$, and x a fixed element of G. Thus the fiber B_x is both a left A-module and a right A-module under the multiplication of $\mathcal{B}$. If we define

$$[r, s] = r^*s \tag{3}$$

and

$$[r, s]' = rs^* \tag{4}$$

$(r, s \in B_x)$, then the ranges of $[\ ,\ ]$ and $[\ ,\ ]'$ are both contained in A; and one checks without difficulty that $\mathcal{L}_x = \langle B_x, [\ ,\ ]', [\ ,\ ] \rangle$ is in fact an A, A imprimitivity bimodule. We refer to $\mathcal{L}_x$ as the A, A *imprimitivity bimodule over x in $\mathcal{B}$.*

6.5. *Example; Hereditary *-subalgebras.* Let A be a *-algebra and B a hereditary *-subalgebra of A (see VII.4.1). We will denote by L the left ideal of A generated by B; that is, L is the linear span of $B \cup AB$. Thus L is a natural left A-module and right B-module (under multiplication in A). As in (3), (4), we define

$$[r, s] = r^*s, \qquad [r, s]' = rs^* \tag{$r, s \in L$}$$

Obviously range$([\ ,\ ]') \subset A$; and the hereditary property of B guarantees that range$([\ ,\ ]) \subset B$. It is now an easy matter to verify that $\mathcal{L} = \langle L, [\ ,\ ]', [\ ,\ ] \rangle$ is an A, B imprimitivity bimodule.

6.6. *Remark.* In the last two examples the riggings $[\ ,\ ]'$ and $[\ ,\ ]$ were derived by means of the formulae (4) and (3) from an "enveloping" *-algebra which contained all of A, B and L. This is by no means an exceptional situation. Indeed, there are enough associativity conditions in the definition of an imprimitivity bimodule to ensure the following result, whose proof we leave to the reader.

Proposition*.   *Let A and B be two *-algebras, and L, [, $]_A$, [, $]_B$ an A, B imprimitivity bimodule satisfying 5.9(6). Then there exist*

(i)   *a *-algebra D,*

(ii)   **-isomorphisms ϕ and ψ of A and B respectively onto *-subalgebras $\phi(A)$ and $\psi(B)$ of D,*

(iii)   *a linear injection $\lambda: L \to D$, such that*

$$\phi(a)\lambda(r) = \lambda(ar), \qquad \lambda(r)\psi(b) = \lambda(rb),$$

$$\phi([r, s]_A) = \lambda(r)(\lambda(s))^*, \qquad \psi([r, s]_B) = (\lambda(r))^*\lambda(s)$$

for all a in A, b in B, and r, s in L.

Further, if $A = B$ we can achieve this with $\psi = \phi$.

6.7.   Remark.   Not every right B-rigged left A-module L, [, $]_B$ can be "completed" to an A, B imprimitivity bimodule L, [, $]_A$, [, $]_B$ by adjoining a suitable rigging [, $]_A$. Take for example the finite group situation studied in 4.18. Let G, H, A, B and ρ be as in 4.18; and put $[f, g]_B = \rho(f^* * g)$ ($f, g \in A$). Then A, [, $]_B$ (A being considered as a left A-module and right B-module under convolution) is a right B-rigged left A-module. Suppose that a rigging [, $]_A$ can be found making A, [, $]_A$, [, $]_B$ an A, B imprimitivity bimodule. Then by 6.2(1)

$$[f, g]_A * h = f * \rho(g^* * h) \qquad (f, g, h \in A). \qquad (5)$$

Putting $f = g = h = 1$ (the unit element of A), we get from (5)

$$[1, 1]_A = 1. \qquad (6)$$

Thus, putting $f = g = 1$ in (5) and letting h remain arbitrary, we obtain from (5) and (6)

$$h = \rho(h) \qquad \text{for all } h \text{ in } A.$$

But this is impossible unless $H = G$.

The Imprimitivity *-algebra

6.8.   Let B be a *-algebra and L, [, $]_B$ any right B-rigged space. It turns out that there is a canonical way to construct from L, [, $]_B$ a new *-algebra E, a left E-module structure for L, and a rigging [, $]_E$, such that L, [, $]_E$,

$[\ , \ ]_B$ becomes an E, B imprimitivity bimodule. This E will be called the imprimitivity *-algebra of L, $[\ , \ ]_B$.

The possibility of this construction should be contrasted with the situation mentioned in Remark 6.7.

6.9. The construction of E is as follows: We form the tensor product $L \otimes \bar{L}$, and introduce into it the following bilinear multiplication:

$$(r \otimes s)(r' \otimes s') = r[s, r']_B \otimes s'. \tag{7}$$

Notice that the right side of (7) is linear in each of its four variables when r, r' are considered as belonging to L and s, s' to $\bar{L}$. The multiplication (7) is associative. Indeed, if r, r', $r'' \in L$ and s, s', $s'' \in \bar{L}$,

$$((r \otimes s)(r' \otimes s'))(r'' \otimes s'') = (r[s, r']_B \otimes s')(r'' \otimes s'')$$
$$= r[s, r']_B[s', r'']_B \otimes s''$$
$$= r[s, r'[s', r'']_B]_B \otimes s''$$
$$= (r \otimes s)(r'[s', r'']_B \otimes s'')$$
$$= (r \otimes s)((r' \otimes s')(r'' \otimes s'')).$$

Thus $L \otimes \bar{L}$ has become an (associative) algebra. Let I be the linear span of

$$\{rb \otimes s - r \otimes sb^* : r \in L, s \in \bar{L}, b \in B\}$$

in $L \otimes \bar{L}$. One checks without difficulty that I is a two-sided ideal of $L \otimes \bar{L}$. We now define E to be the quotient algebra $(L \otimes \bar{L})/I$; and denote by $r \widetilde{\otimes} s$ the image of $r \otimes s$ in E ($r \in L$; $s \in \bar{L}$). We have by (7)

$$(r \widetilde{\otimes} s)(r' \widetilde{\otimes} s') = r[s, r']_B \widetilde{\otimes} s'; \tag{8}$$

and, in view of the definition of I,

$$rb \widetilde{\otimes} s = r \widetilde{\otimes} sb^* \qquad (r \in L; s \in \bar{L}; b \in B). \tag{9}$$

We next observe that the equation

$$(r \otimes s)^* = s \otimes r \qquad\qquad (r, s \in L)$$

defines * as a conjugate-linear mapping of order 2 of $L \otimes \bar{L}$ onto itself. Since * leaves I stable, it generates a conjugate-linear mapping (also called *) of E onto itself of order 2:

$$(r \widetilde{\otimes} s)^* = s \widetilde{\otimes} r \qquad\qquad (r, s \in L). \tag{10}$$

If $r, r' \in L$ and $s, s' \in \bar{L}$, we have by (8), (9), and (10):

$$((r \widetilde{\otimes} s)(r' \widetilde{\otimes} s'))^* = s' \widetilde{\otimes} r[s, r']_B$$

$$= s'[r', s]_B \widetilde{\otimes} r$$

$$= (r' \widetilde{\otimes} s')^*(r \widetilde{\otimes} s)^*.$$

Thus E, with the operations (8) and (10), is a *-algebra.

Definition.   This E is called the *imprimitivity *-algebra* of $B, L, [\ , \ ]_B$.

6.10.   Continuing the notation of 6.9, we shall give to L the structure of a left E-module. First observe that the equation

$$(r \otimes s)t = r[s, t]_B \qquad (r, t \in L; s \in \bar{L}) \qquad (11)$$

defines a left $(L \otimes \bar{L})$-module structure for L. Indeed,

$$\{(r \otimes s)(r' \otimes s')\}t = (r[s, r']_B \otimes s')t = r[s, r']_B[s', t]_B$$

$$= r[s, r'[s', t]_B]_B = (r \otimes s)\{r'[s', t]_B\}$$

$$= (r \otimes s)\{(r' \otimes s')t\}.$$

By 4.3(2) $\zeta t = 0$ whenever $\zeta \in I$ and $t \in L$. Hence the action (11) generates a left E-module structure for L:

$$(r \widetilde{\otimes} s)t = r[s, t]_B \qquad (r, t \in L; s \in \bar{L}). \qquad (12)$$

As a left E-module, L satisfies:

$$(\zeta t)b = \zeta(tb) \qquad (\zeta \in E; t \in L; b \in B). \qquad (13)$$

We shall now write $[r, s]_E$ for $r \widetilde{\otimes} s$ $(r, s \in L)$. Notice that $[\ , \ ]_E$, as a function on $L \times L$ to E, is linear in the first variable and conjugate-linear in the second.

Proposition. $L, [\ , \ ]_E, [\ , \ ]_B$ *is an* E, B *imprimitivity bimodule.*

Proof.   It remains to observe the following relations:

$$[r, s]_E^* = [s, r]_E, \qquad (14)$$

$$[\zeta r, s]_E = \zeta[r, s]_E, \qquad (15)$$

$$[rb, s]_E = [r, sb^*]_E, \qquad (16)$$

$$[\zeta r, s]_B = [r, \zeta^* s]_B, \qquad (17)$$

$$[r, s]_E t = r[s, t]_B \qquad (18)$$

$(r, s, t \in L; \zeta \in E; b \in B)$. To prove (15) and (17) it is enough to suppose that $\zeta = r' \widetilde{\otimes} s'$ $(r', s' \in L)$. The five relations then follow from (8), (9), (10), (12), 4.3(iii), and 4.3(2).   ∎

6.11.   *The Finite Group Case.*   Let us see what the imprimitivity *-algebra becomes in the context of finite groups.

Let H be a subgroup of the finite group G (with unit e); let A and B be the group *-algebras $\mathscr{L}(G)$ and $\mathscr{L}(H)$ of G and H respectively; and let $[f, g]_B = \rho(f^* * g)$ $(f, g \in A)$, where $\rho: f \mapsto f|H$ is the restriction mapping of A onto B. Thus (as in 6.7) A, $[,]_B$ is a right B-rigged space (A being considered as a right B-module under convolution).

Under these conditions we claim that *the imprimitivity *-algebra E of B, A, $[,]_B$ is essentially the same as the G, G/H transformation *-algebra of* VIII.7.9.

To see this, let the G, G/H transformation algebra be denoted by F. By VIII.7.9 F consists of all complex functions on $G \times (G/H)$, with convolution and involution given by

$$(f * g)(x, m) = \sum_{y \in G} f(y, m)g(y^{-1}x, y^{-1}m),$$

$$f^*(x, m) = \overline{f(x^{-1}, x^{-1}m)}$$

$(f, g \in F; x \in G; m \in G/H)$. We now define a linear transformation

$$\Phi: A \otimes \bar{A} \to F$$

as follows:

$$\Phi(\phi \otimes \psi)(x, yH) = \sum_{h \in H} \phi(yh)\psi^*(h^{-1}y^{-1}x) \qquad (19)$$

$(\phi, \psi \in A; x, y \in G)$. Evidently the right side of (19) depends on y only through yH; so (19) makes sense. One easily verifies that

$$\Phi((\phi * b) \otimes \psi) = \Phi(\phi \otimes (\psi * b^*))$$

$(\phi, \psi \in A; b \in B)$. Thus Φ vanishes on the ideal I of 6.9, and so gives rise to a linear map $\tilde{\Phi}: E \to F$:

$$\tilde{\Phi}(\phi \widetilde{\otimes} \psi) = \Phi(\phi \otimes \psi) \qquad (\phi, \psi \in A).$$

The above claim now finds precise expression in the following proposition:

Proposition. $\tilde{\Phi}$ *is a *-isomorphism of E onto F.*

Proof.   A routine calculation shows that $\tilde{\Phi}$ is a *-homomorphism.

Notice that any function in $\mathscr{L}(G \times G)$ belongs to the linear span of the set of functions of the form $\langle x, y \rangle \mapsto \phi(y)\psi^*(y^{-1}x)$, where $\phi, \psi \in A$. On the other hand every function in F is of the form $\langle x, yH \rangle \mapsto \sum_{h \in H} \gamma(x, yh)$ for some γ in $\mathscr{L}(G \times G)$. These two facts together imply that Φ, and hence also $\tilde{\Phi}$, is onto F.

Finally, we must show that $\mathrm{Ker}(\Phi)$ is the I of 6.9. If n and p are the orders of G and H respectively, $A \otimes \bar{A}$ ($= \mathrm{domain}(\Phi)$) and F ($= \mathrm{range}(\Phi)$) have dimensions n^2 and $p^{-1}n^2$; so $\mathrm{Ker}(\Phi)$ has dimension $n^2 - p^{-1}n^2$. On the other hand, if Γ is a transversal for G/H, the $(p^{-1}n)n(p-1)$ elements $\delta_{xh} \otimes \delta_y - \delta_x \otimes \delta_{yh^{-1}}$, where x, y and h run over Γ, G and $H \setminus \{e\}$ respectively, are linearly independent and belong to I. Thus I has dimension at least $(p^{-1}n)n(p-1) = n^2 - p^{-1}n^2$. Since $I \subset \mathrm{Ker}(\Phi)$, these facts imply that $I = \mathrm{Ker}(\Phi)$.   ■

Remark.   We have seen that A is a left E-module under the action defined by (12). Thus, by the preceding proposition, the equation

$$f : \phi = (\tilde{\Phi}^{-1}(f))\phi \qquad (20)$$

defines a left F-module structure : for A. An easy calculation gives the following explicit formula for this structure:

$$(f : \phi)(x) = \sum_{y \in G} f(y, xH)\phi(y^{-1}x)$$

$$(f \in F; \phi \in A; x \in G). \qquad (21)$$

The Abstract Imprimitivity Theorem

6.12.   Now let A and B be any two *-algebras; and let $\mathscr{L} = \langle L, [\ , \ ]_A, [\ , \ ]_B \rangle$ be a *strict* A, B imprimitivity bimodule. Given a *-representation S of B, we can ask whether S is inducible to A (via $L, [\ , \ ]_B$) in the sense of 4.9. For this to be so it is necessary that S be positive with respect to $L, [\ , \ ]_B$, that is:

$$S_{[r,r]_B} \geq 0 \qquad (r \in L). \qquad (22)$$

In fact, because of the strictness of $\mathscr{L}$, this turns out to be also sufficient, and we have:

Proposition.   *Let the *-representation S of B be positive with respect to L, $[\ , \ ]_B$. Thus (by 4.5) $V : \langle r, s \rangle \mapsto V_{[s,r]_B}$ is an operator inner product on L. Let W be the operator inner product on $\bar{L}$ deduced from V as in 2.6. The *-representation S induces (via $L, [\ , \ ]_B$) a *-representation T of A; and in fact T is given by*

$$T_{[s,t]_A} = W_{t,s} \qquad (s, t \in L). \qquad (23)$$

Proof. We denote $X(S)$ by X and by Y the deduced Hilbert space $\mathscr{X}(V)$ (see 1.8.). As usual $r \widetilde{\otimes} \xi$ $(r \in L; \xi \in X)$ is the image of $r \otimes \xi$ in Y. By 4.9 S will induce a *-representation T of A provided that for each a in A there is a bounded linear operator T_a on Y such that

$$T_a(r \widetilde{\otimes} \xi) = ar \widetilde{\otimes} \xi \qquad\qquad (r \in L; \xi \in X).$$

Since the range of $[\ ,\ ]_A$ spans A, it is enough to show the existence of such a T_a when $a = [s, t]_A$ $(s, t \in L)$. To do this, and at the same time to establish (23), it is enough to verify that

$$W_{t, s}(r \widetilde{\otimes} \xi) = ([s, t]_A r) \widetilde{\otimes} \xi. \qquad\qquad (24)$$

for all r, s, t in L and ξ in X. But we have

$$([s, t]_A r) \widetilde{\otimes} \xi = (s[t, r]_B) \widetilde{\otimes} \xi \qquad\qquad \text{(by (1))}$$

$$= s \widetilde{\otimes} S_{[t, r]_B} \xi \qquad\qquad \text{(by 4.5(7))}$$

$$= s \widetilde{\otimes} V_{r, t} \xi$$

$$= W_{t, s}(r \widetilde{\otimes} \xi) \qquad\qquad \text{(by 2.2(4)),}$$

proving (24). ■

Remark. Notice that the above *-representation T of A is non-degenerate. Indeed, W is non-degenerate by 2.3, and by (23) range(T) contains all the $W_{r, s}$.

Notation. The above T will sometimes be denoted by $\mathrm{Ind}_{B \uparrow A}^{\mathscr{L}}(S)$, to indicate its dependence on the imprimitivity bimodule $\mathscr{L}$.

Remark. The above proposition depended heavily, of course, on the assumption that $\mathscr{L}$ was strict.

6.13. Proposition 6.12 can of course be applied to the complex conjugate $\overline{\mathscr{L}}$ (see 6.3) of the strict A, B imprimitivity bimodule $\mathscr{L}$. Let T be a *-representation of A which is positive with respect to $\overline{L}, [\ ,\ ]_A$, that is,

$$T_{[r, r]_A} \geq 0 \qquad \text{for all } r \text{ in } L. \qquad\qquad (25)$$

By Proposition 6.12 applied to $\overline{\mathscr{L}}$, T induces (via $\overline{L}, [\ ,\ ]_A$) a non-degenerate *-representation $S = \mathrm{Ind}_{A \uparrow B}^{\overline{\mathscr{L}}}(T)$ of B satisfying

$$S_{[r, s]_B} = V_{s, r} \qquad\qquad (r, s \in L), \qquad (26)$$

where V is the operator inner product on L deduced from the operator inner product $W: \langle r, s \rangle \mapsto T_{[s, r]_A}$ on $\bar{L}$.

It will be convenient to refer to a *-representation S of B satisfying (22), and also to a *-representation T of A satisfying (25), as being *positive with respect to $\mathscr{L}$, or $\mathscr{L}$-positive.*

6.14. The next proposition is just a specialization of Proposition 2.6 to the present setting.

Proposition. *Let $\mathscr{L}$ be a strict A, B imprimitivity bimodule, and S a *-representation of B which is positive with respect to $\mathscr{L}$. Then $T = \operatorname{Ind}_{B\uparrow A}^{\mathscr{L}}(S)$ is positive with respect to $\bar{\mathscr{L}}$, so that we can form $\tilde{S} = \operatorname{Ind}_{A\uparrow B}^{\bar{\mathscr{L}}}(T)$. If S is non-degenerate, then $\tilde{S} \cong S$.*

Proof. Let W be the operator inner product on $\bar{L}$ deduced (2.6) from the operator inner product $V: \langle r, s \rangle \mapsto S_{[s, r]_B}$. The positivity of T then follows from (23), and the first statement is proved.

Now assume that S is non-degenerate, so that V is likewise non-degenerate. Let $\tilde{V}$ be the operator inner product on L deduced from W. By (26)

$$\tilde{S}_{[r, s]_B} = \tilde{V}_{s, r} \qquad\qquad (r, s \in L). \qquad (27)$$

Since V is non-degenerate, Proposition 2.6 says that V and $\tilde{V}$ are unitarily equivalent under an isometry F. By (27), the definition of V, and the strictness of $\mathscr{L}$, F must intertwine S and $\tilde{S}$. ∎

6.15. Proposition 6.14 gives rise immediately to the following abstract Imprimitivity Theorem, which is the crux of the present section.

Abstract Imprimitivity Theorem. *Let A, B be two *-algebras, and $\mathscr{L} = \langle L, [\ ,\ ]_A, [\ ,\ ]_B \rangle$ a strict A, B imprimitivity bimodule. Let $\mathscr{S}$ and $\mathscr{T}$ be the collections of all unitary equivalence classes of non-degenerate *-representations of B and A respectively which are positive with respect to $\mathscr{L}$. Then $\Phi: S \mapsto \operatorname{Ind}_{B\uparrow A}^{\mathscr{L}}(S)$ (considered as a map of unitary equivalence classes) is a bijection of $\mathscr{S}$ onto $\mathscr{T}$. Its inverse Φ^{-1} is the map $T \mapsto \operatorname{Ind}_{A\uparrow B}^{\bar{\mathscr{L}}}(T)$.*

6.16. Let $\mathscr{L}$, S, T, V and W be as in Proposition 6.14 and its proof. The strictness of $\mathscr{L}$ implies that the commuting algebras of S and V coincide, and likewise the commuting algebras of T and W. Hence 2.12 gives the following very important result, which should be considered as an integral part of the abstract Imprimitivity Theorem:

Theorem. *The mapping Φ of 6.15 preserves the *-isomorphism class of the commuting algebras. More precisely, let S be a non-degenerate *-representation of B which is positive with respect to $\mathscr{L}$, and let $T = \mathrm{Ind}^{\mathscr{L}}_{B\uparrow A}(S)$. Then the commuting algebras $\mathscr{I}(S)$ and $\mathscr{I}(T)$ are *-isomorphic under the mapping $F \mapsto \tilde{F}$ defined as in 2.12:*

$$\tilde{F}(r \tilde{\otimes} \xi) = r \tilde{\otimes} F\xi \qquad\qquad (r \in L; \xi \in X(S)).$$

In particular T is irreducible if and only if S is irreducible.

6.17. Remark. Suppose in 6.15 that the algebra A has a unit $\mathfrak{1}$ satisfying $\mathfrak{1}r = r$ for all r in L; and suppose also that we can find finitely many elements $s_1, \ldots, s_n$ of L such that

$$\sum_{i=1}^{n} [s_i, s_i]_A = \mathfrak{1}.$$

Then for $r \in L$ we have

$$[r, r]_B = [r, \mathfrak{1}r]_B$$

$$= \sum_{i=1}^{n} [r, [s_i, s_i]_A r]_B$$

$$= \sum_i [r, s_i[s_i, r]_B]_B \qquad\qquad \text{(by (1))}$$

$$= \sum_i b_i^* b_i \qquad\qquad \text{(by 4.3(ii), (iii)),} \qquad (28)$$

where $b_i = [s_i, r]_B$. If S is any *-representation of B we get from (28)

$$S_{[r,r]_B} = \sum_i S_{b_i^* b_i} = \sum_i (S_{b_i})^*(S_{b_i}) \geq 0.$$

Therefore, under these hypotheses, *every* *-representation of B is positive with respect to $\mathscr{L}$.

Similarly, suppose that B has a unit $\mathfrak{1}'$ satisfying $r\mathfrak{1}' = r$ $(r \in L)$, and that there are elements $t_1, \ldots, t_m$ of L such that

$$\sum_{i=1}^{m} [t_i, t_i]_B = \mathfrak{1}'.$$

Then every *-representation of A is positive with respect to $\mathscr{L}$.

6.18. The Case of Finite Groups. Let H be a subgroup of a finite group G (with unit e), and let F be the $G, G/H$ transformation algebra (see 6.11). We have seen in 6.10 and 6.11 that $\mathscr{L}(G)$ becomes an $F, \mathscr{L}(H)$ imprimitivity bimodule $\mathcal{L}$ when equipped with the module structures and riggings $[\ ,\ ]_F$ and $[\ ,\ ]_{\mathcal{L}(H)}$ defined as follows:

$$[\phi, \psi]_F(x, yH) = \sum_{h \in H} \phi(yh)\psi^*(h^{-1}y^{-1}x), \tag{29}$$

$$[\phi, \psi]_{\mathcal{L}(H)}(h) = (\phi^* * \psi)(h), \tag{30}$$

$$(f : \phi)(x) = \sum_{y \in G} f(y, xH)\phi(y^{-1}x) \qquad \text{(see (21))},$$

$$\phi b = \phi * b$$

$(f \in F; \phi, \psi \in \mathcal{L}(G); b \in \mathcal{L}(H); x, y \in G; h \in H)$. Let δ_x be the function $y \mapsto \delta_{xy}$. Then δ_e is the unit of $\mathcal{L}(H)$; and of course $[\delta_e, \delta_e]_{\mathcal{L}(H)} = \delta_e$. The unit of F is the function $u : \langle x, yH \rangle \mapsto \delta_{xe}$; and, if Γ is a transversal for G/H, we check that

$$u = \sum_{x \in \Gamma} [\delta_x, \delta_x]_F. \tag{31}$$

It follows from this and 6.17 that every *-representation of $\mathcal{L}(H)$, and likewise of F, is positive with respect to $\mathcal{L}$. Also, (31) implies that $\mathcal{L}$ is strict.

Now the non-degenerate *-representations of $\mathcal{L}(H)$ are just the unitary representations of H. The non-degenerate *-representations of F are the systems of imprimitivity for G over G/H (see VIII.18.7, VIII.18.22). Thus the Φ of 6.15, in the present context, sets up a one-to-one correspondence between unitary representations of the subgroup H and systems of imprimitivity for G over G/H. This is the content of the classical Imprimitivity Theorem for finite groups.

For finite groups the classical Imprimitivity Theorem has an easy direct proof, which we shall sketch in XII.1.7. By obtaining it here as a special case of the (somewhat cumbersome) abstract Imprimitivity Theorem, we hope to help the reader to understand how the same abstract theorem will give rise in 14.17 to the (far less superficial) Imprimitivity Theorem for Banach *-algebraic bundles over arbitrary locally compact groups.

6.19. To conclude this section we shall mention some other properties of representations which are preserved by the correspondence Φ of 6.15.

Proposition. *Let $\mathcal{S}, \mathcal{T}$ and Φ be as in 6.15. Then Φ is a homeomorphism with respect to the regional topologies of $\mathcal{S}$ and $\mathcal{T}$.*

Proof. Since Φ and Φ^{-1} are both inducing maps (see 6.15), they are continuous by 5.3. ∎

6.20. Proposition. *The mapping Φ of 6.15 preserves Hilbert direct sums. More precisely, let $\{S^i\}$ be an indexed collection of *-representations of B which belong to $\mathscr{S}$ and whose Hilbert direct sum $S = \sum_i^\oplus S^i$ exists (and so belongs to $\mathscr{S}$). Then*

$$\Phi(S) \cong \sum_i^\oplus \Phi(S^i).$$

This follows from 5.1.

6.21. Definition. A *-representation S of a *-algebra B is *compact* if S_b is a compact operator for every b in B.

Proposition. *Let $\mathscr{S}$, $\mathscr{T}$ and Φ be as in 6.15. If $S \in \mathscr{S}$, then S is compact if and only if $\Phi(S)$ is compact.*

Proof. This follows from 2.15 when we recall (23). ∎

6.22. Remark. An example of a property of *-representations which is *not* preserved by the correspondence Φ of 6.15 is cyclicity. See Remark 2 of 7.5.

7. Topological Versions of the Abstract Imprimitivity Theorem

7.1. The abstract Imprimitivity Theorem 6.15 as it stands is unsatisfactory for topological contexts. In many interesting situations the A and B of 6.15 are Banach *-algebras, and the imprimitivity bimodule $\mathscr{L}$ is not strict, but merely topologically strict in the following sense:

Definition. Let A and B be normed *-algebras. An A, B imprimitivity bimodule L, $[\ ,\ ]_A$, $[\ ,\ ]_B$ is *topologically strict* if the linear spans of range($[\ ,\ ]_A$) and range($[\ ,\ ]_B$) are dense in A and B respectively.

7.2. In the situation of 7.1 the abstract Imprimitivity Theorem continues to hold.

Theorem. *Let A and B be Banach *-algebras and $\mathscr{L} = \langle L, [\ ,\ ]_A, [\ ,\ ]_B \rangle$ a topologically strict A, B imprimitivity bimodule. Let $\mathscr{S}$ and $\mathscr{T}$ be the spaces of all unitary equivalence classes of non-degenerate *-representations of B and A respectively which are positive with respect to $\mathscr{L}$. Then:*

(I) *Every element S of $\mathscr{S}$ is inducible (via $L, [\ ,\ ]_B$) to an element $\Phi(S)$ of $\mathscr{T}$;*

(II) *every element T of $\mathcal{T}$ is inducible (via $\bar{L}$, [, $]_A$) to an element $\Psi(T)$ of S;*

(III) *$\Phi: S \mapsto \Phi(S)$ (as a mapping of unitary equivalence classes) is a bijection of $\mathcal{S}$ onto $\mathcal{T}$ whose inverse is $T \mapsto \Psi(T)$. Furthermore, if $S \in \mathcal{S}$ then S and $\Phi(S)$ have *-isomorphic commuting algebras. In particular S is irreducible if and only if $\Phi(S)$ is.*

Note. Positivity with respect to $\mathcal{L}$ means of course that 6.12(22) or 6.13(25) holds.

Proof. Let A_0 and B_0 denote the linear spans of the ranges of [, $]_A$ and [, $]_B$ respectively. Thus A_0 and B_0 are dense *-subalgebras (in fact *-ideals; see 4.2) of A and B respectively; and $\mathcal{L}$ can be regarded as a *strict* A_0, B_0 imprimitivity bimodule. Thus, if $\mathcal{S}_0$ and $\mathcal{T}_0$ are the families of all unitary equivalence classes of non-degenerate *-representations of A_0 and B_0 respectively which are positive with respect to $\mathcal{L}$, 6.15 says that the inducing map $\Phi_0: S \mapsto \mathrm{Ind}_{B_0 \uparrow A_0}(S)$ is a bijection of $\mathcal{S}_0$ onto $\mathcal{T}_0$ whose inverse is the inducing map $\Psi_0: T \mapsto \mathrm{Ind}_{A_0 \uparrow B_0}(T)$.

Now each element of $\mathcal{S}$, being continuous on B (by VI.3.8), is determined by its restriction to B_0. So $\mathcal{S}$ can be regarded as a subset of $\mathcal{S}_0$; and similarly $\mathcal{T}$ can be regarded as a subset of $\mathcal{T}_0$. By 4.12 every element of $\mathcal{S}_0$ is inducible not merely to A_0 but to A. From this it follows immediately that $\Phi_0(\mathcal{S}_0) \subset \mathcal{T} \subset \mathcal{T}_0$. Likewise $\Psi_0(\mathcal{T}_0) \subset \mathcal{S} \subset \mathcal{S}_0$. Since Φ_0 and Ψ_0 are bijections onto $\mathcal{T}_0$ and $\mathcal{S}_0$ respectively, it follows that $\mathcal{S}_0 = \mathcal{S}$ and $\mathcal{T}_0 = \mathcal{T}$. With this identification it is clear that Φ and Ψ coincide with Φ_0 and Ψ_0 respectively. Therefore $\Phi: \mathcal{S} \to \mathcal{T}$ is a bijection, with inverse Ψ. ■

Remark 1. In the above proof we have actually shown that $\mathcal{S}_0 = \mathcal{S}$ and $\mathcal{T}_0 = \mathcal{T}$; that is, *every *-representation of B_0 [A_0] which is positive with respect to $\mathcal{L}$ is continuous in the norm of B [A] and so extendable to a *-representation of B [A]*.

Remark 2. The above theorem is stronger than 6.15 in that only topological strictness of $\mathcal{L}$ is required. On the other hand, A and B must be Banach *-algebras in order that 4.12 be applicable.

7.3. The Propositions 6.19, 6.20, and 6.21 apply without any change to the correspondence Φ of 7.2; and we shall not rewrite them here. This is proved, of course, by applying 6.19, 6.20, and 6.21 as they stand to the correspondence Φ_0 of the proof of 7.2. In the case of 6.19 we must in addition recall VII.1.18.

Examples

7.4. The principal application of the abstract Imprimitivity Theorem will be to the Imprimitivity Theorem for Banach *-algebraic bundles, (see 14.17), which generalizes Mackey's classical Imprimitivity Theorem for locally compact groups. This will require considerable preparation. Before embarking on that, we shall present a few simpler examples of the usefulness of 7.2.

7.5. *The Compact Operators.* Let L be any non-zero Hilbert space (with inner product (,)), B the complex field $\mathbb{C}$, and A the C^*-algebra $\mathcal{O}_C(L)$ of all compact operators on L. Thus L is a natural left A-module and a natural right B-module (under scalar multiplication). We define B- and A-valued riggings on L as follows:

$$[\xi, \eta]_B = (\eta, \xi), \tag{1}$$

$$[\xi, \eta]_A(\zeta) = (\zeta, \eta)\xi \tag{2}$$

$(\xi, \eta, \zeta \in L)$. One verifies easily that $\mathscr{L} = \langle L, [\ ,\ ]_A, [\ ,\ ]_B \rangle$ is then an A, B imprimitivity bimodule. The range of $[\ ,\ ]_A$ is the *-algebra $\mathcal{O}_F(L)$ of all bounded linear operators on L of finite rank, and so is dense in A by VI.15.7; thus $\mathscr{L}$ is topologically strict. Notice that if ξ is a unit vector in L we have $[\xi, \xi]_B \geq 0$ and also

$$[\xi, \xi]_A = p_\xi = p_\xi^* p_\xi, \tag{3}$$

where p_ξ is projection onto $\mathbb{C}\xi$. Therefore all *-representations of A and of B are positive with respect to $\mathscr{L}$.

Thus we conclude from 7.2 (and 7.3) the existence of a natural one-to-one correspondence Φ between the non-degenerate *-representations of B and of A, which preserves direct sums. Now obviously a non-degenerate *-representation of B $(=\mathbb{C})$ is just a direct sum of copies of the unique trivial one-dimensional irreducible *-representation of B. Hence, applying Φ, we see that A $(=\mathcal{O}_C(L))$ has to within unitary equivalence exactly one irreducible *-representation T, and that every non-degenerate *-representation of A is a Hilbert direct sum of copies of T. Thus we have recovered by a more abstract route the result obtained in VI.15.12.

Remark 1. The argument of (3) shows that every *-representation of $\mathcal{O}_F(L)$ is positive with respect to $\mathscr{L}$. Hence by Remark 1 of 7.2 every *-representation of $\mathcal{O}_F(L)$ is continuous with respect to the operator norm, and so extends to a *-representation of $\mathcal{O}_C(L)$.

Remark 2.   Assume that $\dim(L) > 1$; and let T be the unique irreducible *-representation of A. Then $T \oplus T$ will be cyclic; but the corresponding *-representation of B (consisting of the scalar operators acting on a two-dimensional space) will not be cyclic. This shows that cyclicity is one property of *-representations which is *not* preserved by the Φ of 7.2 or 6.15.

Remark 3.   The operator inner product defined in 1.10 comes of course from the unique irreducible *-representation of the A of the present number.

7.6.   *Hereditary *-subalgebras.*   Let A be a Banach *-algebra, and B a hereditary *-subalgebra of A (not necessarily closed). As in 6.5 we define L to be the left ideal of A generated by B, and set $[r, s] = r^*s$, $[r, s]' = rs^*$ $(r, s \in L)$, so that $\mathscr{L} = \langle L, [\ ,\ ]', [\ ,\ ]\rangle$ is an A, B imprimitivity bimodule. Of course, $\mathscr{L}$ is not in general strict. However, if we denote by A_0 and B_0 the linear spans of the ranges of $[\ ,\ ]'$ and $[\ ,\ ]$ respectively, then $\mathscr{L}$ considered as an A_0, B_0 imprimitivity bimodule *is* strict. We recall that A_0 and B_0 are *-ideals of A and B respectively. In fact, since $B \subset L$, we have $BB \subset B_0$.

Let $\mathscr{S}_0$ and $\mathscr{T}_0$ be the spaces of all unitary equivalence classes of non-degenerate *-representations of B_0 and A_0 respectively which are positive with respect to $\mathscr{L}$. Further let $\mathscr{T}$ be the space of all unitary equivalence classes of *-representations T of A such that $T|A_0$ is non-degenerate. Since an element T of $\mathscr{T}$ is determined by $T|A_0$ and is automatically $\mathscr{L}$-positive, we can identify $\mathscr{T}$ with a subset of $\mathscr{T}_0$. Also, let $\mathscr{S}$ be the space of all unitary equivalence classes of non-degenerate $\mathscr{L}$-positive *-representations of B. Since $BB \subset B_0$, elements of $\mathscr{S}$ are automatically non-degenerate on the *-ideal B_0 (by the second Remark V.1.7); so $\mathscr{S}$ can be identified with a subset of $\mathscr{S}_0$.

Let $\Phi: \mathscr{S}_0 \to \mathscr{T}_0$ be the bijection set up by the inducing process as in Theorem 6.15.

We are going to show that $\mathscr{S} = \mathscr{S}_0$ and $\mathscr{T} = \mathscr{T}_0$; so that Φ is a bijection of $\mathscr{S}$ onto $\mathscr{T}$.

To see this, two observations are in order. The first is that since A is a Banach *-algebra Proposition 4.12 permits us to induce any element S of $\mathscr{S}_0$ via $\mathscr{L}$ to all of A; and by the first Remark 6.12 the result is non-degenerate on A_0. It follows that

$$\Phi(\mathscr{S}_0) \subset \mathscr{T}. \tag{4}$$

The second observation is that any *-representation T of A can be induced via $\mathscr{L}$ to B. Indeed, an easy calculation (which we leave to the reader) shows that $S = \operatorname{Ind}_{A \uparrow B}^{\overline{\mathscr{Z}}}(T)$ is nothing but the non-degenerate part of $T|B$. From this it follows that

$$\Phi^{-1}(\mathscr{T}) \subset \mathscr{S}. \tag{5}$$

Combining (4), (5) and the bijective character of Φ we conclude that $\mathscr{S} = \mathscr{S}_0$ and $\mathscr{T} = \mathscr{T}_0$.

Putting these facts together with 5.3, 6.15, 6.16, we obtain:

Proposition.   *Let A be a Banach *-algebra and B a hereditary *-subalgebra of A (not necessarily closed). Let $\mathscr{T}$ be the family of all unitary equivalence classes of *-representations T of A with the property that $X(T)$ is the closed linear span of $\{T_{ab}\xi : a \in A, \ b \in B, \ \xi \in X(T)\}$. For each $T \in \mathscr{T}$, let $\Psi(T)$ be the non-degenerate part of $T|B$. Then Ψ is a bijection from $\mathscr{T}$ onto the family $\mathscr{S}$ of all those non-degenerate *-representations S of B such that*

$$S_{r^*r} \geq 0 \qquad for \ all \ r \ in \ the \ linear \ span \ of \ B \cup AB. \tag{6}$$

The bijection Ψ is a homeomorphism with respect to the regional topologies of $\mathscr{T}$ and $\mathscr{S}$.

If T and S correspond under Ψ, their commuting algebras are *-isomorphic. In particular T is irreducible if and only if S is irreducible.

The inverse of Ψ is the inducing map $S \mapsto T = \operatorname{Ind}_{B \uparrow A}^{\mathscr{L}}(S)$, where $\mathscr{L}$ is as in 6.5.

Remark 1.   This Proposition is a strengthened form of VII.4.2.

Note.   The definition of $\mathscr{T}$ given in this Proposition is clearly equivalent to that given earlier in 7.6.

Remark 2.   It follows that every element of $\mathscr{S}$ is continuous with respect to the norm of A.

Remark 3.   Suppose that A is a C*-algebra and that B is closed. Then, since an element of B is positive with respect to B if it is positive with respect to A (VI.7.3), condition (6) holds automatically; and the range of Ψ is just the family of *all* non-degenerate *-representations of B.

Remark 4.   In the general context of the above Proposition, condition (6) is by no means superfluous. See Remark VII.4.5.

Now restricting Ψ to irreducible representations, we obtain:

Corollary. *Let A, B, Ψ be as in the Proposition; and let $\hat{A}_B$ be the family of those irreducible *-representations T of A such that $B \not\subset \mathrm{Ker}(T)$. Then $\Psi|\hat{A}_B$ is a homeomorphism (with respect to the regional topologies of $\hat{A}$ and $\hat{B}$) of $\hat{A}_B$ onto the family of all those S in $\hat{B}$ which satisfy* (6).

7.7. Suppose that the B of 7.6 is a *-ideal of the Banach *-algebra A (not necessarily closed). Then B is certainly hereditary, and (6) holds automatically. So from 7.6 (see especially Remark 2) it follows that *every nondegenerate *-representation S of B is continuous with respect to the norm of A, and can be extended to a *-representation of A acting in the same space as S.*

Thus we have recovered VI.19.11 by a different route.

7.8.   *Groups with Compact Subgroups.* As an illustration of the situation of 7.6, consider a locally compact group G (with unit e and left Haar measure λ), a compact subgroup K of G (with normalized Haar measure ν) and an irreducible unitary representation D of K (finite-dimensional by IX.8.6). We shall denote by p the self-adjoint idempotent element w_D of $\mathscr{L}(K)$ as defined in IX.7.1:

$$p(k) = \dim(D)\overline{\mathrm{Trace}(D_k)} \qquad\qquad (k \in K). \qquad (7)$$

Thus, by IX.7.4, if T is any unitary representation of G, range(T_p) is the D-subspace of $X(T)$ (with respect to $T|K$).

Let A stand for the $\mathscr{L}_1$ group algebra $\mathscr{L}_1(\lambda)$ of G. Then p, considered as the measure $p\,d\nu$ on G, can be identified with a self-adjoint idempotent multiplier of A (as in VIII.1.23). Denoting by B_D the hereditary *-subalgebra $p * A * p$ of A, and applying 7.6 to A and B_D, we obtain the following useful result:

Proposition. *There is a natural one-to-one correspondence Φ_D between the family of those S in $(B_D)\hat{\ }$ which satisfy*

$$S_{p*f^* *f*p} \geq 0 \qquad\qquad (f \in A) \qquad (8)$$

and the family of those irreducible unitary representations T of G such that $T_p \neq 0$, that is, such that D occurs in $T|K$.

Remark 1. Given any irreducible unitary representation T of G, by IX.8.10 there will be some irreducible unitary representation D of K occurring in $T|K$; and so T will belong to the range of the correspondence Φ_D of the preceding proposition. Thus, roughly speaking, the above proposition reduces the study of $\hat{G}$ to the study of the different $(B_D)\hat{\ }$ (D running over the space $\hat{K}$).

This reduction would be very advantageous if the algebras B_D turned out to have easily accessible structure spaces—for example, if they turned out to be commutative, or even if they satisfied the weaker condition that all their irreducible *-representations were finite-dimensional. Let us make the following definition:

Definition. We shall say that the compact subgroup K is *large* in G if, for every irreducible unitary representation D of K, the irreducible *-representations of B_D are all finite-dimensional.

It is a remarkable fact that every semisimple Lie group G with finite center has a compact subgroup which is large in G. This fact (which of course we cannot prove here) is the foundation of Harish–Chandra's theory of the representations of semisimple Lie groups.

Remark 2. If the group G is finite, then of course A, algebraically speaking, is a C^*-algebra; and so by Remark 3 of 7.6 the condition (8) is superfluous. In general however it is by no means superfluous; see Remark VII.4.5.

Remark 3. Suppose that the compact subgroup K is large in G. The fact that condition (8) may fail for some D and S is a symptom of something rather surprising, namely, that in the representation theory of such groups G the restriction to unitary representations is unnatural. In fact the above proposition remains true when condition (8) is eliminated, provided we work in the context, not of irreducible unitary representations of G, but of totally irreducible integrable locally convex representations of G. That is the reason why Harish–Chandra's monumental series of papers on the representation theory of semisimple Lie groups deals almost exclusively with Banach space representations, and only incidentally with unitary representations.

7.9. *Conjugation of Representations in a Saturated Banach *-algebraic Bundle.* Let $\mathscr{B} = \langle B, \pi, \cdot, * \rangle$ be a saturated Banach *-algebraic bundle over a topological group G (with unit e); and denote the unit fiber algebra of $\mathscr{B}$ by A.

Given an element x of G and a *-representation S of A, is it possible to define an "x-conjugate" of S, analogous to the process of conjugating a unitary representation of a normal subgroup of a group? It turns out that there does exist a natural such conjugation process, and that it is a special case of the abstract Rieffel inducing process. We shall mention it only cursorily here, since it will be treated at much greater length and in greater generality in §16.

So let x be a fixed element of G; and let $\mathscr{L}_x$ be the A, A imprimitivity bimodule over x in $\mathscr{B}$, introduced in 6.4. Since $\mathscr{B}$ is assumed to be saturated, $\mathscr{L}_x$ is topologically strict. So, if S is a non-degenerate *-representation of A satisfying

$$S_{b^*b} \geq 0 \qquad \text{for all } b \in B_x, \tag{9}$$

then by Theorem 7.2 $T = \operatorname{Ind}_{A \uparrow A}^{\mathscr{L}}(S)$ is a non-degenerate *-representation of A satisfying

$$T_{bb^*} \geq 0 \qquad \text{for all } b \in B_x. \tag{10}$$

We call T the x-conjugate of S, and denote it by ^xS; thus the existence of ^xS is equivalent to (9).

Evidently ^eS exists always; and $^eS \cong S$ if S is non-degenerate.

We leave it to the reader at this point to verify the following consequence of the "stages" Theorem 5.6: Suppose that $x, y \in G$, and S is a non-degenerate *-representation of A such that ^xS exists. Then $^y(^xS)$ exists if and only if $^{(yx)}S$ exists, and in that case the two are equivalent:

$$^y(^xS) \cong {}^{(yx)}S. \tag{11}$$

Now let us say that a *-representation S of A is $\mathscr{B}$-positive if $S_{b^*b} \geq 0$ for all $b \in B$, that is, if ^xS exists for all $x \in G$. It follows from (11) that, if S is $\mathscr{B}$-positive, then ^xS is $\mathscr{B}$-positive for all x. Thus, combining Theorem 7.2 with (11) we obtain a large portion of the following statement (to be proved in §16):

*Let $\mathscr{T}$ be the family of all unitary equivalence classes of non-degenerate $\mathscr{B}$-positive *-representations of A, equipped with the regional topology. Then the operation of conjugation*:

$$\langle x, S \rangle \mapsto {}^xS \qquad (x \in G; S \in \mathscr{T})$$

*is a (continuous) left action of G on $\mathscr{T}$. This action preserves the commuting algebra of the *-representation S; so for example ^xS is irreducible if and only if S is. In particular, then, the operation of conjugation generates a continuous left action of G on $\mathscr{T} \cap \hat{A}$.*

7.10. Example. As a simple example of the preceding proposition, consider the following finite-dimensional situation (which is a very special case of the construction in VIII.16.15):

Let A be a finite-dimensional C^*-algebra, that is, a finite direct sum of total matrix *-algebras (see VI.3.14). Thus we may write $A = \sum_{i=1}^{\oplus n} \mathcal{O}(X_i)$, where the $X_1, \ldots, X_n$ are non-zero finite-dimensional Hilbert spaces. In fact, putting $X = \sum_{i=1}^{\oplus n} X_i$ (Hilbert direct sum), we shall identify A with the *-subalgebra of $\mathcal{O}(X)$ consisting of those operators which leave each X_i stable. The family $\hat{A}$ of all irreducible *-representations of A is in one-to-one correspondence with the set $\{1, 2, \ldots, n\}$, the element of $\hat{A}$ corresponding to $i\ (= 1, 2, \ldots, n)$ being $S^i: a \mapsto a|X_i\ (a \in A)$.

Now let G be a finite group acting (to the left) on $\hat{A} \cong \{1, \ldots, n\}$; that is, we are given a homomorphism $x \mapsto \pi_x$ of G into the group of permutations of $\{1, \ldots, n\}$. For each x in G let B_x be the linear subspace of $\mathcal{O}(X)$ consisting of those b such that $b(X_i) \subset X_{\pi_x(i)}$ for all $i = 1, 2, \ldots, n$. One verifies easily that

$$B_x B_y \subset B_{xy}, \qquad (B_x)^* = B_{x^{-1}}$$

(the multiplication and adjoint operators being those of $\mathcal{O}(X)$). Thus, when each B_x is equipped with the operator norm, the $\{B_x\}\ (x \in G)$ form the fibers of a C^*-algebraic bundle $\mathscr{B}$ over G whose unit fiber *-algebra B_e coincides A. One checks that $\mathscr{B}$ is saturated. Furthermore, we leave it to the reader to verify that, for each $x \in G$ and each $i = 1, \ldots, n$, the x-conjugate of S^i relative to $\mathscr{B}$ is unitarily equivalent to $S^{\pi_x(i)}$. Thus the operation of x-conjugation in $\mathscr{B}$ coincides (on $\hat{A}$) with the originally given action of G on $\hat{A}$.

Remark 1. Since the different S^i may vary in dimension, this example shows that *the operation of conjugation in $\mathscr{B}$ may change the dimension of the representation*. As we shall see in Proposition 16.16, this implies that $\mathscr{B}$ need not have enough unitary multipliers (see VIII.3.8). In fact it is easy to see directly that, in the above example, $\mathscr{B}$ has enough unitary multipliers if and only if the operation of conjugation preserves the dimension of the representation (i.e. $\dim(S^{\pi_x(i)}) = \dim(S^i)$ for all $x \in G$ and all i).

Remark 2. Notice that the dimension of the fiber B_x is given by

$$\dim(B_x) = \sum_{i=1}^{n} d_i\, d_{\pi_x(i)} \tag{12}$$

(where $d_i = \dim(X_i)$). Thus by Schwarz's Inequality

$$\dim B_x \le \sum_{i=1}^{n} d_i^2 = \dim(B_e), \tag{13}$$

so that B_e has the biggest dimension of any of the fibers. Note that equality holds in (13) if and only if there is a number λ such that $d_{\pi_x(i)} = \lambda\, d_i$ for all i. Since $\sum_{i=1}^{n} d_i^2 = \sum_{i=1}^{n} d_{\pi_x(i)}^2$, such a λ must be 1. So we conclude that $\dim(B_x) = \dim(B_e)$ if and only if the action of x on $\{1, 2, \ldots, n\}$ leaves the dimension of the S^i unaltered. Thus we have another condition equivalent to the possession by $\mathscr{B}$ of enough unitary multipliers (compare Remark 1)—namely, that all the fibers of $\mathscr{B}$ have the same dimension.

As we shall see in XII.6.20, (12) and (13) hold under much more general conditions.

Remark 3.   The above construction of $\mathscr{B}$ introduces a question about which we shall have more to say in 16.31: Given a C^*-algebra A, a locally compact group G, and an action of G on the structure space $\hat{A}$, is it always possible to construct a saturated C^*-algebraic bundle $\mathscr{B}$ over G, with unit fiber algebra A, such that the action of G on $\hat{A}$ by conjugation in $\mathscr{B}$ is the same as the given action? The above construction shows that the answer is affirmative if G is finite and A is finite-dimensional.

Norm Estimates for Imprimitivity Bimodules

7.11.   Suppose that A and B are two *normed* *-algebras and that $\mathscr{L} = \langle L, [\ ,\ ]_A, [\ ,\ ]_B \rangle$ is a strict A, B imprimitivity bimodule; and let Φ be the correspondence of 6.15 between the non-degenerate $\mathscr{L}$-positive *-representations of A and those of B. Let A_c and B_c be the Banach *-algebra completions of A and B. If the actions of A and B on L can be extended to actions of A_c and B_c on L in such a way that $\mathscr{L}$ becomes a (necessarily topologically strict) A_c, B_c imprimitivity bimodule, then by Remark 1 of 7.2 the *-representations in the domain and range of Φ are all automatically continuous with respect to the norms of A and B. However, it may happen that no such extension is possible. In that case it is useful to know a relationship between the continuity of S and the continuity of $\Phi(S)$. The following proposition provides such a relationship.

By a *-*seminorm* on a *-algebra D we mean a seminorm σ on D such that $\sigma(c^*) = \sigma(c)$ and $\sigma(cd) \le \sigma(c)\sigma(d)$ for all c, d in D.

Proposition.   *Let A and B be two *-algebras, and σ a *-seminorm on A. Let $\mathscr{L} = \langle L, [\ ,\ ]_A, [\ ,\ ]_B \rangle$ be a strict A, B imprimitivity bimodule, S an $\mathscr{L}$-positive non-degenerate *-representation of B, and T the induced *-representation $\operatorname{Ind}_{B \uparrow A}^{\mathscr{L}} (S)$ of A. Then the following two conditions are equivalent:*

(I) $\|T_a\| \le \sigma(a) \quad$ *for all a in A.*   (14)

(II) *For each pair r, s of elements of L, there is a positive number k such that*

$$\|S_{[r,\, as]_B}\| \le k\sigma(a) \quad\quad \text{for all } a \text{ in } A.$$ (15)

Proof. Assume (I); and take elements r, s of L. By 6.12(23) and 2.2(5) (applied to $\mathscr{L}$ instead of $\mathscr{L}$) we get for $a \in A$

$$\|S_{[r, as]_B}\|^2 \le \|T_{[r, r]_A}\| \, \|T_{[as, as]_A}\|. \tag{16}$$

Now $[as, as]_A = a[s, s]_A a^*$ by 4.2. So by (I) and (16)

$$\|S_{[r, as]_B}\|^2 \le k^2 \|T_a\|^2 \le k^2 \sigma(a)^2,$$

where $k^2 = \|T_{[r, r]_A}\| \, \|T_{[s, s]_A}\|$. This establishes (II); and we have shown that (I) $\Rightarrow$ (II).

Now assume (II); and let us adopt the notation of 6.12, so that $X = X(S)$, $Y = X(T)$, $V_{r,s} = S_{[s, r]_B}$, $W_{r,s} = T_{[s, r]_A}$. As usual the image of $r \otimes \xi$ in Y ($r \in L$; $\xi \in X$) is written $r \widetilde{\otimes} \xi$. To prove (I) it is enough (in view of VI.8.14) to fix a non-zero vector ξ in X and an element a of A and to show that

$$\|ar \widetilde{\otimes} \xi\|_Y \le \sigma(a)\|r \widetilde{\otimes} \xi\|_Y \qquad (r \in L). \tag{17}$$

For any s, t in L let us define $\{s, t\} = (s \widetilde{\otimes} \xi, t \widetilde{\otimes} \xi)_Y = (S_{[t, s]_B} \xi, \xi)$. Notice that $\{ \, , \, \}$ is conjugate-bilinear, and that

$$\{cs, t\} = \{s, c^*t\}, \tag{18}$$

$$\{s, s\} \ge 0 \tag{19}$$

($s, t \in L$; $c \in A$). Applying the ordinary Schwarz Inequality to $\{ \, , \, \}$ (by (19)), and writing $N(r)$ for $\{r, r\}^{1/2}$, we get by (18)

$$(N(ar))^2 = \{a^*ar, r\} \le N(a^*ar)N(r). \tag{20}$$

Replacing a by a^*a in (20) we obtain $N(a^*ar) \le N((a^*a)^2 r)^{1/2} N(r)^{1/2}$. Combining this with (20) gives

$$(N(ar))^2 \le N((a^*a)^2 r)^{1/2} N(r)^{1+(1/2)}.$$

Iterating this argument, we find for each $n = 1, 2, \ldots$ and each r in L:

$$(N(ar))^2 \le N((a^*a)^{2^n} r)^{2^{-n}} (N(r))^{2 - 2^{-n}}. \tag{21}$$

Now fix $r \in L$. By (II) there is a positive number k (depending of course on r) such that for all c in A

$$(N(cr))^2 = (S_{[r, c^*cr]_B} \xi, \xi)$$

$$\le \|\xi\|^2 k^2 \sigma(c^*c) \le \|\xi\|^2 k^2 \sigma(c)^2. \tag{22}$$

Substituting (22) in (21) with $c = (a^*a)^{2^n}$, we find

$$(N(ar))^2 \le (\|\xi\| k)^{2^{-n}} (\sigma((a^*a)^{2^n}))^{2^{-n}} (N(r))^{2 - 2^{-n}}. \tag{23}$$

In view of the seminorm properties of σ,

$$(\sigma((a^*a)^{2^n}))^{2^{-n}} \le \sigma(a^*a) \le \sigma(a)^2.$$

So (23) becomes

$$(N(ar))^2 \leq (\|\xi\|k)^{2^{-n}}(\sigma(a))^2(N(r))^{2-2^{-n}}. \tag{24}$$

Now (24) holds for all positive integers n. Passing to the limit $n \to \infty$ in (24) we obtain

$$(N(ar))^2 \leq (\sigma(a))^2(N(r))^2.$$

By the definition of N this is just (17). So we have shown that (II) $\Rightarrow$ (I).   ∎

7.12. Corollary. *In Proposition 6.15 suppose that A and B are normed *-algebras, and that, for every pair r, s of elements of L, the linear map $a \mapsto [r, as]_B$ of A into B is continuous. Then, for each $\mathscr{L}$-positive *-representation of B which is norm-continuous on B, the induced *-representation $\mathrm{Ind}_{B\uparrow A}^{\mathscr{L}}(S)$ of A is norm-continuous on A.*

Proof. Apply 7.11 with $\sigma(a) = \|a\|$.   ∎

7.13. On replacing $\mathscr{L}$ by $\bar{\mathscr{L}}$ the roles of A and B become reversed in 7.11 and 7.12. Thus one obtains:

Corollary. *In Proposition 6.15 suppose that A and B are normed *-algebras, and that for each pair r, s of elements of L the two linear maps $a \mapsto [r, as]_B$ of A into B and $b \mapsto [rb, s]_A$ of B into A are continuous. Then, under the correspondence Φ of 6.15, the norm-continuous elements of $\mathscr{S}$ correspond exactly with the norm-continuous elements of $\mathscr{T}$.*

7.14. Here is a topological analogue of 6.17.

In Proposition 6.15 suppose that B is a normed *-algebra, and that there exists a net $\{a_\alpha\}$ of elements of A such that

  (i) each a_α is of the form $\sum_{i=1}^n [s_i, s_i]_A$, where $s_1, \ldots, s_n \in L$ (and n may depend on α),

  (ii) $[r, a_\alpha r]_B \underset{\alpha}{\to} [r, r]_B$ in B for each r in L.

Then we claim that every norm-continuous *-representation S of B is $\mathscr{L}$-positive.

To see this, observe that, if $a_\alpha = \sum_{i=1}^n [s_i, s_i]_A$, then

$$[r, a_\alpha r]_B = \sum_i [r, s_i[s_i, r]_B]_B$$

$$= \sum_i [s_i, r]_B^* [s_i, r]_B;$$

so

$$S_{[r, a_\sigma r]_B} = \sum_i (S_{[s_i, r]_B})^* S_{[s_i, r]_B} \geq 0. \tag{25}$$

By assumption (ii) above, and the norm-continuity of S, (24) implies that $S_{[r, r]_B} \geq 0$ for all r in L. Thus S is $\mathscr{L}$-positive.

Replacing $\mathscr{L}$ by $\mathscr{T}$, we obtain a similar statement with the roles of A and B reversed.

7.15. We conclude this section with a useful application of the preceding theory.

Let X be a non-zero Hilbert space; and let G be a locally compact group with unit e, left Haar measure λ, and modular function Δ. Let B be the compacted group algebra $\mathscr{L}(G)$; and let A be the *-algebra $\mathscr{L}(G; \mathcal{O}_c(X))$, with operations formally identical with those of $\mathscr{L}(G)$:

$$(f * g)(x) = \int f(y)g(y^{-1}x)d\lambda y,$$

$$f^*(x) = \Delta(x^{-1})(f(x^{-1}))^*$$

($f, g \in A$; $x \in G$). Let L be the linear space $\mathscr{L}(G; X)$. We shall consider L as a left A-module and a right B-module under the actions:

$$(fr)(x) = \int f(y)r(y^{-1}x)d\lambda y, \tag{26}$$

$$(r\phi)(x) = \int r(y)\phi(y^{-1}x)d\lambda y \tag{27}$$

($r \in L$; $f \in A$; $\phi \in B$; $x \in G$). Further, let us define riggings $[\ ,\ ]_A: L \times L \to A$ and $[\ ,\ ]_B: L \times L \to B$:

$$[r, s]_A(x)(\xi) = \int \Delta(x^{-1}y)(\xi, s(x^{-1}y))r(y)d\lambda y, \tag{28}$$

$$[r, s]_B(x) = \int (s(yx), r(y))d\lambda y \tag{29}$$

($r, s \in L$; $x \in G$; $\xi \in X$).

Let $\|\ \|_1$ denote the L_1 norms in A, B, and L.

One now verifies that with the module structures (26), (27), and the riggings (28), (29), the system $\mathscr{L} = \langle L, [\ ,\ ]_A, [\ ,\ ]_B \rangle$ is an A, B imprimitivity bimodule. Further, the linear spans of the ranges of $[\ ,\ ]_A$ and $[\ ,\ ]_B$

are dense in A and B respectively with respect to their inductive limit topologies. The following inequalities hold for $r, s \in L$, $f \in A$, $\phi \in B$:

$$\|[fr, s]_B\|_1 \leq \|f\|_1 \|r\|_1 \|s\|_1,$$

$$\|[r\phi, s]_A\|_1 \leq \|\phi\|_1 \|r\|_1 \|s\|_1.$$

We leave the verification of these facts as an exercise to the reader.

Remark. To motivate the formulae (26)–(29), recall that X becomes an $\mathcal{O}_c(X)$, $\mathbb{C}$ imprimitivity bimodule as in 7.5; and that B is an obvious B, B imprimitivity bimodule (with the module structures of left and right convolution and the riggings $\langle \phi, \psi \rangle \mapsto \phi * \psi^*$ and $\langle \phi, \psi \rangle \mapsto \phi^* * \psi$). Now there is an obvious way to take the tensor product of two imprimitivity bimodules. Applied to the two just mentioned, this tensor product operation makes $B \otimes X$ into a $B \otimes \mathcal{O}_c(X)$, B imprimitivity bimodule. Identifying $B \otimes X$ and $B \otimes \mathcal{O}_c(X)$ with dense subspaces of $L = \mathcal{L}(G; X)$ and $A = \mathcal{L}(G; \mathcal{O}_c(X))$ respectively, we verify that the structure of this product imprimitivity bimodule $B \otimes X$ is given by (26)–(29).

7.16. In the context of 7.15 the results 7.13, 7.14, and 6.19 imply the following proposition.

Proposition*.   *For any unitary representation S of G, the integrated form of S is inducible to A via $\mathcal{L}$; and $\mathrm{Ind}_{B \uparrow A}^{\mathscr{L}}(S)$ is $\| \ \|_1$-continuous and is unitarily equivalent to the *-representation T of A described as follows:*

$$X(T) = X(S) \otimes X,$$

$$T_f \zeta = \int_G (S_x \otimes f(x))(\zeta) d\lambda x \qquad (\zeta \in X(T); f \in A). \qquad (30)$$

Conversely, every non-degenerate $\| \ \|_1$-continuous *-representation T' of A is unitarily equivalent to the T constructed in (30) from some unitary representation S of G; and S is determined by T' to within unitary equivalence.

   Furthermore, the above correspondence $S \mapsto T$ is a homeomorphism with respect to the regional topologies of the spaces of *-representations of A and B.

Remark. The most interesting part of the above proposition is the last statement, about regional homeomorphism. This will be useful in XII.5.5.

8.   Positivity of *-Representations with Respect to a Banach *-Algebraic Bundle

8.1.   In this and the next few sections we shall see how the classical Mackey–Blattner construction of induced representations of locally compact groups (presented in §IX.10 for compact groups) emerges as a special case of the abstract inducing process of 4.9. In fact we shall show how the latter includes a natural generalization of the Mackey–Blattner construction to Banach *-algebraic bundles over locally compact groups.

8.2.   Throughout this section G is a locally compact group with unit element e, left Haar measure λ, and modular function Δ; and $\mathscr{B} = \langle B, \pi, \cdot, * \rangle$ is a Banach *-algebraic bundle over G. By Appendix C $\mathscr{B}$ automatically has enough continuous cross-sections. As usual, $\mathscr{L}(\mathscr{B})$ denotes the *-algebra of all continuous cross-sections of $\mathscr{B}$ with compact support, with the operations $*$ and * defined in VIII.5.2(1) and VIII.5.6(7). Every *-representation T of $\mathscr{B}$ gives rise to a *-representation T' of $\mathscr{L}(\mathscr{B})$, its integrated form (see VIII.11.4). We shall usually write T_f instead of T'_f (for $f \in \mathscr{L}(\mathscr{B})$).

We also consider a fixed closed subgroup H of G, with left Haar measure ν and modular function δ. Thus $\mathscr{B}_H$, the reduction of $\mathscr{B}$ to H, is a Banach *-algebraic bundle over H (see VIII.3.1); and every *-representation of $\mathscr{B}_H$ has an integrated form which is a *-representation of $\mathscr{L}(\mathscr{B}_H)$.

8.3.   We shall often want to specialize our results to the group context treated by Mackey, Blattner, and others. For brevity the phrase "the group case" will be used to refer to the situation when $\mathscr{B}$ is the group bundle (VIII.4.4). In that case $\mathscr{L}(\mathscr{B}) = \mathscr{L}(G)$; and non-degenerate *-representations of $\mathscr{B}$ are essentially unitary representations of G (see VIII.9.5).

8.4.   In the context of the Banach *-algebraic bundle $\mathscr{B}$, the inducing process will consist in taking a (non-degenerate) *-representation S of $\mathscr{B}_H$ and constructing from it a (non-degenerate) *-representation T of $\mathscr{B}$. To fit this into the pattern of the Rieffel inducing process 4.9, it will be necessary to set up a right $\mathscr{L}(\mathscr{B}_H)$-rigged left $\mathscr{L}(\mathscr{B})$-module $\mathscr{M}$. Then T will be defined as that *-representation of $\mathscr{B}$ (if such exists) whose integrated form is induced via $\mathscr{M}$ from the integrated form of S.

The example of finite groups in 4.18 suggests that $\mathscr{M}$ is going to come from an $\mathscr{L}(\mathscr{B})$, $\mathscr{L}(\mathscr{B}_H)$ conditional expectation p; and that p will essentially be the operation of restricting a function on G to H. With this in mind, we make the following definition:

Definition.   Let $p: \mathscr{L}(\mathscr{B}) \to \mathscr{L}(\mathscr{B}_H)$ be the linear map given by

$$p(f)(h) = f(h)(\Delta(h))^{1/2}(\delta(h))^{-1/2} \qquad (f \in \mathscr{L}(\mathscr{B}); h \in H). \tag{1}$$

Except for the factor $\Delta(h)^{1/2}\delta(h)^{-1/2}$, p is just the operation of restriction to H. The insertion of this factor makes p self-adjoint; that is, we have

$$p(f^*) = (p(f))^* \qquad (f \in \mathscr{L}(\mathscr{B})). \tag{2}$$

Indeed: If $h \in H$ and $f \in \mathscr{L}(\mathscr{B})$,

$$\begin{aligned}
p(f^*)(h) &= f(h^{-1})^*\Delta(h^{-1})\Delta(h)^{1/2}\delta(h)^{-1/2} \\
&= f(h^{-1})^*\Delta(h^{-1})^{1/2}\delta(h^{-1})^{-1/2}\delta(h)^{-1} \\
&= (p(f)(h^{-1}))^*\delta(h^{-1}) \\
&= (p(f))^*(h).
\end{aligned}$$

So (2) holds.

In order to speak of an $\mathscr{L}(\mathscr{B})$, $\mathscr{L}(\mathscr{B}_H)$ conditional expectation, we must make $\mathscr{L}(\mathscr{B}_H)$ act to the right on $\mathscr{L}(\mathscr{B})$. This we do as follows:

$$(f\phi)(x) = \int_H f(xh)\phi(h^{-1})\Delta(h)^{1/2}\delta(h)^{-1/2} \, dvh \tag{3}$$

$(f \in \mathscr{L}(\mathscr{B}); \ \phi \in \mathscr{L}(\mathscr{B}_H); \ x \in G)$. The integrand on the right of (3) is continuous on H to B_x with compact support; so the right side exists as a B_x-valued integral. The cross-section $f\phi$ of $\mathscr{B}$ defined by (3) has compact support, and is continuous by II.15.19. So $f\phi \in \mathscr{L}(\mathscr{B})$. Obviously $f\phi$ is linear in f and in ϕ. We claim that (3) makes the linear space underlying $\mathscr{L}(\mathscr{B})$ into a right $\mathscr{L}(\mathscr{B}_H)$-module, that is,

$$(f\phi_1)\phi_2 = f(\phi_1 * \phi_2) \qquad (f \in \mathscr{L}(\mathscr{B}); \ \phi_1, \phi_2 \in \mathscr{L}(\mathscr{B}_H)). \tag{4}$$

Indeed: By (3) and II.5.7

$$((f\phi_1)\phi_2)(x) = \int_H \int_H f(xh^{-1}k^{-1})\phi_1(k)\phi_2(h)(\delta(kh)\Delta(kh))^{-1/2} \, dvk \, dvh.$$

Making the substitutions $h \mapsto h^{-1}$ and $k \mapsto kh$, and using Fubini's Theorem II.16.3, we get

$$\begin{aligned}
((f\phi_1)\phi_2)(x) &= \int_H \int_H f(xk^{-1})\phi_1(kh)\phi_2(h^{-1})(\delta(k)\Delta(k))^{-1/2} \, dvh \, dvk \\
&= (f(\phi_1 * \phi_2))(x);
\end{aligned}$$

and (4) is proved.

We are now in a position to assert the crucial relationship:

$$p(f\phi) = p(f) * \phi \qquad (f \in \mathscr{L}(\mathscr{B}); \phi \in \mathscr{L}(\mathscr{B}_H)). \qquad (5)$$

This is easily verified. Relations (2) and (5) now show that p is an $\mathscr{L}(\mathscr{B})$, $\mathscr{L}(\mathscr{B}_H)$ conditional expectation in the sense of 4.13.

A calculation similar to that which led to (4) shows that the right action (3) of $\mathscr{L}(\mathscr{B}_H)$ commutes with left multiplication by elements of $\mathscr{L}(\mathscr{B})$:

$$f * (g\phi) = (f * g)\phi \qquad (f, g \in \mathscr{L}(\mathscr{B}), \phi \in \mathscr{L}(\mathscr{B}_H)). \qquad (6)$$

We also notice for later use that $f\phi$ is separately continuous in f and ϕ in the inductive limit topologies (II.14.3) of $\mathscr{L}(\mathscr{B})$ and $\mathscr{L}(\mathscr{B}_H)$.

8.5. Remark. In the group case, when $\mathscr{L}(\mathscr{B}) = \mathscr{L}(G)$ and $\mathscr{L}(\mathscr{B}_H) = \mathscr{L}(H)$, we can form the convolution $f * \phi$ according to III.11.8(12) whenever $f \in \mathscr{L}(G)$ and $\phi \in \mathscr{L}(H)$; and (3) becomes, not $f\phi = f * \phi$ as one might have expected, but

$$f\phi = f * \alpha(\phi), \qquad (7)$$

where α is the automorphism (not in general a *-automorphism) of $\mathscr{L}(H)$ given by

$$\alpha(\phi) = \phi(h)\Delta(h)^{1/2}\delta(h)^{-1/2}. \qquad (8)$$

The presence of the α in (7) is necessary in order to ensure (5).

8.6. We now make a very important definition:

Definition. A *-representation S of $\mathscr{B}_H$ is said to be *positive with respect to $\mathscr{B}$ (or $\mathscr{B}$-positive)* if the integrated form of S is positive with respect to the conditional expectation p, that is, if

$$S_{p(f^* * f)} \geq 0 \qquad \text{for all } f \text{ in } \mathscr{L}(\mathscr{B}).$$

Remark 1. A *-representation of $\mathscr{B}$ itself is obviously $\mathscr{B}$-positive.

Remark 2. If $H = \{e\}$, so that $\mathscr{B}_H = B_e$, we shall see in 8.11 that the above definition of $\mathscr{B}$-positivity coincides with that of 7.9.

Remark 3. In general S need not be positive with respect to $\mathscr{B}$ (see 8.15). However there are important classes of Banach *-algebraic bundles (including the group bundles) for which $\mathscr{B}$-positivity always holds. See for example 8.14, 11.9, 11.10.

Remark 4.   Suppose that S is $\mathscr{B}$-positive. Let $\mathscr{X}$ be the Hilbert space deduced from the operator inner product V: $\langle f, g \rangle \mapsto S_{p(g^* * f)}$ on $\mathscr{L}(\mathscr{B})$; and let $f \widetilde{\otimes} \xi$ be the image of $f \otimes \xi$ in $\mathscr{X}$. It was observed in 1.8 that $\xi \mapsto f \widetilde{\otimes} \xi$ is continuous for each fixed f in $\mathscr{L}(\mathscr{B})$. We claim that for each fixed ξ in $X(S)$, $f \mapsto f \widetilde{\otimes} \xi$ is continuous on $\mathscr{L}(\mathscr{B})$ to $\mathscr{X}$ with respect to the inductive limit topology of $\mathscr{L}(\mathscr{B})$. Indeed: Since $f \mapsto f \widetilde{\otimes} \xi$ is linear, it is enough to take a net $\{f_i\}$ of elements of $\mathscr{L}(\mathscr{B})$ converging to 0 uniformly on G and all vanishing outside the same compact set, and show that $\| f_i \widetilde{\otimes} \xi \| \to 0$. But our assumption implies that the $\{f_i^* * f_i\}$ converge to 0 uniformly on G and all vanish outside the same compact set; and this in turn implies that

$$\| f_i \widetilde{\otimes} \xi \|^2 = (S_{p(f^* * f)} \xi, \xi) \to 0.$$

Equivalent Conditions for $\mathscr{B}$-Positivity

8.7.   Our next main goal is as follows: Starting from an arbitrary $\mathscr{B}$-positive *-representation S of $\mathscr{B}_H$, we shall show (in 9.26) that the integrated form of S is inducible via p to a *-representation T of $\mathscr{L}(\mathscr{B})$, and that T is the integrated form of a (unique) *-representation, also called T, of $\mathscr{B}$. This T will be called the *-representation of $\mathscr{B}$ induced from S.

It turns out that the Hilbert space $X(T)$ in which T acts has a natural presentation as the $\mathscr{L}_2$ space of a Hilbert bundle over the coset space G/H. From this presentation we shall be able to deduce easily (in §10) that, in the group case, T coincides with the induced representation of G as constructed by Mackey and Blattner. To obtain the required Hilbert bundle over G/H one needs a different formulation of the definition of $\mathscr{B}$-positivity. Most of the rest of this section is devoted to obtaining this equivalent formulation.

8.8.   Recall that, for each coset α in G/H, $\mathscr{B}_\alpha$ denotes the (Banach bundle) reduction of $\mathscr{B}$ to the closed subset α of G. The underlying set of $\mathscr{B}_\alpha$ is $B_\alpha = \bigcup_{x \in \alpha} B_x$; and $\mathscr{L}(\mathscr{B}_\alpha)$ is of course the linear space of all continuous cross-sections of $\mathscr{B}_\alpha$ with compact support.

Let us fix a coset α in G/H. It is very easy to give to $\mathscr{L}(\mathscr{B}_\alpha)$ the structure of a right $\mathscr{L}(\mathscr{B}_H)$-rigged space. Indeed, one defines a right $\mathscr{L}(\mathscr{B}_H)$-module structure for $\mathscr{L}(\mathscr{B}_\alpha)$ as follows:

$$(f\phi)(x) = \int_H f(xh)\phi(h^{-1})\,dvh \tag{9}$$

$(f \in \mathscr{L}(\mathscr{B}_\alpha); \ \phi \in \mathscr{L}(\mathscr{B}_H); \ x \in \alpha)$, and an $\mathscr{L}(\mathscr{B}_H)$-valued rigging $[\ , \ ]_\alpha$ as follows:

$$[f, g]_\alpha(h) = \int_H (f(xk))^* g(xkh) dvk \tag{10}$$

$(f, g \in \mathscr{L}(\mathscr{B}_\alpha); \ h \in H)$, where x is a fixed element of α. The left-invariance of v shows that the right side of (10) is actually independent of the choice of x. One verifies immediately that $\mathscr{L}(\mathscr{B}_\alpha), [\ , \ ]_\alpha$ satisfies the postulates for a right $\mathscr{L}(\mathscr{B}_H)$-rigged space.

8.9. Proposition. *Let S be a *-representation of $\mathscr{B}_H$. The following three conditions are equivalent:*

(I)   *S is $\mathscr{B}$-positive in the sense of 8.6.*

(II) *For every coset α in G/H, S is positive with respect to $\mathscr{L}(\mathscr{B}_\alpha), [\ , \ ]_\alpha;$ that is,*

$$S_{[f, f]_\alpha} \geq 0 \qquad \text{for all } \alpha \text{ in } G/H \text{ and } f \text{ in } \mathscr{L}(\mathscr{B}_\alpha). \tag{11}$$

(III) *For every coset α in G/H, every positive integer n, all $b_1, \ldots, b_n$ in B_α, and all $\xi_1, \ldots, \xi_n$ in $X(S)$,*

$$\sum_{i, j=1}^{n} (S_{b_j^* b_i} \xi_i, \ \xi_j) \geq 0. \tag{12}$$

Note. (12) makes sense since $b_j^* b_i \in B_H$ whenever b_i and b_j both belong to the same B_α.

Proof. The first step is to express $(S_{p(g^* * f)} \xi, \eta)$ in a new form. Choose a continuous everywhere positive H-rho function ρ on G (see III.13.2 and III.14.5); and let $\rho^\#$ be the corresponding regular Borel measure on G/H (III.13.10). We claim that for all f, g in $\mathscr{L}(\mathscr{B})$ and ξ, η in $X(S)$

$(S_{p(g^* * f)} \xi, \eta)$

$$= \int_{G/H} d\rho^\#(xH) \int_H \int_H (\rho(xh)\rho(xk))^{-1/2} (S_{(g(xh))^* f(xk)} \xi, \eta) dvh \, dvk. \tag{13}$$

Indeed: If $h \in H$ we have $(g^* * f)(h) = \int_G g(x)^* f(xh) d\lambda x$ (B_h-valued integral); so by II.5.7

$$(S_{(g^* * f)(h)} \xi, \eta) = \int_G (S_{g(x)^* f(xh)} \xi, \eta) d\lambda x.$$

Thus

$$(S_{\rho(g^* * f)} \xi, \eta) = \int_H \Delta(h)^{1/2} \delta(h)^{-1/2} (S_{(g^* * f)(h)} \xi, \eta) dvh$$

$$= \int_H \int_G \Delta(h)^{1/2} \delta(h)^{-1/2} (S_{g(x)^* f(xh)} \xi, \eta) d\lambda x \, dvh$$

$$= \int_H \int_{G/H} I(xH, h) d\rho^\#(xH) dvh \qquad \text{(by III.13.10(4)),}$$

where $I(xH, h) = \Delta(h)^{1/2} \delta(h)^{-1/2} \int_H \rho(xk)^{-1} (S_{g(xk)^* f(xkh)} \xi, \eta) dvk$. Now I is continuous on $(G/H) \times H$ with compact support. So, applying Fubini's Theorem, we get

$(S_{\rho(g^* * f)} \xi, \eta)$

$$= \int_{G/H} d\rho^\#(xH) \int_H \int_H \rho(xk)^{-1} \Delta(h)^{1/2} \delta(h)^{-1/2} (S_{g(xk)^* f(xkh)} \xi, \eta) dvk \, dvh$$

$$= \int_{G/H} d\rho^\#(xH) \int_H \int_H \rho(xk)^{-1} \Delta(k^{-1}h)^{1/2}$$

$$\delta(k^{-1}h)^{-1/2} (S_{g(xk)^* f(xh)} \xi, \eta) dvh \, dvk$$

$$= \int_{G/H} d\rho^\#(xH) \int_H \int_H (\rho(xk)\rho(xh))^{-1/2} (S_{g(xk)^* f(xh)} \xi, \eta) dvh \, dvk$$

(using the rho-function identity III.13.2(1)). But this is (13), and the claim is proved.

Now assume (II); and let $\xi \in X(S)$, $f \in \mathscr{L}(\mathscr{B})$. If $x \in \alpha \in G/H$, we obtain from (10) and (11) (on replacing the f in (11) by $(\rho^{-1/2}f)|\alpha$):

$$\int_H \int_H (\rho(xh)\rho(xk))^{-1/2} (S_{f(xh)^* f(xk)} \xi, \xi) dvh \, dvk \geq 0.$$

This and (13) together imply that

$$(S_{\rho(f^* * f)} \xi, \xi) \geq 0$$

for all ξ in $X(S)$ and f in $\mathscr{L}(\mathscr{B})$. So (II) $\Rightarrow$ (I).

Conversely, we shall show that (I) $\Rightarrow$ (II). Let g be in $\mathscr{L}(\mathscr{B})$ and ξ in $X(S)$; and let σ be any continuous complex function on G which is constant on each coset xH. Assuming (I), we have by (13):

$$0 \le (S_{p((\sigma g)^* * (\sigma g))}\xi, \xi)$$

$$= \int_{G/H} d\rho^\#(xH)|\sigma(x)|^2 \int_H \int_H (\rho(xh)\rho(xk))^{-1/2}(S_{g(xh)^*g(xk)}\xi, \xi)dvh\,dvk.$$

$$(14)$$

Since g is continuous with compact support, the inner double integral in (14) is continuous as a function on G/H. So by (14) and the arbitrariness of σ one deduces that

$$\int_H \int_H (\rho(xh)\rho(xk))^{-1/2}(S_{g(xh)^*g(xk)}\xi, \xi)dvh\,dvk \ge 0$$

for all x in G. Replacing g by $\rho^{1/2}g$ in this inequality, we obtain by (10)

$$S_{[f, f]_\alpha} \ge 0 \qquad (15)$$

for all α in G/H and all f in $\mathscr{L}(\mathscr{B}_\alpha)$ which are of the form $g|\alpha$ for some g in $\mathscr{L}(\mathscr{B})$. But by II.14.7 every f in $\mathscr{L}(\mathscr{B}_\alpha)$ is of this form. So (15) holds for all f in $\mathscr{L}(\mathscr{B}_\alpha)$; and we have shown that (I) $\Rightarrow$ (II).

We shall now show that (III) $\Rightarrow$ (II). Assume (III); and let $x \in \alpha \in G/H$; $f \in \mathscr{L}(\mathscr{B}_\alpha)$; $\xi \in X(S)$. Then

$$(S_{[f, f]_\alpha}\xi, \xi) = \int_H \int_H (S_{f(xk)^*f(xh)}\xi, \xi)dvh\,dvk. \qquad (16)$$

Now the right side of (16) can be approximated by finite sums of the form

$$\sum_{r,s=1}^n v(E_r)v(E_s)(S_{f(xh_s)^*f(xh_r)}\xi, \xi), \qquad (17)$$

where the $E_1, \ldots, E_n$ are Borel subsets of H (with compact closure), and $h_r \in E_r$. By (III) the summation (17) is non-negative. So the right side of (16) is non-negative, and (II) holds.

Finally we must verify that (II) $\Rightarrow$ (III). Fix $x \in \alpha \in G/H$; $\xi_1, \ldots, \xi_n \in X(S)$; $b_1, \ldots, b_n \in B_\alpha$. We can then choose $f'_1, \ldots, f'_n$ in $\mathscr{L}(\mathscr{B}_\alpha)$ and $h_1, \ldots, h_n$ in H so that $f'_i(xh_i) = b_i$. Now let $f_i = \tau_i f'_i$, where τ_i is a non-negative element of $\mathscr{L}(\alpha)$ which vanishes outside a small neighborhood U_i of xh_i and for which

$\int_H \tau_i(xh)dvh = 1$. Assuming (II), we know from 4.5 that $\langle f, g \rangle \mapsto S_{[g,f]_\alpha}$ is an operator inner product on $\mathscr{L}(\mathscr{B}_\alpha)$, and hence that

$$0 \le \sum_{i,j=1}^n (S_{[f_j, f_i]_\alpha}\xi_i, \xi_j)$$

$$= \sum_{i,j=1}^n \int_H \int_H (S_{f_j(xh)^* f_i(xk)}\xi_i, \xi_j)dvh\, dvk$$

$$= \sum_{i,j=1}^n \int_H \int_H \tau_j(xh)\tau_i(xk)(S_{f_j(xh)^* f_i(xk)}\xi_i, \xi_j)dvh\, dvk.$$

As the U_i shrink down to xh_i, the last expression approaches

$$\sum_{i,j=1}^n (S_{f_j(xh_j)^* f_i(xh_i)}\xi_i, \xi_j) = \sum_{i,j=1}^n (S_{b_j^* b_i}\xi_i, \xi_j).$$

So the latter is non-negative, and (III) is proved.

We have now completely proved the equivalence of (I), (II), and (III).   ∎

8.10.   Remark.   Let G_d be the same group as G only with the discrete topology, and $\mathscr{B}^d$ the Banach *-algebraic bundle over G_d coinciding with $\mathscr{B}$ except for its topology. In view of Proposition 4.5, condition (III) of 8.9 asserts that for each α in G/H S is positive with respect to the right $\mathscr{L}((\mathscr{B}^d)_H)$-rigged space $\mathscr{L}((\mathscr{B}^d)_\alpha)$, $[\ ,\ ]^d_\alpha$, constructed like the $\mathscr{L}(\mathscr{B}_\alpha)$, $[\ ,\ ]_\alpha$ of 8.8 except that we start from $\mathscr{B}^d$ instead of $\mathscr{B}$.

Thus the concept of $\mathscr{B}$-positivity of a *-representation of $\mathscr{B}_H$ is not altered when the topology of G is replaced by the discrete topology.

8.11.   Remark.   Applying either 8.9(II) or 8.9(III) to the case $H = \{e\}$, we see that, for *-representations of B_e, $\mathscr{B}$-positivity in the sense of 8.6 is the same as $\mathscr{B}$-positivity in the sense of 7.9.

8.12.   Corollary.   *For any *-representation T of $\mathscr{B}$, the restriction of T to $\mathscr{B}_H$ is positive with respect to $\mathscr{B}$. More generally, if M is another closed subgroup of G with $H \subset M$, and if T is a *-representation of $\mathscr{B}_M$ which is positive with respect to $\mathscr{B}$, then $T|\mathscr{B}_H$ is positive with respect to $\mathscr{B}$.*

The proof is an obvious application of 8.9(III).

8.13.   Corollary.   *Let M be a closed subgroup of G with $H \subset M$; and let S be a *-representation of $\mathscr{B}_H$ which is positive with respect to $\mathscr{B}$. Then S is positive with respect to $\mathscr{B}_M$.*

This follows immediately from 8.9(II) or 8.9(III).

8.14. Corollary. *If $\mathscr{B}$ has enough unitary multipliers, then every *-representation of $\mathscr{B}_H$ is positive with respect to $\mathscr{B}$. This is the case if $\mathscr{B}$ is a semidirect product bundle (see VIII.4.3), in particular if $\mathscr{B}$ is the group bundle of G.*

Proof. Let S be a *-representation of $\mathscr{B}_H$. Take an element x of G, elements $b_1, \ldots, b_n$ of B_{xH}, and vectors $\xi_1, \ldots, \xi_n$ in $X(S)$. By hypothesis there is a unitary multiplier u of $\mathscr{B}$ of order x. Setting $c_i = u^*b_i$, we have $c_i \in B_H$ and $b_i = uc_i$. By the associative laws for multipliers, $b_j^*b_i = (c_j^*u^*)(uc_i) = c_j^*(u^*u)c_i = c_j^*c_i$. Thus

$$\sum_{i,j}(S_{b_j^*b_i}\xi_i, \xi_j) = \sum_{i,j}(S_{c_j^*c_i}\xi_i, \xi_j)$$

$$= \sum_{i,j}(S_{c_i}\xi_i, S_{c_j}\xi_j) \geq 0.$$

So by 8.9(III) S is $\mathscr{B}$-positive. ∎

8.15. Remark. If $\mathscr{B}$ is merely saturated, the conclusion of 8.14 need no longer hold. For example let $\mathscr{B}$ be the saturated Banach *-algebraic bundle of VIII.3.15 over the two-element group $\{1, -1\}$. The one-dimensional representation $\chi: \langle r, r \rangle \mapsto r$ of B_1 is not $\mathscr{B}$-positive, since for $b = \langle 1, -1 \rangle \in B_{-1}$ we have $\chi(b^*b) = -1$.

8.16. Remark. In 11.9 we shall see that if $\mathscr{B}$ is a saturated C^*-algebraic bundle, every *-representation of $\mathscr{B}_H$ is $\mathscr{B}$-positive. (The example of VIII.3.16 shows that a saturated C^*-algebraic bundle need not have enough unitary multipliers.)

$\mathscr{B}$-positivity and Weak Containment

8.17. It follows easily from 1.14 that a Hilbert direct sum of $\mathscr{B}$-positive *-representations of $\mathscr{B}_H$ is again $\mathscr{B}$-positive.

8.18. The set $\mathscr{S}^+$ of all $\mathscr{B}$-positive *-representations of $\mathscr{B}_H$ is closed in the regional topology.

Indeed: Let W belong to the regional closure of $\mathscr{S}^+$; and let $f \in \mathscr{L}(\mathscr{B})$, $\xi \in X(W)$. By the definition VIII.21.2 of the regional topology, given $\varepsilon > 0$ we can find a *-representation S in $\mathscr{S}^+$ and a vector ξ' in $X(S)$ such that

$$|(W_{p(f^* * f)}\xi, \xi) - (S_{p(f^* * f)}\xi', \xi')| < \varepsilon. \tag{18}$$

Since $S \in \mathscr{S}^+$, $(S_{p(f^* * f)} \xi', \xi') \geq 0$; and so by (18) $(W_{p(f^* * f)} \xi, \xi) > -\varepsilon$. This is true for any $\varepsilon > 0$; hence $(W_{p(f^* * f)} \xi, \xi) \geq 0$ for every f in $\mathscr{L}(\mathscr{B})$ and ξ in $X(W)$. Consequently $W \in \mathscr{S}^+$.

8.19.   Combining 8.17 and 8.18, we obtain:

Proposition.   *If W is a *-representation of $\mathscr{B}_H$ which is weakly contained in the set of all $\mathscr{B}$-positive *-representations of $\mathscr{B}_H$, then W is $\mathscr{B}$-positive.*

8.20.   For Proposition 8.21 we shall need the following general fact:

Proposition.   *Let D be any Banach *-algebra, and D_0 a dense subset of D. Let T be a non-degenerate *-representation of D, and $\mathscr{S}$ a collection of non-degenerate *-representations of D. Then the following two conditions are equivalent:*

(I)   *T is weakly contained in $\mathscr{S}$.*
(II)   *We have*

$$\|T_a\| \leq \sup\{\|S_a\| : S \in \mathscr{S}\} \tag{19}$$

for every a in D_0.

Proof.   Let us assume (I). Then T belongs to the regional closure of the set $\mathscr{S}_f$ of all finite direct sums of elements of $\mathscr{S}$. Thus by VII.1.15

$$\|T_a\| \leq \sup\{\|S_a\| : S \in \mathscr{S}_f\} \tag{20}$$

for all a in D. Since for each a the right hand sides of (19) and (20) coincide, (II) must hold.

Conversely, assume (II). Let D_c, $\| \ \|_c$ be the C^*-completion of D (see VI.10.4). We shall denote corresponding *-representations of D and of D_c (see VI.10.5) by the same letters. We claim that (19) now holds for all a in D_c. Indeed: Since D_0 is dense in D it is dense in D_c. Given a in D_c and $\varepsilon > 0$, we choose b in D_0 such that $\|a - b\|_c < \varepsilon$. Then by (II) $\sup\{\|S_a\| : S \in \mathscr{S}\} \geq \sup\{\|S_b\| : S \in \mathscr{S}\} - \varepsilon \geq \|T_b\| - \varepsilon > \|T_a\| - 2\varepsilon$. Thus by the arbitrariness of ε (19) holds for a, and the claim is proved.

The above claim now implies that

$$\bigcap \{\mathrm{Ker}(S) : S \in \mathscr{S}\} \subset \mathrm{Ker}(T), \tag{21}$$

where the kernels in (21) are the kernels in D_c. From this and the Equivalence Theorem VII.5.5 it follows that (I) holds.   ■

8.21. Proposition. *Let S and W be two *-representations of $\mathscr{B}_H$ such that*

(i) *W is $\mathscr{B}$-positive,*

(ii) *$\|S_f\| \le \|W_f\|$ for all f in $\mathscr{L}(\mathscr{B}_H)$.*

Then S is $\mathscr{B}$-positive.

Proof. Taking the D of 8.20 to be the $\mathscr{L}_1$ cross-sectional algebra of $\mathscr{B}_H$ and the D_0 to be $\mathscr{L}(\mathscr{B}_H)$, we conclude from (ii) and 8.20 that S is weakly contained in W. From this and (i) the conclusion follows by 8.19. ∎

9. Induced Hilbert Bundles and Induced Representations of a Banach *-Algebraic Bundle

9.1. We maintain the notation of 8.2. In addition we will choose once for all a continuous everywhere positive H-rho function ρ on G (see III.14.5), and denote by $\rho^{\#}$ the regular Borel measure on G/H constructed from ρ (as in III.13.10).

We now fix a non-degenerate $\mathscr{B}$-positive *-representation S of $\mathscr{B}_H$, and write X for X(S). Our first goal is to construct from S a Hilbert bundle $\mathscr{Y}$ over G/H. To this end the first step is to construct a Hilbert space Y_α for each coset α in G/H. Let $x \in \alpha \in G/H$. We form the algebraic tensor product $\mathscr{L}(\mathscr{B}_\alpha) \otimes X$, and introduce into it the conjugate-bilinear form $(\ , \ )_\alpha$ given by

$$(f \otimes \xi, g \otimes \eta)_\alpha = \int_H \int_H (\rho(xh)\rho(xk))^{1/2}(S_{g(xk)^*f(xh)}\xi, \eta)dvh\,dvk \qquad (1)$$

$(f, g \in \mathscr{L}(\mathscr{B}_\alpha); \xi, \eta \in X)$. Notice that the right side of (1) depends only on α (not on x). In fact, if $f \mapsto f'$ is the linear automorphism of $\mathscr{L}(\mathscr{B}_\alpha)$ given by $f'(y) = \rho(y)^{-1/2}f(y)$ ($y \in \alpha$), (1) can be written in the form

$$(f \otimes \xi, g \otimes \eta)_\alpha = (S_{[g', f']_\alpha}\xi, \eta) \qquad (2)$$

(recall 8.8(10)). It therefore follows from 8.9(II) and 4.5 that $(\ , \ )_\alpha$ is positive. Let Y_α be the Hilbert space completion of the pre-Hilbert space

$$(\mathscr{L}(\mathscr{B}_\alpha) \otimes X)/N_\alpha$$

(N_α being the null space of $(\ , \ )_\alpha$). This Y_α is going to be the fiber of $\mathscr{Y}$ over α. We write $(\ , \ )_\alpha$ also for the inner product in Y_α, and $\| \ \|_\alpha$ for the norm in Y_α; and we denote by κ_α the quotient map of $\mathscr{L}(\mathscr{B}_\alpha) \otimes X$ into Y_α.

One easily verifies:

Proposition. *$\kappa_\alpha(f \otimes \xi)$ is separately continuous in f (with the inductive limit topology of $\mathscr{L}(\mathscr{B}_\alpha)$) and in ξ.*

9.2.   Let Y be the disjoint union of the Y_α ($\alpha \in G/H$).

Proposition.   *There is a unique topology for Y making $\mathcal{Y} = \langle Y, \{Y_\alpha\} \rangle$ a Hilbert bundle over G/H such that for each f in $\mathcal{L}(\mathcal{B})$ and each ξ in X the cross-section*

$$\alpha \mapsto \kappa_\alpha((f|\alpha) \otimes \xi) \qquad (3)$$

of $\mathcal{Y}$ is continuous.

Proof.   In order to apply II.13.18 to the linear span of the family of cross-sections (3) and so complete the proof, it is enough to verify the following two facts:

(I) If $f, g \in \mathcal{L}(\mathcal{B})$ and $\xi, \eta \in X$, then $\alpha \mapsto ((f|\alpha) \otimes \xi, (g|\alpha) \otimes \eta)_\alpha$ is continuous on G/H;

(II) for each α in G/H, $\{\kappa_\alpha((f|\alpha) \otimes \xi): f \in \mathcal{L}(\mathcal{B}), \xi \in X\}$ has dense linear span in Y_α.

The first of these facts results from a simple uniform continuity argument based on (1). The second is an immediate consequence of II.14.8. ∎

Definition.   The Hilbert bundle $\mathcal{Y}$ whose existence has just been established is called the *Hilbert bundle over G/H induced by S*.

9.3.   Remark.   Strictly speaking $\mathcal{Y}$ depends on the particular choice of ρ in 9.1, and so should be referred to as $\mathcal{Y}^\rho$. But the dependence of $\mathcal{Y}^\rho$ on ρ is not very serious. Indeed, let ρ' be another everywhere positive continuous H-rho function. Then by III.13.2(1) there is a continuous positive-valued function σ on G/H such that $\rho'(x) = \sigma(xH)\rho(x)$ ($x \in G$). If $(\ ,\ )'_\alpha$ is constructed as in (1) from ρ', we have

$$(\zeta, \eta)'_\alpha = \sigma(\alpha)^{-1}(\zeta, \eta)_\alpha \qquad (\zeta, \eta \in \mathcal{L}(\mathcal{B}_\alpha) \otimes X). \qquad (4)$$

It follows that the completed spaces $Y_\alpha^{\rho'}$ and Y_α^ρ are the same except that their inner products differ by the positive multiplicative constant $\sigma(\alpha)^{-1}$. The topologies of $\mathcal{Y}^{\rho'}$ and $\mathcal{Y}^\rho$ derived from Proposition 9.2 are the same.

In the future we shall usually write $\mathcal{Y}$ rather than $\mathcal{Y}^\rho$.

9.4.   The fiber Y_H of $\mathcal{Y}$ over the coset H is essentially the same as the space X of S. Indeed, if $\phi, \psi \in \mathcal{L}(\mathcal{B}_H)$ and $\xi, \eta \in X$,

$$(\phi \otimes \xi, \psi \otimes \xi)_H = \int_H \int_H (\rho(h)\rho(k))^{-1/2}(S_{\phi(h)}\xi, S_{\psi(k)}\eta)dvh\, dvk$$

$$= \rho(e)^{-1}(S_{\phi'}\,\xi, S_{\psi'}\,\eta), \qquad (5)$$

where we have set $\phi'(h) = \Delta(h)^{1/2}\delta(h)^{-1/2}\phi(h)$ (and similarly for ψ'). It follows that the equation

$$F(\kappa_H(\phi \otimes \xi)) = S_{\phi'}\xi \qquad (\phi \in \mathcal{L}(\mathcal{B}_H); \xi \in X) \qquad (6)$$

defines a linear map $F: Y_H \to X$ such that $\rho(e)^{-1/2}F$ is an isometry. Since S is non-degenerate, its integrated form is also non-degenerate, and hence F is *onto* X.

Notice that F is independent of the choice of ρ.

9.5. *An Alternative Construction of Y_α.* One can also build Hilbert spaces Y'_α over each coset α by starting from 8.9(III) instead of 8.9(II). To be specific, take $\alpha \in G/H$; let Z_α be the algebraic direct sum $\sum_{x \in \alpha}^{\oplus}(B_x \otimes X)$; and introduce into Z_α the conjugate-bilinear form $(\ ,\ )'_\alpha$ given by:

$$(b \otimes \xi, c \otimes \eta)'_\alpha = (\rho(x)\rho(y))^{-1/2}(S_{c^*b}\xi, \eta) \qquad (7)$$

$(x, y \in \alpha; b \in B_x; c \in B_y; \xi, \eta \in X)$. By 8.9(III) $(\ ,\ )'_\alpha$ is positive. So one can form a Hilbert space Y'_α by factoring out from Z_α the null space of $(\ ,\ )'_\alpha$ and completing. Let $\kappa'_\alpha: Z_\alpha \to Y'_\alpha$ be the quotient map.

We claim that Y_α and Y'_α are canonically isomorphic. Indeed: From the continuity of S on $\mathcal{B}_H$ it is easy to see that $\kappa'_\alpha(b \otimes \xi)$ is continuous in b on B_α. So if $f \in \mathcal{L}(\mathcal{B}_\alpha)$ and $\xi \in X$, the right side of the definition

$$F_\alpha(f \otimes \xi) = \int_H \kappa'_\alpha(f(xh) \otimes \xi)dvh \qquad (8)$$

(where $x \in \alpha$) exists as a Y'_α-valued integral. For $f, g \in \mathcal{L}(\mathcal{B}_\alpha)$ and $\xi, \eta \in X$, it follows from (1), (7), and (8) that

$$(F_\alpha(f \otimes \xi), F_\alpha(g \otimes \eta))'_\alpha = (\kappa_\alpha(f \otimes \xi), \kappa_\alpha(g \otimes \eta))_\alpha.$$

So there is a linear isometry $\tilde{F}_\alpha: Y_\alpha \to Y'_\alpha$ satisfying

$$\tilde{F}_\alpha(\kappa_\alpha(f \otimes \xi)) = F_\alpha(f \otimes \xi) = \int_H \kappa'_\alpha(f(xh) \otimes \xi)dvh. \qquad (9)$$

We claim that $\tilde{F}_\alpha$ is *onto* Y'_α. To see this it is enough to show that $\kappa'_\alpha(b \otimes \xi)$ belongs to the closure of range($\tilde{F}_\alpha$) whenever $x \in \alpha$, $b \in B_x$, and $\xi \in X$. This is proved by a standard argument based on (9), using cross-sections f that "peak" around x at the value b.

The isometry $\tilde{F}_\alpha$ is clearly independent of ρ.

In the future we shall identify Y'_α and Y_α by means of the isometry $\tilde{F}_\alpha$, writing $\kappa_\alpha(b \otimes \xi)$ instead of $\tilde{F}_\alpha^{-1}(\kappa'_\alpha(b \otimes \xi))$ ($b \in B_\alpha$; $\xi \in X$). Thus by (7) we have

$$(\kappa_\alpha(b \otimes \xi), \kappa_\alpha(c \otimes \eta))_\alpha = (\rho(x)\rho(y))^{-1/2}(S_{c^*b}\xi, \eta) \qquad (10)$$

$(x, y \in \alpha; b \in B_x; c \in B_y; \xi, \eta \in X)$.

Remark. If S is non-zero and the bundle $\mathscr{B}$ is *saturated*, then $Y_\alpha \neq \{0\}$ for each α in G/H. Indeed: if $x \in \alpha$, the linear span of $\{c^*b : c, b \in B_x\}$ is dense in B_e (by saturation); hence we can choose $c, b \in B_x$ and $\xi, \eta \in X$ so that $(S_{c^*b}\xi, \eta) \neq 0$. Thus the inner product (10) in Y_α is not identically 0.

Notice the important identity

$$\kappa_\alpha(bd \otimes \xi) = \Delta(h)^{1/2}\delta(h)^{-1/2}\kappa_\alpha(b \otimes S_d\xi) \tag{11}$$

($\alpha \in G/H$; $b \in B_\alpha$; $h \in H$; $d \in B_h$; $\xi \in X$). Indeed, we need only to show that both sides of (11) have the same inner product with $\kappa_\alpha(c \otimes \eta)$ ($c \in B_\alpha$; $\eta \in X$); but this follows from (10) and the rho-function identity III.13.2 (1).

Remark. In 9.20 we shall give a simple description of the topology of Y in terms of the spaces Y'_α.

9.6. In the case of the coset H, composing the map F of 9.4 with the identification of Y_H and Y'_H in 9.5, we obtain

$$F(\kappa_H(b \otimes \xi)) = \Delta(h)^{1/2}\delta(h)^{-1/2}S_b\xi \qquad (h \in H; b \in B_h; \xi \in X). \tag{12}$$

The $\mathscr{L}_2$ Cross-Sectional Space of the Induced Hilbert Bundle

9.7. Having constructed $\mathscr{Y}$, one can form the cross-sectional Hilbert space $\mathscr{L}_2(\rho^\#; \mathscr{Y})$ of $\mathscr{Y}$ (see II.15.12).

It is worth noticing that the elements of $\mathscr{L}_2(\rho^\#; \mathscr{Y})$ and their norms are quite independent of ρ. Indeed, let ρ' be another everywhere positive continuous rho-function, related to ρ as in 9.3 by the function σ on G/H. Then $d\rho'^\#\alpha = \sigma(\alpha)d\rho^\#\alpha$ (see III.14.6). So in view of (4) the change from $\rho^\#$ to $\rho'^\#$ exactly compensates the difference in the inner products of the fibers of $\mathscr{Y}^{\rho'}$ and $\mathscr{Y}^\rho$. The details of the argument are left to the reader.

9.8. Remembering the original definition 8.6 of the $\mathscr{B}$-positivity of S (in terms of the conditional expectation p), let us denote by V the operator inner product on $\mathscr{L}(\mathscr{B})$ constructed from S and p:

$$V_{f,g} = S_{p(g^* * f)} \qquad (f, g \in \mathscr{L}(\mathscr{B})). \tag{13}$$

As in 1.8, let $\mathscr{X}(V)$ be the Hilbert space deduced from V. We are going to show that $\mathscr{X}(V)$ and $\mathscr{L}_2(\rho^\#; \mathscr{Y})$ are essentially the same Hilbert space. Thus the *-representation of $\mathscr{B}$ induced from S, as tentatively defined in 8.7, will (if it exists) act on $\mathscr{L}_2(\rho^\#; \mathscr{Y})$.

To see the identity of $\mathscr{X}(V)$ and $\mathscr{L}_2(\rho^\#; \mathscr{Y})$, recall from the definition 9.2 of the topology of $\mathscr{Y}$ that, for each f in $\mathscr{L}(\mathscr{B})$ and each ξ in X, the cross-section

(3) of $\mathscr{Y}$ is continuous and has compact support; call this cross-section $\kappa(f \otimes \xi)$. Since $\kappa(f \otimes \xi)$ is linear in f and in ξ, we have defined a linear map $\kappa: \mathscr{L}(\mathscr{B}) \otimes X \to \mathscr{L}(\mathscr{Y}) \subset \mathscr{L}_2(\rho^{\#}; \mathscr{Y})$. Let $(\ ,\ )_0$ and $(\ ,\ )_2$ be the inner products of $\mathscr{X}(V)$ and $\mathscr{L}_2(\rho^{\#}; \mathscr{Y})$ respectively; and let $f \widetilde{\otimes} \xi$ be the image of $f \otimes \xi$ in $\mathscr{X}(V)$. Then, for f, g in $\mathscr{L}(\mathscr{B})$ and ξ, η in X,

$$(f \widetilde{\otimes} \xi, g \widetilde{\otimes} \eta)_0 = (S_{p(g^* * f)} \xi, \eta)$$

$$= \int_{G/H} d\rho^{\#}(xH) \int_H \int_H (\rho(xh)\rho(xk))^{-1/2} (S_{g(xh)^* f(xk)} \xi, \eta) dvh\, dvk$$

(by 8.9(13))

$$= \int_{G/H} ((f|\alpha) \otimes \xi, (g|\alpha) \otimes \eta)_\alpha\, d\rho^{\#}\alpha \qquad \text{(by (1))}$$

$$= (\kappa(f \otimes \xi), \kappa(g \otimes \eta))_2. \tag{14}$$

This equality says that the equation

$$E(f \widetilde{\otimes} \xi) = \kappa(f \otimes \xi) \qquad (f \in \mathscr{L}(\mathscr{B}); \xi \in X) \tag{15}$$

defines a linear isometry E of $\mathscr{X}(V)$ into $\mathscr{L}_2(\rho^{\#}; \mathscr{Y})$.

Proposition. *E is onto $\mathscr{L}_2(\rho^{\#}; \mathscr{Y})$.*

Proof. It must be shown that the cross-sections $\kappa(f \otimes \xi)$ span a dense subspace Γ of $\mathscr{L}_2(\rho^{\#}; \mathscr{Y})$.

The set of all $\kappa(f \otimes \xi)$ $(f \in \mathscr{L}(\mathscr{B}); \xi \in X)$ is clearly closed under multiplication by continuous complex functions on G/H. Also, by II.14.8 for each fixed α in G/H the linear span of $\{(\kappa(f \otimes \xi))(\alpha): f \in \mathscr{L}(\mathscr{B}), \xi \in X\}$ coincides with the linear span of $\{\kappa_\alpha(g \otimes \xi): g \in \mathscr{L}(\mathscr{B}_\alpha), \xi \in X\}$, and so is dense in Y_α. We have thus verified that Γ satisfies the hypotheses of II.15.10. So by the latter proposition Γ is dense in $\mathscr{L}_2(\rho^{\#}; \mathscr{Y})$. ∎

Thus E maps $\mathscr{X}(V)$ linearly and isometrically onto $\mathscr{L}_2(\rho^{\#}; \mathscr{Y})$. If we wish, E can be considered as identifying the two spaces $\mathscr{X}(V)$ and $\mathscr{L}_2(\rho^{\#}; \mathscr{Y})$.

The Action of $\mathscr{B}$ on $\mathscr{Y}$

9.9. In order to show that S is inducible to $\mathscr{L}(\mathscr{B})$ via p (see 8.7), we are going to construct an explicit *-representation T of $\mathscr{B}$ on $\mathscr{L}_2(\rho^{\#}; \mathscr{Y})$. Transferring T to $\mathscr{X}(V)$ by means of E, we will obtain a *-representation $T': b \mapsto E^{-1}T_b E$ of $\mathscr{B}$ whose integrated form will turn out to be just the result of inducing S via p. The *-representation T will be derived from an action τ of $\mathscr{B}$ on the Hilbert bundle $\mathscr{Y}$, which we next proceed to construct.

9.10.   Lemma.   *Let α be a coset in G/H, and c any element of B. Let $b_1, \ldots, b_n$ be elements of B_α and let $\xi_1, \ldots, \xi_n$ be vectors in X. Then*

$$\sum_{i,j=1}^{n} (S_{b_j^* c^* c b_i} \xi_i, \xi_j) \le \|c\|^2 \sum_{i,j=1}^{n} (S_{b_j^* b_i} \xi_i, \xi_j). \tag{16}$$

Proof.   Write S as a Hilbert direct sum $\sum_i^\oplus S^i$ of cyclic *-representations. Since S is $\mathscr{B}$-positive so is each S^i; and it is easy to see that, if each S^i satisfies (16), so does S. Therefore it is enough to prove (16) for each S^i, that is, to assume from the beginning that S is cyclic, with cyclic vector η.

Also it is enough by Remark 8.10 to assume that G carries the discrete topology. Let $\mathscr{L}_1(\mathscr{B})$ and $\mathscr{L}_1(\mathscr{B}_H)$ be the $\mathscr{L}_1$ cross-sectional algebras of $\mathscr{B}$ and $\mathscr{B}_H$ (with respect to counting measure on G and H).

Since G is discrete, the restriction map $p: \mathscr{L}_1(\mathscr{B}) \to \mathscr{L}_1(\mathscr{B}_H)$ (sending f into $f|H$) is continuous; and so the functional $w: f \mapsto (S_{p(f)} \eta, \eta)$ on $\mathscr{L}_1(\mathscr{B})$ is continuous on $\mathscr{L}_1(\mathscr{B})$. From this and the original definition 8.6 of $\mathscr{B}$-positivity it follows that w is a positive linear functional on the Banach *-algebra $\mathscr{L}_1(\mathscr{B})$. Thus by VI.19.5 w satisfies Condition (R), and hence generates a *-representation U of $\mathscr{L}_1(\mathscr{B})$. Let $f \mapsto \tilde{f}$ be the canonical quotient map of $\mathscr{L}_1(\mathscr{B})$ into $X(U)$, so that

$$(\tilde{f}, \tilde{g})_{X(U)} = w(g^* * f). \tag{17}$$

By the continuity of the operators S_b, it is sufficient to prove (16) when the ξ_i range only over a dense subspace of X. We can therefore assume that $\xi_i = S_{\phi_i} \eta$, where $\phi_i \in \mathscr{L}_1(\mathscr{B}_H)$. Recalling that $\mathscr{B}$ and $\mathscr{L}_1(\mathscr{B}_H)$ are both inside $\mathscr{L}_1(\mathscr{B})$, we then have (omitting the convolution symbol $*$):

$$\sum_{i,j=1}^{n} (S_{b_j^* c^* c b_i} \xi_i, \xi_j) = \sum_{i,j} (S_{\phi_j^* b_j^* c^* c b_i \phi_i} \eta, \eta)$$

$$= \sum_{i,j} w(\phi_j^* b_j^* c^* c b_i \phi_i)$$

$$= \sum_{i,j} ((c b_i \phi_i)^{\tilde{}}, (c b_j \phi_j)^{\tilde{}}) \qquad \text{(by (17))}$$

$$= \sum_{i,j} (U_{c b_i} \tilde{\phi}_i, U_{c b_j} \tilde{\phi}_j)$$

$$= \left\| \sum_i U_{c b_i} \tilde{\phi}_i \right\|^2$$

$$= \left\| U_c \sum_i U_{b_i} \tilde{\phi}_i \right\|^2$$

$$\leq \| U_c \|^2 \left\| \sum_i U_{b_i} \tilde{\phi}_i \right\|^2$$

$$\leq \| c \|^2 \sum_{i,j} (U_{b_i} \tilde{\phi}_i, U_{b_j} \tilde{\phi}_j)$$

$$= \| c \|^2 \sum_{i,j} (S_{b_j^* b_i} \xi_i, \xi_j)$$

(reversing the first four steps of this calculation).   ∎

9.11. Corollary. *For each x in G, c in B_x, and α in G/H, there is a unique continuous linear map $\tau_c^{(\alpha)} : Y_\alpha \to Y_{x\alpha}$ satisfying*

$$\tau_c^{(\alpha)}(\kappa_\alpha(b \otimes \xi)) = \kappa_{x\alpha}(cb \otimes \xi) \tag{18}$$

for all b in B_α and ξ in X.

Proof. The uniqueness follows from the denseness of the linear span of $\{\kappa_\alpha(b \otimes \xi)\}$. To prove the existence of $\tau_c^{(\alpha)}$, we take $\xi_1, \ldots, \xi_n$ in X, $y_1, \ldots, y_n$ in α, and $b_i \in B_{y_i}$, and argue as follows:

$$\left\| \sum_{i=1}^n \kappa_{x\alpha}(cb_i \otimes \xi_i) \right\|^2$$

$$= \sum_{i,j} (\rho(xy_i)\rho(xy_j))^{-1/2} (S_{b_j^* c^* c b_i} \xi_i, \xi_j) \tag{by (7)}$$

$$\leq \| c \|^2 \sum_{i,j} (\rho(xy_i)\rho(xy_j))^{-1/2} (S_{b_j^* b_i} \xi_i, \xi_j) \tag{19}$$

(by (16), replacing ξ_i by $\rho(xy_i)^{-1/2}\xi_i$). Let y be a fixed element of α. By the rho-function identity III.13.2(1)

$$\rho(y_i)^{-1} \rho(xy_i) = \rho(y)^{-1} \rho(xy).$$

So

$$(\rho(xy_i)\rho(xy_j))^{-1/2} = (\rho(y_i)\rho(y_j))^{-1/2} \rho(xy)^{-1} \rho(y).$$

Substituting this into (19) and again using (7) we get

$$\left\| \sum_{i=1}^n \kappa_{x\alpha}(cb_i \otimes \xi_i) \right\|^2 \leq \| c \|^2 \rho(xy)^{-1} \rho(y) \left\| \sum_{i=1}^n \kappa_\alpha(b_i \otimes \xi) \right\|^2. \tag{20}$$

From this follows the existence of the continuous map $\tau_c^{(\alpha)}$ satisfying (18).   ∎

In view of III.13.2(1) the quantity $\rho(xy)^{-1}\rho(y)$ occurring in (20) depends only on x and yH; call it $\sigma(x, yH)$. Thus σ is a positive-valued continuous function on $G \times G/H$; and by (20)

$$\|\tau_c^{(\alpha)}\|^2 \leq \|c\|^2 \sigma(x, \alpha). \tag{21}$$

9.12. $\tau_c^{(\alpha)}$ was defined in 9.11 in terms of the description 9.5 of Y_α. In terms of the original description 9.1, we have

$$\tau_c^{(\alpha)}(\kappa_\alpha(f \otimes \xi)) = \kappa_{x\alpha}(cf \otimes \xi) \tag{22}$$

$(x \in G; c \in B_x; \alpha \in G/H; f \in \mathcal{L}(\mathcal{B}_\alpha); \xi \in X)$. Here cf is the element of $\mathcal{L}(\mathcal{B}_{x\alpha})$ given by $(cf)(y) = cf(x^{-1}y)$ $(y \in x\alpha)$.

To prove (22), recall that, in view of the identification $Y'_\alpha \cong Y_\alpha$ of 9.5, equation (9) becomes

$$\kappa_\alpha(f \otimes \xi) = \int_H \kappa_\alpha(f(yh) \otimes \xi)dvh \tag{23}$$

(where $y \in \alpha$). Applying $\tau_c^{(\alpha)}$ to both sides of this, we have by II.5.7

$$\tau_c^{(\alpha)}(\kappa_\alpha(f \otimes \xi)) = \int \tau_c^{(\alpha)}(\kappa_\alpha(f(yh) \otimes \xi))dvh$$

$$= \int \kappa_{x\alpha}(cf(yh) \otimes \xi)dvh \qquad \text{(by (18))}$$

$$= \kappa_{x\alpha}(cf \otimes \xi) \qquad \text{(by (23))}.$$

So (22) is proved.

9.13.   Definition.   If $c \in B$, let τ_c be the union of the maps $\tau_c^{(\alpha)}$ $(\alpha \in G/H)$. That is, $\tau_c : Y \to Y$ is the map whose restriction to each fiber Y_α is $\tau_c^{(\alpha)}$. Thus, if $c \in B_x$,

$$\tau_c(Y_\alpha) \subset Y_{x\alpha}. \tag{24}$$

9.14.   Notice that the definition of τ is independent of ρ.

We know that τ_c is linear on each fiber of $\mathcal{Y}$, and is linear in c on each fiber of $\mathcal{B}$. Evidently

$$\tau_b \tau_c = \tau_{bc} \qquad (b, c \in B). \tag{25}$$

As regards involution, we claim that for each $x \in G$, $b \in B_x$, $\alpha \in G/H$, $\zeta \in Y_\alpha$, and $\zeta' \in Y_{x\alpha}$,

$$(\tau_b \zeta, \zeta')_{x\alpha} = \sigma(x, \alpha)(\zeta, \tau_{b^*}\zeta')_\alpha. \tag{26}$$

Indeed: By continuity and linearity it is enough to prove (26) assuming that $\zeta = \kappa_\alpha(c \otimes \xi)$, $\zeta' = \kappa_{x\alpha}(d \otimes \eta)$ $(\xi, \eta \in X; c \in B_\alpha; d \in B_{x\alpha})$. Now $\pi(d) = x\pi(c)h$ for some h in H; and so

$$\rho(x\pi(c))\rho(\pi(d))$$

$$= \rho(\pi(c))\rho(x^{-1}\pi(d))[(\rho(\pi(c))^{-1}\rho(x\pi(c))][(\rho(\pi(c)h))^{-1}\rho(x\pi(c)h)]$$

$$= \rho(\pi(c))\rho(x^{-1}\pi(d))\sigma(x, \alpha)^{-2}.$$

Therefore by (7)

$$(\tau_b\zeta, \zeta')_{x\alpha} = (\rho(x\pi(c))\rho(\pi(d)))^{-1/2}(S_{d*bc}\xi, \eta)$$

$$= \sigma(x, \alpha)(\rho(\pi(c))\rho(x^{-1}\pi(d)))^{-1/2}(S_{d*bc}\xi, \eta)$$

$$= \sigma(x, \alpha)(\zeta, \tau_{b*}\zeta')_\alpha;$$

and (26) is proved.

9.15. Recall from 9.4 the definition of the bijection $F: Y_H \to X$.

Proposition. *If $\xi \in X$, $\alpha \in G/H$, and $b \in B_\alpha$,*

$$\tau_b(F^{-1}\xi) = \kappa_\alpha(b \otimes \xi). \tag{27}$$

Proof. Suppose that $h \in H$, $c \in B_h$, $d \in B_\alpha$, and $\eta \in X$; then by (7) and III.13.2.(1)

$$(bc \otimes \xi, d \otimes \eta)_\alpha = \Delta(h)^{1/2}\delta(h)^{-1/2}(b \otimes S_c\xi, d \otimes \eta)_\alpha.$$

By the arbitrariness of d and η this implies

$$\kappa_\alpha(bc \otimes \xi) = \Delta(h)^{1/2}\delta(h)^{-1/2}\kappa_\alpha(b \otimes S_c\xi). \tag{28}$$

Now, if b is fixed, both sides of (27) are continuous in ξ. So by 9.4 it is enough to prove (27) assuming that $\xi = F(\kappa_H(c \otimes \eta))$, where $\eta \in X$, $h \in H$, $c \in B_H$. But then

$$\kappa_\alpha(b \otimes \xi) = \Delta(h)^{1/2}\delta(h)^{-1/2}\kappa_\alpha(b \otimes S_c\eta) \qquad \text{(by (12))}$$

$$= \kappa_\alpha(bc \otimes \eta) \qquad \text{(by (28))}$$

$$= \tau_b\kappa_H(c \otimes \eta) = \tau_b(F^{-1}\xi);$$

and (27) is proved. ∎

Remark. If in the above proposition $b \in B_h$ $(h \in H)$, then by (27) and (12)

$$\tau_b(F^{-1}\xi) = \Delta(h)^{1/2}\delta(h)^{-1/2}F^{-1}(S_b\xi). \tag{29}$$

9.16.   Corollary.   *For each α in G/H, the linear span of $\{\tau_b\zeta: b\in B_\alpha, \zeta\in Y_H\}$ is dense in Y_α.*

Proof.   This follows from (27) and the definition of Y_α.   ∎

9.17.   In this connection we notice:

Proposition.   *If $\mathscr{B}$ is saturated, and if $\alpha\in G/H$ and $x\in G$, then the linear span L of $\{\tau_b\zeta: b\in B_x, \zeta\in Y_\alpha\}$ is dense in $Y_{x\alpha}$.*

Proof.   We have observed already that $c\mapsto\kappa_{x\alpha}(c\otimes\xi)$ is continuous on $B_{x\alpha}$ for each ξ (see 9.5). So it is enough to fix ξ and show that L contains $\{\kappa_{x\alpha}(c\otimes\xi): c\in M\}$ for some dense subset M of $B_{x\alpha}$. But this follows from (18) and the saturation of $\mathscr{B}$, when we take $M=\{bd: b\in B_x, d\in B_\alpha\}$.   ∎

9.18.   We come now to the very important continuity property of τ:

Proposition.   *The map $\langle b, \zeta\rangle\mapsto\tau_b\zeta$ is continuous on $B\times Y$ to Y.*

Proof.   We shall first show that, if $f\in\mathscr{L}(\mathscr{B})$ and $c_i\to c$ in B, then

$$c_i f \to cf \qquad \text{uniformly on } G. \tag{30}$$

(Here, as usual, $(cf)(y) = cf(\pi(c)^{-1}y)$, $(c_i f)(y) = c_i f(\pi(c_i)^{-1}y)$.)

Indeed: $\|(cf)(y) - (df)(y)\| = \|cf(\pi(c)^{-1}y) - df(\pi(d)^{-1}y)\|$ is continuous in c, d and y, and vanishes when $c=d$. So given $\varepsilon>0$ and $c\in B$, there is a neighborhood U of c such that $\|(cf)(y) - (df)(y)\| < \varepsilon$ for all d in U and all y in any preassigned compact set $K\subset G$. Taking K to be so large that cf and $c_i f$ vanish outside K for large enough i, we obtain (30).

The next step is to show that, for fixed ξ in X and f in $\mathscr{L}(\mathscr{B})$, the function

$$\langle c, \beta\rangle\mapsto\kappa_\beta((cf\,|\,\beta)\otimes\xi) \tag{31}$$

is continuous on $B\times G/H$ to Y. This follows from two facts: First, by the definition of the topology of Y (9.2), (31) is continuous in β when c is fixed. Secondly, by (30), as $c_i\to c$ in B, $\kappa_\beta((c_i f\,|\,\beta)\otimes\xi)\to\kappa_\beta((cf\,|\,\beta)\otimes\xi)$ uniformly for β in G/H.

We are now ready to prove the continuity of τ. Let $b_r\to b$ in B and $\zeta_r\to\zeta$ in Y. It must be shown that

$$\tau_{b_r}\zeta_r \to \tau_b\zeta. \tag{32}$$

Put $x = \pi(b)$, $x_r = \pi(b_r)$; and suppose $\zeta\in Y_\alpha$, $\zeta_r\in Y_{\alpha_r}$.

Let $\varepsilon > 0$. By assertion (II) in the proof of 9.2, there are elements $f_1, \ldots, f_n$ in $\mathscr{L}(\mathscr{B})$ and $\xi_1, \ldots, \xi_n$ in X such that

$$\left\| \zeta - \sum_{i=1}^{n} \kappa_\alpha((f_i|\alpha) \otimes \xi_i) \right\|_\alpha < \varepsilon. \tag{33}$$

By continuity this implies that

$$\left\| \zeta_r - \sum_{i=1}^{n} \kappa_{\alpha_r}((f_i|\alpha_r) \otimes \xi_i) \right\|_{\alpha_r} < \varepsilon \tag{34}$$

for all large enough r. If k is a positive constant majorizing $\|b\|\sigma(x, \alpha)^{1/2}$ and all the $\|b_r\|\sigma(x_r, \alpha_r)^{1/2}$, (33) and (34) imply by (21) and (22) that

$$\left\| \tau_b \zeta - \sum_{i=1}^{n} \kappa_{x\alpha}((bf_i|x\alpha) \otimes \xi_i) \right\|_{x\alpha} < k\varepsilon, \tag{35}$$

$$\left\| \tau_{b_r} \zeta_r - \sum_{i=1}^{n} \kappa_{x_r\alpha_r}((b_r f_i|x_r\alpha_r) \otimes \xi_i) \right\|_{x_r\alpha_r} < k\varepsilon. \tag{36}$$

Now (32) follows from (35), (36), and the continuity of (31). ∎

9.19. Corollary. *The map $\langle b, \xi \rangle \mapsto \kappa_{\pi(b)H}(b \otimes \xi)$ is continuous on $B \times X$ to Y.*

Proof. Combine 9.15 and 9.18. ∎

9.20. From 9.19 we deduce the following simple description of the topology of Y in terms of the spaces Y'_α of 9.5.

Proposition. *The topology of Y defined in 9.2 is the unique topology which (i) makes $\mathscr{Y}$ a Banach bundle, and (ii) for each fixed ξ in X makes the map $b \mapsto \kappa_{\pi(b)H}(b \otimes \xi)$ (of B into Y) continuous.*

Proof. By 9.19 the topology of Y defined in 9.2 satisfies (i) and (ii). Hence it is enough to show its uniqueness, that is, to take two topologies $\mathscr{T}_1$ and $\mathscr{T}_2$ for Y satisfying (i) and (ii) and to prove that they are the same. By symmetry it is obviously sufficient to show that the identity map is continuous from Y, $\mathscr{T}_1$ to Y, $\mathscr{T}_2$.

So assume that $\zeta_r \to \zeta$ in $Y, \mathcal{T}_1$; and let $\zeta_r \in Y_{\alpha_r}$, $\zeta \in Y_\alpha$; thus $\alpha_r \to \alpha$ in G/H. Fix any $\varepsilon > 0$; and choose a positive integer n, elements $b_1, \ldots, b_n$ in B for which $\pi(b_i) = x_i \in \alpha$ $(i = 1, \ldots, n)$, and $\xi_1, \ldots, \xi_n \in X$, so that

$$\left\| \zeta - \sum_{i=1}^{n} \kappa_\alpha(b_i \otimes \xi_i) \right\| < \varepsilon \tag{37}$$

(see 9.5). For each $i = 1, \ldots, n$ take a continuous cross-section ϕ_i of $\mathcal{B}$ such that $\phi_i(x_i) = b_i$. Since $G \to G/H$ is open and $\alpha_r \to \alpha$ in G/H, we can pass to a subnet (without change of notation) and choose for each $i = 1, \ldots, n$ a net $\{y_i^r\}_r$ in G such that

$$y_i^r \underset{r}{\to} x_i \quad \text{in} \quad G, \qquad \pi(y_i^r) = \alpha_r.$$

Then by the continuity of ϕ_i,

$$\phi_i(y_i^r) \underset{r}{\to} \phi_i(x_i) = b_i \quad \text{in} \quad B.$$

So by hypothesis (ii) and the continuity of addition in a Banach bundle,

$$\sum_{i=1}^{n} \kappa_{\alpha_r}(\phi_i(y_i^r) \otimes \xi_i) \underset{r}{\to} \sum_{i=1}^{n} \kappa_\alpha(b_i \otimes \xi_i) \tag{38}$$

in both $Y, \mathcal{T}_1$ and $Y, \mathcal{T}_2$. Now the fact that $\zeta_r \to \zeta$ in $Y, \mathcal{T}_1$, together with (37) and (38) (for $\mathcal{T}_1$), implies (by the continuity of norm and subtraction in a Banach bundle) that

$$\left\| \zeta_r - \sum_{i=1}^{n} \kappa_{\alpha_r}(\phi_i(y_i^r) \otimes \xi_i) \right\| < \varepsilon \qquad \text{for all large } r. \tag{39}$$

But now (37), (38) (for $\mathcal{T}_2$) and (39) imply (by the arbitrariness of ε) that $\zeta_r \to \zeta$ in $Y, \mathcal{T}_2$.

Since $\{\zeta_r\}$ was an arbitrary convergent net in $\mathcal{T}_1$, this shows that the identity map from $Y, \mathcal{T}_1$ to $Y, \mathcal{T}_2$ is continuous. ∎

The Definition of the Induced Representation

9.21. We shall now use the action τ of $\mathcal{B}$ on $\mathcal{Y}$ to obtain a *-representation of $\mathcal{B}$ acting on $\mathcal{L}_2(\rho^*; \mathcal{Y})$.

Fix $b \in B_x$ $(x \in G)$. If f is any cross-section (continuous or not) of $\mathcal{Y}$, then

$$T_b' f : \alpha \mapsto \tau_b f(x^{-1}\alpha) \qquad (\alpha \in G/H) \tag{40}$$

is a cross-section of $\mathcal{Y}$ in virtue of (24).

Lemma. *If the cross-section f of $\mathcal{Y}$ is continuous, then so is $T'_b f$. If f is locally $\rho^\#$-measurable, then so is $T'_b f$. If $f \in \mathcal{L}_2 (\rho^\# ; \mathcal{Y})$, then so is $T'_b f$, and*

$$\|T'_b f\|_2 \le \|b\| \, \|f\|_2 \tag{41}$$

(where $\| \ \|_2$ is the norm in $\mathcal{L}_2(\rho \# ; \mathcal{Y})$).

Proof. The first statement follows from 9.18.

To prove the second we shall use the criterion II.15.4 for local $\rho^\#$-measurability. Let K be a compact subset of G/H, and assume that f is locally $\rho^\#$-measurable. Thus by II.15.4 there is a sequence $\{g_n\}$ of continuous cross-sections of $\mathcal{Y}$ such that $g_n(\alpha) \to f(\alpha)$ $\rho^\#$-almost everywhere on $x^{-1}K$ (where $x = \pi(b)$). Recalling from III.14.7 that $\rho^\#$ is quasi-invariant, we see from (40) that $(T'_b g_n)(\alpha) \to (T'_b f)(\alpha)$ $\rho^\#$-almost everywhere on K. By the first statement of the lemma the $T'_b g_n$ are continuous. So by II.15.4 $T'_b f$ is locally $\rho^\#$-measurable.

To prove the last statement it is enough, in view of the preceding paragraph, to verify the norm inequality (41). To begin with we recall from III.14.7(3) that

$$d\rho^\#(x^{-1}\alpha) = \sigma(x, x^{-1}\alpha)d\rho^\# \alpha. \tag{42}$$

Therefore, if $b \in B_x$ and $f \in \mathcal{L}_2(\rho^\# ; \mathcal{Y})$,

$$\int_{G/H} \|(T'_b f)(\alpha)\|_\alpha^2 \, d\rho^\# \alpha = \int \|\tau_b f(x^{-1}\alpha)\|_\alpha^2 \, d\rho^\# \alpha$$

$$\le \|b\|^2 \int \sigma(x, x^{-1}\alpha)\|f(x^{-1}\alpha)\|_{x^{-1}\alpha}^2 \, d\rho^\# \alpha \qquad \text{(by (21))}$$

$$= \|b\|^2 \int \|f(\alpha)\|_\alpha^2 \, d\rho^\# \alpha \qquad \text{(by (42))}.$$

But this is (41). ∎

9.22. Definition. If $b \in B$, we shall define T_b as the restriction to $\mathcal{L}_2 (\rho^\# ; \mathcal{Y})$ of the map T'_b of (40). By Lemma 9.21, T_b is a bounded linear operator on $\mathcal{L}_2(\rho^\# ; \mathcal{Y})$, with

$$\|T_b\| \le \|b\|. \tag{43}$$

9.23. Proposition. *The $T: b \mapsto T_b$ of 9.22 is a non-degenerate *-representation of $\mathcal{B}$ on $\mathcal{L}_2 (\rho^\# ; \mathcal{Y})$.*

Proof. T is obviously linear on each fiber. That $T_b T_c = T_{bc}$ follows immediately from (25). If $b \in B_x$ and $f, g \in \mathcal{L}_2(\rho^*; \mathcal{Y})$,

$$(T_{b*}f, g)_2 = \int (\tau_{b*}f(x\alpha), g(\alpha))_\alpha \, d\rho^* \alpha$$

$$= \int \sigma(x^{-1}, x\alpha)(f(x\alpha), \tau_b g(\alpha))_{x\alpha} \, d\rho^* \alpha \qquad \text{(by (26))}$$

$$= \int (f(x\alpha), \tau_b g(\alpha))_{x\alpha} \, d\rho^* (x\alpha) \qquad \text{(by (42))}$$

$$= (f, T_b g)_2.$$

Therefore $(T_b)^* = T_{b*}$.

Thus T will be a *-representation of $\mathcal{B}$ if we can show that $b \mapsto T_b f$ is continuous on B for each f in $\mathcal{L}_2(\rho^*; \mathcal{Y})$. By (43) it is enough to prove this when f belongs to the dense subspace $\mathcal{L}(\mathcal{Y})$ of $\mathcal{L}_2(\rho^*; \mathcal{Y})$ (see II.15.9). But if $b_i \to b$ in B and $f \in \mathcal{L}(\mathcal{Y})$, it follows from 9.18 (by the same argument that was used for (30)) that $T_{b_i} f \to T_b f$ uniformly on G/H with uniform compact support, and hence that $T_{b_i} f \to T_b f$ in $\mathcal{L}_2(\rho^*; \mathcal{Y})$. Therefore T is a *-representation.

To see that T is non-degenerate, let Γ be the linear span (in $\mathcal{L}(\mathcal{Y})$) of $\{T_b f : f \in \mathcal{L}(\mathcal{Y}), b \in B\}$. If $\phi \in \mathscr{C}(G/H)$, $b \in B_x$, and $f \in \mathcal{L}(\mathcal{Y})$, we verify that $\phi(T_b f) = T_b g$, where $g(\alpha) = \phi(x\alpha)f(\alpha)$. Therefore Γ is closed under multiplication by continuous complex functions on G/H. Also, it follows from 9.16 that $\{g(\alpha) : g \in \Gamma\}$ is dense in Y_α for each fixed α. Consequently Γ is dense in $\mathcal{L}_2(\rho^*; \mathcal{Y})$ by II.15.10; and T is non-degenerate.   ∎

9.24.   Definition.   The T of 9.23 is called *the *-representation of $\mathcal{B}$ concretely induced by* the $\mathcal{B}$-positive non-degenerate *-representation S of $\mathcal{B}_H$, and will be denoted by $\mathrm{Ind}_{\mathcal{B}_H \uparrow \mathcal{B}}(S)$.

9.25.   We have used the term "concretely" in 9.24 to distinguish T from the formally different Rieffel abstract inducing process 4.9. To fulfill the promise made in 9.9, however, we must now show that the integrated form of the *-representation $b \mapsto E^{-1} T_b E$ of $\mathcal{B}$ on $\mathscr{X}(V)$ (see 9.9) is just the result of abstractly inducing S to $\mathcal{L}(\mathcal{B})$ via p. That is, keeping the notation of 9.8, we must prove:

Proposition.   *Let T^0 denote the integrated form of the T of 9.23. Then, if $f, g \in \mathcal{L}(\mathcal{B})$ and $\xi \in X$, we have*

$$E((f * g) \widetilde{\otimes} \xi) = T^0_f(E(g \widetilde{\otimes} \xi)). \qquad (44)$$

Proof. As usual let $(bg)(y) = bg(x^{-1}y)$ $(x, y \in G; b \in B_x)$. From (22) and the definition of $\kappa(g \otimes \xi)$ in 9.8 we check that

$$T_b(\kappa(g \otimes \xi)) = \kappa(bg \otimes \xi). \tag{45}$$

Now by VIII.12.5

$$f * g = \int_G (f(x)g)d\lambda x, \tag{46}$$

the right side being an $\mathscr{L}(\mathscr{B})$-valued integral with respect to the inductive limit topology. Also, for fixed ξ, the linear map

$$\phi \mapsto \kappa(\phi \otimes \xi) \qquad (\phi \in \mathscr{L}(\mathscr{B})) \tag{47}$$

is continuous from $\mathscr{L}(\mathscr{B})$ to $\mathscr{L}_2(\rho^*; \mathscr{Y})$ in the inductive limit topology of the former. So, applying (47) to both sides of (46) (and invoking II.6.3) we get

$$\kappa((f * g) \otimes \xi) = \int_G \kappa(f(x)g \otimes \xi)d\lambda x$$

$$= \int T_{f(x)}\kappa(g \otimes \xi)d\lambda x \qquad \text{(by (45))}$$

$$= T_f^0 \kappa(g \otimes \xi).$$

This combined with (15) gives (44). ∎

9.26. In view of 9.25 we have now finally proved the following result.

Theorem. *Let S be a $\mathscr{B}$-positive non-degenerate *-representation of $\mathscr{B}_H$. Then:*

(I) *The integrated form of S is (abstractly) inducible to $\mathscr{L}(\mathscr{B})$ via the conditional expectation p of 8.4.*

(II) *$\text{Ind}^p_{\mathscr{L}(\mathscr{B}_H)\uparrow\mathscr{L}(B)}(S)$ is the integrated form of a (unique) non-degenerate *-representation $\tilde{T}$ of $\mathscr{B}$.*

(III) *$\tilde{T}$ is unitarily equivalent with the concretely induced *-representation $T = \text{Ind}_{\mathscr{B}_H\uparrow\mathscr{B}}(S)$ of 9.24.*

Remark. The hypothesis that S is non-degenerate is of no importance. If S is degenerate, with non-degenerate part S^0, both the abstract and the concrete inducing processes give the same result when applied to S^0 as when applied to S.

9.27. In view of 9.26(III) *we shall usually omit the word "concretely" in Definition 9.24. The latter will be called the bundle formulation of the inducing construction, as contrasted with the (abstract) Rieffel formulation of 8.7.*

9.28. If $H = G$ we have the following simple result:

Proposition. *If S is a non-degenerate *-representation of $\mathscr{B}$, then*

$$\operatorname*{Ind}_{\mathscr{B} \uparrow \mathscr{B}}(S) \cong S.$$

Proof. In this case $\mathscr{L}_2(\rho^\#; \mathscr{Y}) = Y_G = Y_H$ and $T_b = \tau_b$; and the result follows from (29) (since $\Delta(h)^{1/2} \delta(h)^{-1/2}$ is now 1). ∎

9.29. Proposition. *If the bundle space B of $\mathscr{B}$ is second-countable and X $(= X(S))$ is separable, then $\mathscr{L}_2(\rho^\#; \mathscr{Y})$ $(= X(\operatorname{Ind}_{\mathscr{B}_H \uparrow \mathscr{B}}(S)))$ is separable.*

Proof. By II.14.8 there is a countable subset Γ of $\mathscr{L}(\mathscr{B})$ which is dense in $\mathscr{L}(\mathscr{B})$ in the inductive limit topology. Let X_c be a countable dense subset of X. Keeping the notation of 9.8, let $\tilde{\Gamma}$ be the countable subfamily of $\mathscr{X}(V)$ consisting of all finite sums of elements of the form $f \widetilde{\otimes} \xi$, where $f \in \Gamma$ and $\xi \in X_c$. Since by Remark 4 of 8.6 $f \widetilde{\otimes} \xi$ is separately continuous in f (with the inductive limit topology) and ξ, the $\mathscr{X}(V)$-closure of $\tilde{\Gamma}$ contains all $g \widetilde{\otimes} \eta$ $(g \in \mathscr{L}(\mathscr{B}); \eta \in X)$ and hence is equal to $\mathscr{X}(V)$. So $\mathscr{X}(V)$ is separable. Now apply 9.8. ∎

Extension of Induced Representations to the Multiplier Bundle

9.30. By VIII.15.3 the induced representation T of $\mathscr{B}$ can be extended to a *-representation T'' of the multiplier bundle of $\mathscr{B}$. What does T'' look like? The answer, as we shall see below, is a very natural one.

Let us return to Lemma 9.10 and Corollary 9.11, and suppose that the c of Lemma 9.10, instead of being an element of B, is a multiplier of $\mathscr{B}$ of some order x.

The reader will verify that all steps of the proofs of Lemma 9.10 and Corollary 9.11 are still valid. (Since c can be regarded as a multiplier of the "discrete" cross-sectional algebra $\mathscr{L}_1(\mathscr{B})$, the existence of the operator U_c of the proof of Lemma 9.10 follows from VIII.1.15 applied to the non-degenerate part of U). Hence by Corollary 9.11, for each $\alpha \in G/H$ there is a unique continuous linear map $\tau_c^{(\alpha)}: Y_\alpha \to Y_{x\alpha}$ satisfying:

$$\tau_c^{(\alpha)} \kappa_\alpha(b \otimes \xi) = \kappa_{x\alpha}(cb \otimes \xi) \qquad (b \in B_\alpha, \xi \in X), \qquad (48)$$

with

$$\|\tau_c^{(\alpha)}\|^2 \leq \|c\|^2 \sigma(x, \alpha) \tag{49}$$

(see (21)). As in 9.13 we define $\tau_c: Y \to Y$ to be the union of the maps $\tau_c^{(\alpha)}$ ($\alpha \in G/H$). The proof of 9.18 can be repeated (or modified) to show that $\tau_c: Y \to Y$ is continuous on Y. Next we define a linear operator T_c'' on $\mathscr{L}_2(\rho^{\#}; \mathscr{Y})$ as in (40):

$$(T_c''f)(\alpha) = \tau_c f(x^{-1}\alpha) \qquad (f \in \mathscr{L}_2(\rho^{\#}; \mathscr{Y}); \alpha \in G/H). \tag{50}$$

As in 9.21 (see (49)) we prove that T_c'' is a bounded linear operator on $\mathscr{L}_2(\rho^{\#}; \mathscr{Y})$ with norm $\leq \|c\|$. The argument of 9.23 now shows that $c \mapsto T_c''$ is a *-representation of the multiplier bundle of $\mathscr{B}$. Since it clearly extends T, we have:

Proposition. *The *-representation T'' of the multiplier bundle of $\mathscr{B}$ which extends the induced representation $T = \mathrm{Ind}_{\mathscr{B}_H \uparrow \mathscr{B}}(S)$ is given by (50) and (48).*

10. Mackey's and Blattner's Formulations of Induced Representations of Groups

10.1. Suppose in §9 that $\mathscr{B}$ is the group bundle of G. Then $\mathscr{B}_H$ is the group bundle of H; and the non-degenerate *-representation S of $\mathscr{B}_H$ with which we started §§8, 9 is just a unitary representation of H. By 8.14 S is automatically $\mathscr{B}$-positive. So we can form $T = \mathrm{Ind}_{\mathscr{B}_H \uparrow \mathscr{B}}(S)$; and by 9.23 this is a non-degenerate *-representation of $\mathscr{B}$, that is, a unitary representation of G. Thus, in the group case (see 8.3), the inducing operation carries an arbitrary unitary representation S of the closed subgroup H of G into a unitary representation T of the whole group G. *In this case we shall write* $\mathrm{Ind}_{H \uparrow G}(S)$ *for* T, instead of the more cumbersome $\mathrm{Ind}_{\mathscr{B}_H \uparrow \mathscr{B}}(S)$.

Since the group case is by far the most important special case of the inducing operation, we shall now show in detail how the general formulation of the inducing operation is given in §9 reduces in the group case to the more classical formulations of Mackey and Blattner. (For finite groups this was done in 4.18.)

10.2. We assume now that $\mathscr{B}$ is the group bundle $\mathbb{C} \times G$ of the locally compact group G (see VIII.4.4). Thus an element of B_x is of the form $\langle t, x \rangle$, where $t \in \mathbb{C}$. If T is a non-degenerate *-representation of $\mathscr{B}$, the unitary representation of G with which T is identified is just $x \mapsto T_{\langle 1, x \rangle}$ (see VIII.8.5(3)).

10.3. Fix a unitary representation S of the closed subgroup H of G, acting in the Hilbert space X. We shall continue to use all the notation of §9 without explicit reference.

Our basic tool will be the map $D: G \times X \to Y$ given by

$$D(x, \xi) = \kappa_{xH}(\langle 1, x \rangle \otimes \xi) \qquad (x \in G; \xi \in X). \qquad (1)$$

By 9.14 this can be written in the form

$$D(x, \xi) = \tau_{\langle 1, x \rangle}(F^{-1}\xi). \qquad (2)$$

Denoting by D_x the map $\xi \mapsto D(x, \xi)$, we notice that

$$D_x(X) \subset Y_{xH}. \qquad (3)$$

10.4. We claim that $D_x: X \to Y_{xH}$ is a bijection and that $\rho(x)^{1/2}D_x$ is an isometry. To see this, one checks from 9.5(7) that

$$(D_x(\xi), D_x(\eta))_{xH} = \rho(x)^{-1}(\xi, \eta) \qquad (\xi, \eta \in X; x \in G), \qquad (4)$$

whence $\rho(x)^{1/2}D_x$ is an isometry. By 9.15(29)

$$D_{xh} = \Delta(h)^{1/2}\delta(h)^{-1/2}D_x S_h \qquad (x \in G; h \in H). \qquad (5)$$

By the definition of Y_α (see 9.5), the latter is spanned by $\{D_x(\xi): x \in \alpha, \xi \in X\}$. So (5) implies that, for each fixed x in α, $D_x(X)$ is dense in Y_{xH}. Since $\rho(x)^{1/2}D_x$ is an isometry, $D_x(X)$ is closed and hence must be equal to Y_{xH}. So the claim is proved.

10.5. Let f be any cross-section of $\mathcal{Y}$. Since each $D_x: X \to Y_{xH}$ is a bijection, we can construct from f a function $\phi: G \to X$ as follows:

$$\phi(x) = D_x^{-1}(f(xH)) \qquad (x \in G). \qquad (6)$$

If $x \in G$ and $h \in H$, then by (5)

$$\phi(xh) = D_{xh}^{-1} f(xH) = \delta(h)^{1/2}\Delta(h)^{-1/2}S_{h^{-1}}D_x^{-1} f(xH),$$

or

$$\phi(xh) = \delta(h)^{1/2}\Delta(h)^{-1/2}S_{h^{-1}}(\phi(x)) \qquad (x \in G; h \in H). \qquad (7)$$

Conversely, let $\phi: G \to X$ be any map satisfying (7). Combining (5) and (7) we find that $D_{xh}\phi(xh) = D_x\phi(x)$ for all x in G and $h \in H$. Hence ϕ generates a cross-section f of $\mathcal{Y}$:

$$f(xH) = D_x(\phi(x)) \qquad (x \in G). \qquad (8)$$

Obviously the constructions $f \mapsto \phi$ and $\phi \mapsto f$ described in (6) and (8) are each other's inverses. So (6) and (8) set up a one-to-one correspondence $f \leftrightarrow \phi$ between cross-sections f of $\mathcal{Y}$ and functions $\phi: G \to X$ satisfying (7).

Notice how the action of G behaves under this correspondence. Let $y \in G$; and let T'_y stand for the map $T'_{\langle 1, y \rangle}$ of 9.21(40), sending the cross-section f of $\mathcal{Y}$ into the cross-section $\alpha \mapsto \tau_{\langle 1, y \rangle} f(y^{-1}\alpha)$. If $f \leftrightarrow \phi$, $f' = T'_y f$, and $f' \leftrightarrow \phi'$, then for all x in G

$$D_x \phi'(x) = f'(xH) \qquad\qquad \text{(by (8))}$$

$$= \tau_{\langle 1, y \rangle} f(y^{-1}xH)$$

$$= \tau_{\langle 1, y \rangle} D_{y^{-1}x} \phi(y^{-1}x) \qquad\qquad \text{(by (8))}$$

$$= \tau_{\langle 1, y \rangle} \kappa_{y^{-1}xH}(\langle 1, y^{-1}x \rangle \otimes \phi(y^{-1}x)) \qquad \text{(by (1))}$$

$$= \kappa_{xH}(\langle 1, x \rangle \otimes \phi(y^{-1}x))$$

$$= D_x(\phi(y^{-1}x)) \qquad\qquad \text{(by (1))}.$$

Since D_x is one-to-one, this implies that

$$\phi'(x) = \phi(y^{-1}x) \qquad\qquad (x \in G). \qquad (9)$$

Thus, under the correspondence (6), (8), the action of T'_y goes into simple y-translation of the functions ϕ.

10.6.   Does the correspondence $f \leftrightarrow \phi$ of (6), (8) preserve interesting properties of functions? In particular what is the image under $f \mapsto \phi$ of the important space $\mathcal{L}_2(\rho^*; \mathcal{Y})$?

As a first remark in this direction, we notice:

Proposition.   *If $f \leftrightarrow \phi$ as in (6), (8), then f vanishes locally ρ^*-almost everywhere if and only if ϕ vanishes locally λ-almost everywhere.*

Proof.   This follows from III.14.8.   ■

10.7.   Proposition.   *Let $f \leftrightarrow \phi$ as in (6), (8). Then: (I) f is a continuous cross-section of $\mathcal{Y}$ if and only if ϕ is continuous on G. (II) f is locally ρ^*-measurable if and only if ϕ is locally λ-measurable.*

Proof.   (I) By (2) we can write (6) and (8) in the form

$$\phi(x) = F(\tau_{\langle 1, x^{-1} \rangle} f(xH)), \qquad f(xH) = \tau_{\langle 1, x \rangle}(F^{-1}(\phi(x))).$$

From this and 9.18 it follows that the continuity of either f of ϕ implies that of the other.

(II) Assume that f is locally ρ^*-measurable; and let K be any compact subset of G. Denoting by K' the image of K in G/H, by II.15.4 we can find a sequence $\{f_n\}$ of continuous cross-sections of $\mathscr{Y}$ such that $f_n(\alpha) \to f(\alpha)$ for all α in $K' \setminus N'$, where N' is a $\rho^\#$-null set. Let ϕ_n correspond to f_n as in (6). By (I) the ϕ_n are continuous. Since the D_x^{-1} are continuous, (6) implies that $\phi_n(x) \to \phi(x)$ for all x in $K \setminus N$, where N is the inverse image of N' in G. But by III.14.8 $K \cap N$ is λ-null. Thus, by II.15.4 and the arbitrariness of K, ϕ is locally λ-measurable.

Conversely, we shall assume that ϕ is locally λ-measurable and shall prove that f is locally $\rho^\#$-measurable. This turns out to be more intricate than one might expect. Let Q be any compact subset of G/H. By III.2.5 there is a compact subset C of G such that the image in G/H of the interior of C contains Q. By II.15.4 there is a sequence $\{\psi_n\}$ of continuous functions on G to X, and a λ-null subset N of C such that

$$\psi_n(x) \to \phi(x) \qquad \text{in } X \text{ for all } x \text{ in } C \setminus N. \tag{10}$$

Now for each n put $Z_n = \{D_x(\psi_n(x)): x \in C\}$. By (2) and 9.18 D is continuous on $G \times X$. Hence by the compactness of C each Z_n is a compact subset of Y. By III.14.16 there is a $\rho^\#$-null subset P of G/H such that, if $xH \notin P$, then $\{h \in H: xh \in N\}$ is ν-null. Since each coset in Q has non-void intersection with the interior of C, this implies that

$$\alpha \in Q \setminus P \Rightarrow \alpha \cap (C \setminus N) \neq \emptyset.$$

Thus, if $\alpha \in Q \setminus P$, we can take an element x of $\alpha \cap (C \setminus N)$; and by (10) we then have $D_x(\psi_n(x)) \to D_x(\phi(x)) = f(xH)$. Since $D_x(\psi_n(x)) \in Z_n \cap Y_\alpha$, we have shown that

$$\alpha \in Q \setminus P \Rightarrow f(\alpha) \in \left(\bigcup_n (Z_n \cap Y_\alpha) \right)^-. \tag{11}$$

Furthermore let g be any locally $\rho^\#$-measurable cross-section of $\mathscr{Y}$, corresponding by (6) to $\psi: G \to X$. We have seen above that ψ is locally λ-measurable. Therefore $x \mapsto \|\phi(x) - \psi(x)\|$ is a locally λ-measurable numerical function. Since by (4) $\|f(xH) - g(xH)\| = \rho(x)^{-1/2}\|\phi(x) - \psi(x)\|$ $(x \in G)$, it follows from III.14.18 that

$$\alpha \mapsto \|f(\alpha) - g(\alpha)\|_\alpha \qquad \text{is locally } \rho^\#\text{-measurable.} \tag{12}$$

Now (11) and (12) are just the hypotheses (II) and (I) of II.15.6. Therefore II.15.6 asserts that f is locally $\rho^\#$-measurable; and the proof is complete. ■

The Blattner Formulation

10.8. Let $f \leftrightarrow \phi$ as in (6), (8). By (4)

$$\| f(xH) \|_{xH}^2 = \rho(x)^{-1} \| \phi(x) \|^2 \qquad (x \in G). \qquad (13)$$

In particular the right side of (13) depends only on the coset xH. If ϕ is locally $\lambda^\#$-measurable, then by (13) and 10.7 $\rho(x)^\# \| \phi(x) \|^2$ is locally ρ-measurable as a function of xH, and will be $\rho^\#$-summable if and only if $f \in \mathcal{L}_2(\rho^\#; \mathcal{Y})$. Combining this fact with 10.7(II), 10.6, and (9), we obtain the following theorem, which is essentially the Blattner formulation of the construction of induced representations of groups.

Theorem.   *Let S be a unitary representation of H, acting on the Hilbert space X. Define $\mathcal{X}$ to be the linear space of all those functions $\phi : G \to X$ such that: (i) ϕ is locally λ-measurable, (ii) the relation (7) holds for all x in G and h in H, and (iii) the numerical function $xH \mapsto \rho(x)^{-1} \| \phi(x) \|^2$ on G/H is $\rho^\#$-summable. Then, if we identify functions in $\mathcal{X}$ which differ only on a locally λ-null set, $\mathcal{X}$ becomes a Hilbert space under the inner product*

$$(\phi, \psi)_{\mathcal{X}} = \int_{G/H} \rho(x)^{-1} (\phi(x), \psi(x))_X \, d\rho^\#(xH). \qquad (14)$$

(By (13) the integrand in (14) depends only on xH.) For each y in G the left translation operator T_y:

$$(T_y \phi)(x) = \phi(y^{-1}x) \qquad (\phi \in \mathcal{X}; x \in G) \qquad (15)$$

is a unitary operator on $\mathcal{X}$; and $y \mapsto T_y$ $(y \in G)$ is a unitary representation of G on $\mathcal{X}$, unitarily equivalent (under the correspondence $f \leftrightarrow \phi$ of 10.5) with $\mathrm{Ind}_{H \uparrow G}(S)$.

Remark.   In view of 9.7, the definition of $\mathcal{X}$ and of its inner product (14) is quite independent of the particular choice of the rho-function ρ.

The Mackey Formulation

10.9.   Set $\sigma'(y, xH) = \rho(x)^{-1}\rho(y^{-1}x)$ $(x, y \in G)$. Thus σ' is a continuous function on $G \times (G/H)$. By III.13.12 and III.14.4, $\sigma(y, \cdot)$ is the Radon–Nikodym derivative of $y\rho^\#$ (the y-translate $W \mapsto \rho^\#(y^{-1}W)$ of $\rho^\#$) with respect to $\rho^\#$; that is, if $y \in G$,

$$d(y\rho^\#)m = \sigma'(y, m)d\rho^\# m.$$

To obtain the Mackey formulation from the Blattner formulation 10.8, we have only to apply the transformation which sends the function $\phi : G \to X$

into the function $x \mapsto \rho(x)^{-1/2}\phi(x)$. Under this transformation Theorem 10.8 becomes:

Theorem.   *Let S be a unitary representation of H acting on the Hilbert space X. Define $\mathscr{W}$ to be the linear space of all those functions $\phi: G \to X$ such that:* (i) *ϕ is locally λ-measurable,* (ii) *for all x in G and h in H*

$$\phi(xh) = S_{h^{-1}}(\phi(x)), \tag{16}$$

and (iii) *the numerical function $xH \mapsto \|\phi(x)\|^2$ (which is well defined since by (16) $\|\phi(x)\|^2$ depends only on xH) is $\rho^\#$-summable. Then, if functions in $\mathscr{W}$ which differ only on a locally λ-null set are identified, $\mathscr{W}$ becomes a Hilbert space under the inner product*

$$(\phi, \psi)_{\mathscr{W}} = \int_{G/H} (\phi(x), \psi(x))_X \, d\rho^\#(xH). \tag{17}$$

For each y in G the equation

$$(T_y\phi)(x) = (\sigma'(y, xH))^{1/2}\phi(y^{-1}x) \qquad (\phi \in \mathscr{W}; x \in G) \tag{18}$$

defines a unitary operator T_y on $\mathscr{W}$; and $y \mapsto T_y$ is a unitary representation of G on $\mathscr{W}$, unitarily equivalent to $\mathrm{Ind}_{H\uparrow G}(S)$.

10.10.   We shall refer to Theorem 10.9 as the *Mackey formulation* of the inducing construction, as contrasted with the *Blattner formulation* in Theorem 10.8, and the (more general) *bundle formulation* of 9.24.

The Mackey formulation has the advantage that it contains explicit reference only to the quasi-invariant measure $\rho^\#$ on G/H, not to the rho-function ρ itself. Its disadvantage lies in the Radon–Nikodym factor $\sigma'(y, xH)^{1/2}$ occurring in (18), and also in the fact that different choices of ρ lead to formally different Hilbert spaces $\mathscr{W}$ in Theorem 10.9.

Remark.   For convenience we have always assumed that ρ is continuous, and have therefore obtained the Mackey formulation, strictly speaking, only for those quasi-invariant measures $\rho^\#$ which come from a continuous ρ. As Mackey has shown, this is quite inessential (at least for second-countable groups): Formulae (16), (17), and (18) define the induced representation $\mathrm{Ind}_{H\uparrow G}(S)$ no matter what quasi-invariant measure $\rho^\#$ we start with.

10.11.   Suppose that there exists a non-zero G-invariant regular Borel measure μ on G/H. By III.13.16 this amounts to saying that $\Delta(h) \equiv \delta(h)$ $(h \in H)$, that is, that we can take $\rho(x) \equiv 1$. In this case $\rho^\# = \mu$ (if μ is properly normalized); and there is no difference between the Mackey and the Blattner descriptions of $T = \mathrm{Ind}_{H \uparrow G}(S)$. According to both, the Hilbert space $X(T)$ consists of all functions $\phi: G \to X$ such that (i) ϕ is locally λ-measurable, (ii) $\phi(xh) = S_{h^{-1}}(\phi(x))$ $(x \in G; h \in H)$, and (iii) $\|\phi\|^2 = \int_{G/H} \|\phi(x)\|^2 \, d\mu(xH) < \infty$. The operators of T are those of left translation: $(T_y \phi)(x) = \phi(y^{-1}x)$ $(\phi \in X(T); x, y \in G)$.

Miscellaneous Observations

10.12. Suppose that $H = \{e\}$ and that S is the trivial one-dimensional representation of $\{e\}$. By 10.11 we have:

Proposition. $\mathrm{Ind}_{\{e\} \uparrow G}(S)$ *is just the left-regular representation of* G.

Remark. We shall see in 12.13 that the generalized regular representations of VIII.14.8 are also induced representations.

10.13. It is evident from 10.11 that, if G is compact, the $\mathrm{Ind}_{H \uparrow G}(S)$ defined in this chapter is identical with the construction of IX.10.4.

10.14. Let us now make the additional assumption that there exists a closed subgroup K of G which is complementary to H in the following sense:

(i) $KH = G$;
(ii) $K \cap H = \{e\}$;
(iii) the bijection $\langle k, h \rangle \mapsto kh$ of $K \times H$ onto G is a homeomorphism.

(Notice from III.3.14 that hypothesis (iii) can be omitted if one of K and H is σ-compact, in particular if G is second-countable.) Then the Blattner description of $\mathrm{Ind}_{H \uparrow G}(S)$ takes a somewhat simpler form.

The equation

$$\rho(kh) = \delta(h)\Delta(h)^{-1} \qquad (k \in K; h \in H) \qquad (19)$$

defines a continuous everywhere positive H-rho function ρ on G. Identifying G/H with K (via the homeomorphism $k \to kH$), we notice, since ρ is invariant under left translation by elements of K, that $\rho^{\#}$ is just left Haar measure μ on K.

Let S be a unitary representation of H acting on the Hilbert space X. In view of (7), a function ϕ belonging to the space $\mathscr{L}$ of Theorem 10.8 is determined by $\phi|K$. Furthermore $\phi|K$ is locally μ-measurable; and $k \mapsto \|\phi(k)\|^2$ is μ-summable. Conversely, every locally μ-measurable function

$\psi: K \to X$ such that $k \to \|\psi(k)\|^2$ is μ-summable can be extended by means of (7) to a function in $\mathcal{L}$. An easy calculation concerning the action of the operators of translation on G now gives the following useful description of $\mathrm{Ind}_{H \uparrow G}(S)$:

Proposition.   *Under these conditions, $\mathrm{Ind}_{H \uparrow G}(S)$ is unitarily equivalent to the unitary representation T' of G on $\mathcal{L}_2(\mu; X)$ given by:*

$$(T'_x \phi)(k) = \Delta(h')^{1/2} \delta(h')^{-1/2} S_{h'}(\phi(k')) \tag{20}$$

($x \in G$; $\phi \in \mathcal{L}_2(\mu; X)$; $k \in K$), where h' and k' are the elements of H and K respectively satisfying

$$x^{-1} k = k' h'^{-1}.$$

10.15.   Notice in particular that if $x = k_0 \in K$, (20) becomes

$$(T'_{k_0} \phi)(k) = \phi(k_0^{-1} k). \tag{21}$$

Thus we obtain from 10.14 the following corollary.

Corollary.   *Assume that H has a complementary closed subgroup K in the sense of 10.14. If χ is a one-dimensional unitary representation of H, then $\mathrm{Ind}_{H \uparrow G}(\chi)|K$ is unitarily equivalent to the regular representation of K.*

10.16.   As one might expect, there are Blattner and Mackey descriptions not merely of induced representations of groups, but also of induced representations of more general classes of Banach *-algebraic bundles. Let us see how the Blattner description works for semidirect product bundles.

Let A be a Banach *-algebra, and $\mathcal{B}$ the τ-semidirect product $A \underset{\tau}{\times} G$ as in VIII.4.2. Thus $\mathcal{B}_H$ is the $(\tau|H)$-semidirect product of A and H. We shall take a non-degenerate *-representation S of $\mathcal{B}_H$, and recall from 8.14 that S is automatically $\mathcal{B}$-positive. By VIII.15.6, S gives rise to a unitary representation W of H and a non-degenerate *-representation R of A (both acting on $X(S)$) such that

$$W_h R_a W_h^{-1} = R_{\tau_h(a)},$$

$$S_{\langle a, h \rangle} = R_a W_h$$

($a \in A$; $h \in H$).

We now form the induced unitary representation $V = \mathrm{Ind}_{H \uparrow G}(W)$, using the Blattner description 10.8 of V. Thus V acts by left translation on the space $\mathcal{L}$ defined in 10.8. Furthermore, for each a in A, we define an operator Q_a on $\mathcal{L}$ as follows:

$$(Q_a \phi)(x) = R_{\tau_{x^{-1}(a)}}(\phi(x)) \qquad (\phi \in \mathcal{L}; x \in G). \tag{22}$$

One can then prove the following result which constitutes the Blattner description of $\mathrm{Ind}_{\mathscr{B}_H \uparrow \mathscr{B}}(S)$:

Proposition*. *Equation* (22) *defines a non-degenerate *-representation* $Q: a \mapsto Q_a$ *of A on $\mathscr{L}$ satisfying*

$$V_x Q_a V_x^{-1} = Q_{\tau_x(a)} \qquad\qquad (x \in G; a \in A).$$

Thus (see VIII.15.6) *the equation*

$$T_{\langle a, x \rangle} = Q_a V_x \qquad\qquad (x \in G; a \in A)$$

*defines a non-degenerate *-representation T of $\mathscr{B}$ acting on $\mathscr{L}$. This T is unitarily equivalent with* $\mathrm{Ind}_{\mathscr{B}_H \uparrow \mathscr{B}}(S)$.

Thus the operators of T are given by

$$
\begin{aligned}
(T_{\langle a, x \rangle} \phi)(y) &= R_{\tau_{y^{-1}}(a)} \phi(x^{-1} y) \\
&= S_{\langle \tau_{y^{-1}}(a), e \rangle} \phi(x^{-1} y)
\end{aligned}
\qquad (23)
$$

$(\phi \in \mathscr{L}; a \in A; x, y \in G)$.

Remark. For a "Blattner description" of induced representations of the more general so-called homogeneous Banach *-algebraic bundles, see Fell [14], §11.

11. $\mathscr{B}$-Positive Representations and the C^*-Completion of $\mathscr{B}$

11.1. As a first application of the inducing construction, we propose to establish an important connection (11.5) between $\mathscr{B}$-positive *-representations and the C^*-completion of a Banach *-algebraic bundle $\mathscr{B}$. This will lead to Theorem 11.11 on $\mathscr{B}$-positive representations in saturated bundles, which has an important bearing on the generalized Mackey analysis to be developed in the next chapter (see the proof of XII.5.2).

11.2. Throughout this section we fix a Banach *-algebraic bundle $\mathscr{B} = \langle B, \pi, \cdot, * \rangle$ over a locally compact group G with unit e and left Haar measure λ. The unit fiber algebra B_e will be denoted by A.

Let $\mathscr{C} = \langle C, \pi', \cdot, * \rangle$ be the bundle C^*-completion of $\mathscr{B}$ (with norm $\| \ \|_c$), and $\rho: B \to C$ the canonical quotient map (see VIII.16.7). In Remark VIII.16.9 we observed that C_e is not in general the C^*-completion of A; in other words, not every *-representation S of A is obtained by composing ρ with a *-representation of C_e. The S which we obtain in this way are, as we shall soon show, just the $\mathscr{B}$-positive ones.

11.3.   Lemma.   *Let H be a closed subgroup of G and S a $\mathscr{B}$-positive non-degenerate *-representation of $\mathscr{B}_H$; and put $T = \mathrm{Ind}_{\mathscr{B}_H \uparrow \mathscr{B}}(S)$. Then*

$$\|S_a\| \le \|T_a\| \text{for all a in A.} \tag{1}$$

Proof.   We adopt the notation of §9 for the Hilbert bundle $\mathscr{Y}$ over G/H induced by S and the action τ of $\mathscr{B}$ on $\mathscr{Y}$. For $a \in A$ we have

$$(T_a f)(\alpha) = \tau_a(f(\alpha))   (f \in \mathscr{L}_2(\rho^{\#}; \mathscr{Y}); \alpha \in G/H). \tag{2}$$

Identifying Y_{eH} with $X(S)$ by means of the mapping F of 9.4, we have by 9.15(29)

$$\tau_a \xi = S_a \xi   (a \in A; \xi \in X(S)). \tag{3}$$

Take a non-zero vector ξ in $X(S)$; and choose a continuous cross-section ϕ of $\mathscr{Y}$ with $\phi(eH) = \xi$. Let $\{u_i\}$ be an "approximate unit" on G/H, that is, a net of elements of $\mathscr{L}_+(G/H)$ such that (i) $\int u_i \, d\rho^{\#} = 1$ for all i, and (ii) for any (G/H)-neighborhood W of the coset eH, u_i vanishes outside W for all large enough i. We now set $\phi_i(\alpha) = u_i(\alpha)^{1/2}\phi(\alpha)$ $(\alpha \in G/H)$. Thus $\phi_i \in \mathscr{L}_2(\rho^{\#}; \mathscr{Y})$; and one verifies that

$$\|\phi_i\|_2 \xrightarrow[i]{} \|\phi(eH)\| = \|\xi\|. \tag{4}$$

Let a be an element of A. Then $\alpha \mapsto \tau_a \phi(\alpha)$ is continuous; and by (2) and (3), the same argument that led to (4) gives:

$$\|T_a \phi_i\|_2 \to \|S_a \xi\|. \tag{5}$$

Combining (4) and (5) we find that $\|T_a\| \ge \|\xi\|^{-1}\|S_a \xi\|$. Since ξ was an arbitrary non-zero vector in $X(S)$, this implies (1).   ∎

11.4.   Combining 11.3 with 8.20 we obtain the following useful fact:

Proposition.   *Let H be a closed subgroup of G and S a non-degenerate $\mathscr{B}$-positive *-representation of $\mathscr{B}_H$; and put $T = \mathrm{Ind}_{\mathscr{B}_H \uparrow \mathscr{B}}(S)$. Then $S|A$ is weakly contained in $T|A$.*

11.5.   Proposition.   *For any *-representation S of A, the following two conditions are equivalent:*

(I)   *S is $\mathscr{B}$-positive.*

(II)   *There is a (unique) *-representation S' of C_e such that $S_a = S'_{\rho(a)}$ for all a in A.*

Proof.   Clearly (II) amounts to asserting that

$$\|S_a\| \le \|\rho(a)\|_c   \text{for all a in A.} \tag{6}$$

We may as well suppose that S is non-degenerate.

Assume (I). We can then form the induced *-representation $T = \text{Ind}_{A\uparrow\mathscr{B}}(S)$ of $\mathscr{B}$. By the definition of $\mathscr{C}$, $\|T_b\| \leq \|\rho(b)\|_c$ for all b in B. On the other hand, by 11.3 $\|S_a\| \leq \|T_a\|$ for all a in A. The last two facts imply (6), and hence (II).

Conversely, assume (II). Thus (6) holds. By the definition of $\mathscr{C}$ there is a *-representation W of $\mathscr{B}$ such that $\|W_b\| = \|\rho(b)\|_c$ $(b \in B)$. Combining this with (6) we find:

$$\|S_a\| \leq \|W_a\| \qquad\qquad (a \in A). \qquad (7)$$

Now by 8.12 $W|A$ is $\mathscr{B}$-positive. Hence by (7) and 8.21 S is $\mathscr{B}$-positive; and we have shown that (II) $\Rightarrow$ (I). ∎

11.6. Corollary. *The map $H: A_c \to C_e$ of Remark VIII.16.9 is one-to-one (hence an isometric *-isomorphism) if and only if every *-representation of A is $\mathscr{B}$-positive.*

By 8.14, this is the case if $\mathscr{B}$ has enough unitary multipliers.

11.7. Corollary. *If $\mathscr{B}$ is a C*-algebraic bundle, then every *-representation of A is $\mathscr{B}$-positive.*

Proof. By 11.6 and VIII.16.10. ∎

11.8. Proposition 11.5 has the following generalization:

Proposition. *Let H be any closed subgroup of G, and S a *-representation of $\mathscr{B}_H$. Then the following two conditions are equivalent:*

(I) *$S|A$ is $\mathscr{B}$-positive.*
(II) *There is a (unique) *-representation S' of $\mathscr{C}_H$ such that $S_b = S'_{\rho(b)}$ for all b in $\mathscr{B}_H$.*

If (I) *and* (II) *hold, then S is $\mathscr{B}$-positive if and only if S' is $\mathscr{C}$-positive.*

Proof. (II) $\Rightarrow$ (I) by 11.5.

Let us assume (I). Then by 11.5

$$\|S_a\| \leq \|\rho(a)\|_c \qquad \text{for all } a \text{ in } A. \qquad (8)$$

If $b \in \mathscr{B}_H$ then $b^*b \in A$, and so by (8)

$$\|S_b\|^2 = \|S_{b^*b}\| \leq \|\rho(b^*b)\|_c = \|\rho(b)\|_c^2. \qquad (9)$$

From this it follows that S gives rise to a map $S': \mathscr{C}_H \to \mathcal{O}(X(S))$ which is linear on each fiber and satisfies

$$S'_{\rho(b)} = S_b \qquad\qquad (b \in \mathscr{B}_H), \qquad (10)$$

$$\|S'_d\| \leq \|d\|_c \qquad\qquad (d \in \mathscr{C}_H). \qquad (11)$$

We claim that S' is a *-representation of $\mathscr{C}_H$. Indeed: An easy continuity argument shows that S' preserves multiplication and *. It remains only to prove the strong continuity of S'. Let $d_i \to d$ in $\mathscr{C}_H$; and put $x_i = \pi'(d_i)$, $x = \pi'(d)$. Given $\varepsilon > 0$, choose a continuous cross-section ϕ of $\mathscr{B}$ such that

$$\|\rho(\phi(x)) - d\|_c < \varepsilon. \tag{12}$$

Since $\rho: B \to C$ is continuous (VIII.16.7), the map $y \mapsto \rho(\phi(y))$ is continuous, and so by (12)

$$\|\rho(\phi(x_i)) - d_i\|_c < \varepsilon \qquad \text{for all large enough } i. \tag{13}$$

Since S is strongly continuous on B_H,

$$S_{\phi(y)} \to S_{\phi(x)} \qquad \text{strongly as } y \to x \text{ in } H. \tag{14}$$

Combining (10), (11), (12), (13), and (14), we obtain for any unit vector ξ in $X(S)$,

$$\|S'_{d_i}\xi - S'_d\xi\| \le \|d_i - \rho(\phi(x_i))\|_c + \|d - \rho(\phi(x))\|_c$$

$$+ \|S_{\phi(x_i)}\xi - S_{\phi(x)}\xi\|$$

$$< 3\varepsilon \qquad \text{for all large enough } i.$$

So $S'_{d_i} \to S'_d$ strongly; and we have shown that S' is a *-representation of $\mathscr{C}_H$.

Thus by (10) (II) holds; and we have shown that (I) $\Rightarrow$ (II). Consequently (I) $\Leftrightarrow$ (II).

If we assume (I) and (II), the last statement of the proposition follows from 8.9(III) by an easy density argument.   ∎

11.9.   The most interesting results of this section concern saturated bundles. We begin with a lemma on saturated C^*-algebraic bundles.

Lemma.   *Suppose that $\mathscr{B}$ is a saturated C^*-algebraic bundle over G; and let x be any element of G. There exists an approximate unit $\{u_r\}$ of $\mathscr{B}$ such that each u_r is of the form*

$$\sum_{i=1}^{n} b_i^* b_i, \tag{15}$$

where $n = 1, 2, \dots$ and the $b_1, \dots, b_n$ are in B_x.

Proof.   Let A as usual be the unit fiber C^*-algebra B_e; and let Q be the subset of A consisting of all elements of the form (15) (for fixed x). Thus Q is a cone, and by the definition of a C^*-algebraic bundle the elements of Q are positive in A. Replacing the b_i in (15) by $b_i c$, we see that

$$a \in Q, c \in A \Rightarrow c^* a c \in Q. \tag{16}$$

It follows that the norm-closure $\bar{Q}$ of Q also has property (16). Also, by the polarization identity, the linear span of Q in A is the same as the linear span of $B_{x^{-1}}B_x$; and by saturation this is dense in A. Hence by VI.8.17 (applied to $\bar{Q}$), $\bar{Q}$ contains all positive elements of A. From this and the existence of positive approximate units in A (see VI.8.4), we conclude that A has an approximate unit $\{u_i\}$ such that $u_i \in Q$ for all i. By the proof of VIII.16.3 this $\{u_i\}$ is an approximate unit of $\mathscr{B}$.   ■

11.10. Theorem. *Let $\mathscr{B}$ be a saturated C^*-algebraic bundle over the locally compact group G; and let H be any closed subgroup of G. Then every *-representation of $\mathscr{B}_H$ is positive with respect to $\mathscr{B}$.*

Proof. Let S be a *-representation of $\mathscr{B}_H$. Let $x \in \alpha \in G/H$; and take elements $c_1, \ldots, c_n$ of B_α and vectors $\xi_1, \ldots, \xi_n$ in $X(S)$. By Lemma 11.9 (applied to x^{-1} instead of x) there is an approximate unit $\{u_r\}$ of $\mathscr{B}$ such that for each r

$$u_r = \sum_{t=1}^{m_r} (b_t^r)^* b_t^r, \tag{17}$$

where each b_t^r belongs to $B_{x^{-1}}$.

Now for each i, j we have $c_j^* c_i = \lim_r c_j^* u_r c_i$. So by (17)

$$\sum_{i,j=1}^{n} (S_{c_j^* c_i} \xi_i, \xi_j) = \lim_r \sum_{t=1}^{m_r} \sum_{i,j=1}^{n} (S_{c_j^*(b_t^r)^* b_t^r c_i} \xi_i, \xi_j). \tag{18}$$

Since $b_i^r c_i \in B_{x^{-1}} B_\alpha \subset B_H$, the last summation in (18) becomes

$$\sum_{i,j=1}^{n} (S_{b_i^r c_i} \xi_i, S_{b_i^r c_j} \xi_j) = \left\| \sum_i S_{b_i^r c_i} \xi_i \right\|^2 \geq 0.$$

Therefore by (18) $\sum_{i,j}(S_{c_j^* c_i} \xi_i, \xi_j) \geq 0$; and so by 8.9(III) S is $\mathscr{B}$-positive.   ■

Remark. Does the above theorem remain true when the saturation hypothesis is omitted? We do not know. If it does remain true, then the hypothesis of saturation can also be omitted in the following Theorem 11.11.

11.11. Theorem. *Let $\mathscr{B}$ be any saturated Banach *-algebraic bundle over the locally compact group G; and let H be any closed subgroup of G. For any *-representation S of $\mathscr{B}_H$, the following two conditions are equivalent:*

(I)   *S is $\mathscr{B}$-positive;*
(II) *$S|B_e$ is $\mathscr{B}$-positive.*

Proof.   (I) $\Rightarrow$ (II) by 8.12.

Assume (II). Form the bundle C^*-completion $\mathscr{C}$ of $\mathscr{B}$ as in VIII.16.7; and let $\rho: B \to C$ be the canonical quotient map. By (II) and 11.8 there is a $*$-representation S' of $\mathscr{C}_H$ such that $S_b = S'_{\rho(b)}$ ($b \in \mathscr{B}_H$). By VIII.16.8 $\mathscr{C}$ is saturated. Hence by 11.10 S' is $\mathscr{C}$-positive. Therefore by the last statement of 11.8 S is $\mathscr{B}$-positive.   ∎

11.12.   Remark.   Keeping the assumptions of 11.2 we might make the following conjecture: Let H be any closed subgroup of G and S a $*$-representation of $\mathscr{B}_H$. Then S is $\mathscr{B}$-positive if and only if S is weakly contained in $\{T | \mathscr{B}_H : T$ is a $*$-representation of $\mathscr{B}\}$.

It follows from 11.4 that the conjecture is true if $H = \{e\}$. More generally it is true if H is normal in G (see 12.8). We do not know whether the conjecture holds in general.

12.   Elementary Properties of Induced Representations of Banach *-Algebraic Bundles

12.1.   Again throughout this section $\mathscr{B} = \langle B, \pi, \cdot, * \rangle$ is a Banach $*$-algebraic bundle over a locally compact group G (with unit e, left Haar measure λ, and modular function Δ).

Direct Sums

12.2.   Proposition.   *Let H be a closed subgroup of G; and let $\{S^i\}$ ($i \in I$) be an indexed collection of $*$-representations of $\mathscr{B}_H$, with Hilbert direct sum $S = \sum_{i \in I}^{\oplus} S^i$. Then S is $\mathscr{B}$-positive if and only if each S^i is $\mathscr{B}$-positive; and in that case*

$$\underset{\mathscr{B}_H \uparrow \mathscr{B}}{\mathrm{Ind}}\,(S) \cong \sum_{i \in I}^{\oplus}\ \underset{\mathscr{B}_H \uparrow \mathscr{B}}{\mathrm{Ind}}\,(S^i). \tag{1}$$

Proof.   By 1.14 and 5.1.   ∎

Remark.   One can be more explicit about the equivalence (1). For each i let $\mathscr{Y}^i$ be the Hilbert bundle over G/H induced by S^i (as in 9.2). Then the Hilbert bundle $\mathscr{Y}$ induced by S can be naturally identified with the direct sum $\sum_{i \in I}^{\oplus} \mathscr{Y}^i$ of the Hilbert bundles $\mathscr{Y}^i$, as defined in II.15.14; and the corresponding identification of $\mathscr{L}_2(\rho^\#; \mathscr{Y})$ with $\sum_{i \in I}^\oplus \mathscr{L}_2(\rho^\#; \mathscr{Y}^i)$ (see II.15.15) is the unitary equivalence which realizes (1). The details are left to the reader.

12.3. Let S be a non-zero non-degenerate $\mathscr{B}$-positive *-representation of $\mathscr{B}_H$, and $\mathscr{Y}$ the Hilbert bundle induced by S. Then at least some of the fiber Hilbert spaces Y_{xH} are non-zero; for example $Y_H \neq \{0\}$ by 9.4. It follows that $\mathscr{L}_2(\rho^\#; \mathscr{Y}) \neq \{0\}$; and $\mathrm{Ind}_{\mathscr{B}_H \uparrow \mathscr{B}}(S)$ is non-zero.

Combining this remark with 12.2 we obtain:

Corollary. *Let H be a closed subgroup of G, and S a non-degenerate $\mathscr{B}$-positive *-representation of $\mathscr{B}_H$. In order for $\mathrm{Ind}_{\mathscr{B}_H \uparrow \mathscr{B}}(S)$ to be irreducible it is necessary that S be irreducible.*

Remark. The irreducibility of S is by no means sufficient for the irreducibility of $\mathrm{Ind}_{\mathscr{B}_H \uparrow \mathscr{B}}(S)$. For example, the regular representation of G is induced from the one-dimensional representation of $\{e\}$ (10.12), but is not itself irreducible (see for example IX.9.1).

It would be very helpful to find a neat necessary and sufficient condition for the irreducibility of $\mathrm{Ind}_{\mathscr{B}_H \uparrow \mathscr{B}}(S)$. Unfortunately no such condition is known, even in the group case. (However, for some partial results see Quigg [1, 2].)

Regional Continuity of the Inducing Operation

12.4. Proposition. *Let H be a closed subgroup of G; and let $\mathscr{S}$ be the space of all non-degenerate $\mathscr{B}$-positive *-representations of $\mathscr{B}_H$. Then the inducing map $S \mapsto \mathrm{Ind}_{\mathscr{B}_H \uparrow \mathscr{B}}(S)$ is continuous on $\mathscr{S}$ with respect to the regional topologies of $\mathscr{S}$ and the space of *-representations of $\mathscr{B}$.*

Proof. This follows from 5.3 and the description 9.25 of $\mathrm{Ind}_{\mathscr{B}_H \uparrow \mathscr{B}}(S)$ in terms of the Rieffel inducing process. ∎

12.5. In the group case this becomes:

Corollary. *The inducing map $S \mapsto \mathrm{Ind}_{H \uparrow G}(S)$ is continuous, with respect to the regional topologies, from the space of unitary representations of H to the space of unitary representations of G.*

Dense Embedding of Bundles

12.6.   Suppose that $\mathscr{C} = \langle C, \pi', \cdot, * \rangle$ is another Banach *-algebraic bundle over the same group G, and that $\Phi: B \to C$ satisfies the following conditions:

(i)   For each x in G, $\Phi|B_x$ is a linear map of B_x into C_x;
(ii) $\Phi(b_1 b_2) = \Phi(b_1)\Phi(b_2)$ $(b_1, b_2 \in B)$;
(iii) $\Phi(b^*) = (\Phi(b))^*$ $(b \in B)$;
(iv) $\Phi: B \to C$ is continuous;
(v) $\Phi(B_x)$ is dense in C_x for each x in G.

Proposition.   *Let H be a closed subgroup of G and S a non-degenerate *-representation of $\mathscr{C}_H$. Then $S': b \mapsto S_{\Phi(b)}$ $(b \in \mathscr{B}_H)$ is a non-degenerate *-representation of $\mathscr{B}_H$, and is $\mathscr{B}$-positive if and only if S is $\mathscr{C}$-positive. If S is $\mathscr{C}$-positive, then*

$$\operatorname*{Ind}_{\mathscr{B}_H \uparrow \mathscr{B}} (S') \cong \left(\operatorname*{Ind}_{\mathscr{C}_H \uparrow \mathscr{C}} (S) \right) \circ \Phi. \tag{2}$$

Proof.   Conditions (i)–(iv) show that S' is a *-representation, and condition (v) implies that S' is non-degenerate. The equivalence of the positivity of S and of S' follows from (v) and version 8.9(III) of the definition of positivity.

Assume now that S and S' are positive. To prove (2) we shall use the Rieffel description 9.25 of the inducing process. Let $\Psi: \mathscr{L}(\mathscr{B}) \to \mathscr{L}(\mathscr{C})$ be the linear map given by $\Psi(f)(x) = \Phi(f(x))$ $(f \in \mathscr{L}(\mathscr{B}); \ x \in G)$, and $\Psi_H: \mathscr{L}(\mathscr{B}_H) \to \mathscr{L}(\mathscr{C}_H)$ the corresponding map for $H: \Psi_H(\phi)(h) = \Phi(\phi(h))$ $(\phi \in \mathscr{L}(\mathscr{B}_H); \ h \in H)$. Conditions (i)–(iv) imply that Ψ and Ψ_H are *-homomorphisms of the cross-sectional algebras. Notice that

$$S'_\phi = S_{\Psi_H(\phi)} \qquad\qquad (\phi \in \mathscr{L}(\mathscr{B}_H)). \tag{3}$$

Let p denote the conditional expectation 8.4(1) both on $\mathscr{L}(\mathscr{B})$ and on $\mathscr{L}(\mathscr{C})$. We shall write X for $X(S)$, and T and T' for $\operatorname{Ind}_{\mathscr{C}_H \uparrow \mathscr{C}}(S)$ and $\operatorname{Ind}_{\mathscr{B}_H \uparrow \mathscr{B}}(S')$ respectively.

Now $X(T')$ is the completion of $\mathscr{L}(\mathscr{B}) \otimes X$ with respect to the positive form $(\ ,\ )_0$ given by:

$$(f' \otimes \xi, g' \otimes \eta)_0 = (S'_{p(g'^* * f')}\xi, \eta) \qquad (f', g' \in \mathscr{L}(\mathscr{B}); \xi, \eta \in X); \tag{4}$$

and $X(T)$ is the completion of $\mathscr{L}(\mathscr{C}) \otimes X$ with respect to the positive form $(\ ,\ )_{00}$ given by:

$$(f \otimes \xi, g \otimes \eta)_{00} = (S_{p(g^* * f)}\xi, \eta) \qquad (f, g \in \mathscr{L}(\mathscr{C}); \xi, \eta \in X). \tag{5}$$

Observe that, if ξ, $\eta \in X$, f', $g' \in \mathcal{L}(\mathcal{B})$, $f = \Psi(f')$, and $g = \Psi(g')$, then $p(g^* * f) = p(\Psi(g'^* * f')) = \Psi_H(p(g'^* * f'))$, and so by (3)

$$(S_{p(g^* * f)}\xi, \eta) = (S'_{p(g'^* * f')}\xi, \eta).$$

From this and (4), (5) we conclude that the map F sending the image $f' \widetilde{\otimes} \xi$ of $f' \otimes \xi$ in $X(T')$ into the image $\Psi(f') \widetilde{\otimes} \xi$ of $\Psi(f') \otimes \xi$ in $X(T)$ is a linear isometry of $X(T')$ into $X(T)$. Now by (v) and II.14.6 $\Psi(\mathcal{L}(\mathcal{B}))$ is dense in $\mathcal{L}(\mathcal{C})$ in the inductive limit topology. Also we have seen in Remark 4 of 8.6 that the linear map $f \mapsto f \widetilde{\otimes} \xi$ of $\mathcal{L}(\mathcal{C})$ into $X(T)$ is continuous in the inductive limit topology. From the last two facts it follows that the range of F is dense in $X(T)$ and hence equal to $X(T)$.

Thus $F: X(T') \to X(T)$ is a linear isometric surjection. If f', $g' \in \mathcal{L}(\mathcal{B})$ and $\xi \in X$ we have

$$\begin{aligned} T_{\Psi(f')}F(g' \widetilde{\otimes} \xi) &= T_{\Psi(f')}(\Psi(g') \widetilde{\otimes} \xi) \\ &= (\Psi(f') * \Psi(g')) \widetilde{\otimes} \xi \\ &= \Psi(f' * g') \widetilde{\otimes} \xi \\ &= F((f' * g') \widetilde{\otimes} \xi) \\ &= FT'_{f'}(g' \widetilde{\otimes} \xi). \end{aligned}$$

This shows that F intertwines T' and $T \circ \Psi$ (as *-representations of $\mathcal{L}(\mathcal{B})$). Since $T \circ \Psi$ is clearly the integrated form of $T \circ \Phi$, it follows that $T' \cong T \circ \Phi$.　■

Remark. Thus, in the context of this proposition, it makes no difference whether we induce S within $\mathcal{B}$ or within $\mathcal{C}$.

Remark. In particular, this proposition holds when Φ is an isomorphism of the bundles $\mathcal{B}$ and $\mathcal{C}$ covariant with the identity automorphism of G (in the sense of VIII.3.3).

Remark. The hypotheses of this proposition also hold when $\mathcal{C}$ is the bundle C^*-completion of $\mathcal{B}$ and Φ is the canonical quotient map ρ of VIII.16.7.

Partial Cross-Sectional Bundles and Induced Representations

12.7. Let N be a closed normal subgroup of G and let G' be the quotient group G/N. Let H' be a closed subgroup of G' and H the inverse image of H' in G. We can then construct the $\mathcal{L}_1$ partial cross-sectional bundle $\mathcal{B}'$ over G'

derived from $\mathscr{B}$, as in VIII.6.5; and $\mathscr{B}'_{H'}$ will be the $\mathscr{L}_1$ partial cross-sectional bundle over H' derived from $\mathscr{B}_H$.

Now we recall from VIII.15.9 that the *-representations of $\mathscr{B}$ are in natural correspondence with those of $\mathscr{B}'$, and *-representations of $\mathscr{B}_H$ with those of $\mathscr{B}'_{H'}$. Suppose that S is a non-degenerate *-representation of $\mathscr{B}_H$, and S' the (non-degenerate) *-representation of $\mathscr{B}'_{H'}$ corresponding to S as in VIII.15.9. We want to show that the process of inducing S' within $\mathscr{B}'$ corresponds exactly to the process of inducing S within $\mathscr{B}$.

Proposition.   *In the above context, S' is $\mathscr{B}'$-positive if and only if S is $\mathscr{B}$-positive. If this is the case, $\mathrm{Ind}_{\mathscr{B}'_{H'}\uparrow\mathscr{B}'}(S')$ is unitarily equivalent to the *-representation of $\mathscr{B}'$ corresponding via VIII.15.9 with the *-representation $\mathrm{Ind}_{\mathscr{B}_H\uparrow\mathscr{B}}(S)$ of $\mathscr{B}$.*

Proof.   We shall again proceed by way of the Rieffel formulation of the inducing process. Let Δ, δ, Δ', δ' be the modular functions of G, H, G', H' respectively. If $f \in \mathscr{L}(\mathscr{B})$ let f' denote the element $xN \mapsto f|xN$ of $\mathscr{L}(\mathscr{B}')$. As in 8.4 we introduce the conditional expectations $p: \mathscr{L}(\mathscr{B}) \to \mathscr{L}(\mathscr{B}_H)$ and $p': \mathscr{L}(\mathscr{B}') \to \mathscr{L}(\mathscr{B}'_{H'})$:

$$p(f)(h) = \Delta(h)^{1/2}\delta(h)^{-1/2}f(h) \qquad (f \in \mathscr{L}(\mathscr{B}); h \in H),$$

$$p'(\phi)(h') = \Delta'(h')^{1/2}\delta'(h')^{-1/2}\phi(h') \qquad (\phi \in \mathscr{L}(\mathscr{B}'); h' \in H').$$

We claim that, for all f, g in $\mathscr{L}(\mathscr{B})$,

$$S'_{p'(g'^**f')} = S_{p(g^**f)}. \tag{6}$$

Indeed: Let τ, ν, λ, ν', λ' be left Haar measures on N, H, G, H', G' respectively, normalized so that the integration formula III.13.17(8) holds for H, N, H' and also for G, N, G'. Applying III.13.20 to the pairs G, N and H, N, we conclude that

$$\Delta(h)\delta(h)^{-1} = \Delta'(hN)\delta'(hN)^{-1} \qquad (h \in H). \tag{7}$$

Therefore, since by VIII.6.7 the map $\Phi: f \mapsto f'$ is a *-isomorphism of $\mathscr{L}(\mathscr{B})$ into $\mathscr{L}(\mathscr{B}')$, we have for f, $g \in \mathscr{L}(\mathscr{B})$ and ξ, $\eta \in X(S)$:

$$(S_{p(g^**f)}\xi, \eta) = \int_H \Delta(h)^{1/2}\delta(h)^{-1/2}(S_{(g^**f)(h)}\xi, \eta)d\nu h$$

$$= \int_{H'} d\nu'(hN)\int_N \Delta(hn)^{1/2}\delta(hn)^{-1/2}(S_{(g^**f)(hn)}\xi, \eta)d\tau n$$

$$\text{(by III.13.17(8))}$$

$$= \int_{H'} \Delta'(h')^{1/2} \delta'(h')^{-1/2} (S'_{(g^* * f)'(h')} \xi, \eta) dv'h'$$

(by (7) and VIII.15.9(6))

$$= \int_{H'} \Delta'(h')^{1/2} \delta'(h')^{-1/2} (S'_{(g'^* * f')(h')} \xi, \eta) dv'h'$$

(since $f \mapsto f'$ is a *-homomorphism)

$$= (S'_{p'(g'^* * f')} \xi, \eta).$$

This proves (6).

Suppose now that S' is $\mathscr{B}'$-positive. Then by (6)

$$(S_{p(f^* * f)} \xi, \xi) = (S'_{p'(f'^* * f')} \xi, \xi) \ge 0 \qquad \text{for all } f \text{ in } \mathscr{L}(\mathscr{B}) \text{ and } \xi \text{ in } X(S);$$

whence S is $\mathscr{B}$-positive. Conversely, suppose S is $\mathscr{B}$-positive. Then by (6)

$$S'_{p'(\phi^* * \phi)} \ge 0 \qquad \text{for all } \phi \text{ in } M, \tag{8}$$

where $M = \{f': f \in \mathscr{L}(\mathscr{B})\}$. Let ψ be any element of $\mathscr{L}(\mathscr{B}')$. By II.14.6 there is a net $\{\phi_i\}$ of elements of M converging to ψ in the inductive limit topology. Now M is clearly closed under multiplication by continuous complex functions on G'; so, choosing a function w in $\mathscr{L}(G')$ such that $w(x') = 1$ whenever $\psi(x') \ne 0$, and replacing ϕ_i by $w\phi_i$, we may as well assume that the ϕ_i all vanish outside the same compact subset of G'. But in that case, the fact that $\phi_i \to \psi$ implies that $\phi_i^* * \phi_i \to \psi^* * \psi$, whence $p'(\phi_i^* * \phi_i) \to p'(\psi^* * \psi)$ in the inductive limit topology. Therefore by (8) $S'_{p'(\psi^* * \psi)} \ge 0$; and S' is $\mathscr{B}'$-positive.

We have shown that S is $\mathscr{B}$-positive if and only if S' is $\mathscr{B}'$-positive. Assume that this is the case; and put $T = \text{Ind}_{\mathscr{B}_H \uparrow \mathscr{B}}(S)$, $T' = \text{Ind}_{\mathscr{B}_{H'} \uparrow \mathscr{B}'}(S')$. By (6) the equation

$$F(f \widetilde{\otimes} \xi) = f' \widetilde{\otimes} \xi \qquad (f \in \mathscr{L}(\mathscr{B}); \xi \in X(S))$$

defines a linear isometry of $X(T)$ onto the closure in $X(T')$ of $M \widetilde{\otimes} X(S)$. The argument of the preceding paragraph concerning the denseness of M in $\mathscr{L}(\mathscr{B}')$ shows that F is *onto* $X(T')$. If $f, g \in \mathscr{L}(\mathscr{B})$ and $\xi \in X(S)$ we have

$$FT_f(g \widetilde{\otimes} \xi) = F((f * g) \widetilde{\otimes} \xi)$$

$$= (f * g)' \widetilde{\otimes} \xi$$

$$= (f' * g') \widetilde{\otimes} \xi$$

$$= T'_{f'}(g' \widetilde{\otimes} \xi) = T'_{f'} F(g \widetilde{\otimes} \xi).$$

This shows that F intertwines the integrated form of T and the *-representation $W: f \mapsto T'_{f'}$ of $\mathscr{L}(\mathscr{B})$. But by the definition of the correspondence Ξ at the beginning of the proof of VIII.15.9, W is just the integrated form of the *-representation $\tilde{T}$ of $\mathscr{B}$ corresponding under Ξ with T'. So $T \cong \tilde{T}$; and the proposition is proved.   ∎

12.8. Corollary. *Let N be a closed normal subgroup of G, H a closed subgroup of G containing N, and S a non-degenerate $\mathscr{B}$-positive *-representation of $\mathscr{B}_H$. Denote $\mathrm{Ind}_{\mathscr{B}_H \uparrow \mathscr{B}}(S)$ by T. Then $T | \mathscr{B}_N$ weakly contains $S | \mathscr{B}_N$.*

Proof. Let $\mathscr{B}'$ be the partial cross-sectional bundle over G/N derived from $\mathscr{B}$. By 12.7 we can induce S within $\mathscr{B}'$ instead of $\mathscr{B}$; and in this setting 11.4 can be applied to give the required result. The details are left to the reader.   ∎

12.9. Corollary. *Let $\mathscr{B}$ be a saturated C^*-algebraic bundle over G, and N a closed normal subgroup of G. Let $\mathscr{D}$ be the $\mathscr{L}_1$ partial cross-sectional bundle over G/N derived from $\mathscr{B}$. Then the condition of 11.6 holds for the unit fiber of $\mathscr{D}$.*

Proof. By 11.9 and 12.7.   ∎

Positive Functionals and Induced Representations

12.10. In this number we will suppose that $\mathscr{B}$ has an approximate unit.

Let H be a closed subgroup of G, with left Haar measure ν and modular function δ. Let $p: \mathscr{L}(\mathscr{B}) \to \mathscr{L}(\mathscr{B}_H)$ be the conditional expectation defined in 8.4.

Suppose we are given a positive linear functional q on the *-algebra $\mathscr{L}(\mathscr{B}_H)$ which is continuous in the inductive limit topology. By VIII.13.4 q generates a *-representation S of $\mathscr{B}_H$; and S is non-degenerate by VIII.13.5 (since $\mathscr{B}$ and hence $\mathscr{B}_H$ has an approximate unit). In this situation the following theorem is of considerable interest.

Theorem. *The following two conditions are equivalent:*

   (I) *The linear functional $q \circ p$ on $\mathscr{L}(\mathscr{B})$ is positive;*
  (II) *S is $\mathscr{B}$-positive.*

*If (I) and (II) hold, the *-representation T of $\mathscr{B}$ generated by $q \circ p$ is unitarily equivalent to $\mathrm{Ind}_{\mathscr{B}_H \uparrow \mathscr{B}}(S)$.*

Remark. Since p is continuous in the inductive limit topology, so is $q \circ p$. Hence (I) and VIII.13.4 imply that T is well defined.

Proof.   This theorem is a topological application of the Proposition 5.6 on "inducing in stages."

We are going to apply 5.6 taking A and B to be $\mathscr{L}(\mathscr{B})$ and $\mathscr{L}(\mathscr{B}_H)$; L, [,] to be $\mathscr{L}(\mathscr{B})$ together with the mapping $\langle f, g \rangle \mapsto p(f^* * g)$ and the module structures defined by multiplication in $\mathscr{L}(\mathscr{B})$ and 8.4(3); and K and V to be $\mathscr{L}(\mathscr{B}_H)$ and $\langle \phi, \psi \rangle \mapsto q(\psi^* * \phi)$. As in 5.6 let $P = \mathscr{L}(\mathscr{B}) \otimes \mathscr{L}(\mathscr{B}_H)$; and let Γ be the conjugate-bilinear form on P given by

$$\Gamma(f \otimes \phi, g \otimes \psi) = q(\psi^* * p(g^* * f) * \phi)$$

$$= (q \circ p)[(g\psi)^* * (f\phi)]. \tag{9}$$

(The last equality follows from 8.4(2), 8.4(5), 8.4(6).) Proposition 5.6 then asserts that Γ is positive if and only if S is $\mathscr{B}$-positive; and that, if this is the case, the integrated form of $\mathrm{Ind}_{\mathscr{B}_H \uparrow \mathscr{B}}(S)$ is unitarily equivalent to the *-representation of $\mathscr{L}(\mathscr{B})$ generated by Γ.

Let N be the linear span in $\mathscr{L}(\mathscr{B})$ of $\{ f\phi : f \in \mathscr{L}(\mathscr{B}), \phi \in \mathscr{L}(\mathscr{B}_H) \}$. In view of (9) and 5.8, the conclusion of the last paragraph can be rephrased as follows: S is $\mathscr{B}$-positive if and only if $q \circ p$ is positive on N; and, if this is the case, the integrated form of $\mathrm{Ind}_{\mathscr{B}_H \uparrow \mathscr{B}}(S)$ is unitarily equivalent to the *-representation of $\mathscr{L}(\mathscr{B})$ generated by the restriction of $q \circ p$ to N.

To prove the theorem, then, it remains only to justify the replacement of N by $\mathscr{L}(\mathscr{B})$ in the last statement. Since $\mathscr{B}$ has an approximate unit, the argument of VIII.5.13 shows that there is a net $\{ \phi_i \}$ of elements of $\mathscr{L}(\mathscr{B}_H)$ such that (i) $f\phi_i \to f$ in the inductive limit topology for every f in $\mathscr{L}(\mathscr{B})$, and (ii) the compact supports of the ϕ_i are all contained in the same compact subset of H.

Now fix an element f of $\mathscr{L}(\mathscr{B})$, and put $f_i = f\phi_i$ for each index i. It follows from (i) and (ii) that $f_i \to f$ in the inductive limit topology, and that all the f_i vanish outside the same compact set. So $f_i^* * f_i \to f^* * f$ in the inductive limit topology; whence

$$(q \circ p)(f_i^* * f_i) \to (q \circ p)(f^* * f). \tag{10}$$

Now the f_i belong to N. Thus, if $q \circ p$ is positive on N, it follows from (10) that $(q \circ p)(f^* * f) \geq 0$, whence $q \circ p$ is positive on $\mathscr{L}(\mathscr{B})$. We have shown that $q \circ p$ is positive (on $\mathscr{L}(\mathscr{B})$) if and only if its restriction to N is positive.

Assume that $q \circ p$ is positive; and let f and f_i be as in the last paragraph. Then $f_i^* * f_i \to f^* * f, f_i^* * f \to f^* * f$, and $f^* * f_i \to f^* * f$ in the inductive limit topology. Denoting by $\tilde{f}$ the image of f in the Hilbert space generated by $q \circ p$, it follows that

$$\| \tilde{f}_i - \tilde{f} \|^2 = (q \circ p)(f_i^* * f_i - f_i^* * f - f^* * f_i + f^* * f) \underset{i}{\to} 0.$$

Thus the Hilbert space generated by $(q \circ p)|N$ is dense in, and so coincides with, that generated by $q \circ p$. Now $q \circ p$ generates a *-representation T of $\mathscr{L}(\mathscr{B})$ in virtue of VIII.13.4. By the preceding sentence $(q \circ p)|N$ generates the same *-representation T.

Assembling the facts proved above, we find that S is $\mathscr{B}$-positive if and only if $q \circ p$ is positive, and that, if this is the case, $\mathrm{Ind}_{\mathscr{B}_H \uparrow \mathscr{B}}(S)$ is unitarily equivalent to T. ∎

12.11. Corollary. *Suppose that $\mathscr{B}$ has an approximate unit and enough unitary multipliers. Let H be a closed subgroup of G, and q a positive linear functional on $\mathscr{L}(\mathscr{B}_H)$ which is continuous in the inductive limit topology; and let S be the (non-degenerate) *-representation of $\mathscr{B}_H$ generated by q (in virtue of VIII.13.4, VIII.13.5). Let $p: \mathscr{L}(\mathscr{B}) \to \mathscr{L}(\mathscr{B}_H)$ be the conditional expectation defined in 8.4. Then $q \circ p$ is a positive linear functional on $\mathscr{L}(\mathscr{B})$; and the *-representation of $\mathscr{B}$ generated by $q \circ p$ is unitarily equivalent to $\mathrm{Ind}_{\mathscr{B}_H \uparrow \mathscr{B}}(S)$.*

Proof. By 8.14 S is automatically $\mathscr{B}$-positive. So the required conclusion follows from 12.10. ∎

12.12. Suppose in 12.11 that $\mathscr{B}$ is the group bundle of G. A regular complex Borel measure μ on H is said to be *of positive type* if the corresponding linear functional $\phi \mapsto \int \phi \, d\mu$ on $\mathscr{L}(\mathscr{B}_H)$ is positive. (This definition was made in X.2.7 in the commutative context.) By the Riesz Theorem II.8.12 the q of 12.11 can now be regarded as a regular complex Borel measure on H of positive type; and $q \circ p$ can then be identified with the injection onto G (see II.8.5) of the measure $\Delta(h)^{1/2}\delta(h)^{-1/2} \, dqh$ on H. Thus in this case 12.11 becomes:

Corollary. *Let q be a regular complex Borel measure of positive type on H. Then the injection $\tilde{q}$ of $\Delta(h)^{1/2}\delta(h)^{-1/2} \, dqh$ onto G is a regular complex Borel measure of positive type on G. If S and T are the unitary representations of H and G generated by q and $\tilde{q}$ respectively, we have $T \cong \mathrm{Ind}_{H \uparrow G}(S)$.*

12.13. Example. Assume that $\mathscr{B}$ has an approximate unit and enough unitary multipliers; and let q be a continuous positive linear functional on B_e. The *-representation R^q of $\mathscr{B}$ defined in VIII.14.6 was referred to in VIII.14.8 as a *generalized regular representation* of $\mathscr{B}$. By 12.11 (together with VI.19.9) these generalized regular representations are exactly those of the form $\mathrm{Ind}_{B_e \uparrow \mathscr{B}}(S)$ where S is a cyclic *-representation of B_e.

12.14. *Example.* Adopting the context of X.4.19 and X.4.20, let us take a character τ of the closed subgroup N of H, and form the unitary representation T^τ of H as in X.4.20. Combining 12.11 with 12.7, we conclude that T^τ is unitarily equivalent to $\mathrm{Ind}_{N\uparrow H}(\tau)$. Thus Proposition X.4.20 gives us the harmonic analysis of $\mathrm{Ind}_{N\uparrow H}(\tau)$.

Inducing in Stages

12.15. **Theorem.** *Let H and L be two closed subgroups of G with $H \subset L$; and let S be a non-degenerate *-representation of $\mathscr{B}_H$. Then the following two conditions are equivalent:*

(I) S *is $\mathscr{B}$-positive;*
(II) S *is $\mathscr{B}_L$-positive and $\mathrm{Ind}_{\mathscr{B}_H\uparrow\mathscr{B}_L}(S)$ is $\mathscr{B}$-positive.*

If these two conditions hold, then

$$\underset{\mathscr{B}_H\uparrow\mathscr{B}}{\mathrm{Ind}}\ (S) \cong \underset{\mathscr{B}_L\uparrow\mathscr{B}}{\mathrm{Ind}}\left(\underset{\mathscr{B}_H\uparrow\mathscr{B}_L}{\mathrm{Ind}}\ (S)\right). \tag{11}$$

Proof. To begin with we shall suppose that $\mathscr{B}$ has an approximate unit. Afterwards this assumption will be dropped.

Let us write $S = \sum_i^{\oplus} S^i$, where each S^i is a cyclic *-representation of $\mathscr{B}_H$. By 12.2 the theorem will hold for S provided it holds for each S^i. Hence we may, and shall, assume from the beginning that S is cyclic, with cyclic vector ξ.

Let δ_H and δ_L be the modular functions of H and L respectively; and define the three conditional expectations $p: \mathscr{L}(\mathscr{B}) \to \mathscr{L}(\mathscr{B}_H)$, $p_1: \mathscr{L}(\mathscr{B}) \to \mathscr{L}(\mathscr{B}_L)$, $p_2: \mathscr{L}(\mathscr{B}_L) \to \mathscr{L}(\mathscr{B}_H)$ as in 8.4:

$$p(f)(h) = \Delta(h)^{1/2}\delta_H(h)^{-1/2}f(h),$$

$$p_1(f)(m) = \Delta(m)^{1/2}\delta_L(m)^{-1/2}f(m),$$

$$p_2(\phi)(h) = \delta_L(h)^{1/2}\delta_H(h)^{-1/2}\phi(h)$$

$(f \in \mathscr{L}(\mathscr{B}); \phi \in \mathscr{L}(\mathscr{B}_L); m \in L; h \in H)$. We notice that $p_2 \circ p_1 = p$.

Now by 8.13 conditions (I) and (II) both imply that S is $\mathscr{B}_L$-positive. So we shall assume this. Put $q(\phi) = (S_\phi\xi, \xi)$ $(\phi \in \mathscr{L}(\mathscr{B}_H))$. Then q is a positive linear functional on $\mathscr{L}(\mathscr{B}_H)$ which generates S. Since S is assumed to be $\mathscr{B}_L$-positive, Theorem 12.10 (applied to $\mathscr{B}_L$) tells us that $q \circ p_2$ is positive on $\mathscr{L}(\mathscr{B}_L)$ and generates $V = \mathrm{Ind}_{\mathscr{B}_H\uparrow\mathscr{B}_L}(S)$. Thus Theorem 12.10 (applied this time to $\mathscr{B}$) tells us that $q \circ p$ $(= (q \circ p_2) \circ p_1)$ is positive on $\mathscr{L}(\mathscr{B})$ if and only if V is $\mathscr{B}$-positive, and that in that case $q \circ p$ generates $\mathrm{Ind}_{\mathscr{B}_L\uparrow\mathscr{B}}(V)$. On the other hand, a third application of 12.10 shows that $q \circ p$ is positive on $\mathcal{L}(\mathcal{B})$ if and only if S is $\mathcal{B}$-positive, in which case $q \circ p$ generates $\mathrm{Ind}_{\mathcal{B}_H \uparrow \mathcal{B}}(S)$. Putting these facts together we find that V is $\mathcal{B}$-positive if and only if S is $\mathcal{B}$-positive, and that in that case $\mathrm{Ind}_{\mathcal{B}_H \uparrow \mathcal{B}}(S) \cong \mathrm{Ind}_{\mathcal{B}_L \uparrow \mathcal{B}}(V)$. This is the required result.

We now discard the assumption that $\mathcal{B}$ has an approximate unit (which was needed in the above argument in order to be able to apply 12.10). Let $\mathcal{C}$ be the bundle C^*-completion of $\mathcal{B}$ and $\rho: B \to C$ the canonical quotient map (see VIII.16.7). Since $\mathcal{C}$ automatically has an approximate unit (VIII.16.3), we shall try to obtain the required result for $\mathcal{B}$ by applying the preceding part of the proof to $\mathcal{C}$.

First we claim that the conditions (I) and (II) both imply that $S|B_e$ is $\mathcal{B}$-positive. Indeed: Condition (I) implies this by 8.12. Now assume Condition (II); and set $V = \mathrm{Ind}_{\mathcal{B}_H \uparrow \mathcal{B}_L}(S)$, $T = \mathrm{Ind}_{\mathcal{B}_L \uparrow \mathcal{B}}(V)$. By 11.3 we have $\|S_a\| \leq \|V_a\|$ and also $\|V_a\| \leq \|T_a\|$ for all a in B_e; therefore

$$\|S_a\| \leq \|T_a\| \qquad\qquad (a \in B_e). \qquad (12)$$

Since $T|B_e$ is $\mathcal{B}$-positive by 8.12, it follows from (12) and 8.21 that $S|B_e$ is $\mathcal{B}$-positive. This establishes the claim.

In view of this claim we may as well assume from the beginning that $S|B_e$ is $\mathcal{B}$-positive. Thus by 11.7 there is a *-representation S' of $\mathcal{C}_H$ such that

$$S_b = S'_{\rho(b)} \qquad \text{for } b \in B_H. \qquad (13)$$

Applying the preceding part of the proof to $\mathcal{C}$ (which has an approximate unit), we conclude that the following two conditions are equivalent: (I') S' is $\mathcal{C}$-positive; (II') S' is $\mathcal{C}_L$-positive and $V' = \mathrm{Ind}_{\mathcal{C}_H \uparrow \mathcal{C}_L}(S')$ is $\mathcal{C}$-positive. If these conditions hold, then

$$\underset{\mathcal{C}_H \uparrow \mathcal{C}}{\mathrm{Ind}\,(S')} \cong \underset{\mathcal{C}_L \uparrow \mathcal{C}}{\mathrm{Ind}\,(V')}. \qquad (14)$$

On the other hand, by (13) and 12.6, (I') holds if and only if S is $\mathcal{B}$-positive; and (II') holds if and only if S is $\mathcal{B}_L$-positive and $V = \mathrm{Ind}_{\mathcal{B}_H \uparrow \mathcal{B}_L}(S)$ is $\mathcal{B}$-positive. This proves the equivalence of (I) and (II) of the theorem. If (I) and (II) hold, that is, if (I') and (II') hold, then (14) and Proposition 12.6 imply (11). This completes the proof. ■

12.16.   Just as we obtained 12.11 and 12.12 from 12.10 so from 12.15 we obtain as special cases the next two results.

Corollary. *Suppose that $\mathscr{B}$ has enough unitary multipliers. Let H and L be two closed subgroups of G with $H \subset L$. If S is a non-degenerate *-representation of $\mathscr{B}_H$, we have*

$$\underset{\mathscr{B}_H \uparrow \mathscr{B}}{\mathrm{Ind}}\,(S) \cong \underset{\mathscr{B}_L \uparrow \mathscr{B}}{\mathrm{Ind}} \left(\underset{\mathscr{B}_H \uparrow \mathscr{B}_L}{\mathrm{Ind}}\,(S) \right).$$

12.17. In the group case this becomes:

Corollary. *Let H and L be two closed subgroups of G with $H \subset L$. If S is a unitary representation of H, then*

$$\underset{H \uparrow G}{\mathrm{Ind}}(S) \cong \underset{L \uparrow G}{\mathrm{Ind}} \left(\underset{H \uparrow L}{\mathrm{Ind}}(S) \right).$$

12.18. The results 12.15–12.17 have of course many noteworthy special cases. Here is one:

Corollary. *Let H be any closed subgroup of G; and let R^H and R^G be the regular representations of H and G respectively. Then*

$$R^G \cong \underset{H \uparrow G}{\mathrm{Ind}}(R^H).$$

Proof. Apply 12.17 to the two subgroups $\{e\}$ and H. ■

Lifting and Induced Representations

12.19. Let G' be another locally compact group, and $p: G' \to G$ a continuous open homomorphism of G' onto G. Thus we can form the retraction $\mathscr{B}' = \langle B', \pi', \cdot, * \rangle$ of $\mathscr{B}$ by p (see VIII.3.16); this is a Banach *-algebraic bundle over G'. Let $P: B' \to B$ be the map given by

$$P(\langle x', b \rangle) = b \qquad\qquad (\langle x', b \rangle \in B').$$

If T is a *-representation of $\mathscr{B}$, then $T \circ P$ is a *-representation of $\mathscr{B}'$, called the *lift of T to $\mathscr{B}'$*.

Proposition. *Let H be a closed subgroup of G, and set $H' = p^{-1}(H)$. Let S be a non-degenerate *-representation of $\mathscr{B}_H$, and $S' = S \circ (P|B'_{H'})$ the lift of S to $\mathscr{B}'_{H'}$. Then:*

(I) *S is $\mathscr{B}$-positive if and only if S' is $\mathscr{B}'$-positive; and*

(II) *if this is the case, $\mathrm{Ind}_{\mathscr{B}'_{H'} \uparrow \mathscr{B}'}(S')$ is equivalent to the lift of $\mathrm{Ind}_{\mathscr{B}_H \uparrow \mathscr{B}}(S)$ to $\mathscr{B}'$.*

Proof.   (I) follows immediately from 8.9(III).

Assume that S and S' are positive; and let $\mathcal{Y}$ and $\mathcal{Y}'$ be the Hilbert bundles which they induce. We shall show that $\mathcal{Y}$ and $\mathcal{Y}'$ are essentially the same.

It follows from III.13.20 that if ρ is a continuous everywhere positive H-rho function on G, then $\rho' = \rho \circ p$ is an H'-rho function on G'. Now let α be a coset in G/H, and $\alpha' = p^{-1}(\alpha)$ the corresponding coset in G'/H'. Let κ_α and $\kappa'_{\alpha'}$ be the quotient maps into Y_α and $Y'_{\alpha'}$ respectively defined in 9.5. If $\xi, \eta \in X(S)$, b', $c' \in B'_{\alpha'}$, and $\pi'(b') = x'$, $\pi'(c') = y'$, we have by 9.5(7)

$$(\kappa'_{\alpha'}(b' \otimes \xi), \kappa'_{\alpha'}(c' \otimes \eta)) = (\rho'(x')\rho'(y'))^{-1/2}(S'_{c'*b'}\xi, \eta)$$

$$= (\rho(x)\rho(y))^{-1/2}(S_{c*b}\xi, \eta)$$

$$= (\kappa_\alpha(b \otimes \xi), \kappa_\alpha(c \otimes \eta)),$$

where $b = P(b')$, $c = P(c')$, $x = \pi(b) = p(x')$, and $y = \pi(c) = p(y')$. It follows that the equation

$$F_\alpha(\kappa'_{\alpha'}(b' \otimes \xi)) = \kappa_\alpha(P(b') \otimes \xi) \qquad (b' \in B'_{\alpha'}; \xi \in X(S))$$

defines a linear isometry F_α of $Y'_{\alpha'}$ onto Y_α. The map $F: \mathcal{Y}' \to \mathcal{Y}$ which coincides with F_α on each $Y'_{\alpha'}$ is thus a bijection; and it follows from 9.20 that F is a homeomorphism. Thus F identifies $\mathcal{Y}'$ with $\mathcal{Y}$, and so $\mathcal{L}_2(\rho'^\#; \mathcal{Y}')$ with $\mathcal{L}_2(\rho^\#; \mathcal{Y})$. Under this identification the action of b' on $\mathcal{Y}'$ ($b' \in B'$) coincides with the action of $P(b')$ on $\mathcal{Y}$. Hence the same is true of $\mathcal{L}_2(\rho'^\#; \mathcal{Y}')$ and $\mathcal{L}_2(\rho^\#; \mathcal{Y})$; and the proof is complete. ∎

12.20.   In the group case, of course, Proposition 12.19 becomes:

Corollary.   *Let G, G', p, H, H' be as in 12.19. Let V be a unitary representation of H, and $V' = V \circ (p|H')$ the lifted unitary representation of H'. Then*

$$\operatorname*{Ind}_{H' \uparrow G'}(V') \cong \left(\operatorname*{Ind}_{H \uparrow G}(V) \right) \circ p.$$

12.21.   In particular, 12.20 holds when $G' = G$ and p is an automorphism of G. Suppose in fact that p is the inner automorphism $x \mapsto yxy^{-1}$ of G. Then $T \circ p \cong T$ for any unitary representation T of G; and 12.20 gives:

Corollary.   *Let H be a closed subgroup of G, V a unitary representation of H, and y an element of G. Let $H' = y^{-1}Hy$; and let V' be the unitary representation $x \mapsto V_{yxy^{-1}}$ of H'. Then*

$$\operatorname*{Ind}_{H' \uparrow G}(V') \cong \operatorname*{Ind}_{H \uparrow G}(V).$$

13. Restriction and Tensor Products of Induced Representations of Banach *-Algebraic Bundles

13.1. Throughout this section $\mathscr{B}$ is a Banach *-algebraic bundle over the locally compact group G with unit e, left Haar measure λ, and modular function Δ.

Restriction of Representations

13.2. Proposition. *Let H and K be two closed subgroups of G such that $KH = G$. Thus the map $\Phi: k(H \cap K) \mapsto kH$ $(k \in K)$ is a bijection of $K/(H \cap K)$ onto G/H; and we shall assume that this bijection is a homeomorphism. We shall also assume that for each x in G the linear span of $\bigcup \{B_k B_h : k \in K, h \in H, kh = x\}$ is dense in B_x. Now let S be a $\mathscr{B}$-positive non-degenerate *-representation of $\mathscr{B}_H$. Then $S' = S|\mathscr{B}_{H \cap K}$ is non-degenerate and $\mathscr{B}_K$-positive, and*

$$\operatorname*{Ind}_{\mathscr{B}_{H \cap K} \uparrow \mathscr{B}_K} (S') \cong \operatorname*{Ind}_{\mathscr{B}_H \uparrow \mathscr{B}} (S)|\mathscr{B}_K. \tag{1}$$

Proof. S' is non-degenerate by VIII.9.4. It is $\mathscr{B}$-positive by 8.12, and hence $\mathscr{B}_K$-positive by 8.13. So the induced representation on the left side of (1) can be formed.

As in §9 let ρ be a continuous everywhere positive H-rho function on G. By III.14.14,

$$\rho|K \text{ is an } (H \cap K)\text{-rho function on } K. \tag{2}$$

Let us identify the topological spaces G/H and $K/(H \cap K)$ by means of the homeomorphism Φ. Then the Hilbert bundles $\mathscr{Y}$ and $\mathscr{Z}$ induced by S and S' (using ρ and $\rho|K$) respectively are both over G/H. We shall now show that $\mathscr{Y}$ and $\mathscr{Z}$ are essentially the same Hilbert bundle.

Let $\alpha \in G/H$. If $b \in B_{\alpha \cap K}$ and $\xi \in X(S)$, we shall denote by $\kappa_\alpha(b \otimes \xi)$ and $\kappa'_\alpha(b \otimes \xi)$ the images of $b \otimes \xi$ in Y_α and Z_α respectively. If in addition $c \in B_{\alpha \cap K}$ and $\eta \in X(S)$, it follows from 9.5(7) that

$$(\kappa_\alpha(b \otimes \xi), \kappa_\alpha(c \otimes \eta))_{Y_\alpha} = (\kappa'_\alpha(b \otimes \xi), \kappa'_\alpha(c \otimes \eta))_{Z_\alpha}. \tag{3}$$

Hence the equation

$$F_\alpha(\kappa'_\alpha(b \otimes \xi)) = \kappa_\alpha(b \otimes \xi) \tag{4}$$

$(b \in B_{\alpha \cap K}; \xi \in X)$ defines a linear isometry $F_\alpha: Z_\alpha \to Y_\alpha$, whose range R_α is the closure of the linear span of $\{\kappa_\alpha(b \otimes \xi): b \in B_{\alpha \cap K}, \xi \in X(S)\}$. We claim that

$R_\alpha = Y_\alpha$. Indeed: If $c \in B_H$, $b \in B_{\alpha \cap K}$ and $\xi \in X(S)$, it follows from 9.15(28) that $\kappa_\alpha(bc \otimes \xi) \in R_\alpha$. By the denseness hypothesis of this proposition, such products bc span a dense subspace of B_x for each x in α. Hence, by the continuity assertion 9.19, R_α contains $\kappa_\alpha(d \otimes \xi)$ for all d in B_α and ξ in $X(S)$, and the claim is proved. So $F_\alpha : Z_\alpha \to Y_\alpha$ is an isometric surjection.

Thus the map $F : \mathscr{Z} \to \mathscr{Y}$ which coincides on each Z_α with F_α is a norm-preserving bijection. It follows from (4) and 9.20 that F is a homeomorphism. In particular F gives rise to an isometric bijection $\tilde{F} : \mathscr{L}_2(\rho^\# ; \mathscr{Z}) \to \mathscr{L}_2(\rho^\# ; \mathscr{Y})$.

Comparing (4) with the definition 9.11(18) of the action of $\mathscr{B}$ on $\mathscr{Y}$, we see that F carries the action of $\mathscr{B}_K$ on $\mathscr{Z}$ into the restriction to $\mathscr{B}_K$ of the action of $\mathscr{B}$ on $\mathscr{Y}$. From this and the definition 9.21(40) of the operators of the induced representations, it follows that $\tilde{F}$ intertwines the two sides of (1). So the latter are unitarily equivalent. ∎

Remark 1. If K is σ-compact, it follows from Theorem III.3.10 (applied with G and M replaced by K and G/H) is automatically a homeomorphism.

Remark 2. The denseness hypothesis in the above proposition certainly holds if $\mathscr{B}$ is saturated, in particular if $\mathscr{B}$ is the group bundle. Here is a simple example showing that in general it cannot be omitted. Let $G = \{e, x, y, z\}$ be the four-element non-cyclic group; and let H, K, L be its two-element subgroups $\{e, x\}$, $\{e, y\}$, $\{e, z\}$ respectively. Let $\mathscr{B}$ be the Banach *-algebraic bundle over G such that $\mathscr{B}_H$ is the group bundle of H, while $B_y = \{0\}$, $B_z = \{0\}$. Let χ be the one-dimensional *-representation of $\mathscr{B}_K$ sending $\langle t, e \rangle$ into t ($t \in \mathbb{C}$). Then χ is $\mathscr{B}$-positive and $\mathrm{Ind}_{\mathscr{B}_K \uparrow \mathscr{B}}(\chi)$ is two-dimensional. On the other hand $LK = G$, $L \cap K = \{e\}$, and $\mathrm{Ind}_{B_e \uparrow \mathscr{B}_L}(\chi | B_e)$ is one-dimensional.

Remark 3. This proposition generalizes Corollary 10.15.

Inducing and Outer Tensor Products

13.3. We remind the reader of the notion of the outer tensor product, defined in VIII.21.25.

Let K be another locally compact group; and let P be the projection of $K \times G$ onto $G : P(k, x) = x$. The retraction $K \times \mathscr{B}$ of $\mathscr{B}$ by P is then a Banach *-algebraic bundle over $K \times G$, whose bundle space is $K \times B$ (see VIII.3.16). If T is a non-degenerate *-representation of $\mathscr{B}$ and U is a unitary representation of K, the equation

$$(U \underset{o}{\otimes} T)_{k, b} = U_k \otimes T_b \tag{5}$$

defines a non-degenerate *-representation $U \underset{\circ}{\otimes} T$ of $K \times \mathscr{B}$, called the *outer tensor product* of U and T.

We are going to show that the operation of inducing commutes with the formation of outer tensor products. First we need a lemma.

13.4.   Lemma.   *Keep the notation of 13.3; and define* $\Phi: \mathscr{L}(K) \otimes \mathscr{L}(\mathscr{B}) \to \mathscr{L}(K \times \mathscr{B})$ *as follows:* $\Phi(\phi \otimes f)(k, x) = \phi(k)f(x)$ $(\phi \in \mathscr{L}(K); \; f \in \mathscr{L}(\mathscr{B}); \; k \in K; \; x \in G)$. *Then* Φ *is a *-isomorphism of the tensor product *-algebra* $\mathscr{L}(K) \otimes \mathscr{L}(\mathscr{B})$ *into* $\mathscr{L}(K \times \mathscr{B})$; *and its range is dense in* $\mathscr{L}(K \times \mathscr{B})$ *in the inductive limit topology. Further, if* U *and* T *are as in 13.3,*

$$(U \underset{\circ}{\otimes} T)_{\Phi(\phi \otimes f)} = U_\phi \otimes T_f \tag{6}$$

for all ϕ *in* $\mathscr{L}(K)$ *and* f *in* $\mathscr{L}(\mathscr{B})$.

Sketch of Proof.   If Y and Z are any linear spaces and M and N are subsets of $Y^{\#}$ and $Z^{\#}$ which separate the points of Y and Z respectively, then the subset $\{\alpha \otimes \beta : \alpha \in M, \; \beta \in N\}$ of $(Y \otimes Z)^{\#}$ separates the points of $Y \otimes Z$. From this general fact follows the one-to-oneness of Φ.

That Φ is a *-homomorphism is easily checked.

If $\mathscr{B}$ is the group bundle of G, the denseness of range(Φ) follows from the Stone-Weierstrass Theorem. From this it is easy to see that, in the general case, the inductive limit closure R of range(Φ) is closed under multiplication by functions in $\mathscr{C}(K \times G)$. From this and II.14.6 we deduce the denseness of R, hence of range(Φ).

An easy calculation verifies equation (6).   ■

13.5.   Let K and $K \times \mathscr{B}$ be as in 13.3. In addition let H and M be closed subgroups of G and K respectively; and take a unitary representation V of M and a non-degenerate $\mathscr{B}$-positive *-representation S of $\mathscr{B}_H$. We can then form:

(i)   the (non-degenerate) *-representation $V \underset{\circ}{\otimes} S$ of $M \times \mathscr{B}_H = (K \times \mathscr{B})_{M \times H}$,

(ii)   the induced unitary representation $U = \mathrm{Ind}_{M \uparrow K}(V)$ of K, and

(iii)   the induced *-representation $T = \mathrm{Ind}_{\mathscr{B}_H \uparrow \mathscr{B}}(S)$ of $\mathscr{B}$.

Proposition. $V \underset{\circ}{\otimes} S$ *is positive with respect to* $K \times \mathscr{B}$; *and*

$$\underset{M \times \mathscr{B}_H \uparrow K \times \mathscr{B}}{\mathrm{Ind}} (V \underset{\circ}{\otimes} S) \cong U \underset{\circ}{\otimes} T. \tag{7}$$

Proof. This will be proved by combining 5.2 with the Rieffel description of the inducing operation on bundles.

Let Δ, δ, Δ', δ' be the modular functions of G, H, K, M respectively. By III.9.3 the modular functions Δ'' and δ'' of $K \times G$ and $M \times H$ respectively are given by:

$$\Delta''(k, x) = \Delta'(k)\Delta(x), \qquad \delta''(m, h) = \delta'(m)\delta(h). \tag{8}$$

Let $\quad p: \mathscr{L}(\mathscr{B}) \to \mathscr{L}(\mathscr{B}_H), \quad p': \mathscr{L}(K) \to \mathscr{L}(M), \quad$ and $\quad p'': \mathscr{L}(K \times \mathscr{B}) \to \mathscr{L}(M \times \mathscr{B}_H)$ be the conditional expectations introduced in 8.4:

$$p(f)(h) = \Delta(h)^{1/2}\delta(h)^{-1/2}f(h),$$

$$p'(\phi)(m) = \Delta'(m)^{1/2}\delta'(m)^{-1/2}\phi(m),$$

$$p''(F)(m, h) = \Delta''(m, h)^{1/2}\delta''(m, h)^{-1/2}F(m, h)$$

$(f \in \mathscr{L}(\mathscr{B}); \phi \in \mathscr{L}(K), F \in \mathscr{L}(K \times \mathscr{B}); h \in H; m \in M)$. In view of 13.4 we can identify $\mathscr{L}(K) \otimes \mathscr{L}(\mathscr{B})$ with a dense *-subalgebra of $\mathscr{L}(K \times \mathscr{B})$, and $\mathscr{L}(M) \otimes \mathscr{L}(\mathscr{B}_H)$ with a dense *-subalgebra of $\mathscr{L}(M \times \mathscr{B}_H)$. By (8) we then have:

$$p''(\phi \otimes f) = p'(\phi) \otimes p(f) \qquad (\phi \in \mathscr{L}(K); f \in \mathscr{L}(\mathscr{B})). \tag{9}$$

Let p_0'' be the restriction of p'' to $\mathscr{L}(K) \otimes \mathscr{L}(\mathscr{B})$. It follows from 5.2 and (9) (and the analogue of (6) for $V \underset{\circ}{\otimes} S$) that the integrated form of $V \underset{\circ}{\otimes} S$ is inducible via p_0'' to $\mathscr{L}(K) \otimes \mathscr{L}(\mathscr{B})$, and that

$$\mathrm{Ind}^{p_0''}(V \underset{\circ}{\otimes} S) \cong \mathrm{Ind}^{p'}(V) \otimes \mathrm{Ind}^{p}(S). \tag{10}$$

In particular $V \underset{\circ}{\otimes} S$ is positive with respect to p_0''. On the other hand, by 9.25, $\mathrm{Ind}^{p'}(V)$ and $\mathrm{Ind}^{p}(S)$ are the integrated forms $\tilde{U}$ and $\tilde{T}$ of U and T respectively. So (10) becomes:

$$\mathrm{Ind}^{p_0''}(V \underset{\circ}{\otimes} S) \cong \tilde{U} \otimes \tilde{T}. \tag{11}$$

Now by (6) the right side of (11) is the restriction to $\mathscr{L}(K) \otimes \mathscr{L}(\mathscr{B})$ of the integrated form of $U \otimes T$. So, by the denseness of $\mathscr{L}(K) \underset{\circ}{\otimes} \mathscr{L}(\mathscr{B})$ (13.4), the proposition will be proved if we show that (i) $V \otimes S$ is positive with respect to p'' (i.e., $(K \times \mathscr{B})$-positive), and that (ii) the left side of (11) is the restriction to $\mathscr{L}(K) \otimes \mathscr{L}(\mathscr{B})$ of the integrated form of the left side of (7).

For this we argue as in the proof of 12.10. Take any element F of $\mathscr{L}(K \times \mathscr{B})$. By 13.4 F is the limit, in the inductive limit topology, of a net $\{F_r\}$ of elements of $\mathscr{L}(K) \otimes \mathscr{L}(\mathscr{B})$; and since the latter is closed under multiplication by complex functions of the form $\langle k, x \rangle \mapsto \phi(k)\psi(x)$ (where $\phi \in \mathscr{C}(K)$, $\psi \in \mathscr{C}(G)$), we may as well assume that the F_r all vanish outside the same compact set. But then $F_r^* * F_r \to F^* * F$ in the inductive limit topology, so that, for any ζ in $X(V) \otimes X(S)$,

$$((V \underset{\circ}{\otimes} S)_{p''(F_r^* * F_r)} \zeta, \zeta) \to ((V \underset{\circ}{\otimes} S)_{p''(F^* * F)} \zeta, \zeta). \tag{12}$$

Since $F_r \in \mathscr{L}(K) \otimes \mathscr{L}(\mathscr{B})$ and $V \underset{\circ}{\otimes} S$ is positive with respect to p_0'', the left side of (12) is non-negative for each r; hence the right side is non-negative. So $V \underset{\circ}{\otimes} S$ is $(K \times \mathscr{B})$-positive; and the left side of (7) has meaning; call it Q.

If ζ, F, and $\{F_r\}$ are as in the preceding paragraph, we show (as in the corresponding step of the proof of 12.10) that

$$F_r \underset{\sim}{\widetilde{\otimes}} \zeta \to F \widetilde{\otimes} \zeta \qquad \text{in } X(Q).$$

This implies that the space of $\text{Ind}^{p_0''}(V \underset{\circ}{\otimes} S)$ is dense in, and hence equal to, the space of Q. Hence $\text{Ind}^{p_0''}(V \underset{\circ}{\otimes} S)$ is the restriction of Q to $\mathscr{L}(K) \otimes \mathscr{L}(\mathscr{B})$; and the proof is finished. ∎

Remark. Let $\mathscr{Y}$, $\mathscr{Y}'$, and $\mathscr{Y}''$ be the Hilbert bundles induced by S, V, and $V \underset{\circ}{\otimes} S$ in the above discussion. One can verify without difficulty that $\mathscr{Y}''$ is the Hilbert bundle tensor product, in the sense of II.15.16, of $\mathscr{Y}$ and $\mathscr{Y}'$, and that the unitary equivalence (7) is realized by the isometry Φ of II.15.17.

13.6. In the group case, of course, 13.5 becomes:

Corollary. *Let G, H, K, M be as in 13.5; and let V and S be unitary representations of M and H respectively. Then*

$$\underset{M \times H \uparrow K \times G}{\text{Ind}} (V \underset{\circ}{\otimes} S) \cong \underset{M \uparrow K}{\text{Ind}}(V) \underset{\circ}{\otimes} \underset{H \uparrow G}{\text{Ind}}(S).$$

Inducing and Inner Tensor Products

13.7. The reader will recall from VIII.9.16 the definition of the tensor product $U \otimes T$ of a unitary representation U of G and a *-representation T of $\mathscr{B}$. This $U \otimes T$ is often called the *inner tensor product*, to distinguish it from the outer tensor products discussed in 13.5.

13.8. Proposition. *Assume that $\mathscr{B}$ has an approximate unit. Let H be a closed subgroup of G, and S a non-degenerate $\mathscr{B}$-positive *-representation of $\mathscr{B}_H$; and let U be a unitary representation of G. Then the inner tensor product *-representation $(U|H) \otimes S$ of $\mathscr{B}_H$ is $\mathscr{B}$-positive; and*

$$\underset{\mathscr{B}_H \uparrow \mathscr{B}}{\text{Ind}} ((U|H) \otimes S) \cong U \otimes \underset{\mathscr{B}_H \uparrow \mathscr{B}}{\text{Ind}} (S). \tag{13}$$

Proof. By 13.5 $U \underset{\circ}{\otimes} S$ (as a *-representation of $G \times \mathscr{B}_H = (G \times \mathscr{B})_{G \times H}$) is positive with respect to $G \times \mathscr{B}$, and we have (recalling 9.28):

$$\underset{G \times \mathscr{B}_H \uparrow G \times \mathscr{B}}{\text{Ind}} \ (U \underset{\circ}{\otimes} S) \cong U \underset{\circ}{\otimes} \underset{\mathscr{B}_H \uparrow \mathscr{B}}{\text{Ind}} (S). \tag{14}$$

Let D be the diagonal subgroup $\{\langle x, x \rangle : x \in G\}$ of $G \times G$. In the Restriction Theorem 13.2 let us replace the G, H, K, and $\mathscr{B}$ of that theorem by our present $G \times G$, $G \times H$, D, and $G \times \mathscr{B}$. Then the hypotheses of 13.2 are satisfied; the denseness hypothesis results from the presence of an approximate unit in $\mathscr{B}$. By the preceding paragraph we can replace the S of 13.2 by $U \underset{\circ}{\otimes} S$. Thus, combining (14) with 13.2(1) we find that the restriction T of the right side of (14) to $(G \times \mathscr{B})_D$ is equivalent to the result of inducing to $(G \times \mathscr{B})_D$ the restriction W of $U \underset{\circ}{\otimes} S$ to $(G \times \mathscr{B})_{(G \times H) \cap D}$.

Now $(G \times \mathscr{B})_D$ is isometrically isomorphic with $\mathscr{B}$ itself under the map $\langle \pi(b), b \rangle \mapsto b$ (which is covariant with the isomorphism $\langle x, x \rangle \mapsto x$ of D with G). When $(G \times \mathscr{B})_D$ and $\mathscr{B}$ (and D and G) are identified via these isomorphisms, T becomes the inner tensor product *-representation $U \otimes \text{Ind}_{\mathscr{B}_H \uparrow \mathscr{B}}(S)$ of $\mathscr{B}$; and W becomes the inner tensor product *-representation $(U|H) \otimes S$ of $\mathscr{B}_H$. Therefore the last sentence of the preceding paragraph becomes just (13).   ∎

13.9. The next result differs from 13.8 only in that the roles of the *-representation and of the unitary representation are interchanged.

Proposition. *Assume that $\mathscr{B}$ is saturated; and let H be a closed subgroup of G. Let T be a non-degenerate *-representation of $\mathscr{B}$, and V a unitary representation of H. Then $V \otimes (T|\mathscr{B}_H)$ is $\mathscr{B}$-positive, and*

$$\underset{\mathscr{B}_H \uparrow \mathscr{B}}{\text{Ind}} (V \otimes (T|\mathscr{B}_H)) \cong \underset{H \uparrow G}{\text{Ind}}(V) \otimes T. \tag{15}$$

The proof follows from 13.5 and 13.2 in the same way as did the proof of 13.8. The only difference is that in the present context one needs saturation of $\mathscr{B}$ in order to satisfy the denseness hypothesis of 13.2. We leave the details to the reader.

13.10. In the group case both 13.8 and 13.9 reduce to the following:

Corollary. *Let H be a closed subgroup of G; and let T and V be unitary representations of G and H respectively. Then*

$$\underset{H \uparrow G}{\text{Ind}}(V \otimes (T|H)) \cong \underset{H \uparrow G}{\text{Ind}}(V) \otimes T. \tag{16}$$

13.11. In the special case when $H = \{e\}$ and V is trivial, $\text{Ind}_{H \uparrow G}(V)$ is just the regular representation R of G. By 12.2 $\text{Ind}_{H \uparrow G}(V \otimes (T|H))$ is the direct sum $\dim(T) \cdot R$ of $\dim(T)$ copies of R, and we obtain:

Corollary. *For any unitary representation T of G,*

$$R \otimes T \cong \dim(T) \cdot R. \tag{17}$$

Remark. (17) says that, among unitary representations of G, R behaves somewhat like the zero element with respect to the operation of (inner) tensor product.

14. The Imprimitivity Theorem for Banach *-Algebraic Bundles

14.1. In this section, as usual, $\mathscr{B}$ is a Banach *-algebraic bundle over the locally compact group G with unit e, left Haar measure λ, and modular function Δ. We shall also fix a closed subgroup H of G, with left Haar measure ν and modular function δ. We choose once for all an everywhere positive continuous H-rho function ρ on G, and denote by $\rho^{\#}$ the corresponding measure on G/H (see III.13.10).

14.2. Our goal in this section is to prove the Imprimitivity Theorem for Banach *-algebraic bundles. The approach is as follows: In VIII.7.8 we defined the compacted transformation algebra E derived from $\mathscr{B}$ and the G-space G/H; and in VIII.18.22 we established a one-to-one correspondence between the (continuous) non-degenerate *-representations of E and the systems of imprimitivity for $\mathscr{B}$ over G/H. Furthermore, in 8.4 of this chapter we made $\mathscr{L}(\mathscr{B})$ into a right $\mathscr{L}(\mathscr{B}_H)$-rigged space (with a rigging $[\ ,\ ]$ derived from an $\mathscr{L}(\mathscr{B})$, $\mathscr{L}(\mathscr{B}_H)$ conditional expectation p). In this section we are going to define a left E-module structure for $\mathscr{L}(\mathscr{B})$, and also an E-valued rigging $[\ ,\ ]_E$ on $\mathscr{L}(\mathscr{B})$, such that $\mathscr{L}(\mathscr{B})$, $[\ ,\ ]_E$, $[\ ,\ ]$ *will become an* E, $\mathscr{L}(\mathscr{B}_H)$ *imprimitivity bimodule*. In fact, if $\mathscr{B}$ is saturated, this imprimitivity bimodule will turn out to be topologically strict in a certain sense, and so, by the abstract Imprimitivity Theorem (§6), will generate an isomorphism between the *-representation theory of $\mathscr{L}(\mathscr{B}_H)$—that is, of $\mathscr{B}_H$—and the *-representation theory of E—that is, the systems of imprimitivity for $\mathscr{B}$ over G/H. In other words, *if $\mathscr{B}$ is saturated, we shall have a one-to-one correspondence, set up by the inducing process, between *-representations of $\mathscr{B}_H$ and systems of imprimitivity for $\mathscr{B}$ over G/H.* This is the Imprimitivity Theorem in the bundle context. In particular, in the group case we will get a one-to-one correspondence between unitary representations of H and systems of imprimitivity for G over G/H. The important direction of this correspondence is expressed by the classical Imprimitivity Theorem of Mackey: *Every system of imprimitivity for G over the G-space G/H is obtained by the inducing process from a unique unitary representation of H.*

Remark.   We have already seen in 6.11 how this $E, \mathscr{L}(\mathscr{B}_H)$ imprimitivity bimodule should be constructed in the group case when G is finite. The construction of 6.11 serves as an illuminating guide for the much more general construction of this section.

Induced Systems of Imprimitivity

14.3.   We shall begin by observing that the inducing process, in its bundle formulation, directly generates not merely a *-representation but a system of imprimitivity for $\mathscr{B}$.

Let S be a non-degenerate $\mathscr{B}$-positive *-representation of $\mathscr{B}_H$, and $\mathscr{Y} = \{Y_\alpha\}$ the Hilbert bundle over G/H induced by S. Thus $\mathscr{L}_2(\rho^\#; \mathscr{Y})$ is the space $X(T)$ of the induced *-representation $T = \text{Ind}_{\mathscr{B}_H \uparrow \mathscr{B}}(S)$ of $\mathscr{B}$. By II.15.13 the equation

$$(P(W)f)(\alpha) = \text{Ch}_W(\alpha) f(\alpha) \tag{1}$$

($f \in X(T)$; W a Borel subset of G/H; $\alpha \in G/H$) defines a regular $X(T)$-projection-valued Borel measure P on G/H. If $f \in X(T)$, $b \in B_x$, and W is a Borel subset of G/H, then for α in G/H

$$
\begin{aligned}
(T_b P(W)f)(\alpha) &= \tau_b[(P(W)f)(x^{-1}\alpha)] &&\text{(see 9.21(40))} \\
&= \text{Ch}_W(x^{-1}\alpha)\tau_b(f(x^{-1}\alpha)) \\
&= \text{Ch}_{xW}(\alpha)(T_b f)(\alpha) \\
&= (P(xW)T_b f)(\alpha),
\end{aligned}
$$

whence

$$T_b P(W) = P(xW)T_b.$$

Thus, according to VIII.18.7, $\langle T, P \rangle$ is a system of imprimitivity for $\mathscr{B}$ over G/H.

Definition. P and $\langle T, P \rangle$ are called respectively the *projection-valued measure induced by S* and the *system of imprimitivity induced by S*.

By the uniqueness statement in II.12.8, a regular projection-valued Borel measure P on G/H is uniquely determined by the corresponding *-representation $P': \phi \mapsto \int_{G/H} \phi \, dP$ of $\mathscr{C}_0(G/H)$. We call P' the *integrated form of P*.

In this connection notice from Proposition II.15.13 that, for any bounded continuous complex function ϕ on G/H, the spectral integral $\int \phi \, dP$ is just the operator of pointwise multiplication of cross-sections of $\mathscr{Y}$ by the function ϕ.

Remark. Suppose that $\mathscr{B}$ is the group bundle, so that S is a unitary representation of H. Let T be the Blattner formulation (acting on $\mathscr{L}$, as in 10.8) of the induced unitary representation $\mathrm{Ind}_{H \uparrow G}(S)$ of G. It is easy to see that in that case the corresponding description of P is almost the same as (1), namely:

$$(P(W)f)(x) = \mathrm{Ch}_W(xH)f(x) \tag{2}$$

$(f \in \mathscr{L}; W$ a Borel subset of $G/H; x \in G)$.

14.4. Like T, the P of 14.3 admits another description, in terms of the Rieffel inducing process. Let P' be the integrated form of P; and let us adopt the notation of 9.8.

Notice that $\mathscr{L}(\mathscr{B})$ is a left $\mathscr{C}_0(G/H)$-module under the action:

$$(\phi f)(x) = \phi(xH)f(x) \tag{3}$$

$(\phi \in \mathscr{C}_0(G/H); f \in \mathscr{L}(\mathscr{B}); x \in G)$. One easily checks that with this left action, and with the right $\mathscr{L}(\mathscr{B}_H)$-module structure of 8.4(3) and the rigging $[\ ,\ ] : \langle f, g \rangle \mapsto p(f^* * g)$ derived from the conditional expectation p of 8.4, $\mathscr{L}(\mathscr{B})$ becomes a right $\mathscr{L}(\mathscr{B}_H)$-rigged left $\mathscr{C}_0(G/H)$-module. Further, if $\phi \in \mathscr{C}_0(G/H), f \in \mathscr{L}(\mathscr{B})$, and $\xi \in X(S)$, the cross-section $\kappa(\phi f \otimes \xi)$ of $\mathscr{Y}$ (see 9.8) is the same as $P'_\phi(\kappa(f \otimes \xi))$. From this and Proposition 9.8 we see that S induces a *-representation of $\mathscr{C}_0(G/H)$ via $\mathscr{L}(\mathscr{B}), [\ ,\ ]$ (in the abstract sense of 4.9); and that this induced *-representation is unitarily equivalent to P' (under the E of 9.8).

Remark. This fact should be placed side by side with the analogous fact proved in 9.25 for the induced *-representation T of $\mathscr{B}$. It shows that we had every right to describe P in 14.3 as *induced by S*.

14.5. For the reader's convenience we recall from VIII.7.7 the definition of the compacted transformation *-algebra E derived from $\mathscr{B}$ and the G-space G/H: E consists of all continuous functions $u: G \times (G/H) \to B$ satisfying

$u(x, \alpha) \in B_x$ $(x \in G; \alpha \in G/H)$ and having compact support; and its multiplication and involution are given by:

$$(u * v)(x, \alpha) = \int_G u(y, \alpha)v(y^{-1}x, y^{-1}\alpha)d\lambda y, \qquad (4)$$

$$u^*(x, \alpha) = \Delta(x^{-1})[u(x^{-1}, x^{-1}\alpha)]^* \qquad (5)$$

$(u, v \in E; x \in G; \alpha \in G/H)$. In fact E is a normed *-algebra under the norm:

$$\|u\| = \int_G \sup\{\|u(x, \alpha)\| : \alpha \in G/H\}d\lambda x. \qquad (6)$$

14.6.   Our next step is to make $\mathcal{L}(\mathcal{B})$ into a left E-module. This we do by means of the following action of E on $\mathcal{L}(\mathcal{B})$:

$$(uf)(x) = \int_G u(y, xH)f(y^{-1}x)d\lambda y \qquad (7)$$

$(u \in E; f \in \mathcal{L}(\mathcal{B}); x \in G)$.

Notice that (7) is the obvious generalization of 6.11(21).

Since the integrand in (7), as a function of y, is B_x-valued and continuous with compact support, the right side of (7) exists as a B_x-valued integral. Thus uf as defined by (7) is a cross-section of $\mathcal{B}$ with compact support; and it is continuous by Proposition II.15.19. So it belongs to $\mathcal{L}(\mathcal{B})$. Obviously uf is linear in u and in f. Using II.5.7 and the Fubini Theorem II.16.3, we have:

$$((u * v)f)(x) = \int (u * v)(y, xH)f(y^{-1}x)d\lambda y$$

$$= \iint u(z, xH)v(z^{-1}y, z^{-1}xH)f(y^{-1}x)d\lambda z \, d\lambda y$$

$$= \iint u(z, xH)v(y, z^{-1}xH)f(y^{-1}z^{-1}x)d\lambda y \, d\lambda z$$

$$= \int u(z, xH)(vf)(z^{-1}x)d\lambda z$$

$$= (u(vf))(x)$$

$(u, v \in E; f \in \mathcal{L}(\mathcal{B}); x \in G)$. So

$$(u * v)f = u(vf),$$

and (7) does define $\mathcal{L}(\mathcal{B})$ as a left E-module.

We next observe that, with the $\mathscr{L}(\mathscr{B}_H)$-valued rigging $[\ ,\ ]$ of 14.4, and the module structures (7) and 8.4(3), $\mathscr{L}(\mathscr{B})$ becomes a right $\mathscr{L}(\mathscr{B}_H)$-rigged left E-module. Indeed, it is only necessary to verify that

$$[uf, g] = [f, u^*g] \qquad (f, g \in \mathscr{L}(\mathscr{B}); u \in E).$$

This results from a routine calculation which we leave to the reader.

Remark. Suppose that the element u of E is of the following special form:

$$u(x, \alpha) = \chi(\alpha)g(x) \qquad (x \in G; \alpha \in G/H), \qquad (8)$$

where $g \in \mathscr{L}(\mathscr{B})$ and $\chi \in \mathscr{L}(G/H)$. Then (7) becomes

$$uf = \chi(g * f) \qquad (9)$$

(the left action of χ on $\mathscr{L}(\mathscr{B})$ being given by (3)).

14.7. As in 14.3, let $\langle T, P \rangle$ be the system of imprimitivity (for $\mathscr{B}$ over G/H) induced by the $\mathscr{B}$-positive non-degenerate *-representation S of $\mathscr{B}_H$. Let R be the E-integrated form of $\langle T, P \rangle$ (defined in VIII.18.6). Thus R is a non-degenerate *-representation of E. The following proposition is the analogue for $\langle T, P \rangle$ of the fact observed for P in 14.4.

Proposition. *S induces a* *-representation R' of E via the right $\mathscr{L}(\mathscr{B}_H)$-rigged left E-module $\mathscr{L}(\mathscr{B})$ (defined in 14.6); and $R' \cong R$.*

Proof. Let E_0 be the linear span in E of the set of those u which are of the form (8). It follows from II.14.6 that E_0 is dense in E in the inductive limit topology of E.

Recall the isometry

$$f \widetilde{\otimes} \xi \mapsto \kappa(f \otimes \xi) \qquad (10)$$

defined in 9.8(15). We shall adopt the notation of 9.8 (except that (10) cannot now be denoted by E). Let R' stand for the *-representation of E on $X(V)$ which corresponds to R via (10). It is enough to show that R' is the *-representation of E abstractly induced via $\mathscr{L}(\mathscr{B})$. This means, after transforming by (10), that it is enough to show that

$$R_u(\kappa(f \otimes \xi)) = \kappa(uf \otimes \xi) \qquad (11)$$

for $u \in E, f \in \mathscr{L}(\mathscr{B})$, and $\xi \in X(S)$.

Now, applying (10) to Remark 4 of 8.6, we see that, for fixed $\xi, g \mapsto \kappa(g \otimes \xi)$ is continuous in g with respect to the inductive limit topology. Also it is easy to verify that, for fixed f, $u \mapsto uf$ is continuous in u with respect to the inductive limit topologies. Combining the last two facts with the first paragraph of this proof, one concludes that it is enough to verify (11) in the special case that u is of the form (8).

Thus, assume that $\chi \in \mathcal{L}(G/H)$, g, $f \in \mathcal{L}(\mathcal{B})$, and $\xi \in X(S)$; and let u be given by (8). In that case we have by VIII.18.17

$$R_u = P'_\chi T_g;$$

and so by (9) together with 14.4 and 9.25(44)

$$\kappa(uf \otimes \xi) = \kappa(\chi(g * f) \otimes \xi)$$
$$= P'_\chi(\kappa((g * f) \otimes \xi))$$
$$= P'_\chi T_g(\kappa(f \otimes \xi))$$
$$= R_u(\kappa(f \otimes \xi)).$$

Thus we have verified (11) for the required special case. ∎

Remark. This proposition shows that it is very appropriate to describe $\langle T, P \rangle$, or its E-integrated form, as *induced by S*. Compare Remark 14.4.

Construction of the Imprimitivity Bimodule

14.8. In the preceding numbers we have made $\mathcal{L}(\mathcal{B})$ into a left module for three different *-algebras, namely $\mathcal{L}(\mathcal{B})$ itself (under left convolution), $\mathscr{C}_0(G/H)$, and E. Of these three we shall next show that E is distinguished by the following important property: By defining a suitable E-valued rigging $[\ ,\ ]_E$ on $\mathcal{L}(\mathcal{B})$, we can make $\mathcal{L}(\mathcal{B})$, $[\ ,\ ]_E$, $[\ ,\ ]$ an E, $\mathcal{L}(\mathcal{B}_H)$ imprimitivity bimodule.

14.9. Motivated by the formula 6.11(19), we make the following definition:

Definition. If $f, g \in \mathcal{L}(\mathcal{B})$, let $[f, g]_E$ be the function on $G \times G/H$ to B given by:

$$[f, g]_E(x, yH) = \int_H f(yh)g^*(h^{-1}y^{-1}x)dvh \qquad (x, y \in G). \qquad (12)$$

To begin with, we note that the right side of (12) exists as a B_x-valued integral, and depends only on x and yH; by II.15.19 it is continuous as a function of $\langle x, y \rangle$, and hence as a function of $\langle x, yH \rangle$; and it vanishes outside a compact subset of $G \times G/H$. So $[f, g]_E \in E$.

Evidently $[f, g]_E$ is linear in f and conjugate-linear in g.

We observe for later use that $[f, g]_E$ is separately continuous in f and g in the inductive limit topologies of $\mathscr{L}(\mathscr{B})$ and E.

14.10. As before we denote by $[\ ,\ ]$ the $\mathscr{L}(\mathscr{B}_H)$-valued rigging $\langle f, g \rangle \mapsto p(f^* * g)$ on $\mathscr{L}(\mathscr{B})$ derived from the conditional expectation p of 8.4. We consider $\mathscr{L}(\mathscr{B})$ as a right $\mathscr{L}(\mathscr{B}_H)$-module by means of 8.4(3), and as a left E-module by means of (7).

Proposition. $\mathscr{L}(\mathscr{B}), [\ ,\ ]_E, [\ ,\ ]$ is an $E, \mathscr{L}(\mathscr{B}_H)$ imprimitivity bimodule.

Proof. The identities that remain to be verified are:

$$[f, g]_E^* = [g, f]_E,$$

$$u * [f, g]_E = [uf, g]_E,$$

$$[f\phi, g]_E = [f, g\phi^*]_E,$$

$$[f, g]_E q = f[g, q] \tag{13}$$

$(f, g, q \in \mathscr{L}(\mathscr{B}); u \in E; \phi \in \mathscr{L}(\mathscr{B}_H))$. These verifications are of quite a routine nature. We shall content ourselves with checking (13). By Fubini's Theorem and II.5.7,

$$(f[g, q])(x) = \int_H f(xh^{-1})[g, q](h)(\delta(h)\Delta(h))^{-1/2}\,dvh$$

$$= \int_H \int_G \delta(h)^{-1} f(xh^{-1})g^*(y)q(y^{-1}h)d\lambda y\,dvh$$

$$= \int_H \int_G \delta(h^{-1}) f(xh^{-1})g^*(hx^{-1}y)q(y^{-1}x)d\lambda y\,dvh$$

$$= \int_H \int_G f(xh)g^*(h^{-1}x^{-1}y)q(y^{-1}x)d\lambda y\,dvh$$

$$= \int_G [f, g]_E(y, xH)q(y^{-1}x)d\lambda y$$

$$= ([f, g]_E q)(x).$$

So (13) holds. ∎

Definition. This $E, \mathscr{L}(\mathscr{B}_H)$ imprimitivity bimodule $\mathscr{L}(\mathscr{B}), [\ ,\ ]_E, [\ ,\ ]$ will be called the *canonical imprimitivity bimodule derived from $\mathscr{B}$ and H*. We shall denote it by $\mathscr{I}(\mathscr{B}; H)$, or simply by $\mathscr{I}$.

The Imprimitivity Theorem

14.11. Let E' and B' denote the linear spans in E and $\mathscr{L}(\mathscr{B}_H)$ of range($[\ ,\ ]_E$) and range($[\ ,\ ]$) respectively. Considered as an E', B' imprimitivity bimodule, $\mathscr{I}$ is of course strict, and so by the abstract Imprimitivity Theorem gives rise to a one-to-one correspondence between the $\mathscr{I}$-positive *-representations of E' and of B'. However, it is not primarily the *-representations of E' and B' that interest us, but the norm-continuous *-representations of the normed *-algebras E (with norm (6)) and $\mathscr{L}(\mathscr{B}_H)$ (with the $\mathscr{L}_1$ norm). In order to pass from the conclusions about E' and B' to the conclusions that interest us, we shall have to ask the following four questions:

(I)   Is B' dense in $\mathscr{L}(\mathscr{B}_H)$?

(II) Is E' dense in E?

(III) Which *-representations of E are positive with respect to $\mathscr{I}$?

(IV) In the correspondence $S \leftrightarrow R$ between *-representations S of B' and *-representations R of E', set up by the abstract Imprimitivity Theorem, how is the norm-continuity of S related to the norm-continuity of R?

Remark. In connection with question (IV), notice that the normed *-algebras E and $\mathscr{L}(\mathscr{B}_H)$ are not complete. We know of no way of extending $\mathscr{I}$ to an imprimitivity bimodule for the completions of E and $\mathscr{L}(\mathscr{B}_H)$. So the strong result of 7.2 is not available here.

14.12. As regards question (I) we have:

Proposition. *If $\mathscr{B}$ has an approximate unit, then B' is dense in $\mathscr{L}(\mathscr{B}_H)$ in the inductive limit topology.*

Proof. By the definition of $[\ ,\ ]$, B' is the linear span of $\{p(f * g): f, g \in \mathscr{L}(\mathscr{B})\}$. Since p is continuous in the inductive limit topologies and $p(\mathscr{L}(\mathscr{B})) = \mathscr{L}(\mathscr{B}_H)$ (by II.14.7), we have only to show that the linear span of $\{f * g: f, g \in \mathscr{L}(\mathscr{B})\}$ is dense in $\mathscr{L}(\mathscr{B})$ in the inductive limit topology. But this follows from the hypothesis of an approximate unit and Remark VIII.5.12.   ∎

14.13. The answer to question 14.11(II) is "no" in general (see 14.24); but it is "yes" if $\mathscr{B}$ is saturated.

Proposition. *If $\mathscr{B}$ is saturated, E' is dense in E in the inductive limit topology.*

Proof. Let $\mathscr{C}' = \langle C', \{C'_{x,y}\}\rangle$ be the Banach bundle over $G \times G$ which is the retraction of $\mathscr{B}$ by the continuous map $\langle x, y\rangle \mapsto xy$ of $G \times G$ onto G (see II.13.7). For each pair f, g of elements of $\mathscr{L}(\mathscr{B})$ we form the element $f \times g : \langle x, y\rangle \mapsto \langle x, y, f(x)g(y)\rangle$ of $\mathscr{L}(\mathscr{C}')$. We claim that the linear span M' of $\{f \times g : f, g \in \mathscr{L}(\mathscr{B})\}$ is dense in $\mathscr{L}(\mathscr{C}')$ in the inductive limit topology. Indeed: By the saturation of $\mathscr{B}$, $\{F(x, y) : F \in M'\}$ is dense in $C'_{x,y}$ for each $\langle x, y\rangle$ in $G \times G$. Furthermore, an easy Stone-Weierstrass argument with complex-valued functions shows that the inductive limit closure of M' is closed under multiplication by arbitrary complex functions on $G \times G$. Therefore the claim follows from II.14.6.

Let us now transform $\mathscr{C}'$ and M' by the "shear transformation" $\langle x, y\rangle \mapsto \langle x, xy\rangle$ of the base space $G \times G$. Under this transformation $\mathscr{C}'$ goes into the Banach bundle $\mathscr{C} = G \times \mathscr{B}$ over $G \times G$ which is the retraction of $\mathscr{B}$ by the map $\langle x, y\rangle \mapsto y$; and M' goes into the linear span M of the set of all cross-sections of $\mathscr{C}$ of the form $\langle x, y\rangle \mapsto f(x)g(x^{-1}y)$ ($f, g \in \mathscr{L}(\mathscr{B})$). By the above claim, M is dense in $\mathscr{L}(\mathscr{C})$ in the inductive limit topology.

We now point out another general fact: For each u in $\mathscr{L}(\mathscr{C})$ let u^0 be the element of E given by:

$$u^0(y, xH) = \int_H u(xh, y)dvh \qquad (x, y \in G).$$

If N is any linear subspace of $\mathscr{L}(\mathscr{C})$ which is dense in $\mathscr{L}(\mathscr{C})$ in the inductive limit topology, we claim that $N^0 = \{u^0 : u \in N\}$ is dense in E in the inductive limit topology. This follows from the two easily verified facts that the linear map $u \mapsto u^0$ is onto E and that it is continuous in the inductive limit topologies.

Applying the last paragraph to M, we conclude that M^0 is inductively dense in E. But M^0 is the linear span in E of the set of all functions v of the form

$$v(y, xH) = \int f(xh)g(h^{-1}x^{-1}y)dvh \qquad (x, y \in G),$$

that is (by (12)),

$$v = [f, g^*]_E,$$

where $f, g \in \mathscr{L}(\mathscr{B})$. So $M^0 = E'$; and E' is dense in E in the inductive limit topology. ∎

Remark.   Suppose that $H = G$. Then G/H is the one-element space, $E \cong \mathcal{L}(\mathcal{B})$, and (12) becomes:

$$[f, g]_E = f * g^*.$$

Thus the above proposition has the following useful special case: *If $\mathcal{B}$ is saturated, the linear span of $\{f * g : f, g \in \mathcal{L}(\mathcal{B})\}$ is dense in $\mathcal{L}(\mathcal{B})$ in the inductive limit topology.*

14.14.   To answer question 14.11(III), we need a lemma.

Lemma.   *Assume that $\mathcal{B}$ has an approximate unit. Then there exists a net $\{g_i\}$ of elements of $\mathcal{L}(\mathcal{B})$ such that*

$$f[g_i, g_i] \to f \qquad \text{for all } f \text{ in } \mathcal{L}(\mathcal{B}) \tag{14}$$

in the inductive limit topology.

Proof.   Since the involution operation is continuous on $\mathcal{L}(\mathcal{B})$ in the inductive limit topology, it is enough to obtain

$$(f[g_i, g_i])^* \to f^*. \tag{15}$$

A simple calculation shows that

$$(f[g_i, g_i])^*(x) = \int_H (g_i^* * g_i)(h) f^*(h^{-1}x) dvh \tag{16}$$

$(f, g_i \in \mathcal{L}(\mathcal{B}); \ x \in G)$.

Now let $\{w_r\}$ be an approximate unit of $\mathcal{B}$. By VIII.3.4,

$$\{w_r^* w_r\} \text{ is an approximate unit of } \mathcal{B}. \tag{17}$$

For each r choose χ_r in $\mathscr{C}(\mathcal{B})$ such that $\chi_r(e) = w_r$. Next, let $\{\sigma_\alpha'\}$ be an approximate unit on G (in the sense of III.11.17); and for each α let

$$\gamma_\alpha = k_\alpha(\sigma_\alpha'^* * \sigma_\alpha')|H,$$

where k_α is such a positive constant that $\int_H \gamma_\alpha \, dv = 1$. Thus, setting $\sigma_\alpha = k_\alpha^{1/2}\sigma_\alpha'$, we have

$$\gamma_\alpha = (\sigma_\alpha^* * \sigma_\alpha)|H. \tag{18}$$

Since the supports of the γ_α shrink down to e, $\{\gamma_\alpha\}$ is an approximate unit on H.

Now let $\phi_{r,\alpha}(h) = \gamma_\alpha(h)\chi_r(e)^*\chi_r(h)$ $(h \in H)$, so that $\phi_{r,\alpha} \in \mathcal{L}(\mathcal{B}_H)$. Defining the convolution $\phi_{r,\alpha} * f$ as in Remark VIII.5.13, we conclude from VIII.5.13 (and (17)) that

$$\lim_r \limsup_\alpha \|\phi_{r,\alpha} * f - f\|_\infty = 0 \tag{19}$$

for every f in $\mathscr{L}(\mathscr{B})$.

We now propose to show that (19) remains true when $\phi_{r,\alpha}$ is replaced by

$$\psi_{r,\alpha} = ((\sigma_\alpha \chi_r)^* * (\sigma_\alpha \chi_r))|H.$$

To see this, we observe from (18) that for $h \in H$

$$\psi_{r,\alpha}(h) - \phi_{r,\alpha}(h) = \int_G \sigma_\alpha(y)\sigma_\alpha(yh)[\chi_r(y)^*\chi_r(yh) - \chi_r(e)^*\chi_r(h)]d\lambda y. \quad (20)$$

We claim that for each r

$$\lim_\alpha \int_H \|\psi_{r,\alpha}(h) - \phi_{r,\alpha}(h)\|dvh = 0. \quad (21)$$

Indeed: Fix r. The bundle element $\chi_r(y)^*\chi_r(yh) - \chi_r(e)^*\chi_r(h)$ is continuous in $\langle y, h \rangle$ and vanishes when $y = e$. So, given $\varepsilon > 0$, we can find a neighborhood U of e such that

$$\|\chi_r(y)^*\chi_r(yh) - \chi_r(e)^*\chi_r(h)\| < \varepsilon \qquad \text{when } y, h \in U. \quad (22)$$

Now choose an index α_0 so large that, if $\alpha \succ \alpha_0$, $\sigma_\alpha(y)\sigma_\alpha(yh)$ vanishes unless y, $h \in U$. Then, by (20) and (22), for all h and all $\alpha \succ \alpha_0$ $\|\psi_{r,\alpha}(h) - \phi_{r,\alpha}(h)\|$ is majorized by $\varepsilon(\sigma_\alpha^* * \sigma_\alpha)(h)$; and therefore by (18)

$$\int_H \|\psi_{r,\alpha}(h) - \phi_{r,\alpha}(h)\|dvh \le \varepsilon$$

for all $\alpha \succ \alpha_0$. This proves (21).

Now for any r and α, any f in $\mathscr{L}(\mathscr{B})$, and any x in G,

$$\|(\psi_{r,\alpha} * f - \phi_{r,\alpha} * f)(x)\| = \left\| \int_H (\psi_{r,\alpha}(h) - \phi_{r,\alpha}(h))f(h^{-1}x)dvh \right\|$$

$$\le \|f\|_\infty \int_H \|\psi_{r,\alpha}(h) - \phi_{r,\alpha}(h)\|dvh.$$

So by (21), for each r

$$\lim_\alpha \|\psi_{r,\alpha} * f - \phi_{r,\alpha} * f\|_\infty = 0. \quad (23)$$

Combining (23) with (19), we deduce that

$$\lim_r \limsup_\alpha \|\psi_{r,\alpha} * f - f\|_\infty = 0. \quad (24)$$

Since the $\{\psi_{r,\alpha}\}$ have their compact supports contained in a single compact set, (24) implies that we can find a net $\{\psi^i\}$ such that each ψ^i is one of the $\psi_{r,\alpha}$, and such that for every f in $\mathscr{L}(\mathscr{B})$

$$\psi^i * f \underset{i}{\to} f \qquad (25)$$

in the inductive limit topology. On the other hand, by (16) and the definition of $\psi_{r,\alpha}$,

$$(\psi_{r,\alpha} * f)(x) = \int ((\sigma_\alpha \chi_r)^* * (\sigma_\alpha \chi_r))(h) f(h^{-1}x) dvh$$

$$= (f^*[\sigma_\alpha \chi_r, \sigma_\alpha \chi_r])^*(x). \qquad (26)$$

It follows from (25) and (26) that (15) holds provided we take g_i to be $\sigma_\alpha \chi_r$, where α and r are so chosen that $\psi^i = \psi_{r,\alpha}$.   ∎

14.15. Proposition.   *Assume that $\mathscr{B}$ has an approximate unit. Then every *-representation of E' which is continuous with respect to the norm $\| \ \|$ (6) of E is positive with respect to $\mathscr{I}$.*

Proof.   By 7.14 (applied with A and B reversed) it is enough to find a net $\{g_i\}$ of elements of $\mathscr{L}(\mathscr{B})$ such that

$$\|[f, f[g_i, g_i]]_E - [f, f]_E\| \to 0 \qquad (27)$$

for all f in $\mathscr{L}(\mathscr{B})$.

Take $\{g_i\}$ to be as in Lemma 14.14. We have observed in 14.9 that $[f, g]_E$ is separately continuous in f and g in the inductive limit topologies. Evidently $\| \ \|$ is continuous with respect to the inductive limit topology of E. In view of these facts, (27) is implied by the defining property (14) of $\{g_i\}$.   ∎

14.16.   We shall now answer question 14.11(IV) by showing that, if $\mathscr{B}$ has an approximate unit, the $\mathscr{I}$-positive non-degenerate *-representations of B' which are $\mathscr{L}_1$-continuous correspond exactly to the $\mathscr{I}$-positive non-degenerate *-representations of E' which are continuous with respect to the norm of E.

Assume that $\mathscr{B}$ has an approximate unit.

By 14.12 B' is dense in $\mathscr{L}(\mathscr{B}_H)$. Hence by VIII.13.2 the non-degenerate $\mathscr{L}_1$-continuous *-representations S' of B' are just the restrictions to B' of the integrated forms of non-degenerate *-representations S of $\mathscr{B}_H$. Further, positivity of S' with respect to $\mathscr{I}$ is the same as $\mathscr{B}$-positivity of the corresponding *-representation S of $\mathscr{B}_H$.

Let S be a non-degenerate $\mathscr{B}$-positive *-representation of $\mathscr{B}_H$, and S' the ($\mathscr{I}$-positive) restriction to B' of the integrated form of S. We shall suppose that, under the correspondence of 6.15 applied to $\mathscr{I}$, S' corresponds to the non-degenerate *-representation R' of E'. But, by the discussion of 14.7, R' is just the restriction to E' of the E-integrated form of the system of imprimitivity over G/H induced by S. Therefore R' is continuous with respect to the norm of E (see VIII.18.16).

Thus we have established one direction of the equivalence asserted in the following proposition:

Proposition. *Let S' and R' be $\mathscr{I}$-positive non-degenerate *-representations of E' and B' respectively which correspond under the correspondence of 6.15 (applied to $\mathscr{I}$). Then R' is continuous in the norm of E if and only if S' is continuous in the $\mathscr{L}_1$-norm.*

Proof. To prove the other direction we assume that R' is continuous in the norm of E.

From given elements ξ of $X(R')$ and f of $\mathscr{L}(\mathscr{B})$ we obtain an element of $X(S')$, namely the quotient image $\zeta = f \widetilde{\otimes} \xi$ of $f \otimes \xi$. From the definition of the inducing process $\mathrm{Ind}_{E' \uparrow B'}$, we have $S'_\phi \zeta = f\phi^* \widetilde{\otimes} \xi$, and so

$$(S'_\phi \zeta, \zeta) = (R'_{[f, f\phi^*]_E} \xi, \xi) \qquad (\phi \in B'). \qquad (28)$$

Now for fixed f it follows from continuity observations in 8.4 and 14.9 that $\phi \mapsto [f, f\phi^*]_E$ is continuous with respect to the inductive limit topologies of $\mathscr{L}(\mathscr{B}_H)$ and E. Since R' is continuous by hypothesis, (28) implies that $\phi \mapsto (S'_\phi \zeta, \zeta)$ is continuous on B' with respect to the inductive limit topology. Further, B' is inductive-limit dense in $\mathscr{L}(\mathscr{B}_H)$; and it is easy to check that, for each b in B_H, B' is closed under the left action of the multiplier m_b on $\mathscr{L}(\mathscr{B}_H)$ (see VIII.12.3). Finally, notice that the vectors ζ of the above form span a dense subspace of $X(S')$.

We have thus verified all the hypotheses of Theorem VIII.13.8 as applied to B' and S'. So VIII.13.8 implies that S' is $\mathscr{L}_1$-continuous. ∎

14.17. We are now in a position to derive the chief result of this chapter—the Imprimitivity Theorem for Banach *-algebraic bundles.

Let us say that a system of imprimitivity $\mathscr{T}$ for $\mathscr{B}$ over G/H is *non-degenerate on E'* if the restriction to E' of the E-integrated form of $\mathscr{T}$ is non-degenerate. (Recall that E' is the linear span in E of range($[\ ,\ ]_E$).)

In this theorem we assume merely the standing hypotheses of 14.1. In particular $\mathscr{B}$ need not have an approximate unit.

Imprimitivity Theorem for Banach *-algebraic Bundles. (I) *If S is a non-degenerate $\mathscr{B}$-positive *-representation of $\mathscr{B}_H$, then the system of imprimitivity $\mathscr{T}$ for $\mathscr{B}$ over G/H induced by S is non-degenerate on E'.*

(II) *Conversely, let $\mathscr{T}$ be any system of imprimitivity for $\mathscr{B}$ over G/H which is non-degenerate on E'. Then there is a non-degenerate $\mathscr{B}$-positive *-representation S of $\mathscr{B}_H$ such that $\mathscr{T}$ is unitarily equivalent to the system of imprimitivity induced by S; and this S is unique to within unitary equivalence.*

Proof. To begin with we suppose that $\mathscr{B}$ has an approximate unit.

(I) If S' and R' are the restrictions of the integrated forms of S and $\mathscr{T}$ to B' and E' respectively, we have seen in the first part of 14.16 that S' is non-degenerate and that R' is the *-representation of E' corresponding to S' via 6.15 applied to $\mathscr{I}$. So R' is non-degenerate.

(II) Let R be the (necessarily continuous) E-integrated form of $\mathscr{T}$, and R' its (non-degenerate) restriction to E'. By 14.15 R' is positive with respect to $\mathscr{I}$, and so corresponds via 6.15 with some non-degenerate $\mathscr{I}$-positive *-representation S' of B'. By Proposition 14.16 S' is continuous with respect to the $\mathscr{L}_1$-norm, and so is the restriction to B' of the integrated form of a non-degenerate *-representation S of $\mathscr{B}_H$ which is positive with respect to $\mathscr{B}$. Let $\mathscr{T}^0$ be the system of imprimitivity induced by S, and R^0 the E-integrated form of T^0. By 14.7

$$R^0 | E' \cong R' = R | E'. \tag{29}$$

Since E' is a *-ideal of E and the two sides of (29) are non-degenerate, it follows from (29) that $R^0 \cong R$. By VIII.18.18 this implies that $\mathscr{T}^0 \cong \mathscr{T}$. This shows the existence of an S having the required property.

The uniqueness of S follows from the biuniqueness of the correspondence 6.15. To be more specific, S is determined by the restriction S' of its integrated form to B'; and S' is determined to within unitary equivalence by the representation R' of E' to which it corresponds by 6.15. On the other hand R' is the restriction to E' of the system of imprimitivity $\mathscr{T}$ induced by S. So S is determined to within unitary equivalence by $\mathscr{T}$. This completes the proof if $\mathscr{B}$ has an approximate unit.

We now discard the assumption that $\mathscr{B}$ has an approximate unit.

Let $\mathscr{C} = \langle C, \{C_x\} \rangle$ be the bundle C^*-completion of $\mathscr{B}$, and $\rho: B \to C$ the canonical quotient map (VIII.16.7). Let F be the compacted transformation *-algebra for $\mathscr{C}$ and G/H; let $[\ ,\ ]_F$ be the F-valued rigging on $\mathscr{L}(\mathscr{C})$ defined as in (12) (with $\mathscr{B}$ replaced by $\mathscr{C}$); and let F' be the linear span of range($[\ ,\ ]_F$). The map ρ gives rise to *-homomorphisms $\rho': \mathscr{L}(\mathscr{B}) \to \mathscr{L}(\mathscr{C})$ and $\tilde{\rho}: E \to F$:

$$\rho'(f)(x) = \rho(f(x)), \qquad \tilde{\rho}(u)(x, \alpha) = \rho(u(x, \alpha));$$

and we have

$$\tilde{\rho}([f, g]_E) = [\rho'(f), \rho'(g)]_F \quad (f, g \in \mathscr{L}(\mathscr{B})). \tag{30}$$

By II.14.6 $\rho'(\mathscr{L}(\mathscr{B}))$ and $\tilde{\rho}(E)$ are dense in $\mathscr{L}(\mathscr{C})$ and F respectively in the inductive limit topologies. Hence by (30) and the separate continuity of $[\ ,\ ]_F$ (see 14.9), $\tilde{\rho}(E')$ is dense in F'.

By VIII.16.3 $\mathscr{C}$ has an approximate unit; so our theorem holds for $\mathscr{C}$ by the first part of the proof.

To prove (I) for $\mathscr{B}$, let S be a non-degenerate $\mathscr{B}$-positive *-representation of $\mathscr{B}_H$. By 11.7 there is a non-degenerate $\mathscr{C}$-positive *-representation S^0 of $\mathscr{C}_H$ such that $S = S^0 \circ (\rho | B_H)$. If R and R^0 are the integrated forms of the systems of imprimitivity induced by S and S^0 respectively, one verifies (see 12.6) that

$$R_u = R^0_{\tilde{\rho}(u)} \qquad (u \in E). \tag{31}$$

Now by (I) of the present theorem applied to $\mathscr{C}$, $R^0 | F'$ is non-degenerate. So (31) and the denseness of $\tilde{\rho}(E')$ in F' imply that $R|E'$ is non-degenerate; and (I) is proved for $\mathscr{B}$.

Since S^0 is determined to within unitary equivalence by R^0, the above argument also shows that S is determined to within unitary equivalence by R. Thus the uniqueness statement in (II) holds for $\mathscr{B}$.

Finally, let $\mathscr{T} = \langle T, P \rangle$ be a system of imprimitivity for $\mathscr{B}$ over G/H. By the definition of $\mathscr{C}$, we have $T = T^0 \circ \rho$ for some *-representation T^0 of $\mathscr{C}$; and $\mathscr{T}^0 = \langle T^0, P \rangle$ is evidently a system of imprimitivity for $\mathscr{C}$ over G/H. By (II) applied to $\mathscr{C}$, $\mathscr{T}^0$ is induced by some non-degenerate *-representation S^0 of $\mathscr{C}_H$. Hence, in view of 12.6, $\mathscr{T}$ is induced by $S = S^0 \circ (\rho | B_H)$. This completes the proof of (II) for $\mathscr{B}$. ∎

14.18. If $\mathscr{B}$ is saturated, then by 14.13 E' is dense in E in the inductive limit topology. In that case *every* system of imprimitivity for $\mathscr{B}$ over G/H is non-degenerate on E'; and Theorem 14.17 becomes:

Theorem. *Assume that $\mathscr{B}$ is saturated. Then, given any system of imprimitivity $\mathscr{T}$ for $\mathscr{B}$ over G/H, there is a non-degenerate $\mathscr{B}$-positive *-representation S of $\mathscr{B}_H$ such that $\mathscr{T}$ is unitarily equivalent to the system of imprimitivity induced by S. Further, the S having this property is unique to within unitary equivalence.*

14.19. In the group case $\mathscr{B}$ (being the group bundle) is automatically saturated; and 14.18 becomes the Imprimitivity Theorem of Mackey:

Theorem. *Given any system of imprimitivity $\mathscr{T}$ for G over G/H, there is a unitary representation S of H (unique to within unitary equivalence) such that the system of imprimitivity induced by S is unitarily equivalent to $\mathscr{T}$.*

The Case of Discrete G/H

14.20. Suppose that G/H is discrete. In that case the recovery of S from the system of imprimitivity $\mathscr{T}$ induced by S is very much simpler than in the general case. Also, the rather mysterious condition of non-degeneracy on E' which appears in Theorem 14.17 becomes much more transparent if G/H is discrete.

14.21. As regards the recovery of S from $\mathscr{T}$ we have:

Proposition. *Assume that G/H is discrete. Let S be a $\mathscr{B}$-positive non-degenerate *-representation of $\mathscr{B}_H$, and $\langle T, P \rangle$ the system of imprimitivity for $\mathscr{B}$ over G/H induced by S. Then S is unitarily equivalent to the subrepresentation of $T|\mathscr{B}_H$ which acts on $\mathrm{range}(P(\{H\}))$.*

Proof. Let $\mathscr{Y} = \{Y_\alpha\}$ ($\alpha \in G/H$) be the Hilbert bundle over G/H induced by S. Since G/H is discrete, $\mathrm{range}(P(\{H\}))$ is just the fiber Y_H of $\mathscr{Y}$. Also, since G/H being discrete has a G-invariant measure, it follows from III.13.16 that $\delta(h) = \Delta(h)$ for $h \in H$. Hence, identifying Y_H with $X(S)$ by means of the F of 9.4, we have from 9.15(29):

$$T_b \xi = \tau_b \xi = S_b \xi \qquad\qquad (b \in B_H;\ \xi \in X(S)).$$

This shows that $T|\mathscr{B}_H$ coincides on $\mathrm{range}(P(\{H\}))$ with S.   ∎

14.22. Proposition. *Suppose that G/H is discrete; and let $\langle T, P \rangle$ be a system of imprimitivity for $\mathscr{B}$ over G/H. Then $\langle T, P \rangle$ is non-degenerate on E' (in the sense of 14.17) if and only if $\mathrm{range}(P(\{H\}))$ generates $X(T)$ under T, that is, if and only if $\{T_b \xi : b \in B,\ \xi \in \mathrm{range}(P(\{H\}))\}$ spans a dense subspace of $X(T)$.*

Proof. Since G/H is discrete, H is open in G; and we may as well assume that ν coincides on H with λ. As usual, we denote by R the integrated form of $\langle T, P \rangle$.

Suppose that $f, g \in \mathscr{L}(\mathscr{B})$ and that f vanishes outside the coset zH ($z \in G$). Then, for $x, y \in G$,

$$[f, g]_E(x, yH) = \int_H f(yh)g^*(h^{-1}y^{-1}x)d\lambda h$$

$$= \begin{cases} 0 & \text{if } yH \neq zH \\ (f * g^*)(x) & \text{if } yH = zH. \end{cases}$$

It follows that, for $\xi \in X(T)$,

$$R_{[f, g]_E}\xi = \int_G \sum_{\alpha \in G/H} P(\{\alpha\})T_{[f, g]_E(x, \alpha)}\xi \, d\lambda x \qquad \text{(by VIII.18.16(21))}$$

$$= P(\{zH\}) \int_G T_{(f * g^*)(x)}\xi \, d\lambda x$$

$$= P(\{zH\})T_{f * g^*}\xi,$$

or

$$R_{[f, g]_E} = P(\{zH\})T_f(T_g)^*. \tag{32}$$

Now by VIII.18.7(6), for $x \in zH$ we have $T_{f(x)}P(\{H\}) = P(\{zH\})T_{f(x)}$. Integrating this with respect to x (over zH) gives

$$T_f P(\{H\}) = P(\{zH\})T_f. \tag{33}$$

Together, (32) and (33) imply that

$$R_{[f, g]_E} = T_f P(\{H\})(T_g)^*. \tag{34}$$

Now any f in $\mathscr{L}(\mathscr{B})$ is a sum of elements of $\mathscr{L}(\mathscr{B})$ each of which vanishes outside some coset zH. Hence by linearity (34) holds for all f, g in $\mathscr{L}(\mathscr{B})$.

Now $\langle T, P \rangle$ is non-degenerate on E' if and only if $\{R_{[f, g]_E}: f, g \in \mathscr{L}(\mathscr{B})\}$ acts non-degenerately on $X(T)$. By (34) and the non-degeneracy of the integrated form of T, this happens if and only if the linear span of $\{T_f(\text{range}(P(\{H\}))): f \in \mathscr{L}(\mathscr{B})\}$ is dense in $X(T)$. ∎

14.23. Remark. The last two propositions make it very simple to prove the Imprimitivity Theorem in case G/H is discrete. Indeed, given a system of imprimitivity $\langle T, P \rangle$ for $\mathscr{B}$ over G/H, all we have to do in that case is to define S in accordance with 14.21, that is, as the subrepresentation of $T|\mathscr{B}_H$ acting on range$P(\{H\})$, and then to verify (using the equivalent condition in Proposition 14.22) that $\langle T, P \rangle$ is canonically equivalent to the system of imprimitivity induced by S. We suggest that the reader carry out this verification for this own instruction.

If G/H is not discrete, then $P(\{H\}) = 0$, and the above approach breaks down. In fact the main purpose of this chapter has been to provide machinery strong enough to handle the case that G/H is not discrete.

14.24.   Remark.   Proposition 14.22 suggests almost trivial examples of the failure of the property of non-degeneracy on E'. Suppose for instance that G is the two-element group $\{e, u\}$, and that $\mathcal{B}$ is a Banach *-algebraic bundle over G with $B_u = \{0\}$. Let T be a non-zero non-degenerate *-representation of $\mathcal{B}$ (i.e., of B_e), and P the trivial $X(T)$-projection-valued measure on G given by $P_{\{e\}} = 0$, $P_{\{u\}} = 1$. Then $\langle T, P \rangle$ is a system of imprimitivity for $\mathcal{B}$ over G which by 14.22 obviously fails to be non-degenerate on E'.

Thus, in general E' will not be dense in the compacted transformation *-algebra E.

The Commuting Algebra of a System of Imprimitivity

14.25.   Let S be a non-degenerate $\mathcal{B}$-positive *-representation of $\mathcal{B}_H$, and $\mathcal{T} = \langle T, P \rangle$ the system of imprimitivity induced by S. Let $\mathcal{I}(S)$ and $\mathcal{I}(\mathcal{T})$ be the commuting algebras of S and $\mathcal{T}$ respectively (that is, the *-algebras of all S, S and all $\mathcal{T}$, $\mathcal{T}$ intertwining operators respectively). We are going to set up a canonical *-isomorphism between $\mathcal{I}(S)$ and $\mathcal{I}(\mathcal{T})$. As in 6.16, this result should be considered as an integral part of the Imprimitivity Theorem.

Let $\mathcal{Y} = \langle Y, \{Y_\alpha\} \rangle$ be the Hilbert bundle over G/H induced by S; and let us readopt the rest of the notation of §9.

Fix an element γ of $\mathcal{I}(S)$. By 8.9(III) and 1.11, for each α in G/H the equation

$$\gamma_\alpha(\kappa_\alpha(b \otimes \xi)) = \kappa_\alpha(b \otimes \gamma(\xi)) \tag{35}$$

$(\xi \in X(S); b \in B_\alpha)$ defines a bounded linear operator γ_α on Y_α satisfying

$$\|\gamma_\alpha\| \leq \|\gamma\|. \tag{36}$$

It follows from (35) and 9.5(9) that

$$\gamma_\alpha(\kappa_\alpha(\phi \otimes \xi)) = \kappa_\alpha(\phi \otimes \gamma(\xi)) \tag{37}$$

for $\phi \in \mathcal{L}(\mathcal{B}_\alpha)$, $\xi \in X(S)$. Let $\tilde{\gamma} : Y \to Y$ be the map coinciding with γ_α on Y_α (for each α). If f is a cross-section of $\mathcal{Y}$ of the form

$$\alpha \mapsto \kappa_\alpha((\phi|\alpha) \otimes \xi) \qquad (\phi \in \mathcal{L}(\mathcal{B}); \xi \in X(S)),$$

then by (37) $(\tilde{\gamma} \circ f)(\alpha) = \kappa_\alpha((\phi|\alpha) \otimes \gamma(\xi))$. From this, (36), and the definition of the topoplogy of $\mathcal{Y}$, we deduce that $\tilde{\gamma}$ is continuous. This implies (by (36) and II.15.4) that, if f is a locally $\rho^{\#}$-measurable cross-section of $\mathscr{Y}$, then so is $\tilde{\gamma} \circ f$. Thus, again using (36), we see that $\gamma^0: f \mapsto \tilde{\gamma} \circ f$ $(f \in \mathscr{L}_2(\rho^{\#}; \mathscr{Y}))$ is a bounded linear operator on $\mathscr{L}_2(\rho^{\#}; \mathscr{Y})$ satisfying $\|\gamma^0\| \leq \|\gamma\|$. It is evident that γ^0 commutes with all T_b $(b \in B)$ and all $P(W)$ (W a Borel subset of G/H), and so belongs to $\mathscr{I}(\mathscr{T})$.

Theorem. *The map $\gamma \mapsto \gamma^0$ just defined is a *-isomorphism of $\mathscr{I}(S)$ onto $\mathscr{I}(\mathscr{T})$.*

Proof. Let B', E' be as in 14.11. We first claim that the restriction S' to B' of the integrated form of S is non-degenerate. Indeed: Let $\mathscr{C}$ be the bundle C^*-completion of $\mathscr{B}$. As in the latter part of the proof of 14.17, S is lifted from a non-degenerate *-representation S^0 of $\mathscr{C}_H$. Since $\mathscr{C}$ has an approximate unit, the analogue C' of B' in $\mathscr{C}$ is dense in $\mathscr{L}(\mathscr{C}_H)$ in the inductive limit topology by 14.12; and so $S^0|C'$ is non-degenerate. On the other hand, by an argument similar to one found in 14.17, the image of B' in $\mathscr{L}(\mathscr{C}_H)$ is dense in C'. Therefore $S' = S|B'$ is non-degenerate; and the claim is proved.

As in 14.17 let R' be the restriction to E' of the integrated form R of the system of imprimitivity $\mathscr{T} = \langle T, P \rangle$ induced by S. By 14.17(I) R', like S', is non-degenerate; and by 14.7 R' corresponds to S' under the correspondence of 6.15 applied to the strict E', B' imprimitivity bimodule $\mathscr{I}$ of 14.10. Therefore by 6.16 the commuting algebras of S' and R' are *-isomorphic under the *-isomorphism $\Phi: F \mapsto \tilde{F}$ of 6.16. Passing from the Rieffel to the bundle description of the induced representation via the isometry of 9.8, we check that the Φ of 6.16 becomes just the mapping $\gamma \mapsto \gamma^0$.

To complete the proof we need only to show that the commuting algebras of S and S' are the same, and that those of R and R' are the same. This follows from the following general fact whose verification was essentially carried out in VIII.15.4:

If I is a *-ideal of a *-algebra A, and T is any *-representation of A such that $T|I$ is non-degenerate, then T and $T|I$ have the same commuting algebras. ∎

14.26. Corollary. *Let S and $\mathscr{T} = \langle T, P \rangle$ be as in 14.25. Then $\mathscr{T}$ is irreducible if and only if S is irreducible.*

Remark. Thus, if S is irreducible, $X(T)$ is irreducible under the combined action of T and P, though it is not in general irreducible under the action of T alone. Compare Remark 12.3.

Compact Induced Representations

14.27.   Definition.   A *-representation T of $\mathscr{B}$ is said to be *compact* if the integrated form of T is compact in the sense of 6.21, that is, if T_f is a compact operator for every f in $\mathscr{L}(\mathscr{B})$.

Theorem.   *Assume that $\mathscr{B}$ is saturated; and let S be a compact non-degenerate $\mathscr{B}$-positive *-representation of $\mathscr{B}_H$. Then the integrated form R of the system of imprimitivity $\langle T, P \rangle$ induced by S is a compact *-representation of E. In particular, the product $P(\phi)T_f$ is compact whenever $\phi \in \mathscr{L}(G/H)$ and $f \in \mathscr{L}(\mathscr{B})$. (Here of course $P(\phi)$ is the spectral integral $\int \phi \, dP$.)*

Proof.   Let B', E' be as in 14.11. Applying 6.21 to $\mathscr{I}$ considered as a strict E', B' imprimitivity bimodule, we conclude that R_u is compact whenever $u \in E'$. But by 14.13 this implies that R_u is compact for all u in E. Taking u to be of the special form $\langle x, \alpha \rangle \mapsto \phi(\alpha)f(x)$, where $\phi \in \mathscr{L}(G/H)$ and $f \in \mathscr{L}(\mathscr{B})$, we have $R_u = P(\phi)T_f$ by VIII.18.17; and from this the last statement of the theorem follows.   ∎

14.28.   If G/H is compact, we can take $\phi \equiv 1$ in the last statement of 14.27, in which case $P(\phi)T_f = T_f$. Thus the following interesting corollary emerges:

Corollary.   *Assume that $\mathscr{B}$ is saturated, and that G/H is compact. Then, if S is a compact non-degenerate $\mathscr{B}$-positive *-representation of $\mathscr{B}_H$, $\mathrm{Ind}_{\mathscr{B}_H \uparrow \mathscr{B}}(S)$ is a compact *-representation of $\mathscr{B}$.*

Remark.   It follows from this and VI.23.2 that, under the hypotheses of the above corollary, $\mathrm{Ind}_{\mathscr{B}_H \uparrow B}(S)$ is automatically discretely decomposable.

15.   A Generalized Mackey–Stone–von Neumann Theorem

15.1.   One of Mackey's beautiful applications of his Imprimitivity Theorem for groups was his derivation of a generalization of the Stone–von Neumann Theorem on the essential uniqueness of the operators representing the position and momentum observables of quantum mechanics. From our more general Imprimitivity Theorem we can easily obtain a generalization of Mackey's version.

15.2.   Fix a saturated Banach *-algebraic bundle $\mathscr{B} = \langle B, \pi, \cdot, * \rangle$ over a locally compact *Abelian* group G with unit e and Haar measure λ. Let $\hat{G}$ be the (locally compact Abelian) character group of G (defined in X.1.8).

We now construct a new Banach *-algebraic bundle $\mathcal{D} = \langle D, \pi', \cdot, * \rangle$ over $G \times \hat{G}$ as follows: As a Banach bundle, $\mathcal{D}$ is the retraction of $\mathcal{B}$ by the projection $\langle x, \chi \rangle \mapsto x$ of $G \times \hat{G}$ onto G. Thus $D \cong B \times \hat{G}$; and $\pi'\langle b, \chi \rangle = \langle \pi(b), \chi \rangle$ ($\chi \in \hat{G}$; $b \in B$). We now introduce multiplication and involution into D as follows:

$$\langle b, \chi \rangle \langle b', \chi' \rangle = \langle \chi(\pi(b'))bb', \chi\chi' \rangle, \qquad \langle b, \chi \rangle^* = \langle \overline{\chi(\pi(b))}b^*, \chi^{-1} \rangle$$

($b, b' \in B$; $\chi, \chi' \in \hat{G}$). With these definitions one checks that $\mathcal{D}$ becomes a saturated Banach *-algebraic bundle over $G \times \hat{G}$.

15.3.   What do the *-representations of $\mathcal{D}$ look like?

Suppose that S is a non-degenerate *-representation of $\mathcal{B}$ and V a unitary representation of $\hat{G}$ acting in the same space as S and satisfying

$$V_\chi S_b = \chi(\pi(b)) S_b V_\chi \qquad (\chi \in \hat{G}; b \in B). \qquad (1)$$

Then one verifies immediately that the equation

$$T_{\langle b, \chi \rangle} = S_b V_\chi \qquad (\langle b, \chi \rangle \in D) \qquad (2)$$

defines a non-degenerate *-representation T of $\mathcal{D}$. Conversely, we have:

Proposition. *Every non-degenerate *-representation T of $\mathcal{D}$ is of the form (2), where S is a unique non-degenerate *-representation of $\mathcal{B}$ acting in $X(T)$ and V is a unique unitary representation of $\hat{G}$ acting in $X(T)$ and satisfying (1).*

Proof. Notice that each χ in $\hat{G}$ gives rise to a unitary multiplier u_χ of $\mathcal{D}$ as follows:

$$u_\chi \langle b', \chi' \rangle = \langle \chi(\pi(b'))b', \chi\chi' \rangle,$$

$$\langle b', \chi' \rangle u_\chi = \langle b', \chi'\chi \rangle$$

($\langle b', \chi' \rangle \in D$); and $\chi \mapsto u_\chi$ is a continuous homomorphism of $\hat{G}$ into the group of unitary multipliers, with respect to the strong topology of the latter (see VIII.15.2). Hence, if T' is the extension of T to $\mathcal{W}(\mathcal{D})$ described in VIII.15.3, $V: \chi \mapsto T'_{u_\chi}$ must be a unitary representation of $\hat{G}$. Since $\langle b, \chi \rangle = \langle b, 1 \rangle u_\chi$, we have

$$T_{\langle b, \chi \rangle} = T_{\langle b, 1 \rangle} V_\chi \qquad (\langle b, \chi \rangle \in D),$$

which gives (2) when we define $S_b = T_{\langle b, 1\rangle}$ (1 being the unit of $\hat{G}$). Equation (1) holds in view of the identity $u_\chi\langle b, 1\rangle = \chi(\pi(b))\langle b, 1\rangle u_\chi$.

Thus the existence of S and V has been proved. Their uniqueness is evident.   ■

15.4.   Thus non-degenerate *-representations of $\mathscr{D}$ are essentially just pairs S, V satisfying (1). Here is a way of constructing such pairs.

Take a non-degenerate *-representation Q of B_e which is positive with respect to $\mathscr{B}$. Thus Q induces a Hilbert bundle $\mathscr{Y} = \{Y_x\}$ over G (as in 9.2); and the induced *-representation $S = \mathrm{Ind}_{B_e \uparrow \mathscr{B}}(Q)$ acts on $\mathscr{L}_2(\lambda; \mathscr{Y})$. For $\chi \in \hat{G}$ let V_χ be the operator on $\mathscr{L}_2(\lambda; \mathscr{Y})$ of pointwise multiplication by χ. So $V: \chi \mapsto V_\chi$ is a unitary representation of $\hat{G}$. Notice from 14.3 that

$$V_\chi = \int_G \chi \, dP, \tag{3}$$

where P is the projection-valued measure on G induced by Q. Further, S and V satisfy (1); for, if $\chi \in \hat{G}$, $b \in B_x$, and $f \in \mathscr{L}_2(\lambda; \mathscr{Y})$, we have:

$$
\begin{aligned}
(V_\chi S_b f)(y) &= \chi(y)\tau_b(f(x^{-1}y)) &&(\tau_b \text{ being as in 9.14})\\
&= \chi(x)\tau_b[\chi(x^{-1}y)f(x^{-1}y)]\\
&= \chi(x)(S_b V_\chi f)(y).
\end{aligned}
$$

So $V_\chi S_b = \chi(x)S_b V_\chi$.

Thus every non-degenerate $\mathscr{B}$-positive *-representation Q of B_e gives rise to a pair S, V satisfying (1), and hence via (2) to a non-degenerate *-representation T of $\mathscr{D}$.

15.5.   The interesting fact is that every such pair, S, V satisfying (1), that is (by 15.3), every non-degenerate *-representation T of $\mathscr{D}$, arises as in 15.4 from a unique $\mathscr{B}$-positive non-degenerate *-representation Q of B_e. This is our generalized version of the Mackey–Stone–von Neumann Theorem.

Theorem.   *Let T^0 be a non-degenerate *-representation of $\mathscr{D}$. Then there exists a unique (to within unitary equivalence) non-degenerate *-reprensenta- tion Q of B_e which is positive with respect to $\mathscr{B}$, and such that T^0 is unitarily equivalent to the *-representation T of $\mathscr{D}$ obtained from Q as in 15.4.*

Proof. Let S^0, V^0 be the pair satisfying (1) and corresponding to T^0 as in 15.3.

By Pontryagin Duality (X.3.11), the character group of $\hat{G}$ is G. Hence the spectral measure (see X.2.1) of V^0 is a regular $X(T^0)$-projection-valued Borel measure P^0 on G satisfying

$$V_\chi^0 = \int_G \chi(x)dP^0 x \qquad\qquad (\chi \in \hat{G}). \qquad (4)$$

We claim that S^0, P^0 is a system of imprimitivity for $\mathscr{B}$ based on G. Indeed: Let M be the set of all bounded continuous complex functions ϕ on G such that

$$S_b^0 P^0(\phi) = P^0(x\phi)S_b^0 \qquad (5)$$

whenever $x \in G$ and $b \in B_x$. (Here $(x\phi)(y) = \phi(x^{-1}y)$, and $P^0(\phi) = \int \phi\, dP^0$.) If $\chi \in \hat{G}$, we have by (4) $P^0(x\chi) = \int \chi(x^{-1}y)dP^0 y = \chi(x^{-1})V_\chi^0$; so by (1) and (4) $S_b^0 P^0(\chi) = \chi(x^{-1})V_\chi^0 S_b^0 = P^0(x\chi)S_b^0$, showing that $\hat{G} \subset M$. Thus M contains the linear span of $\hat{G}$; and the latter is a *-algebra of bounded complex functions on G which separates points of G and contains the constant functions. Our claim now follows from Proposition VIII.18.8.

Now since $\mathscr{B}$ is saturated, Theorem 14.18, says that S^0, P^0 is equivalent to the system of imprimitivity S, P induced by some $\mathscr{B}$-positive non-degenerate *-representation Q of B_e. Comparing (3) and (4), we see from this that the pair S^0, V^0 is unitarily equivalent to the pair S, V obtained from Q as in 15.4. This proves the existence of Q.

The uniqueness of Q is obtained by combining the uniqueness assertion in 14.17 with the fact (X.2.1) that a unitary representation of $\hat{G}$ uniquely determines its spectral measure. ∎

15.6. Theorem 15.5 says that there is a one-to-one correspondence between the set of all unitary equivalence classes of non-degenerate *-representations Q of B_e which are positive with respect to $\mathscr{B}$, and the set of all unitary equivalence classes of non-degenerate *-representations T of $\mathscr{D}$.

It is easily seen that this correspondence preserves direct sums. It also preserves the isomorphism class of the commuting algebra; that is:

Proposition. *If Q and T correspond as above, the commuting algebras of Q and of T are *-isomorphic. In particular Q is irreducible if and only if T is irreducible.*

Proof.   Proposition VIII.18.8 (applied with $x = e$) shows that the commuting algebra of a unitary representation of $\hat{G}$ is the same as the commuting algebra of its spectral measure. Combining this fact with 14.25 we obtain the first statement.   ∎

The Group Case

15.7.   Let us specialize the theory developed above to the case where $\mathscr{B}$ is the group bundle of G. In that case $B_e = \mathbb{C}$, and the *-representations of $\mathbb{C}$ are just multiples of the identity representation. We thus obtain Mackey's generalization of the Stone–von Neumann Theorem:

Mackey's Theorem.   *Let G be a locally compact Abelian group. Then there exists to within unitary equivalence just one pair S, V with the properties:*

   (i)   *S is a unitary representation of G;*
   (ii) *V is a unitary representation of $\hat{G}$ acting on the same space as S;*
   (iii) *$V_\chi S_x = \chi(x)S_x V_\chi$ for all x in G and χ in $\hat{G}$;*
   (iv) *$X(S)$ is irreducible under the combined actions of S and V.*

The proof follows immediately from 15.5 and 15.6.

It is worth recalling from 15.4 the form which the pair S, V in the above theorem must take: It must be defined, up to unitary equivalence, as follows:

$$X(S) = X(V) = \mathscr{L}_2(\lambda),$$

$$(S_x f)(y) = f(x^{-1}y)$$

(i.e., S is the regular representation),

$$(V_\chi f)(y) = \chi(y)f(y)$$

$(f \in \mathscr{L}_2(\lambda); \ x, y \in G; \ \chi \in \hat{G})$.

Remark.   If condition (iv) is removed from the above theorem, the only extra flexibility allowed to S, V is that it may be a direct sum of copies of the irreducible pair just defined.

Application to Quantum Mechanics

15.8.   We should like to end this section with a brief mention of the relevance of 15.7 to quantum mechanics. Of course, this will hardly be comprehensible unless the reader already has at least a rudimentary acquaintance with quantum mechanics.

Consider a physical system with n degrees of freedom, whose position observables are $q_1, \ldots, q_n$ and whose corresponding momentum observables are $p_1, \ldots, p_n$. Quantum mechanics says that $q_1, \ldots, q_n, p_1, \ldots, p_n$ are to be identified with certain unbounded self-adjoint operators $Q_1, \ldots, Q_n$, $P_1, \ldots, P_n$ respectively on some Hilbert space X, and that Q_j and P_j must satisfy the Heisenberg commutation relations (here stated "formally", without a precise specification of the domains of the operators):

$$Q_j Q_k - Q_k Q_j = 0, \qquad P_j P_k - P_k P_j = 0, \tag{6}$$

$$P_j Q_k - Q_k P_j = i\delta_{jk}\mathbf{1} \tag{7}$$

($i = \sqrt{-1}$, and $\mathbf{1}$ is the identity operator on X). We also require that X be irreducible under the action of $Q_1, \ldots, Q_n, P_1, \ldots, P_n$. The question is: What operators $Q_1, \ldots, Q_n, P_1, \ldots, P_n$ can satisfy these conditions?

We shall transform the problem so that only bounded operators appear. For $j = 1, \ldots, n$ and real t, put

$$U_j(t) = e^{itQ_j}, \qquad V_j(t) = e^{itP_j}. \tag{8}$$

Thus U_j and V_j are the unitary representations of the real line corresponding as in X.2.2 to Q_j and P_j. Expanding e^{itQ_j} and e^{itP_j} formally as power series in t, one can verify that (6) and (7) correspond formally to the following conditions on the U_j and V_j:

$$U_j(t)U_k(s) = U_k(s)U_j(t), \tag{9}$$

$$V_j(t)V_k(s) = V_k(s)V_j(t), \tag{10}$$

$$U_j(t)V_k(s) = \delta_{jk} e^{ist} V_k(s)U_j(t). \tag{11}$$

We shall in fact consider (6) and (7) as replaced by the precise conditions (9), (10), (11), and the irreducibility condition on the $\{Q_j\}$, $\{P_j\}$ as replaced by the condition that X is irreducible under the combined action of all the $U_j(t)$ and $V_j(t)$.

In view of (9) and (10) we can define two unitary representations U and V of the additive group $\mathbb{R}^n$:

$$U(\tau) = U_1(\tau_1) \ldots U_n(\tau_n),$$

$$V(\tau) = V_1(\tau_1) \ldots V_n(\tau_n)$$

($\tau \in \mathbb{R}^n$). Condition (11) says that U and V are connected by the relation

$$U(\tau)V(\sigma) = e^{i(\sigma \cdot \tau)} V(\sigma)U(\tau) \tag{12}$$

($\sigma, \tau \in \mathbb{R}^n$), where $\sigma \cdot \tau = \sum_{j=1}^{n} \sigma_j \tau_j$.

Now the character group of $\mathbb{R}^n$ is isomorphic to $\mathbb{R}^n$ under the isomorphism which sends each σ in $\mathbb{R}^n$ into the character $\tau \mapsto e^{i(\sigma \cdot \tau)}$. So we can think of V and U as unitary representations of $\mathbb{R}^n$ and $(\mathbb{R}^n)\hat{\ }$ respectively; and (12) becomes just a special case of condition 15.7(iii). Theorem 15.7 now states that the pair U, V is essentially unique; to within unitary equivalence it must be defined as follows:

$$X = \mathscr{L}_2(\mathbb{R}^n) \qquad \text{(with respect to Lebesgue measure),}$$

$$(V(\tau)f)(\sigma) = f(\sigma - \tau),$$

$$(U(\tau)f)(\sigma) = e^{i(\sigma \cdot \tau)}f(\sigma)$$

($f \in X; \sigma, \tau \in \mathbb{R}^n$). This is the original Stone-von Neumann Theorem.

From these formulae together with (8) we deduce the forms of the original unbounded operators Q_j and P_j acting on $\mathscr{L}_2(\mathbb{R}^n)$:

$$(Q_j f)(\sigma) = \sigma_j f(\sigma), \qquad (P_j f)(\sigma) = i\frac{\partial f}{\partial \sigma_j}.$$

These are the familiar forms of the quantum-mechanical position and momentum operators. Granting the legitimacy of the replacement of (6), (7) by (9), (10), (11), we have shown that they are the only unbounded self-adjoint operators which act irreducibility and satisfy (6) and (7).

16.   Conjugation of Representations

16.1.   Let G be a group, H a subgroup of G, and S a unitary representation of H. For each element x of G, the formula

$$S'_h = S_{x^{-1}hx} \qquad\qquad (h \in xHx^{-1}) \qquad (1)$$

defines a unitary representation S' of the conjugate subgroup xHx^{-1}. We call S' the representation *conjugate to S under x*, and denote it by $^x S$.

The formation of conjugates of representations is very important in the theory of group representations, especially in the Mackey normal subgroup analysis. For the bundle generalization of the latter, to be presented in Chapter XII, we shall require a bundle generalization of the process of conjugation of representations. Given a Banach *-algebraic bundle $\mathscr{B}$ over G, a closed subgroup H of G, a *-representation S of $\mathscr{B}_H$, and an element x of G, is there a natural way to construct a "conjugate" *-representation S' of $\mathscr{B}_{xHx^{-1}}$? Formula (1) is of no use as it stands, since xhx^{-1} has no meaning for elements h of the bundle space. This question was answered affirmatively in

7.9 in case $H = \{e\}$ and S is $\mathscr{B}$-positive, as an illustration of the abstract inducing process. In the present section we shall answer it affirmatively for arbitrary H, and make a more systematic study of the process of conjugation. The most satisfactory results are obtained when $\mathscr{B}$ is saturated, and we shall restrict ourselves to this case.

16.2. Throughout this section $\mathscr{B} = \langle B, \pi, \cdot, * \rangle$ is a saturated Banach *-algebraic bundle over the locally compact group G with unit e, left Haar measure λ, and modular function Δ. We fix a closed subgroup H of G, with left Haar measure ν and modular function δ, and choose once for all an everywhere positive continuous H-rho-function ρ on G.

16.3. Take a $\mathscr{B}$-positive non-degenerate *-representation S of $\mathscr{B}_H$. Let $\mathscr{Y} = \langle Y, \{Y_\alpha\} \rangle$ be the Hilbert bundle over G/H induced by S as in 9.2; and let us adopt without further ado the rest of the notation of §9. In particular τ_b is the action of b on Y defined in 9.13.

Fix an element x of G. If $c \in B_{xHx^{-1}}$, so that the group element $h = x^{-1}\pi(c)x$ belongs to H, then by 9.13(24)

$$\tau_c(Y_{xH}) \subset Y_{\pi(c)xH} = Y_{xhH} = Y_{xH};$$

and the equation

$$S'_c(\zeta) = \delta(h)^{1/2}\Delta(h)^{-1/2}\tau_c(\zeta) \qquad (\zeta \in Y_{xH}) \qquad (2)$$

defines a bounded linear operator S'_c on Y_{xH}.

Proposition. $S' : c \mapsto S'_c$ is a non-degenerate *-representation of $B_{xHx^{-1}}$ on the Hilbert space Y_{xH}.

Proof. S' is obviously linear on fibers and is multiplicative by 9.14(25). It is continuous in the strong operator topology by 9.18. To show that it preserves adjoints, let $\xi, \eta \in Y_{xH}$, $c \in B_{xHx^{-1}}$, and put $h = x^{-1}\pi(c)x \in H$. By (2) and 9.14(26)

$$(S'_c\xi, \eta)_{Y_{xH}} = \delta(h)^{1/2}\Delta(h)^{-1/2}(\tau_c\xi, \eta)_{Y_{xH}}$$

$$= \delta(h)^{1/2}\Delta(h)^{-1/2}\sigma(\pi(c), xH)(\xi, \tau_{c^*}\eta)_{Y_{xH}}$$

$$= \delta(h)^{1/2}\Delta(h)^{-1/2}\rho(x)\rho(xh)^{-1}(\xi, \tau_{c^*}\eta)_{Y_{xH}}$$

$$= \delta(h^{-1})^{1/2}\Delta(h^{-1})^{-1/2}(\xi, \tau_{c^*}\eta)_{Y_{xH}}$$

$$= (\xi, S'_{c^*}\eta)_{Y_{xH}}.$$

Thus $(S'_c)^* = S'_{c^*}$; and we have shown that S' is a *-representation. Its non-degeneracy follows from 9.17. ∎

Definition.   The above S' is called the *-representation conjugate to S under x, or simply the x-conjugate of S, and is denoted by ^xS.

Remark 1. ^xS evidently depends only on the H-coset xH to which x belongs.

Remark 2.   If $H = \{e\}$, the ^xS defined here coincides with that of 7.9.

Remark 3.   It follows from (2) and 9.15(29) that

$$^eS \cong S.$$

Remark 4.   If S is not the zero representation, the ^xS is also not the zero representation. This follows from Remark 9.5.

Remark 5. ^xS can be constructed as above, and will be non-degenerate, even if $\mathscr{B}$ is not saturated. But Propositions 16.9 and 16.11 will then fail to hold.

16.4.   From Remarks 1 and 3 of 16.3 applied to the case that $H = G$, we obtain the following useful fact:

Proposition.   *If T is a non-degenerate *-representation of $\mathscr{B}$, then*

$$^xT \cong T \quad \text{for all } x \text{ in } G.$$

16.5.   Returning to 16.3, let α denote the coset xH, and recall the description of Y_α given in 9.5. When the Z_α of 9.5 is subjected to the linear automorphism $b \otimes \xi \mapsto \rho(y)^{-1/2} b \otimes \xi$ ($\xi \in X$; $b \in B_y$; $y \in \alpha$), the inner product 9.5(7) is transformed into

$$(b \otimes \xi, c \otimes \eta)''_\alpha = (S_{c^*b}\xi, \eta);$$

and one verifies that formula (2) takes the simpler form

$$S''_c(\kappa''_\alpha(b \otimes \xi)) = \kappa''_\alpha(cb \otimes \xi)$$

(where κ''_α is the quotient map with respect to the transformed inner product). Thus the latter formula defines an equivalent version S'' of ^xS.

Notice that in this new formulation of the operation of conjugation there is no reference to modular functions and rho-functions on G. We conclude that *the definition of the conjugation operation is unaltered when the topology of G is replaced by the discrete topology.*

16.6.   Proposition.   *Let K be another closed subgroup of G with $K \subset H$. Let S be a $\mathscr{B}$-positive non-degenerate *-representation of $\mathscr{B}_H$; and let $x \in G$. Then $^x(S|\mathscr{B}_K)$ and $(^xS)|\mathscr{B}_{xKx^{-1}}$ are unitarily equivalent *-representations of $\mathscr{B}_{xKx^{-1}}$.*

Proof. Let $\mathscr{L}$ be the Hilbert bundle over G/K induced by $S|\mathscr{B}_K$; and put $\alpha = xH$, $\beta = xK$. On using the description 16.5 of Y_α and Z_β, we see that $Z_\beta \subset Y_\alpha$ and that $^x(S|\mathscr{B}_K)$ is the same as the subrepresentation of $(^xS)|\mathscr{B}_{xKx^{-1}}$ acting on Z_β. Hence the proposition will be established if we show that $Z_\beta = Y_\alpha$.

Now Z_β certainly contains $\kappa''_\alpha(B_x \otimes X(S))$. Furthermore the definition of $(\ ,\ )''$ in 16.5 implies that

$$\kappa''_\alpha(b \otimes S_d\xi) = \kappa''_\alpha(bd \otimes \xi)$$

for all $b \in B_x$, $d \in B_H$, and $\xi \in X(S)$. It follows that Z_β contains $\kappa''_\alpha(B_x B_H \otimes X(S))$. But by the saturation of $\mathscr{B}$, the linear span of $B_x B_h$ is dense in B_{xh} for every h in H. Therefore Z_β contains, and so is equal to, Y_α. ∎

16.7. Applying 16.6 with H and K replaced by G and H, and using 16.4, we obtain the following corollary:

Corollary. *If $x \in G$ and T is a non-degenerate *-representation of $\mathscr{B}$, then $^x(T|\mathscr{B}_H)$ is unitarily equivalent to $T|\mathscr{B}_{xHx^{-1}}$.*

Remark. It is useful to observe the explicit form of the unitary equivalence claimed in this corollary. Let $\alpha = xH \in G/H$; put $S = T|\mathscr{B}_H$; and adopt the notation of §9. Then the equation

$$F(\kappa_\alpha(b \otimes \xi)) = \rho(\pi(b))^{-1/2}T_b\xi \qquad (b \in B_\alpha; \xi \in X(T))$$

defines F as a linear isometry of $X(^xS)$ onto $X(T)$ which intertwines ^xS and $T|\mathscr{B}_{xHx^{-1}}$.

Conjugation and Imprimitivity Bimodules

16.8. As one might guess from the discussion in 7.9 the operation of conjugation can also be described as an inducting operation by means of an $\mathscr{L}(\mathscr{B}_{xHx^{-1}})$, $\mathscr{L}(\mathscr{B}_H)$ imprimitivity bimodule.

Fix an element x of G. We shall make $\mathscr{L}(\mathscr{B}_{xH})$ into a left $\mathscr{L}(\mathscr{B}_{xHx^{-1}})$-module and a right $\mathscr{L}(\mathscr{B}_H)$-module by means of the actions : described as follows:

$$(\phi:f)(xh) = \int_H \delta(k)^{1/2}\Delta(k)^{-1/2}\phi(xkx^{-1})f(xk^{-1}h)dvk, \qquad (3)$$

$$(f:\psi)(xh) = \int_H \Delta(k)^{1/2}\delta(k)^{-1/2}f(xhk)\psi(k^{-1})dvk \qquad (4)$$

$(f \in \mathscr{L}(\mathscr{B}_{xH}); \phi \in \mathscr{L}(\mathscr{B}_{xHx^{-1}}); \psi \in \mathscr{L}(\mathscr{B}_H); h \in H)$. We also equip $\mathscr{L}(\mathscr{B}_{xH})$ with an $\mathscr{L}(\mathscr{B}_{xHx^{-1}})$-valued rigging $[\ ,\ ]'_x$ and an $\mathscr{L}(\mathscr{B}_H)$-valued rigging $[\ ,\ ]_x$ defined by:

$$[f, g]'_x(xhx^{-1}) = \Delta(h)^{1/2}\delta(h)^{-1/2} \int_H \Delta(k)f(xhk)g(xk)^* dvk, \qquad (5)$$

$$[f, g]_x(h) = \Delta(h)^{1/2}\delta(h)^{-1/2} \int_H \Delta(k)\delta(k)^{-1}f(xk)^*g(xkh)dvk \qquad (6)$$

$(f, g \in \mathscr{L}(\mathscr{B}_{xH}); h \in H)$. We leave to the reader the routine verification of the fact that, when $\mathscr{L}(\mathscr{B}_{xH})$ is equipped with the module structures (3) and (4) and the riggings (5) and (6) it becomes an $\mathscr{L}(\mathscr{B}_{xHx^{-1}})$, $\mathscr{L}(\mathscr{B}_H)$ imprimitivity bimodule. This imprimitivity bimodule will be denoted throughout this section by $\mathscr{I}_x$.

Remark. It is assumed above that in defining the convolution in $\mathscr{L}(\mathscr{B}_{xHx^{-1}})$ one uses the left Haar measure v' on xHx^{-1} obtained by x-conjugation of v: $v'(W) = v(x^{-1}Wx)$.

16.9. Notice that the $\mathscr{L}(\mathscr{B}_H)$-valued rigging (6) on $\mathscr{L}(\mathscr{B}_{xH})$ differs from the rigging 8.8(10) only by the linear automorphism $f \mapsto f'$ of $\mathscr{L}(\mathscr{B}_{xH})$, where $f'(xh) = \Delta(h)^{1/2}\delta(h)^{-1/2}f(xh)$. It follows therefore from 9.1(2) that the integrated form of a $\mathscr{B}$-positive *-representation of $\mathscr{B}_H$ is positive with respect to $\mathscr{I}_x$.

16.10.   Proposition. *The linear spans of the ranges of* $[\ ,\ ]_x$ *and* $[\ ,\ ]'_x$ *are dense in* $\mathscr{L}(\mathscr{B}_H)$ *and* $\mathscr{L}(\mathscr{B}_{xHx^{-1}})$ *respectively in the inductive limit topologies.*

Proof. We shall first show that range $([\ ,\ ]'_x)$ has dense linear span in $\mathscr{L}(\mathscr{B}_{xHx^{-1}})$.

Let us denote by M the inductive limit closure of the linear span of range $([\ ,\ ]'_x)$, and by L the linear span in $\mathscr{L}(\mathscr{B}_{xHx^{-1}})$ of the set of all cross-sections of $\mathscr{B}_{xHx^{-1}}$ of the form

$$xhx^{-1} \mapsto b\phi(h)c^* \qquad (h \in H), \qquad (7)$$

where $\phi \in \mathscr{L}(\mathscr{B}_H)$ and $b, c \in B_x$. We claim that

$$L \subset M. \qquad (8)$$

To prove (8), take $b, c \in B_x$ and $\psi, \chi \in \mathscr{L}(\mathscr{B}_H)$. Then the equations

$$f(xh) = \delta(h)^{1/2}\Delta(h)^{-1/2}b\psi(h),$$

$$g(xh) = \delta(h)^{1/2}\Delta(h)^{-1/2}c\chi(h)$$

$(h \in H)$ define elements f and g of $\mathscr{L}(\mathscr{B}_{xH})$; and, for $h \in H$, we have by (5)

$$[f, g]_x'(xhx^{-1}) = \int b\psi(hk)\chi(k)^*c^*\delta(k)dvk$$

$$= b\left[\int \psi(hk)\chi(k)^*\delta(k)dvk\right]c^*$$

$$= b(\psi * \chi^*)(h)c^*.$$

It follows from this that (7) belongs to M whenever ϕ is of the form $\psi * \chi^*$ $(\psi, \chi \in \mathscr{L}(\mathscr{B}_H))$. By Remark 14.13 such ϕ span a dense subset of $\mathscr{L}(\mathscr{B}_H)$. Since the map carrying ϕ into (7) is clearly continuous in the inductive limit topologies of $\mathscr{L}(\mathscr{B}_H)$ and $\mathscr{L}(\mathscr{B}_{xHx^{-1}})$, the last two sentences imply that (7) belongs to M for all ϕ in $\mathscr{L}(\mathscr{B}_H)$ and all b, c in B_x. Thus (8) is established.

Now L is clearly closed under multiplication by continuous complex functions on xHx^{-1}. Also the saturation of $\mathscr{B}$ implies that $\{f(k): f \in L\}$ is dense in B_k for every k in xHx^{-1}. So by II.14.6 L is dense in $\mathscr{L}(\mathscr{B}_{xHx^{-1}})$ in the inductive limit topology. This and (8) show that $M = \mathscr{L}(\mathscr{B}_{xHx^{-1}})$. Hence, by the definition of M, the linear span of range($[\ ,\ ]_x'$) is dense in $\mathscr{L}(\mathscr{B}_{xHx^{-1}})$ in the inductive limit topology.

The proof that the linear span of range($[\ ,\ ]_x$) is dense in $\mathscr{L}(\mathscr{B}_H)$ is very similar, and is left to the reader.   ■

16.11.   The next proposition shows that x-conjugation of a *-representation of $\mathscr{B}_H$ is the same as inducing with respect to $\mathscr{I}_x$.

Proposition.   *Let x be an element of G, and S a non-degenerate $\mathscr{B}$-positive *-representation of $\mathscr{B}_H$. Then the integrated form of ^xS is unitarily equivalent to the *-representation of $\mathscr{L}(\mathscr{B}_{xHx^{-1}})$ induced from S via $\mathscr{I}_x$.*

Proof.   Denote ^xS and its integrated form by S'. Writing α for xH, and adopting the notation of §9, we have only to show that

$$S_\phi'(\kappa_\alpha(f \otimes \xi)) = \kappa_\alpha(\phi : f \otimes \xi) \tag{9}$$

whenever $\phi \in \mathscr{L}(\mathscr{B}_{xHx^{-1}})$, $f \in \mathscr{L}(\mathscr{B}_\alpha)$, and $\xi \in X(S)$. But for any ζ in Y_α we have

$$(S'_\phi \kappa_\alpha(f \otimes \xi), \zeta)_{Y_\alpha} = \int (S'_{\phi(xkx^{-1})}\kappa_\alpha(f \otimes \xi), \zeta)_{Y_\alpha} \, dvk$$

$$= \int \delta(k)^{1/2}\Delta(k)^{-1/2}(\tau_{\phi(xkx^{-1})}\kappa_\alpha(f \otimes \xi), \zeta)_{Y_\alpha} \, dvk \qquad \text{(by (2))}$$

$$= \int \delta(k)^{1/2}\Delta(k)^{-1/2}(\kappa_\alpha(\phi(xkx^{-1})f \otimes \xi), \zeta)_{Y_\alpha} \, dvk$$

$$\text{(by 9.12(22))} \qquad (10)$$

(where $(\phi(xkx^{-1})f)(xh) = \phi(xkx^{-1})f(xk^{-1}h))$. Now, just as in VIII.12.5, the definition (3) of $\phi{:}f$ can be written in the form

$$\phi{:}f = \int \delta(k)^{1/2}\Delta(k)^{-1/2}(\phi(xkx^{-1})f)dvk, \qquad (11)$$

where the right side of (11) is an $\mathscr{L}(\mathscr{B}_\alpha)$-valued integral with respect to the inductive limit topology. Applying to both sides of (11) the linear functional $g \mapsto (\kappa_\alpha(g \otimes \xi), \zeta)_{Y_\alpha}$ (which is continuous on $\mathscr{L}(\mathscr{B}_\alpha)$ in the inductive limit topology by Proposition 9.1), we find

$$(\kappa_\alpha(\phi{:}f \otimes \xi), \zeta)_{Y_\alpha} = \int \delta(k)^{1/2}\Delta(k)^{-1/2}(\kappa_\alpha(\phi(xkx^{-1})f \otimes \xi), \zeta)_{Y_\alpha} \, dvk. \quad (12)$$

Combining (10) and (12), by the arbitrariness of ζ one obtains (9).   ■

Properties of Conjugation

16.12.   The process of conjugation preserves Hilbert direct sums.

Proposition.   *Let x be an element of G, and $\{S^i\}$ an indexed collection of non-degenerate $\mathscr{B}$-positive *-representations of $\mathscr{B}_H$. Then*

$$x\left(\sum_i{}^\oplus S^i\right) \cong \sum_i{}^\oplus {}^x(S^i).$$

Proof.   Combine 16.11 with 5.1.   ■

16.13.   Proposition.   *Let x be an element of G, and S a non-degenerate $\mathscr{B}$-positive *-representation of $\mathscr{B}_H$. The commuting algebras of S and ${}^xS$ are *-isomorphic under the *-isomorphism $F \mapsto \tilde{F}$, where*

$$\tilde{F}(\kappa_{xH}(f \otimes \xi)) = \kappa_{xH}(f \otimes F(\xi)) \qquad (13)$$

$(f \in \mathscr{L}(\mathscr{B}_{xH}); \xi \in X(S))$. *In particular ^xS is irreducible if and only if S is irreducible.*

Proof. Let B and E be the linear spans of the ranges of $[\ ,\ ]_x$ and $[\ ,\ ]'_x$ respectively (see (5), (6)); and let T and T' be the restrictions of S and ^xS to B and E respectively. Thus T and T' are non-degenerate by 16.10, and by 16.11 T' is induced from T via $\mathscr{J}_x$ when the latter is considered as a strict E, B imprimitivity bimodule. Hence by Theorem 6.16 the commuting algebras of T and T' are *-isomorphic under (13). But, again by 16.10, the commuting algebras of S and T are the same, and likewise those of ^xS and T'. This completes the proof. ■

16.14. Corollary. *Suppose that B_e has the property that all its irreducible *-representations are finite-dimensional. If S is a finite-dimensional $\mathscr{B}$-positive *-representation of $\mathscr{B}_H$, then every fiber of the Hilbert bundle $\mathscr{Y}$ induced by S is finite-dimensional.*

Proof. Let $x \in G$. The fiber Y_{xH} of $\mathscr{Y}$ is the space of ^xS.

Now $S|B_e \cong \sum_{i \in I}^{\oplus} R^i$, where I is a finite set and each R^i is an irreducible (finite-dimensional) *-representation of B_e. So by 16.6 and 16.12 $(^xS)|B_e \cong{}^x(S|B_e) \cong \sum_{i \in I}^{\oplus}{}^x(R^i)$. By 16.13 each $^x(R^i)$ is irreducible, and so by the hypothesis of the corollary is finite-dimensional. It follows that $(^xS)|B_e$ is finite-dimensional, whence by the opening remark Y_{xH} is finite-dimensional. ■

16.15. Corollary. *In addition to the hypotheses of 16.14 suppose that G/H is finite. Then $\mathrm{Ind}_{\mathscr{B}_H \uparrow \mathscr{B}}(S)$ is finite-dimensional.*

This follows immediately from 16.14.

Remark. Without the hypothesis that the elements of $(B_e)\hat{\ }$ are all finite-dimensional, the conclusions of 16.14 and 16.15 would be false. Indeed, we will see from the example of 16.32 that in general the process of conjugation in $\mathscr{B}$ may carry a finite-dimensional *-representation of B_e into an infinite-dimensional one.

16.16. In the group case one would like to know that the conjugation defined in 16.3 coincides (at least up to unitary equivalence) with the more obvious conjugation given by (1). More generally we will prove:

Proposition. *Let S be a non-degenerate $\mathscr{B}$-positive *-representation of $\mathscr{B}_H$, and x an element of G; and suppose that $\mathscr{B}$ has an approximate unit and that there exists a unitary multiplier u of $\mathscr{B}$ of order x. Then ${}^xS$ is unitarily equivalent to the *-representation T of $\mathscr{B}_{xHx^{-1}}$ defined by*

$$T_b = S_{u^*bu} \qquad\qquad (b \in \mathscr{B}_{xHx^{-1}}). \qquad (14)$$

Remark. We have assumed an approximate unit in $\mathscr{B}$ in order that $\mathscr{B}$ should have no annihilators. All associative laws between elements and multipliers of $\mathscr{B}$ then conveniently hold (see VIII.2.14). Thus $b \mapsto u^*(bu) = (u^*b)u$ is a *-isomorphism of $\mathscr{B}_{xHx^{-1}}$ onto $\mathscr{B}_H$; and the T of (14) is indeed a *-representation of $\mathscr{B}_{xHx^{-1}}$. Suppose however that $\mathscr{B}$ has non-zero annihilators. Even then, the above proposition holds in modified form. For, though $u^*(bu)$ and $(u^*b)u$ may then be unequal, it follows from the non-degeneracy of S that $S_{u^*(bu)} = S_{(u^*b)u}$; and one verifies that the modification

$$T_b = S_{u^*(bu)} = S_{(u^*b)u}$$

of equation (14) still defines a *-representation T of $\mathscr{B}_{xHx^{-1}}$. The proof below continues to be valid in this slightly more general context, and shows that $T \cong {}^xS$.

Proof. In this argument we will use the description 9.5 of Y_{xH}.

Let ξ, η be in $X(S)$ and b, c in B_{xH}; and put $\pi(b) = y$, $\pi(c) = z$, $\alpha = xH$ Then

$$(\kappa_\alpha(b \otimes \xi), \kappa_\alpha(c \otimes \eta))_{Y_\alpha} = (\rho(y)\rho(z))^{-1/2}(S_{c^*b}\xi, \eta) \qquad \text{(by 9.5(10)}$$

$$= (\rho(y)\rho(z))^{-1/2}(S_{(u^*c)^*u^*b}\xi, \eta)$$

$$= (\rho(y)\rho(z))^{-1/2}(S_{u^*b}\xi, S_{u^*c}\eta)$$

(since u^*b, u^*c are in B_H). It follows that the equation

$$F(\kappa_\alpha(b \otimes \xi)) = (\rho(\pi(b)))^{-1/2}S_{u^*b}\xi$$

$(b \in B_{xH}; \; \xi \in X(S))$ defines a linear isometry of $X({}^x S)$ into $X(S)$. Since $u^*(B_{xh}) = B_h$ for each h in H and S is non-degenerate, F is *onto* $X(S)$. If $\xi \in X(S)$, h, $k \in H$, $b \in B_{xh}$, and $c \in B_{xkx^{-1}}$, then

$$F[({}^x S)_c \kappa_\alpha(b \otimes \xi)] = \delta(k)^{1/2}\Delta(k)^{-1/2}F(\kappa_\alpha(cb \otimes \xi)) \qquad \text{(by (2) and 9.11(18))}$$

$$= \delta(k)^{1/2}\Delta(k)^{-1/2}\rho(xkh)^{-1/2}S_{u^*cb}\xi$$

$$= \rho(xh)^{-1/2}S_{u^*cb}\xi$$

$$= \rho(\pi(b))^{-1/2}S_{u^*cu}S_{u^*b}\xi$$

$$= T_c F(\kappa_\alpha(b \otimes \xi)) \qquad \text{(by (14))}.$$

This shows that F intertwines ${}^x S$ and T, completing the proof. ∎

In particular, in the group case the conjugation defined by 16.3 does coincide, up to unitary equivalence, with that defined by (1).

16.17. We come now to a very important result (which is certainly to be expected by analogy with (1)).

Proposition. *Let S be a non-degenerate $\mathcal{B}$-positive *-representation of $\mathcal{B}_H$, and let x, y be two elements of G. Then*

(I) ${}^x S$ *is $\mathcal{B}$-positive (so that we can form ${}^y({}^x S)$); and*
(II) ${}^y({}^x S) \cong {}^{(yx)}S$.

Proof. Put $H' = xHx^{-1}$; and notice that the formula

$$\rho'(z) = \rho(zx) \qquad (z \in G) \qquad (15)$$

defines ρ' as a continuous everywhere positive H'-rho-function on G.

Take vectors $\xi_1, \ldots, \xi_n$ in $X = X(S)$, elements $b_1, \ldots, b_n$ of B_{xH}, and elements $c_1, \ldots, c_n$ of $B_{yH'}$. Thus $\pi(b_i) = xh_i$ and $\pi(c_i) = yxk_ix^{-1}$, where h_i and k_i are in H. Putting $\alpha = xH$, we have by (2), 9.11(18), and 9.5(10)

$$\sum_{i,j=1}^{n} (\rho'(\pi(c_i))\rho'(\pi(c_j)))^{-1/2}(({}^x S)_{c_j^* c_i}\kappa_\alpha(b_i \otimes \xi_i), \kappa_\alpha(b_j \otimes \xi_j))$$

$$= \sum_{i,j}(\rho(yxk_i)\rho(yxk_j))^{-1/2}\delta(k_j^{-1}k_i)^{1/2}\Delta(k_j^{-1}k_i)^{-1/2}(\kappa_\alpha(c_j^* c_i b_i \otimes \xi_i), \kappa_\alpha(b_j \otimes \xi_j))$$

$$= \sum_{i,j=1}^{n} N_{ij}(S_{b_j^* c_j^* c_i b_i}\xi_i, \xi_j), \qquad (16)$$

where

$$N_{ij} = (\rho(yxk_i)\rho(yxk_j))^{-1/2}\delta(k_j^{-1}k_i)^{1/2}\Delta(k_j^{-1}k_i)(\rho(xk_j^{-1}k_ih_i)\rho(xh_j))^{-1/2}$$

$$= \rho(x)^{-1}(\rho(yxk_ih_i)\rho(yxk_jh_j))^{-1/2}$$

$$= \rho(x)^{-1}(\rho(\pi(c_ib_i))\rho(\pi(c_jb_j)))^{-1/2}.$$

Therefore the right side of (16) becomes

$$\rho(x)^{-1}\sum_{i,\,j=1}^{n}(\kappa_{y\alpha}(c_ib_i\otimes\xi_i),\kappa_{y\alpha}(c_jb_j\otimes\xi_j)). \tag{17}$$

Now (17) is non-negative by the $\mathscr{B}$-positivity of S. So the left side of (16) is non-negative. Since the $\kappa_\alpha(b\otimes\xi)$ span a dense subspace of Y_α, this implies (by the arbitrariness of y) that ^xS is $\mathscr{B}$-positive. So statement (I) of the proposition is proved.

Thus we can form the Hilbert bundle $\mathscr{Y}' = \langle Y', \{Y'_\beta\}\rangle$ over G/H' generated by $S' = {}^xS$ (using the H'-rho-function ρ'). Let κ'_β ($\beta\in G/H'$) denote the canonical quotient map of $\mathscr{L}(\mathscr{B}_\beta)\otimes X(S')$ into Y'_β. The equality of (16) and (17) asserts that there is a linear isometry $F_{yH'}: Y'_{yH'}\to Y_{yxH}$ satisfying

$$F_\beta(\kappa'_\beta(c\otimes\kappa_\alpha(b\otimes\xi))) = \rho(x)^{-1/2}\kappa_{y\alpha}(cb\otimes\xi) \tag{18}$$

($\xi\in X$; $b\in B_\alpha$; $c\in B_\beta$; $\beta = yH'$). Since $\mathscr{B}$ is saturated, the set of all such products cb spans a dense subspace of each fiber within $B_{y\alpha}$. Therefore by (18) the range of F_β is dense in $Y_{y\alpha}$ and so equal to $Y_{y\alpha}$. Using (2) and (18) one verifies easily that F_β intertwines $^y(^*S)$ (whose space is Y'_β) and ^{yx}S (whose space is $Y_{y\alpha}$). So statement (II) is proved.   ∎

16.18.   Proposition 16.17 (together with Remark 3 of 16.3) asserts that G acts as a left transformation group on the set of all pairs $\langle K, S\rangle$, where K is a closed subgroup of G and S is a unitary equivalence class of non-degenerate $\mathscr{B}$-positive *-representations of $\mathscr{B}_K$; the element x of G sends $\langle K, S\rangle$ into $\langle xKx^{-1}, {}^xS\rangle$. In particular, if H is *normal*, $\langle x, S\rangle\mapsto {}^xS$ is a left action of G on the space of all unitary equivalence classes of non-degenerate $\mathscr{B}$-positive *-representations of $\mathscr{B}_H$. We shall see in 16.23 that this action is continuous (with respect to the regional topology of the space of *-representations of $\mathscr{B}_H$). In view of 16.13 *this action leaves stable the space of all equivalence classes of irreducible $\mathscr{B}$-positive *-representations of $\mathscr{B}_H$.*

16.19.   The proof of 16.17 gives further interesting information. Thus we can prove:

Proposition. *Let x be an element of G, and S a non-degenerate $\mathscr{B}$-positive
-representation of $\mathscr{B}_H$. Then

$$\underset{\mathscr{B}_{xHx^{-1}}\uparrow\mathscr{B}}{\text{Ind}}\ (^xS) \cong \underset{\mathscr{B}_H\uparrow\mathscr{B}}{\text{Ind}}\ (S). \tag{19}$$

Proof. We adopt without further ado all the notation of the proof of 16.17.
Notice that the correspondence $\beta \mapsto \beta x$ $(\beta \in G/H')$ is a homeomorphism of
G/H' onto G/H.

Let $F: Y' \to Y$ be the bijection coinciding on each Y'_β with the bijection
$F_\beta: Y'_\beta \to Y_{\beta x}$ constructed in the proof of 16.17. Observe that (17) amounts to

$$F_\beta(\kappa'_\beta(c \otimes \zeta)) = \rho(x)^{-1/2}\tau_c(\zeta) \tag{20}$$

$(\beta \in G/H'; c \in B_\beta; \zeta \in Y_{xH})$, where τ of course is the action of $\mathscr{B}$ on $\mathscr{Y}$ (9.13).

We claim that $F: Y' \to Y$ is a homeomorphism. Indeed: Suppose $\mathscr{T}$ stands
for that topology of Y' which makes F a homeomorphism. Then, equipped
with $\mathscr{T}$ instead of its original topology, $\mathscr{Y}'$ is still a Hilbert bundle over G/H';
and by (20) and the continuity of τ (9.18), for each fixed ζ in $X(^xS)$, the map
$c \mapsto \kappa'_{\pi(c)H'}(c \otimes \zeta)$ is continuous on B with respect to $\mathscr{T}$. By 9.20 this implies
that $\mathscr{T}$ is the same as the original topology of $\mathscr{Y}'$. So F is a homeomorphism.

Thus, to prove the proposition, it remains only to show that F carries the
action τ of $\mathscr{B}$ on $\mathscr{Y}$ into the corresponding action τ' of $\mathscr{B}$ on $\mathscr{Y}'$. If β, c, ζ are as
in (20) and $d \in B_z$, we have by (20)

$$\begin{aligned}
F[\tau'_d\kappa'_\beta(c \otimes \zeta)] &= F[\kappa'_{z\beta}(dc \otimes \zeta)] \\
&= \rho(x)^{-1/2}\tau_{dc}(\zeta) = \rho(x)^{-1/2}\tau_d\tau_c(\zeta) \\
&= \tau_d F[\kappa'_\beta(c \otimes \zeta)].
\end{aligned}$$

So $F \circ \tau'_d = \tau_d \circ F$; and the proof is complete. ∎

Remark. Combining this proposition with 16.16, we obtain in the group
case another proof of 12.21.

16.20. Here is an interesting and useful application of the conjugation
process:

Suppose that $\mathscr{B}$ is a saturated C^*-algebraic bundle (over a locally compact
group G with unit e) with unit fiber $A = B_e$; and let u be a fixed multiplier of
A which is central, that is, $ua = au$ for all $a \in A$. It then follows (from VIII.1.15
and VI.14.1) that, for each $D \in \hat{A}$, the *-representation of the multiplier
*-algebra of A which extends D sends u into a scalar operator, which we shall
denote by $\lambda(D) \cdot 1$.

By Proposition VIII.3.8 u has a unique extension to a multiplier u_0 of $\mathscr{B}$ of order e.

Notice that every element of $\hat{A}$ is $\mathscr{B}$-positive (by 11.10); hence the process of conjugation gives rise to a left action of G on *all* of $\hat{A}$.

Proposition. *With the above notation, the following two conditions are equivalent:*

(i) u_0 *is central in the sense that* $u_0 b = b u_0$ *for all* $b \in B$.

(ii) $\lambda(D) = \lambda(^x D)$ *for all* $D \in \hat{A}$ *and* $x \in G$.

Proof. (i) $\Rightarrow$ (ii). Assume (i), and let $D \in \hat{A}$, $x \in G$. Preserving the notation of §9 (with $H = \{e\}$, $S = D$), we see from 9.16 that $\{\kappa_x(ba \otimes \xi): a \in A, b \in B_x, \xi \in X(D)\}$ is dense in $X(^x D)$. Hence to prove (ii) it is enough to verify that

$$(^x D)_{u_0} \kappa_x(ba \otimes \xi) = \lambda(D)\kappa_x(ba \otimes \xi)$$

($a \in A$; $b \in B_x$; $\xi \in X(D)$). But in view of 9.30

$$
\begin{aligned}
(^x D)_{u_0} \kappa_x(ba \otimes \xi) &= \kappa_x(u_0 ba \otimes \xi) \\
&= \kappa_x(bu_0 a \otimes \xi) && \text{(by (i))} \\
&= \kappa_x(b(ua) \otimes \xi) \\
&= \kappa_x(b \otimes D_{ua}\xi) && \text{(by 9.5 (11))} \\
&= \kappa_x(b \otimes D_u D_a \xi) \\
&= \lambda(D)\kappa_x(b \otimes D_a \xi) \\
&= \lambda(D)\kappa_x(ba \otimes \xi) && \text{(by 9.5 (11))}.
\end{aligned}
$$

(ii) $\Rightarrow$ (i). We first remark that for each $b \neq 0$ in B there exists an element D of $\hat{A}$ such that $(\mathrm{Ind}_{A \uparrow \mathscr{B}}(D))_b \neq 0$. Indeed: It is enough to find $D \in \hat{A}$ such that $(\mathrm{Ind}_{A \uparrow \mathscr{B}}(D))_{b^*b} \neq 0$ and this is possible by Lemma 11.3 (since $0 \neq b^*b \in A$). Now assume (ii). In view of the above remark, (i) will be proved if we show that

$$(\mathrm{Ind}(D))_{u_0 b} = (\mathrm{Ind}(D))_{bu_0} \qquad \text{for all } D \in \hat{A};$$

and this in turn will follow if we can show that $(\mathrm{Ind}(D))_{u_0}$ is the scalar operator $\lambda(D) \cdot 1$ for each $D \in \hat{A}$. But the latter is a consequence of (ii) together with Proposition 9.30. ∎

An Equivalent Definition of Conjugation

16.21. A closer look at the proof of 16.19, combined with the Imprimitivity Theorem 14.18, suggests a useful equivalent definition of the process of conjugation.

Fix an element x of G and write $H' = xHx^{-1}$; and let us denote by Φ the homeomorphism $\beta \mapsto \beta x$ of G/H' onto G/H observed in the proof of 16.19. Of course Φ is an isomorphism of G/H' and G/H as left G-spaces. Thus if $\langle T, P \rangle$ is any system of imprimitivity for $\mathcal{B}$ over G/H, and if we write P' for the regular $X(T)$-projection-valued Borel measure $W \mapsto P(\Phi(W))$ on G/H', then clearly $\langle T, P' \rangle$ is a system of imprimitivity for $\mathcal{B}$ over G/H'. Let us refer to $\langle T, P' \rangle$ as the *x-conjugate* of $\langle T, P \rangle$.

Proposition. *Let S be a non-degenerate $\mathcal{B}$-positive *-representation of $\mathcal{B}_H$, and x an element of G. Let $\langle T, P \rangle$ and $\langle T', P' \rangle$ be the systems of imprimitivity for $\mathcal{B}$ over G/H and $G/(xHx^{-1})$ induced by S and ^xS respectively. Then $\langle T', P' \rangle$ is equivalent to the x-conjugate of $\langle T, P \rangle$.*

Proof. The map F of the proof of 16.19 is an isometric isomorphism of $\mathcal{Y}$ and $\mathcal{Y}'$ covariant with Φ (see II.13.8), and carries the action τ of $\mathcal{B}$ on $\mathcal{Y}$ into the action τ' of $\mathcal{B}$ on $\mathcal{Y}'$. From this the result follows immediately. ∎

In view of the uniqueness statement in 14.18, the preceding proposition leads to the following equivalent description of ^xS.

Let S be a non-degenerate $\mathcal{B}$-positive *-representation of $\mathcal{B}_H$, and $\langle T, P \rangle$ the system of imprimitivity for $\mathcal{B}$ over G/H induced by S. Let x be an element of G, and $\langle T, P' \rangle$ the x-conjugate of $\langle T, P \rangle$. Then ^xS is that non-degenerate $\mathcal{B}$-positive *-representation of $\mathcal{B}_{xHx^{-1}}$ (unique to within unitary equivalence) such that the system of imprimitivity induced by ^xS is equivalent to $\langle T, P' \rangle$.

Conjugation and Partial Cross-Sectional Bundles

16.22. It is often useful to observe that the conjugation process described in this section is essentially unaltered on passing to partial cross-sectional bundles.

To state this more precisely, let N be a closed normal subgroup of G contained in the closed subgroup H; and form the partial cross-sectional bundle $\mathcal{C} = \langle C, \{C_\alpha\} \rangle$ over $G' = G/N$ derived from $\mathcal{B}$ (as in VIII.6.5). Since $\mathcal{B}$ is saturated the same is true of $\mathcal{C}$ (by VIII.6.8). Let H' be the (closed) image of H in G'. Evidently $\mathcal{C}_{H'}$ is the partial cross-sectional bundle over H' ($= H/N$) derived from $\mathcal{B}_H$. Thus the natural correspondence of VIII.15.9 holds between *-representations of $\mathcal{B}_H$ and of $\mathcal{C}_{H'}$.

Proposition.   *Let S be a non-degenerate $\mathscr{B}$-positive *-representation of $\mathscr{B}_H$, and S' the *-representation of $\mathscr{C}_H$ corresponding to S via VIII.15.9. Then S' is $\mathscr{C}$-positive. If $x \in G$ and $x' = xN \in G'$, then $^{x'}(S')$ is unitarily equivalent to the *-representation of $\mathscr{C}_{x'H'x'^{-1}}$ corresponding via VIII.15.9 with the *-representation ^xS of $\mathscr{B}_{xHx^{-1}}$.*

Proof.   It was shown in 12.7 that S' is $\mathscr{C}$-positive.

Let $V = {}^xS$, and let V' be the *-representation of $\mathscr{C}_{x'H'x'^{-1}}$ corresponding (via VIII.15.9) with V. We shall make the natural identification of G'/H' with G/H. Then, if $\langle T, P \rangle$ is the system of imprimitivity induced by S, the proof of 12.7 shows us that the system of imprimitivity induced by S' coincides (to within unitary equivalence) with $\langle T', P \rangle$, where T' corresponds to T via VIII.15.9. By 16.21 the system of imprimitivity induced by V is (to within equivalence) the x-conjugate $\langle T, Q \rangle$ of $\langle T, P \rangle$; and so, again by the proof of 12.7, the system of imprimitivity induced by V' is $\langle T', Q \rangle$ (where T' as before corresponds to T). But $\langle T', Q \rangle$ is the x'-conjugate of $\langle T', P \rangle$. Hence by 16.21 V' is equivalent to the x'-conjugate of S'.   ■

The Continuity of the Conjugation Operation

16.23.   We shall now show that, if H is normal, the process of conjugation is continuous in the regional topology.

Proposition.   *Suppose that H is normal; and let $\mathscr{S}$ be the set of all non-degenerate $\mathscr{B}$-positive *-representations of $\mathscr{B}_H$. Then the conjugation map $\langle x, S \rangle \mapsto {}^xS$ is continuous on $G \times \mathscr{S}$ to $\mathscr{S}$ with respect to the regional topology of $\mathscr{S}$.*

Proof.   We have seen in 16.17 that the range of the conjugation map lies in $\mathscr{S}$.

It is sufficient to prove the proposition for the case that $H = \{e\}$. Indeed: Let $\mathscr{C} = \langle C, \{C_\alpha\} \rangle$ be the partial cross-sectional bundle over G/H derived from $\mathscr{B}$. If $S \in \mathscr{S}$, let S' be the corresponding *-representation of $C_e = \mathscr{L}_1(v; \mathscr{B}_H)$; that is, S' is the integrated form of S. By the definition VIII.21.2 of the regional topology the map $S \mapsto S'$ is a homeomorphism. By 16.22 this map carries conjugation in $\mathscr{S}$ by the element x into conjugation of *-representations of C_e by xN. If we have proved that the proposition holds for $H = \{e\}$, then it holds for $\mathscr{C}$ and the one-element subgroup of G/H; that is, $\langle x, S' \rangle \mapsto {}^{xN}(S')$ is continuous. It then follows from what we have said that $\langle x, S \rangle \mapsto {}^xS$ is continuous.

Thus we may and shall assume from the beginning that $H = \{e\}$. We shall write A for B_e. Suppose that $S^i \to S$ in $\mathscr{S}$ (in the regional topology) and $x^i \to x$ in G. We must show that $^{x^i}(S^i) \to {}^x S$ regionally. For this it is enough to show that, for each U in a basis of regional neighborhoods of $^x S$, some subnet of $\{^{x^i}(S^i)\}$ eventually lies in U.

As usual let κ_x be the canonical quotient map of $B_x \otimes X(S)$ into $X(^xS)$ (the x-fiber of the Hilbert bundle induced by S).

Choose finitely many vectors $\zeta^1, \ldots, \zeta^n$ in $X(^xS)$ of the form

$$\zeta^r = \kappa_x(b^r \otimes \xi^r) \quad (b^r \in B_x; \xi^r \in X(S)). \tag{21}$$

Passing to a subnet (and recalling VII.1.8), we can find vectors ξ^r_i in $X(S^i)$ such that for $r, s = 1, \ldots, n$ and $a \in A$

$$(\xi^r_i, \xi^s_i) \underset{i}{\to} (\xi^r, \xi^s), \tag{22}$$

$$(S^i_a \xi^r_i, \xi^s_i) \underset{i}{\to} (S_a \xi^r, \xi^s). \tag{23}$$

Now pick continuous cross-sections γ^r of $\mathscr{B}$ such that $\gamma^r(x) = b^r$ $(r = 1, \ldots, n)$, and set

$$\zeta^r_i = \kappa_{x^i}(\gamma^r(x^i) \otimes \xi^r_i) \in X(^{x^i}(S^i)).$$

From (23) and the fact that

$$\|\gamma^s(x^i)^* \gamma^r(x^i) - (b^s)^* b^r \| \underset{i}{\to} 0$$

we obtain

$$|(S^i_{\gamma^s(x^i)^* \gamma^r(x^i)} \xi^r_i, \xi^s_i) - (S_{(b^s)^* b^r} \xi^r, \xi^s)| \leq |(S^i_{(b^s)^* b^r} \xi^r_i, \xi^s_i) - (S_{(b^s)^* b^r} \xi^r, \xi^s)|$$

$$+ |(S^i_{(\gamma^s(x^i)^* \gamma^r(x^i) - (b^s)^* b^r)} \xi^r_i, \xi^s_i)|$$

$$\underset{i}{\to} 0,$$

whence

$$(\zeta^r_i, \zeta^s_i) = (S^i_{\gamma^s(x^i)^* \gamma^r(x^i)} \xi^r_i, \xi^s_i)$$

$$\underset{i}{\to} (S_{(b^s)^* b^r} \xi^r, \xi^s)$$

$$= (\zeta^r, \zeta^s). \tag{24}$$

Further, if $a \in A$, a similar argument yields

$$((^{x^i}(S^i))_a \zeta_i^r, \zeta_i^s) = (S^i_{\gamma^s(x^i)^* a \gamma^r(x^i)} \xi_i^r, \xi_i^s)$$

$$\xrightarrow[i]{} (S_{(b^s)^* a b^r} \xi^r, \xi^s)$$

$$= ((^x S)_a \zeta^r, \zeta^s). \tag{25}$$

Now suppose that U is a regional neighborhood of $^x S$ of the form VII.1.3(1), using vectors in $X(^x S)$ of the form (21). It follows from (24) and (25) that $^{x^i}(S^i)$ is i-eventually in U. By VII.1.4 such U form a basis of regional neighborhoods of $^x S$. So, as we observed earlier, it has been established that the original net $\{^{x^i}(S^i)\}$ converges to $^x S$. ∎

16.24. Corollary. *Suppose that H is normal, and that $\mathscr{S}$ is the set of all unitary equivalence classes of elements of $\mathscr{S}$. For each x in G, the conjugation map $S \mapsto {}^x S$ is a homeomorphism of $\mathscr{S}$ onto itself.*

16.25. Corollary. *Suppose that H is normal, and let $\mathscr{S}$ be as in 16.23. If $x \in G$, $S \in \mathscr{S}$, and $\mathscr{W} \subset \mathscr{S}$, then $^x S$ is weakly contained in $\{{}^x T : T \in \mathscr{W}\}$ if and only if S is weakly contained in $\mathscr{W}$.*

Proof. By 16.24 and 16.12 (and the definition VII.1.21 of weak containment). ∎

16.26. By 16.23 and 16.18, if H is *normal*, the $\mathscr{B}$-positive part of the structure space $(\mathscr{B}_H)\hat{\ }$ of $\mathscr{B}_H$ (defined in VIII.21.2) is a left topological G-space under the action $\langle x, S \rangle \mapsto {}^x S$ of G by conjugation. This action of G on the $\mathscr{B}$-positive part of $(\mathscr{B}_H)\hat{\ }$ will be extremely important in Chapter XII.

Suppose that H is normal and that $\mathscr{B}$ is a saturated C^*-*algebraic bundle*. Then by Theorem 11.10 the $\mathscr{B}$-positive part of $(\mathscr{B}_H)\hat{\ }$ is in fact all of $(\mathscr{B}_H)\hat{\ }$, so that G acts by conjugation as a left topological transformation group on $(\mathscr{B}_H)\hat{\ }$ itself.

Inducing Followed by Restriction

16.27. What happens when the inducing operation is followed by restriction to another subgroup? The result is expressed in the proposition below by means of conjugation.

In this number we shall suppose that G/H is *discrete*. Let K be another closed subgroup of G. A subset of G of the form KxH ($x \in G$) is called a K, H *double coset*. The K, H double cosets are pairwise disjoint, and their union is of course G. Let us pick a subset M of G containing exactly one element from each K, H double coset. Thus $\{xH : x \in M\}$ contains exactly one element from each orbit in G/H under the action of K on G/H.

Proposition. *Let S be a non-degenerate $\mathscr{B}$-positive *-representation of $\mathscr{B}_H$. For each x in M let V^x be the restriction of ^xS to $\mathscr{B}_{(xHx^{-1} \cap K)}$; and let W^x be the result of inducing V^x up to $\mathscr{B}_K$. Then*

$$\left(\underset{\mathscr{B}_H \uparrow \mathscr{B}}{\operatorname{Ind}} (S) \right) \Big| \mathscr{B}_K \cong \sum_{x \in M}^{\oplus} W^x. \tag{26}$$

Proof. Since ^xS is $\mathscr{B}$-positive by 16.17, V^x is $\mathscr{B}$-positive by 8.12 and hence $\mathscr{B}_K$-positive by 8.13. So the definition of W^x makes sense.

Let $\mathscr{Y} = \langle Y, \{Y_\alpha\}\rangle$ be the Hilbert bundle induced by S; and put $T = \operatorname{Ind}_{\mathscr{B}_H \uparrow \mathscr{B}}(S)$, so that $X(T) = \mathscr{L}_2(\mathscr{Y}) = \sum_{\alpha \in G/H}^{\oplus} Y_\alpha$ (since G/H is discrete). For each x in M let θ_x be the K-orbit in G/H containing xH; and let Y^x be the Hilbert direct sum $\sum_{\alpha \in \theta_x}^{\oplus} Y_\alpha$, so that

$$X(T) = \sum_{x \in M}^{\oplus} Y^x. \tag{27}$$

Fix an element x of M. Since θ_x is K-stable, Y^x is stable under $T | \mathscr{B}_K$ (in virtue of the identity VIII.18.7(8) for systems of imprimitivity). Let the subrepresentation of $T | \mathscr{B}_K$ acting on Y^x be called R^x; and let P^x be the restriction to subsets of θ_x of the projection-valued measure P induced by S. By VIII.9.4 $T | \mathscr{B}_K$ is non-degenerate; hence so is R^x, and $\langle R^x, P^x \rangle$ is a system of imprimitivity for $\mathscr{B}_K$ over θ_x. Since θ_x is transitive under K, and since the stability subgroup of K for the point xH in θ_x is $L_x = xHx^{-1} \cap K$, the map $\Phi : K/L_x \to \theta_x$ given by $\Phi(kL_x) = kxH$ $(k \in K)$ is an isomorphism of K-spaces. Therefore, if $Q^x(W) = P^x(\Phi(W))$ $(W \subset K/L_x)$, $\langle R^x, Q^x \rangle$ is a system of imprimitivity for $\mathscr{B}_K$ over K/L_x. By the Imprimitivity Theorem 14.18, R^x is the result of inducing up to $\mathscr{B}_K$ from some *-representation V'^x of $\mathscr{B}_{L_x}$. In fact, since K/L_x is discrete, 14.21 tells us that V'^x is equivalent to the subrepresentation of $R^x | \mathscr{B}_{L_x}$ acting on the range of $Q^x(\{L_x\})$, i.e., to the subrepresentation of $T | \mathscr{B}_{L_x}$ acting on the range of $P(\{xH\})$. Now the subrepresentation of $T | \mathscr{B}_{xHx^{-1}}$ acting on range$(P(\{xH\}))$ is exactly ^xS. Hence the subrepresentation of $T | \mathscr{B}_{L_x}$ acting on range$(P(\{xH\}))$ is what we called V^x; that is, we have shown that $R^x \cong \operatorname{Ind}_{\mathscr{B}_{L_x} \uparrow \mathscr{B}_K}(V^x) = W^x$. This and (27) now imply (26). ∎

Remark. The above proposition and 13.2 clearly must have a common generalization. In fact, in the context of second-countable bundles, one can presumably invoke direct integral theory to set up a generalization of both of these which makes no mention of discreteness at all, and which reduces to Mackey's Restriction Theorem in the group case (Mackey [5], Theorem 12.1).

16.28. Let us apply 16.27 to the case that K is normal and contained in H. By 16.27 and 16.6 we obtain:

Corollary. *Assume that G/H is discrete; and let K be a closed normal subgroup of G with $K \subset H$. Let S be a non-degenerate $\mathscr{B}$-positive *-representation of $\mathscr{B}_H$. Then*

$$\underset{\mathscr{B}_H \uparrow \mathscr{B}}{\operatorname{Ind}} (S)|\mathscr{B}_K \cong \sum_{x \in M}{}^{\oplus} {}^x(S|\mathscr{B}_K),$$

where M is any transversal for G/H.

16.29. Examples. We have already mentioned in 7.10 an interesting finite-dimensional example of the action of G by conjugation in $\mathscr{B}$ on the structure space of the unit fiber of $\mathscr{B}$. This example was a very special case of the class of saturated C^*-algebraic bundles $\mathscr{B}$ over a group G which we explicitly constructed in VIII.16.15–17. We shall see in XII.6.3 that the action of G on $(B_e)\hat{}$ by conjugation works out in this more general case to be again the action of G on $\hat{A} = (B_e)\hat{}$ with which the construction VIII.16.15 began.

In this example we see that in general the dimensions of D and ${}^x D$ $(D \in (B_e)\hat{} \; ; x \in G)$ will be different. In fact, there is no reason why one should not be finite and the other infinite.

General Remarks on Conjugation in Saturated C^*-Algebraic Bundles

16.30. Let A and B be two Banach *-algebras. We have seen in 7.2 that a topologically strict A, B imprimitivity bimodule $\mathscr{L}$ gives rise to an "isomorphism" Φ of the "$\mathscr{L}$-positive parts" of the *-representation theories of A and B. It turns out that, in a modified von Neumann algebra context (the details of which we cannot describe here), a converse of this result can be proved: Given an isomorphism Φ of the *-representation theories of A and B (in a suitable precise sense of the term "isomorphism"), one can canonically construct an A, B imprimitivity bimodule $\mathscr{L}_\Phi$ from which Φ is derived more or less as in 7.2. See Rieffel [6]. The basic idea behind the construction of $\mathscr{L}_\Phi$ is not hard to describe. Suppose we are given an imprimitivity bimodule

$\mathscr{L} = \langle L, [\ ,\ ]_A, [\ ,\ ]_B \rangle$ such that Φ is derived from it by 7.2; and let us ask ourselves how we might reconstruct $\mathscr{L}$ from Φ. Take a non-degenerate $\mathscr{L}$-positive *-representation S of B; and put $T = \Phi(S) = \mathrm{Ind}_{B\uparrow A}^{\mathscr{L}}(S)$, $X = X(S)$, $Y = X(T)$. We recall from 6.16 that the map $\Phi^c: F \mapsto \tilde{F}$ of 2.12 is a *-isomorphism of the commuting algebras $\mathscr{I}(S)$ and $\mathscr{I}(T)$. Now each element r of L gives rise to a bounded linear operator $U_r: Y \to X$ as in 2.2(1), and one verifies without difficulty that this U_r belongs to the space M of all bounded linear operators $U: Y \to X$ satisfying

$$F \circ U = U \circ \Phi^c(F) \qquad \text{for all } F \text{ in } \mathscr{I}(S). \tag{28}$$

If S is "large enough" one will expect the mapping $r \mapsto U_r$ to be one-to-one and to map L onto a dense subset of M (hence onto M if M is finite-dimensional). It is now quite easy to see how the rest of the structure of $\mathscr{L}$ is reflected in the image of A, B and L under T, S and $r \mapsto U_r$.

Now suppose that $A = B$. Then the abstractly given Φ of the preceding paragraph is an "automorphism" of the *-representation theory of A. Suppose now that we are given not just one such "automorphism," but a group G acting as a group of "automorphisms" $x \mapsto \Phi_x$ $(x \in G)$ on the *-representation theory of A. The preceding paragraph suggests that under suitable conditions each x in G will give rise to an A, A imprimitivity bimodule $\mathscr{L}_{\Phi_x}$ from which Φ_x is derived as in 7.2. In fact, *if A is a C^*-algebra, we might conjecture the possibility of constructing a saturated C^*-algebraic bundle $\mathscr{B}$ over G with unit fiber C^*-algebra A, such that for each x in G, $\mathscr{L}_{\Phi_x}$ is just the A, A imprimitivity bimodule over x in $\mathscr{B}$, so that Φ_x coincides with the process of x-conjugation in $\mathscr{B}$.* The precise conditions under which such a conjecture holds true will undoubtedly be a topic for future research.

Remark. Suppose that the preceding action $x \mapsto \Phi_x$ of G (as a group of automorphisms of the *-representation theory of the C^*-algebra A) happens to come from an action $x \mapsto \tau_x$ of G as a continuous group of *-automorphisms of A itself. Then by 16.16 the semidirect product C^*-algebraic bundle $\mathscr{B} = A \times_\tau G$ (formed as in §VIII.4) verifies the preceding conjecture for this case.

16.31. The first step in giving precise form to the conjecture of 16.30 would be of course to define appropriately the notion of "an action Φ of a group G by automorphisms of the *-representation theory of a C^*-algebra A." We shall not attempt such a definition here, but merely comment that such an action would certainly imply in particular a continuous action σ of G on the structure space $\hat{A}$ of A. This suggests the following related question which is now perfectly precise:

Question. *Suppose we are given* (i) *a C*-algebra A,* (ii) *a locally compact group G, and* (iii) *a continuous left action σ of G on Â. Does there exist a saturated C*-algebraic bundle ℬ over G whose unit fiber is A, such that the action of G on Â by conjugation in ℬ is the same as the action σ?*

16.32. We have seen (Remark 3 of 7.10) that the answer to the preceding Question is "yes" when A is finite-dimensional and G is finite.

Notice that the data of the Question will not in general determine ℬ uniquely (to within isomorphism). Indeed, if $A = \mathbb{C}$ (and G is finite), then of course σ is trivial, and every cocycle bundle over G will satisfy the conditions on ℬ.

16.33. More generally, let ℋ be a Hilbert bundle (whose fibers are all non-zero) over a locally compact Hausdorff space M, and let A be the cross-sectional C*-algebra of the bundle $\mathcal{O}_c(\mathcal{H})$ of elementary C*-algebras generated by ℋ (see VII.8.4, VII.8.19). By Proposition VII.8.14, $\hat{A}$ can be identified both setwise and topologically with M (via the correspondence sending each point m of M into the *-representation $\phi \mapsto \phi(m)$ of A). Let G be a locally compact group, and σ a continuous left action of G on $M \cong \hat{A}$. With these ingredients A, G, σ, the answer to Question 16.31 is "yes." The construction of a corresponding ℬ is a natural, though technically tedious, generalization of the construction of 7.10 (See Exercise 55).

If the M of the preceding paragraph is discrete, then A is just an arbitrary C*-algebra of compact type. In that case, not only is the answer to Question 16.31 affirmative, but in XII.6.19 we shall completely classify all the ℬ which satisfy the conditions of the Question: They are indexed by the cocycle bundles over the stability subgroups for the various elements of $\hat{A}$ (compare the last statement of 16.32).

16.34. *In general the answer to Question* 16.31 *is negative.* In Exercise 56, we give an example of a negative answer in a situation where $\hat{A}$ is compact and Hausdorff and all the elements of $\hat{A}$ are finite-dimensional.

Remark. This negative answer does not of course negate the vague conjecture of 16.30, since it is not clear that every group of automorphisms of $\hat{A}$ will necessarily come from a group of automorphisms of the *-representation theory of A (whatever the appropriate definition of the latter may turn out to be).

17. Non-Involutory Induced Representations

17.1. In this final section we shall briefly describe (without proofs) a generalization of the abstract inducing process, and of the Imprimitivity Theorem, to the non-involutory context. As one would expect, the results are much weaker, since statements about unitary equivalence must in general be replaced by statements about Naimark-relatedness (see V.3.2). There is however this advantage—that in the non-involutory context the inducing process can always be performed; no special conditions (such as positivity with respect to the rigged module; see 4.9) need be imposed on the representation being induced.

Dual Representations

17.2. The non-involutory inducing process will be described in terms of dual representations. The notion of a dual representation is a reformulation of the idea of a locally convex representation on a space carrying the weak topology—a reformulation in which the space and its adjoint play symmetrical roles.

Definition. Let X_1 and X_2 be two linear spaces. A *duality for X_1 and X_2* is a (complex-valued) bilinear form $(\ |\ )$ on $X_1 \times X_2$ which is non-degenerate in the sense that (i) if $\xi_1 \in X_1$ and $(\xi_1|\xi_2) = 0$ for all ξ_2 in X_2 then $\xi_1 = 0$, and (ii) if $\xi_2 \in X_2$ and $(\xi_1|\xi_2) = 0$ for all ξ_1 in X_1 then $\xi_2 = 0$.

Definition. A *dual system* is a triple $X = \langle X_1, X_2, (\ |\ )\rangle$ consisting of a pair of linear spaces X_1, X_2 and a duality $(\ |\ )$ for X_1 and X_2.

Usually we omit explicit reference to $(\ |\ )$, and refer simply to a dual system $X = \langle X_1, X_2 \rangle$. The duality enables us to identify X_2 and X_1 with total linear subspaces of $X_1^\#$ and $X_2^\#$ respectively. The weak topologies for X_1 and X_2 generated by X_2 and X_1 are called $\sigma(X_1, X_2)$ and $\sigma(X_2, X_1)$ respectively.

17.3. Two dual systems $\langle X_1, X_2, (\ |\ )\rangle$ and $\langle X_1', X_2', (\ |\ )'\rangle$ are *isomorphic under* an isomorphism $\langle F_1, F_2 \rangle$ if $F_i: X_i \to X_i'$ is a linear bijection (for $i = 1, 2$) and

$$(F_1(\xi)|F_2(\eta))' = (\xi|\eta)$$

for all ξ in X_1 and η in X_2.

17.4.   Every LCS X gives rise to a dual system $\langle X_0, X^* \rangle$ (where X_0 denotes the linear space underlying X), the duality being of course given by $(\xi|\alpha) = \alpha(\xi)$. We call $\langle X_0, X^* \rangle$ the dual system *associated with X*.

17.5.   Definition.   Let $X = \langle X_1, X_2 \rangle$ be a dual system. Let X'_1 be a $\sigma(X_1, X_2)$-dense linear subspace of X_1; and let X'_2 be a $\sigma(X_2, X_1)$-dense linear subspace of X_2. Then $X' = \langle X'_1, X'_2 \rangle$, equipped with the restriction to $X'_1 \times X'_2$ of the duality of X, is a dual system. A dual system X' constructed in this way is called a *dense contraction* of the original X.

Remark.   A dense contraction of a dense contraction of X need *not* be a dense contraction of X. We leave it to the reader to construct an example showing this.

17.6.   Now fix an algebra A; and let $\tilde{A}$ denote the *reverse algebra of A*, that is, the algebra having the same underlying linear space as A but with multiplication reversed:

$$(ab)_{\text{in } \tilde{A}} = (ba)_{\text{in } A}.$$

Definition.   Let $X = \langle X_1, X_2 \rangle$ be a dual system. By a *dual representation* of the algebra A *on X* we mean a pair $T = \langle T^1, T^2 \rangle$, where (i) T^1 is an algebraic representation of A on X_1, (ii) T^2 is an algebraic representation of $\tilde{A}$ on X_2, and (iii) the "adjointness" identity

$$(T_a^1 \xi | \eta) = (\xi | T_a^2 \eta) \tag{1}$$

holds for all ξ in X_1, η in X_2 and a in A.

The T^1 and T^2 in this definition are called the *first* and *second members of* T. The space X_i of T^i is sometimes written $X_i(T)$, and the dual system $X = \langle X_1(T), X_2(T) \rangle$ is sometimes referred to as $X(T)$, the *dual system on which T acts*.

Notice that each of T^1 and T^2 determines the other through (1).

17.7.   Definition.   Two dual representations T and T' of A, acting on dual systems X and X' respectively, are *equivalent* if there is an isomorphism $F = \langle F_1, F_2 \rangle$ of X and X' such that $F_i \circ T_a^i = T_a'^i \circ F_i$ for $i = 1, 2$ and all a in A.

Sometimes, when no ambiguity can arise, we may deliberately confuse a dual representation with the equivalence class to which it belongs.

17.8. Every locally convex representation S of A gives rise to a dual representation $T = \langle T^1, T^2 \rangle$ of A acting on the dual system associated with $X(S)$:

$$T^1_a = S_a, \qquad (T^2_a \beta)(\xi) = \beta(S_a \xi)$$

$(a \in A; \xi \in X(S); \beta \in X(S)^*)$. This T is the *dual representation associated with S*.

Conversely, if $T = \langle T^1, T^2 \rangle$ is a dual representation of A, acting on $\langle X_1, X_2 \rangle$, then T^1, acting on X_1 equipped with the weak topology $\sigma(X_1, X_2)$, is a locally convex representation S of A which we call the *locally convex version* of T. (The continuity of the T^1_a with respect to $\sigma(X_1, X_2)$ follows immediately from (1).)

Under the correspondences $S \mapsto T$ and $T \mapsto S$ of the preceding two paragraphs, the family of all equivalence classes of dual representations T of A is in one-to-one correspondence with the family of all homeomorphic equivalence classes (see V.1.4) of locally convex representations S which are *weakly topologized*, i.e., for which $X(S)$ carries the weak topology generated by $X(S)^*$.

17.9. Fix a dual representation $T = \langle T^1, T^2 \rangle$ of A, acting on the dual system $X = \langle X_1, X_2 \rangle$. In this number we shall think of T^1 and T^2 as locally convex representations of A and $\tilde{A}$ respectively (their spaces X_1 and X_2 being equipped, of course, with the weak topologies $\sigma(X_1, X_2)$ and $\sigma(X_2, X_1)$).

We recall from V.1.7, V.1.8, V.1.10, and V.3.5 the definitions of non-degeneracy, irreducibility, total irreducibility, and local finite-dimensionality of locally convex representations.

Proposition*. T^1 *is non-degenerate [irreducible, totally irreducible, locally finite-dimensional] if and only if T^2 is non-degenerate [irreducible, totally irreducible, locally finite-dimensional].*

Definition. T *is said to be non-degenerate [irreducible, totally irreducible, locally finite-dimensional] if T^1 (and hence also T^2) has the corresponding property.*

Notice that any locally convex representation S of A is non-degenerate [irreducible, totally irreducible, locally finite-dimensional] if and only if its associated dual representation has the corresponding property.

17.10. Definition. Let $T = \langle T^1, T^2 \rangle$ be a dual representation of A acting on the dual system $X = \langle X_1, X_2 \rangle$. Let $X' = \langle X_1', X_2' \rangle$ be a dense contraction of X (see 17.5) such that X_i' is T^i-stable $(i = 1, 2)$; and set

$$(T')_a^1 = T_a^1 | X_1', \qquad (T')_a^2 = T_a^2 | X_2' \qquad\qquad (a \in A).$$

Thus T' is a dual representation of A acting on X'. A dual representation T' of A constructed in this way is called a *dense contraction* of T.

If Y is an LCS with associated dual system X, it is convenient to refer to a dense contraction of X simply as a *dense contraction of Y*. Similarly, if S is a locally convex representation of A, with associated dual representation T, a dense contraction of T is called simply a *dense contraction of S*.

17.11. The importance of Definition 17.10 arises from the following fact (proved in Fell [10], Lemma 8).

Proposition*. *Two locally convex representations S and T of A are Naimark-related if and only if S and T have equivalent dense contractions, that is, if and only if there are dense contractions S' and T' of S and T respectively such that S' and T' are equivalent.*

17.12. Combining 17.11 with VI.13.16 we obtain the following interesting fact:

Proposition*. *Suppose that A is a *-algebra, and that T is a dual representation of the algebra underlying A. Then there is, to within unitary equivalence, at most one *-representation S of A such that T is equivalent to some dense contraction of S.*

Definition. If such an S exists, it is called the *involutory expansion of T*.

The Non-Involutory Abstract Inducing Process

17.13. The non-involutory abstract inducing process is best described in terms of dual representations. To induce from a dual representation of one algebra B to a dual representation of another algebra A we make use of the notion of a non-involutory B-rigged A-module system.

17.14. Fix two algebras A and B.

Definition. A (*non-involutory*) *B-rigged A-module system* is a triple $\langle L, M, [\ ,\ ]\rangle$ where (i) L is a right B-module and a left A-module, (ii) M is a left B-module and a right A-module, and (iii) $[\ ,\ ]: M \times L \to B$ is a bilinear map satisfying

$$[bs, r] = b[s, r], \tag{2}$$

$$[s, rb] = [s, r]b, \tag{3}$$

$$[sa, r] = [s, ar] \tag{4}$$

for all r in L, s in M, a in A, and b in B.

17.15. Let $\mathscr{L} = \langle L, M, [\ ,\ ]\rangle$ be a fixed B-rigged A-module system; and let $S = \langle S^1, S^2\rangle$ be a dual representation of B, acting on a dual system $X = \langle X_1, X_2, (\ |\ )\rangle$. To construct the induced dual representation T of A, we begin by setting up the dual system Y on which T will act, as follows: Since the expression $(S^1_{[s, r]}\xi | \eta)$ is linear in each of r, s, ξ and η, the equation

$$\sigma(r \otimes \xi, \eta \otimes s) = (S^1_{[s, r]}\xi | \eta)$$

$(r \in L; s \in M; \xi \in X_1; \eta \in X_2)$ defines σ as a bilinear form on $(L \otimes X_1) \times (X_2 \otimes M)$. Let

$$I_1 = \{\alpha \in L \otimes X_1 : \sigma(\alpha, \beta) = 0 \text{ for all } \beta \text{ in } X_2 \otimes M\},$$

$$I_2 = \{\beta \in X_2 \otimes M : \sigma(\alpha, \beta) = 0 \text{ for all } \alpha \text{ in } L \otimes X_1\};$$

and set $Y_1 = (L \otimes X_1)/I_1$, $Y_2 = (X_2 \otimes M)/I_2$. Then $Y = \langle Y_1, Y_2\rangle$ is evidently a dual system when equipped with the quotient duality $(\ |\ )_0$:

$$(r \widetilde{\otimes} \xi | \eta \widetilde{\otimes} s)_0 = (S^1_{[s, r]}\xi | \eta) \tag{5}$$

$(r \in L; s \in M; \xi \in X_1; \eta \in X_2)$. (Here $r \widetilde{\otimes} \xi$ and $\eta \widetilde{\otimes} s$ are the quotient images of $r \otimes \xi$ and $\eta \otimes s$ in Y_1 and Y_2 respectively.) If $a \in A$ and $\sum_{i=1}^n r_i \widetilde{\otimes} \xi_i = 0$, then for all η and s

$$\sigma\left(\sum_{i=1}^n ar_i \otimes \xi_i, \eta \otimes s\right) = \sum_{i=1}^n (S^1_{[s, ar_i]}\xi_i | \eta)$$

$$= \sum_{i=1}^n (S_{[sa, r_i]}\xi_i | \eta) \qquad \text{by (4)}$$

$$= \sigma\left(\sum_{i=1}^n r_i \otimes \xi_i, \eta \otimes sa\right) \tag{6}$$

$$= 0,$$

whence $\sum_{i=1}^{n} ar_i \widetilde{\otimes} \xi_i = 0$. It follows that the equation

$$T_a^1(r \widetilde{\otimes} \xi) = ar \widetilde{\otimes} \xi \qquad (a \in A; r \in L; \xi \in X_1) \qquad (7)$$

defines T^1 as an algebraic representation of A on Y_1. Similarly the equation

$$T_a^2(\eta \widetilde{\otimes} s) = \eta \widetilde{\otimes} sa \qquad (a \in A; s \in M; \eta \in X_2) \qquad (8)$$

defines T^2 as an algebraic representation of the reverse algebra $\tilde{A}$ on Y_2. The calculation (6) verifies the identity (1), showing that $T = \langle T^1, T^2 \rangle$ is a dual representation of A, acting on Y.

Definition.   This T is said to be *induced by S via $\mathscr{L}$*, and is denoted by $\text{Indual}_{B\uparrow A}^{\mathscr{L}}(S)$, or simply $\text{Indual}_{B\uparrow A}(S)$.

If S' is a locally convex representation of B, with associated dual representation S (see 17.8), by $\text{Indual}_{B\uparrow A}(S')$ we shall mean the same as $\text{Indual}_{B\uparrow A}(S)$.

17.16.   The construction of $\text{Indual}_{B\uparrow A}^{\mathscr{L}}(S)$ is interesting less for its own sake than as a stepping-stone to further definitions. Thus, suppose in addition that A is a *-algebra. There are then two distinct situations in which the construction of 17.15 might lead to a *-representation of A:

(I)   It might happen that $\text{Indual}_{B\uparrow A}(S)$ has an involutory expansion T' (see 17.12). In this case T' is unique up to unitary equivalence by 17.12, and can be regarded as induced by S in a generalized sense.

(II)   Suppose that $T = \text{Indual}_{B\uparrow A}(S)$ turns out to be locally finite-dimensional, and furthermore that there exists a locally finite-dimensional *-representation T'' of A which is Naimark-related with T (or, more precisely, with the locally convex version of T; see 17.8). By V.3.6 and VI.13.16, T'' is unique up to unitary equivalence, and so again can be regarded as induced by S in a generalized sense.

17.17.   *Example: The involutory special case.*   Let us observe how the involutory inducing process of 4.9 fits into the more general non-involutory scheme of this section.

Let A and B be *-algebras; and suppose that $\mathscr{L} = \langle L, [\ ,\ ] \rangle$ is a right B-rigged left A-module (in the sense of 4.7). Let M be the complex-conjugate space $\bar{L}$. Since L is a right B-module and a left A-module, M is a left B-module and a right A-module under the complex-conjugate module structures : defined in 4.4. As a map of $M \times L$ into B, $[\ ,\ ]$ is bilinear; and the postulates of 4.7 imply (recalling 4.3(2)) that $[s, rb] = [s, r]b$, $[b:s, r] = b[s, r]$, and $[s:a, r] = [s, ar]$ $(r \in L; s \in M; a \in A; b \in B)$. That is, $\langle L, M, [\ ,\ ] \rangle$ is a non-involutory B-rigged A-module system. We refer to $\langle L, M, [\ ,\ ] \rangle$ as the *non-involutory B-rigged A-module system underlying $\mathscr{L}$*, and denote it by $\mathscr{L}^0$.

Now let S be a (non-degenerate) *-representation of B which is inducible to A via $\mathscr{L}$ in the sense of 4.9. On comparing (5) with 4.9(12), one sees that $\mathrm{Indual}_{B\uparrow A}^{\mathscr{L}^0}(S)$ is a dense contraction of $\mathrm{Ind}_{B\uparrow A}^{\mathscr{L}}(S)$, and hence that $\mathrm{Ind}_{B\uparrow A}^{\mathscr{L}}(S)$ is the involutory expansion of $\mathrm{Indual}_{B\uparrow A}^{\mathscr{L}^0}(S)$.

Are there situations in which S fails to be both involutory and inducible, and yet $\mathrm{Indual}_{B\uparrow A}^{\mathscr{L}^0}(S)$ has an involutory expansion? We do not know. (See Remark 2 of 17.18).

17.18. Example: The supplementary series of representations of $SL(2, \mathbb{R})$. Let G be the locally compact group $SL(2, \mathbb{R})$ of all 2×2 matrices with real entries and determinant 1; and let H be the triangular subgroup of G consisting of all

$$h = \begin{pmatrix} a & b \\ 0 & a^{-1} \end{pmatrix} \quad (a, b \text{ real}, a \neq 0). \tag{9}$$

We shall denote by $\mathscr{L}$ the right $\mathscr{L}(H)$-rigged left $\mathscr{L}(G)$-module constructed in §8 from H and the group bundle of G. That is, $\mathscr{L}$ consists of $\mathscr{L}(G)$ (considered as a left $\mathscr{L}(G)$-module by left multiplication and a right $\mathscr{L}(H)$-module by 8.4(3)) together with the $\mathscr{L}(H)$-valued rigging $[\ ,\ ]:\langle f, g\rangle \mapsto p(f^* * g)$ on $\mathscr{L}(G) \times \mathscr{L}(G)$ (p being the conditional expectation 8.4(1)).

Now fix a number r satisfying $0 < r < 1$; and let χ be the one-dimensional (non-unitary) representation of H given by

$$\chi(h) = |a|^r \qquad (h \text{ being as in (9)}).$$

Let χ also denote the integrated form of χ (which is a one-dimensional representation of $\mathscr{L}(H)$); and form the dual induced representation

$$T^r = \underset{\mathscr{L}(H)\uparrow\mathscr{L}(G)}{\mathrm{Indual}}{}^{\mathscr{L}^0}(\chi)$$

of $\mathscr{L}(G)$ ($\mathscr{L}^0$ being as before the non-involutory rigged module system underlying $\mathscr{L}$). It turns out (though we shall not take the space to prove it) that T^r is locally finite-dimensional, and that there is a unitary representation V^r of G whose integrated form is locally finite-dimensional and Naimark-related to T^r. As we pointed out in 17.16(II), this V^r is uniquely determined up to unitary equivalence.

Remark 1. As a matter of fact, as r runs over the open interval $]0, 1[$, the V^r form a family of pairwise inequivalent irreducible unitary representations of $SL(2, \mathbb{R})$—the so-called *supplementary series* (see Gelfand and Naimark [5]). Such supplementary series, obtained by non-involutory induction from non-unitary characters of a subgroup, are found to occur for all semisimple Lie groups.

Remark 2. It can be shown that the integrated form of V^r is *not* an involutory expansion of T^r. Thus the "non-unitary induction process" leading to the supplementary series of unitary representations of $SL(2, \mathbb{R})$ is an example of 17.16(II) but not of 17.16(I).

Non-Involutory Imprimitivity Bimodules

17.19. On the basis of 17.15 one can formulate a non-involutory abstract Imprimitivity Theorem.

Let A and B be two fixed algebras.

Definition. A *non-involutory A, B, imprimitivity bimodule* is a system $\mathscr{L} = \langle L, M, [\ , \ ]_A, [\ , \ ]_B \rangle$, where (i) L is a right B-module and a left A-module, (ii) M is a left B-module and a right A-module, (iii) $\langle L, M, [\ , \ ]_B \rangle$ is a non-involutory B-rigged A-module system, (iv) $\langle M, L, [\ , \ ]_A \rangle$ is a non-involutory A-rigged B-module system, and (v) the associative relations

$$r[s, r']_B = [r, s]_A r', \qquad s[r, s']_A = [s, r]_B s'$$

hold for all r, r' in L and s, s' in M.

17.20. Suppose that A and B are *-algebras and that $\mathscr{L} = \langle L, [\ , \ ]_A, [\ , \ ]_B \rangle$ is an A, B imprimitivity bimodule (in the sense of 6.2). Then the complex-conjugate space $\bar{L}$, which we shall denote by M, is a left B-module and a right A-module under the complex-conjugate actions of 4.4; and $\langle L, M, [\ , \ ]_A, [\ , \ ]_B \rangle$ is easily seen to be a non-involutory A, B imprimitivity bimodule in the sense of 17.19. Analogously with 17.17, we call $\langle L, M, [\ , \ ]_A, [\ , \ ]_B \rangle$ the *non-involutory A, B imprimitivity bimodule underlying $\mathscr{L}$.*

17.21. Returning to the case of arbitrary algebras A and B, suppose we are given a non-involutory A, B imprimitivity bimodule $\mathscr{L} = \langle L, M, [\ , \ ]_A, [\ , \ ]_B \rangle$ which is *strict* in the sense of 6.2, that is, the linear spans of the ranges of $[\ , \ ]_A$ and $[\ , \ ]_B$ are all of A and B respectively. If S is any dual representation of B, we can form the dual representation T of A induced from S via $L, M, [\ , \ ]_B$; this T will be denoted by $\mathrm{Indual}^{\mathscr{L}}_{B\uparrow A}(S)$. In the opposite direction, if T is any dual representation of A, we can form the dual representation S of B induced from T via $M, L, [\ , \ ]_A$; this S will be called $\mathrm{Indual}^{\mathscr{L}}_{A\uparrow B}(T)$.

In this context the abstract Imprimitivity Theorem takes the following form:

Theorem*. *Let S be a non-degenerate dual representation of B, acting on the dual system* $X = \langle X_1, X_2 \rangle$. *Then* $T = \text{Indual}^{\mathscr{L}}_{B \uparrow A}(S)$ *is non-degenerate; and* $S' = \text{Indual}^{\mathscr{L}}_{A \uparrow B}(T)$ *is equivalent to the dense contraction of S acting on* $X' = \langle X'_1, X'_2 \rangle$, *where* X'_i *is the linear span of* $\{S^i_b \xi : b \in B, \xi \in X_i\}$.

This is proved by the same sort of calculations that led to 6.14.

Remark. The weakness of this theorem lies of course in the fact that S' is equivalent merely to a dense contraction of S, not to S itself. This weakness disappears if the S^i are *algebraically* non-degenerate, so that $X'_i = X_i$.

17.22. In this connection, suppose that S^1 is any *algebraic* representation of B. Then we can "enlarge" S^1 to a dual representation $S = \langle S^1, S^2 \rangle$ of B having S^1 as its first member, and form the induced dual representation $T = \langle T^1, T^2 \rangle = \text{Indual}^{\mathscr{L}}_{B \uparrow A}(S)$. It is easy to see that the algebraic representation T^1 of A depends only on S^1; let us write $T^1 = \text{Ind}^{\mathscr{L}}_{B \uparrow A}(S^1)$. Similarly, any algebraic representation T^1 of A gives rise to an induced algebraic representation $S^1 = \text{Ind}^{\mathscr{L}}_{A \uparrow B}(T^1)$ of B. Removing the superscripts 1, one can now easily obtain the following purely algebraic non-involutory abstract Imprimitivity Theorem:

Theorem*. *Let A, B, $\mathscr{L}$ be as in 17.21; and let S be an (algebraically) non-degenerate algebraic representation of B. Then:*

(I) $T = \text{Ind}^{\mathscr{L}}_{B \uparrow A}(S)$ *is algebraically non-degenerate, and*
(II) $S' = \text{Ind}^{\mathscr{L}}_{A \uparrow B}(T)$ *is algebraically equivalent to S.*

Because of the symmetry of A and B in $\mathscr{L}$, this result establishes a one-to-one correspondence between the family of all algebraic equivalence classes of (algebraically) non-degenerate algebraic representations of B and the similar family for A.

17.23. Proposition*. *Let A, B, $\mathscr{L}$, S be as in 17.22. Then* $T = \text{Ind}^{\mathscr{L}}_{B \uparrow A}(S)$ *is algebraically irreducible if and only if S is algebraically irreducible.*

Thus, under the correspondence of 17.22, the algebraically irreducible representations of B correspond exactly to the algebraically irreducible representations of A.

18.   Exercises for Chapter XI

1. Prove Proposition 1.15.

2. Let $\{V^\alpha : \alpha \in I\}$ be an indexed collection of operator inner products on L whose direct sum V exists. For each α let W^α be dual to V^α. Prove that the dual of V is unitarily equivalent to the direct sum of the W^α (see 2.10).

3. Let L, M, P, V, W, U be as in 1.16. Prove that if V', W', U' are dual to V, W, U respectively, then U' is unitarily equivalent to the tensor product of V' and W' (see 2.11).

4. Find a generalization of Theorem 3.14 to *-algebras without a unit. Also prove a uniqueness result as indicated in Remark 3.14.

5. In the context of Definition 4.9 of the Rieffel inducing process, give an example in which the *-representation S of B satisfies (i) (i.e., is positive with respect to $\mathscr{L}$) but not (ii), so that S is not inducible to A via $\mathscr{L}$.

6. Show that the linear isometry F defined in 4.25 by $F(b \widetilde{\otimes} \xi) = S_b \xi$, $(b \in B; \xi \in X(S))$, intertwines $T|B$ and S.

7. Prove Propositions 5.2 and 5.8.

8. The following result on "inducing in stages" is a variant on Proposition 5.6:

Let A, B, C be three algebras. Let $\mathscr{L} = \langle L, [\ , \ ] \rangle$ be a right B-rigged left A-module, and let $\mathscr{K} = \langle K, [\ , \ ]' \rangle$ be a right C-rigged left B-module. Thus $P = L \otimes K$ is a natural right C-module and left A-module:

$$a(r \otimes k) = ar \otimes k, (r \otimes k)c = r \otimes kc \qquad (r \in L; k \in K; a \in A, c \in C).$$

(I) Show that $\mathscr{P} = \langle P, [\ , \ ]'' \rangle$ is a right C-rigged left A-module, where $[\ , \ ]''$ is determined by the equation:

$$[r \otimes k, s \otimes m]'' = [[s, r]k, m]' \qquad (r, s \in L; k, m \in K).$$

We call $\mathscr{P}$ the *composition* of $\mathscr{L}$ and $\mathscr{K}$.

(II) Let W be a *-representation of C which is inducible to B via $\mathscr{K}$, and set $S = \text{Ind}_{C \uparrow B}^{\mathscr{K}}(W)$. Show that the following two conditions are equivalent:
(i) S is inducible to A via $\mathscr{L}$;
(ii) W is inducible to A via $\mathscr{P}$.
Show that, if (i) and (ii) hold, then

$$\underset{B \uparrow A}{\text{Ind}^{\mathscr{L}}}(S) = \underset{C \uparrow A}{\text{Ind}^{\mathscr{P}}}(W).$$

(III) Assume that the above modules L and K satisfy the associative laws: $(ar)b = a(rb), (bk)c = b(kc)$ $(r \in L; k \in K; a \in A; b \in B; c \in C)$. Now let N be the linear span in P of $\{rb \otimes k - r \otimes bk : r \in L, k \in K, b \in B\}$; let P_B be the quotient space P/N; and let $r \otimes_B k$ be the image of $r \otimes k$ under the quotient map $P \to P_B$. Show that:

(i) P_B inherits from P a left A-module and right C-module structure:

$$a(r \otimes_B k) = ar \otimes_B k, \qquad (r \otimes_B k)c = r \otimes_B kc.$$

(ii) $[\ , \ ]''$ lifts to a C-rigging $[\ , \ ]''_B$ of P_B:

$$[r \otimes_B k, \ s \otimes_B m]''_B = [[s, r]k, m]'.$$

(iii) $\mathscr{P}_B = \langle P_B, [\ , \ ]''_B \rangle$ is a right C-rigged left A-module. We call it the *reduced composition* of $\mathscr{L}$ and $\mathscr{K}$.

(iv) The process of inducing from C to A via $\mathscr{P}_B$ is identical with that via $\mathscr{P}$. Hence statement (II) above continues to hold when $\mathscr{P}$ is replaced (in both occurrences) by $\mathscr{P}_B$.

  9. Give direct proofs of 13.6, 13.10 and 13.11 for compact groups (starting from Definition IX.10.4).

  10. Prove Proposition 6.6.

  11. Verify that the mapping Φ defined by equation (19) of 6.11 satisfies

$$\Phi((\phi * b) \otimes \psi) = \Phi(\phi \otimes (\psi * b^*))$$

for $\phi, \psi \in A$, $b \in B$. Show also that $\hat{\Phi}$ in Proposition 6.11 is a *-homomorphism.

  12. Verify formula (21) in Remark 6.11. Show that with the left F-module structure for A given by (21), and with the F-rigging $[\ , \ ]_F$ given by

$$[\phi, \psi]_F(x, yH) = \sum_{h \in H} \phi(yh)\psi^*(h^{-1}y^{-1}x)$$

for $\phi, \psi \in A$; $x, y \in G$, the triple $(A, [\ , \ ]_F, [\ , \ ]_B)$ becomes an F, B imprimitivity bimodule.

  13. Verify equation (31) in 6.18 and show that (31) implies that the imprimitivity bimodule $\mathscr{L}$ is strict.

  14. Let $A, B, \mathscr{L}$ be as in 6.12; and suppose in addition that B is a normed *-algebra, and that there exists a net $\{a_i\}$ of elements of A such that (i) each a_i is of the form $\sum_{j=1}^n [s_j, s_j]_A$, where $s_1, \ldots, s_n \in L$ (and of course n depends on a_i), and (ii) $[r, a_i r]_B \underset{i}{\to} [r, r]_B$ in B for each r in L. Prove that every norm-continuous *-representation S of B is $\mathscr{L}$-positive.

  15. Show that $\mathscr{L} = \langle L, [\ , \ ]_A, [\ , \ ]_B \rangle$ in 7.5, with the riggings on L given by (1) and (2), is an A, B imprimitivity bimodule.

  16. Verify the following statements made in Example 7.10.

  (a) $B_x B_y \subset B_{xy}$, $(B_x)^* = B_{x^{-1}}$ for $x, y \in G$.

  (b) The C^*-algebraic bundle $\mathscr{B}$ over G is saturated.

  (c) For each $x \in G$ and each $i = 1, \ldots, n$, the x-conjugate of S^i relative to $\mathscr{B}$ is unitarily equivalent to $S^{\pi_x(i)}$.

  17. Verify, in 7.6, that $\mathrm{Ind}_{A \uparrow B}^{\mathscr{L}}(T)$ exists and is the non-degenerate part of $T|B$.

  18. Prove Remark 2 of 17.18.

  19. Show that equation (12) in 7.10 is true.

20. Verify the statements in 7.15 which were left to the reader.

21. Give a complete detailed proof of Proposition 7.16.

22. Prove equations (5) and (6) in 8.4. Show also that $f\phi$ is separately continuous in f and ϕ in the inductive limit topologies of $\mathcal{L}(\mathcal{B})$ and $\mathcal{L}(\mathcal{B}_H)$.

23. Verify equation (7) in 8.5.

24. Show that $\mathcal{L}(\mathcal{B}_a)$, $[\ ,\ ]_a$ in 8.8 satisfies the postulates for a right $\mathcal{L}(\mathcal{B}_H)$-rigged space.

25. Show that a Hilbert direct sum of $\mathcal{B}$-positive *-representations of $\mathcal{B}_H$ is again $\mathcal{B}$-positive (see 8.17).

26. Prove Proposition 9.1.

27. Verify the two facts (I) and (II) left to the reader in the proof of Proposition 9.2.

28. Prove that the linear isometry $\tilde{F}_a: Y_a \to Y'_a$ described in 9.5 (see (9)) is surjective.

29. Verify the details of 9.7.

30. Check that equation (45) in 9.25 holds.

31. In the context of 9.30, verify that all steps of the proofs of Lemma 9.10 and Corollary 9.11 are still valid.

32. Write out the details which show that Theorem 10.9 follows from Theorem 10.8.

33. Prove the statements and the proposition in 10.14.

34. Prove Proposition 10.16.

35. Verify relations (4) and (5) in the proof of Lemma 11.3.

36. Verify in the proof of Proposition 11.8 that the map S' preserves multiplication and *.

37. Let $\mathcal{B} = \{B_x\}_{x \in G}$ be a C^*-algebraic bundle over a discrete group G. Show that, for each $x \in G$, there is a (unique) continuous linear map $\pi'_x: C^*(\mathcal{B}) \to B_x$ with the properties:

(i) π'_x is the identity map on B_x, and

(ii) $\pi'_x(B_y) = 0$ for $y \neq x$. (Here we identify the B_x with subspaces of $C^*(\mathcal{B})$ as in VIII.16.11.)

[Hint: It is enough to show that $\|a\|_{C^*(\mathcal{B})} \geq \|a(x)\|$ for all $a \in \mathcal{L}_1(\mathcal{B})$ and $x \in G$. Let S be a faithful *-representation of B_e. Then S is $\mathcal{B}$-positive (by 11.7), so we can form the *-representation $T = \mathrm{Ind}_{B_e \uparrow \mathcal{B}}(S)$. Now show that $\|T_a\| \geq \|a(x)\|$.]

38. Exercise 37 has a bearing on VIII.16.12. It says that, in the context of VIII.16.12, a *necessary* condition for the identification of E with $C^*(\mathcal{B})$ is that there should be "continuous projections onto each B_x", i.e., that for each $x \in G$ there should be a continuous linear map $\pi'_x: E \to B_x$ which annihilates B_y for each $y \neq x$ and coincides with the identity on B_x.

Here are two examples which illustrate the situation:

(i) Let $0 < u < v < \pi$. Let $E = \mathscr{C}([u, v])$ (continuous complex functions on $[u, v]$ with pointwise operations and the supremum norm). For each $n \in \mathbb{Z}$, let $B_n = \mathbb{C}\phi_n$, where $\phi_n(t) = e^{int}$ ($t \in [u, v]$). Show that the B_n satisfy (i), (ii), (iii) of VIII.16.2, but that there are no "continuous projections onto the B_n". Thus E cannot be identified with $C^*(\mathcal{B})$ in this case.

(ii) Let G be a discrete group which is not amenable (see Greenleaf [3]). Let R be the left regular representation of G, and E the closure of the linear span of the range of R; and for each $x \in G$ let $B_x = \mathbb{C}R_x$. Show that VIII.16.12 (i), (ii), (iii) hold, and that there are continuous projections onto the B_x. The non-amenability of G, however, implies that the natural *-homomorphism $C^*(\mathscr{B}) \to E$ is *not* one-to-one. So E cannot be identified with $C^*(\mathscr{B})$ in this case.

Thus the condition that there be continuous projections onto the B_x is necessary but not sufficient for the identification of E with $C^*(\mathscr{B})$.

39. Let $\mathscr{B}$ be a C^*-algebraic bundle over the locally compact group G (with $B_e \neq \{0\}$). Show that, if the cross-sectional C^*-algebra $C^*(\mathscr{B})$ has a unit element 1, then G must be discrete. [Hint: Suppose G is not discrete. Since $C^*(\mathscr{B})$ has a unit, there is $g \in \mathscr{L}(\mathscr{B})$ such that $\|T_g - 1_{X(T)}\| < \frac{1}{3}$ for every non-degenerate *-representation of $\mathscr{B}$. Let S be a faithful *-representation of B_e, and put $T = \text{Ind}(S)$; and let $\mathscr{Y}$ be the Hilbert bundle induced by S. Construct a net $\{\zeta_i\}$ of elements of $\mathscr{L}_2(\mathscr{Y}) \,(= X(T))$, of norm near to 1, and with supports shrinking down to e; and verify that $(T_g \zeta_i, \zeta_i) \to 0$. It follows that $\|T_g \zeta_i - \zeta_i\|$ is eventually $> \frac{1}{3}$, a contradiction.]

40. Verify the statements made in Remark 12.2.

41. Supply the missing details in the proof of Corollary 12.8.

42. Give a complete proof of Lemma 13.4.

43. Show in Remark 13.5 that $\mathscr{Y}''$ is the Hilbert bundle tensor product of $\mathscr{Y}$ and $\mathscr{Y}'$, and that the described mapping Φ gives the desired unitary equivalence.

44. Write out a complete proof of Proposition 13.9.

45. Verify formula (2) of Remark 14.3.

46. Show, in 14.6, that $[uf, g] = [f, u^*g]$ for $f, g \in \mathscr{L}(\mathscr{B})$, $u \in E$.

47. Verify the remaining identities (other than (13)) in the proof of Proposition 14.10.

48. Show, in the proof of Proposition 14.13 that the set $N^0 = \{u^0 : u \in N\}$ is dense in E in the inductive limit topology.

49. This exercise is concerned with the explicit form of the Imprimitivity Theorem for *semidirect product bundles*. Let A be a saturated Banach *-algebra; and suppose that $\mathscr{C} = A \underset{\tau}{\times} G$ is a τ-semidirect product of A and G as in III.4.4. Thus $\mathscr{C}$ is saturated.

Let (V, P) be a system of imprimitivity for the group G over G/H; and let Q be a non-degenerate *-representation of A, acting on the same space as V and P, satisfying:

(i) $Q_a P_\phi = P_\phi Q_a$ $(a \in A; \phi \in \mathscr{C}_0(G/H))$,

(ii) $V_x Q_a V_x^{-1} = Q_{\tau_x(a)}$ $(a \in A; x \in G)$.

Prove that there exists a unitary representation W of H and a non-degenerate *-representation R of A (acting in the same space as W) such that:

(a) $W_h R_a W_h^{-1} = R_{\tau_h(a)}$ $(a \in A; h \in H)$;

(b) The triple (Q, V, P) is unitarily equivalent to (Q', V', P'), where (V', P') is the Blattner formulation of the system of imprimitivity induced by W and Q' is the Q of 10.16(22).

(c) Show, furthermore, that the pair (W, R) is unique to within unitary equivalence.

50. Verify Remark 14.23.

51. Show that the object $\mathscr{D}$ constructed in 15.2 is a saturated Banach *-algebraic bundle over $G \times \hat{G}$.

52. Check that the T defined in (2) of 15.3 is a non-degenerate *-representation of the Banach *-algebraic bundle $\mathscr{D}$ constructed in 15.2.

53. Show that, in 16.8, $\mathscr{L}(\mathscr{B}_{xH})$ is an $\mathscr{L}(\mathscr{B}_{xHx^{-1}})$, $\mathscr{L}(\mathscr{B}_H)$ imprimitivity bimodule when furnished with the module structures (3) and (4) and the riggings (5) and (6).

54. Assume that $H = \{e\}$ in Proposition 16.17, and give a proof of it using the abstract "stages" Theorem 5.6.

55. Let M be a locally compact Hausdorff space and $\mathscr{H}$ a Hilbert bundle over M (whose fibers H_m are all non-zero). Let G be a locally compact group acting continuously to the left on M (the action being denoted by $\langle x, m \rangle \mapsto xm$). In this exercise we shall construct a saturated C^*-algebraic bundle $\mathscr{B}$ over G such that (i) the unit fiber A of $\mathscr{B}$ is $\mathscr{C}_0(\mathscr{O}_c(\mathscr{H}))$, and (ii) the action of G on $\hat{A} \cong M$ by conjugation in $\mathscr{B}$ is the same as the original action $\langle x, m \rangle \mapsto xm$. (See 16.33).

First, we denote by $\mathscr{H}^1$ and $\mathscr{H}^2$ the retractions of $\mathscr{H}$ to Hilbert bundles over $M \times M$ given by the maps $\langle m_1, m_2 \rangle \mapsto m_1$ and $\langle m_1, m_2 \rangle \mapsto m_2$ respectively. Now (recalling VIII.8.17) we consider $\mathscr{O}_c(\mathscr{H}^1, \mathscr{H}^2)$; this is a Banach bundle over $M \times M$ whose fiber at $\langle m_1, m_2 \rangle$ consists of all compact operators from H_{m_1} to H_{m_2}. Now let $\mathscr{E}$ be the retraction of $\mathscr{O}_c(\mathscr{H}^1, \mathscr{H}^2)$ to a Banach bundle over $G \times M$ via the map $\langle x, m \rangle \mapsto \langle m, xm \rangle$; so $\mathscr{E}$ is a Banach bundle over $G \times M$ whose fiber at $\langle x, m \rangle$ consists of all compact operators from H_m to H_{xm}. Finally, let $\mathscr{B}$ be the C_0 partial cross-sectional bundle over G derived from $\mathscr{E}$ and the (continuous open) projection $\langle x, m \rangle \mapsto x$ of $G \times M$ onto G (see II.14.9). Thus $\mathscr{B}$ is a Banach bundle over G whose fiber B_x at x is the space (with the supremum norm) of all continuous functions ϕ on M to $\mathscr{E}$ such that (i) for each $m \in M$, $\phi(m)$ is a compact operator on H_m to H_{xm}, and (ii) ϕ vanishes at infinity on M.

We shall now introduce a multiplication and involution into $\mathscr{B}$:

If $x, y \in G$, $\phi \in B_x$, and $\psi \in B_y$, we define $\phi \cdot \psi$ and ϕ^* as functions on M as follows:

$$(\phi \cdot \psi)(m) = \phi(ym) \cdot \psi(m), \tag{1}$$

$$\phi^*(m) = (\phi(x^{-1}m))^* \tag{2}$$

($m \in M$). Since $\psi(m): H_m \to H_{ym}$, $\phi(ym): H_{ym} \to H_{xym}$, and $\phi(x^{-1}m): H_{x^{-1}m} \to H_m$, we have $(\phi \cdot \psi)(m): H_m \to H_{xym}$ and $\phi^*(m): H_m \to H_{x^{-1}m}$.

(i)   If x, y, ϕ, ψ are fixed as above, prove that $\phi \cdot \psi$ and ϕ^* are continuous functions from M to $\mathscr{E}$ and also vanish at infinity on M; so $\phi \cdot \psi$ and ϕ^* belong to B_{xy} and $B_{x^{-1}}$ respectively.

(ii) Prove that under the operations (1) and (2) of multiplication and involution, $\mathscr{B}$ becomes a saturated C^*-algebraic bundle, whose unit fiber A is $\mathscr{C}_0(\mathscr{O}_c(\mathscr{H}))$.

(iii) Identifying $\hat{A}$ with M as in 16.33, prove that the action of G on $\hat{A}$ by conjugation in $\mathscr{B}$ is the same as the original action $\langle x, m \rangle \mapsto xm$ of G on M.

[Hint: For (i) use Proposition VII.8.18.]

56. In this exercise we construct ingredients A, G, σ such that the answer to Question 16.31 is *negative*.

Let E be the C^*-algebra of all continuous functions on the closed interval $[-1, 1]$ to M_2 (the 2×2 total matrix $*$-algebra), with pointwise multiplication and involution and the supremum norm. Let A be the closed $*$-subalgebra of E consisting of those $a \in E$ such that

$$a(-1) = \begin{pmatrix} \lambda & 0 \\ 0 & \lambda \end{pmatrix} \quad \text{for some complex } \lambda, \tag{1}$$

and

$$a(1) = \begin{pmatrix} 0 & 0 \\ 0 & \mu \end{pmatrix} \quad \text{for some complex } \mu. \tag{2}$$

If $-1 < t < 1$, let T^t be the 2-dimensional element $a \mapsto a(t)$ of $\hat{A}$. Also, let $T_a^{-1} = \lambda$ and $T_a^1 = \mu$, where λ and μ are as in (1) and (2) above; thus T^{-1} and T^1 are one-dimensional elements of $\hat{A}$.

   (i) Prove that $t \mapsto T^t$ $(-1 \le t \le 1)$ is a homeomorphism of $[-1, 1]$ onto $\hat{A}$.

   (ii) Show that there does *not* exist a saturated C^*-algebraic bundle $\mathscr{B}$ over the two-element group G such that: (a) the unit fiber C^*-algebra of $\mathscr{B}$ is A and (b) the action of the non-unit element of G on $\hat{A}$ by conjugation in $\mathscr{B}$ is given by

$$T^t \mapsto T^{-t} \quad (-1 \le t \le 1).$$

57. Show by an example that a dense contraction of a dense contraction of a dual system X need *not* be a dense contraction of X (see 17.5).

58. Prove Proposition 17.9.

59. Prove Proposition 17.11.

60. The following construction is a non-involutory version of the Gelfand–Naimark–Segal construction of $*$-representations from positive functionals (see VI.19.3).

Let A be any algebra, and $\phi: A \to \mathbb{C}$ any fixed linear functional. Let us define:

$$I_1 = \{a \in A : \phi(ba) = 0 \quad \text{for all } b \in A\},$$

$$I_2 = \{b \in A : \phi(ba) = 0 \quad \text{for all } a \in A\},$$

$$X_1 = A/I_1, \qquad X_2 = A/I_2.$$

Note that I_1 and I_2 are a left and right ideal of A respectively.

Show that $X = \langle X_1, X_2 \rangle$ is a dual system under the duality $(\ |\ )$ given by

$$(a + I_1 | b + I_2) = \phi(ba) \qquad (a, b \in A).$$

Show also that the equations

$$T_c^1(a + I_1) = ca + I_1,$$

$$T_c^2(b + I_2) = bc + I_2$$

$(a, b \in A)$ define a dual representation $T = \langle T^1, T^2 \rangle$ of A on X. This T is called the *dual representation of A generated by ϕ.*

61. Suppose that A is a *-algebra, A_0 the algebra (without involution) underlying A, and $\phi: A_0 \to \mathbb{C}$ any fixed linear functional. We shall say that ϕ *generates the *-representation S of A* if the dual representation T of A_0 generated by ϕ (in the sense of the preceding Exercise) has involutory expansion S (in the sense of 17.12).

Show that, if ϕ is a positive linear functional on A satisfying Condition (R) (see VI.19.3), then the *-representation of A generated by ϕ in the sense of VI.19.3 is also generated (to within unitary equivalence) by ϕ in the sense of the preceding paragraph.

Give an example of a linear functional ϕ on A which is *not* positive, and yet which generates a *-representation of A in the sense of this Exercise (see 17.18).

62. Let A be an algebra and $T = \langle T^1, T^2 \rangle$ a dual representation of A on $X = \langle X_1, X_2 \rangle$. A linear functional ϕ on A will be said to be *affiliated* with T if it belongs to the linear span (in $A^\#$) of $\{\psi_{\xi,\eta} : \xi \in X_1, \eta \in X_2\}$, where $\psi_{\xi,\eta}(a) = (T_a^1 \xi | \eta)$ $(a \in A)$.

Show that if T^1 and T^2 are both (algebraically) totally irreducible, and if ϕ is affiliated with T, then the dual representation of A generated by ϕ is equivalent to a direct sum of finitely many copies of T. (Here we make the obvious definition of a finite direct sum of dual representations.)

63. Prove Theorem 17.22.

Notes and Remarks

For a cursory account of the history and development of the theory of induced representations and the imprimitivity theorem, together with the principal contributors, we refer the reader to the general introduction to this second volume. Here are a few additional historical remarks.

§§1 and 2 of this chapter owe a great deal to the ground-breaking article of Stinespring [1].

The key concept of this chapter, the Rieffel inducing process, appears first in Rieffel [5]. It is formulated there in the context of C^*-algebras; but our version of it is essentially the same thing.

The important Corollary XI.12.12 was first observed by Blattner [5].

It is worth mentioning that Moscovici [1] developed an interesting generalization of Mackey's inducing process on groups. Moscovici's construction is a special case of the Rieffel inducing process, but in a different direction from the bundle construction of the second half of this chapter.

Bennett [1] generalized the notion of a B-rigged A-module (in the C^*-algebra context) by eliminating the right B-module structure and replacing it with a complete positivity condition on the rigging. His construction thus lies, roughly speaking, halfway in generality between rigged modules and our operator inner products.

Imprimitivity bimodules and the abstract Imprimitivity Theorem of §6 have been familiar to algebraists for some time in the purely algebraic non-involutory context (see Morita [1]). In fact, Morita's work shows that under suitable conditions a converse holds: Given two algebras A and B having "isomorphic representation theories," there is a canonical A, B imprimitivity bimodule (in a non-involutory sense) which implements the isomorphism of their representation theories. In the C^*-algebraic and W^*-algebraic contexts this converse has been developed by Rieffel [6].

The Imprimitivity Theorem for *homogeneous* Banach *-algebraic bundles was proved by Fell [14], using the classical methods of Loomis and Blattner. Philip Green [2] proved it for his "twisted covariant systems," which (as we have pointed out in the Notes and Remarks to Chapter VIII) are basically the same as homogeneous C^*-algebraic bundles; his proof uses the same Morita–Rieffel approach to imprimitivity on which the present chapter is based.

The idea of a dual representation, on which §17 is based, is due to Mackey [10]. This notion has also been developed by Bonsall and Duncan [1, §§27–29].

To obtain an overview of the entire field of group representations, from which the present chapter took its rise, the reader would do well to consult the survey articles of Mackey [10, 18, 22, 23].

Additional references which are relevant to the material discussed in this chapter and which may be of interest to the reader are the following: Aarnes [6], Abellanas [1], Armacost [1], Backhouse and Gard [1], Bagchi, Mathew, and Nadkarni [1], Baggett [10], Baggett and Ramsay [1], Baris [1], Barut and Raczka [1], Bennett [1, 2], Bernstein and Zelevinsky [1], Bruhat [1], Busby and Schochetman [1], Busby, Schochetman, and Smith [1], Carey [2, 3], Castrigiano and Henrichs [1], Cattaneo [5], Coleman [1, 2], Combes [1], Combes and Zettle [1], Corwin [1, 2], Corwin and Greenleaf [1], Deliyannis [1], Fabec [1, 2], Fakler [1, 2, 3, 4], Fontenot and Schochetman [1], Gaal [1], Gootman [5], Green [1, 2, 3, 4, 5], Guichardet [8], Havenschild, Kaniuth, and Kumar [1], Herb [1], Hulanicki and Pytlik [1, 2, 3], Jacquet and Shalika [1], Kehlet [2], Kirillov [1, 6], Kleppner [1, 2, 4], Koppinen and Neuvonen [1], Koornwinder and Van der Meer [1], Kraljević [1], Kunze [3], Langworthy [1], Lipsman [1, 3, 4], Mathew and Nadkarni

[1], Maurin and Maurin [1], Moore [1, 2, 3, 4, 5], Moore and Repka [1], Mueller–Roemer [1], Okamoto [1], Ørsted [1], N. Pedersen [1], Penney [2, 3, 4], Quigg [1, 2], Rieffel [2, 7, 9, 12, 13], Rigelhof [1], Rousseau [3, 7, 8, 10, 11], Sankaran [2], Schochetman [5, 6, 7, 9, 10], Scutaru [1], Seda [4, 5], Sen [1], Sund [1], Szmidt [1], Thieleker [1], Ward [1], G. Warner [1, 2], Wawrzyńczyk [1, 2, 3], Wigner [1], and Zimmer [2].

> In scientific work, it is not enough to be able to solve one's problems. One must also turn these problems around and find out what problems one has solved. It is frequently the case that, in solving a problem, one has automatically given the answer to another, which one has not even considered in the same connection.
>
> —Norbert Wiener

XII The Generalized Mackey Analysis

This final chapter, on the generalized Mackey analysis, is the climax toward which the earlier chapters have been pointing. Indeed, our choice of the material to be included in earlier chapters has been determined to a considerable extent by the requirements of the present chapter. Our interest in Banach *-algebraic bundles, especially saturated ones, is largely due to the fact that they seem to form the most general natural setting for the Mackey analysis.

Given a locally compact group G, let us consider the problem of classifying all possible irreducible unitary representations of G. Roughly speaking, two general methods have been developed for doing this. In both, one begins by choosing a suitable closed subgroup N of G; one then tries to classify the irreducible representations of G by analyzing what their behavior can be when restricted to N. In the first of these methods the chosen subgroup N is *compact*. A brief suggestion of the approach taken when N is compact is to be found in XI.7.8. This method, as applied to the representations of semisimple Lie groups, has been developed with great power by Harish-Chandra (see also Godement [9]). For a thorough exposition of Harish-Chandra's work the reader is referred to G. Warner [1, 2]. The second of these methods is the so-called Mackey normal subgroup analysis. This assumes that N is *normal* in G. Its crucial tools are the notion of induced representation and the Imprimitivity Theorem, developed (in greater generality, of course) in Chapter XI.

1243

In §1 of this chapter, for pedagogical reasons, we develop the Mackey normal subgroup analysis for finite groups, where its algebraic essentials appear uncluttered with topology and measure theory. In this context the method is much older than Mackey's work. Indeed, surely nothing in §1 would have been unfamiliar to Frobenius (who invented induced representations of finite groups in his paper [5] in 1898). The first explicit presentation of the method, however, seems to occur long after Frobenius, in Clifford's 1937 paper [1]. In this paper, which is still purely algebraic in character, Clifford sets down an algorithm for determining the finite-dimensional representations of an arbitrary (perhaps infinite) group G in terms of the finite-dimensional representations of a normal subgroup N and of certain subgroups of G/N.

The technical difficulties of carrying through Clifford's method become substantial when we attempt to extend the method to infinite-dimensional representations. In 1939, in his epoch-making paper [1], Wigner analyzed the irreducible unitary representations of one particular continuous group—namely the physically important Poincaré group (see 8.14–8.18)—following an analytic version of Clifford's method. In 1947 Gelfand and Naimark [2] did the same thing for the "$ax + b$" group (see 8.3). But it was Mackey who first systematically extended Clifford's method to the context of infinite-dimensional unitary representations of arbitrary second-countable locally compact groups. The representation-theoretic tools which he needed for this purpose were, first, the general definition of an induced unitary representation of a (second-countable) locally compact group, secondly the Imprimitivity Theorem, and thirdly the direct integral decomposition of unitary representations into irreducible "parts". The first two of these stem from his paper [2], and the third was developed by him in [7]. His final version of his extension of Clifford's method is contained in his important 1958 paper [8]. In view of these achievements, the infinite-dimensional generalization of the normal subgroup analysis of Clifford has come to be known by the name "Mackey normal subgroup analysis" (or, more briefly, the "Mackey machine").

The present chapter embodies two further developments which have taken place since Mackey's work. In the first place the contributions of Blattner [1], [6] now enable us to dispense entirely with the assumptions of second countability which are always adopted in Mackey's work. In particular, Blattner pointed out that for the purposes of the Mackey normal subgroup analysis one does not need the full apparatus of direct integral decomposition theory (for which separability assumptions seem to be essential), but only the limited portion of it contained in the "spectral measure" introduced by

Glimm [5] (see our §VII.9), which is valid without any separability requirements.

The second development, which has governed not only this chapter but Chapter XI also, is the recognition that the entire Mackey normal subgroup analysis (including the theory of induced representations and the Imprimitivity Theorem) is valid in the context not merely of locally compact groups but of *saturated* Banach *-algebraic bundles over locally compact groups. For the more restrictive class of *homogeneous* Banach *-algebraic bundles, this development was given in Fell [14] and in Leptin [5] (the latter reference being concerned with objects called generalized L_1 algebras, which are extremely closely related to homogeneous Banach *-algebraic bundles). As far as the authors know, the present chapter is the first place in which the Mackey normal subgroup analysis for *saturated* bundles has appeared in print.

Further fruitful generalizations of the Mackey normal subgroup analysis are of course to be expected in future. Possibly it can be given a purely algebraic formulation, bearing somewhat the same relation to the present chapter as the abstract Imprimitivity Theorem XI.6.15 bore to the Imprimitivity Theorem XI.14.18 for saturated Banach *-algebraic bundles. For another direction of possible generalization see 5.13 of this chapter.

The titles of the sections of this chapter are almost self-explanatory. The Mackey normal subgroup analysis, as developed for finite groups in §1, can be divided for convenience into three steps (see 1.28). §§2, 3, and 4 are devoted to carrying out these three steps respectively, in the most general context of a saturated Banach *-algebraic bundle over a locally compact group. In §5 these three steps are tied together, and the final results of the Mackey analysis in its generalized form are formulated.

§6 is somewhat of a digression from the mainstream of this chapter. Making essential use of the Mackey obstruction developed in §4, we are able to give a complete description of the structure of an arbitary saturated C^*-algebraic bundle (over a locally compact group G) whose unit fiber C^*-algebra is of compact type. It turns out that such objects have a more restricted structure than one might expect. For example, if the unit fiber A is finite-dimensional, then all the other fibers are finite-dimensional; and their dimensions are completely known when A and the action of G on $\hat{A}$ by conjugation are known (see 6.20).

In §7 these results are applied to the case of a saturated Banach *-algebraic bundle $\mathscr{B}$ over a *compact* group. If in addition the unit fiber *-algebra B_e of $\mathscr{B}$ is of finite type (i.e., $(B_e)\hat{\ }$ is discrete and every element of $(B_e)\hat{\ }$ is finite-dimensional), then it turns out that all the main features of the unitary representation theory of compact groups (as developed in Chapter IX) hold for the *-representations of $\mathscr{B}$: Its irreducible *-representations are all finite-dimensional; all its *-representations are discretely decomposable; and the Frobenius Reciprocity Theorem has a perfectly satisfying generalization to $\mathscr{B}$.

Finally, §§8 and 9 are devoted to specific examples of the working out of the Mackey analysis (mostly in the group context). In §8 we give straightforward examples in which all the hypotheses of that analysis are satisfied. In §9 we discuss some examples of theoretical interest, which show what happens when one or other of the hypotheses of the Mackey analysis fails.

1. The Mackey Normal Subgroup Analysis for Finite Groups

1.1. To help the reader to grasp the essentials of the Mackey normal subgroup analysis, we begin by carrying it out in the context of finite groups, where it appears unencumbered with topological and measure-theoretic technicalities.

Certain arguments and results presented in earlier chapters in greater generality will be repeated here in the context of finite groups, in order to make this section more self-contained.

1.2. Throughout this section G is a finite group (with unit e) and N is a fixed normal subgroup of G. By 'representation' in this section we mean a finite-dimensional unitary representation. As in Chapter VIII, if H is a finite group, $\hat{H}$ is the (finite) family of all unitary equivalence classes of irreducible representations of H.

The action of G on $\hat{N}$

1.3. Our first step is to make G act (to the left) as a group of transformations of $\hat{N}$. Let D be an irreducible representation of N, and x an arbitrary element of G. Composing D with the inner automorphism $m \mapsto x^{-1}mx$ of N, we obtain a new irreducible representation D' of N:

$$D'_m = D_{x^{-1}mx} \qquad\qquad (m \in N).$$

This D' is denoted by $^x D$. The unitary equivalence class of D' depends only on the unitary equivalence class Δ of D, and so can be denoted by $^x\Delta$. Evidently $^e D = D$ and $^{yx} D = {}^y({}^x D)$; so

$$^e\Delta = \Delta \quad \text{and} \quad {}^{yx}\Delta = {}^y({}^x\Delta) \qquad\qquad (x, y \in G).$$

Thus $\langle x, \Delta \rangle \mapsto {}^x\Delta$ is a left action of G on $\hat{N}$.

Notice that, if $x \in N$, then $(^xD)_m = D_x^{-1}D_mD_x$; so $^x\Delta = \Delta$. Thus N acts trivially on $\hat{N}$; and the action of G on $\hat{N}$ is actually the result of lifting to G an action of G/N on $\hat{N}$.

As we saw in XI.16.16, this left action of G on $\hat{N}$ is a very special case of the general conjugation operation discussed in §XI.16.

As usual we will often not distinguish notationally between a representation and the equivalence class to which it belongs.

1.4. Now let T be any representation of G. The restriction $T|N$ of T to N is a representation of N, and, being finite-dimensional, is discretely decomposable (see VI.14.6). So for each D in $\hat{N}$ we can form the D-subspace X_D of $X(T)$. Let $p(D)$ be the projection onto X_D. Thus the $p(D)$ $(D \in \hat{N})$ are orthogonal and their sum is $\mathbb{1}_{X(T)}$.

Proposition. *For $x \in G$ and $D \in \hat{N}$,*

$$T_x p(D)T_x^{-1} = p(^xD). \tag{1}$$

Proof. Let Y be a $(T|N)$-stable subspace of $X(T)$ on which $T|N$ acts equivalently to D. From the relation

$$T_m(T_x\xi) = T_x(T_{x^{-1}mx}\xi) \qquad (\xi \in Y; m \in N), \tag{2}$$

we see, first, that T_xY is $(T|N)$-stable, and secondly that $T|N$ acts on T_xY equivalently to ^xD. So $T_xY \subset X_{(^xD)}$. Since X_D is the sum of all such Y, we have $T_xX_D \subset X_{(^xD)}$. Replacing x by x^{-1} and D by ^xD we obtain $T_x^{-1}X_{(^xD)} \subset X_D$, or $X_{(^xD)} \subset T_xX_D$. Thus $T_xX_D = X_{(^xD)}$. Translated into a statement about projections this becomes (1). ∎

1.5. Equation (1) says that T together with the $p(D)$ $(D \in \hat{N})$ is a system of imprimitivity for the group G over $\hat{N}$. This fact will be of great importance in a short while. In the meantime notice what happens if the representation T of G is irreducible. Let θ be any one of the orbits in $\hat{N}$ under G. Then by (1), putting $p(\theta) = \sum_{D \in \theta} p(D)$, we have

$$T_x p(\theta) = \sum_{D \in \theta} T_x p(D) = \sum_{D \in \theta} p(^xD)T_x = p(\theta)T_x$$

(since ^xD runs over θ when D does). Thus $p(\theta)$ commutes with all T_x. Since T is being assumed irreducible, for each orbit θ the projection $p(\theta)$ must be either 0 or $\mathbb{1}$. Now the $p(\theta)$ are mutually orthogonal and add up to $\mathbb{1}$. It follows that $p(\theta) = \mathbb{1}$ *for exactly one orbit* θ, and that $p(\theta') = 0$ *for all other orbits* θ'.

By (1), if $p(D) = 0$ for some D in this orbit θ, then $p(^xD) = 0$ for all x in G, whence $p(\theta)$ would be 0. But $p(\theta) = 1$. It follows that $p(D) \neq 0$, i.e., D occurs in $T|N$, for all D in θ. Since $p(\theta') = 0$ for orbits $\theta' \neq \theta$, no element of $\hat{N} \setminus \theta$ occurs in $T|N$.

We have proved:

Proposition.   *For each irreducible representation T of G, the set of those D in $\hat{N}$ which occur in $T|N$ forms an orbit θ under the action of G on $\hat{N}$.*

Definition.   In this situation we say that θ is *associated with T*, or that T is *associated with θ*.

1.6.   According to Proposition 1.5 each T in $\hat{G}$ is associated with exactly one orbit θ in $\hat{N}$. So the problem of classifying the irreducible representations of G splits into two parts: (A) Classify the orbits θ in $\hat{N}$ under the action of G; (B) given an orbit θ, classify those T in $\hat{G}$ which are associated with θ. We shall now attack part (B). The main tool for this is the Imprimitivity Theorem, whose statement and proof we shall repeat here in the simple context of finite groups (following the pre-Rieffel argument).

Systems of Imprimitivity

1.7.   Let M be a finite left G-space. For our purposes, a *system of imprimitivity for G over M* is a pair T, p, where

(i) T is a representation of G;
(ii) $p: m \mapsto p(m)$ assigns to each m in M a non-zero projection $p(m)$ on $X(T)$ such that $p(m)p(n) = 0$ if $m \neq n$ and $\sum_{m \in M} p(m) = 1_{X(T)}$;
(iii)   for all x in G and m in M

$$T_x p(m) T_x^{-1} = p(xm). \tag{3}$$

Let H be any subgroup of G, S a non-zero representation of H, and T the induced representation $\mathrm{Ind}_{H \uparrow G}(S)$ of G. Thus T acts by left translation on the space of all functions $f: G \to X(S)$ which satisfy the identity

$$f(xh) = S_{h^{-1}}(f(x)) \qquad (x \in G; h \in H). \tag{4}$$

If we set $p(\alpha)(f) = \mathrm{Ch}_\alpha f$ $(\alpha \in G/H; f \in X(T))$, we verify easily that T, p is a system of imprimitivity for G over G/H (compare XI.14.3); it is called the system of imprimitivity *induced by S*.

Conversely we have:

Imprimitivity Theorem. *Let H be a subgroup of G and T,p a system of imprimitivity for G over G/H. Then there is a representation S of H such that T,p is unitarily equivalent to the system of imprimitivity induced by S; and S is unique to within unitary equivalence.*

Proof. We shall first prove the existence of S. Let $Y = \text{range}(p(eH))$. Thus by (3) Y is stable under $T|H$. Let S stand for the subrepresentation of $T|H$ acting on Y; and write T', p' for the system of imprimitivity induced by S. We shall show that T, p and T', p' are unitarily equivalent.

Choose a transversal Γ for G/H; and let $F: X(T') \to X(T)$ be the linear map defined by

$$F(f) = \sum_{x \in \Gamma} T_x(f(x)) \qquad (f \in X(T')). \qquad (5)$$

From the fact that $T_{xh}(f(xh)) = T_x T_h(S_{h^{-1}}(f(x))) = T_x(f(x))$ it follows that F is independent of Γ. Hence, if $y \in G$ and $f \in X(T')$,

$$F(T'_y f) = \sum_{x \in \Gamma} T_x(f(y^{-1}x))$$

$$= T_y \sum_{x \in y^{-1}\Gamma} T_x(f(x))$$

$$= T_y F(f) \qquad (6)$$

(since $y^{-1}\Gamma$ is also a transversal for G/H). So F intertwines T' and T. Also, if $f \in X(T')$ and $\alpha \in G/H$,

$$F(p'(\alpha)f) = \sum_{x \in \Gamma} Ch_\alpha(x) T_x(f(x))$$

$$= T_{x_\alpha}(f(x_\alpha)) \qquad (\text{where } x_\alpha \in \Gamma \cap \alpha)$$

$$= p(\alpha)\left(\sum_{x \in \Gamma} T_x(f(x)) \right) \qquad (\text{since } T_x(f(x)) \in \text{range}(p(xH)))$$

$$= p(\alpha)(F(f)); \qquad (7)$$

so F intertwines p' and p.

Since $T_x(Y) = \text{range}(p(xH))$ (by (3)), F maps $X(T')$ onto $X(T)$. Furthermore, if $f, g \in X(T')$,

$$(F(f), F(g))_{X(T)} = \sum_{x, y \in \Gamma} (T_{y^{-1}x}(f(x)), g(y))$$

$$= \sum_{x \in \Gamma} (f(x), g(x)) \qquad (\text{by (3)})$$

$$= (f, g)_{X(T')};$$

so F is a linear isometric bijection. This together with (6) and (7) shows that T, p and T', p' are unitarily equivalent. Thus S has the property required by the theorem.

The uniqueness of S results from the following easily verified fact: If T', p' is the system of imprimitivity induced by the unitary representation S of H, then S is equivalent to the subrepresentation of $T' | H$ acting on range($p'(eH)$) (see XI.14.21).   ∎

The Elements of $\hat{G}$ Associated with an Orbit

1.8. We now take up the question (B) of 1.6.

Fix an orbit θ in $\hat{N}$ and an element E of θ. Let H be the stability subgroup for E, i.e., $H = \{x \in G : {}^{x}E \cong E\}$. We notice that $N \subset H$, since N acts trivially on $\hat{N}$. As a G-space, θ is isomorphic with the coset space G/H under the bijection $w : xH \mapsto {}^{x}E$.

Now let T be an element of $\hat{G}$ associated with θ; form the $p(D)$ ($D \in \hat{N}$) as in 1.4; and for $\alpha \in G/H$ put $p'(\alpha) = p(w(\alpha))$. By 1.4 the $p'(\alpha)$ ($\alpha \in G/H$) are orthogonal projections on $X(T)$ whose sum is 1, and

$$T_x p'(\alpha) T_x^{-1} = p'(x\alpha) \qquad\qquad (x \in G; \alpha \in G/H).$$

Thus T, $\{p'(\alpha)\}$ is a system of imprimitivity for G over G/H. Applying the Imprimitivity Theorem 1.7 we find that T, $\{p'(\alpha)\}$ is unitarily equivalent to the system of imprimitivity induced by some unique representation S of H.

What properties must S have? By its definition

$$\operatorname{Ind}_{H \uparrow G}(S) \cong T.$$

Since T is irreducible, S is irreducible (see XI.12.3). Furthermore, we claim that

$$S | N \text{ is equivalent to a direct sum of copies of } E. \qquad (8)$$

Indeed: By the last statement of the proof of 1.7, S is equivalent to the subrepresentation of $T | H$ acting on range($p'(eH)$) ($=$ range($p(E)$)). So $S | N$ is equivalent to the subrepresentation of $T | N$ acting on $X_E = $ range($p(E)$). But by the definition of X_E the latter is a direct sum of copies of E. Thus (8) holds.

1.9. The next interesting fact is the converse: Whenever S is an irreducible representation of H satisfying (8), the induced representation $T = \operatorname{Ind}_{H \uparrow G}(S)$ is irreducible and associated with θ.

To prove this, take an irreducible representation S of H satisfying (8), and let $\mathcal{F} = \langle T, p \rangle$ be the system of imprimitivity for G over G/H induced by S. We claim that, for each α in G/H,

$$\text{range}(p(\alpha)) \text{ is the } w(\alpha)\text{-subspace for } T|N. \tag{9}$$

Indeed: Since the different $w(\alpha)$-subspaces are orthogonal, and since $\sum_{\alpha \in G/H} p(\alpha) = 1$, it is enough to prove that, for all α,

$$\text{range}(p(\alpha)) \subset \text{the } w(\alpha)\text{-subspace for } T|N. \tag{10}$$

In fact, it is enough to prove (10) for some *one* α. Indeed, assume that (10) holds for some α, and take any x in G. By 1.4 T_x carries the right side of (10) onto the $w(x\alpha)$-subspace for $T|N$; and by the definition of induced representations $T_x(\text{range}(p(\alpha))) = \text{range}(p(x\alpha))$. So (10) holds with α replaced by $x\alpha$. Since x is arbitrary and G acts transitively on G/H, (10) must hold for all α in G/H.

Now by the last sentence of the proof of 1.7 $T|N$ acts on $\text{range}(p(eH))$ like $S|N$, which by hypothesis (8) is a multiple of E. So (10) holds for $\alpha = eH$. Combining this with the preceding paragraph we deduce that (10) holds for all α. This in turn establishes (9).

We shall now show that T is irreducible. Let F be any T, T intertwining operator. Then F intertwines $T|N$ with itself, and so by VI.13.14 leaves stable each $w(\alpha)$-subspace for $T|N$. By (9) this implies that F commutes with all the $p(\alpha)$, i.e., that F belongs to the commuting algebra not merely of T but of $\mathcal{F}$. But the irreducibility of S implies that of $\mathcal{F}$ (see XI.14.26); so the commuting algebra of $\mathcal{F}$ consists of the scalar operators only. Therefore all T, T intertwining operators F are scalar, and T must be irreducible.

Condition (9) tells us that T must be associated with the original orbit θ. More than that, it shows that the S constructed from T as in 1.7 coincides with the S with which this number began. In particular, the map $S \mapsto \text{Ind}_{H \uparrow G}(S) = T$ is one-to-one (as a map of unitary equivalence classes) when S runs over the irreducible representations of H satisfying (8).

1.10. Summing up the results of 1.8 and 1.9, we have shown:

Theorem. *Let θ, E, H be as in 1.8. Then the inducing map $S \mapsto \text{Ind}_{H \uparrow G}(S) = T$ is a bijection from the set $\hat{H}_E$ of those elements S of $\hat{H}$ which satisfy (8) onto the set of those T in $\hat{G}$ which are associated with θ.*

1.11. Theorem 1.10 reduces problem (B) of 1.6 to the determination of the set $\hat{H}_E$. As an example of the power of this reduction, suppose for the moment that the action of G/N is free on the orbit θ, that is, $H = N$. In that case $\hat{H}_E$

evidently has just one element, namely E. So Theorem 1.10 says that there is exactly one element of $\hat{G}$ associated with the orbit θ, namely $T = \text{Ind}_{N\uparrow G}(E)$.

However, in general H may be much bigger than N. Indeed, sometimes H is all of G (for example, if E is the trivial representation of N). Also, condition (8) appears at first sight rather difficult to work with. A further analysis of the set $\hat{H}_E$ is therefore desirable.

The Analysis of $\hat{H}_E$; the Mackey Obstruction

1.12. To see how this analysis should go, it is instructive to consider first the special case that E satisfies the following two conditions:

E is one-dimensional;   (11)

E can be extended to a (one-dimensional) representation E' of H.   (12)

One important special situation in which (11) and (12) hold for all E in $\hat{N}$ is that in which N is Abelian and G is the semidirect product of the normal subgroup N and another subgroup K. Indeed: (11) holds then since N is Abelian. To prove (12), notice that since $N \subset H$ each element of H can be written in just one way in the form mk, where $m \in N$ and $k \in H \cap K$. We now define $E'(mk) = E(m)$ $(m \in N; k \in H \cap K)$; and recall from (11) and the definition of H that

$$E(k^{-1}mk) = E(m) \qquad (m \in N; k \in H). \qquad (13)$$

Thus, for m, n in N and h, k in $H \cap K$,

$$
\begin{aligned}
E'((mh)(nk)) &= E'([m(hnh^{-1})]hk) \\
&= E(m(hnh^{-1})) \\
&= E(m)E(hnh^{-1}) \\
&= E(m)E(n) \qquad\qquad\qquad \text{(by (13))} \\
&= E'(mh)E'(nk).
\end{aligned}
$$

So E' is multiplicative on H. Since it clearly extends E, (12) holds in this special situation.

1.13. Assume now that E satisfies (11) and (12). Given an element V of $(H/N)\hat{}$, the equation

$$S_h = E'(h)V_{hN} \qquad\qquad (h \in H) \qquad (14)$$

defines a representation S of H which is clearly irreducible. Since E' extends E, we have $S_m = E(m)\text{1}$ for $m \in N$; and so $S \in \hat{H}_E$. Notice that S determines V

(the extension E' being considered as fixed). Further, we claim that every S in $\hat{H}_E$ is constructed as in (14) from some V in $(H/N)\hat{\ }$. Indeed, let $S \in \hat{H}_E$. Then $S_m = E(m)\mathbb{1}$ for $m \in N$. So the representation $V': h \mapsto (E'(h))^{-1} S_h$ of H satisfies

$$V'_{mh} = (E'(mh))^{-1} S_{mh}$$
$$= E'(h)^{-1} E(m)^{-1} E(m) S_h$$
$$= V'_h$$

for all h in H and m in N, and so is obtained by lifting to H an element V of $(H/N)\hat{\ }$. We now have

$$S_h = E'(h)V'_h = E'(h)V_{hN} \qquad (h \in H),$$

proving the claim.

Thus we have shown:

Proposition. *If the E of 1.8 satisfies (11) and (12), the mapping $V \mapsto S$ defined by (14) is a bijection from $(H/N)\hat{\ }$ onto $\hat{H}_E$.*

1.14. Combining 1.10 and 1.13, we find:

Theorem. *Let θ, E, H be as in 1.8; and assume that E satisfies (11) and (12). Let E' be an extension of E to H; and for each V in $(H/N)\hat{\ }$ let $E' \times V$ stand for the representation S of H defined in (14). Then*

$$V \mapsto \underset{H\uparrow G}{\mathrm{Ind}}(E' \times V)$$

(considered as a map of unitary equivalence classes) is a bijection from $(H/N)\hat{\ }$ onto the set of all those T in $\hat{G}$ which are associated with θ.

1.15. By 1.12, this theorem provides a complete recipe for classifying $\hat{G}$ in case G is the semidirect product of an Abelian normal subgroup N and another subgroup K. Besides $\hat{N}$, the classification will involve only the irreducible representations of subgroups of the quotient group G/N. We will carry out this recipe for specific groups in Examples 1.23, 1.24, 1.25, and 1.27.

Notice another simple case in which Theorem 1.14 is applicable, namely, when E is the trivial representation I of N. In that case $H = G$, and E' can be taken to be the trivial representation I' of G. The theorem states that the elements of $\hat{G}$ associated with the orbit $\{I\}$ in $\hat{N}$ are just those which are lifted to G from elements of $(G/N)\hat{\ }$.

1.16. Suppose that E fails to satisfy (11) and (12). Can we extend the analysis of 1.13 to cover this case? It will turn out that (11) at least is not essential. Whatever the dimension of E, if (12) holds, that is, if E can be extended to a representation E' of H (acting in the same space), Proposition 1.14 will remain true provided (14) is generalized to read:

$$S_h = E'_h \otimes V_{hN} \qquad (h \in H). \qquad (15)$$

So in this case the structure space $(H/N)\hat{\ }$ still indexes the elements of $\hat{G}$ which are associated with θ.

Even if (12) fails, the analysis of 1.13 can still be generalized, provided we are willing to consider not merely (ordinary) representations, but also cocycle representations of H/N (see VIII.10.9). Indeed, we shall show in 1.17 that, whether (12) holds or not, the element E of $\hat{N}$ gives rise to a cocycle class t in $Z(\mathbb{E}, H/N)$ (see III.5.12) which, as it were, measures the obstruction to extending E to H. E can be extended to H if and only if t is the unit element of $Z(\mathbb{E}, H/N)$. The desired generalization of 1.13 will then consist in showing that *the elements of $\hat{G}$ associated with θ are indexed by the set of all equivalence classes of irreducible cocycle representations of H/N with cocycle τ^{-1}, where τ is any element of $C(\mathbb{E}, H/N)$ of class t.*

1.17.   Let E be any element of $\hat{N}$; and as before let $H = \{x \in G: {}^xE \cong E\}$. To realize the program of 1.16, we must construct from E a cocycle in $C(\mathbb{E}, H/N)$ (see III.5.12).

Choose a transversal for H/N, i.e., a subset Γ of H containing exactly one element from each N coset. We shall suppose that $e \in \Gamma$. Since ${}^hE \cong E$ for all h in H, we can select for each h in Γ a unitary operator E'_h on $X(E)$ such that

$$E_{hmh^{-1}} = E'_h E_m E'^{-1}_h \qquad (16)$$

for all m in N. Since E is irreducible, E'_h is determined up to a multiplicative constant in $\mathbb{E}$. We agree to set

$$E'_e = 1_{X(E)}. \qquad (17)$$

So far E' is defined only on Γ. We now extend it to all of H by setting

$$E'_{hm} = E'_h E_m \qquad (h \in \Gamma; m \in N). \qquad (18)$$

It is easy to see that (16) and (18) then hold for all m in N and h in H.

If $h, k \in H$ and $m \in N$, (16) gives

$$E'_h E'_k E_m E'^{-1}_k E'^{-1}_h = E'_h E_{kmk^{-1}} E'^{-1}_h$$

$$= E_{hkmk^{-1}h^{-1}}$$

$$= E'_{hk} E_m E'^{-1}_{hk}. \qquad (19)$$

By the uniqueness of the E'_h up to scalars, we conclude from (19) that there is a number $\sigma(h, k)$ in $\mathbb{E}$ such that

$$E'_{hk} = \sigma(h, k)E'_h E'_k. \tag{20}$$

The map $\sigma: H \times H \to \mathbb{E}$ is a cocycle. Indeed, it follows from (17) that $\sigma(e, h) = \sigma(h, e) = 1$; and the cocycle identity III.5.12(12) is proved by the same calculation as in III.5.13(18).

Next, we claim that $\sigma(h, k)$ $(h, k \in H)$ depends only on the cosets hN and kN. Indeed, let $h, k \in H$ and $m \in N$. We have by (18) and (20):

$$\sigma(h, km)E'_h E'_{km} = E'_{hkm}$$

$$= E'_{hk} E_m$$

$$= \sigma(h, k)E'_h E'_k E_m$$

$$= \sigma(h, k)E'_h E'_{km},$$

whence

$$\sigma(h, km) = \sigma(h, k). \tag{21}$$

Again, notice that by (17) and (20) (applied with $k = h^{-1}$) $E'_{h^{-1}}$ and $(E'_h)^{-1}$ differ only by a scalar; hence by (16)

$$E'^{-1}_h E_m E'_h = E'_{h^{-1}} E_m E'^{-1}_{h^{-1}} = E_{h^{-1}mh} \qquad (h \in H; m \in M). \tag{22}$$

By (18), (20), and (22)

$$\sigma(h, k)E'_{hmk} = \sigma(h, k)\sigma(hm, k)E'_{hm} E'_k$$

$$= \sigma(h, k)\sigma(hm, k)E'_h E_m E'_k$$

$$= \sigma(h, k)\sigma(hm, k)E'_h E'_k E_{k^{-1}mk}$$

$$= \sigma(hm, k)E'_{hk} E_{k^{-1}mk}$$

$$= \sigma(hm, k)E'_{hmk};$$

so

$$\sigma(h, k) = \sigma(hm, k). \tag{23}$$

Equations (21) and (23) establish the claim.

It follows that the equation

$$\tau(hN, kN) = \sigma(h, k) \qquad (h, k \in H) \tag{24}$$

defines a cocycle in $C(\mathbb{E}, H/N)$. Let t be the cocycle class of τ, so that $t \in Z(\mathbb{E}, H/N)$.

There was considerable arbitrariness in the choice of Γ and the E'_h ($h \in \Gamma$); but this arbitrariness, though it affects τ, does not affect the cocycle class t. Indeed, the cocycles τ' which could have been obtained by different choices are easily seen to be precisely those which are cohomologous with τ, i.e., which are in t.

Definition. We call t the *Mackey obstruction* of E. If t is the unit of $Z(\mathbb{E}, H/N)$, the Mackey obstruction of E is said to be *trivial*.

1.18. Proposition. *The Mackey obstruction t of E is trivial if and only if E can be extended to a representation E' of H (acting in the same space as E).*

Proof. If E can be so extended, its extension E' can be used as the E' of 1.17, and (20) then holds with $\sigma(h, k) \equiv 1$; so t is the unit.

Conversely, if t is the unit, then, as we observed at the end of 1.17, the choices of Γ and the E'_h could have been made so that $\sigma(h, k) \equiv 1$. But by (20) this says that E' is a representation of H. By (17) and (18) E' extends E. ∎

Remark. This proposition suggests that t measures the "obstruction to extending E to H"; hence the term "obstruction."

1.19. Let E, E', σ, τ be as in 1.17. Generalizing 1.13, we shall see that $\hat{H}_E$ is in one-to-one correspondence with the set $(H/N)^{\widehat{}\,(\tau^{-1})}$ of all unitary equivalence classes of irreducible τ^{-1}-representations of H/N (see VIII.10.9).

Proposition. *Let V be an irreducible τ^{-1}-representation of H/N. Then the equation*

$$S_h = E'_h \otimes V_{hN} \qquad\qquad (h \in H) \qquad (25)$$

defines an irreducible representation S of H (acting on $X(E) \otimes X(V)$); in fact $S \in \hat{H}_E$. The map $V \mapsto S$ defined by (25), considered as a map of unitary equivalence classes, is a bijection from $(H/N)^{\widehat{}\,(\tau^{-1})}$ onto $\hat{H}_E$.

Proof. From (20) and the fact that V is a τ^{-1}-representation we see that S is an (ordinary) representation of H. Since E' extends E, S reduces on N to a multiple of E. To show that S is irreducible, suppose that F is any operator in its commuting algebra. In particular F commutes with all $S_m = E_m \otimes 1$ ($m \in N$). Since E is irreducible, this implies by VI.14.5 that $F = 1 \otimes \Phi$, where Φ is an operator on $X(V)$. Thus $E_h \otimes \Phi V_{hN} = F S_h = S_h F = E_h \otimes V_{hN} \Phi$ for all h in H, implying that Φ belongs to the commuting algebra of V. But the latter is irreducible; so Φ, and hence also F, is a scalar operator. Thus S is irreducible. We have now shown that $S \in \hat{H}_E$.

Let the S of (25) be called $S^{(V)}$.

Suppose that V and W are two elements of $(H/N)^{\hat{}(\tau^{-1})}$ such that $S^{(V)} \cong S^{(W)}$ under a unitary equivalence P. An argument very similar to the last paragraph shows that $P = 1 \otimes \Psi$, where Ψ is a unitary equivalence of V and W. So $V \mapsto S^{(V)}$ is one-to-one.

It remains to show that every S in $\hat{H}_E$ is of the form $S^{(V)}$. Let $S \in \hat{H}_E$. By the definition of $\hat{H}_E$, we may as well assume that $X(S) = X(E) \otimes Y$, where Y is a finite-dimensional Hilbert space, and that

$$S_m = E_m \otimes 1_Y \qquad (m \in N). \qquad (26)$$

Now we claim that, for each h in H, the operator $(E_h'^{-1} \otimes 1)S_h$ commutes with all $E_m \otimes 1$ $(m \in N)$. Indeed, notice first from (18), (20), and (23) that

$$E_{hmh^{-1}}E_h' = E_{hm}' = E_h'E_m. \qquad (27)$$

By (26) and (27)

$$(E_h'^{-1} \otimes 1)S_h(E_m \otimes 1) = (E_h'^{-1} \otimes 1)S_{hm}$$

$$= (E_h'^{-1} \otimes 1)S_{hmh^{-1}}S_h$$

$$= (E_h'^{-1}E_{hmh^{-1}} \otimes 1)S_h$$

$$= (E_m E_h'^{-1} \otimes 1)S_h$$

$$= (E_m \otimes 1)(E_h'^{-1} \otimes 1)S_h,$$

proving the claim. In view of this, the irreducibility of E, and VI.14.5, each h in H gives rise to a unitary operator W_h on Y satisfying $(E_h'^{-1} \otimes 1)S_h = 1 \otimes W_h$, or

$$S_h = E_h' \otimes W_h. \qquad (28)$$

If $h \in H$ and $m \in N$, we have by (18), (26), and (28),

$$E_{hm}' \otimes W_{hm} = S_{hm} = S_h S_m = E_h' E_m \otimes W_h = E_{hm}' \otimes W_h,$$

whence $W_{hm} = W_h$. It follows that W_h depends only on the coset hN, and one can define

$$V_{hN} = W_h \qquad (h \in H). \qquad (29)$$

If $h, k \in H$, we have by (28) and (20)

$$E_{hk}' \otimes W_{hk} = S_{hk}$$

$$= S_h S_k$$

$$= E_h' E_k' \otimes W_h W_k$$

$$= E_{hk}' \otimes (\sigma(h, k))^{-1} W_h W_k,$$

so that $W_{hk} = (\sigma(h, k))^{-1} W_h W_k$. Applying (24) and (29) to this, we see that V is a τ^{-1}-representation of H/N on Y. Equation (28) becomes

$$S_h = E'_h \otimes V_{hN},$$

which together with the irreducibility of S implies that V is irreducible. Consequently $V \in (H/N)^{\widehat{}(\tau^{-1})}$, and $S = S^{(V)}$. This completes the proof of the proposition.   ∎

1.20.   Combining 1.10 and 1.19, we have (keeping the notation of 1.19):

Theorem.   *The map* $V \mapsto \mathrm{Ind}_{H \uparrow G}(S^{(V)})$ *is a bijection from the set* $(H/N)^{\widehat{}(\tau^{-1})}$ *of all unitary equivalence classes of irreducible* τ^{-1}*-representations of* H/N, *onto the set of all elements of* $\hat{G}$ *which are associated with the orbit* θ *containing* E.

1.21.   *Remark*.   Theorem 1.20 is the final solution obtained by the Mackey normal subgroup analysis for problem (B) of 1.6. It gives a recipe for determining all irreducible representations of G in terms of the representation theories of N and of the stability subgroups of G/N.

Theorem 1.20 has one unpleasant feature: In order to study the (ordinary) representations of G, we are obliged in general to classify more general objects, namely cocycle representations, for subgroups of G/N. However, this feature will not seem so unpleasant in the generalized Mackey analysis to be set forth in the remainder of this chapter, since, as we have seen in §VIII.10, cocycle representations of H/N are just special cases of representations of Banach *-algebraic bundles over H/N.

1.22.   *Remark*.   We see from 1.12 and 1.18 that the Mackey obstruction of E is always trivial when E satisfies (11) and (12) of 1.12. In particular, this is so if N is Abelian and G is the semidirect product of N with another subgroup.

Another situation in which the Mackey obstruction of E is trivial is when H/N is *cyclic*. In fact, letting K be any finite cyclic group, we claim that $Z(\mathbb{E}, K)$ is the one-element group. By III.5.13 this is the same as saying that the only central extension of $\mathbb{E}$ by K is the direct product extension; and the latter fact was proved in III.5.11(II).

This remark has an important application to solvable groups. Assume that G is *solvable*; that is, it has an increasing sequence of subgroups

$$\{e\} = N_0 \subset N_1 \subset \cdots \subset N_{r-1} \subset N_r = G \tag{30}$$

such that, for $i = 1, \ldots, r$, N_{i-1} is normal in N_i and N_i/N_{i-1} is cyclic. The analysis of $\hat{G}$ can then proceed by repeated application of the Mackey analysis: N_1 presents no problem since it is cyclic. So we first catalogue $(N_2)\hat{\ }$ using the normal subgroup N_1 of N_2; then we catalogue $(N_3)\hat{\ }$ using the normal subgroup N_2 of N_3 (whose structure space $(N_2)\hat{\ }$ has just been found); then $(N_4)\hat{\ }$ using N_3; finally reaching $\hat{G}$ after $r-1$ steps. By the above remark, since at each stage all the stability subgroups of the quotients N_i/N_{i-1} are cyclic, *no non-trivial Mackey obstructions will ever arise.*

Whether G is solvable or not, any composition series for G of the form (30) can be used as the basis for an investigation of $\hat{G}$ by repeated application of the Mackey analysis; but it may be impossible to avoid the appearance of a non-trivial Mackey obstruction at some stage. See Example 1.26.

Examples of the Mackey Analysis for Finite Groups

1.23. Example; The Symmetric Group on Three Objects. We begin with the simplest non-Abelian group, the 6-element group G of all permutations of three objects a, b, c. This has a 3-element normal subgroup N consisting of all the even permutations; and G is the semidirect product of N with the two-element subgroup whose non-unit element is the transposition w of a and b. $\hat{N}$ has three elements, I (the trivial character) and two nontrivial characters χ and χ^{-1}; and the action of w on $\hat{N}$ transposes χ and χ^{-1}. So $\hat{N}$ has two orbits, $\{I\}$ and $\theta = \{\chi, \chi^{-1}\}$. The elements of $\hat{G}$ associated with the orbit $\{I\}$ are by 1.15 just the two characters of G lifted from the two characters of G/N (sending even permutations into 1 and odd permutations into 1 and -1 respectively). Since G/N acts freely on θ, 1.11 says that G has just one irreducible representation associated with θ, namely $\mathrm{Ind}_{N\uparrow G}(\chi)$ (or the equivalent representation $\mathrm{Ind}_{N\uparrow G}(\chi^{-1})$). So $\hat{G}$ has just three elements, of dimensions 1, 1, and 2 respectively. (See Exercise 26(A) of Chapter IV.)

1.24. Example; The Dihedral Groups. By the n-th *dihedral group* G_n $(n = 3, 4, \ldots)$ we mean the semidirect product of a normal cyclic subgroup N of order n and a subgroup $K = \{e, w\}$ of order 2, the multiplication being determined by

$$wmw^{-1} = m^{-1} \qquad\qquad (m \in N). \qquad (31)$$

The order of G_n is of course $2n$. G_3 is the group of the preceding Example 1.23.

Geometrically speaking, G_n is the symmetry group of the n-sided regular polygon, reflections being included.

Like N, $\hat{N}$ is a cyclic group of order n. Let χ be a generating element of $\hat{N}$ and I the trivial (unit) element of $\hat{N}$. Since by (31) w sends χ into χ^{-1}, the G-orbits in $\hat{N}$ are the sets $\theta_r = \{\chi^r, \chi^{-r}\}$ (r integral).

Assume first that n is odd. Then there are $\frac{1}{2}(n-1)$ distinct two-element orbits θ_r ($r = 1, \ldots, \frac{1}{2}(n-1)$), and one one-element orbit $\theta_0 = \{I\}$, and no others. By 1.11 and 1.15 there are therefore just two one-dimensional elements of $(G_n)^{\hat{}}$ (those lifted from the two characters of K), $\frac{1}{2}(n-1)$ two-dimensional elements $T^{(r)} = \mathrm{Ind}_{N\uparrow G}(\chi^r)$ ($r = 1, \ldots, \frac{1}{2}(n-1)$) of $(G_n)^{\hat{}}$, and no others. (Note that this conclusion squares with IV.6.4, since $2 \cdot 1^2 + \frac{1}{2}(n-1)2^2 = 2n$.)

Now let n be even. There are then $\frac{1}{2}(n-2)$ two-element orbits θ_r ($r = 1, 2, \ldots, \frac{1}{2}(n-2)$), and two one-element orbits $\{I\}$ and $\{\chi^{(n/2)}\}$. As before, with the two-element orbits are associated $\frac{1}{2}(n-2)$ two-dimensional elements $T^{(r)} = \mathrm{Ind}_{N\uparrow G}(\chi^r)$ ($r = 1, \ldots, \frac{1}{2}(n-2)$) of $(G_n)^{\hat{}}$; and with $\{I\}$ are associated the two one-dimensional representations lifted from characters of K. The orbit $\{\chi^{n/2}\}$ has G_n for its stability group; and by 1.15 there are just two elements of $(G_n)^{\hat{}}$ associated with it, both one-dimensional. So we have altogether four one-dimensional and $\frac{1}{2}(n-2)$ two-dimensional elements of $(G_n)^{\hat{}}$, and no others. (This again squares with IV.6.4.)

These facts were obtained by a different route in IV.6.18.

Notice that the four one-dimensional elements of $(G_n)^{\hat{}}$, in case n is even, are just those lifted from the characters of the four-element Abelian quotient group G_n/M, where M is the $(\frac{1}{2}n)$-element subgroup of N (which is normal in G_n).

1.25. Example; The Quaternion Group.

Let G be the eight-element multiplicative group consisting of the quaternions ± 1, $\pm i$, $\pm j$, $\pm k$. The four-element cyclic subgroup $N = \{\pm 1, \pm i\}$ is normal in G. Let χ be the identity character of N, so that $\hat{N} = \{I, \chi, \chi^2, \chi^{-1}\}$. Clearly j, acting by conjugation, sends each ψ in $\hat{N}$ into ψ^{-1}. There are thus three orbits in $\hat{N}$, namely $\{I\}$, $\{\chi^2\}$, and $\{\chi, \chi^{-1}\}$. As in the previous examples, we check that the first two orbits are each associated with two one-dimensional representations, while $\{\chi, \chi^{-1}\}$ is associated with one two-dimensional element $\mathrm{Ind}_{N\uparrow G}(\chi)$ of $\hat{G}$. So $\hat{G}$ has four one-dimensional and one two-dimensional elements. (Compare IV.6.7.)

Suppose now that we apply the Mackey analysis not to the normal subgroup N but to $M = \{1, -1\}$. $\hat{M}$ of course has two elements, the trivial I and the identity character J. Since M is central in G, both of these have stability group G. Furthermore, M is the commutator subgroup of G; and this implies that any one-dimensional representation of G restricts to I on M. So J cannot be extended to its stability subgroup G, i.e., its Mackey obstruction is non-trivial. As a matter of fact, it is easy to verify that one of the cocycles in the Mackey obstruction of J is the "bicharacter" σ of G/M given by

$$\sigma(\langle r, s \rangle, \langle r', s' \rangle) = \begin{cases} 1 & \text{unless } s = r' = w, \\ -1 & \text{if} \quad s = r' = w. \end{cases}$$

(Here we identify G/M with $\mathbb{Z}_2 \times \mathbb{Z}_2$, where $\mathbb{Z}_2$ is the two-element group $\{e, w\}$.) From Mackey's Theorem XI.15.7 it follows that G/M has just one (two-dimensional) irreducible σ^{-1}-representation. Thus, by the Mackey analysis applied to M, there is just one irreducible representation of G associated with the orbit $\{J\}$ in $\hat{M}$, and it is two-dimensional. As usual, associated with $\{I\}$ there are four elements of $\hat{G}$, namely the four one-dimensional representations lifted from the four characters of G/M. In brief, the Mackey analysis based on M agrees (as of course it must) with what we obtained using N.

1.26. *An Example with Non-Trivial Mackey Obstruction.* Example 1.25 illustrated the general principle mentioned in 1.22, that if G is solvable the investigation of $\hat{G}$ by repeated applications of the Mackey analysis need never involve (non-ordinary) cocycle representations if we take the successive normal subgroups properly, though it may if we do not. In this number we present an example of a non-solvable group G for which non-ordinary cocycle representations cannot be avoided.

Let F be a finite field (with unit 1) of order $n > 3$. We form the finite group $G = SL(2, F)$ of all 2×2 matrices with entries in F and determinant 1. The only non-trivial normal subgroup of G is the center N of G consisting of $e = \begin{pmatrix} 1 & 0 \\ 0 & 1 \end{pmatrix}$ and $-e = \begin{pmatrix} -1 & 0 \\ 0 & -1 \end{pmatrix}$; and the quotient group G/N is simple (see Bourbaki [15, Exercise 14(e), p. 422]). Furthermore, G is its own commutator subgroup (Bourbaki [15, Exercise 13(c), p. 421]); in particular G has no one-dimensional representation other than the trivial one. Thus the non-trivial character of N cannot be extended to its stability subgroup G, and so has non-trivial Mackey obstruction.

1.27. *Example; A Nilpotent Group.* We end this section with one more example of the Mackey analysis, applied to a nilpotent group.

Fix a finite field F (with unit 1) of order n; and let G be the multiplicative group of all 3×3 matrices of the form

$$\begin{pmatrix} 1 & x & z \\ 0 & 1 & y \\ 0 & 0 & 1 \end{pmatrix} \qquad (x, y, z \in F). \qquad (32)$$

The matrix (32) will be denoted by $[x, y, z]$. As our normal subgroup N we take $\{[x, 0, z] : x, z \in F\}$. Then G is the semidirect product of N with the subgroup $K = \{[0, y, 0] : y \in F\}$.

Since $[x, 0, z][x', 0, z'] = [x + x', 0, z + z']$, we have $N \cong F_+ \times F_+$ (F_+ being the additive group of F). So the elements of $\hat{N}$ are of the form

$$\phi \times \psi : [x, 0, z] \mapsto \phi(x)\psi(z),$$

where $\phi, \psi \in (F_+)\hat{\,}$. One verifies easily that the action of G on $\hat{N}$ is given by:

$$[0, y, 0](\phi \times \psi) = \phi\psi_y \times \psi \qquad (y \in F; \phi, \psi \in (F_+)\hat{\,}), \qquad (33)$$

where $\psi_y(x) = \psi(yx)$. From this we can read off the orbits. If γ stands for the trivial (unit) element of $(F_+)\hat{\,}$, then $\{\phi \times \gamma\}$ is a one-element orbit (with stability subgroup G) for each ϕ in $(F_+)\hat{\,}$. The group G/N is Abelian, and $\phi \times \gamma$ extends to G. So the elements of $\hat{G}$ which are associated with $\{\phi \times \gamma\}$ are just the n one-dimensional representations of G of the form

$$\phi \cdot \chi : [x, y, z] \mapsto \phi(x)\chi(y) \qquad (x, y, z \in F),$$

where χ runs over $(F_+)\hat{\,}$.

Let ψ be an element of $(F_+)\hat{\,}$ different from γ. By Remark X.3.16 $\phi\psi_y$ runs (once) over $(F_+)\hat{\,}$ as y runs over F; so the orbit of $\phi \times \psi$ is $(F_+)\hat{\,} \times \psi$; and the stability subgroup of $\phi \times \psi$ is just N. Thus by 1.11 there is exactly one element of $\hat{G}$ associated with $(F_+)\hat{\,} \times \psi$, namely the n-dimensional representation $T^\psi = \mathrm{Ind}_{N \uparrow G}(\gamma \times \psi)$.

Thus $\hat{G}$ has just n^2 one-dimensional elements (namely those lifted from the characters of the Abelian quotient group G/C, where $C = \{[0, 0, z] : z \in F\}$), and $n - 1$ n-dimensional elements; it has no others.

Remark.   Suppose we had applied the Mackey analysis to C instead of N. Notice that C is both the center and the commutator subgroup of G. Thus a non-trivial element ψ of $\hat{C}$ ($\cong (F_+)\hat{\,}$) cannot be extended to its stability subgroup G, that is, it has non-trivial Mackey obstruction. By the above discussion there is only one element of $\hat{G}$ associated with the orbit $\{\psi\}$ in $\hat{C}$, namely T^ψ.

Question. Notice that $SL(2, F)$ acts as a group of automorphisms of G, the action being:

$$u[x, y, z] = [ax + by, cx + dy, z + \tfrac{1}{2}(acx^2 + bdy^2 + 2bcxy)],$$

$(x, y, z \in F; u = \begin{pmatrix} a & b \\ c & d \end{pmatrix} \in SL(2, F))$. Let L be the semidirect product of G with $SL(2, F)$ with respect to this action.

Now T^ψ is that unique irreducible n-dimensional representation of G whose restriction to C is a direct sum of copies of the character ψ. Since each element u of $SL(2, F)$ leaves C pointwise fixed, it follows that T^ψ is fixed under the action of $SL(2, F)$ on $\hat{G}$; that is, the stability subgroup of L for T^ψ is L itself.

Can T^ψ be extended to L? That is, does T^ψ have trivial Mackey obstruction (with reference to L)? We do not know.

Recapitulation

1.28. In preparation for discussing the Mackey analysis in full generality, let us recapitulate in three steps the procedure developed in this section for classifying the structure space of a finite group G in terms of a normal subgroup N of G.

Step I: We make G act as a transformation group on the structure space $\hat{N}$ of N. We then show that each element T of $\hat{G}$ "belongs" to some orbit θ in $\hat{N}$ under this action, in the sense that the set of those D in $\hat{N}$ which occur in the direct sum decomposition of $T|N$ into irreducible parts is exactly θ.

Step II: Fix an orbit θ in $\hat{N}$ and an element E of θ, and denote by H the stability subgroup of G for E. By means of the inducing construction and the Imprimitivity Theorem we classify the elements of $\hat{G}$ which "belong" to θ in terms of the set $\hat{H}_E$ of those elements of $\hat{H}$ whose restriction to N is a direct sum of copies of E.

Step III: Let E and H be as in Step II. We construct from E a well-defined element t_E of the second cohomology group $Z(\mathbb{E}, H/N)$, called the Mackey obstruction of E. If τ is a fixed cocycle belonging to t_E, we classify the set $\hat{H}_E$ of Step II by putting it into one-to-one correspondence with the "τ^{-1}-structure space of H/N," i.e., the set of all irreducible τ^{-1}-representations of H/N.

2.   The Mackey Analysis in the General Case; Step I

2.1.   Having developed the Mackey normal subgroup analysis for finite groups, we shall now attempt to carry through a similar development for saturated Banach *-algebraic bundles over arbitrary locally compact groups.

We start out with a locally compact group G (with unit e), and a saturated Banach *-algebraic bundle $\mathscr{B} = \langle B, \pi, \cdot, * \rangle$ over G. Let N be a closed normal subgroup of G. In analogy with the finite group case, our goal is roughly to classify the irreducible *-representations of $\mathscr{B}$ in terms of the irreducible *-representations of $\mathscr{B}_N$ and the representation theory of subgroups of G/N.

As in VIII.21.2 we denote by $\hat{\mathscr{B}}$ and $(\mathscr{B}_N)\hat{\ }$ the structure spaces of $\mathscr{B}$ and $\mathscr{B}_N$ respectively, equipped with the regional topologies. Let $(\mathscr{B}_N)\hat{\ }^+$ be the subset of $(\mathscr{B}_N)\hat{\ }$ consisting of those irreducible *-representations of $\mathscr{B}_N$ which are $\mathscr{B}$-positive (in the sense of XI.8.6). The reader will recall that in the group case (i.e., when $\mathscr{B}$ is the group bundle of G) we have $\hat{\mathscr{B}} \cong \hat{G}$ and (by XI.8.14) $(\mathscr{B}_N)\hat{\ }^+ = (\mathscr{B}_N)\hat{\ } \cong \hat{N}$.

2.2.   We have seen in §XI.16 that, in view of the saturation of $\mathscr{B}$, G acts by conjugation as a transformation group on $(\mathscr{B}_N)\hat{\ }^+$. By Remark 1 of XI.16.3 the action of an element x of G on $(\mathscr{B}_N)\hat{\ }^+$ depends only on the N-coset to which x belongs; so the action of G on $(\mathscr{B}_N)\hat{\ }^+$ is actually lifted from an action of G/N on $(\mathscr{B}_N)\hat{\ }^+$. By XI.16.23 this action is continuous; so that G (and G/N) become topological transformation groups on $(\mathscr{B}_N)\hat{\ }^+$.

2.3.   It is vital to observe that the preceding situation suffers no loss of generality whatsoever if we assume that $N = \{e\}$. Indeed: Suppose that N is a general closed normal subgroup as before; and form the $\mathscr{L}_1$ partial cross-sectional bundle $\mathscr{B}' = \langle B', \pi', \cdot, * \rangle$ over G/N derived from $\mathscr{B}$, as in VIII.6.5. By VIII.6.8 $\mathscr{B}'$ is saturated. By VIII.6.7 (see VIII.15.9) the *-representation theories of $\mathscr{B}$ and $\mathscr{B}'$ are identical (including their regional topologies); and the same holds for $\mathscr{B}_N$ and B'_e $(= \mathscr{L}_1(\mathscr{B}_N))$. Furthermore, by XI.12.7 an element of $(\mathscr{B}_N)\hat{\ }$ is $\mathscr{B}$-positive if and only if the corresponding element of $(B'_e)\hat{\ }$ is $\mathscr{B}'$-positive; so $(\mathscr{B}_N)\hat{\ }^+$ and $(B'_e)\hat{\ }^+$ can be identified. By XI.16.22 the action of G/N on $(\mathscr{B}_N)\hat{\ }^+$ relative to $\mathscr{B}$ is the same as the action of G/N on $(B'_e)\hat{\ }^+$ relative to $\mathscr{B}'$. Thus nothing is altered if we agree to consider $\mathscr{B}'$ and B'_e rather than $\mathscr{B}$ and $\mathscr{B}_N$.

In view of this, *we shall suppose that the passage from $\mathscr{B}$ to $\mathscr{B}'$ has already been made, and shall assume from now on that $N = \{e\}$. We denote the unit fiber Banach *-algebra B_e by A, and $(B_e)\hat{\ }^+$ by $\hat{A}^+$.*

The goal of the Mackey analysis is now to classify the *-representations of $\mathscr{B}$ in terms of the *-representations of A and the representation theory of subgroups of G.

2.4. It will be useful to work not only with $\mathscr{B}$ but with its bundle C^*-completion.

Let $\mathscr{C} = \langle C, \pi', \cdot, * \rangle$ be the bundle C^*-completion of $\mathscr{B}$, and $\rho: B \to C$ the natural map (with dense range) constructed in VIII.16.7. Let E denote the unit fiber C^*-algebra C_e. By XI.11.5 the correspondence

$$S \mapsto R = S \circ (\rho | A) \tag{1}$$

maps the set of all non-degenerate *-representations S of E biuniquely onto the set of all non-degenerate $\mathscr{B}$-positive *-representations R of A. This correspondence preserves irreducibility and so identifies $\hat{E}$ and $\hat{A}^+$ as sets. In fact, by VII.1.18 it identifies them as topological spaces (with the regional topologies). This identification will be especially useful since in view of VII.5.11 the topology of $\hat{E}$ has the simple "hull-kernel" description.

For simplicity of notation we shall often consider the correspondence (1) as an identification, writing R_a instead of S_a when $a \in E$ and S and R correspond as in (1), and writing $\mathrm{Ker}_E(R)$ for $\{a \in E: S_a = 0\}$.

2.5. Looking back at 1.10, we foresee that, in carrying out Step II of 1.28 in our present more general context, it will be necessary to form the stability subgroup H of G for an element of $\hat{A}^+$, and then to induce certain *-representations of $\mathscr{B}_H$ up to $\mathscr{B}$. Now in general (see Example 9.17) H need not be closed. On the other hand the assumption that H was closed was basic to the definition of induction in §XI.9. To overcome this difficulty one could assume from the beginning that $\hat{A}^+$ is a T_0 space, since then by III.3.8 H would automatically be closed. However, it proves useful to proceed more generally, as follows:

We fix once for all a T_0 topological G-transformation space Z, together with a continuous map $\Phi: \hat{A}^+ \to Z$ which is G-equivariant, that is,

$$\Phi(^xD) = x\Phi(D) \tag{2}$$

$(x \in G; D \in \hat{A}^+)$.

2.6. Such pairs Z, Φ certainly exist. Indeed, the most important such Z is the primitive ideal space $\mathrm{Prim}(E)$ of E, with the hull-kernel topology (see VII.3.9). Along with it we take $\Phi: \hat{A}^+ \to \mathrm{Prim}(E)$ to be the natural surjection which sends D into $\mathrm{Ker}_E(D)$. Now by VII.5.11 the hull-kernel and regional topologies of $\hat{E}$ coincide; and so Φ identifies $\mathrm{Prim}(E)$ with the space of all regional equivalence classes of elements of $\hat{E}$ (see VII.2.2). Thus, since the action of G on $\hat{E}$ ($\cong \hat{A}^+$) is regionally continuous, it generates a continuous action of G on $\mathrm{Prim}(E)$ with respect to which Φ is equivariant. We saw in VII.3.9 that $\mathrm{Prim}(E)$ is always a T_0 space. So the pair $\mathrm{Prim}(E)$, Φ has the properties of 2.5.

Although $\mathrm{Prim}(E)$, Φ is by far the most important pair with the properties of 2.5, it is occasionally useful (as we shall see in Example 9.12) to consider more general such pairs.

2.7. Throughout the rest of our discussion of the generalized Mackey analysis, the reader will notice that it is the G-space Z, rather than $\hat{A}^+$, which will play the role that $\hat{N}$ played in §1.

By III.3.8 the stability subgroup of G for any point of the T_0 space Z is closed.

The Association of Elements of $\hat{B}$ with Orbits in Z

2.8. Now Step I of 1.28 consisted in showing that every irreducible representation T of G was associated with some orbit in $\hat{N}$. This was done by considering the direct sum decomposition of $T|N$ into irreducible parts. In our present more general context, what does it mean for a *-representation T of $\mathscr{B}$ to be "associated with" some G-orbit in Z? In general $T|A$ cannot now be decomposed as a direct sum of irreducible parts. Luckily, however, the spectral measure of $T|A$, developed in §VII.9, provides just the needed generalization of a direct sum decomposition into irreducible parts.

2.9. To see this, suppose that S is any $\mathscr{B}$-positive non-degenerate *-representation of A. Considering S as a *-representation of E (see 2.4), we can form the spectral measure Q_*^S of S as in VII.9.12. This of course is an $X(S)$-projection-valued measure on the Borel σ-field of $\hat{E}$. Let Q^S denote the image of Q_*^S under the Φ of 2.5; that is, Q^S is the $X(S)$-projection-valued measure on the Borel σ-field of Z given by:

$$Q^S(W) = Q_*^S(\Phi^{-1}(W))$$

(for each Borel subset W of Z).

Definition. We refer to Q^S as the *spectral measure of S on Z.*

Suppose that W is a *closed* subset of Z. Then there are useful equivalent descriptions of $Q^S(W)$. Let $I(W)$ be the closed two-sided ideal of E corresponding (under the mapping Γ of VII.5.13) with the closed subset $\Phi^{-1}(W)$ of $\hat{E}$; that is,

$$I(W) = \bigcap\{\mathrm{Ker}_E(R) : R \in \Phi^{-1}(W)\}. \tag{3}$$

In view of VII.9.14,

$$\text{range}(Q^S(W)) = \{\xi \in X(S): S_a\xi = 0 \text{ for all } a \text{ in } I(W)\}. \tag{4}$$

Also, by (4) and VII.5.5, $\text{range}(Q^S(W))$ can be described as the largest closed S-stable subspace Y of $X(S)$ such that the subrepresentation of S acting on Y is weakly contained in $\Phi^{-1}(W)$.

2.10. Our next immediate aim is to prove the generalization 2.11 of 1.4(1). For this one must know the relation between the spectral measure of a *-representation of A and of its conjugates.

Let S and Q^S be as in 2.9, and fix x in G. Let $F \mapsto \tilde{F}$ be the *-isomorphism between the commuting algebras $\mathscr{I}(S)$ and $\mathscr{I}(^xS)$ of S and ^xS respectively which was described in Proposition XI.16.13. We recall from VII.9.13 that $\text{range}(Q^S) \subset \mathscr{I}(S)$. Thus the following assertion makes sense:

Proposition. *For any Borel subset* W *of* Z,

$$(Q^S(W))^{\tilde{}} = Q^{(^xS)}(xW). \tag{5}$$

Proof. Evidently each side of (5) is a projection-valued measure (as a function of W). Since a projection-valued Borel measure is determined by its values on closed sets, it is enough to prove (5) assuming that W is closed.

Let R be the subrepresentation of S acting on $\text{range}(Q^S(W))$. Since W is closed, the last statement of 2.9 says that R is weakly contained in $\Phi^{-1}(W)$. Therefore by XI.16.24 ^xR is weakly contained in $^x(\Phi^{-1}(W)) = \Phi^{-1}(xW)$ (see (2)). Now it is easy to see from the construction in XI.16.13 that ^xR is equivalent to the subrepresentation of ^xS acting on $\text{range}((Q^S(W))^{\tilde{}})$. So by the last sentence of 2.9,

$$(Q^S(W))^{\tilde{}} \leq Q^{(^xS)}(xW). \tag{6}$$

To obtain the reverse inequality, let ^{0} be the *-isomorphism from $\mathscr{I}(^xS)$ onto $\mathscr{I}(^{x^{-1}}(^xS))$ constructed as in XI.16.13 with S and x replaced by ^xS and x^{-1}. When $^{x^{-1}}(^xS)$ is identified with S as in XI.16.17 (and Remark 3 of XI.16.3), $^{\tilde{}}$ and ^{0} become each other's inverses. Hence (6) applied with S, x, W replaced by ^xS, x^{-1}, xW gives just the reverse inequality of (6), proving (5). ∎

2.11. Let T be a non-degenerate *-representation of $\mathscr{B}$; and put $S = T|A$. Thus S is non-degenerate by VIII.9.4 and $\mathscr{B}$-positive by XI.8.12.

Proposition. *Let x be an element of G. For each Borel subset W of Z,* range($Q^S(xW)$) *is equal to the closed linear span of*

$$\{T_b\xi : b \in B_x, \xi \in \text{range}(Q^S(W))\}.$$

Proof. Adopting the notation of §XI.9, we verify (see Remark XI.16.7) that the equation

$$F(\kappa_x(b \otimes \xi)) = T_b\xi \quad (b \in B_x; \xi \in X(T)) \qquad (7)$$

defines a unitary equivalence F of ^xS and S. It follows that

$$F[\text{range}\{Q^{(^xS)}(xW)\}] = \text{range}(Q^S(xW)). \qquad (8)$$

Now by 2.10 range$\{Q^{(^xS)}(xW)\}$ is the closed linear span of $\{\kappa_x(b \otimes \xi) : b \in B_x,$ $\xi \in \text{range}(Q^S(W))\}$. Combining this with (7) and (8) we obtain the required result. ∎

Remark. This proposition asserts in particular that $T, Q^{T|A}$ is a system of imprimitivity for $\mathscr{B}$ over the G-space Z (in the sense of VIII.18.7, except that here Z need not be a locally compact Hausdorff space).

2.12. Keep the notation of 2.11; and suppose that W is a Borel subset of Z such that $xW = W$ for all x in G (i.e., W is a union of orbits). It follows from 2.11 that range($Q^S(W)$) is T-stable, so that $Q^S(W)$ commutes with all T_b. Since by VII.9.13 $Q^S(W)$ lies in the von Neumann algebra generated by range(T), we obtain the following important observation:

Proposition. *Let W be a Borel subset of Z such that xW = W for all x in G; and let T be any non-degenerate *-representation of $\mathscr{B}$. Then $Q^{T|A}(W)$ lies in the center of the von Neumann algebra generated by* range(T). *If in addition T is primary (see VI.24.9), then $Q^{T|A}(W)$ is either 0 or $1_{X(T)}$.*

2.13. We will now answer the question raised in 2.8: What does it mean for a *-representation of $\mathscr{B}$ to be associated with an orbit in Z?

Definition. Let P be any projection-valued measure on a σ-field $\mathscr{Y}$ of subsets of a set Y. If W is a subset of Y (not necessarily belonging to $\mathscr{Y}$), we shall say that P is *concentrated on* W if $P(V) = 0$ whenever $V \in \mathscr{Y}$ and $V \cap W = \emptyset$. (If $W \in \mathscr{Y}$, this means of course that W carries P, that is, $P(Y \setminus W) = 0$.)

As in VII.10.1, P is *concentrated at* a point y of Y if it is concentrated on $\{y\}$.

Definition. Let T be a non-degenerate *-representation of $\mathscr{B}$, and θ an orbit in Z (under the action of G). We say that T is *associated with* θ if $Q^{T|A}$ is concentrated on θ.

Remark 1. By VII.9.16 the family of all *-representations which are associated with a given θ is closed under the formation of Hilbert direct sums and subrepresentations.

Remark 2. It is easy to see that for irreducible representations of finite groups the idea of association defined above coincides with that of 1.5.

2.14. We saw in 1.5 that, in the finite group context, every irreducible representation of G is associated with some orbit in $\hat{N}$. Is it true, in our present more general situation, that every irreducible *-representation of $\mathscr{B}$ is associated with some orbit in Z? In general, as we shall see in Example 9.5, the answer is "no." There are, however, rather broad conditions under which the answer is "yes." These conditions, whatever their precise formulation, assert that the action of G on Z is "smooth" in the sense that (intuitively speaking) the orbits in Z do not get too badly tangled up with each other. In 2.16 and 2.17 we will give two conditions, each of which is sufficient to ensure a "yes" answer to the above question.

2.15. Remark. The notion of "smoothness" referred to in 2.14 is closely related to the concept of ergodicity introduced in VIII.19.3. Let us explore this relationship a little more closely. We shall say tentatively that G *acts smoothly on* Z if every (numerical) measure μ which is defined on the Borel σ-field of Z and which is ergodic with respect to the action of G on Z (see VIII.19.3) is concentrated on some orbit θ in Z (that is, $\mu(V) = 0$ for all Borel subsets V of Z such that $V \cap \theta = \emptyset$). Now, as we observed in VIII.19.3, ergodicity is a measure-theoretic version of transitivity. A G-ergodic measure μ on Z can be thought of as the measure-theoretic analogue of the notion of an orbit; roughly speaking, it describes a conglomeration of actual orbits which are so closely "tangled up together" that it cannot be split up non-trivially into G-stable Borel parts. The smoothness condition now says that the only "measure-theoretic orbits" are actual orbits—that is, an aggregate of more than one orbit can never be so "tangled" that it becomes measure-theoretically transitive.

Let us now briefly indicate the relationship between the smoothness property just defined and an affirmative answer to the question raised in 2.14. Suppose for simplicity that B is second-countable and that $Z = \mathrm{Prim}(E)$ (see

2.6). Let T be an irreducible *-representation of $\mathscr{B}$. Since B is second-countable, $X(T)$ is separable (VIII.9.6); and so it can be shown that $Q^{T|A}$ has a "separating vector"; that is, there is a unit vector ξ in $X(T)$ such that $Q^{T|A}(W)\xi = 0 \Rightarrow Q^{T|A}(W) = 0$ for all Borel subsets W of Z. The equation

$$\mu(W) = \|Q^{T|A}(W)\xi\|^2$$

now defines a numerical measure μ on the Borel σ-field of Z. By 2.12, if W is a Borel set which is a union of orbits, $\mu(W)$ is either 0 or 1. From this, together with the countability assumptions which have been made, it is possible to prove that μ is ergodic with respect to the action of G on Z. Let us now assume that G acts smoothly on Z. Then μ must be confined to a single orbit θ. Since ξ was a separating vector for $Q^{T|A}$, it follows that $Q^{T|A}$ is confined to θ, that is, T is associated with θ.

The above discussion is intended to be merely suggestive. In the numbers which follow we will bypass the notion of smoothness as defined above, and give conditions which are verifiable in concrete cases and which imply an affirmative answer to the question raised in 2.14 not only for irreducible but even for primary *-representations of $\mathscr{B}$.

2.16.   For the moment let M be any left topological G-space. The space of all orbits in M under G is denoted by M/G.

Definition.   We shall say that M/G *satisfies Condition (C)* if there exists a countable family $\mathscr{W}$ of Borel subsets of M such that (i) each W in $\mathscr{W}$ is G-stable (i.e., is a union of orbits), and (ii) $\mathscr{W}$ separates the points of M/G (i.e., given any two distinct G-orbits θ_1 and θ_2, there is a set W in $\mathscr{W}$ which contains one of θ_1 and θ_2 and is disjoint from the other).

Remark.   Let $q: M \to M/G$ be the quotient map sending each point m of M into its orbit Gm. The orbit space M/G carries the natural quotient topology $\mathscr{T}$ derived from M and q; let $\mathscr{S}_1$ be the σ-field of all $\mathscr{T}$-Borel subsets of M/G. The space M/G also carries another σ-field $\mathscr{S}_2$ directly derived from the Borel σ-field of M:

$$S \in \mathscr{S}_2 \Leftrightarrow q^{-1}(S) \quad \text{is a Borel subset of } M.$$

We have $\mathscr{S}_1 \subset \mathscr{S}_2$; but in general $\mathscr{S}_1 \neq \mathscr{S}_2$. Condition (C) asserts that some countable subfamily of $\mathscr{S}_2$ separates the points of M/G. This is weaker than the assertion that some countable subfamily of $\mathscr{S}_1$ separates the points of M/G. In particular, Condition (C) holds if $\mathscr{T}$ is T_0 and second-countable.

Returning to the topological G-space Z, we have:

Proposition. *Suppose that Z/G satisfies Condition (C). Then every primary*
*-representation T of $\mathscr{B}$ is associated with a unique orbit in Z.

Proof. The proof is very similar to that of VII.10.5.

Let $\mathscr{W}$ be the countable family whose existence is assumed in the definition
of Condition (C). Adjoining the complements of sets in $\mathscr{W}$, we may assume
without loss of generality that

$$W \in \mathscr{W} \Rightarrow Z \setminus W \in \mathscr{W}. \tag{9}$$

Let us write Q for $Q^{T|A}$. By 2.12,

$$Q(W) = 0 \text{ or } 1 \quad \text{whenever } W \in \mathscr{W}. \tag{10}$$

Put $\mathscr{W}_1 = \{W \in \mathscr{W} : Q(W) = 1\}$ and $W_1 = \bigcap \mathscr{W}_1$. Since $\mathscr{W}_1$ is countable, the
intersection property of projection-valued measures gives

$$Q(W_1) = 1. \tag{11}$$

Now we claim that W_1 is an orbit. Indeed: W_1 is certainly a union of orbits
(since each W in $\mathscr{W}$ is). Suppose that W_1 contains two distinct orbits θ and θ'.
By Condition (C) and (9) there is an element W_2 of $\mathscr{W}$ containing θ but not θ'.
By (10) either $Q(W_2) = 1$ or $Q(Z \setminus W_2) = 1$; that is, either W_2 or $Z \setminus W_2$
belongs to $\mathscr{W}_1$. In either case W_1 cannot contain both θ and θ'. So W_1 is either
void or a single orbit. It cannot be void by (11); so it is a single orbit, proving
the claim.

Now that W_1 is known to be an orbit, (11) asserts that T is associated with
W_1.

Notice that every orbit θ in Z is a Borel set. Indeed, by Condition (C) and
(9) θ is the intersection of the countable family $\{W \in \mathscr{W} : \theta \subset W\}$ of Borel sets,
and so is itself Borel. This fact implies that the orbit W_1 with which T is
associated is unique. ■

2.17. The hypothesis of Proposition 2.16 will clearly be applicable mostly to
separable situations. The hypothesis of the next result has no ring of
separability about it.

Proposition. *Consider the orbit space Z/G, consisting of all the G-orbits in Z*
and equipped with the quotient topology. If Z/G is almost Hausdorff (in the
sense of VII.10.6), then every primary *-representation T of $\mathscr{B}$ is associated
with some unique orbit in Z.

Proof. Put $Q = Q^{T|A}$. Let $\Psi : Z \to Z/G$ be the quotient mapping; and let P
be the image of Q under Ψ, that is,

$$P(W) = Q(\Psi^{-1}(W))$$

for all Borel subsets W of Z/G. From VII.9.11 and the continuity of Φ and Ψ it follows that

$$P(\bigcup \mathscr{W}) = \sup\{P(W): W \in \mathscr{W}\} \tag{12}$$

for any collection $\mathscr{W}$ of open subsets of Z/G. For any Borel subset W of Z/G, $\Psi^{-1}(W)$ is a union of orbits in Z, and therefore by Proposition 2.12

$$P(W) \text{ is either } 0 \text{ or } 1. \tag{13}$$

In view of (12) and (13), the same proof that we gave for Proposition VII.9.6 now shows that there is a point θ in Z/G such that $\{\theta\}$ is a Borel set and $P(\{\theta\}) = 1$. This of course implies that θ is a Borel subset of Z and $Q(\theta) = 1$. So T is associated with θ.

We leave it to the reader to verify that any one-element subset of an almost Hausdorff space is a Borel set. It follows that (as in 2.16) every orbit in Z is a Borel set. Therefore the orbit θ with which T is associated is unique.   ∎

2.18.   Remark.   In the absence of the hypotheses of 2.16 or 2.17 we do not know whether or not it is possible for a *-representation T of $\mathscr{B}$ to be associated with two distinct orbits in Z.

2.19.   Remark.   If G and E are second-countable and $Z = \mathrm{Prim}(E)$, it can be shown that Z/G satisfies Condition (C) if and only if Z/G is almost Hausdorff. See Glimm [4]. In general, however, the relation between them is obscure.

We shall see in Example 9.18 that the conclusion of Propositions 2.16 and 2.17 sometimes holds even though Condition (C) fails and Z/G is not almost Hausdorff.

2.20.   Even if the hypotheses of Propositions 2.16 and 2.17 fail, it may still be possible to obtain weaker information about the primary representations of $\mathscr{B}$ by passing to a new space $\tilde{Z}$.

Suppose, for instance, that we are able to find a new T_0 topological space $\tilde{Z}$ and a continuous map $\Psi: Z \to \tilde{Z}$ such that

$$\Psi(xz) = \Psi(z) \qquad \text{for all } z \text{ in } Z \text{ and } x \text{ in } G$$

(that is, Ψ is constant on each orbit in Z). Then Propositions 2.16 and 2.17 can be applied to the pair $\tilde{Z}, \Psi \circ \Phi$, and we obtain:

Proposition.   *In the above context suppose that either $\tilde{Z}$ is almost Hausdorff or there exists a countable family of Borel subsets of $\tilde{Z}$ which separates points of $\tilde{Z}$. Then, for each primary *-representation T of $\mathscr{B}$, there is a point $\tilde{z}$ of $\tilde{Z}$ such that $Q^{T|A}$ is carried by the Borel subset $\Psi^{-1}(\tilde{z})$ of Z.*

2.21. *Remark.* Under the hypotheses of the last proposition, each primary *-representation T of $\mathscr{B}$ will have the property that the spectral measure $Q_*^{T|A}$ is carried by a Borel subset of $\hat{E}$ of the form $W = (\Psi \circ \Phi)^{-1}(\tilde{z})$, where $\tilde{z}$ is some point of $\tilde{Z}$. If the set W is closed, so that (by VII.5.13) it corresponds to a closed two-sided ideal I of E, the classification of these T becomes just the classification of the primary *-representations of a smaller bundle obtained by factoring out I. For brevity we formulate the precise statement of this fact only for semidirect product bundles, leaving its proof (and generalization) to the reader.

Proposition*. *Let $\mathscr{B} = D \underset{\tau}{\times} G$ be the bundle τ-semidirect product of the locally compact group G with a C^*-algebra D; and let I be a closed τ-stable two-sided ideal of D. Thus τ generates a strongly continuous action $\tilde{\tau}$ of G on $\tilde{D} = D/I$; and we can form the $\tilde{\tau}$-semidirect product $\tilde{\mathscr{B}} = \tilde{D} \underset{\tau}{\times} G$. The family of all *-representations of $\tilde{\mathscr{B}}$ is in natural one-to-one correspondence with the family of all those *-representations T of $\mathscr{B}$ such that $I \subset \mathrm{Ker}(T|D)$.*

2.22. *Remark.* If the action of G on Z is "non-smooth," the Mackey analysis to be developed in §§3 and 4 will in general describe only *certain* of the primary *-representations of $\mathscr{B}$ (namely those which happen to be associated with some orbit in Z), but by no means all of them.

Actually, using the notion of a virtual subgroup, the Mackey analysis can be generalized so that it classifies *all* primary *-representations of $\mathscr{B}$ even when the action of G on Z is "non-smooth." But this generalization is far less satisfactory than the smooth theory, in that it merely reduces the problem of classifying $\tilde{\mathscr{B}}$ to another problem which appears hopeless. Indeed, there is reason to suppose that in general, when the action of G on Z is not smooth, the space $\tilde{\mathscr{B}}$ is essentially unclassifiable in some metamathematical sense. At any rate, the theory of virtual subgroups is beyond the scope of this work. (For information on virtual groups see, for example, Mackey [10, 13, 21] and Ramsay [1, 2, 3, 4].)

3. The Mackey Analysis in the General Case; Step II

3.1. In this section we shall generalize Step II of 1.28 to the context of bundles over locally compact groups. We keep all the notation and conventions of 2.1–2.5 and 2.9.

Let z be a fixed element of Z and $\theta = Gz$ the orbit in Z containing z; and let H be the stability subgroup of G for z. By III.3.8 H is closed.

3.2. In the finite group context of §1 (where $Z = \hat{N}$), Step II consisted in showing that the inducing map $S \mapsto T = \mathrm{Ind}_{H \uparrow G}(S)$ sets up a one-to-one correspondence between the family of all elements S of $\hat{H}$ such that $S|N$ is a multiple of z (i.e., the spectral measure of $S|N$ is carried by $\{z\}$) and the family of all those elements T of $\hat{G}$ which are associated with θ (i.e., the spectral measure of $T|N$ is carried by θ). Analogously, it is reasonable to hope that, in the general case, *the inducing map* $S \to T = \mathrm{Ind}_{\mathscr{B}_H \uparrow \mathscr{B}}(S)$ *will set up a one-to-one correspondence between the family of all elements S of* $(\mathscr{B}_H)\hat{}$ *such that* $Q^{S|A}$ *is concentrated at z, and the family of all elements T of $\hat{\mathscr{B}}$ which are associated with θ.* This conjecture is false in general (see Example 9.4); but we shall show that it is true under the added general hypothesis that G/H is σ-compact. In particular it is true whenever G is second-countable.

3.3. We begin with two general facts about topologies on coset spaces. From here to 3.6 H can be any closed subgroup of G. We write M for G/H.

Lemma. *Let P be any regular projection-valued Borel measure on M; and let $\mathscr{T}$ be a T_0 topology on M, which is contained in the natural topology, and such that the action $\langle x, \alpha \rangle \mapsto x\alpha$ of G on M is continuous with respect to $\mathscr{T}$. Then* range(P) *is contained in the von Neumann algebra generated by* $\{P(W): W$ is $\mathscr{T}$-open$\}$.

Proof. Let P act on the Hilbert space X. Words referring to the topology of M mean the natural topology unless $\mathscr{T}$ is explicitly mentioned.

By VI.24.2 it suffices to show that range(P) commutes with the commuter of $\{P(W): W$ is $\mathscr{T}$-open$\}$. So it is enough to take a self-adjoint element q of $\mathcal{O}(X)$ which commutes with $P(W)$ for all $\mathscr{T}$-open sets W, and to show that $qP(V) = P(V)q$, in other words

$$P(M \setminus V)qP(V) = 0, \tag{1}$$

for all Borel subsets V of M. Since P is regular, (1) will be proved if we show that

$$P(C)qP(D) = 0 \tag{2}$$

whenever C, D are disjoint compact subsets of M.

To prove (2) it is enough, we claim, to take any two distinct points α, β of M and find disjoint neighborhoods U and V of α and β respectively such that

$$P(U)qP(V) = 0. \tag{3}$$

Indeed, suppose we can achieve (3). Then, given a fixed point β of D, by the compactness of C we can cover C by finitely many open sets $U_1, \ldots, U_r$, and find corresponding open neighborhoods $V_1, \ldots, V_r$ of β, such that $P(U_i)qP(V_i) = 0$ for each i. This implies that $P(C)qP(V) = 0$, where $V = \bigcap_i V_i$. Since V is an open neighborhood of β, another similar argument, making β variable and using the compactness of D, shows that (2) holds. So the claim is proved.

Now take two distinct points α, β of M. Since $\mathcal{T}$ is T_0, we can find a $\mathcal{T}$-closed set S which separates α and β; say $\alpha \notin S$, $\beta \in S$. Now S is also closed in the natural topology of M; so there is a compact neighborhood N of e such that $N^{-1}N\alpha \cap S = \emptyset$. Put $U = N\alpha$, $V = N\beta$. Thus

$$U \cap NS = \emptyset, \qquad V \subset NS. \tag{4}$$

Since N is compact and G acts continuously with respect to $\mathcal{T}$, NS is $\mathcal{T}$-closed (by the same argument as in III.1.6). So by hypothesis $P(M \setminus NS)qP(NS) = 0$. Combining this with (4) we get (3). Since U and V are neighborhoods of α and β, the earlier claim shows that the lemma is proved. ∎

3.4. The next lemma is similar to 3.3; but it has a stronger hypothesis and a stronger conclusion. Let H and M be as in 3.3.

Lemma. *Assume that M is σ-compact. As in 3.3 let $\mathcal{T}$ be a T_0 topology on M which is contained in the natural topology of M and such that the action of G on M is continuous with respect to $\mathcal{T}$. Then every compact G_δ subset C of M belongs to the Borel field Γ generated by $\mathcal{T}$.*

Remark. By a G_δ set we mean an intersection of countably many open sets. Topological words refer to the natural topology of M unless $\mathcal{T}$ is explicitly mentioned. Recall that a subset of M has the *Lindelöf property* if every open covering of it has a countable subcovering.

Proof. Let $C = \bigcap_{n=1}^{\infty} U_n$, where C is a compact subset of M and the U_n are open in M. Since $M \setminus U_n$ is closed, it is σ-compact (because M is). Hence $M \setminus C$ is σ-compact, and so has the Lindelöf property. Now if $\alpha \in C$ and $\beta \in M \setminus C$, then as in the proof of 3.3 we can find open neighborhoods $U_{\alpha\beta}$ and $V_{\alpha\beta}$ of α and β respectively and a subset $S_{\alpha\beta}$ of M such that $U_{\alpha\beta} \subset S_{\alpha\beta}$, $V_{\alpha\beta} \subset M \setminus S_{\alpha\beta}$, and either $S_{\alpha\beta}$ or $M \setminus S_{\alpha\beta}$ is $\mathcal{T}$-open. Fix $\beta \in M \setminus C$. Since C is compact, we can pick finitely many $\alpha_1, \ldots, \alpha_n$ so that $C \subset \bigcup_{i=1}^{n} U_{\alpha_i\beta}$. Set $S_\beta = \bigcup_{i=1}^{n} S_{\alpha_i\beta}$, $V_\beta = \bigcap_{i=1}^{n} V_{\alpha_i\beta}$. Then $S_\beta \in \Gamma$, $C \subset S_\beta$, $V_\beta \subset M \setminus S_\beta$, and V_β is an open neighborhood of β (disjoint from C). Therefore, since $M \setminus C$ has the Lindelöf property, we can find a sequence $\beta_1, \beta_2, \ldots$ of elements of $M \setminus C$

such that $\bigcup_{i=1}^{\infty} V_{\beta_i} = M \setminus C$. If we then set $S = \bigcap_{i=1}^{\infty} S_{\beta_i}$, we conclude that $S \in \Gamma$ and $S = C$. So $C \in \Gamma$. Since C was an arbitrary compact G_δ set, the lemma is proved. ∎

3.5.   Next we need a general fact about induced representations.

Let v be a left Haar measure and δ the modular function of the closed subgroup H of G; and take a $\mathcal{B}$-positive non-degenerate *-representation S of $\mathcal{B}_H$, acting on X. By XI.8.12 Ind$(S)|A$ is $\mathcal{B}$-positive; so we can ask what $(\text{Ind}(S)|A)_a$ looks like when $a \in E$ (see 2.4).

Let $\mathcal{Y} = \{Y_\alpha\}$ be the Hilbert bundle over G/H induced by S, and $\kappa_\alpha : \sum_{x \in \alpha}^\oplus B_x \otimes X \to Y_\alpha \, (\alpha \in G/H)$ the quotient maps, as in XI.9.5. We claim that, if $x \in G$ and $\alpha = xH$, then

$$\kappa_\alpha(B_x \otimes X) \qquad \text{is dense in } Y_\alpha. \tag{5}$$

Indeed: We recall from XI.9.15(28) that

$$\kappa_\alpha(bc \otimes \xi) = \kappa_\alpha(b \otimes \Delta(h)^{1/2}\delta(h)^{-1/2}S_c\xi)$$

whenever $b \in B_x$, $h \in H$, $c \in B_h$, and $\xi \in X$. It follows from this that

$$\kappa_\alpha(B_x \otimes X) \supset \kappa_\alpha([B_x B_h] \otimes X) \tag{6}$$

for all h in H ([ ] denoting linear spans). Now by XI.9.19 $\kappa_\alpha(d \otimes \xi)$ is continuous in d (for $d \in \mathcal{B}_\alpha$). So (6) and the saturation of $\mathcal{B}$ imply that the closure of $\kappa_\alpha(B_x \otimes X)$ contains $\kappa_\alpha(B_y \otimes X)$ for all y in α. This establishes (5).

If $b \in B_x$ and $\alpha \in G/H$ let $\tau_b^{(\alpha)} : Y_\alpha \to Y_{x\alpha}$ be the map defined in XI.9.11. From (5) and the definition of conjugate representations in XI.16.3, we obtain:

Proposition.   *Let $x \in \alpha \in G/H$. Then*

$$t^{(\alpha)} : a \mapsto \tau_a^{(\alpha)} \qquad\qquad (a \in A)$$

*is a *-representation of A on Y_α which is equivalent to the conjugate representation ${}^x(S|A)$. In particular $t^{(\alpha)}$ is itself $\mathcal{B}$-positive.*

The last statement of the proposition follows from XI.16.17.

3.6.   In view of 2.4 and the last proposition, $t_a^{(\alpha)}$ is a well-defined operator on Y_α for each a in E (not merely for each a in A).

Proposition.   *Let $T = \text{Ind}_{\mathcal{B}_H \uparrow \mathcal{B}}(S)$. For each a in E and each f in the space of T,*

$$[(T|A)_a f](\alpha) = t_a^{(\alpha)}(f(\alpha)) \qquad \text{for almost all } \alpha. \tag{7}$$

Remark.   "Almost all α" refers to the unique G-invariant measure class on G/H; see §III.14.

Proof. By the definition of $\mathrm{Ind}(S)$, (7) holds for $a \in A$. Now let $a \in E$; and choose a sequence $\{a_n\}$ of elements of A with $\rho(a_n) \to a$ in E. Then

$$(T \mid A)_{a_n} f \to (T \mid A)_a f \qquad \text{in } X(T) = \mathscr{L}_2(\mathscr{Y}). \tag{8}$$

By (8) and II.3.5 one can pass to a subsequence and assume that

$$((T \mid A)_{a_n} f)(\alpha) \to ((T \mid A)_a f)(\alpha) \tag{9}$$

for almost all α. Since (7) holds for the a_n, and since $\|t_{a_n}^{(\alpha)} - t_a^{(\alpha)}\| \to 0$ for each α, (9) implies that (7) holds for a. ∎

3.7. We now return to the point z of Z. As in 3.1 H is its (closed) stability subgroup $\{x \in G : xz = z\}$. Let $\gamma : G/H \to Z$ be the continuous injection given by $\gamma(xH) = xz$ $(x \in G)$.

Let S be a non-degenerate $\mathscr{B}$-positive *-representation of $\mathscr{B}_H$. We can then form the system of imprimitivity T, P over G/H induced by S (see XI.14.3). As in 2.9 $Q^{T \mid A}$ is the spectral measure of $T \mid A$ on Z.

Proposition. *Assume that $Q^{S \mid A}$ is concentrated at z. Then*

$$Q^{T \mid A}(W) = P(\gamma^{-1}(W)) \tag{10}$$

for every Borel subset W of Z.

Remark. This proposition is the generalized analogue of statement 1.9(9), which, as the reader may recall, was crucial in its own context. The hypothesis that $Q^{S \mid A}$ is concentrated at z is the generalized version of 1.8(8).

Proof. Evidently both sides of (10) are projection-valued Borel measures (as functions of W). Since a projection-valued Borel measure is determined by its values on closed sets (II.11.11), it is enough to prove (10) assuming that W is a closed subset of Z.

Let $\mathscr{X}$ be the space of the induced representation T. In view of 2.9(3), (4), the equality (10) will be proved provided we can show that an element f of $\mathscr{X}$ vanishes almost everywhere outside $\gamma^{-1}(W)$ if and only if $(T \mid A)_a f = 0$ for all a in the $I(W)$ of 2.9(3). On the other hand, by 3.6, $((T \mid A)_a f)(\alpha) = t_a^{(\alpha)}(f(\alpha))$. Consequently, given $f \in \mathscr{X}$ and a closed subset W of Z, we must show that

$$f \text{ vanishes almost everywhere outside } \gamma^{-1}(W) \tag{11}$$

if and only if

$$\text{for each } a \text{ in } I(W), \ t_a^{(\alpha)}(f(\alpha)) = 0 \qquad \text{for almost all } \alpha \text{ in } G/H. \tag{12}$$

Assume (11). To prove (12) it is enough to show that $t_a^{(\alpha)} = 0$ for $a \in I(W)$ and $\alpha \in \gamma^{-1}(W)$, that is (by 3.5),

$$I(W) \subset \operatorname{Ker}_E[^x(S|A)] \qquad \text{for } x \in \alpha \in \gamma^{-1}(W). \tag{13}$$

Thus, let $\alpha = xH \in \gamma^{-1}(W)$, so that $xz \in W$, or $z \in x^{-1}W$. Since $Q^{S|A}$ is concentrated at z, this means that $Q^{S|A}(x^{-1}W) = 1$, whence by 2.10

$$Q^{x(S|A)}(W) = 1.$$

By 2.9(4) this implies (13); so (12) holds.

Conversely, suppose (11) fails. By Lusin's Theorem II.15.5 we can find a compact subset C of G/H, disjoint from $\gamma^{-1}(W)$ and of positive measure, such that f is continuous and never zero on C. Pick a point xH in C with the property that $C \cap U$ is of positive measure for every neighborhood U of xH. Since $xH \notin \gamma^{-1}(W)$, we have $xz \notin W$, or $z \notin x^{-1}W$. Since $Q^{S|A}$ is concentrated at z this says that $Q^{S|A}(x^{-1}W) = 0$, whence by 2.10

$$Q^{x(S|A)}(W) = 0. \tag{14}$$

But $^x(S|A) \cong t^{(xH)}$ (by 3.5); so (14) asserts that no non-zero vector in Y_{xH} (the fiber over xH of the induced Hilbert bundle $\mathscr{Y}$) is annihilated by $t_a^{(xH)}$ for all a in $I(W)$. In particular we can choose a in $I(W)$ so that

$$t_a^{(xH)}(f(xH)) \neq 0. \tag{15}$$

Since f is continuous on C, and since by XI.9.18 $\langle \alpha, \zeta \rangle \mapsto t_a^{(\alpha)}\zeta$ is continuous on $\{\langle \alpha, \zeta \rangle : \zeta \in Y_\alpha\}$, (15) implies that there is a neighborhood U of xH such that

$$t_a^{(\alpha)}(f(\alpha)) \neq 0 \qquad \text{for all } \alpha \text{ in } C \cap U. \tag{16}$$

But $C \cap U$ has positive measure (by the choice of xH). So (16) implies that (12) fails. Thus (12) $\Rightarrow$ (11).

We have now shown that (11) $\Leftrightarrow$ (12); and this, as was observed earlier, proves the proposition. ∎

3.8.   Proposition.   *Keep the notation and hypotheses of Proposition 3.7. Then:*

(I)   *T is associated with the orbit Gz.*

(II)   *The commuting algebras of S and of T are *-isomorphic. In particular T is irreducible [primary] if and only if S is.*

Proof.   Since range$(\gamma) = Gz$, assertion (I) follows immediately from (10) and the definition in 2.13.

To prove (II), recall from XI.14.25 that the commuting algebra of S is *-isomorphic with that of the system of imprimitivity T, P. So (II) will be established if we show that the commuting algebra of T, P coincides with that of T. This is clearly the same as showing that

$$\text{range}(P) \subset \mathscr{A}, \tag{17}$$

where $\mathscr{A}$ is the von Neumann algebra generated by $\text{range}(T)$.

Let $\mathscr{T}$ be that topology of G/H under which $\gamma: G/H \to Z$ is a homeomorphism. Since γ is continuous, $\mathscr{T}$ is weaker than the natural topology; and $\mathscr{T}$ is T_0 since Z is. Since G acts continuously on Z, its action on G/H is continuous with respect to $\mathscr{T}$. Therefore, by 3.3, (17) will be proved if we show that

$$P(U) \in \mathscr{A} \qquad \text{for all } \mathscr{T}\text{-open sets } U. \tag{18}$$

Now $\mathscr{T}$-open sets are those of the form $\gamma^{-1}(W)$, where W is open in Z; and for these we have (10). On the other hand, $Q^{T|A}(W) \in \mathscr{A}$ by VII.9.13. So (18) holds; and (II) is proved. ∎

3.9. **Proposition.** *Suppose in 3.7 that S and S' are two non-degenerate $\mathscr{B}$-positive *-representations of $\mathscr{B}_H$ such that both $Q^{S|A}$ and $Q^{S'|A}$ are concentrated at z. Then $\text{Ind}_{\mathscr{B}_H \uparrow \mathscr{B}}(S') \cong \text{Ind}_{\mathscr{B}_H \uparrow \mathscr{B}}(S)$ if and only if $S' \cong S$.*

Proof. Let $S'' = S \oplus S'$; and let T, P and T', P' and T'', P'' be the systems of imprimitivity over G/H induced by S and S' and S'' respectively (so that $T = \text{Ind}(S)$, $T' = \text{Ind}(S')$). Let q and q' be the projections of the space of T'', P'' onto the spaces of T, P and T', P' respectively (see XI.12.2). Now assume $T \cong T'$. This says that there is an element u of the commuting algebra of T'' satisfying $u^*u = q$, $uu^* = q'$. Now since $Q^{S''|A}$ is concentrated at z by VII.9.16, the proof of 3.8 shows that the commuting algebras of T'' and of T'', P'' are the same. So u lies in the commuting algebra of T'', P'', and is thus an equivalence between T, P and T', P'. The uniqueness clause in the Imprimitivity Theorem XI.14.18 therefore asserts that $S' \cong S$. We have proved the "only if" statement. The "if" statement is trivial. Therefore the proof is finished. ∎

3.10. Now 3.8 says that, from every non-degenerate *-representation S of $\mathscr{B}_H$ whose restriction to A is "essentially confined to z" (i.e., $Q^{S|A}$ is concentrated at z) we obtain by the inducing operation a *-representation T of $\mathscr{B}$ which is associated with the orbit Gz. By 3.9 the map $S \mapsto T$ is one-to-one. What is the range of this map? Is every non-degenerate *-representation of $\mathscr{B}$ which is associated with Gz induced from such an S?

To answer this affirmatively, one has to impose some extra general hypothesis. It turns out that σ-compactness of G/H will do. Another sufficient hypothesis is that the injection $\gamma: G/H \to Z$ is a homeomorphism.

3.11. Proposition. *Let T be any non-degenerate *-representation of $\mathscr{B}$ which is associated with Gz. Assume that either G/H is σ-compact or the injection $\gamma: G/H \to Z$ is a homeomorphism. Then there exists a non-degenerate $\mathscr{B}$-positive *-representation S of $\mathscr{B}_H$ such that (i) $Q^{S|A}$ is concentrated at z, and (ii) $T \cong \mathrm{Ind}_{\mathscr{B}_H \uparrow \mathscr{B}}(S)$.*

Proof. We shall first assume that G/H is σ-compact.

Put $M = G/H$. Let $\mathscr{T}$ be the topology on M making γ a homeomorphism, and Γ the Borel σ-field generated by $\mathscr{T}$. As we saw in the proof of 3.8, the hypotheses on $\mathscr{T}$ needed in 3.4 hold; and in addition M is σ-compact. So by 3.4 every compact G_δ set is in Γ.

Since $Q^{T|A}$ is concentrated on Gz, there is an $X(T)$-projection-valued measure Q_0 on Γ satisfying

$$Q_0(\gamma^{-1}(W)) = Q^{T|A}(W) \tag{19}$$

for all Borel subsets W of Z. Notice that every function in $\mathscr{C}_0(M)$ is Γ-measurable; this follows from the fact that, if $k > 0$ and $0 \le f \in \mathscr{C}_0(M)$, then $\{\alpha \in M: f(\alpha) \ge k\}$ is a compact G_δ and so belongs to Γ. Therefore the spectral integral

$$Q_1(f) = \int_M f \, dQ_0 \qquad (f \in \mathscr{C}_0(M)) \tag{20}$$

defines a *-representation Q_1 of $\mathscr{C}_0(M)$ (see II.11.8). We claim that Q_1 is non-degenerate. Indeed, since M is σ-compact, there exists an increasing *sequence* $\{f_n\}$ of non-negative elements of $\mathscr{C}_0(M)$ such that $f_n(\alpha) \uparrow 1$ for all α in M. It then follows from II.11.8(VII) and the Lebesgue Dominated Convergence Theorem that $Q_1(f_n) \to \int_M dQ_0 = 1$ weakly, proving the claim. So by the Generalized Riesz Theorem II.12.8 there is a regular $X(T)$-projection-valued *Borel* measure Q_2 on M such that

$$Q_1(f) = \int_M f \, dQ_2 \qquad (f \in \mathscr{C}_0(M)). \tag{21}$$

Notice that

$$Q_0(W) = Q_2(W) \qquad \text{for compact } G_\delta \text{ sets } W. \tag{22}$$

Indeed: Suppose W is a compact G_δ subset of M. Then there is a decreasing sequence $\{g_n\}$ of non-negative elements of $\mathscr{L}(M)$ such that $g_n(\alpha) \downarrow \mathrm{Ch}_W(\alpha)$ for all α in M. Thus, applying the Lebesgue Dominated Convergence Theorem to $\int g_n \, dQ_0$ and $\int g_n \, dQ_2$, and using (20) and (21), we obtain (22).

Next, we claim that T, Q_2 is a system of imprimitivity over M. Indeed, to see this it is enough by (21) and VIII.18.8 to show that

$$T_b Q_1(f) = Q_1(xf)T_b \qquad (x \in G; b \in B_x; f \in \mathscr{C}_0(M)).$$

By (20), this in turn will be the case if

$$T_b Q_0(V) = Q_0(xV)T_b \qquad (x \in G; b \in B_x)$$

for each V in Γ; that is, by (19), if

$$T_b Q^{T|A}(W) = Q^{T|A}(xW)T_b \qquad (x \in G; b \in B_x) \qquad (23)$$

for all Borel subsets W of Z. But (23) has already been observed in 2.11. So the claim is true.

Therefore by the Imprimitivity Theorem (XI.14.18) there is a non-degenerative $\mathscr{B}$-positive *-representation S of $\mathscr{B}_H$ such that T, Q_2 can be taken to coincide with the system of imprimitivity induced by S. To prove the proposition it remains only to show that

$$Q^{S|A} \text{ is concentrated at } z. \qquad (24)$$

To prove (24) it is enough to establish the following two statements (I), (II):

(I) If W is a closed subset of Z and $z \in W$, then $Q^{S|A}(W) = 1$;

(II) If W is a closed subset of Z and $z \notin W$, then $Q^{S|A}(W) = 0$.

To see that (I) and (II) imply (24), let $\mathscr{F}$ be the family of all those Borel subsets W of Z such that *either* $z \in W$ and $Q^{S|A}(W) = 1$ *or* $z \notin W$ and $Q^{S|A}(W) = 0$. Clearly $\mathscr{F}$ is a σ-field. If (I) and (II) hold then all closed sets will be in $\mathscr{F}$, whence $\mathscr{F}$ will be the entire Borel field, proving (24).

Let us first prove (I). Take a closed subset W of Z with $z \in W$. Notice that, when N runs over the compact neighborhoods of e in G, the NW are closed subsets of Z whose intersection is W. Therefore, by VII.9.11, (I) will be proved if we show that $Q^{S|A}(NW) = 1$ for each such N. Fix a compact neighborhood N of e in G.

As usual, let $\mathscr{Y} = \{Y_\alpha\}$ be the Hilbert bundle over M induced by S. We denote by σ the quotient map $G \to M$. Choose a compact G_δ neighborhood C of $\sigma(e)$ contained in $\sigma(N)$; and let h be an element of $\mathscr{L}(\mathscr{Y})$ vanishing outside C. Since Q_2 is the second member of the system of imprimitivity induced by S, we have $h \in \text{range}(Q_2(C)) = \text{range}(Q_0(C))$ (by (22)). Now $C \subset \sigma(N) \subset$

$\gamma^{-1}(Nz) \subset \gamma^{-1}(NW)$ (since $z \in W$). So $Q_0(C) \le Q_0(\gamma^{-1}(NW)) = Q^{T|A}(NW)$ (by (19)), whence

$$h \in \text{range}(Q^{T|A}(NW)).   \tag{25}$$

Let $a \in I(NW)$ (see 2.9(3)). It follows from (25) that $(T|A)_a h = 0$. By 3.6 this says that

$$t_a^{(\alpha)}(h(\alpha)) = 0 \qquad \text{for almost all } \alpha \text{ in } M.   \tag{26}$$

But by the continuity of h (and also XI.9.18), the left side of (26) is continuous in α. So (26) implies that $t_a^{(\alpha)}(h(\alpha)) = 0$ for *all* α; in particular,

$$t_a^{\sigma(e)}(h(\sigma(e))) = 0.   \tag{27}$$

Now $\mathscr{Y}$ has enough continuous cross-sections. So, as h runs over its possible values, $h(\sigma(e))$ runs over all of $Y_{\sigma(e)}$. So it follows from (27) that $t_a^{\sigma(e)} = 0$, or (by 3.5) $(S|A)_a = 0$. Since a was an arbitrary element of $I(NW)$, this implies that $Q^{S|A}(NW) = 1$. As we have already noticed, this proves (I).

We must now prove (II). Take a closed subset W of Z with $z \notin W$; and fix an element ξ of range($Q^{S|A}(W)$). It must be shown that $\xi = 0$.

Since $\sigma(e) \notin \gamma^{-1}(W)$, we may choose a compact neighborhood N of e so that $\sigma(N^{-1}N) \cap \gamma^{-1}(W) = \emptyset$, whence

$$\sigma(N) \cap \gamma^{-1}(NW) = \emptyset.   \tag{28}$$

Now choose a compact G_δ neighborhood C of $\sigma(e)$ in M such that $C \subset \sigma(N)$. Let ϕ be any element of $\mathscr{L}(\mathscr{B})$ with support contained in $N \cap \sigma^{-1}(C)$; and denote by g the element of $\mathscr{L}(\mathscr{Y})$ given by:

$$g(\alpha) = \kappa_\alpha((\phi|\alpha) \otimes \xi) \qquad (\alpha \in M)   \tag{29}$$

(see XI.9.2). Thus supp(g) $\subset C$; and, since Q_2 is the second member of the system of imprimitivity induced by S, we have by (22):

$$g \in \text{range}(Q_2(C)) = \text{range}(Q_0(C)).   \tag{30}$$

Since $C \subset \sigma(N) \subset \gamma^{-1}(Z \setminus NW)$ (by (28)), we see from (19) and (30) that

$$g \in \text{range}(Q^{T|A}(Z \setminus NW)).   \tag{31}$$

On the other hand, take any element a of $I(NW)$ (see 2.9(3)). We claim that

$$(T|A)_a g = 0.   \tag{32}$$

Indeed: (29) implies by XI.9.5(8) that, for $x \in G$,

$$g(xH) = \int_H \kappa_{xH}(\phi(xh) \otimes \xi) \, dvh.   \tag{33}$$

By 2.10 and the definition of ξ, we obtain

$$\kappa_{xH}(\phi(xh) \otimes \xi) \in \text{range}(Q^{xh(S|A)}(xhW). \tag{34}$$

Suppose now that $xh \in N$ ($x \in G$; $h \in H$). Then, since $a \in I(NW) \subset I(xhW)$, we have by (34)

$$t_a^{(xH)}(\kappa_{xH}(\phi(xh) \otimes \xi)) = {}^{xh}(S|A)_a(\kappa_{xH}(\phi(xh) \otimes \xi)) = 0. \tag{35}$$

If however $xh \notin N$, then $\phi(xh) = 0$ (since $\text{supp}(\phi) \subset N$); so again the left side of (35) is 0. Thus the left side of (35) is 0 for all x and h, implying by (33) and 3.6 that (32) holds. So the claim is proved.

By the arbitrariness in a, (32) implies that $g \in \text{range}(Q^{T|A}(NW))$. Combining this with (31) we conclude that $g = 0$. By the continuity of g this means that $g(eH) = 0$, that is,

$$\kappa_{eH}((\phi|H) \otimes \xi) = 0 \tag{36}$$

for all ϕ of the form we are considering. Making the identification XI.9.4(6), we can write (36) in the form

$$S_{(\phi|H)'}\xi = 0. \tag{37}$$

Now by II.14.8 any element ψ of $\mathcal{L}(\mathcal{B}_H)$ with compact support contained in a sufficiently small H-neighborhood of e is of the form $(\phi|H)'$. From this and (37) we see that

$$S_\psi \xi = 0 \qquad \text{for all such } \psi. \tag{38}$$

Now for any a in A, $S_a\xi$ can be strongly approximated by vectors of the form $S_\psi\xi$, ψ being as in (38). Hence by (38) $S_a\xi = 0$ for all a in A. Consequently, since S is non-degenerate on $\mathcal{B}_H$ and therefore on A, we obtain $\xi = 0$. This completes the proof of (II), and so of the proposition, provided we assume that G/H is σ-compact.

Instead of assuming that M is σ-compact, let us now assume that γ is a homeomorphism. In that case the Γ of the opening paragraph of the proof is the entire Borel field of M. Since by VII.9.11

$$Q^{T|A}(\bigcup \mathcal{W}) = \sup\{Q^{T|A}(W): W \in \mathcal{W}\}$$

whenever $\mathcal{W}$ is a family of open subsets of Z, the projection-valued measure Q_0 defined by (19) satisfies the similar condition on M, and so by II.11.9 is regular. Thus the Q_2 defined by (20) and (21) coincides with Q_0. The proof now proceeds exactly as before. ∎

3.12. Summarizing the results of this section, we have proved the following generalization of 1.10:

Theorem. *Let z be a point of Z, and H the (closed) stability subgroup $\{x \in G : xz = z\}$ for z; and assume that either G/H is σ-compact or the injection $xH \mapsto xz$ is a homeomorphism of G/H into Z. Then the inducing map*

$$S \mapsto \operatorname*{Ind}_{\mathscr{B}_H \uparrow \mathscr{B}} (S) = T \tag{39}$$

(as a map of unitary equivalence classes) is a bijection from the family of all unitary equivalence classes of non-degenerate $\mathscr{B}$-positive *-representations S of $\mathscr{B}_H$ such that $Q^{S|A}$ is concentrated at z, onto the family of all unitary equivalence classes of non-degenerate *-representations of $\mathscr{B}$ which are associated with the orbit Gz.

If S and T correspond under (39), they have *-isomorphic commuting algebras. In particular, one is irreducible if and only if the other is; and one is primary if and only if the other is.

4. The Mackey Analysis in the General Case; Step III

4.1. We keep the notation of §§2, 3.

In Step I (§2) we were able under favorable circumstances (namely, when the hypotheses of 2.16 or 2.17 hold) to associate each element of $\hat{\mathscr{B}}$ with an orbit in Z. In Step II (§3), given an element z of Z with stability subgroup H, we classified the set of those elements of $\hat{\mathscr{B}}$ which are associated with the fixed orbit $\theta = Gz$, in terms of the set $\mathscr{S}$ of those elements S of $(\mathscr{B}_H)\hat{\ }$ such that the spectral measure of $S|A$ is concentrated at z. As we expect from 1.28, Step III will (under favorable circumstances) classify this set $\mathscr{S}$ in terms of the projective representation theory of H.

4.2. In Step III the advantages of allowing Z to be more general than $\operatorname{Prim}(E)$ (see 2.6) seem to evaporate. So *in this section we shall take Z to be the primitive ideal space $\operatorname{Prim}(E)$, and Φ to be the natural map $D \mapsto \operatorname{Ker}_E(D)$, as in 2.6. Thus the z of 4.1 is now a primitive ideal of E.

4.3. Definition. An element D of $\hat{E}$ is called *semicompact* if D_a is a non-zero compact operator for some a in E.

By VI.15.17 this is equivalent to saying that

$$\mathcal{O}_c(X(D)) \subset \operatorname{range}(D). \tag{1}$$

Remark. Regarded as an element of $\hat{A}$, D is semicompact if and only if the *norm-closure* of $\{D_a : a \in A\}$ contains at least one non-zero compact operator (or, equivalently, all compact operators).

4.4. Let us suppose that the z of 4.1 happens to be the kernel of some semicompact element D of $\hat{E}$. By VI.15.18 D is the *only* element of $\hat{E}$ whose kernel is z. Furthermore, by VII.10.10 a non-degenerate *-representation R of E has its spectral measure concentrated at z if and only if R is unitarily equivalent to a direct sum of copies of D. Thus, in this case the classification of the set $\mathscr{S}$ of 4.1 amounts to the problem of finding all those *-representations of $\mathscr{B}_H$ whose restriction to A is equivalent to a direct sum of copies of D. Here H, the stability subgroup for z, is now equal to the stability subgroup for D:

$$H = \{x \in G : {}^x D \cong D\}. \tag{2}$$

If the group G were discrete, a first step in solving this problem might be, as in the analogous situation in 1.17, to construct from D a cocycle in $C(\mathbb{E}, H)$ and thence a cohomology class in $Z(\mathbb{E}, H)$ (the generalized Mackey obstruction of D). However, cocycles are not really appropriate tools for non-discrete groups, especially if the groups are not second-countable. Now we saw in III.5.13 that, for discrete groups, the cohomology classes in $Z(\mathbb{E}, H)$ correspond with isomorphism classes of central extensions of $\mathbb{E}$ by H. In our present context therefore (in which H may be neither discrete nor second-countable) we will try to construct directly from D an isomorphism class of central extensions γ of $\mathbb{E}$ by H. If this effort is successful (as it always will be if D is semicompact), the result will be called the Mackey obstruction of D. In analogy with 1.19, we shall find that the *-representations S of $\mathscr{B}_H$ which reduce on A to a direct sum of copies of D are in natural correspondence with the γ^{-1}-representations of H (see VIII.10.6), where γ^{-1} is the central extension inverse to γ (see III.5.7). This fact will constitute our solution of Step III in the general case. The arguments involving group extensions, by which it will be proved, are (in spite of their different appearance) a direct generalization of the cocycle arguments of 1.17 and 1.19.

In solving this problem there is clearly no loss of generality in assuming that $H = G$. If $H \neq G$, we simply cut down $\mathscr{B}$ and G to $\mathscr{B}_H$ and H. Then (2) becomes (3) of the following number.

4.5. Let us review the problem before us: $\mathscr{B} = \langle B, \pi, \cdot, * \rangle$ is a saturated Banach *-algebraic bundle over the locally compact group G (with unit e); its unit fiber algebra B_e is denoted by A. Let $\hat{A}^+$ be the space of all $\mathscr{B}$-positive elements of $\hat{A}$, considered as a topological G-space under conjugation. As in 2.4, E is the unit fiber C^*-algebra of the bundle C^*-completion of $\mathscr{B}$; and we identify $\hat{A}^+$ and $\hat{E}$ (as topological G-spaces). Fix an element D of $\hat{E}$ (not necessarily semicompact) acting in X; and assume that

$$^xD \cong D \qquad \text{for all } x \text{ in } G. \tag{3}$$

Our problem is to classify those *-representations of $\mathscr{B}$ whose restrictions to A are equivalent to a direct sum of copies of D.

The Mackey Obstruction

4.6.   Let us fix a central extension

$$\gamma: \mathbb{E} \underset{i}{\to} K \underset{j}{\to} G \tag{4}$$

of $\mathbb{E}$ by G. Let $\mathscr{C}_\gamma$ be the retraction of $\mathscr{B}$ under $j: K \to G$ (see VIII.3.17).

Definition.   A *-representation V of $\mathscr{C}_\gamma$ is called a *natural extension of D to $\mathscr{C}_\gamma$* if $X(V) = X (= X(D))$ and

$$V_{\langle i(u), a\rangle} = uD_a \qquad \text{for all } a \in A, u \in \mathbb{E}. \tag{5}$$

It follows from (5) that

$$V_{\langle i(u)k, b\rangle} = uV_{\langle k, b\rangle} \qquad (\langle k, b\rangle \in \mathscr{C}_\gamma; u \in \mathbb{E}). \tag{6}$$

Indeed: Fixing k, b, and u, and using (5), we have

$$V_{\langle i(u)k, b\rangle}D_a = V_{\langle i(u)k, b\rangle}V_{\langle e, a\rangle}$$

$$= V_{\langle i(u)k, ba\rangle}$$

$$= V_{\langle k, b\rangle}V_{\langle i(u), a\rangle}$$

$$= uV_{\langle k, b\rangle}D_a$$

for all a in A. Since D is non-degenerate, this implies (6).

Remark.   By XI.16.7, condition (3) is *necessary* for the existence of a natural extension of D to $\mathscr{C}_\gamma$.

4.7.   If γ is the direct product extension

$$\mathbb{E} \underset{i}{\to} \mathbb{E} \times G \underset{j}{\to} G,$$

the requirement that D have a natural extension V to $\mathscr{C}_\gamma$ is the same as the requirement that D can be extended to a *-representation V' of $\mathscr{B}$ (acting in the same space as D). Indeed: V and V' are then related by the simple formula

$$V_{\langle u, x, b\rangle} = uV'_b \qquad (b \in B; x = \pi(b); u \in \mathbb{E}).$$

4.8. We shall now show that the existence of a natural extension V of D to $\mathscr{C}_\gamma$ determines both γ and V essentially uniquely.

Proposition. *Let*

$$\gamma: \mathbb{E} \underset{i}{\to} K \underset{j}{\to} G$$

and

$$\gamma': \mathbb{E} \underset{i'}{\to} K' \underset{j'}{\to} G$$

be two central extensions of $\mathbb{E}$ by G; and let V and V' be natural extensions of D to $\mathscr{C}_\gamma$ and $\mathscr{C}_{\gamma'}$ respectively. Then there exists an isomorphism $F: K \to K'$ of the central extensions γ and γ' (see III.5.3) such that

$$V_{\langle k, b\rangle} = V'_{\langle F(k), b\rangle} \qquad (\langle k, b\rangle \in \mathscr{C}_\gamma). \qquad (7)$$

Note. $\langle k, b\rangle \mapsto \langle F(k), b\rangle$ is an isomorphism of the two bundles $\mathscr{C}_\gamma$ and $\mathscr{C}_{\gamma'}$ covariant with F (see VIII.3.3).

Proof. Fix elements k and k' of K and K' respectively such that $j(k) = j'(k') = x$. By (5) we have for any b, c in B_x and any a in A

$$V_{\langle k, b\rangle} D_a V^*_{\langle k, c\rangle} = V_{\langle k, b\rangle} V_{\langle e, a\rangle} V_{\langle k^{-1}, c^*\rangle}$$

$$= D_{bac^*}$$

$$= V'_{\langle k', b\rangle} D_a V'^*_{\langle k', c\rangle}. \qquad (8)$$

Since D is irreducible, by VI.24.7 $\{D_a : a \in A\}$ is strongly dense in $\mathcal{O}(X)$. So by the strong separate continuity of operator multiplication, (8) implies that

$$V_{\langle k, b\rangle} q V^*_{\langle k, c\rangle} = V'_{\langle k', b\rangle} q V'^*_{\langle k', c\rangle} \qquad (9)$$

for all q in $\mathcal{O}(X)$. Now (9) is easily seen to imply that, if the operators $V_{\langle k, b\rangle}$, $V_{\langle k, c\rangle}$, $V'_{\langle k', b\rangle}$, $V'_{\langle k', c\rangle}$ are all non-zero, there must exist a complex constant u (perhaps depending on b, c, k, k') such that

$$V'_{\langle k', b\rangle} = u V_{\langle k, b\rangle}, \qquad V'_{\langle k', c\rangle} = \bar{u}^{-1} V_{\langle k, c\rangle}. \qquad (10)$$

The saturation of $\mathscr{B}$ implies that there do exist b, c in B_x making these four operators non-zero. So, applying (10) with $c = b$, we conclude that $|u| = 1$, and (10) becomes

$$V'_{\langle k', b\rangle} = u V_{\langle k, b\rangle}, \qquad V'_{\langle k', c\rangle} = u V_{\langle k, c\rangle}.$$

Thus there is a single number u in $\mathbb{E}$ (depending only on k and k') such that

$$V'_{\langle k', b\rangle} = u V_{\langle k, b\rangle} \qquad \text{for all } b \text{ in } B_x. \qquad (11)$$

Now let M stand for the topological subgroup $\{\langle k, k' \rangle : j(k) = j'(k')\}$ of $K \times K'$; and for each $\langle k, k' \rangle$ in M let $\phi(k, k')$ be the element u of $\mathbb{E}$ satisfying (11) (where $x = j(k) = j'(k')$). One verifies from (6) that

$$\phi(i(w)k, k') = w^{-1}\phi(k, k'), \qquad \phi(k, i'(w)k') = w\phi(k, k')$$

$(\langle k, k' \rangle \in M; w \in \mathbb{E})$; and from the saturation of $\mathscr{B}$ we see that ϕ is a homomorphism.

We claim that ϕ is continuous. Indeed: Suppose $\langle k_\alpha, k'_\alpha \rangle \underset{\alpha}{\to} \langle k, k' \rangle$ in M. Choose b in $B_{j(k)}$ so that $V_{\langle k, b \rangle} \neq 0$. Since the projection map π of $\mathscr{B}$ is open, we can pass to a subnet and find a net $\{b_\alpha\}$ such that $b_\alpha \to b$ in B and $\pi(b_\alpha) = j(k_\alpha)$. Letting $u = \phi(k, k')$, $u_\alpha = \phi(k_\alpha, k'_\alpha)$, we thus have by (11)

$$V'_{\langle k'_\alpha, b_\alpha \rangle} = u_\alpha V_{\langle k_\alpha, b_\alpha \rangle}. \tag{12}$$

The left side of (12) approaches $V'_{\langle k', b \rangle}$ strongly; and $V_{\langle k_\alpha, b_\alpha \rangle} \to V_{\langle k, b \rangle}$ strongly. Since $V_{\langle k, b \rangle} \neq 0$, it follows from this and (11) and (12) that $u_\alpha \to u$. So the claim is proved.

From the above properties of ϕ we deduce via Remark III.5.3 that $\gamma \cong \gamma'$, the isomorphism being implemented by $F: K \to K'$, where

$$F(k) = (i'(\phi(k, k')))^{-1}k' \qquad (\langle k, k' \rangle \in M). \tag{13}$$

Now (11) and (13) imply (7). So the proof of the proposition is complete. ∎

4.9.   Definition.   Suppose there exists a central extension γ of $\mathbb{E}$ by G and a natural extension of D to $\mathscr{C}_\gamma$. We then refer to the isomorphism class of γ (which by 4.8 depends only on D) as the *Mackey obstruction of D*.

4.10.   If the Mackey obstruction of D exists and is the class of the direct product extension of $\mathbb{E}$ by G, we say that D has *trivial Mackey obstruction*. By 4.7 D has trivial Mackey obstruction if and only if D can be extended to a *-representation of $\mathscr{B}$ (acting in $X(D)$).

Existence of the Mackey Obstruction

4.11.   We do not know whether or not the Mackey obstruction of D exists in all cases. Our next step will be to show that the Mackey obstruction exists if D is semicompact.

As usual, let $\mathscr{Y} = \{Y_x\}$ be the Hilbert bundle over G induced by D (XI.9.2). Let $\kappa_x: B_x \otimes X \to Y_x$ be the quotient map (XI.9.5); and let τ be the "action" of $\mathscr{B}$ on $\mathscr{Y}$ (XI.9.13).

Definition. For each x in G, let K_x denote the set of all norm-preserving bicontinuous bijections $u: \mathcal{Y} \to \mathcal{Y}$ such that:

(i) for each y in G, $u|Y_y$ is linear from Y_y onto $Y_{yx^{-1}}$; and
(ii) $\tau_b u = u\tau_b$ for all b in B.

We notice that, if $u \in K_x$ and $v \in K_y$, then $uv \in K_{xy}$ and $u^{-1} \in K_{x^{-1}}$.

Now define $K = \bigcup_{x \in G} K_x$. By the preceding sentence K is a group under composition; and the map $\sigma: K \to G$ given by $\sigma^{-1}(x) = K_x$ is a homomorphism.

Remark. It follows from XI.9.17 and condition (ii) of the above definition that an element of K is determined by what it does on any one fiber of $\mathcal{Y}$.

4.12. Proposition. *σ is onto G; that is, K_x is non-void for every x in G.*

Proof. Recall from XI.9.4 and XI.9.15 that $\kappa_e(a \otimes \xi) \leftrightarrow D_a\xi$ identifies Y_e with X and $\tau_a^{(e)}$ with D_a $(a \in A)$. (In the present context we of course take the rho-function ρ of XI.9.1 to be identically 1.)

Fix $x, y \in G$. By (3) there is a linear isometric bijection $w: Y_e \cong X \to Y_{x^{-1}}$ such that

$$\tau_a w\xi = w\tau_a\xi \qquad (a \in A; \xi \in X). \qquad (14)$$

If $b, c \in B_y$ and $\xi, \eta \in X$,

$$(\tau_b w(\xi), \tau_c w(\eta))_{Y_{yx^{-1}}} = (\tau_{c^*b} w(\xi), w(\eta))_{Y_{x^{-1}}} \qquad \text{(by XI.9.14(26))}$$

$$= (w(\tau_{c^*b}\xi), w(\eta))_{Y_{x^{-1}}} \qquad \text{(by (14))}$$

$$= (\tau_{c^*b}\xi, \eta) = (\tau_b\xi, \tau_c\eta) \qquad \text{(by XI.9.14(26)).}$$

It follows that there is a unique linear isometry $u_y: Y_y \to Y_{yx^{-1}}$ given by

$$u_y(\tau_b\xi) = \tau_b(w(\xi)), \qquad (15)$$

that is, $u_y(\kappa_y(b \otimes \xi)) = \tau_b w(\xi)$ $(b \in B_y; \xi \in X)$. By XI.9.17 the range of u_y is dense in $Y_{yx^{-1}}$, hence equal to $Y_{yx^{-1}}$.

We shall now let x be fixed and y vary in the above calculation; and define $u: \mathcal{Y} \to \mathcal{Y}$ to coincide with u_y on Y_y (for each y). Thus u is an isometric bijection sending Y_y onto $Y_{yx^{-1}}$.

Notice that u commutes with the τ_c $(c \in B)$. Indeed: If $c \in B_z$, $b \in B_y$, and $\xi \in X$,

$$u\tau_c(\tau_b\xi) = u\tau_{cb}\xi$$

$$= \tau_{cb}w(\xi)$$

$$= \tau_c\tau_bw(\xi)$$

$$= \tau_cu(\tau_b\xi).$$

So $u\tau_c = \tau_cu$, proving the claim.

Next we claim that u is continuous on $\mathscr{Y}$. To see this it is enough (by II.13.12 and the isometric property of u) to take a continuous cross-section ϕ of $\mathscr{Y}$ of the form $y \mapsto \kappa_y(f(y) \otimes \xi)$, where f is a continuous cross-section of $\mathscr{B}$ and $\xi \in X$, and to show that $u \circ \phi$ is continuous. But

$$(u \circ \phi)(y) = \tau_{f(y)}w(\xi),$$

and the right side is continuous in y by XI.9.18. So u is continuous.

Likewise, if f, ϕ, ξ are as in the preceding paragraph and $\psi(y) = \tau_{f(yx)}w(\xi)$, then $(u^{-1} \circ \psi)(y) = \phi(yx)$. Since by XI.9.17 the $\psi(y)$ are continuous in y and span a dense subspace in each fiber B_y (as f and ξ vary), and since the ϕ are also continuous, u^{-1} is continuous on $\mathscr{Y}$ by II.13.12.

We have now shown that u is bicontinuous. Combining this with the preceding remarks, we find that $u \in K_x$. So the proposition is proved.   ■

4.13. Now let us topologize K with the topology of pointwise convergence on $\mathscr{Y}$; that is, $u_i \to u$ in K if and only if $u_i(\zeta) \to u(\zeta)$ in $\mathscr{Y}$ for all ζ in $\mathscr{Y}$.

Notice that $\sigma: K \to G$ is then continuous. For if $u_i \to u$ in K and ξ is a fixed vector in Y_e, we have $\sigma(u_i) = (\pi(u_i\xi))^{-1} \to (\pi(u\xi))^{-1} = \sigma(u)$.

We shall return in 4.19 to the question of whether σ is open.

4.14. Lemma.   *In order that $u_i \to u$ in K it is sufficient that $u_i\xi \to u\xi$ in $\mathscr{Y}$ for some one non-zero vector ξ in Y.*

Proof.   Let $0 \neq \xi \in Y_x$. Since ^xD is irreducible, $\{\tau_a\xi : a \in A\}$ is dense in Y_x. By XI.9.17 it follows that $\{\tau_b\xi : b \in B\}$ is dense in each fiber of $\mathscr{Y}$.

Now suppose that $u_i\xi \to u\xi$ in $\mathscr{Y}$. Then for all b in B, $u_i(\tau_b\xi) = \tau_bu_i\xi \to \tau_bu\xi = u(\tau_b\xi)$ (by the continuity of τ_b and 4.11(ii)). Thus by the preceding paragraph $u_i\zeta \to u\zeta$ for a set of vectors ζ which is dense in each fiber Y_x. From this and the isometric property of the u_i it follows that $u_i\zeta \to u\zeta$ for all ζ in $\mathscr{Y}$, that is, $u_i \to u$ in K.   ■

4.15. Proposition.   *With the topology of 4.13, K is a topological group.*

Proof. We shall first show that $\langle u, \zeta \rangle \mapsto u\zeta$ is continuous on $K \times \mathcal{Y}$. Let $u_i \to u$ in K and $\zeta_i \to \zeta$ in $\mathcal{Y}$; and set $x_i = \pi(\zeta_i)$, $x = \pi(\zeta)$. Given $\varepsilon > 0$, choose continuous cross-sections $f_1, \ldots, f_n$ of $\mathcal{B}$ and vectors $\xi_1, \ldots, \xi_n$ of X so that

$$\left\| \sum_{r=1}^{n} \kappa_x(f_r(x) \otimes \xi_r) - \zeta \right\| < \varepsilon. \tag{16}$$

Since $x_i \to x$, we have by (16)

$$\left\| \sum_{r=1}^{n} \kappa_{x_i}(f_r(x_i) \otimes \xi_r) - \zeta_i \right\| < \varepsilon \qquad \text{for all large } i; \tag{17}$$

whence

$$\left\| \sum_{r=1}^{n} u_i \kappa_{x_i}(f_r(x_i) \otimes \xi_r) - u_i \zeta_i \right\| < \varepsilon \qquad \text{for all large } i. \tag{18}$$

But

$$u_i \kappa_{x_i}(f_r(x_i) \otimes \xi_r) = u_i \tau_{f_r(x_i)} \xi_r = \tau_{f_r(x_i)} u_i \xi_r; \tag{19}$$

and since $u_i \xi_r \to u \xi_r$ for each r, the continuity of τ (XI.9.18) gives

$$\tau_{f_r(x_i)} u_i \xi_r \to \tau_{f_r(x)} u \xi_r. \tag{20}$$

The right side of (20) is $u\tau_{f_r(x)} \xi_r = u\kappa_x(f_r(x) \otimes \xi_r)$. So from (16), (18), (19), (20), and II.13.2 we deduce that $u_i \zeta_i \to u\zeta$. Consequently $\langle u, \zeta \rangle \mapsto u\zeta$ is continuous.

From this we see that multiplication in K is continuous. For if $u_i \to u$ and $v_i \to v$ in K, then for $\zeta \in \mathcal{Y}$ we have $v_i \zeta \to v\zeta$ and so $u_i(v_i \zeta) \to u(v\zeta)$; thus $u_i v_i \to uv$.

To prove that the inverse map is continuous, suppose $u_i \to u$ in K, $\sigma(u_i) = x_i$, $\sigma(u) = x$. Take $\zeta \in Y_y$; and let f be a continuous cross-section of $\mathcal{Y}$ passing through $\eta = u^{-1}\zeta$ ($\in Y_{yx}$). By the first paragraph of the proof

$$u_i(f(yx_i)) \to u(f(yx)) = u\eta = \zeta. \tag{21}$$

Since both sides of (21) are in Y_y, this means that

$$\|u_i(f(yx_i)) - \zeta\| \to 0. \tag{22}$$

Since u_i^{-1} is an isometry, (22) implies

$$\|u_i^{-1}\zeta - f(yx_i)\| \to 0. \tag{23}$$

But $f(yx_i) \to f(yx) = u^{-1}\zeta$. This, (23), and II.13.2 imply that $u_i^{-1}\zeta \to u^{-1}\zeta$, showing that $u_i^{-1} \to u^{-1}$. So the inverse map is continuous. ∎

4.16.   If $z \in \mathbb{E}$, let us denote by $i(z)$ the element of K consisting of scalar multiplication by z on each fiber. Thus $i: \mathbb{E} \to K$ is an isomorphism into the center of K.

Proposition. $\mathrm{Ker}(\sigma)$ $(= K_e)$ *is equal to* $i(\mathbb{E})$.

Proof.   An element u of K_e carries $Y_e \cong X$ onto itself and intertwines D with itself. Since D is irreducible, u is a scalar operator on Y_e. Hence by Remark 4.11 u is the same scalar operator on every fiber of $\mathscr{Y}$.   ∎

Remark.   It follows that two elements of the same K_x can differ only by a scalar multiple of absolute value 1.

4.17.   Since K is a topological group and $\sigma: K \to G$ is a continuous homomorphism, we can form the retraction $\mathscr{C}$ of $\mathscr{B}$ under σ as in VIII.3.16. If $\langle u, b \rangle \in \mathscr{C}$ (i.e., $b \in B$; $u \in K$; $\sigma(u) = \pi(b)$), the equation

$$V_{\langle u, b \rangle} = (u\tau_b) | Y_e = (\tau_b u) | Y_e$$

defines $V_{\langle u, b \rangle}$ as a bounded linear operator on $Y_e \cong X$. Clearly $V_{\langle u, b \rangle}$ is linear in b for each u; and we have (using XI.9.14(26))

$$V_{\langle u, b \rangle} V_{\langle v, c \rangle} = V_{\langle uv, bc \rangle}, \quad (V_{\langle u, b \rangle})^* = V_{\langle u^{-1}, b^* \rangle}, \tag{24}$$

$$V_{\langle i(z), a \rangle} = zD_a \tag{25}$$

($\langle u, b \rangle$, $\langle v, c \rangle \in \mathscr{C}$; $a \in A$; $z \in \mathbb{E}$). Also V is continuous on $\mathscr{C}$ in the strong operator topology; this follows from the continuity of τ (XI.9.18) and the first statement of the proof of 4.15. So V *is a* *-*representation of* $\mathscr{C}$.

Notice from (24) and (25) that

$$(V_{\langle u, b \rangle})^* D_a V_{\langle u, c \rangle} = D_{b^*ac}, \quad V_{\langle u, b \rangle} D_a (V_{\langle u, c \rangle})^* = D_{bac^*}, \tag{26}$$

$$(V_{\langle u, b \rangle})^* V_{\langle u, c \rangle} = D_{b^*c}, \quad V_{\langle u, b \rangle} (V_{\langle u, c \rangle})^* = D_{bc^*} \tag{27}$$

($\langle u, b \rangle$, $\langle u, c \rangle \in \mathscr{C}$; $a \in A$).

In the future we will usually write V_b^u instead of $V_{\langle u, b \rangle}$.

4.18.   Suppose we knew that $\sigma: K \to G$ were *open*. Then from 4.12, 4.13, 4.15, and 4.16 we would conclude that

$$\gamma: \mathbb{E} \underset{i}{\to} K \underset{\sigma}{\to} G$$

is a central extension of $\mathbb{E}$ by G. (Notice that K would then be locally compact by III.2.4.) Further, by (25), V would be a natural extension of D to $\mathscr{C}$ in the sense of 4.6. So we would then know that D has a Mackey obstruction, namely the class of γ. Thus, to show that D has a Mackey obstruction, it is enough to show that σ is open.

If G is discrete, σ is automatically open. So we have:

Proposition. *If the base group G is discrete, then every element of $\hat{E}$ has a Mackey obstruction.*

4.19. We shall obtain a useful condition equivalent to the openness of σ (in the general case of non-discrete G).

If $w \in \mathcal{O}(X)$, let $\tilde{w}$ be the class of w modulo the equivalence relation $\sim$ given by:

$$w_1 \sim w_2 \Leftrightarrow w_2 = cw_1 \qquad \text{for some } c \text{ in } \mathbb{E};$$

and let $\tilde{\mathcal{O}} = \mathcal{O}(X)/\sim$ be the space of all $\sim$-classes. By the strong topology of $\tilde{\mathcal{O}}$ we mean the quotient topology derived from the strong topology of $\mathcal{O}(X)$.

By 4.16, if $x \in G$ and $b \in B_x$, the class $(V_b^u)^{\tilde{}}$ is independent of which u in K_x we take. We shall therefore write $V_b^{\tilde{}}$ instead of $(V_b^u)^{\tilde{}}$.

Proposition. *The following two conditions are equivalent:*

(i) $\sigma: K \to G$ is open;

(ii) the map $b \mapsto V_b^{\tilde{}}$ is continuous from B to $\tilde{\mathcal{O}}$ in the strong topology of the latter.

Proof. Assume (ii). Let $x_i \to x$ in G, and let $u \in K_x$. To prove (i) we must show that, on passing to a subnet, we can find elements u_i of K_{x_i} such that $u_i \to u$ in K. To do this, let us first choose elements ξ of X, b of B_x, and c of $B_{x^{-1}}$ so that $\tau_{cb}\xi \neq 0$ (using the saturation of $\mathcal{B}$). Since $\pi: B \to G$ is open, we can pass to a subnet and choose $b_i \in B_{x_i}$ so that $b_i \to b$ in B. Now, since $V_{b_i}^{\tilde{}} \to V_b^{\tilde{}}$ strongly by (ii), and since the quotient map $\mathcal{O}(X) \to \tilde{\mathcal{O}}$ is clearly open, we can pass to a subnet again and find elements u_i of K_{x_i} so that

$$V_{b_i}^{u_i} \to V_b^u \text{ strongly.} \tag{28}$$

We claim that $u_i \to u$ in K. As we have remarked, this will establish (i). To prove the claim, we again pass to a subnet and choose $c_i \in B_{x_i^{-1}}$ so that $c_i \to c$ (by the openness of π). By the continuity of τ,

$$\tau_{c_i b_i}\xi \to \tau_{cb}\xi. \tag{29}$$

Also, by (28) and the continuity of τ,

$$u_i \tau_{c_i b_i}\xi = \tau_{c_i} V_{b_i}^{u_i}\xi \to \tau_c V_b^u\xi = u\tau_{cb}\xi. \tag{30}$$

Since the $\tau_{c_i b_i}\xi$ and $\tau_{cb}\xi$ are all in $X \cong Y_e$, (29), (30), and the isometric property of the u_i show that $u_i\tau_{cb}\xi \to u\tau_{cb}\xi$. Since $\tau_{cb}\xi \neq 0$, this and 4.14 imply that $u_i \to u$ in K. So (i) holds; and we have shown that (ii) $\Rightarrow$ (i).

Conversely, assume (i); and let $b_i \to b$ in B. Putting $x = \pi(b)$, $x_i = \pi(b_i)$, we have $x_i \to x$. By (i) we can pass to a subnet and choose u_i and u in K_{x_i} and K_x respectively so that $u_i \to u$ in K. For any ξ in X we then have (using the first assertion of the proof of 4.15)

$$V_{b_i}^{u_i}\xi = u_i\tau_{b_i}\xi \to u\tau_b\xi = V_b^u\xi.$$

So $V_{b_i}^{u_i} \to V_b^u$ strongly, implying that $V_{b_i}^{\sim} \to V_b^{\sim}$ strongly. Thus (i) $\Rightarrow$ (ii); and the proof is complete. ■

4.20. In order to use 4.19 to prove the openness of σ in case D is semicompact, we need a lemma.

Lemma. *Let Z be any Hilbert space. Let $\{q_i\}$ be a net of bounded linear operators on Z, and q a bounded linear operator on Z, such that*

$$\|q_i c q_i^* - q c q^*\| \to 0 \tag{31}$$

for all compact operators c on Z. Then there exist numbers λ_i in $\mathbb{E}$ such that

$$\lambda_i q_i \to q \text{ strongly}. \tag{32}$$

Proof. For any vector η in Z, let p_η denote projection onto $\mathbb{C}\eta$. If $\xi \in Z$ and $\|\xi\| = 1$, we verify that

$$qp_\xi q^* = \|q\xi\|^2 p_{q\xi}, \qquad q_i p_\xi q_i^* = \|q_i\xi\|^2 p_{q_i\xi}. \tag{33}$$

We consider separately the cases $q = 0$ and $q \neq 0$.

Suppose first that $q = 0$. In that case, (31) (applied with $c = p_\xi$) and (33) give $\|q_i\xi\|^2 p_{q_i\xi} \to 0$ (in norm), and so $\|q_i\xi\|^2 \to 0$, for all unit vectors ξ. This says that $q_i \to 0 = q$ strongly.

Next, suppose $q \neq 0$; and fix a unit vector ξ such that $q\xi \neq 0$. By (31) (with $c = p_\xi$) and (33) we get

$$\|q_i\xi\|^2 p_{q_i\xi} \to \|q\xi\|^2 p_{q\xi} \text{ in norm}. \tag{34}$$

Taking norms on both sides of (34) we get (since $q\xi \neq 0$)

$$\|q_i\xi\| \to \|q\xi\|. \tag{35}$$

From this and (34) we deduce that

$$p_{q_i\xi} \to p_{q\xi} \text{ in norm}. \tag{36}$$

Now (35) and (36) imply that there exist numbers λ_i in $\mathbb{E}$ such that

$$\lambda_i q_i \xi \to q\xi. \tag{37}$$

With these λ_i we shall prove (32). For convenience let us replace q_i by $\lambda_i q_i$; this makes no difference to (31), and (37) becomes

$$q_i \xi \to q\xi. \tag{38}$$

We shall show that $q_i \to q$ strongly. Choose any vector ζ in Z; and let c be the operator $\tau \mapsto (\tau, \zeta)\zeta$ of rank 1. We then have, putting $\eta = q\xi$,

$$q_i c q_i^* \eta = (\eta, q_i\xi)q_i\zeta, \qquad qcq^*\eta = (\eta, \eta)q\zeta. \tag{39}$$

Now by (31)

$$q_i c q_i^* \eta \to qcq^*\eta; \tag{40}$$

and (38) gives

$$(\eta, q_i\xi) \to (\eta, \eta) \neq 0. \tag{41}$$

Combining (39), (40), and (41), we get $q_i\zeta \to q\zeta$. Since ζ was arbitrary, this says that $q_i \to q$ strongly, proving (32).　∎

4.21. Proposition. *If D is semicompact, it has a Mackey obstruction.*

Proof. Suppose $b_i \to b$ in B; and choose an element u of $K_{\pi(b)}$ and (for each i) an element u_i of $K_{\pi(b_i)}$. Let us denote by $\mathscr{F}$ the set of all those bounded linear operators g on X such that

$$V_{b_i}^{u_i} g (V_{b_i}^{u_i})^* \to V_b^u g (V_b^u)^* \tag{42}$$

in norm. Since the V_b^u are bounded in norm, $\mathscr{F}$ is closed in the norm-topology. Now, if $g = D_a$ ($a \in A$), then by (26) the left and right sides of (42) are $D_{b_i a b_i^*}$ and D_{bab^*} respectively; so, since $b_i a b_i^* \to bab^*$ in A, (42) holds in this case. Thus $\mathscr{F}$ contains $D(A)$ and is norm-closed. Since D is semicompact, it follows that $\mathscr{F}$ contains all the compact operators. So by 4.20 $V_{b_i}^{\sim} \to V_b^{\sim}$ strongly. Thus $b \mapsto V_b^{\sim}$ is strongly continuous; and by 4.19 this implies that σ is open. So by 4.18 D has a Mackey obstruction.　∎

The Classification of the *-Representations that Reduce to a Multiple of D

4.22. *From here to the end of this section D is assumed to have a Mackey obstruction. Let*

$$\gamma: \mathbb{E} \underset{i}{\to} K \underset{\sigma}{\to} G$$

be a central extension of $\mathbb{E}$ by G whose class is the Mackey obstruction of D; and let V be a natural extension of D to the retraction $\mathscr{C}$ of $\mathscr{B}$ under σ. As in previous numbers we write V_b^u for $V_{\langle u, b \rangle}$ ($\langle u, b \rangle \in \mathscr{C}$). If $x \in G$ we put $K_x = \sigma^{-1}(x)$.

The problem proposed in 4.5 will be solved by setting up a natural correspondence between the non-degenerate *-representations T of $\mathscr{B}$ such that

$$T|A \cong \text{a direct sum of copies of } D, \tag{43}$$

and those unitary representations W of K which satisfy

$$W_{i(z)} = \bar{z} 1_{X(W)} \qquad (z \in \mathbb{E}). \tag{44}$$

4.23. To set up this correspondence, take a unitary representation W of K satisfying (44). Let T' be the tensor product $W \otimes V$ (defined as in VIII.9.16); thus T' is a *-representation of $\mathscr{C}$. We shall denote by $\Phi: \mathscr{C} \to \mathscr{B}$ the surjection $\langle u, b \rangle \mapsto b$. In view of (6) and (44), $T'_{\langle u, b \rangle}$ depends only on b, that is, on $\Phi(\langle u, b \rangle)$; so T' can be factored in the form

$$T' = T \circ \Phi,$$

where $T: B \to \mathcal{O}(X(T'))$. From the evident fact that Φ is open and preserves all operations, it follows that T is a *-representation of $\mathscr{B}$. We thus have $X(T) = X(W) \otimes X(D)$, and

$$T_b = W_u \otimes V_b^u \qquad (b \in B; u \in K_{\pi(b)}). \tag{45}$$

For a in A we have

$$T_a = 1_{X(W)} \otimes V_a^1 = 1_{X(W)} \otimes D_a \qquad \text{(by (25))}. \tag{46}$$

So T satisfies (43). In particular T is non-degenerate.

4.24. We shall now investigate some properties of the map

$$W \mapsto T \tag{47}$$

constructed in 4.23.

First, it is clear that (47) preserves direct sums.

Suppose that $W \mapsto T$ and $W' \mapsto T'$ under (47); and let $\mathscr{I}$ and $\mathscr{J}$ be the spaces of intertwining operators for W, W' and T, T' respectively. If $q \in \mathscr{I}$, clearly $q \otimes 1_X \in \mathscr{J}$. We claim that the range of the linear map $q \mapsto q \otimes 1_X$ is all of $\mathscr{J}$. Indeed, let $Q \in \mathscr{J}$. Then in particular, by (46) and VI.14.5, $Q = q \otimes 1_X$ for some bounded linear map $q: X(W) \to X(W')$. The intertwining condition for Q now becomes

$$q W_u \otimes V_b^u = W'_u q \otimes V_b^u. \tag{48}$$

Now for every u in K there is a b in $B_{\sigma(u)}$ for which $V_b^u \neq 0$. Thus (48) implies that $q \in \mathscr{I}$, proving the claim.

We have shown that $q \mapsto q \otimes 1$ is a linear isomorphism of $\mathscr{I}$ onto $\mathscr{J}$. It also preserves composition and adjoints. So we obtain immediately the following two results.

4.25. Proposition.   *If* $W \mapsto T$ *and* $W' \mapsto T'$ *under* (47) *and* $T' \cong T$, *then* $W' \cong W$; *that is,* (47) *is one-to-one as a map of unitary equivalence classes.*

4.26. Proposition.   *If* $W \mapsto T$ *under* (47), *the commuting algebras of* W *and of* T *are* *-*isomorphic under the mapping* $q \mapsto q \otimes 1_X$. *In particular,* T *is irreducible* [*primary*] *if and only if* W *is irreducible* [*primary*].

4.27.   Our main result concerning the mapping (47) is its surjectivity.

Proposition.   *Every* *-*representation* T *of* $\mathscr{B}$ *satisfying* (43) *is unitarily equivalent to the image under* (47) *of a unitary representation* W *of* K *satisfying* (44).

Proof.   By (43) we may as well assume that $X(T) = Y \otimes X$ for some Hilbert space Y, and that

$$T_a = 1_Y \otimes D_a \qquad \text{for } a \in A. \tag{49}$$

Fix $x \in G$, $u \in K_x$. We claim that there exists a unique linear isometry $Z: X(T) \to X(T)$ satisfying

$$Z \circ (1_Y \otimes V_b^u) = T_b \qquad \text{for all } b \text{ in } B_x. \tag{50}$$

Indeed: Since by XI.9.17 the range(V_b^u) ($b \in B_x$) span a dense subspace of X, Z must be unique if it exists. To show that it exists, we put $\zeta = \sum_{i=1}^{n} (1 \otimes V_{b_i}^u)(\eta_i \otimes \xi_i)$, $\zeta' = \sum_{i=1}^{n} T_{b_i}(\eta_i \otimes \xi_i)$ (where $\xi_1, \ldots, \xi_n \in X$; $\eta_1, \ldots, \eta_n \in Y$; $b_1, \ldots, b_n \in B_x$), and show that $\|\zeta'\| = \|\zeta\|$. We have:

$$\|\zeta\|^2 = \sum_{i,j} (V_{b_i}^u \xi_i, V_{b_j}^u \xi_j)(\eta_i, \eta_j)$$

$$= \sum_{i,j} (V_{b_j}^{u^{-1}} V_{b_i}^u \xi_i, \xi_j)(\eta_i, \eta_j) \qquad \text{(by (24))}$$

$$= \sum_{i,j} (D_{b_j^* b_i} \xi_i, \xi_j)(\eta_i, \eta_j) \qquad \text{(by (24), (25))}$$

$$= \sum_{i,j} (T_{b_j^* b_i}(\eta_i \otimes \xi_i), \eta_j \otimes \xi_j) \qquad \text{(by (49))}$$

$$= \sum_{i,j} (T_{b_i}(\eta_i \otimes \xi_i), T_{b_j}(\eta_j \otimes \xi_j)) = \|\zeta'\|^2.$$

In view of this there is a linear isometry Z on $X(T)$ satisfying $Z\zeta = \zeta'$ for all such ξ_i, η_i, b_i. But this implies (50).

Next we claim that

$$Z \circ (\mathbb{1} \otimes D_a) = (\mathbb{1} \otimes D_a) \circ Z \qquad (a \in A). \qquad (51)$$

Indeed: For $a \in A$, $b \in B_x$, we have

$$ZT_a(\mathbb{1} \otimes V_b^u) = Z(\mathbb{1} \otimes V_a^{\mathbb{1}})(\mathbb{1} \otimes V_b^u)$$

$$= Z(\mathbb{1} \otimes V_{ab}^u) \qquad \text{(by (24), (25), (49))}$$

$$= T_{ab} \qquad \text{(by (50))}$$

$$= T_a T_b = T_a Z(\mathbb{1} \otimes V_b^u) \qquad \text{(by (50))}.$$

We pointed out earlier that the ranges of the $\mathbb{1} \otimes V_b^u$ ($b \in B_x$) span a dense subspace. So the last equation implies $ZT_a = T_a Z$ ($a \in A$). By (49) this gives (51).

Since D is irreducible, (51) implies by VI.14.5 that Z is of the form $W_u \otimes \mathbb{1}$, W_u being a bounded linear operator on Y. By (50) this says that

$$T_b = W_u \otimes V_b^u \qquad (x \in G; b \in B_x; u \in K_x). \qquad (52)$$

It remains only to show that $W: u \mapsto W_u$ is a unitary representation of K satisfying (44).

If $b, c \in B$, $u \in K_{\pi(b)}$, $v \in K_{\pi(c)}$, we have by (52)

$$W_{uv} \otimes V_{bc}^{uv} = T_{bc}$$

$$= T_b T_c$$

$$= (W_u \otimes V_b^u)(W_v \otimes V_c^v)$$

$$= W_u W_v \otimes V_{bc}^{uv}. \qquad (53)$$

By XI.9.17 we can choose b and c so that $V_{bc}^{uv} \neq 0$. It then follows from (53) that

$$W_{uv} = W_u W_v. \qquad (54)$$

Now clearly $W_{\mathbb{1}} = \mathbb{1}_Y$. So (54) implies $W_{u^{-1}} = (W_u)^{-1}$. Since Z was an isometry, W_u is likewise an isometry; hence W_u is unitary.

We shall show that W is strongly continuous. Let $u_i \to u$ in K; choose b in $B_{\sigma(u)}$ so that $V_b^u \neq 0$, and (passing to a subnet) choose a net $\{b_i\}$ such that $b_i \to b$ and $b_i \in B_{\sigma(u_i)}$. Fix $\xi \in X$ so that $\|V_b^u \xi\| = 1$. By (52), for each η in Y

$$T_{b_i}(\eta \otimes \xi) = W_{u_i}\eta \otimes V_{b_i}^{u_i}\xi, \qquad (55)$$

$$T_b(\eta \otimes \xi) = W_u\eta \otimes V_b^u\xi. \qquad (56)$$

Since T is strongly continuous,

$$T_{b_i}(\eta \otimes \xi) \to T_b(\eta \otimes \xi). \tag{57}$$

Also, as we have seen before,

$$V_{b_i}^{u_i}\xi = u_i\tau_{b_i}\xi \to u\tau_b\xi = V_b^u\xi. \tag{58}$$

Now

$$\|W_{u_i}\eta - W_u\eta\| = \|(W_{u_i}\eta - W_u\eta) \otimes V_b^u\xi\|$$

$$\leq \|(W_{u_i}\eta \otimes V_{b_i}^{u_i}\xi) - (W_u\eta \otimes V_b^u\xi)\|$$

$$+ \|W_{u_i}\eta \otimes (V_b^u\xi - V_{b_i}^{u_i}\xi)\|. \tag{59}$$

The first summand on the right of (59) approaches 0 (in i) because of (55), (56), (57); the second approaches 0 by (58). So (59) shows that $W_{u_i}\eta \to \Delta_u\eta$, and hence by the arbitrariness of η that $W_{u_i} \to W_u$ strongly. So by (54) W is a unitary representation of K.

Let $z \in \mathbb{E}$, $u = i(z)$, $a \in A$. Then

$$1 \otimes D_a = T_a$$

$$= W_u \otimes V_a^u$$

$$= W_u \otimes zD_a \qquad \text{(by (25))}$$

$$= zW_u \otimes D_a.$$

Since for some a $D_a \neq 0$, this gives $zW_u = 1$, or $W_{i(z)} = \bar{z}1_Y$. So W satisfies (44); and the proof is complete. ∎

4.28. In summary we have the following generalization of 1.19.

Theorem. *Let D be an element of $\hat{E}$ satisfying (3); and assume that D has a Mackey obstruction $\tilde{\gamma}$ (which is automatically the case if D is semicompact or if G is discrete). Let*

$$\gamma: \mathbb{E} \underset{i}{\to} K \underset{\sigma}{\to} G$$

be a central extension of class $\tilde{\gamma}$, and V a natural extension of D to the $\mathscr{C}_\gamma$ of 4.6. Then the map $W \mapsto T$ given by

$$T_b = W_u \otimes V_{\langle u, b\rangle} \qquad (\langle u, b\rangle \in \mathscr{C}_\gamma) \tag{60}$$

*is a one-to-one correspondence between the set of all equivalence classes of unitary representations W of K satisfying (44) and the set of all equivalence classes of *-representations T of $\mathscr{B}$ satisfying (43). If W and T are thus related* by (60), *they have *-isomorphic commuting algebras; in particular, T is irreducible [primary] if and only if W is irreducible [primary].*

Remark 1.   In view of the definitions of III.5.7 and VIII.10.6, the domain of the map $W \mapsto T$ in the above theorem can be described as the set of all γ^{-1}-representations of G, where γ^{-1} is the central extension

$$\gamma^{-1} : \mathbb{E} \underset{i^{-1}}{\to} K \underset{\sigma}{\to} G$$

inverse to γ.

Remark 2.   By 4.7 the Mackey obstruction of D is trivial if and only if D can be extended to a *-representation V' of $\mathscr{B}$ (acting in $X(D)$). In that case the above theorem asserts that the family of all equivalence classes of unitary representations W of G is in one-to-one correspondence with the family of all equivalence classes of *-representations T of $\mathscr{B}$ satisfying (43), under the Hilbert tensor product map defined in VIII.9.16:

$$W \mapsto T = W \otimes V'. \tag{61}$$

Remark 3.   By III.5.11 the Mackey obstruction of D is automatically trivial if G is either $\mathbb{Z}$ or $\mathbb{Z}_p$ (a finite cyclic group) or $\mathbb{R}$ or $\mathbb{E}$.

4.29.   We conclude by abandoning the hypothesis 4.5(3).

Fix any semicompact element D of $\hat{E}$, that is, a semicompact $\mathscr{B}$-positive element of $\hat{A}$. We remarked in 4.4 that the stability subgroup H for D (as an element of $\hat{E} \cong \hat{A}^+$) is the same as for $\mathrm{Ker}_E(D)$ (as an element of $\mathrm{Prim}(E)$). Since $\mathrm{Prim}(E)$ is T_0, H is closed by III.3.8. Thus 4.5(3) holds when $\mathscr{B}$ and G are replaced by $\mathscr{B}_H$ and H; and so by 4.21 there is a central extension γ of $\mathbb{E}$ by H whose isomorphism class $\tilde{\gamma}$ is the Mackey obstruction of D relative to $\mathscr{B}_H$.

Definition. $\tilde{\gamma}$ is called the *Mackey obstruction of D*. If $\tilde{\gamma}$ is the class of the direct product extension, D is said to have *trivial Mackey obstruction*.

Remark.   One easily verifies that this generalizes the cocycle definition 1.17 when the latter is reformulated in terms of group extensions.

4.30.   *A Class of Examples.*   Consider the interesting class of saturated C^*-algebraic bundles $\mathscr{B}$ constructed in VIII.16.15-17. In 6.4 we shall determine the Mackey obstructions of the elements D of the structure spaces of the unit fiber C^*-algebras for these $\mathscr{B}$. It will turn out that *every* possible central extension of $\mathbb{E}$ by the stability subgroup for D occurs as the Mackey obstruction when the $\mathscr{B}$ is suitably chosen.

5. The Mackey Analysis in the General Case; Final Results

5.1. In this section we combine together the results of Steps I, II, and III.

Reviewing the notation, we take $\mathscr{B} = \langle B, \pi, \cdot, * \rangle$ to be a saturated Banach *-algebraic bundle over the locally compact group G (with unit e), and denote by A the unit fiber Banach *-algebra B_e. Let $\hat{A}^+$ be the topological G-space (under the conjugation operation) of all $\mathscr{B}$-positive elements of $\hat{A}$.

The Analysis on a Single Semicompact Orbit

5.2. The following theorem combines Steps II and III.

Theorem. *Let D be a semicompact element of $\hat{A}^+$, and H the stability subgroup of G for D. (H is closed by 4.4.) Assume that either G/H is σ-compact or the injection $xH \mapsto {}^x D$ of G/H into $\hat{A}^+$ is a homeomorphism. Let*

$$\gamma: \mathbb{E} \xrightarrow[i]{} K \xrightarrow[\sigma]{} H$$

be a central extension of $\mathbb{E}$ by H whose class is the Mackey obstruction of D; this exists by 4.21. Then there is a natural one-to-one correspondence

$$W \mapsto T \tag{1}$$

*between the family $\mathscr{W}$ of all equivalence classes of γ^{-1}-representations W of H (see Remark 1 of 4.28) and the family $\mathscr{T}$ of all those non-degenerate *-representations T of $\mathscr{B}$ which are associated with the orbit θ of D (i.e., such that the spectral measure of $T|A$ is concentrated on θ).*

The correspondence (1) is given by

$$T = \mathop{\mathrm{Ind}}_{\mathscr{B}_H \uparrow \mathscr{B}} (S^W). \tag{2}$$

*Here S^W is the *-representation of $\mathscr{B}_H$ defined by*

$$S_b^W = W_u \otimes V_{\langle u, b \rangle} \qquad (b \in \mathscr{B}_H), \tag{3}$$

where $\langle u, b \rangle$ is an element of the retraction $\mathscr{C}$ of $\mathscr{B}_H$ by σ, and V is a fixed natural extension of D to $\mathscr{C}$ (see 4.6).

*If W and T correspond under (1), they have *-isomorphic commuting algebras; thus T is irreducible [primary] if and only if W is irreducible [primary].*

Proof. By 4.4 the non-degenerate *-representations S of $\mathscr{B}_H$ such that the spectral measure of $S|A$ is concentrated at D are just those for which $S|A$ is a direct sum of copies of D; and by 4.28 these are just the S^W defined by (3), where $W \in \mathscr{W}$. By XI.11.11 such S are necessarily $\mathscr{B}$-positive. Hence by 3.12 the elements of $\mathscr{T}$ are just the *-representations of $\mathscr{B}$ of the form $\mathrm{Ind}(S^W)$, where $W \in \mathscr{W}$. Thus the proof is complete, once we recall that the correspondences of 3.12 and 4.28 preserve the *-isomorphism classes of the commuting algebras. ∎

Remark. If the D of the above theorem has trivial Mackey obstruction, then as in Remark 2 of 4.28 we can take $\mathscr{W}$ to be the family of all unitary equivalence classes of unitary representations of H; and (3) takes the simpler form

$$S^W = W \otimes V' \qquad \text{(Hilbert tensor product)}, \qquad (4)$$

where V' is any fixed extension of D to $\mathscr{B}_H$.

5.3.   Corollary.   *Let D be as in Theorem 5.2. Then there exist elements of $\hat{\mathscr{B}}$ which are associated with the orbit of D.*

Proof.   By Theorem 5.2 we have only to show that irreducible γ^{-1}-representations of H exist. But this was proved in VIII.14.13. ∎

5.4.   *Remark on Dimension.*   What is the dimension of the T of 5.2(2)?

Let W and S^W be as in Theorem 5.2. Since $S^W | A$ is the direct sum of $\dim(W)$ copies of D, it follows from XI.16.6 and XI.16.12 that, if $x \in \alpha \in G/H$,

$$\dim(Y_\alpha) = \dim({}^x(S^W)) = \dim(W)\dim({}^x D)$$

(Y_α being the α-fiber of the induced Hilbert bundle), and hence that

$$\dim(T) = \sum_{\alpha \in G/H} \dim(Y_\alpha) = \dim(W) \sum_{E \in \theta} \dim(E),$$

where θ is the orbit of D in $\hat{A}^+$. Thus:

  The T of 5.2(2) is finite-dimensional if and only if (a) W is finite-dimensional, (b) the orbit θ of D in $\hat{A}^+$ is finite, and (c) each E in θ is finite-dimensional. In that case

$$\dim(T) = \dim(W) \sum_{E \in \theta} \dim(E). \qquad (5)$$

5.5.   We have seen in Remark 3 of 4.28 that the D of Theorem 5.2 automatically has trivial Mackey obstruction if H is either $\mathbb{Z}$, $\mathbb{Z}_p$, $\mathbb{R}$, or $\mathbb{E}$. Here is another special circumstance (generalizing 1.12) in which the Mackey obstruction is trivial.

Proposition.   *Suppose in Theorem 5.2 that D is one-dimensional and that $\mathscr{B}_H$ is a semidirect product of A with H. Then the Mackey obstruction of D is trivial.*

Proof. Suppose that $\mathscr{B}_H = A \underset{\tau}{\times} H$ (see VIII.4.2), where τ is a strongly continuous homomorphism of G into the group of isometric *-automorphisms of A. By XI.4.22 $(^x D)(a) = D(\tau_{x^{-1}}(a))$ for any a in A and x in H; hence, since H is the stability group for D,

$$D(\tau_x(a)) = D(a) \qquad \text{for all } a \text{ in } A, x \text{ in } H.$$

In view of this the equation

$$D'_{\langle a, x \rangle} = D_a \qquad\qquad (a \in A; x \in H)$$

evidently defines an extension of D to a one-dimensional *-representation D' of the semidirect product bundle $\mathscr{B}_H$. So by 4.7 the Mackey obstruction of D is trivial. ∎

5.6. It can be shown (though we shall only sketch the proof here) that the correspondence (1), (2) of Theorem 5.2 is a homeomorphism with respect to the regional topologies, provided a certain further assumption holds.

Theorem*. *In addition to the assumptions of Theorem 5.2, let us suppose that (i) the orbit θ of D is open relative to its closure in $\hat{A}^+$ (i.e., $\theta = U \cap C$, where U is open in $\hat{A}^+$ and C is closed in $\hat{A}^+$), and (ii) the natural bijection $\sigma: G/H \to \theta$ (given by $\sigma(xH) = {}^x D$) is a homeomorphism. Then the correspondence $W \mapsto T$ of (1), (2) is a homeomorphism with respect to the regional topologies.*

Sketch of Proof. The correspondence (1) is the composition of

$$W \mapsto S = S^W \tag{6}$$

and

$$S \mapsto T = \text{Ind}(S). \tag{7}$$

So it is sufficient to show that (6) and (7) are both homeomorphisms.

We shall first show that (6) is a homeomorphism.

This is an easy consequence of the following general fact about Hilbert tensor products:

*Let R be any fixed *-representation of $\mathscr{B}$ such that (i) $R|A$ is irreducible, and (ii) the norm-closure of $\{R_a : a \in A\}$ contains $\mathcal{O}_c(X(R))$ (i.e., $R|A$ is semicompact). Then we claim that the Hilbert tensor product map*

$$S \mapsto T = S \otimes R$$

*(see VIII.9.16) is a regional homeomorphism from the space of all unitary representations of G into the space of all *-representations of $\mathscr{B}$.*

Indeed: Replacing $\mathscr{B}$ by its bundle C^*-completion, we may assume from the beginning that $\mathscr{B}$ is a C^*-algebraic bundle. So by (ii) $\mathcal{O}_c(X(R)) \subset \{R_a : a \in A\}$.

For each x in G let $C_x = \{b \in B_x : R_b \text{ is compact}\}$. Then $\bigcup_{x \in G} C_x$, with the relativized topology, norm, and operations of $\mathscr{B}$, is a C^*-algebraic bundle over G which we shall denote by $\mathscr{C}$. (The only non-trivial step in proving this fact is the observation that the bundle projection of $\mathscr{B}$ is still open when restricted to $\mathscr{C}$. To see this, take any element b of C_x; and let u be an element of C_e such that b^*bu is very close to b^*b. Then by a calculation in VIII.16.3 bu is very close to b. So, if f is a continuous cross-section of $\mathscr{B}$ passing through b, $y \mapsto f(y)u$ is a continuous cross-section of $\mathscr{C}$ passing very near to b.)

Let $R' = R|\mathscr{C}$. Thus R' is an irreducible *-representation of $\mathscr{C}$ whose range consists of compact operators; and, if S is a unitary representation of G,

$$S \otimes R' = (S \otimes R)|\mathscr{C}.$$

Since $\mathscr{L}(\mathscr{C})$ is a *-ideal of $\mathscr{L}(\mathscr{B})$, it follows that the correspondence $S \otimes R \mapsto S \otimes R'$ is a regional homeomorphism. So, to prove the claim, it is enough to show that $S \mapsto S \otimes R'$ is a regional homeomorphism.

Notice that, for each x in G,

$$\{R'_c : c \in C_x\} \qquad \text{is norm-dense in } \mathcal{O}_c(X(R)).$$

To prove this, let Z_x be the norm-closure of $\{R'_c : c \in C_x\}$; and put $D = C_e$. Since $C_x D \subset C_x$ and $DC_x \subset C_x$, Z_x is a closed two-sided ideal of $\mathcal{O}_c(X(R))$. Now the latter is topologically simple. Hence Z_x is either $\{0\}$ or $\mathcal{O}_c(X(R))$. But $Z_x \neq \{0\}$ by Lemma XI.11.9. So $Z_x = \mathcal{O}_c(X(R))$.

Let L be the *-algebra $\mathscr{L}(G; \mathcal{O}_c(X(R)))$ of XI.7.15. Referring forward to Lemma 7.8 of this chapter, we see that the equation

$$\Phi(g)(x) = R'_{g(x)} \qquad\qquad (g \in \mathscr{L}(\mathscr{C}); x \in G)$$

defines a *-homomorphism $\Phi : \mathscr{L}(\mathscr{C}) \to L$; and by the preceding paragraph and II.14.6 range(Φ) is dense in L (in the inductive limit topology). So the composition map $T \mapsto T \circ \Phi$ is a regional homeomorphism from the space of *-representations of L to that of $\mathscr{L}(\mathscr{C})$.

Referring to the context of XI.7.16, let S be a unitary representation of G and T the *-representation of L (called A in XI.7.16) defined by XI.7.16(30). Then, for $g \in \mathscr{L}(\mathscr{C})$ and $\zeta \in X(T)$,

$$(T \circ \Phi)_g(\zeta) = \int (S_x \otimes R'_{g(x)})(\zeta) d\lambda x$$

$$= (S \otimes R')_g(\zeta);$$

that is, $T \circ \Phi$ is the integrated form of $S \otimes R'$. Since $S \mapsto T$ is a regional homeomorphism by XI.7.16, and $T \mapsto T \circ \Phi$ is a regional homeomorphism by the preceding paragraph, it follows that

$$S \mapsto T \circ \Phi \cong S \otimes R'$$

is a regional homeomorphism, completing the proof of the italicized claim.

The fact that (6) is a homeomorphism now results from the above claim applied to the $\mathscr{C}$ and V of Theorem 5.2.

Next we must show that (7) is a homeomorphism.

The continuity of (7) follows from XI.12.4. We shall show that its inverse is continuous. Let F denote the compacted transformation algebra derived from $\mathscr{B}$ and the G-space G/H, as in VIII.7.8. For each S in the domain of (7) let T, P be the system of imprimitivity induced by S (see XI.14.3); and let T^0 be the F-integrated form of T, P (see VIII.18.16). By XI.6.19 and the fact that S and T^0 are induced from each other via an imprimitivity bimodule (see §XI.14), the map $S \mapsto T^0$ is a regional homeomorphism. So, to prove the continuity of the inverse of (7), it is enough to show that $T \mapsto T^0$ is regionally continuous.

Let E be the unit fiber C^*-algebra of the bundle C^*-completion of $\mathscr{B}$, as in 2.4. The assumption (i) on θ implies that θ can be identified with the structure space of a C^*-algebra J which is a closed two-sided ideal of a quotient C^*-algebra of E. Given a $*$-representation T in the range of (7), let Q be the spectral measure of $T|E$. Since Q is carried by θ, T generates a non-degenerate $*$-representation $\tilde{T}$ of J (acting on $X(T)$) whose spectral measure is again Q (or rather, the restriction of Q to Borel subsets of θ).

Now let ϕ be a function in $\mathscr{L}(G/H)$. By assumption (ii) we have $\phi = \psi \circ \sigma$, where $\psi \in \mathscr{L}(\theta)$. By the Dauns-Hofmann Theorem VIII.1.23, ψ gives rise to a multiplier m of J; and by VIII.1.24

$$\tilde{T}'_m = \int \psi \, dQ \tag{8}$$

($\tilde{T}'$ being the extension of $\tilde{T}$ to the multiplier algebra of J). On the other hand, the proof of Proposition 3.11 shows that P and Q are each other's transforms under the homeomorphism $\sigma: G/H \to \theta$. Since $\phi = \psi \circ \sigma$, this implies $\int \psi \, dQ = \int \phi \, dP$, whence by (8)

$$\tilde{T}'_m = \int \phi \, dP. \tag{9}$$

If $\phi \in \mathcal{L}(G/H)$ and $f \in \mathcal{L}(\mathcal{B})$, let $\phi \times f$ denote the element $\langle x, m \rangle \mapsto \phi(m)f(x)$ of F. By VIII.18.17

$$T^0_{\phi \times f} = \left(\int \phi \, dP \right) T_f. \tag{10}$$

Combining (9) and (10) we obtain

$$T^0_{\phi \times f} = \tilde{T}'_m T_f. \tag{11}$$

Now the maps $T \mapsto \tilde{T} \mapsto \tilde{T}'$ are regionally continuous. Also, by VIII.18.17 the set of all such $\phi \times f$ is dense in F. Combining these facts with (11) and VII.1.16, it is not hard to show that $T \mapsto T^0$ is continuous; and this, as we have seen, is enough to complete the proof. ∎

The Classification of All Primary Representations of $\mathcal{B}$

5.7. Theorem 5.2 classifies those *-representations of $\mathcal{B}$, in particular those primary *-representations of $\mathcal{B}$, which happen to be associated with the orbit of a semicompact element of $\hat{A}^+$. Now in Step I (§2) we saw that under suitable conditions *every* primary *-representation of $\mathcal{B}$ is associated with an orbit. Combining the three steps, we thus obtain as our final result the following Theorem 5.8, which classifies all primary *-representations (in particular all irreducible *-representations) of $\mathcal{B}$ in terms of the primary (or irreducible) projective representation theory of the stability subgroups of G.

Theorem 5.8 can be regarded as the apex of this entire work.

5.8. Keep the notation of 5.1. For the purposes of the coming theorem we shall assume that

$$\text{every element of } \hat{A}^+ \text{ is semicompact.} \tag{12}$$

This implies (see 4.4) that the map $\Phi : D \mapsto \mathrm{Ker}_E(D)$ of 2.6 is one-to-one, so that $\hat{A}^+$ is T_0 and can be identified with the Z ($= \mathrm{Prim}(E)$) of 2.6.

We shall also assume that G acts on $\hat{A}^+$ "smoothly" in the sense that

$$\hat{A}^+/G \text{ either satisfies Condition } (C) \text{ (see 2.16) or is almost}$$
$$\text{Hausdorff (see 2.17).} \tag{13}$$

Finally, if $D \in \hat{A}^+$ let H_D be the (closed) stability subgroup of G for D. We shall assume that, for each D in $\hat{A}^+$,

$$\text{either } G/H_D \text{ is } \sigma\text{-compact or the injection } xH_D \mapsto {}^x D$$
$$\text{of } G/H_D \text{ into } \hat{A}^+ \text{ is a homeomorphism.} \tag{14}$$

Theorem. *Assume (12), (13), and (14); and let T be any primary *-representation of $\mathscr{B}$. Then there is a unique orbit θ in $\hat{A}^+$ with which T is associated. If D is a fixed element of θ and H is the (closed) stability subgroup of G for D, there is a unique primary γ^{-1}-representation W of H such that*

$$T = \underset{\mathscr{B}_H \uparrow \mathscr{B}}{\text{Ind}} (S^W). \tag{15}$$

(Here γ and S^W have the same meanings as in Theorem 5.2.)

If T and W are related by (15), then T and W have *-isomorphic commuting algebras. In particular T is irreducible if and only if W is irreducible.

Proof. By 2.16, 2.17, and 5.2. ∎

Remark. In the commonest specific applications of this theorem G is second-countable. Then G/H_D is automatically σ-compact, and (14) holds. By Remark 2.19 the two "smoothness" conditions of (13) are then equivalent. Of the two, Condition (C) is usually by far the easier to verify.

The Type I Property

5.9. In VI.24.13 we defined a *-algebra to *have a* Type I *-*representation theory* if all its primary *-representations are of Type I. This very important definition can easily be carried over to Banach *-algebraic bundles. The Banach *-algebraic bundle $\mathscr{B}$ is said *to have a Type I *-representation theory* if every primary *-representation T of $\mathscr{B}$ is of Type I, that is, is unitarily equivalent to a direct sum of copies of a single irreducible *-representation of $\mathscr{B}$. This amounts to saying that $\mathscr{B}$ has a Type I *-representation theory if and only if its $\mathscr{L}_1$ cross-sectional algebra does.

Similarly, suppose that

$$\gamma: \mathbb{E} \to K \to H$$

is a central extension of the circle group $\mathbb{E}$ by a locally compact group H. We shall say that the γ-*representation theory of H is of Type* I if every primary γ-representation of H is of Type I (i.e., is a direct sum of copies of a single irreducible γ-representation of H). In particular, if γ is the direct product extension, the γ-representation theory of H is of Type I if and only if H itself has a Type I unitary representation theory, that is, every primary unitary representation of H is a direct sum of copies of a single irreducible unitary representation.

Theorem 5.8 gives the following very useful criterion for $\mathscr{B}$ to have a Type I *-representation theory.

Theorem.   *Assume conditions (12), (13), and (14) of 5.8. From each orbit θ in $\hat{A}^+$ choose an element D_θ of θ; let H_θ be the (closed) stability subgroup of G for D_θ; and let*

$$\gamma_\theta : E \to K_\theta \to H_\theta \tag{16}$$

be a central extension belonging to the Mackey obstruction of D_θ. Then the following two conditions are equivalent:

(I)   *$\mathscr{B}$ has a Type I *-representation theory.*

(II)   *For every orbit θ in $\hat{A}^+$, the γ_θ^{-1}-representation theory of H_θ is of Type I.*

Proof.   Let T be a primary *-representation of $\mathscr{B}$. By Theorem 5.8 T is associated with some orbit θ, and is obtained by the construction

$$T = \operatorname*{Ind}_{\mathscr{B}_{H_\theta} \uparrow \mathscr{B}} (S^W) \tag{17}$$

from some primary γ_θ^{-1}-representation W of H_θ. Now we saw in VI.24.9 that the property of being primary of Type I depends only on the *-isomorphism class of the commuting algebra of the *-representation. Since by 5.8 the T and W of (17) have *-isomorphic commuting algebras, T is of Type I if and only if W is. This implies the equivalence of (I) and (II).   ∎

Remark 1.   If it is somehow known (see for example 5.5) that every D_θ has trivial Mackey obstruction, then (II) can be replaced by:

(II′)   *For every orbit θ in $\hat{A}^+$, the unitary representation theory of H_θ is of Type I.*

Remark 2.   Suppose that θ is an orbit whose stability subgroup H_θ is compact. Then by III.2.4 the K_θ of (16) is also compact; and so by VI.24.12 and IX.8.10 the γ_θ^{-1}-representation theory of H_θ is of Type I.

Remark 3.   Suppose that H_θ is not compact. Then, even if H_θ has a Type I unitary representation theory—in fact, even if H_θ is Abelian—we canno conclude that the γ_θ^{-1}-representation theory of H_θ is necessarily of Type I. A an example, Mackey has constructed (see Mackey [8], p. 306, Example 2) ar Abelian group H and a central extension γ of $\mathbb{E}$ by H such that the γ representation theory of H is essentially the same as the *-representatior theory of the CAR algebra of VI.17.5. Since the latter is known to hav primary *-representations which are not of Type I (see Glimm [1]), it follows that the γ-representation theory of H is not of Type I.

General Remarks

5.10. *Iteration of the Mackey Analysis.* As in VII.10.11 let us say that a Banach *-algebra C *has a smooth *-representation theory* if every irreducible *-representation of C is semicompact. The Banach *-algebraic bundle $\mathscr{B}$ will be said to *have a smooth *-representation theory* if its $\mathscr{L}_1$ cross-sectional algebra has a smooth *-representation theory.

In VII.10.11 we quoted a deep theorem of Glimm and Sakai to the effect that a *Banach *-algebra has a smooth *-representation theory if and only if it has a Type I *-representation theory.* Thus, if the unit fiber algebra A of $\mathscr{B}$ has a smooth *-representation theory, and if in addition conditions (13), (14), and 5.9(II) hold, we can conclude from Theorem 5.9 and the Glimm-Sakai Theorem that $\mathscr{B}$ also has a smooth *-representation theory.

This is important in situations where the generalized Mackey analysis requires to be *iterated.* For example, suppose that N and M are two closed subgroups of the base group G with $M \subset N$, N being normal in G and M being normal in N. To classify $\hat{\mathscr{B}}$, it might be necessary to proceed in two stages: The first stage would be to classify $(\mathscr{B}_N)\hat{\ }$ by means of the Mackey analysis applied to N and M (instead of G and N; recall 2.3). Then, with the knowledge of $(\mathscr{B}_N)\hat{\ }$ so obtained, the second stage would be to classify $\hat{\mathscr{B}}$ by another application of the Mackey analysis to G and N.

In thinking of this iteration process, one has to ask whether in fact the Mackey analysis as applied in the first stage gives us enough information about $(\mathscr{B}_N)\hat{\ }$ to enable us to proceed with the second stage. Earlier in this number we pointed out that if hypothesis (12) and the other general hypotheses hold in the first stage, we shall indeed be able to conclude that $\mathscr{B}_N$ has a smooth *-representation theory; that is, hypothesis (12) will hold for the second stage. However, there remains a real deficiency; for the verification of hypothesis (13) in the second stage requires a knowledge of $(\mathscr{B}_N)\hat{\ }$ not merely as a set but as a topological space; and the Mackey analysis as we have developed it does not give enough information about the topology of $\hat{\mathscr{B}}$. This deficiency is somewhat mitigated by two circumstances: First, the verification of Condition (C) of 2.16 (which in the second-countable situation is equivalent to (13)) requires only a knowledge of the Borel σ-field of $\hat{A}^{+}$ —a much weaker structure than the topology of $\hat{A}^{+}$. Secondly, once we know enough about the topology (or Borel σ-field) of $(\mathscr{B}_N)\hat{\ }$ to be sure that condition (13) holds, the topology of $(\mathscr{B}_N)\widehat{}$ never again enters the computation of $\hat{\mathscr{B}}$.

It is thus very natural to try to supplement the Mackey analysis so that it will provide complete information on the topology (or at least the Borel σ-field) of $\hat{\mathscr{B}}$. Theorem 5.6 gives us this topology as relativized to the space of those elements of $\hat{\mathscr{B}}$ which are associated with a single orbit (at least under favorable general conditions). But of course we need more than this—we need the topology as it relates elements associated with different orbits. For partial successes so far achieved in this direction we refer the reader to general articles such as Glimm [5], Baggett [4, 10], Fell [8], Green [1]. Important topological results have also been obtained for the structure spaces of specific classes of groups; see for example I. D. Brown [2] on nilpotent groups.

5.11.   *The Effros–Hahn Conjecture.* Suppose that $\mathscr{B}$ satisfies (12) and (14), but that hypothesis (13) fails (i.e., the action of G on $\hat{A}^+$ is not "smooth"). Then there is no reason to suppose that the elements of $\hat{\mathscr{B}}$ obtained by the Mackey analysis (Theorem 5.2) from the different orbits in $\hat{A}^+$ will exhaust all of $\hat{\mathscr{B}}$. However, suppose we moderate our ambition, and try to classify not the elements of $\hat{\mathscr{B}}$ but those of the primitive ideal space $\mathrm{Prim}(\mathscr{B})$ of $\mathscr{B}$ (that is, the primitive ideal space of the C^*-completion of the $\mathscr{L}_1$ cross-sectional algebra of $\mathscr{B}$). In other words, suppose we try to classify $\hat{\mathscr{B}}/\sim$, where $\sim$ is the relation of regional equivalence (see VII.2.2). Effros and Hahn [1], 7.4, conjectured that for this weaker purpose the Mackey analysis *is* indeed adequate even if the action of G on $\hat{A}^+$ is not "smooth" (at least in the limited context considered by them). More precisely, they conjectured that if (12) and (14) hold, then (even if (13) fails) every element of $\hat{\mathscr{B}}$ will be regionally equivalent to some element of $\hat{\mathscr{B}}$ obtained as in Theorem 5.2 from some orbit in $\hat{A}^+$. In 1979 a generalized version of this conjecture (containing the original Effros–Hahn version but still working entirely in the group context) was proved by Gootman and Rosenberg [1]. Their method of proof is based on a "local cross-section" theorem together with a "local" version of Mackey's original proof of the Imprimitivity Theorem, and relies essentially on second countability. Green [5] extended the Gootman–Rosenberg result to the context of homogeneous C^*-algebraic bundles (his "twisted crossed products").

Suppose that we are in a situation where (12) and (14) hold and the assumptions of the Gootman–Rosenberg Theorem are satisfied. Suppose further that $\mathscr{B}$ is known by some independent means to have a Type I *-representation theory. By the theorem of Glimm and Sakai, together with VI.15.18, this implies that $\hat{\mathscr{B}} = \mathrm{Prim}(\mathscr{B})$, that is, regional equivalence and unitary equivalence are the same for irreducible *-representations of $\mathscr{B}$. Therefore, by the Gootman–Rosenberg Theorem, the generalized Mackey analysis yields all the elements of $\hat{\mathscr{B}}$ (up to unitary equivalence), even if (13) fails. See Example 9.18.

5.12.   *The Case that $\mathscr{B}$ is not Saturated.*   The hypothesis that $\mathscr{B}$ is saturated has been fundamental to the whole development of this chapter. If $\mathscr{B}$ were not saturated we could not even define the action of G on $\hat{A}^{+}$. However, if $\mathscr{B}$ is not saturated, all is not lost. Indeed, suppose that $\mathscr{B}$ is a $C*$-algebraic bundle which is *not* saturated; and let us form its saturated part $\mathscr{D}$ as in VIII.16.6. Now $\mathscr{D}$ is a "bundle ideal" of $\mathscr{B}$; that is, every element of $\mathscr{B}$ gives rise to a multiplier of $\mathscr{D}$. Therefore by VIII.15.3 every irreducible *-representation of $\mathscr{D}$ extends to an irreducible *-representation of $\mathscr{B}$ (acting on the same space). Conversely, as in V.1.18, the restriction to $\mathscr{D}$ of an irreducible *-representation of $\mathscr{B}$ is either zero or irreducible. Hence $\hat{\mathscr{D}}$ can be identified with the subset $\{T \in \hat{\mathscr{B}} : T_d \neq 0 \text{ for some } d \text{ in } \mathscr{D}\}$ of $\hat{\mathscr{B}}$. Since $\mathscr{D}$ is saturated, we can try to apply to it the Mackey analysis. If the attempt is successful—that is, if $\mathscr{D}$ satisfies hypotheses (12), (13), and (14)—then we shall have classified at least the elements of the well-defined subset $\hat{\mathscr{D}}$ of $\hat{\mathscr{B}}$.

5.13.   *The Analysis in Terms of Non-Normal Subgroups.*   The Mackey analysis of $\hat{\mathscr{B}}$ began in 2.1 with the fixing of a closed normal subgroup N of G; and it proceeded to classify the elements of $\hat{\mathscr{B}}$ in terms of their behavior when restricted to $\mathscr{B}_N$. Now there is no reason why one should not attempt the same program even if N is not normal. Indeed, in XI.7.8 we obtained a result somewhat similar to Step II of the Mackey analysis valid (in the group context) when N is *compact*. One might expect that both the "compact subgroup analysis" of XI.7.8 and the normal subgroup analysis of this chapter will eventually turn out to be special cases of a more general analysis which starts from an *arbitrary* closed subgroup N of G, and analyzes the *-representations of $\hat{\mathscr{B}}$ in terms of their restrictions to $\mathscr{B}_N$. Attempts to develop a more general analysis of this sort have so far been unsuccessful.

6.   Saturated *C*-Algebraic Bundles with Unit Fibers of Compact Type*

6.1.   As the reader may recall from VI.23.3, a $C*$-algebra A is of *compact type* if it is a $C*$-direct sum of elementary $C*$-algebras. For such an A the space $\hat{A}$ is discrete, and for each D in $\hat{A}$ the range of D consists of all the compact operators on $X(D)$ (see VII.5.24).

In this section we are going to use the theory of §4 to describe in detail the structure of a saturated C^*-algebraic bundle $\mathscr{B}$ (over a locally compact group) whose unit fiber C^*-algebra is of compact type. It will turn out that any such $\mathscr{B}$ is a C^*-direct sum of C^*-algebraic bundles of the canonical type constructed in VIII.16.15-17.

By far the greater part of the proof of this fact will be the analysis of those $\mathscr{B}$ which are of transitive type in the following sense:

Definition. A saturated C^*-algebraic bundle $\mathscr{B}$ over a locally compact group G, with unit fiber C^*-algebra A, will be said to be *of transitive type* if the action of G on $\hat{A}$ by conjugation (see §XI.16) is transitive.

6.2. Our first step must be to determine the conjugating action of the group (6.3) and also the Mackey obstruction (6.4) for the C^*-algebraic bundles $\mathscr{B}$ of the canonical type constructed in VIII.16.15-17.

Let us fix ingredients A, G, D_0, K, γ as in VIII.16.15. That is, A is a C^*-algebra of compact type; G is a locally compact group (with unit e) acting *transitively* and continuously to the left on the discrete space $\hat{A}$ (the action being denoted by $\langle x, D \rangle \mapsto xD$); D_0 is a fixed element of $\hat{A}$; K is the stability subgroup of G for D_0; and

$$\gamma: \mathbb{E} \underset{\sigma}{\to} L \underset{\rho}{\to} K$$

is a central extension of the circle group $\mathbb{E}$ by K. Let $\mathscr{Q} = \langle Q, \pi', \cdot, * \rangle$ be the cocycle bundle over K determined by γ as in VIII.16.15; and let $\mu: G \times \hat{A} \to K$ be the function constructed as in VIII.16.16. We construct canonically from these ingredients the saturated C^*-algebraic bundle $\mathscr{B} = \langle B, \pi, \cdot, * \rangle$ over G as in VIII.16.17, having A as its unit fiber C^*-algebra.

6.3. Proposition. *The action* $\langle x, D \rangle \mapsto {}^x D$ *of* G *on* $\hat{A}$ *by conjugation in* $\mathscr{B}$ *is identical with the initially given action* $\langle x, D \rangle \mapsto xD$ *of* G *on* $\hat{A}$.

Proof. Fix $D \in \hat{A}$, $x \in G$, $0 \neq \xi \in X(D)$, and $q \in Q_{\mu_{x,D}}$ with $\|q\| = 1$ (so that $q^*q = 1 \in Q_e$).

Let $a \in \mathcal{O}_c(X(D), X(xD))$, and consider the element $b = a \otimes q$ of B_x (see VIII.16.17); thus

$$b^*b = a^*a \otimes 1 \in A.$$

So (recalling from §XI.16 the definition of ${}^x D$ and κ_x) we have

$$\|\kappa_x(b \otimes \xi)\|^2 = (D_{b^*b}\xi, \xi) = (D_{a^*a}\xi, \xi) = \|a\xi\|^2.$$

Thus the map

$$a\xi \mapsto \kappa_x(b \otimes \xi) \qquad (a \in \mathcal{O}_c(X(D), X(xD))) \qquad (1)$$

is a linear isometry of $X(xD)$ into $X(^xD)$. If $c \in A = B_e$, then clearly $cb = ca \otimes q$; so (1) is an intertwining operator for the two representations xD and ^xD. Since (1) is an isometry and since ^xD is irreducible, the range of (1) must be all of ^xD; so xD and ^xD are unitarily equivalent under (1). ∎

In particular, then, the $\mathscr{B}$ of VIII.16.17 are of transitive type.

6.4. What is the Mackey obstruction for D_0?

Let $\mathscr{Y} = \{Y_x\}_{x \in G}$ be the Hilbert bundle induced as in §XI.9 by D_0; and let τ be the "action" of $\mathscr{B}$ on $\mathscr{Y}$ as defined in §XI.9. Let $\mathscr{B}_K$ and $\mathscr{Y}_K$ be as usual the reductions of $\mathscr{B}$ and $\mathscr{Y}$ to K; and for $b \in \mathscr{B}_K$ let τ_b^K denote the restriction of τ_b to $\mathscr{Y}_K$.

We have seen in 4.11–4.21 that the Mackey obstruction for D_0 is the central extension

$$\gamma_0: \mathbb{E} \xrightarrow{\sigma_0} H \xrightarrow{\rho_0} K \qquad (2)$$

constructed as follows: For each $k \in K$, H_k consists of all those norm-preserving bicontinuous bijections $\phi: \mathscr{Y}_K \to \mathscr{Y}_K$ such that (i) for each $m \in K$, $\phi | Y_m$ carries Y_m linearly onto $Y_{mk^{-1}}$, and (ii) ϕ commutes with τ_b^K for all $b \in \mathscr{B}_K$. The union $H = \bigcup_{k \in K} H_k$ is a topological group under the operation of composition and the topology of pointwise convergence on $\mathscr{Y}_K$. The homomorphism ρ_0 of (2) is given by $\rho_0^{-1}(k) = H_k$ $(k \in K)$; and σ_0 is the homomorphism sending each $u \in \mathbb{E}$ into scalar multiplication by u on each fiber. (As we saw in 4.11–4.21, this description of the Mackey obstruction is valid in a context far more general than the present $\mathscr{B}$.)

Proposition. *In the context of 6.2, the Mackey obstruction of D_0 is the central extension class inverse to γ.*

Proof. We begin by remarking that each fiber Q_h of $\mathcal{Q}$ can be regarded as a one-dimensional Hilbert space under the inner product

$$(p, q) = q^*p \in Q_e \cong \mathbb{C}.$$

Fix $k \in K$. The fiber Y_k of $\mathscr{Y}_K$ is spanned by the $\kappa_k(b \otimes \xi)$ $(b \in B_k$; $\xi \in X(D_0))$; but, since $kD_0 = D_0$, it is clear that components other than the D_0-component of b do not affect $\kappa_k(b \otimes \xi)$. Hence (recalling by VIII.16.16

that $\mu_{k,D_0} = k$) we see that Y_k is the completion of $\mathcal{O}_c(X(D_0)) \otimes Q_k \otimes X(D_0)$ under the inner product

$$(a \otimes q \otimes \xi, a' \otimes q' \otimes \xi') = (q'^*q)(a'^*a\xi, \xi')_{X(D_0)}$$

$$= (q'^*q)(a\xi, a'\xi')_{X(D_0)}.$$

Thus the equation

$$\Phi_k(\kappa_k(b \otimes \xi)) = b_{D_0}\xi \otimes q_{D_0} \tag{3}$$

(where $b \in B_k$, $b = \sum_{D \in A}^{\oplus} b_D \otimes q_D$; $\xi \in X(D_0)$) defines a linear surjective isometry $\Phi_k \colon Y_k \to X(D_0) \otimes Q_k$ of Hilbert spaces.

Let Φ stand for the union of the Φ_k ($k \in K$), so that Φ is a surjective isometry, linear on each fiber, of $\mathscr{Y}_K$ onto $X(D_0) \otimes \mathscr{Q}$. (Here $X(D_0) \otimes \mathscr{Q}$ is the tensor product Hilbert bundle over K defined in II.15.16). We claim that Φ is bicontinuous. Indeed: By Proposition II.13.17 it is enough to show that Φ is continuous. To prove this it is enough (by II.13.16) to choose a family Γ of continuous cross-sections of $\mathscr{Y}_K$ such that

$$\text{for each } k \in K, \{\phi(k) \colon \phi \in \Gamma\} \text{ has dense linear span in } Y_k, \tag{4}$$

and show that $\Phi \circ \phi$ is continuous for each $\phi \in \Gamma$. As the family Γ let us take the collection of all $\phi \colon k \mapsto \kappa_k(\psi(k) \otimes \xi)$, where $\xi \in X(D_0)$ and ψ is a continuous cross-section of $\mathscr{B}_K$ of the form $k \mapsto \alpha \otimes q(k)$, α being in $\mathcal{O}_c(X(D_0))$ and q a continuous cross-section of $\mathscr{Q}$. For such a ϕ we have by (3)

$$\Phi_k(\phi(k)) = \alpha(\xi) \otimes q(k) \qquad (k \in K). \tag{5}$$

Since the right side of (5) exhausts $X(D_0) \otimes Q_k$ and Φ_k is an isometry, it follows from (5) that the $\phi(k)$ (k fixed, $\phi \in \Gamma$) exhaust Y_k; hence the family Γ has property (4). The continuity of $\Phi \circ \phi$ ($\phi \in \Gamma$) is clear from (5). So Φ is continuous, hence bicontinuous.

Thus Φ identifies $\mathscr{Y}_K$ with $X(D_0) \otimes \mathscr{Q}$.

If $b \in \mathscr{B}_K$, let $\check{\tau}_b$ be the image of τ_b^K under Φ, that is:

$$\check{\tau}_b \circ \Phi = \Phi \circ \tau_b^K.$$

Now suppose k, $m \in K$, $c \in B_m$, $c = \sum_{D \in A}^{\oplus} c_D \otimes p_D$ ($c_D \in \mathcal{O}_c(X(D), X(mD))$, $p_D \in Q_{\mu_m, D}$); and let ϕ be as (5) above. Then by (5)

$$\check{\tau}_c(\alpha(\xi) \otimes q(k)) = \check{\tau}_c\Phi(\phi(k))$$

$$= \Phi\tau_c(\phi(k))$$

$$= \Phi(\tau_c(\kappa_k(\psi(k) \otimes \xi)))$$

$$= \Phi(\kappa_{mk}(c\psi(k) \otimes \xi)).$$

Since $c\psi(k) = c_{D_0}\alpha \otimes p_{D_0}q(k)$, we obtain (again by (5))

$$\check{\tau}_c(\alpha(\xi) \otimes q(k)) = c_{D_0}\alpha(\xi) \otimes p_{D_0}q(k). \qquad (6)$$

That is to say, $\check{\tau}_c$ acts on $X(D_0) \otimes Q_k$ by multiplying to the left by c_{D_0} in the first factor, and multiplying to the left by p_{D_0} in the second factor.

Thus, translating the definition of (2) by Φ, we can describe H_k as consisting of all those norm-preserving bicontinuous bijections w of $X(D_0) \otimes \mathcal{Q}$ onto itself such that (i) for each $m \in K$, w carries $X(D_0) \otimes Q_m$ linearly onto $X(D_0) \otimes Q_{mk^{-1}}$, and (ii) w commutes with the $\check{\tau}_c$ of (6) for all $c \in \mathcal{B}_K$.

Let $h \in L$ (see 6.2); and recall from VIII.16.15 that h can be identified with the element $\langle 1, h \rangle^\sim$ of $\mathcal{Q}$; and that this identification preserves multiplication, and sends inverse into involution and projection into projection. Making this identification let us define β_h as the map of $X(D_0) \otimes \mathcal{Q}$ into itself given by

$$\beta_h(\xi \otimes q) = \xi \otimes qh^{-1} \qquad (\xi \in X(D_0), q \in \mathcal{Q}). \qquad (7)$$

Thus, if $k = \rho(h)$, β_h carries $X(D_0) \otimes Q_m$ into $X(D_0) \otimes Q_{mk^{-1}}$. By II.13.16 β_h is continuous. By VIII.16.15 (14) β_h is an isometry. By (6) above β_h commutes with $\check{\tau}_c$ for all $c \in \mathcal{B}_K$. Evidently

$$\beta_{h_1}\beta_{h_2} = \beta_{h_1 h_2} \qquad (h_1, h_2 \in L), \qquad (8)$$

so that in particular $\beta_{h^{-1}} = (\beta_h)^{-1}$, and each β_h is *onto* $X(D_0) \otimes \mathcal{Q}$. Putting these facts together we conclude that $\beta_h \in H_k$. By (8) $h \mapsto \beta_h$ is a homomorphism of L into H, in fact a continuous homomorphism (see Lemma 4.14).

Notice that the diagram

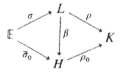

commutes. (Here $\bar{\sigma}_0(\lambda) = \sigma_0(\bar{\lambda})$ for $\lambda \in \mathbb{E}$.) Indeed: If $\lambda \in \mathbb{E}$ we have by (7)

$$\beta_{\sigma(\lambda)}(\xi \otimes q) = \xi \otimes q\sigma(\bar{\lambda})$$

$$= \bar{\lambda}(\xi \otimes q).$$

So $\beta_{\sigma(\lambda)}$ is scalar multiplication by $\bar{\lambda}$, and this is $\bar{\sigma}_0(\lambda)$; whence $\beta \circ \sigma = \bar{\sigma}_0$. Also, if $h \in L$, $\rho_0(\beta_h) = \rho(h)$.

Since the diagram commutes and since $h \mapsto \beta_h$ is continuous, it follows from Remark III.5.3 that the original γ of 6.2 is isomorphic to the central extension

$$\mathbb{E} \underset{\bar{\sigma}_0}{\to} H \underset{\rho_0}{\to} K.$$

But the latter is the inverse (see III.5.7) of the Mackey obstruction of D_0. So the Proposition is proved. ■

The Structure of an Arbitrary Saturated C*-Algebraic Bundle with Unit Fiber of Compact Type

6.5.   To fulfill the promise made in 6.1, we now fix an arbitrary saturated C^*-algebraic bundle $\mathscr{B} = \langle B, \pi, \cdot, * \rangle$ over a locally compact group G (with unit e) such that the unit fiber C^*-algebra $A = B_e$ is of compact type. Throughout 6.5–6.18 we shall assume that $\mathscr{B}$ is *of transitive type*, so that the action $\langle x, D \rangle \mapsto {}^x D$ of G on $\hat{A}$ by conjugation is transitive. Fix an element D_0 of $\hat{A}$; and let K be the stability subgroup of G for D_0.

Let $\mathscr{Y} = \{Y_x\}_{x \in G}$ be the Hilbert bundle over G induced by D_0, and τ the "action" of $\mathscr{B}$ on $\mathscr{Y}$ (see §XI.9). We construct the Mackey obstruction

$$\gamma_0 : \mathbb{E} \underset{\sigma_0}{\to} H \underset{\rho_0}{\to} K \tag{9}$$

as in 6.4 (keeping in particular the same definition of H_k as in 6.4).

6.6.   The following technical fact is vital.

Proposition.   *Let $k \in K$. Each u in H_k can be extended in exactly one way to a norm-preserving bicontinuous bijection $\breve{u} : \mathscr{Y} \to \mathscr{Y}$ such that:*

(i)   *For all $x \in G$, $\breve{u} | Y_x$ carries Y_x linearly onto $Y_{xk^{-1}}$, and*
(ii)   *$\breve{u}$ commutes with τ_b for all $b \in B$.*

Proof.   Fix $k \in K$, $u \in H_k$; and for each $x \in G$ define $v_x : B_x \otimes X(D_0) \to Y_{xk^{-1}}$ as follows:

$$v_x(b \otimes \xi) = \tau_b u(\xi) \qquad (b \in B_x; \xi \in X(D_0) \cong Y_e).$$

Thus, for $b, c \in B_x$ and $\xi, \eta \in X(D_0)$,

$$(v_x(b \otimes \xi), v_x(c \otimes \eta))_{Y_{xk^{-1}}} = (\tau_b u(\xi), \tau_c u(\eta))_{Y_{xk^{-1}}}$$

$$= (\tau_{c^*b}(\xi), u(\eta))_{Y_{k^{-1}}} \qquad \text{(by XI.9.14 (26))}$$

$$= (u\tau_{c^*b}(\xi), u(\eta))_{Y_{k^{-1}}}$$

$$\text{(commutativity property of } H_k\text{)}$$

$$= (\tau_{c^*b}\xi, \eta)_{Y_e} \qquad\qquad \text{(since } u \text{ is an isometry)}$$

$$= (\kappa_x(b \otimes \xi), \kappa_x(c \otimes \eta))_{Y_x} \qquad \text{(by XI.9.15 (29))}.$$

It follows that v_x defines an isometry $\breve{u}_x: Y_x \to Y_{xk^{-1}}$ such that

$$\breve{u}_x(\kappa_x(b \otimes \xi)) = \tau_b u(\xi) \qquad (b \in B_x; \xi \in X(D_0)), \qquad (10)$$

that is (see XI.9.15 (27))

$$\breve{u}_x \tau_b \xi = \tau_b u(\xi) \qquad (b \in B_x; \xi \in X(D_0) \cong Y_e). \qquad (11)$$

Let $\breve{u}$ be the union of the $\breve{u}_x$ ($x \in G$), so that $\breve{u}: \mathcal{Y} \to \mathcal{Y}$ is an isometry carrying each Y_x linearly into $Y_{xk^{-1}}$. From (10) and the fact that u commutes with τ_b^K for $b \in \mathcal{B}_K$ it follows that $\breve{u}$ extends u.

Let $b \in B_x$. If $c \in B_y$ and $\xi \in X(D_0)$,

$$\breve{u}\tau_b \kappa_y(c \otimes \xi) = \breve{u}\tau_{bc}\xi = \tau_{bc}u(\xi) \qquad\qquad \text{(by (11))}$$

$$= \tau_b \tau_c u(\xi) = \tau_b \breve{u}\kappa_y(c \otimes \xi) \qquad \text{(by (10))}.$$

By the arbitrariness of c and ξ this implies

$$\breve{u}\tau_b = \tau_b \breve{u} \qquad \text{for all } b \in \mathcal{B}. \qquad (12)$$

A similar argument shows that

$$(uv)^{\smile} = \breve{u}\breve{v} \qquad \text{for all } u, v \text{ in } H. \qquad (13)$$

Evidently $\breve{u}$ is the identity on $\mathcal{B}$ if u is the identity on $\mathcal{B}_K$. Hence by (13) $(u^{-1})^{\smile} = (\breve{u})^{-1}$, so that $\breve{u}: \mathcal{Y} \to \mathcal{Y}$ is surjective, hence an isometric bijection.

Next we claim that $\breve{u}$ is continuous. To see this, it is enough by II.13.16 to take a continuous cross-section γ of $\mathcal{B}$ and a vector $\xi \in X(D_0)$, form the cross-section $\phi: x \mapsto \kappa_x(\gamma(x) \otimes \xi)$ of $\mathcal{Y}$ (continuous by XI.9.2), and show that $\breve{u} \circ \phi$ is continuous. But $\breve{u}(\phi(x)) = \tau_{\gamma(x)}u(\xi)$; and this is continuous in x by the continuity of τ (XI.9.18).

Applying the preceding paragraph to $(\breve{u})^{-1} = (u^{-1})^{\smile}$, we see that $\breve{u}$ is in fact bicontinuous.

Thus, recalling (12), we conclude that $\breve{u}$ is an extension of u with the properties required in the Proposition.

It remains only to show that the extension is unique. But this follows immediately from (11).   ∎

6.7. Let us topologize the set M of all maps of $\mathcal{Y}$ into $\mathcal{Y}$ with the topology of pointwise convergence; that is, $q_i \to q$ in M if and only if $q_i(\zeta) \to q(\zeta)$ in $\mathcal{Y}$ for all ζ in $\mathcal{Y}$. Then we have:

Proposition.   *The correspondence $u \mapsto \breve{u}$, defined in 6.6, is bicontinuous.*

Proof.   Recall (from 6.4) that the topology of H is also that of pointwise convergence; so, since $\breve{u}$ extends u, the inverse map $\breve{u} \mapsto u$ is certainly continuous. To prove the continuity of $u \mapsto \breve{u}$, assume that $u_i \to u$; we shall show that $(u_i)^{\smile} \to \breve{u}$. By the linear and isometric properties of $\breve{u}$, together with II.13.12, it is enough to show that

$$(u_i)^{\smile}(\kappa_x(b \otimes \xi)) \to \breve{u}(\kappa_x(b \otimes \xi)) \tag{14}$$

for $x \in G$, $b \in B_x$, $\xi \in X(D_0)$. But $(u_i)^{\smile}(\kappa_x(b \otimes \xi)) = \tau_b u_i(\xi)$ and $\breve{u}(\kappa_x(b \otimes \xi)) = \tau_b u(\xi)$: so (14) follows from the continuity of τ_b (XI.9.18).   ∎

6.8.   Our approach to the structure theory of $\mathscr{B}$ will consist of two steps: First we shall show that the entire structure of $\mathscr{B}$ is determined (to within isomorphism) once we know the isomorphism class of the induced bundle $\mathscr{Y}$ together with the collection $\{\breve{u} : u \in H\}$ of isometries of $\mathscr{Y}$; this will occupy 6.9–6.13. Secondly, we shall show in 6.14–6.16 that $\mathscr{Y}$ together with the collection of all $\breve{u}$ is determined (to within isomorphism) by A, the action of G on $\hat{A}$, and the Mackey obstruction of an element D_0 of $\hat{A}$.

6.9.   Recalling that the action of G on $\hat{A}$ is transitive, for each $D \in \hat{A}$ we choose a fixed element ξ_D of G such that ${}^{(\xi_D)}D_0 \cong D$ (taking in particular $\xi_{D_0} = e$). It will be convenient to identify each D in $\hat{A}$ with the (equivalent) conjugated representation ${}^{(\xi_D)}D_0$, so that the space $X(D)$ of D is just Y_{ξ_D}.

Now for each $D \in \hat{A}$ let $\mathscr{X}^D$ be the Hilbert bundle over G with constant fiber $X(D)$; and let $\mathscr{Y}^D$ be the Hilbert bundle retraction of $\mathscr{Y}$ by the map $x \mapsto x\xi_D$ of G onto G (so that $\mathscr{Y}^D$ is a Hilbert bundle over G whose x-fiber is $Y_{x\xi_D}$). Let $\mathscr{W}^D = \mathcal{O}_c(\mathscr{X}^D, \mathscr{Y}^D)$ be the Banach bundle over G whose fiber at x consists of all compact operators from $X(D)$ to $Y_{x\xi_D}$ (the bundle space being topologized as in VII.8.17). Finally, let

$$\mathscr{W} = \sum_{D \in A}^{\oplus 0} \mathscr{W}^D$$

be the C_0 direct sum Banach bundle defined in II.13.20.

6.10.   We shall set up a map $\tau^0 : \mathscr{B} \to \mathscr{W}$ as follows: If $b \in B_x$ ($x \in G$), let $\tau^0(b)$ be the element of $W_x = \sum_{D \in \hat{A}}^{\oplus 0} \mathcal{O}_c(X(D), Y_{x\xi_D})$ given by

$$\tau^0(b) = \sum_{D \in \hat{A}}^{\otimes 0} (\tau_b | X(D)). \tag{15}$$

We must of course verify that $\tau^0(b)$ is indeed in W_x. Since τ_b carries $X(D) = Y_{\xi_D}$ into $Y_{x\xi_D}$, and since $\tau_b^* \tau_b = \tau_{b^*b}$ restricted to $X(D)$ is a compact operator (coinciding with D_{b^*b}), it follows that $\tau_b|X(D) \in \mathcal{O}_c(X(D), Y_{x\xi_D})$ for each D. Furthermore $\|\tau_b|X(D)\|^2 = \|D_{b^*b}\| \to 0$ as $D \to \infty$ in $\hat{A}$ (since A is of compact type). Therefore $\tau^0(b) \in W_x$.

Evidently τ^0 is linear on each fiber B_x.

Also τ^0 is an isometry; for, if $b \in B_x$,

$$\|\tau^0(b)\|^2 = \sup_{D \in \hat{A}} \|\tau_b|X(D)\|^2$$

$$= \sup_{D \in \hat{A}} \|D_{b^*b}\|$$

$$= \|b^*b\| = \|b\|^2.$$

Proposition. $\tau^0: \mathcal{B} \to \mathcal{W}$ is a surjection.

Proof. Since τ^0 is an isometry, $\tau^0(B_x)$ is closed in W_x; hence it is enough to fix $x \in G$ and show that $\tau^0(B_x)$ is dense in W_x. Since the elements of W_x of finite total rank are dense in W_x, it is enough to fix an element α of W_x of finite total rank, and show that $\alpha \in \tau^0(B_x)$.

Such an α is of the form $\sum_{D \in \hat{A}}^{\oplus} \alpha_D$, where (i) $\alpha_D: X(D) \to Y_{x\xi_D}$ is linear and of finite rank for each $D \in \hat{A}$, and (ii) all but finitely many of the α_D are 0. Since $A \cong \sum_{D \in \hat{A}}^{\oplus 0} \mathcal{O}_c(X(D))$, this implies that there is an element $p \in A$ such that

$$\alpha_D \tau_p = \alpha_D \quad \text{on each } X(D). \tag{16}$$

Now by Lemma XI.11.9 there is an approximate unit $\{u_r\}$ of $\mathcal{B}$, and a constant k, such that

$$\|u_r\| \le k \quad \text{for all } r, \tag{17}$$

and each u_r is of the form

$$u_r = \sum_{j=1}^{n_r} (b_r^j)^* b_r^j, \quad \text{where each } b_r^j \in B_x. \tag{18}$$

For each D, j and r, $\alpha_D \tau_{(b_r^j)^*}: Y_{x\xi_D} \to Y_{\xi_D} \to Y_{x\xi_D}$; and $\alpha_D \tau_{(b_r^j)^*}$ is of finite rank (since α_D is) and vanishes for all but finitely many D. Now, x being fixed, ^xD runs over $\hat{A}$ when D runs over $\hat{A}$; hence, given a collection $\{\beta_D\}_{D \in \hat{A}}$ such that $\beta_D \in \mathcal{O}_c(Y_{x\xi_D})$ and $\lim_{D \to \infty} \|\beta_D\| = 0$, there exists an element c of A such that $(^xD)_c = \beta_D$ for all $D \in \hat{A}$. In particular, then, for each j and r there is an element c_r^j of A such that

$$\alpha_D \tau_{(b_r^j)^*} = \tau_{c_r^j} \quad \text{on } Y_{x\xi_D} \text{ for each } D \in \hat{A}. \tag{19}$$

It follows from (18) and (19) that, on each $Y_{\xi_D} = X(D)$,

$$\alpha_D \tau_{u_r} = \sum_{j=1}^{n_r} \alpha_D \tau_{(b_r^j)^* b_r^j} = \sum_{j=1}^{n_r} \tau_{c_r^j} \tau_{b_r^j}$$

$$= \tau_{v_r},$$

where $v_r = \sum_{j=1}^{n_r} c_r^j b_r^j \in B_x$. This and (16) imply that

$$\alpha_D \tau_{pu_r} = \tau_{v_r} \qquad \text{on each } X(D). \tag{20}$$

Now $\{u_r\}$ is an approximate identity of A, so that $\|pu_r - p\| \to 0$. It follows that $\|\tau_{pu_r}^0 - \tau_p^0\| \to 0$, whence by (16) and (20)

$$\alpha = \lim_r \tau^0(v_r) \qquad \text{in } W_x.$$

But, since $\tau^0(B_x)$ is closed in W_x, this implies that $\alpha \in \tau^0(B_x)$, which was to be proved.   ■

6.11.   We have shown that $\tau^0 : \mathscr{B} \to \mathscr{W}$ is an isometric bijection (linear on each fiber). We shall now show that it is bicontinuous, and hence that $\mathscr{B}$ and $\mathscr{W}$ are isometrically isomorphic as Banach bundles.

Proposition. τ^0 *is bicontinuous.*

Proof.   We shall first show that τ^0 is continuous. Let us take an element D of $\hat{A}$, vectors ξ_0, η_0 in $X(D)$, and a continuous cross-section ϕ of $\mathscr{B}$; and define u as the element of A (of rank 1) given by

$$D_u \xi = (\xi, \xi_0)\eta_0 \qquad\qquad (\xi \in X(D)).$$

$$D'_u = 0 \qquad \text{for } D \neq D' \in \hat{A}.$$

By II.13.16 and the denseness of $B_x A$ in B_x, it is enough to show that the cross-section

$$x \mapsto \tau^0(\phi(x)u) \tag{21}$$

of $\mathscr{W}$ is continuous. But notice that $\tau^0(\phi(x)u)$ vanishes in components other than D, and that if $\xi \in X(D)$

$$\tau^0(\phi(x)u)\xi = \tau_{\phi(x)u}\xi$$

$$= \tau_{\phi(x)}\tau_u \xi$$

$$= \tau_{\phi(x)}D_u \xi$$

$$= (\xi, \xi_0)\tau_{\phi(x)}\eta_0.$$

From this, the continuity of τ, and the definition of the topology of $\mathscr{W}$ (VII.8.17), it follows that (21) is continuous. So τ^0 is continuous.

The continuity of the inverse of τ^0 now follows from II.13.17.   ∎

6.12. We have seen in 6.11 that $\mathscr{B}$ is determined *as a Banach bundle* by $\mathscr{W}$, and hence (in view of the definition of $\mathscr{W}$) by $\mathscr{Y}$. What about the multiplication and involution operations on $\mathscr{B}$?

We recall from 6.5 and 6.6 the definition of H and of $\breve{u}$ ($u \in H$).

Lemma. *For each* $b \in B$, τ_b *is the unique map* $\tau' : \mathscr{Y} \to \mathscr{Y}$ *with the following properties*: (i) τ' *commutes with* $\breve{u}$ *for all* u *in* H; (ii) $\tau' | X(D) = \tau_b | X(D)$ *for all* $D \in \hat{A}$ (*i.e.* $\tau^0(b) = \sum_{D \in \hat{A}}^{\oplus} (\tau' | X(D))$).

Proof. Fix $x \in G$; we must show that τ' and τ_b coincide on Y_x.

Let $D \in \hat{A}$ and $k \in K$ be such that $\xi_D k^{-1} = x$; and choose an element u of H_k. By 6.6 $\breve{u}$ carries $X(D) = Y_{\xi_D}$ linearly and isometrically onto Y_x. Thus, if $\zeta \in Y_x$ and $\xi = \breve{u}^{-1}(\zeta)$,

$$\tau'(\zeta) = \tau' \breve{u}(\xi) = \breve{u} \tau'(\xi) \qquad \text{(by (i))}$$

$$= \breve{u} \tau_b(\xi) \qquad \text{(by (ii))}$$

$$= \tau_b \breve{u}(\xi) = \tau_b(\zeta) \qquad \text{(by (ii), Proposition 6.6).}$$

So τ' and τ_b coincide on Y_x.   ∎

6.13. Let $\mathscr{W}$ be equipped with the multiplication $\cdot$ and involution $*$ which are covariant under τ^0 with the multiplication and involution in $\mathscr{B}$. Thus (see 6.11) $\mathscr{W}$ is isometrically isomorphic with $\mathscr{B}$ as a C*-algebraic bundle. In view of the above Lemma, and the fact that $b \mapsto \tau_b$ is multiplicative and involutive, the multiplication and involution in $\mathscr{W}$ can be described as follows:

Proposition. *Given* $\alpha = \sum_{D \in \hat{A}}^{\oplus} \alpha_D \in \mathscr{W}$, *let* $\alpha' : \mathscr{Y} \to \mathscr{Y}$ *be the map* (*existing and unique by Lemma 6.12*) *such that* (i) α' *commutes with* $\breve{u}$ *for all* $u \in H$, *and* (ii) *for each* $D \in \hat{A}$, α' *and* α_D *coincide on* $X(D)$. *Then, if* $\alpha_1, \alpha_2, \alpha \in \mathscr{W}$,

$$(\alpha_1 \cdot \alpha_2)' = \alpha_1' \alpha_2' \qquad (composition \ of \ maps), \tag{22}$$

$$(\alpha^*)' = (\alpha')^* \qquad (involution \ in \ the \ Hilbert \ space \ fiber \ Y_x) \tag{23}$$

Proof. By 6.12, $\alpha' = \tau_{(\tau^0)^{-1}(\alpha)}$.   ∎

Thus $\mathscr{W}$ (and hence also $\mathscr{B}$) is determined to within isometric isomorphism *as a Banach *-algebraic bundle* by (22) and (23), once $\mathscr{Y}$ and the collection $\{\breve{u} : u \in H\}$ are known.

6.14. Since by 6.13 $\mathscr{B}$ is essentially determined by $\mathscr{Y}$ together with $\{\breve{u} : u \in H\}$, it is important to investigate just what $\mathscr{Y}$ and the $\breve{u}$ really depend upon. We shall next obtain a canonical form for $\mathscr{Y}$ and $\{\breve{u} : u \in H\}$ which depends only on (a) A, (b) $\hat{A}$ as a transitive G-space, and (c) the Mackey obstruction (9) of the element D_0 of $\hat{A}$.

We begin by forming the central extension

$$\tilde{\gamma}_0 : \mathbb{E} \underset{\sigma_0}{\to} \tilde{H} \underset{\tilde{\rho}_0}{\to} K \tag{24}$$

inverse to the γ_0 of (9). ($\tilde{H}$ is the reverse group of H, and $\tilde{\rho}_0(h) = \rho_0(h^{-1})$; see Remark III.5.7.) From $\tilde{\gamma}_0$ we form the corresponding cocycle bundle $\mathscr{Q} = \{Q_k\}_{k \in K}$ is in VIII.4.7 (see also VIII.16.15). Thus $\mathscr{Q}$ is a cocycle bundle over K. The unit fiber Q_e is identified with $\mathbb{C}$ under the correspondence $\langle \lambda, \sigma_0(u) \rangle^\sim \leftrightarrow \lambda u$ ($\lambda \in \mathbb{C}$, $u \in \mathbb{E}$) (see VIII.16.15).

Observe that $\mathscr{Q}$ can also be considered as a Hilbert bundle, with inner products

$$(\alpha, \beta)_{Q_k} = \beta^* \alpha \in Q_e \cong \mathbb{C} \qquad (\alpha, \beta \in Q_k). \tag{25}$$

6.15. Regarding $\mathscr{Q}$ as a Hilbert bundle, for each $D \in \hat{A}$ we form the tensor product Hilbert bundle $X(D) \otimes \mathscr{Q}$ over K (see II.15.16); and we let $\mathscr{Z}^D$ be the Hilbert bundle reatraction of $X(D) \otimes \mathscr{Q}$ by the map $\xi_D k \mapsto k$ of $\xi_D K$ onto K. Thus $\mathscr{Z}^D$ is a Hilbert bundle over the coset $\xi_D K$. Now let $\mathscr{Z}$ be the Hilbert bundle over G consisting of the disjoint union of the $\mathscr{Z}^D$; that is, $Z_x = Z_x^D$ for $x \in \xi_D K$, each $\mathscr{Z}^D$ is open in $\mathscr{Z}$, and $\mathscr{Z}$ coincides over $\xi_D K$ with $\mathscr{Z}^D$. (Recall that each coset $\xi_D K$ is open in G.)

For each u in H let $\breve{u}$ be the extension of u defined in 6.6.

Proposition. $\mathscr{Z}$ and $\mathscr{Y}$ are isometrically isomorphic as Hilbert bundles over G, under the isometric isomorphism $\Phi : \mathscr{Z} \to \mathscr{Y}$ given by

$$\Phi(\xi \otimes \alpha) = \lambda \breve{u}(\xi) \tag{26}$$

(where $D \in \hat{A}$, $\xi \in X(D) = Y_{\xi_D}$; $k \in K$, $\alpha \in Q_k$, $\alpha = \langle \lambda, u \rangle^\sim$, $\lambda \in \mathbb{C}$, $u \in (\tilde{H})_k = H_{k^{-1}}$).

Proof. It is clear that the right side of (26) depends only on $\xi \otimes \alpha$.

Since $u \in H_{k^{-1}}$, $\breve{u}$ carries each $\mathscr{Y}$-fiber Y_x linearly and isometrically onto Y_{xk}. Thus $\breve{u}(\xi) \in Y_{\xi_D k}$, and so $\Phi(Z_{\xi_D k}) \subset Y_{\xi_D k}$. In fact $\Phi(Z_{\xi_D k}) = Y_{\xi_D k}$, and

$$\|\Phi(\xi \otimes \alpha)\| = |\lambda| \|\xi\| = \|\alpha\| \|\xi\| = \|\xi \otimes \alpha\|;$$

so $\Phi : \mathscr{Z} \to \mathscr{Y}$ is an isometric bijection.

Let us show that Φ is continuous. For this it is enough (by II.13.16) to take a continuous cross-section ϕ of $\mathscr{Q}$, an element D of $\hat{A}$, and a vector ξ in $X(D)$, and show that

$$k \mapsto \Phi(\xi \otimes \phi(k)) \in Y_{\xi_D k} \qquad (27)$$

is continuous on K to $\mathscr{Y}$. Suppose then that $k_i \to k$ in K. It is enough to replace $\{k_i\}$ by a subnet (without change of notation) and show that some subnet of $\{\Phi(\xi \otimes \phi(k_i))\}$ converges to $\Phi(\xi \otimes \phi(k))$. Now, since ϕ is continuous, $\phi(k_i) = \langle \lambda_i, u_i \rangle^\sim \to \phi(k) = \langle \lambda, u \rangle^\sim$ in $\mathscr{Q}$. By the openness of the quotient map $\sim$ which defines $\mathscr{Q}$, we can pass to a subnet and assume $\lambda_i \to \lambda$ in $\mathbb{C}$ and $u_i \to u$ in $\tilde{H}$. But then

$$\Phi(\xi \otimes \phi(k_i)) = \lambda_i(u_i)^\vee(\xi) \qquad \text{(by (26))}$$

$$\to \lambda \check{u}(\xi) \qquad \text{(by 6.7)}$$

$$= \Phi(\xi \otimes \phi(k)).$$

So (27) is continuous, whence Φ is continuous.

The continuity of Φ^{-1} now follows from II.13.17.

So Φ, being bicontinuous, is an isometric isomorphism of the Hilbert bundles $\mathscr{X}$ and $\mathscr{Y}$. ∎

6.16. We have now to determine the form taken by the isometries $\check{u}$ of 6.6 after they are transferred to $\mathscr{X}$ via the isomorphism Φ of 6.15.

Proposition. *If* $v \in (\tilde{H})_{m^{-1}} = H_m$ $(m \in K)$, *and if* D, ξ, k, α, λ, u *are as in Proposition* 6.15, *then*

$$\check{v}(\Phi(\xi \otimes \alpha)) = \Phi(\xi \otimes \alpha v). \qquad (28)$$

Note. As in VIII.16.15 we identify v with the element $\langle 1, v \rangle^\sim$ of $Q_{m^{-1}}$; and the product αv on the right of (28), being taken in $\mathscr{Q}$, lies in $Q_{km^{-1}}$.

Proof. By (26) the left side of (28) is $\lambda \check{v} \check{u}(\xi)$. Since $\alpha v = \langle \lambda, u \rangle^\sim \langle 1, v \rangle^\sim = \langle \lambda, (uv)_{\tilde{H}} \rangle^\sim = \langle \lambda, (vu)_H \rangle^\sim$, the right side of (28) is $\lambda(vu)_H^\vee(\xi)$. (Here the subscripts H and $\tilde{H}$ mean that the products are taken in H and $\tilde{H}$ respectively.) But $(vu)_H^\vee = \check{v}\check{u}$ by (13). So the two sides of (28) are equal. ∎

Thus the $\tilde{u}$, transferred to $\mathscr{X}$ via Φ, operate on $\mathscr{X}$ by right multiplication in the second factor (by elements of norm 1).

Since the construction of $\mathscr{X}$ in 6.15 used only the three entities (a), (b), (c) of 6.14, we have obtained the canonical form promised in 6.14 for $\mathscr{Y}$ and the $\{\check{u} : u \in H\}$.

6.17. Collecting these results, we arrive at the following qualitative proposition, which asserts that the C^*-algebraic bundles under consideration are essentially uniquely determined by the three items (a), (b), (c) of 6.14.

Proposition. *Suppose that $\mathscr{B}$ and $\mathscr{B}'$ are two saturated C^*-algebraic bundles over the same locally compact group G, having unit fiber C^*-algebras A and A' respectively of compact type. Suppose that A and A' are $*$-isomorphic under a $*$-isomorphism $F: A \to A'$; and let $\hat{F}: \hat{A} \to \hat{A}'$ be the bijection generated by F. Assume further that*

$$\hat{F}(^xD) = {}^x(\hat{F}(D)) \qquad\qquad (x \in G, D \in \hat{A}),$$

that is, $\hat{F}$ carries the action of G on $\hat{A}$ by $\mathscr{B}$-conjugation into the action of G on $\hat{A}'$ by $\mathscr{B}'$-conjugation; and that these actions are transitive. Let D_0 be any element of $\hat{A}$, and K the stability subgroup of G for D_0 (under $\mathscr{B}$-conjugation), and hence also for $\hat{F}(D_0)$ (under $\mathscr{B}'$-conjugation). Finally, assume that the Mackey obstructions

$$\gamma_0: \mathbb{E} \underset{\sigma_0}{\to} H \underset{\rho_0}{\to} K$$

of D_0 (in $\mathscr{B}$) and

$$\gamma_0': \mathbb{E} \underset{\sigma_0'}{\to} H' \underset{\rho_0'}{\to} K$$

of $\hat{F}(D_0)$ (in $\mathscr{B}'$) are isomorphic.

Then $\mathscr{B}$ and $\mathscr{B}'$ are isometrically isomorphic as Banach $*$-algebraic bundles over G.

Proof. From the isomorphism of γ_0 and γ_0' we immediately deduce the isomorphism of the $\mathscr{D}$ and $\mathscr{D}'$ constructed from them as in 6.14 (simultaneously as C^*-algebraic bundles and as Hilbert bundles). Hence the Hilbert bundles $\mathscr{X}$ and $\mathscr{X}'$ constructed as in 6.15 (from $\mathscr{B}$ and $\mathscr{B}'$ respectively) are isometrically isomorphic under an isomorphism which carries the collection of all right multiplications by elements of H into the collection of all right multiplications by elements of H'. From this fact and 6.15 and 6.16 it follows that the Hilbert bundle $\mathscr{Y}$ induced by D_0 (in $\mathscr{B}$) is isometrically isomorphic with the Hilbert bundle $\mathscr{Y}'$ induced by $\hat{F}(D_0)$ (in $\mathscr{B}'$), under an isomorphism which carries $\{\breve{u}: u \in H\}$ into $\{(u')\check{\ }: u' \in H'\}$. But by 6.13 this implies that $\mathscr{B}$ and $\mathscr{B}'$ are themselves isometrically isomorphic as Banach $*$-algebraic bundles over G. ■

6.18. Now in VIII.16.17 we have explicitly constructed a class of saturated C^*-algebraic bundles from precisely the ingredients (a), (b), (c) considered in 6.14. Applying 6.17 to these, we obtain the following final description of an arbitrary saturated C^*-algebraic bundle $\mathscr{B}$ of transitive type with unit fiber of compact type.

Theorem. *Let $\mathscr{B}$ be a saturated C^*-algebraic bundle over the locally compact group G, with unit fiber C^*-algebra A. Assume that A is of compact type and that $\mathscr{B}$ is of transitive type. Let D_0 be an element of $\hat{A}$, K the stability subgroup for D_0, and γ the Mackey obstruction for D_0. Then $\mathscr{B}$ is isometrically isomorphic to the saturated C^*-algebraic bundle $\mathscr{B}'$ canonically constructed as in VIII.16.15–17 from the ingredients A, G (acting on $\hat{A}$ by $\mathscr{B}$-conjugation), D_0, K, $\tilde{\gamma}$ (where $\tilde{\gamma}$ is the central extension inverse to γ).*

Proof. By 6.3 the actions of G on $\hat{A}$ by conjugation in $\mathscr{B}$ and in $\mathscr{B}'$ are the same. By 6.4 the Mackey obstruction of D_0 in $\mathscr{B}'$ is $\tilde{\tilde{\gamma}} = \gamma$, hence the same as that of D_0 in $\mathscr{B}$. Consequently, by 6.17, $\mathscr{B}$ and $\mathscr{B}'$ are isometrically isomorphic. ∎

6.19. We can now remove the restriction that G operates *transitively* on $\hat{A}$, obtaining the final result of this section.

Theorem. *Let $\mathscr{B}$ be a saturated C^*-algebraic bundle over a locally compact group (with unit e), whose unit fiber C^*-algebra A is of compact type. Let Θ be the family of all the orbits in $\hat{A}$ under the action of G on $\hat{A}$ by $\mathscr{B}$-conjugation. For each $\theta \in \Theta$, let A_θ be the closed two-sided ideal $\{a \in A: D_a = 0 \text{ for all } D \notin \theta\}$ of A, so that $(A_\theta)\hat{} = \Theta$; let D_θ be some fixed element of θ, K_θ the stability subgroup of G for D_θ, and γ_θ the Mackey obstruction for D_θ; and define $\mathscr{B}^\theta$ to be the saturated C^*-algebraic bundle over G canonically constructed (as in VIII.16.15–17) from the ingredients A_θ, G (acting transitively on θ by $\mathscr{B}$-conjugation), D_θ, K_θ and $(\gamma_\theta)\tilde{}$ (the class inverse to γ_θ). Then*

$$\mathscr{B} \cong \sum_{\theta \in \Theta}^{\oplus 0} \mathscr{B}^\theta \qquad (C^*\text{-direct sum; see VIII.16.13}). \qquad (29)$$

Proof. For each $\theta \in \Theta$ let u_θ be the (central) multiplier of A given by

$$D_{u_\theta a} = \begin{cases} D_a & \text{if } D \in \theta \\ 0 & \text{if } D \notin \theta \end{cases} \qquad \text{for all } a \in A.$$

If $D \in \hat{A}$ and $\theta \in \Theta$, D_{u_θ} is the scalar operator 1 or 0 according as D does or does not belong to θ. Now by VIII.3.8 u_θ can be regarded as a multiplier of $\mathscr{B}$ of order e; and by the preceding sentence u_θ satisfies condition (ii) of Proposition XI.16.20. Hence by XI.16.20 u_θ is central in $\mathscr{B}$.

Now clearly the $\{u_\theta\}$ satisfy conditions (i), (ii) of the end of VIII.16.13. We have just seen that they also satisfy (iii). Hence by Proposition VIII.16.14

$$\mathscr{B} \cong \sum_{\theta \in \Theta}^{\oplus 0} \mathscr{F}^\theta, \tag{30}$$

where $\mathscr{F}^\theta$ is the saturated C^*-algebraic bundle over G whose bundle space is $u_\theta \mathscr{B}$.

Now $F_e^\theta = A_\theta$; and we leave the reader to check that the actions of G on θ by conjugation in $\mathscr{F}^\theta$ and in $\mathscr{B}$ are the same. Hence $\mathscr{F}^\theta$ is a saturated C^*-algebraic bundle of transitive type whose unit fiber is A_θ. We also leave to the reader the verification that, if $D_\theta \in \theta$, the Mackey obstructions of D_θ in $\mathscr{B}$ and in $\mathscr{F}^\theta$ are the same. Hence by Theorem 6.18 $\mathscr{F}^\theta \cong \mathscr{B}^\theta$ for each θ. Thus (30) implies (29). ■

This theorem fulfills the promise of 6.1.

6.20.   Theorem 6.19 shows that the structure of a saturated C^*-algebraic bundle is surprisingly restrictive. This restrictiveness is illustrated by the following Corollary:

Corollary.   *Let $\mathscr{B}$ be a saturated C^*-algebraic bundle over a locally compact group G, whose unit fiber C^*-algebra A is finite-dimensional. Then every fiber of $\mathscr{B}$ is finite-dimensional. In fact*

$$\dim(B_x) = \sum_{D \in \hat{A}} \dim(D)\,\dim({}^x D) \tag{31}$$

for all $x \in G$ (${}^x D$ being as usual the x-conjugate of D in $\mathscr{B}$). In particular

$$\dim(B_x) \le \dim(A) \qquad \text{for all } x \in G. \tag{32}$$

Proof.   Theorem 6.19 and formula (16) of VIII.16.17 imply (31).

The Schwarz Inequality applied to (31) gives

$$(\dim(B_x))^2 \le \sum_{D \in \hat{A}} (\dim(D))^2 \sum_{D \in \hat{A}} (\dim({}^x D))^2$$

$$= (\dim(A))^2,$$

which is (32). ■

Remark.   A very special class of bundles illustrating the above (31) and (32) was constructed in XI.7.10. In the light of this section, the bundles of XI.7.10 can be characterized as saturated C^*-algebraic bundles over a finite group, with finite-dimensional unit fiber, and with trivial Mackey obstructions.

6.21. Another interesting special case of 6.19 is that in which the unit fiber C*-algebra is elementary. In that case 6.19 evidently gives:

Corollary.   *Let $\mathscr{B}$ be a saturated C*-algebraic bundle over a locally compact group G, whose unit fiber C*-algebra A is elementary, and hence has only one irreducible *-representation D_0. Let γ be the Mackey obstruction of D_0 (thus γ is a central extension of $\mathbb{E}$ by G); and let $\mathscr{Q}$ be the cocycle bundle over G constructed from the inverse $\tilde{\gamma}$ of γ as in 6.14. Then $\mathscr{B}$ is isometrically isomorphic (as a Banach *-algebraic bundle) with the tensor product $A \otimes \mathscr{Q}$ (see VIII.16.18(24)).*

6.22. Finally, let us consider the special case of 6.19 in which the stability subgroups K_θ are all trivial, that is, G acts freely on $\hat{A}$. (This implies of course that G is discrete.) In that case, bearing in mind Remark 2 of VIII.16.16, we deduce from 6.19:

Corollary.   *Assume in 6.19 that the stability subgroups K_θ are all trivial. Let X denote the Hilbert space direct sum $\sum_{D \in \hat{A}} X(D)$. Then (to within isometric isomorphism) $\mathscr{B}$ is obtained as follows: For each $x \in G$, B_x is the closed subspace of $\mathcal{O}_c(X)$ consisting of those α such that*

$$\alpha(X(D)) \subset X({}^x D) \quad \text{for all } D \in \hat{A};$$

and the operations of multiplication and involution in $\mathscr{B}$ are just those of $\mathcal{O}_c(X)$ restricted to $\mathscr{B}$.

7. Saturated Bundles Over Compact Groups

*The Case That B_e has a Smooth *-Representation Theory*

7.1. Suppose that the group G of §5 is compact. Then 5.8(14) holds automatically. By Remark 2 of 5.9, condition (II) of Theorem 5.9 also holds in that case. So, to conclude from Theorem 5.9 that $\mathscr{B}$ has a Type I *-representation theory, it is only necessary to know 5.8(12), (13). As we shall see in 7.2, condition 5.8(12) together with the compactness of G implies 5.8(13) provided that G and B_e are second-countable (and perhaps even without this proviso).

Unfortunately, in order to prove Theorem 7.2 we shall have to invoke a result of Glimm which has not been proved in this work. However, Theorem 7.2 will not be required in what follows.

7.2. Theorem*. *Let $\mathscr{B}$ be a saturated Banach *-algebraic bundle over a second-countable compact group G; and assume that the unit fiber *-algebra B_e of $\mathscr{B}$ is separable and has a smooth *-representation theory (see 5.10). Then $\mathscr{B}$ has a Type I *-representation theory.*

Sketch of Proof. Let us denote B_e by A. Thus $\hat{A}$ is a second-countable T_0 space; and G and $\hat{A}^+$ satisfy the hypotheses of Theorem 1 of Glimm [4]. Now by Lemma 17.2 of Fell [14], for each D in $\hat{A}^+$ the natural injection $xH_D \mapsto {}^x D$ of G/H_D into $\hat{A}^+$ (H_D being the stability subgroup for D) is a homeomorphism. Hence by the equivalence of conditions (3) and (6) in Theorem 1 of Glimm [4], the orbit space $\hat{A}^+/G$ satisfies Condition (C) of 2.16. Thus 5.8(13) holds; and Theorem 5.9 gives the required result. ∎

Remark 1. It seems likely that the above theorem remains true even if we omit the hypothesis that G and B_e are second-countable. To prove this it would have to be shown that if G is any compact group acting as a topological transformation group on the structure space $\hat{E}$ of a Type I C^*-algebra E, then the quotient orbit space $\hat{E}/G$ is almost Hausdorff. Notice that this is true (even in the absence of second countability) if $\hat{E}$ is Hausdorff or if G is finite (see the proof of Proposition 17.5 of Fell [14]).

Remark 2. In view of the theorem of Glimm and Sakai quoted in 5.10, the conclusion of the above theorem amounts to saying that $\mathscr{B}$, like B_e, has a smooth *-representation theory.

The Case That B_e is of Compact Type

7.3. Definition. A Banach *-algebra A will be said to be *of compact type* if (a) the structure space $\hat{A}$ is discrete and (b) every irreducible *-representation of A is compact (see XI.6.21).

Being of compact type is of course a much more special property than having a smooth *-representation theory.

Clearly A is of compact type if and only if its C^*-completion is of compact type.

Remark 1. For C^*-algebras this definition coincides with that of VI.23.3. Indeed: Condition (II) of VI.23.3 implies the above (a) and (b) in virtue of VII.5.21. Conversely, VII.5.22 shows that (a) and (b) imply (II) of VI.23.3.

Remark 2. If A is separable, then by VII.5.20 (a) $\Rightarrow$ (b) in the above definition. If A is not separable, the question of whether or not (a) $\Rightarrow$ (b) is equivalent to the question of whether Rosenberg's Theorem holds in the non-separable case (see VI.23.1).

7.4. ***Definition.*** A Banach *-algebra A is *of finite type* if $\hat{A}$ is discrete and all the irreducible *-representations of A are finite-dimensional.

If A is of finite type it is obviously of compact type.

7.5. ***Definition.*** A Banach *-algebraic bundle $\mathscr{B}$ over a locally compact group G is *of compact type* [*of finite type*] if its $\mathscr{L}_1$ cross-sectional algebra is of compact type [of finite type].

Thus $\mathscr{B}$ is of compact type if and only if (a) $\hat{\mathscr{B}}$ is discrete (in the regional topology) and (b) T_f is a compact operator for every f in $\mathscr{L}(\mathscr{B})$ and every T in $\hat{\mathscr{B}}$. Similarly, $\mathscr{B}$ is of finite type if and only if $\hat{\mathscr{B}}$ is discrete and every irreducible *-representation of $\mathscr{B}$ is finite-dimensional.

If G is a compact group, its group bundle is of finite type by IX.4.7 and IX.8.6.

Remark. The converse of the last statement is true. Indeed, it has been shown by Stern [1] (see also Baggett [5]) that a locally compact group whose structure space is discrete must be compact.

7.6. **Proposition.** *If A is a Banach *-algebra of compact type, then every non-degenerate *-representation of A is discretely decomposable as a Hilbert direct sum of irreducible *-representations of A.*

Proof. We may as well assume that A is a C^*-algebra. Let T be a non-degenerate *-representation of A, and P its spectral measure. For each D in $\hat{A}$ let T^D be the subrepresentation of T acting on $X_D = \text{range}(P(\{D\}))$. Since $\hat{A}$ is discrete, $X(T) = \sum_{D \in \hat{A}}^{\oplus} X_D$. By VII.9.16 the spectral measure of T^D is concentrated at D; and so by VII.10.10 T^D is equivalent to a direct sum of copies of D. ∎

Remark. The conclusion of this proposition holds under weaker hypotheses (at least if A is separable). Thus, if A is a Banach *-algebra with a smooth *-representation theory and for which $\hat{A}$ is countable, the preceding proof remains valid, and we can conclude that every non-degenerate *-representation of A is a Hilbert direct sum of irreducible *-representations.

7.7. Let $\mathscr{B}$ be a saturated Banach *-algebraic bundle over a compact group G such that the unit fiber B_e is a Banach *-algebra of compact type. Our next aim is to extend to $\mathscr{B}$ as much as possible of the representation theory of compact groups (see Chapter IX). In particular we will show that every irreducible *-representation of $\mathscr{B}$ is compact (in fact finite-dimensional if B_e is of finite type), and that every non-degenerate *-representation of $\mathscr{B}$ is discretely decomposable. We will also show that the Frobenius Reciprocity Theorem holds in $\mathscr{B}$.

7.8. We first need a general lemma on *-representations of arbitrary Banach *-algebraic bundles.

Lemma. *Let $\mathscr{B}$ be a Banach *-algebraic bundle over an arbitrary topological group G; and let T be a *-representation of $\mathscr{B}$ such that T_b is a compact operator for every b in $\mathscr{B}$. Then $T: b \mapsto T_b$ is continuous on $\mathscr{B}$ with respect to the norm-topology of operators on $X(T)$.*

Proof. We begin with the following observation, whose verification is left to the reader: If $\{\alpha_i\}$ is a norm-bounded net of elements of $\mathcal{O}(X(T))$, and if $\beta \in \mathcal{O}_C(X(T))$, then

$$\alpha_i \to 0 \text{ strongly} \Rightarrow \|\alpha_i \beta\| \to 0. \tag{1}$$

Now suppose that $b_i \to b$ in $\mathscr{B}$. We wish to show that $T_{b_i} \to T_b$ in the norm topology. We have $b_i^* \to b^*$, so that

$$(T_{b_i})^* \to (T_b)^* \text{ strongly}. \tag{2}$$

Also $b_i^* b_i \to b^* b$ in B_e; so, since $T|B_e$ is norm-continuous,

$$(T_{b_i})^* T_{b_i} = T_{b_i^* b_i} \to T_{b^* b} = (T_b)^* T_b \text{ in norm}. \tag{3}$$

Now

$$(T_{b_i} - T_b)^*(T_{b_i} - T_b) = (T_{b_i})^* T_{b_i} + (T_b)^* T_b - (T_{b_i})^* T_b - ((T_{b_i})^* T_b)^*. \tag{4}$$

By (3), (2), and observation (1), the right side of (4) approaches $2(T_b)^* T_b - 2(T_b)^* T_b = 0$ in norm. So by (4) $\|T_{b_i} - T_b\|^2 = \|(T_{b_i} - T_b)^*(T_{b_i} - T_b)\| \to 0$. ∎

7.9. In connection with 7.8 it is important to remark that, if T is a *-representation of $\mathscr{B}$ such that T_a is compact for all a in the unit fiber *-algebra B_e, then T_b is compact for all b in $\mathscr{B}$.

Indeed: Let $b \in \mathscr{B}$. Then T_b has a polar decomposition $T_b = UP$ (see VI.13.5), where $U \in \mathcal{O}(X(T))$ and $P = ((T_b)^* T_b)^{1/2} = (T_{b^* b})^{1/2}$. Now $b^* b \in B_e$

and so T_{b*b} is compact; hence, since $\mathcal{O}_c(X(T))$ is a C*-algebra, $P = (T_{b*b})^{1/2}$ is compact. Therefore $T_b = UP$ is compact.

7.10.   Proposition. *Let $\mathcal{B}$ be a Banach *-algebraic bundle over a locally compact group G; and let T be a *-representation of $\mathcal{B}$ such that T_a is compact for all a in the unit fiber *-algebra B_e. Then T is compact in the sense of* XI.14.27, *that is, the range of the integrated form of T consists entirely of compact operators.*

Proof. We must show that T_f is compact for all f in $\mathcal{L}(\mathcal{B})$. Let λ be left Haar measure on G; and let $f \in \mathcal{L}(\mathcal{B})$. By 7.8 and 7.9, $x \mapsto T_{f(x)}$ is norm-continuous on G to $\mathcal{O}_c(X(T))$, and of course has compact support. So $S = \int T_{f(x)} \, d\lambda x$ exists as an $\mathcal{O}_c(X(T))$-valued integral (see II.15.18). Clearly $S = T_f$; so the latter is compact. ∎

Remark. The hypothesis that T_a is compact for all a in B_e is of course not *necessary* for T to be compact in the sense of XI.14.27. Indeed, take $\mathcal{B}$ to be the group bundle of an infinite compact group G, and observe (using IX.9.1) that the regular representation of G is infinite-dimensional but compact.

7.11.   Proposition. *Let $\mathcal{B}$ be a saturated Banach *-algebraic bundle over the compact group G; and denote the unit fiber *-algebra of $\mathcal{B}$ by A. We shall suppose that (a) the action of G on $\hat{A}^+$ (by conjugation) satisfies 5.8(13), and (b) every irreducible *-representation of A is compact. Then all the irreducible *-representations of $\mathcal{B}$ are compact (in the sense of* XI.14.27).

Proof. All the hypotheses of Theorem 5.8 are satisfied. So by Theorem 5.8 one constructs the most general irreducible *-representation T of $\mathcal{B}$ as follows: Take an irreducible *-representation D in $\hat{A}^+$, and let H be its stability subgroup in G. Let

$$\gamma: \mathbb{E} \underset{\iota}{\to} K \underset{\sigma}{\to} H \tag{5}$$

be a central extension belonging to the Mackey obstruction of D (see 4.21); and let V be the natural extension of D to $\mathscr{C}_\gamma$ as in 4.6. Further, let W be an irreducible γ^{-1}-representation of H. From these ingredients we form the *-representation S^W of $\mathcal{B}_H$ as in 5.2(3), and put

$$T = \underset{\mathcal{B}_H \uparrow \mathcal{B}}{\mathrm{Ind}} (S^W). \tag{6}$$

By Theorem 5.8 this is the form of the most general irreducible *-representation T of $\mathcal{B}$.

Since the K of (5) is compact (by III.2.4), the above W is finite-dimensional (by IX.8.6). From this, the compactness of D, and condition 4.6(5) on V, it follows that $(S^W)_a$ is compact for all a in A. Hence by 7.10 S^W is a compact *-representation of $\mathscr{B}_H$. From this and XI.14.28 applied to (6) we conclude that T is compact. ∎

7.12. Hypothesis (a) of 7.11 certainly holds if $\hat{A}$ is discrete. Hence the conclusion of 7.11 holds if A is of compact type.

Suppose that the A of 7.11 is of finite type. Then the V and the W in the proof of 7.11 are both finite-dimensional; hence S^W is finite-dimensional. Further, since $\hat{A}$ is discrete, the stability subgroup H in the proof of 7.11 is open in G, and G/H is finite. Therefore, by XI.16.15, the induced *-representation T given by (6) is finite-dimensional. We thus obtain:

Proposition.   *Let $\mathscr{B}$ be a saturated Banach *-algebraic bundle over a compact group, such that the unit fiber *-algebra of $\mathscr{B}$ is of compact type [of finite type]. Then every irreducible *-representation of $\mathscr{B}$ is compact [finite-dimensional].*

7.13.   Proposition.   *Let $\mathscr{B}$ be a saturated Banach *-algebraic bundle over a compact group G, such that the unit fiber *-algebra A of $\mathscr{B}$ is of compact type. Then every non-degenerate *-representation of $\mathscr{B}$ is discretely decomposable.*

Proof.   Let T be a non-degenerate *-representation of $\mathscr{B}$; and let P be the spectral measure of $T|A$. By XI.8.12 any element of $\hat{A}$ that occurs in $T|A$ must be $\mathscr{B}$-positive; so P is carried by $\hat{A}^+$.

For each G-orbit θ in $\hat{A}^+$ let $X_\theta = \mathrm{range}(P(\theta))$. By 2.12 each such X_θ is T-stable. Denoting by T^θ the subrepresentation of T acting on X_θ, and noting that $X(T) = \sum_\theta^\oplus X_\theta$ (since $\hat{A}^+$ is discrete), we have

$$T = \sum_\theta^\oplus T^\theta \qquad \text{(Hilbert direct sum)}.$$

Thus it is sufficient to show that each T^θ is discretely decomposable.

Fix a G-orbit θ. Then T^θ is associated with θ. Choose $D \in \theta$; and let H be the stability subgroup for D. By Theorem 5.2 we have (using the notation of that theorem)

$$T^\theta = \mathop{\mathrm{Ind}}_{\mathscr{B}_H \uparrow \mathscr{B}} (S^W), \qquad (7)$$

where W is a γ^{-1}-representation of H. By 5.2 the T^θ and W of (7) have *-isomorphic commuting algebras. Now the discrete decomposability of any *-representation Z depends only on the *-isomorphism type of the commuting algebra $\mathscr{I}(Z)$ of Z. (Indeed, Z is discretely decomposable if and only if there exists a maximal family $\mathcal{F}$ of non-zero orthogonal projections in $\mathcal{I}(Z)$ with the further property that each p in $\mathcal{F}$ is minimal as a non-zero projection in $\mathcal{I}(Z)$.) Therefore T^θ will be discretely decomposable if and only if W is. But W, being a unitary representation of a compact group, is discretely decomposable by IX.8.10. Therefore T^θ is discretely decomposable, completing the proof. ∎

Remark. This result would follow immediately from 7.6 and 7.16. We prefer, however, to prove it independently of 7.16.

7.14. Our next goal is to prove the bundle generalization of the classical Frobenius Reciprocity Theorem (IX.10.8). For this we need a lemma which says in effect that the correspondence 4.24(47) commutes with the inducing operation.

Let $\mathcal{B}$ be a saturated Banach *-algebraic bundle over a compact group G, such that the unit fiber *-algebra A of $\mathcal{B}$ is of compact type. Let D be an element of $\hat{A}$, with stability subgroup H; and let

$$\gamma: \mathbb{E} \xrightarrow{i} K \xrightarrow{\sigma} H$$

be a central extension belonging to the Mackey obstruction of D. Let V be the natural extension of D to the retraction $\mathcal{C}$ of $\mathcal{B}_H$ by σ.

Furthermore let L be a closed subgroup of H, and $L_0 = \sigma^{-1}(L)$ the corresponding subgroup of K. Thus

$$\gamma_0: \mathbb{E} \xrightarrow{i} L_0 \xrightarrow{\sigma|L_0} L$$

is a central extension of $\mathbb{E}$ by L. Let R be a γ_0^{-1}-representation of L; and define the non-degenerate *-representation S^R of $\mathcal{B}_L$ as in 4.23(45):

$$(S^R)_b = R_u \otimes V_{\langle u, b \rangle} \qquad (b \in \mathcal{B}_L; \langle u, b \rangle \in \mathcal{C}_{L_0}). \qquad (8)$$

Lemma. *Set* $W = \mathrm{Ind}_{L_0 \uparrow K}(R)$. *Then* W *is a* γ^{-1}-*representation of* H; *and*

$$S^W \cong \underset{\mathcal{B}_L \uparrow \mathcal{B}_H}{\mathrm{Ind}} (S^R) \qquad (9)$$

*(S^W being the *-representation of $\mathcal{B}_H$ defined by 4.23(45)).*

Proof. R can be considered as a unitary representation of L_0 satisfying

$$R_{i(z)} = \bar{z}\mathbf{1} \qquad (z \in \mathbb{E}).$$

Since $i(\mathbb{E})$ is central in K, a routine calculation (based on the Mackey form of the definition of induced representations of groups) shows that

$$W_{i(z)} = \bar{z}\mathbf{1} \qquad (z \in \mathbb{E}).$$

So W is a γ^{-1}-representation of H.

Now let $\Phi: \mathcal{C} \to \mathcal{B}$ be the map given by $\Phi(\langle u, b \rangle) = b$ (as in 4.23). Then by the definition of S^R and S^W in 4.23,

$$S^R \circ (\Phi | \mathcal{C}_{L_0}) = R \otimes (V | \mathcal{C}_{L_0}), \tag{10}$$

$$S^W \circ \Phi = W \otimes V \tag{11}$$

(the right sides of (10) and (11) being the tensor products of VIII.9.16). Thus

$$\left[\operatorname*{Ind}_{\mathcal{B}_L \uparrow \mathcal{B}_H} (S^R) \right] \circ \Phi \cong \operatorname*{Ind}_{\mathcal{C}_{L_0} \uparrow \mathcal{C}} (S^R \circ (\Phi | \mathcal{C}_{L_0})) \qquad \text{(by XI.12.19)}$$

$$\cong \operatorname*{Ind}_{\mathcal{C}_{L_0} \uparrow \mathcal{C}} (R \otimes (V | \mathcal{C}_{L_0})) \qquad \text{(by (10))}$$

$$\cong W \otimes V \qquad \text{(by XI.13.9)}$$

$$= S^W \circ \Phi \qquad \text{(by (11))}.$$

Since Φ is surjective, this implies (9).   ∎

7.15.   Bundle Frobenius Reciprocity Theorem.   *Let $\mathcal{B}$ be a saturated Banach *-algebraic bundle over the compact group G; and assume that the unit fiber *-algebra A of $\mathcal{B}$ is of compact type. Let M be a closed subgroup of G. Let T be an irreducible *-representation of $\mathcal{B}$, and V an irreducible $\mathcal{B}$-positive *-representation of $\mathcal{B}_M$. Then the multiplicity of V in $T | \mathcal{B}_M$ is finite and equal to the multiplicity of T in $\operatorname{Ind}_{\mathcal{B}_M \uparrow \mathcal{B}}(V)$.*

Remark.   In view of 7.13 both $T | \mathcal{B}_M$ and $\operatorname{Ind}(V)$ are discretely decomposable. The multiplicities referred to in the theorem are of course to be understood in the sense of VI.14.14 (having transposed the latter to the context of *-representations of bundles in the natural way).

Proof.   Let m be the multiplicity of V in $T | \mathcal{B}_M$, and m' the multiplicity of T in $\operatorname{Ind}(V)$.

Let θ be the G-orbit in $\hat{A}^+$ with which T is associated, and ϕ the M-orbit in $\hat{A}^+$ with which V is associated. (Notice that $V | A$ is $\mathcal{B}$-positive by XI.8.12.) Thus either

$$\phi \subset \theta \tag{12}$$

or

$$\phi \cap \theta = \emptyset. \tag{13}$$

Assume that (13) holds. Then we claim that $m = m' = 0$. Indeed: If V occurred in $T | \mathcal{B}_M$, then any element of $\hat{A}$ occurring in $V | A$ would also occur in $T|A$, contradicting (13). So $m = 0$. Furthermore, by XI.16.28 applied to V, $\text{Ind}(V)$ is associated with the G-orbit containing ϕ, which by (13) is disjoint from θ. So T, being associated with θ, cannot occur in $\text{Ind}(V)$, and $m' = 0$. This proves the claim. So the conclusion of the theorem is valid if (13) holds.

Thus it is sufficient to prove the theorem assuming (12). Let D be an element of ϕ (hence also of θ); let H be the stability subgroup of G for D; and put $L = H \cap M$ (so that L is the stability subgroup of M for D). With these ingredients define γ and γ_0 as in 7.14. Thus by Theorem 5.2 there are an irreducible γ_0^{-1}-representation R of L and an irreducible γ^{-1}-representation W of H such that

$$V = \underset{\mathscr{B}_L \uparrow \mathscr{B}_M}{\text{Ind}} (S^R), \tag{14}$$

$$T = \underset{\mathscr{B}_H \uparrow \mathscr{B}}{\text{Ind}} (S^W) \tag{15}$$

(S^R and S^W being defined as in 5.2 for each case).

Let K and L_0 be as in 7.14. We have

$$\text{Ind}(V) = \underset{\mathscr{B}_M \uparrow \mathscr{B}}{\text{Ind}} \underset{\mathscr{B}_L \uparrow \mathscr{B}_M}{\text{Ind}} (S^R) \qquad \text{(by (14))}$$

$$= \underset{\mathscr{B}_H \uparrow \mathscr{B}}{\text{Ind}} \underset{\mathscr{B}_L \uparrow \mathscr{B}_H}{\text{Ind}} (S^R) \qquad \text{(by XI.12.15 used twice)}$$

$$= \underset{\mathscr{B}_H \uparrow \mathscr{B}}{\text{Ind}} (S^Q) \qquad \text{(by 7.14),} \quad (16)$$

where $Q = \text{Ind}_{L_0 \uparrow K}(R)$. Now Q, being a γ^{-1}-representation of H, is a direct sum

$$Q \cong \sum_i{}^{\oplus} Q^i$$

of irreducible γ^{-1}-representations Q_i of H; and so by (16)

$$\text{Ind}(V) \cong \sum_i{}^{\oplus} \underset{\mathscr{B}_H \uparrow \mathscr{B}}{\text{Ind}} (S^{Q^i}). \tag{17}$$

By 5.2 the terms $\text{Ind}(S^{Q^i})$ on the right of (17) are all irreducible *-representations of $\mathscr{B}$; and $\text{Ind}(S^{Q^i}) \cong \text{Ind}(S^{Q^j})$ if and only if $Q^i \cong Q^j$. From this, (17), and (15) it follows that

$$m' = \text{multiplicity of } W \text{ in } Q. \tag{18}$$

Since $Q = \text{Ind}_{L_0 \uparrow K}(R)$, the classical Frobenius Reciprocity Theorem IX.10.8 applied in the group K shows that the right side of (18) is finite and equal to the multiplicity of R in $W|L_0$. Hence it is enough to verify that

$$m = \text{multiplicity of } R \text{ in } W|L_0. \tag{19}$$

Let N be a subset of G containing exactly one element from each M, H double coset (and suppose $e \in N$). By (15) and XI.16.27

$$T|\mathcal{B}_M \cong \sum_{x \in N}^{\oplus} Z^x, \tag{20}$$

where Z^x is the result of inducing $^x(S^W)|\mathcal{B}_{xHx^{-1} \cap M}$ up to $\mathcal{B}_M$. Let m_x be the multiplicity of V in Z^x. By (20)

$$m = \sum_{x \in N} m_x. \tag{21}$$

Now we claim that $m_x = 0$ if $x \in N$ and $x \neq e$. Indeed: Since $S^W|A$ is concentrated at D, by XI.16.6 $^x(S^W)|A$ is concentrated at ^xD; and so by XI.16.28 $Z^x|A$ is concentrated on the M-orbit of ^xD. Now if $x \in N$ and $x \neq e$, i.e., if $x \notin MH$, then the M-orbit of ^xD is disjoint from the M-orbit of D; so the multiplicity of V in Z^x must be 0, and the claim is proved. Applying this to (21) we obtain

$$m = m_e. \tag{22}$$

Now

$$Z^e = \underset{\mathcal{B}_L \uparrow \mathcal{B}_M}{\text{Ind}} (S^W|\mathcal{B}_L). \tag{23}$$

Writing $W|L_0$ as a direct sum of irreducible parts R^j:

$$W|L_0 \cong \sum_j^{\oplus} R^j,$$

we obtain from (23)

$$Z^e \cong \underset{\mathcal{B}_L \uparrow \mathcal{B}_M}{\text{Ind}} \left(\sum_j^{\oplus} S^{R^j} \right)$$

$$\cong \sum_j^{\oplus} \underset{\mathcal{B}_L \uparrow \mathcal{B}_M}{\text{Ind}} (S^{R^j}). \tag{24}$$

Now by 5.2 (applied in $\mathcal{B}_M$) each term on the right side of (24) is irreducible, and $\text{Ind}(S^{R^j}) \cong \text{Ind}(S^{R^k})$ if and only if $R^j \cong R^k$. From this and (14) it follows that m_e equals the number of indices j for which $R \cong R^j$; that is,

$$m_e = \text{multiplicity of } R \text{ in } W|L_0.$$

Combining this with (22) we obtain (19); and this, as we have seen, completes the proof of the theorem. ■

7.16. The following proposition generalizes IX.4.7.

Proposition. *Let $\mathscr{B}$ be a saturated Banach *-algebraic bundle over a compact group whose unit fiber *-algebra A is of compact type. Then the structure space of $\mathscr{B}$ is discrete.*

Proof. For each G-orbit θ in $\hat{A}^+$, let $\hat{\mathscr{B}}_\theta$ be the set of those elements of $\hat{\mathscr{B}}$ which are associated with θ. We claim that $\hat{\mathscr{B}}_\theta$ is an open subset of $\hat{\mathscr{B}}$.

As in the proof of 7.11 we have

$$\hat{\mathscr{B}} = \bigcup_\theta \hat{\mathscr{B}}_\theta. \tag{25}$$

Let θ_0 be an orbit. To prove the claim it is enough by (25) to show that $\hat{\mathscr{B}}_{\theta_0}$ does not intersect the closure of $\bigcup_{\theta \neq \theta_0} \hat{\mathscr{B}}_\theta$. If the latter statement were false, it would follow from the regional continuity of the restriction operation $T \mapsto T|A$ (see VIII.21.20) that θ_0 would intersect the closure of $\hat{A} \setminus \theta_0$, contradicting the discreteness of $\hat{A}$. So the claim is true.

In view of this claim, together with (25), it is enough to show that $\hat{\mathscr{B}}_\theta$ is discrete for each orbit θ. Fix an element D of an orbit θ, and adopt the notation of Theorems 5.2 and 5.5. By Theorem 5.5 $\hat{\mathscr{B}}_\theta$ is homeomorphic to the space of all irreducible γ^{-1}-representations of H; and the latter space is discrete by IX.4.7. ■

7.17. From 7.16 and 7.12 we obtain:

Theorem. *Let $\mathscr{B}$ be a saturated Banach *-algebraic bundle over a compact group. If the unit fiber *-algebra of $\mathscr{B}$ is of compact type [of finite type], then $\mathscr{B}$ is of compact type [of finite type].*

Question. Does this theorem hold without the hypothesis of saturation? We do not know.

Weak Frobenius Reciprocity and Amenability

7.18. Much fruitful effort has been expended on obtaining partial generalizations of the Frobenius Reciprocity Theorem to non-compact situations (see the bibliographical notes). There is a difficulty even in conjecturing such generalizations; for a *-representation of a Banach *-algebraic bundle which is not of compact type will not in general be discretely decomposable, and hence its structure cannot be described by discrete multiplicities. We shall mention here one such generalized formulation, in which the notion of "occurrence as a subrepresentation" is replaced by the notion of weak containment. It will take the form of a "Weak Frobenius Reciprocity Property," which fails in general but whose range of validity constitutes an interesting subject for research.

Definition. Let $\mathscr{B}$ be a Banach *-algebraic bundle over a locally compact group G, and H a closed subgroup of G. We shall say that the pair $\mathscr{B}$, $\mathscr{B}_H$ satisfies *Weak Frobenius Reciprocity* if the following condition (WFR) holds: If T is any irreducible *-representation of $\mathscr{B}$ and S is any irreducible $\mathscr{B}$-positive *-representation of $\mathscr{B}_H$, then T is weakly contained in $\text{Ind}_{\mathscr{B}_H \uparrow \mathscr{B}}(S)$ if and only if S is weakly contained in $T|\mathscr{B}_H$.

In the group case, of course, we would say that the pair G, H (H being a closed subgroup of G) satisfies (WFR) if, given irreducible unitary representations T and S of G and H respectively, the statement that $\text{Ind}_{H \uparrow G}(S)$ weakly contains T and the statement that $T|H$ weakly contains S are either both true or both false.

7.19. Suppose that $\mathscr{B}$ is a saturated Banach *-algebraic bundle over a compact group G, and that the unit fiber *-algebra of $\mathscr{B}$ is of compact type. Thus by 7.17 $\mathscr{B}$ itself is of compact type. Hence, if $T \in \hat{\mathscr{B}}$ and V is any non-degenerate *-representation of $\mathscr{B}$, it follows from VII.9.15 and VII.10.10 that V weakly contains T if and only if T occurs as a subrepresentation of V. Therefore it follows from the Frobenius Reciprocity Theorem 7.15 that $\mathscr{B}$, $\mathscr{B}_H$ satisfies Weak Frobenius Reciprocity for every closed subgroup H.

7.20. Another situation in which Weak Frobenius Reciprocity holds is provided by §X.4.

Let $\mathscr{B}$ be a saturated commutative Banach *-algebraic bundle over the locally compact Abelian group G (with unit e), satisfying the further condition X.4.1(I) (i.e., every element of $(B_e)\hat{}$ is $\mathscr{B}$-positive). It is not hard to show from X.4.18 (see also XI.12.14) that the pair $\mathscr{B}$, B_e satisfies Weak Frobenius Reciprocity. Further, if H is any closed subgroup of G, we can apply this fact to the partial cross-sectional bundle over G/H, and conclude that $\mathscr{B}$, $\mathscr{B}_H$ satisfies Weak Frobenius Reciprocity. Details are left to the reader.

In particular, then, Weak Frobenius Reciprocity holds for any locally compact Abelian group G and any closed subgroup of G.

7.21.   *Remark*.   In the context of compact groups (or, more generally, of the Banach *-algebraic bundles of 7.15), the Frobenius Reciprocity Theorem enables us to determine the structure of an induced representation up to unitary equivalence provided we know the structure of the restrictions of representations. An example of the usefulness of this was worked out in IX.12.5. Weak Frobenius Reciprocity in the non-compact context, however, even when valid, tells us merely the spectrum of an induced representation —information which is very much weaker than its unitary equivalence class.

Notice that Weak Frobenius Reciprocity, as we have formulated it, makes no distinction between different non-zero multiplicities. A little thought should persuade the reader that there is no natural way to distinguish different multiplicities as long as "containment" is defined by means of regional closure.

7.22.   We have seen that Weak Frobenius Reciprocity holds for G, H (H being a closed subgroup of the locally compact group G) whenever G is either compact or Abelian. It fails in general, however, if G is neither compact nor Abelian. For a simple example see 8.5.

7.23.   An important special case of Weak Frobenius Reciprocity for groups is the notion of amenability.

***Definition*.**   A locally compact group G (with unit e) is *amenable* if G, $\{e\}$ satisfies the Weak Frobenius Reciprocity property.

If I is the one-dimensional representation of $\{e\}$, then $\mathrm{Ind}_{\{e\}\uparrow G}(I)$ is the regular representation and $T|\{e\}$ contains I for every T in $\hat{G}$. Therefore G is amenable if and only if every element of $\hat{G}$ is weakly contained in the regular representation of G.

As a matter of fact, for G to be amenable it is necessary and sufficient that the regular representation R of G should weakly contain the trivial one-dimensional representation J of G. Indeed: Suppose that R weakly contains J and that T is any element of $\hat{G}$. Then by VIII.21.24 $R \otimes T$ weakly contains $J \otimes T \cong T$. But by XI.13.11 $R \otimes T$ is a multiple of R. So R weakly contains T; and G is amenable.

For a large body of results concerning amenability we refer the reader to Greenleaf [3] and Pier [1].

It is well known (see Greenleaf [3] or Pier [1]) that there are many locally compact groups which are not amenable. Such groups provide further examples of the failure of Weak Frobenius Reciprocity.

8.   Examples

8.1.   In this and the final section we shall illustrate the Mackey analysis with various specific examples. Roughly speaking, these examples are of two kinds: First there are those in which the hypotheses listed in 5.8 hold, and in which all the steps of the Mackey analysis are therefore valid. Secondly, there are those of a more pathological nature, through which we observe the effect of the failure of one or other of these hypotheses. In this section we present some useful examples of the first kind, mostly in the group context, that is, dealing with the classification of the irreducible unitary representations of certain important infinite groups. The reader will of course notice the essential similarity between the method followed in these examples and that of the finite group examples of §1. The final section, §9, will consist largely of examples of the second kind.

8.2.   *Remark.*   Let G be a locally compact group and N a closed normal subgroup of G; and let $\mathscr{C} = \{C_\alpha\}$ ($\alpha \in G/N$) be the corresponding (saturated) group extension bundle over G/N (see VIII.6.6). Thus $\hat{G} \cong \hat{\mathscr{C}}$. In accordance with 2.3, the Mackey analysis of $\hat{G}$ in this context amounts to the application of the theory of §§2–5 to the group extension bundle $\mathscr{C}$. However we can avoid explicit mention of $\mathscr{C}$. Specifically, the unit fiber *-algebra C_{eH} of $\mathscr{C}$ is $\mathscr{L}_1(N)$, and so $(C_{eH})^{\hat{}}$ becomes just $\hat{N}$. We shall usually speak of the action of G (rather than of G/N) on $\hat{N}$; this action is given by the classical formula XI.16.1(1). Given an element D of $\hat{N}$, by the stability subgroup H we shall mean the subgroup H of G consisting of those x for which $^{x}D \cong D$, rather than the image H' of H in G/N. Thus the non-degenerate *-representations of $\mathscr{C}_{H'}$, which figure so largely in Steps II and III (§§3 and 4), are just unitary representations of H.

The "ax + b" Group

8.3.   Let G be the multiplicative group of all 2×2 real matrices of the form

$$\{a, b\} = \begin{pmatrix} a & b \\ 0 & 1 \end{pmatrix} \qquad (a, b \in \mathbb{R}; a > 0). \qquad (1)$$

This G is called the *"ax + b" group* (since $\{a, b\}$ can be interpreted as the transformation $x \mapsto ax + b$ of $\mathbb{R}$). Evidently

$$\{a, b\}\{c, d\} = \{ac, ad + b\}, \qquad \{a, b\}^{-1} = \{a^{-1}, -a^{-1}b\}.$$

G is locally compact, but neither Abelian nor compact. Let N be the closed normal subgroup $\{\{1, b\}: b \in \mathbb{R}\}$ of G. Our goal is to determine the structure space $\hat{G}$ of G by applying the Mackey analysis to G, N.

Since N is Abelian and isomorphic to the additive group $\mathbb{R}$ of the reals, $\hat{N}$ is also isomorphic with $\mathbb{R}$ under the isomorphism $t \mapsto \phi_t$, where (for real t) ϕ_t is the character of N given by

$$\phi_t(\{1, b\}) = e^{itb} \qquad\qquad (b \in \mathbb{R}).$$

Under the conjugating action of the element $\{c, d\}$ of G, the element ϕ_t of $\hat{N}$ goes into

$$\{1, b\} \mapsto \phi_t(\{c, d\}^{-1}\{1, b\}\{c, d\}),$$

that is, into $\phi_{c^{-1}t}$. Since c varies over all positive real numbers, there are thus three orbits in $\hat{N}$ under the action of G—the "trivial" orbit $\theta_0 = \{\phi_0\}$ and the two orbits $\theta_+ = \{\phi_t: t > 0\}$ and $\theta_- = \{\phi_t: t < 0\}$. This collection of orbits certainly satisfies Condition (C) of 2.16. So, since the elements of $\hat{N}$ are all one-dimensional and since G is second-countable, all the hypotheses of 5.8 hold. In particular, by 5.8 every primary unitary representation of G is associated with either θ_0, θ_+, or θ_-.

The stability subgroup for ϕ_0 is of course G. Hence the primary unitary representations of G associated with θ_0 are just those whose restriction to N is a multiple of ϕ_0, that is, whose kernels contain N. They are the results of lifting to G the primary unitary representations of G/N. But $G/N \cong R$ is Abelian; and so by VI.24.14 its primary unitary representations are of Type I. The elements of $\hat{G}$ associated with θ_0 are just the results of lifting to G the characters of G/N, that is, the χ_s (s real) given by

$$\chi_s(\{a, b\}) = a^{is} \qquad\qquad (\{a, b\} \in G).$$

The typical element ϕ_1 of θ_+ has stability subgroup N (so that its stability subgroup in G/N is the one-element group). It follows from Step II (3.12) that

$$T^+ = \underset{N \uparrow G}{\mathrm{Ind}}(\phi_1) \qquad\qquad (2)$$

is an irreducible unitary representation of G, and that every primary unitary representation of G associated with θ_+ is a direct sum of copies of T^+ (hence of Type I). In particular, T^+ is the only irreducible unitary representation of G associated with θ_+.

Similarly, every primary unitary representation of G associated with θ_- is a direct sum of copies of the unique irreducible unitary representation

$$T^- = \underset{N \uparrow G}{\mathrm{Ind}}(\phi_{-1}) \tag{3}$$

of G associated with θ_-.

We have now proved:

Proposition.   *The "$ax + b$" group G has a Type I unitary representation theory. Apart from the one-dimensional unitary representations χ_s: $\{a, b\} \mapsto a^{is}$ of G (s real), $\hat{G}$ has only two other inequivalent elements, namely the infinite-dimensional unitary representations T^+ and T^- defined by (2) and (3).*

Remark.   Suppose that in the definition (1) of the "$ax + b$" group we were to replace "$a > 0$" by "$a \neq 0$". The results would then be very similar. Instead of two non-trivial orbits in $\hat{N}$ there would then only be one. Hence there would only be *one* irreducible unitary representation of G other than those lifted from characters of G/N.

Remark.   Let F be any non-discrete locally compact field (see III.6.2); and let G be the "$ax + b$" group over F (that is, the group defined as in (1) with $\mathbb{R}$ replaced by F and "$a > 0$" by "$a \neq 0$"). In view of X.3.16, the same Mackey analysis as above can be applied to show that the preceding Remark remains valid for this new G.

8.4.   The regional topology of the structure space

$$\hat{G} = \{\chi_s : s \in \mathbb{R}\} \cup \{T^+, T^-\}$$

of the "$ax + b$" group G is completely described by the following proposition:

Proposition*.   (I) *The map $s \mapsto \chi_s$ is a homeomorphism of $\mathbb{R}$ onto a closed subset of $\hat{G}$.* (II) *The closure of $\{T^+\}$ is $\{T^+\} \cup \{\chi_s : s \in \mathbb{R}\}$.* (III) *The closure of $\{T^-\}$ is $\{T^-\} \cup \{\chi_s : s \in \mathbb{R}\}$.*

Sketch of Proof.   (I) follows from VIII.21.22.

To prove (II), we first note that the closure of $\{T^+\}$ does not contain T^-. For if it did, $T^+ | N$ would weakly contain $T^- | N$; and this would imply that $\{\phi_t : t > 0\}$ would weakly contain $\{\phi_t : t < 0\}$, an impossibility. Thus (II) will be proved if we show that the closure of $\{T^+\}$ contains all the χ_s. But, by the continuity of the inducing operation, $T^+ = \mathrm{Ind}_{N \uparrow G}(\phi_t) \to \mathrm{Ind}_{N \uparrow G}(\phi_0)$ regionally as t approaches 0 through positive values. So T^+ weakly contains

Ind(ϕ_0). But Ind(ϕ_0) is weakly equivalent to $\{\chi_s : s \in \mathbb{R}\}$. Therefore T^+ weakly contains all the χ_s.

(III) is proved similarly. ∎

Remark. The above description of its topology shows that $\hat{G}$ is a T_0 space but not a T_1 space.

Remark. In view of the preceding proposition and VII.5.20, the integrated forms of T^+ and T^- are not compact. However it is easy to pick out elements f of $\mathscr{L}(G)$ such that T_f^+ and T_f^- are compact. Indeed, let A be the C^*-completion of the $\mathscr{L}_1$ group algebra of G, and I the intersection of the kernels in A of the integrated forms of all the χ_s ($s \in \mathbb{R}$). Since $\{\chi_s : s \in \mathbb{R}\}$ is closed in $\hat{G}$, $\hat{I}$ is homeomorphic (by VII.4.6) with the open subspace $\{T^+, T^-\}$ of $\hat{G}$, which by the above proposition is Hausdorff. So by VII.5.20 $T^+ | I$ and $T^- | I$ are compact representations. Now an element f of $\mathscr{L}(G)$ belongs to I provided

$$\int_{-\infty}^{\infty} f(a, b)db = 0 \qquad \text{for all } a > 0. \tag{4}$$

Therefore T_f^+ and T_f^- are compact operators whenever f is an element of $\mathscr{L}(G)$ satisfying (4).

8.5. Remark. Notice that Weak Frobenius Reciprocity (see 7.18) *fails* for the pair G, N of 8.3. Indeed, consider the trivial representation ϕ_0 of N and the irreducible representation T^+ of G. Since T^+ is associated with θ_+, $T^+ | N$ is weakly equivalent to θ_+ and so weakly contains ϕ_0. On the other hand, as we have already pointed out (in the proof of 8.4), $\text{Ind}_{N \uparrow G}(\phi_0)$ is weakly equivalent to $\{\chi_s : s \in \mathbb{R}\}$, and so does not weakly contain T^+.

The Euclidean Groups

8.6. Let E^2 be the multiplicative group of all 2×2 complex matrices of the form

$$\{u, b\} = \begin{pmatrix} u & b \\ 0 & 1 \end{pmatrix} \qquad (u, b \in \mathbb{C}; |u| = 1).$$

This E^2 is called the *proper Euclidean group of the plane* (since $\{u, b\}$ can be interpreted as the distance-preserving transformation $z \mapsto uz + b$ of the complex plane). It is locally compact, but not Abelian or compact. We have

$$\{u, b\}\{u', b'\} = \{uu', ub' + b\}, \qquad \{u, b\}^{-1} = \{u^{-1}, -u^{-1}b\}.$$

Let N be the closed normal Abelian "translation" subgroup $\{\{1, b\} : b \in \mathbb{C}\}$ of E^2. Thus N is isomorphic to the additive group $\mathbb{C}$; and $\hat{N}$ is also isomorphic to $\mathbb{C}$ under the correspondence $\lambda \mapsto \phi_\lambda$ $(\lambda \in \mathbb{C})$, where

$$\phi_\lambda(\{1, b\}) = e^{i Re(\lambda b)} \qquad\qquad (b \in \mathbb{C}). \qquad (5)$$

The conjugating action of $\{u, a\}$ sends ϕ_λ into the character

$$\{1, b\} \mapsto \phi_\lambda(\{u, a\}^{-1}\{1, b\}\{u, a\}) = e^{i Re(\lambda^{-1} ub)},$$

that is, into $\phi_{u^{-1}\lambda}$. It follows that the orbits in $\hat{N}$ under E^2 are just the circles

$$\theta_r = \{\phi_\lambda : |\lambda| = r\},$$

where $r \geq 0$. The trivial orbit $\theta_0 = \{\phi_0\}$ has E^2 for its stability subgroup. As in 8.3, the primary representations of E^2 associated with θ_0 are of Type I; and the irreducible representations of E^2 associated with θ_0 are just the one-dimensional unitary representations χ_n $(n \in \mathbb{Z})$ lifted from characters of $E^2/N \cong \mathbb{E}$:

$$\chi_n(\{u, b\}) = u^n. \qquad (6)$$

If $r > 0$, the stability subgroup for any element of θ_r is just N. So by 3.12, if $r > 0$ there is only one irreducible unitary representation of E^2 associated with θ_r, namely

$$T^r = \underset{N \uparrow G}{\operatorname{Ind}}(\phi_r); \qquad (7)$$

and every unitary representation of E^2 associated with θ_r is a direct sum of copies of T^r. The orbit space $\hat{N}/E^2 = \{\theta_r : r \geq 0\}$ satisfies the Condition (C) of 2.16. Hence by 5.8 we have:

Proposition.   *The proper Euclidean group E^2 of the plane has a Type I unitary representation theory. Apart from the one-dimensional unitary representations χ_n of (6), its only irreducible unitary representations (to within unitary equivalence) are the T^r of (7) $(r > 0)$. If r, s are distinct positive real numbers, then $T^r \not\cong T^s$.*

8.7.   Keep the notation of 8.6. The regional topology of the structure space

$$(E^2)^{\hat{}} = \{\chi_n : n \in \mathbb{Z}\} \cup \{T^r : r > 0\}$$

of E^2 is completely described by the following proposition:

Proposition*.   (I) *$\{\chi_n : n \in \mathbb{Z}\}$ is a discrete closed subset of $(E^2)^{\hat{}}$. (II) For any fixed integer n, the map $\tau : [0, \infty[\rightarrow (E^2)^{\hat{}}$, sending 0 into χ_n and r into T^r for $r > 0$, is a homeomorphism.*

The proof of this proceeds similarly to that of 8.4.

Remark.   In view of this proposition, $(E^2)\hat{\ }$ is a T_1 space but is not Hausdorff. (Indeed, if $\{r_p\}$ is a sequence of positive numbers converging to 0, then $T^{r_p}_{\ p} \to \chi_n$ for every n.) From this and VII.5.20 it follows that *every element of* $(E^2)\hat{\ }$ *is compact*. This last statement also follows from 7.11.

8.8.   Generalizing 8.6, for any integer $n > 1$ let us define E^n to be the semidirect product (see III.4.4) of the additive group $\mathbb{R}^n$ with the (compact) special orthogonal group $SO(n)$:

$$E^n = \mathbb{R}^n \underset{\tau}{\times} SO(n),$$

the action τ of $SO(n)$ on $\mathbb{R}^n$ being the usual linear action of $n \times n$ matrices on $\mathbb{R}^n$. The group E^n is called the *proper Euclidean group of n-space*. (It differs from the group defined in III.4.8(B) by not including reflections.)

For $n = 2$, this definition gives a group isomorphic to that of 8.6.

The map $y \mapsto \phi_y$ $(y \in \mathbb{R}^n)$, where ϕ_y is the character $x \mapsto \exp[i(\sum_{j=1}^n x_j y_j)]$ of $\mathbb{R}^n$, identifies $(\mathbb{R}^n)\hat{\ }$ with $\mathbb{R}^n$. Regarding $\mathbb{R}^n$ as a closed normal subgroup of E^n, we see that the action of $SO(n)$ by inner automorphisms on $\mathbb{R}^n$ generates the conjugating action $u\phi_y = \phi_{uy}$ $(u \in SO(n); y \in \mathbb{R}^n)$ of $SO(n)$ on $(\mathbb{R}^n)\hat{\ }$. Thus the orbits in $(\mathbb{R}^n)\hat{\ }$ are the spheres

$$\theta_r = \left\{ \phi_y : y \in \mathbb{R}^n, \ \sum_{j=1}^n y_j^2 = r^2 \right\},$$

where $r \geq 0$. As usual, associated with θ_0 are the finite-dimensional irreducible unitary representations of E^n lifted from elements of $(SO(n))\hat{\ }$. Let $r > 0$. Then a typical element of θ_r is $\phi'_r = \phi_{0,\dots,0,r}$; and the stability subgroup of $SO(n)$ for ϕ'_r is the subgroup H consisting of all those rotations which leave $\langle 0, \dots, 0, r \rangle$ fixed. This H can be identified with $SO(n-1)$. By 5.2 (see 5.5), the set of irreducible unitary representations of E^n associated with the orbit θ_r is in one-to-one correspondence with $(SO(n-1))\hat{\ }$. Indeed, each D in $(SO(n-1))\hat{\ }$ gives rise to a unitary representation $S^{r,D}$ of the subgroup $H_0 = \mathbb{R}^n \underset{\tau}{\times} SO(n-1)$ of E^n:

$$S^{r,D}_{\langle x, u \rangle} = \phi'_r(x)D_u \qquad\qquad (\langle x, u \rangle \in H_0);$$

and the corresponding irreducible unitary representation of E^n is

$$T^{r,D} = \underset{H_0 \uparrow G}{\mathrm{Ind}} (S^{r,D}). \qquad\qquad (8)$$

Thus, arguing as before, we have by 5.8:

Proposition. *E^n has a Type I representation theory. The finite-dimensional elements of $(E^n)\hat{\ }$ are just those lifted from the irreducible unitary representations of $E^n/\mathbb{R}^n \cong SO(n)$. The infinite-dimensional elements of $(E^n)\hat{\ }$ are the $T^{r,D}$ of (8), where $r > 0$ and $D \in (SO(n-1))\hat{\ }$. We have $T^{r',D'} \cong T^{r,D}$ if and only if $r' = r$ and $D' \cong D$.*

The Heisenberg Group

8.9. For our purposes we understand by a *nilpotent Lie group* a locally compact group which, for some positive integer n, is isomorphic to a closed subgroup of the multiplicative group

$$\{a \in SL(n, \mathbb{C}): a_{ij} = 0 \text{ for } i > j, a_{ii} = 1 \text{ for all } i\}.$$

8.10. Consider the multiplicative group H of all 3×3 real matrices of the form

$$\{r, s, t\} = \begin{pmatrix} 1 & r & t \\ 0 & 1 & s \\ 0 & 0 & 1 \end{pmatrix} \qquad (r, s, t \in \mathbb{R}).$$

Thus H is a nilpotent Lie group; we call it the *Heisenberg group*. Its operations are evidently given by

$$\{r, s, t\}(r', s', t'\} = \{r + r', s + s', t + t' + rs'\},$$

$$\{r, s, t\}^{-1} = \{-r, -s, rs - t\}.$$

Let N be the Abelian closed normal subgroup of H consisting of all $\{0, s, t\}$ $(s, t \in \mathbb{R})$. Thus $\hat{N}$ consists of all

$$\phi_{p,q}: \{0, s, t\} \mapsto e^{i(ps + qt)}$$

(where $p, q \in \mathbb{R}$); and one verifies that the conjugating action of the element $\{r, s, t\}$ of H sends $\phi_{p,q}$ into $\phi_{p-qr,q}$. There are thus two kinds of orbits in $\hat{N}$ (depending on whether q is 0 or not): For each real p we have a one-element orbit $\{\phi_{p,0}\}$; and for each non-zero real q we have the orbit $\theta_q = \{\phi_{p,q}: p \in \mathbb{R}\}$. This collection of orbits obviously satisfies Condition (C) of 2.16; and so 5.8(12), (13), (14) all hold.

The stability subgroup for $\phi_{p,0}$ is H; in fact, $\phi_{p,0}$ is the restriction to N of any one of the one-dimensional unitary representations

$$\chi_{p,m}: \{r, s, t\} \mapsto e^{i(mr + ps)} \tag{9}$$

of H (where $m \in \mathbb{R}$). The stability subgroup of the typical element $\phi_{0,q}$ of θ_q $(q \neq 0)$ is N. Therefore by 5.8 and 5.9 we have:

Proposition. *H has a Type I unitary representation theory. Apart from the one-dimensional unitary representations $\chi_{p,m}$ of (9) ($p, m \in \mathbb{R}$), the only irreducible unitary representations of H are the infinite-dimensional representations*

$$T^q = \underset{N \uparrow H}{\mathrm{Ind}}(\phi_{0,q}) \qquad\qquad (0 \neq q \in \mathbb{R}). \qquad (10)$$

If q', q are distinct non-zero real numbers then $T^{q'} \not\cong T^q$.

Remark. This analysis of $\hat{H}$ is identical in form with 1.27.

Remark. The center C of H is the subgroup consisting of all $\{0, 0, t\}$ ($t \in \mathbb{R}$). It is easy to check that, if $0 \neq q \in \mathbb{R}$,

$$T^q_{\{0,0,t\}} = e^{iqt}\mathbb{1} \qquad\qquad (t \in \mathbb{R}). \qquad (11)$$

In view of (11), the character $\{0, 0, t\} \mapsto e^{iqt}$ of C is called the *central character* of T^q. By the above proposition every non-trivial character of C is the central character of exactly one element of $\hat{H}$. (Of course the $\chi_{p,m}$ of (9) have trivial central character.)

8.11. The regional topology of the structure space

$$\hat{H} = \{\chi_{p,m} : p, m \in \mathbb{R}\} \cup \{T^q : 0 \neq q \in \mathbb{R}\}$$

of the Heisenberg group is determined by the following proposition (similar in form to 8.7):

Proposition*. (I) *The map $\langle p, m \rangle \mapsto \chi_{p,m}$ ($p, m \in \mathbb{R}$) is a homeomorphism of $\mathbb{R}^2$ onto a closed subset of $\hat{H}$.* (II) *For each pair p, m of real numbers, the map $\tau : \mathbb{R} \to \hat{H}$ sending 0 into $\chi_{p,m}$ and q into T^q (for each real $q \neq 0$) is a homeomorphism.*

The proof is similar to the proofs of 8.4 and 8.7.

Remark. As in 8.7, $\hat{H}$ is a T_1 space but not Hausdorff. (Indeed, if $\{q_n\}$ is a sequence of non-zero real numbers converging to 0, then $T^{q_n} \underset{n}{\to} \chi_{p,m}$ for all real p, m.) Therefore by VII.5.20 *every element of $\hat{H}$ is compact.*

It can be shown that every connected nilpotent Lie group G shares the property which we have established for H, namely that each element of $\hat{G}$ is compact. See Dixmier [5], Fell [6].

8.12. *Remark.* Let M be the (discrete) central subgroup $\{\{0, 0, m\}: m \in \mathbb{Z}\}$ of the Heisenberg group H; and put $H_0 = H/M$. Then $(H_0)\hat{\ }$ is regionally homeomorphic with the subspace $\mathscr{W}$ of $\hat{H}$ consisting of those T whose kernels contain M. Evidently

$$\mathscr{W} = \{\chi_{p, m}: p, m \in \mathbb{R}\} \cup \{T^{2\pi n}: 0 \neq n \in \mathbb{Z}\}.$$

By 8.11 $\mathscr{W}$ is Hausdorff. Thus H_0 is an example of a locally compact group whose structure space is Hausdorff.

The most obvious class of locally compact groups having Hausdorff structure spaces is the class of direct products of compact groups and Abelian groups. The above H_0 does not belong to this class (since, unlike groups of this class, it has infinite-dimensional irreducible representations).

8.13. *Remark.* The term "Heisenberg group" comes from the connection between this group and the Heisenberg commutation relations of quantum mechanics. Indeed, we saw in XI.15.8 that the search for realizations of the Heisenberg commutation relations for systems with one degree of freedom is the same as the search for pairs U, V of unitary representations of $\mathbb{R}$ (acting in the same Hilbert space X) such that

$$U_t V_s = e^{ist} V_s U_t \qquad (s, t \in \mathbb{R}),$$

and such that the combined action of U and V on X is irreducible. Now it is easy to see that, if U and V have these properties, then the equation

$$T_{\{r, s, t\}} = e^{it} V_s U_r$$

defines an irreducible unitary representation T of the Heisenberg group, with central character $\{0, 0, t\} \mapsto e^{it}$. By the remark following Proposition 8.10, this implies that $T \cong T^1$ (as defined in (10)). Thus the U and V are determined to within unitary equivalence; and we have reproved the uniqueness of the Heisenberg commutation relations, at least for systems with one degree of freedom. (Of course the present proof is not really different from that given in §XI.15, since both §XI.15 and the Mackey analysis rest squarely on the Imprimitivity Theorem of §XI.14.) See Howe [6] for further discussion of the Heisenberg group.

The Poincaré Group

8.14. Let $L(4)$ be the multiplicative group of all those 4×4 real matrices a which (considered as linear operators on $\mathbb{R}^4$) leave invariant the form $x_1^2 + x_2^2 + x_3^2 - x_4^2$, that is,

$$(ax)_1^2 + (ax)_2^2 + (ax)_3^2 - (ax)_4^2 = x_1^2 + x_2^2 + x_3^2 - x_4^2$$

for all x in $\mathbb{R}^4$. Equipped with the usual topology, $L(4)$ is called the *Lorentz group*.

Consider the semidirect product

$$P = \mathbb{R}^4 \underset{\tau}{\times} L(4),$$

where τ is the usual action of 4×4 matrices on vectors in $\mathbb{R}^4$. We call P the *Poincaré group*. Let us denote by N the Abelian closed normal subgroup $\{\langle u, 1 \rangle : u \in \mathbb{R}^4\}$ of P (1 being the unit element of $L(4)$), and by H the complementary closed subgroup $\{\langle 0, a \rangle : a \in L(4)\}$.

Now $\hat{N}$ consists of all ϕ_λ ($\lambda \in \mathbb{R}^4$), where

$$\phi_\lambda(\langle u, 1 \rangle) = e^{i(\lambda \cdot u)} \qquad\qquad (u \in \mathbb{R}^4).$$

(Here $\lambda \cdot u = \lambda_1 u_1 + \lambda_2 u_2 + \lambda_3 u_3 - \lambda_4 u_4$.) One verifies that the conjugating action of an element $\langle u, a \rangle$ of P on $\hat{N}$ sends ϕ_λ into $\phi_{a\lambda}$. Thus, for each λ in $\mathbb{R}^4$, the orbit in $\hat{N}$ containing ϕ_λ is just $\{\phi_{a\lambda} : a \in L(4)\}$. Since $(a\lambda) \cdot (a\lambda) = \lambda \cdot \lambda$ for $a \in L(4)$, the orbit that contains ϕ_λ is itself contained in the set $\{\phi_\mu : \mu \in \mathbb{R}^4, \mu \cdot \mu = \lambda \cdot \lambda\}$.

8.15. Lemma. *For each real $r \neq 0$, $\{\phi_\lambda : \lambda \in \mathbb{R}^4, \lambda \cdot \lambda = r\}$ is an orbit under the action of P; call it θ_r. In addition, $\{\phi_\lambda : 0 \neq \lambda \in \mathbb{R}^4, \lambda \cdot \lambda = 0\}$ is an orbit under the action of P; call it θ_0.*

Proof. This lemma follows easily from the fact that $L(4)$ contains all matrices of the form

$$a = \begin{pmatrix} a_{11} & a_{12} & a_{13} & 0 \\ a_{21} & a_{22} & a_{23} & 0 \\ a_{31} & a_{32} & a_{33} & 0 \\ 0 & 0 & 0 & 1 \end{pmatrix}, \tag{12}$$

where

$$a' = \begin{pmatrix} a_{11} & a_{12} & a_{13} \\ a_{21} & a_{22} & a_{23} \\ a_{31} & a_{32} & a_{33} \end{pmatrix} \in O(3); \tag{13}$$

and also all matrices of the form

$$\lambda_t = \begin{pmatrix} 1 & 0 & 0 & 0 \\ 0 & 1 & 0 & 0 \\ 0 & 0 & \cosh t & \sinh t \\ 0 & 0 & \sinh t & \cosh t \end{pmatrix} \qquad (t \in \mathbb{R}). \qquad \blacksquare$$

Thus the orbits in $\hat{N}$ under P are the θ_r (r any real number) and the trivial orbit $\{\phi_0\}$. Clearly conditions 5.8(12), (13), (14) hold for P, N.

8.16.   What are the elements of $\hat{P}$ associated with these orbits?

(I) As in previous examples, the elements of $\hat{P}$ associated with $\{\phi_0\}$ are those of the form

$$T: \langle u, a \rangle \mapsto S_a, \qquad \text{where } S \in (L(4))\hat{}. \tag{14}$$

(II) Let $r < 0$. A typical element of θ_r is then $\phi_{\langle 0, 0, 0, |r|^{1/2} \rangle}$. The stability subgroup K_r for this element consists of those $\langle u, a \rangle$ in P for which

$$a \begin{pmatrix} 0 \\ 0 \\ 0 \\ 1 \end{pmatrix} = \begin{pmatrix} 0 \\ 0 \\ 0 \\ 1 \end{pmatrix};$$

that is, $K_r \cap H = O(3)$. (Here $O(3)$ is regarded as a subgroup of H by identifying each element a' in $O(3)$ with $\langle 0, a \rangle$ (see (12), (13)).) Thus by Theorem 5.2 (see also 5.5) the elements T of $\hat{P}$ associated with θ_r are constructed from elements S of $(O(3))\hat{}$ as follows:

$$\left. \begin{aligned} T = \underset{K_r \uparrow P}{\text{Ind}}(S'), \qquad \text{where } S' \in (K_r)\hat{}, \\ S'_{\langle u, a \rangle} = \exp(-iu_4 |r|^{1/2}) S_a \qquad (u \in \mathbb{R}^4, a \in O(3)). \end{aligned} \right\} \tag{15}$$

(III) Let $r > 0$. A typical element of θ_r is then $\phi_{\langle r^{1/2}, 0, 0, 0 \rangle}$. Its stability subgroup K_r consists of those $\langle u, a \rangle$ in P such that

$$a \begin{pmatrix} 1 \\ 0 \\ 0 \\ 0 \end{pmatrix} = \begin{pmatrix} 1 \\ 0 \\ 0 \\ 0 \end{pmatrix};$$

that is, $K_r \cap H$ consists of all $\langle 0, a \rangle$ such that

$$a = \begin{pmatrix} 1 & 0 & 0 & 0 \\ \hline 0 & & & \\ 0 & & a' & \\ 0 & & & \end{pmatrix}, \tag{16}$$

where a' belongs to the group $L(3)$ of all 3×3 real matrices which leave invariant the form $x_2^2 + x_3^2 - x_4^2$. Thus the elements T of $\hat{P}$ associated with θ_r are constructed from elements S of $(L(3))\hat{\ }$ as follows:

$$\left.\begin{aligned} T &= \underset{K_r \uparrow P}{\mathrm{Ind}}(S'), \qquad \text{where } S' \in (K_r)\hat{\ }, \\ S'_{\langle u, a \rangle} &= \exp(iu_1 r^{1/2}) S_{a'} \qquad (u \in \mathbb{R}^4; a, a' \text{ as in (16)}). \end{aligned}\right\} \quad (17)$$

(IV) Now consider the orbit θ_0. A typical element of this orbit is $\phi_{\langle 0, 0, 1, 1 \rangle}$. Denoting by K_0 the stability subgroup for $\phi_{\langle 0, 0, 1, 1 \rangle}$, we claim that $K_0 \cap H$ is isomorphic to the Euclidean group E_r^2 of the plane, generated by the proper Euclidean group E^2 together with all reflections about lines in the plane (see III.4.8(B)). To prove this claim, let us write $e_1, e_2,$ and w for $\langle 1, 0, 0, 0 \rangle$, $\langle 0, 1, 0, 0 \rangle$, and $\langle 0, 0, 1, 1 \rangle$ respectively. Suppose that $\langle 0, a \rangle \in K_0 \cap H$. Then one verifies that

$$\left.\begin{aligned} ae_1 &= pe_1 + qe_2 + \lambda w, \\ ae_2 &= re_1 + se_2 + \mu w, \end{aligned}\right\} \quad (18)$$

where p, q, r, s, λ, μ are real, and $\alpha = \begin{pmatrix} p & r \\ q & s \end{pmatrix}$ belongs to the group $O(2)$ of all rotations and reflections in the space W generated by e_1 and e_2. Now E_r^2 is the semidirect product of $O(2)$ with $\mathbb{R}^2$, and can be considered as the group of all pairs $\langle \alpha, m \rangle$, where $\alpha \in O(2)$ and $m = (\lambda, \mu)$ is a two-component row vector, multiplication being given by

$$\langle \alpha, m \rangle \langle \alpha', m' \rangle = \langle \alpha\alpha', m\alpha' + m' \rangle.$$

Thus

$$\Phi: \langle 0, a \rangle \mapsto \left\langle \begin{pmatrix} p & r \\ q & s \end{pmatrix}, (\lambda, \mu) \right\rangle$$

$(a, p, q, r, s, \lambda, \mu$ being as in (18)) maps the group $K_0 \cap H$ into E_r^2. It is easy to see that Φ is an injective homomorphism. In fact Φ is onto E_r^2; this follows immediately from the fact that the following matrices are in $K_0 \cap H$ (for all real t):

$$\begin{pmatrix} 1 & 0 & 0 & 0 \\ 0 & 1 & -t & t \\ 0 & t & 1 - \frac{1}{2}t^2 & \frac{1}{2}t^2 \\ 0 & t & -\frac{1}{2}t^2 & 1 + \frac{1}{2}t^2 \end{pmatrix},$$

$$\begin{pmatrix} 1 & 0 & -t & t \\ 0 & 1 & 0 & 0 \\ t & 0 & 1 - \frac{1}{2}t^2 & \frac{1}{2}t^2 \\ t & 0 & -\frac{1}{2}t^2 & 1 + \frac{1}{2}t^2 \end{pmatrix},$$

$$\begin{pmatrix} \cos t & \sin t & 0 & 0 \\ -\sin t & \cos t & 0 & 0 \\ 0 & 0 & 1 & 0 \\ 0 & 0 & 0 & 1 \end{pmatrix}.$$

Thus the claim is proved; and $K_0 \cap H$ is isomorphic under Φ with E_r^2.

So the elements T of $\hat{P}$ associated with θ_0 are constructed from elements S of $(E_r^2)\hat{\ }$ as follows:

$$\left.\begin{aligned} T &= \operatorname*{Ind}_{K_0 \uparrow P}(S'), \qquad \text{where } S' = (K_0)\hat{\ }, \\ S'_{\langle u, a \rangle} &= e^{i(u_3 - u_4)} S_{\Phi(a)} \qquad (\langle u, a \rangle \in K_0). \end{aligned}\right\} \tag{19}$$

8.17.   We have now shown:

Proposition.   *Every element T of $\hat{P}$ is, to within unitary equivalence, of one of the forms* (14), (15), (17), *or* (19).

Remark.   Thus the investigation of the irreducible unitary representations of P has been reduced to the same investigation for four smaller groups, namely $L(4)$, $O(3)$, $L(3)$, and E_r^2. Of these, $(O(3))\hat{\ }$ and $(E_r^2)\hat{\ }$ have already been "almost" classified in previous examples. (Indeed, E^2 and $SO(3)$ are normal subgroups of index 2 in E_r^2 and $O(3)$ respectively; and $(E^2)\hat{\ }$ and $(SO(3))\hat{\ }$ were classified in 8.6 and IX.12.3 respectively. The passage from $(E^2)\hat{\ }$ to $(E_r^2)\hat{\ }$ is an unusually simple application of the Mackey analysis, which we leave to the reader; and the same is true of the passage from $(SO(3))\hat{\ }$ to $(O(3))\hat{\ }$.) The remaining two groups $L(3)$ and $L(4)$ are intimately related to $SL(2, \mathbb{R})$ and $SL(2, \mathbb{C})$ respectively; and their structure spaces are well known, although we shall not attempt to obtain them here. These groups do not have sufficiently many non-trivial normal subgroups for the Mackey analysis to be an effective tool in the study of their structure spaces.

By 5.9 we shall be able to conclude that the Poincaré group has a Type I unitary representation theory provided the same is true of the four groups

$L(4)$, $O(3)$, $L(3)$, and E_r^2. Of these $O(3)$ is compact; and the discussion of the last paragraph, together with 8.6, shows that E_r^2 has a Type I unitary representation theory. It can be shown from the theory of semisimple Lie groups that $L(4)$ and $L(3)$ also have Type I unitary representation theories. Hence the Poincaré group has a Type I unitary representation theory.

It is a curious fact that the stability subgroups for the non-trivial orbits—namely $O(3)$, $L(3)$, and E_r^2—are just the symmetry groups of two-dimensional Riemannian, Lobachevskian, and Euclidean geometry respectively.

8.18. *Remark.* The Poincaré group P is the symmetry group of special relativity, and therefore its unitary representation theory has great relevance to quantum mechanics. See §8 of the Introduction to Volume 1.

Ergodic States of Physical Systems

8.19. In this example we shall show how the Mackey analysis applied to semidirect product bundles helps us to classify the ergodic positive linear functionals discussed in VI.20.7.

Suppose we are given a C^*-algebra A and a homomorphism τ of the additive group $\mathbb{R}$ of the reals into the group of all $*$-automorphisms of A. We assume that τ is strongly continuous in the sense that, for each a in A, the map $t \mapsto \tau_t(a)$ is continuous from $\mathbb{R}$ to A.

As in VI.20.7, a non-zero positive linear functional p on A is called *ergodic* if (a) p is $\mathbb{R}$-invariant (i.e, $p(\tau_t(a)) = p(a)$ for all a in A and t in $\mathbb{R}$), and (b) the only $\mathbb{R}$-invariant positive linear functionals q on A which are subordinate to p (see VI.20.2) are the non-negative multiples of p.

We have outlined in VI.20.8 the physical significance of ergodicity. In the next numbers we will apply the Mackey analysis to construct all possible ergodic positive linear functionals (at least under favorable conditions).

8.20. By VI.20.7, the search for ergodic positive linear functionals p on A is the same as the search for triples U, S, ξ such that (i) S is a $*$-representation of A, (ii) U is a unitary representation of $\mathbb{R}$ acting in the same space X as S and satisfying

$$U_t S_a U_t^{-1} = S_{\tau_t(a)} \qquad (a \in A; t \in \mathbb{R}), \qquad (20)$$

(iii) X is irreducible under the combined action of U and S, and (iv) ξ is a non-zero vector in X satisfying

$$U_t \xi = \xi \qquad \text{for all } t \text{ in } \mathbb{R}. \qquad (21)$$

Such a triple U, S, ξ will be called an *ergodic triple*. By VI.20.7 the most general ergodic positive linear functional p on A is obtained from an ergodic triple U, S, ξ by setting

$$p(a) = (S_a \xi, \xi) \qquad\qquad (a \in A). \qquad (22)$$

8.21. From A and τ let us construct the (saturated) semidirect product bundle $\mathscr{B} = A \underset{\tau}{\times} \mathbb{R}$ as in VIII.4.2. By VIII.15.6 and VIII.15.7 the pairs U, S satisfying (i), (ii), (iii) of 8.20 correspond naturally to the irreducible *-representations of $\mathscr{B}$.

The action τ of $\mathbb{R}$ on A generates of course a conjugating action $\langle t, D \rangle \mapsto {}^t D$ of $\mathbb{R}$ on $\hat{A}$:

$$({}^t D)_a = D_{\tau_{-t}(a)} \qquad\qquad (t \in \mathbb{R}; D \in \hat{A}; a \in A).$$

*We shall assume that A has a smooth *-representation theory, and that the conjugating action of $\mathbb{R}$ on $\hat{A}$ satisfies one of the "smoothness" conditions 5.8(13).* (Notice from XI.8.14 that here $\hat{A}^+ = \hat{A}$.) Thus the hypotheses of our main Theorem 5.8 are all valid for $\mathscr{B}$.

8.22. Now let U, S, ξ be an ergodic triple, and T the irreducible *-representation of $\mathscr{B}$ corresponding to U, S. By 5.8 T is associated with some orbit θ in $\hat{A}$. Take an element D of θ; and let H be the stability subgroup (of $\mathbb{R}$) for D. Since H is either $\{0\}$ or $\mathbb{R}$ or is isomorphic to $\mathbb{Z}$, the Mackey obstruction of D is trivial (by III.5.11). This says that D can be extended to a *-representation R of $\mathscr{B}_H$; that is, there is a unitary representation V of H, acting in $X(D)$, such that

$$V_t D_a V_t^{-1} = D_{\tau_t(a)} \qquad\qquad (t \in H; a \in A). \qquad (23)$$

(D, V is the pair corresponding to R by VIII.15.6.) Now, given any character χ of H, let us form the *-representation R^χ of $\mathscr{B}_H$ which will play the role of the S^W of 5.8; that is, R^χ corresponds (by VIII.15.6) to the pair D, $V \otimes \chi$. Theorem 5.8 now tells us that

$$T \cong \underset{\mathscr{B}_H \uparrow \mathscr{B}}{\mathrm{Ind}} (R^\chi) \qquad\qquad (24)$$

for some character χ of H. Applying the Mackey-Blattner description XI.10.16 of induced representations of semidirect product bundles, we conclude from (24) that

$$U \cong \underset{H \uparrow \mathbb{R}}{\mathrm{Ind}} (V \otimes \chi). \qquad\qquad (25)$$

Now U, S, ξ was an ergodic triple; so ξ is a non-zero vector in $X(U)$ satisfying (21). The reader will verify that the space of the induced group representation U can contain a non-zero U-invariant vector ξ if and only if (a) $\mathbb{R}/H$ is compact (that is, either $H = \mathbb{R}$ or $H \cong \mathbb{Z}$), and (b) there is a non-zero $(V \otimes \chi)$-invariant vector η in $X(V) = X(D)$. Suppose this is the case. Then (using the Mackey description of the right side of (25)) we have

$$\xi(t) = \eta \qquad \text{for all } t \text{ in } \mathbb{R}. \tag{26}$$

In view of (23), the $(V \otimes \chi)$-invariance of η implies that

$$(D_{\tau_{t(a)}}\eta, \eta) = (D_a\eta, \eta) \qquad (a \in A; t \in H). \tag{27}$$

Conversely, by (23) combined with the argument of VI.20.7, the assumption (27) implies that η is $(V \otimes \chi)$-invariant for some character χ of H, and so can be used in (26) to exhibit a vector ξ invariant under the right side of (25). Applying the equation XI.10.16(22) to the induced representation (24) of $\mathscr{B}$, we obtain the following expression for the ergodic positive linear functional (22):

$$p(a) = (S_a\xi, \xi) = \int_{\mathbb{R}/H} (D_{\tau_{-t(a)}}\eta, \eta)d\tilde{t}$$

(where $\tilde{t} = t + H$ and $d\tilde{t}$ is normalized Haar measure on the compact quotient group $\mathbb{R}/H$).

8.23. Maintaining the assumptions of 8.21, we can thus describe the most general ergodic positive linear functional on A.

Proposition. *Let p be an ergodic positive linear functional on A. Then either (I) p is itself indecomposable, or (II) there exist an element D of $\hat{A}$, a smallest positive number s such that $^sD \cong D$, and a vector η in $X(D)$, such that*

$$(D_{\tau_{s(a)}}\eta, \eta) = (D_a\eta, \eta) \qquad (a \in A)$$

and

$$p(a) = \int_0^s (D_{\tau_{t(a)}}\eta, \eta)dt \qquad (a \in A).$$

Remark. Alternative (I) corresponds to the case that $H = \mathbb{R}$ (in 8.22), while alternative (II) corresponds to the case that $H = \mathbb{Z}s \cong \mathbb{Z}$.

8.24. *Remark.* Let us reflect for a moment on this proposition in the light of the interpretation of p, given in VI.20.8, as the "macroscopic" state of a thermodynamical system. Alternative (I) says that p is actually a "microscopic" state, which is not physically reasonable. Alternative (II) says that p

comes from a microscopic state $a \mapsto (D_a \eta, \eta)$ of the system which is periodic in time with period s. This too is unreasonable, since the random fluctuations of a thermodynamical system should preclude any precise return to a former microscopic state. Nevertheless macroscopic states do physically exist. The fact that they do not fall within the purview of Proposition 8.23 indicates that a thermodynamical system must fail to satisfy some one of the hypotheses of the proposition. It is 5.8(13) which breaks down for macroscopic thermodynamical systems. The action of time on the C^*-algebra A of all microscopic physical observables of a thermodynamical system cannot be "smooth" in the sense of 5.8(13). We shall see specific mathematical examples of the failure of 5.8(13) in the next section. It is very interesting that the phenomena associated with the breakdown of 5.8(13), which from the standpoint of the Mackey analysis seem merely pathological, are thus of great physical importance for the mathematical understanding of the macroscopic systems of thermodynamics.

9. Further Examples

9.1.   Most of the examples to be presented in this final section illustrate the phenomena which can occur when one of the hypotheses 5.8(12), (13), (14) of the Mackey analysis breaks down.

Transformation Bundles with Transitive Action

9.2.   We begin with the simple case of a transformation bundle built from a transitive group action.

Let G be any locally compact group, H a closed subgroup of G, and $\mathscr{B}$ the transformation bundle for the natural action of G on G/H (see VIII.7.3). Thus $\mathscr{B}$ is the semidirect product of $A = \mathscr{C}_0(G/H)$ and G:

$$\mathscr{B} = A \underset{\tau}{\times} G,$$

the action τ of G on A being that derived from the action of G on G/H:

$$\tau_x(f)(m) = f(x^{-1}m) \qquad (x \in G; f \in A; m \in G/H).$$

Notice that $\hat{A} \cong G/H$; and the conjugating action of G on $\hat{A}$ is just its natural action on G/H. Thus (taking the Z of 2.5 to be $\hat{A}$), we see that there is only one orbit in $\hat{A}$; and so every non-degenerate *-representation of $\mathscr{B}$ is associated with that orbit. Since the natural bijection $G/H \to \hat{A}$ is a homeomorphism, we can apply 5.2 (and 5.5) to conclude that the non-degenerate *-representations of $\mathscr{B}$ are in natural one-to-one correspondence with the unitary representations of the stability subgroup H of the typical element eH of G/H. Since by VIII.18.22 the non-degenerate *-representations of $\mathscr{B}$ are essentially just systems of imprimitivity for G over G/H, this is nothing more than we already knew from the Imprimitivity Theorem XI.14.18.

9.3. Take the special case that $H = \{e\}$, so that $\hat{A} \cong G/H = G$. Then the above paragraph says that $\mathscr{B}$ has (to within unitary equivalence) exactly one irreducible *-representation T, and that every non-degenerate *-representation of $\mathscr{B}$ is a direct sum of copies of T. (Compare VIII.7.10) By XI.14.27 T must be a compact *-representation. (If $\mathscr{B}$, and hence $\mathscr{L}_1(\mathscr{B})$, is second-countable, this also follows from Rosenberg's Theorem VI.23.1.)

9.4. In 9.2 and 9.3 the Mackey analysis was completely applicable. But a simple modification of 9.3 shows that the fundamental Proposition 3.11 (and hence also 3.12 and 5.2) would fail if we simply omitted the hypothesis that either G/H is σ-compact or $\gamma: G/H \to Z$ is a homeomorphism.

Indeed, suppose that the locally compact group G is second-countable but of uncountable cardinality; and let G_d coincide with G as an abstract group and carry the discrete topology. Let $\mathscr{B}_d$ be the transformation bundle (over G_d) for G_d acting to the left by multiplication on G. Thus $\mathscr{B}_d$ coincides in all respects with the $\mathscr{B}$ of 9.3 except that its topology is bigger. Hence the unique irreducible *-representation T of $\mathscr{B}$ is also an irreducible *-representation of $\mathscr{B}_d$. On the other hand T acts in a separable space (since G is second-countable), while by the uncountability of G_d any *-representation of $\mathscr{B}_d$ which is induced from a non-zero *-representation of $B_e \cong \mathscr{C}_0(G)$ must act in a Hilbert space of uncountable dimension. So T, though associated with the unique G_d-orbit in G, cannot be obtained by the inducing recipe of 3.11 as applied in $\mathscr{B}_d$. Hypothesis 5.8(14) evidently fails in this situation.

Transformation Bundles with Non-Smooth Actions

9.5. In this example we will show that if hypothesis 5.8(13) fails, then not every irreducible *-representation of $\mathscr{B}$ need be associated with an orbit.

Choose a real number α such that $\pi^{-1}\alpha$ is irrational; and consider the action

$$\langle n, u \rangle \mapsto n : u = e^{in\alpha}u \qquad (n \in \mathbb{Z}; u \in \mathbb{E}) \qquad (1)$$

of $\mathbb{Z}$ on the circle group $\mathbb{E}$. The transformation bundle over $\mathbb{Z}$ for this action is the semidirect product

$$\mathscr{B} = \mathscr{C}(\mathbb{E}) \underset{\tau}{\times} \mathbb{Z}$$

(see VIII.7.3); and the conjugating action of $\mathbb{Z}$ on $(\mathscr{C}(\mathbb{E}))\hat{} \cong \mathbb{E}$ generated by τ is just (1).

9.6. Now the action (1) of $\mathbb{Z}$ on $\mathbb{E}$ is not "smooth," that is, it does not satisfy the hypothesis 5.8(13). In 9.8 we will establish this by showing that the conclusion of 5.8 fails for $\mathscr{B}$. Meanwhile, however, it will be illuminating to show directly why Condition (C) of 2.16 fails in the present context. Let us proceed by contradiction and assume that Condition (C) holds; that is, we assume that there exists a countable family $\mathscr{W}$ of Borel subsets of $\mathbb{E}$ such that (i) each set in $\mathscr{W}$ is a union of $\mathbb{Z}$-orbits (under (1)), and (ii) $\mathscr{W}$ separates the set of all $\mathbb{Z}$-orbits. Without loss of generality we can assume that $\mathscr{W}$ is closed under complementation. Now let ν be Haar measure on $\mathbb{E}$. By IX.6.5(I) the subgroup $\{e^{in\alpha} : n \in \mathbb{Z}\}$ is dense in $\mathbb{E}$; so by VIII.19.4 (and VIII.19.5) every set in $\mathscr{W}$ either is ν-null or has ν-null complement. Let $\mathscr{V} = \{W \in \mathscr{W} : \nu(W) = 0\}$. Being a countable union of ν-null sets,

$$\bigcup \mathscr{V} \text{ is } \nu\text{-null.} \tag{2}$$

Since each $\mathbb{Z}$-orbit is countable, hence ν-null, (2) implies that there are at least two distinct orbits θ_1 and θ_2 disjoint from $\bigcup \mathscr{V}$. Since $\mathscr{W}$ separates orbits and is closed under complementation, there must be an element W of $\mathscr{W}$ such that $\theta_1 \subset W, \theta_2 \subset \mathbb{E} \setminus W \in \mathscr{W}$. This implies that neither W nor $\mathbb{E} \setminus W$ is contained in $\bigcup \mathscr{V}$; hence neither belongs to $\mathscr{V}$; hence both W and $\mathbb{E} \setminus W$ have ν-null complements, an impossibility. So Condition (C) fails.

9.7. Let u be any point of $\mathbb{E}$; and let θ_u be the $\mathbb{Z}$-orbit $\{e^{in\alpha}u : n \in \mathbb{Z}\}$ containing u. The stability subgroup of $\mathbb{Z}$ for u is $\{0\}$ (since $\pi^{-1}\alpha$ is irrational). Hence by 5.2 (or 3.12) there is exactly one irreducible *-representation of $\mathscr{B}$ associated with the orbit θ_u, namely

$$V^u = \underset{B_0 \uparrow \mathscr{B}}{\operatorname{Ind}}(\phi_u)$$

(where ϕ_u is the evaluation map $f \mapsto f(u)$ on $\mathscr{C}(\mathbb{E}) = B_0$). Furthermore, since the orbits θ_u are pairwise disjoint Borel sets and $V^u | B_0$ is concentrated on θ_u, we have

$$V^u \cong V^v \Leftrightarrow \theta_u = \theta_v \Leftrightarrow v = e^{in\alpha} \quad \text{for some } n \in \mathbb{Z}. \tag{3}$$

9.8. Now it is easy to see that there are other irreducible *-representations of $\mathscr{B}$ which are not associated with any orbit. Indeed, let μ be any regular Borel measure on $\mathbb{E}$ which is ergodic with respect to the action (1) of $\mathbb{Z}$ (see VIII.19.3); and let $\mathscr{T}^{\mu}$ be the system of imprimitivity for $\mathbb{Z}$ over $\mathbb{E}$ constructed from μ as in VIII.19.6. By VIII.19.7 $\mathscr{T}^{\mu}$ is irreducible. Thus *the *-representation T^{μ} of $\mathscr{B}$ corresponding to $\mathscr{T}^{\mu}$ (by VIII.18.22) is irreducible.*

It is clear from the way in which $\mathscr{T}^{\mu}$ was constructed that the spectral measure of $T^{\mu}|B_0$ has the same null sets as μ. Hence, if μ and μ' are two Borel measures on $\mathbb{E}$ which are ergodic with respect to the action (1)

$$T^{\mu'} \cong T^{\mu} \Leftrightarrow \mu' \text{ and } \mu \text{ have the same null sets.} \tag{4}$$

Thus the unitary equivalence class of T^{μ} is determined by the measure-theoretic equivalence class of μ.

Now each orbit θ_u (see 9.7) determines an equivalence class of measures μ_u ergodic under the action (1), namely those measures which are carried by θ_u and assign positive measure to each point of θ_u. Furthermore T^{μ_u} clearly coincides with the V^u of 9.7. Thus, to show that there are irreducible *-representations of $\mathscr{B}$ not associated with any orbit θ_u, by (4) we need only exhibit a Borel measure μ on $\mathbb{E}$ which is ergodic under the action (1) and assigns measure 0 to each one-element set (and so is not equivalent to any of the μ_u); then T^{μ} will be a *-representation of the required sort. But by VIII.19.4 the Haar measure ν of $\mathbb{E}$ is ergodic under (1), and certainly $\nu(\{u\}) = 0$ for $u \in \mathbb{E}$. So T^{ν} is an irreducible *-representation of $\mathscr{B}$ not associated with any orbit.

Thus we have shown that *the failure of 5.8(13) implies in general the failure of the first conclusion of Theorem 5.8.*

9.9. *Remark.* Keep the notation of 9.5–9.8. In view of 9.8 one is tempted to conjecture that *every* irreducible *-representation of $\mathscr{B}$ is unitarily equivalent to one of the form T^{μ} (μ being a Borel measure on $\mathbb{E}$ ergodic with respect to the action (1)). This however is false, as the following proposition shows:

Proposition*. *For each u in $\mathbb{E}$ let χ_u be the character $n \mapsto u^n$ of $\mathbb{Z}$. Then*

$$u, v \in \mathbb{E}, u \neq v \Rightarrow T^{\nu} \otimes \chi_u \not\cong T^{\nu} \otimes \chi_v.$$

We leave the proof of this to the reader. (VIII.19.2 will be helpful!) $T^{\nu} \otimes \chi_u$ is of course the tensor product representation of $\mathscr{B}$.

Suppose that $1 \neq u \in \mathbb{E}$. By the above proposition $T^{\nu} \otimes \chi_u \not\cong T^{\nu} \otimes \chi_1 = T^{\nu}$. Also $(T^{\nu} \otimes \chi_u)|B_0 = T^{\nu}|B_0$, which cannot be equivalent to $T^{\mu}|B_0$ unless μ and ν are equivalent (see 9.8). Therefore $T^{\nu} \otimes \chi_u$ is an irreducible

*-representation of $\mathscr{B}$ distinct from all the T^μ constructed in 9.8. Furthermore there are at least a continuum of such $T^\nu \otimes \chi_u$ (corresponding to the continuum of possible choices of u).

9.10.   Remark.   Although, as we have seen in 9.8, the V^u do not exhaust all of $\hat{\mathscr{B}}$, nevertheless the Effros–Hahn conjecture (see 5.11) holds for this $\mathscr{B}$. Indeed, the C^*-completion C of $\mathscr{L}_1(\mathscr{B})$ turns out to be simple, so that the primitive ideal space of C has only a single point. See Effros and Hahn [1], Corollary 5.16 and 7.3.

It should be pointed out that the *-representation theory of this $\mathscr{B}$ is not of Type I. Unfortunately we are not in a position to prove this, since we have never given examples of non-Type I representations.

An Example of a Non-Trivial Reduction in the Non-Smooth Case.

9.11.   The next example is also "non-smooth." However we shall see that by a judicious choice of the pair Z, Φ of 2.5 we can make a non-trivial reduction of the problem of determining *all* the non-degenerate *-representations.

9.12.   Consider the action of $\mathbb{R}$ on the torus $\mathbb{E}^2$ given as follows:

$$t:\langle u, v \rangle = \langle e^{it}u, e^{i\beta t}v \rangle \qquad (t \in \mathbb{R}; \langle u, v \rangle \in \mathbb{E}^2), \qquad (5)$$

where $\beta\cdot$ is a fixed irrational number. Notice incidentally that $\{\langle e^{it}, e^{i\beta t} \rangle : t \in \mathbb{R}\}$ is dense in $\mathbb{E}^2$ (by IX.6.5(II)), so that by VIII.19.4 Haar measure on $\mathbb{E}^2$ is ergodic with respect to the action (5).

Let $\mathscr{B} = \mathscr{C}(\mathbb{E}^2) \underset{\tau}{\times} \mathbb{R}$ be the transformation bundle for the action (5) of $\mathbb{R}$ on $\mathbb{E}^2$ (see VIII.7.3). As usual we identify the structure space of $B_0 = \mathscr{C}(\mathbb{E}^2)$ with $\mathbb{E}^2$ itself, so that the conjugating action of $\mathbb{R}$ on $(B_0)^\wedge$ becomes just (5).

Now, instead of taking the Z of 2.5 to be $\mathbb{E}^2$ (and Φ to be the identity map), similarly to what we have done in past examples, let us take Z to be $\mathbb{E}$, considered as an $\mathbb{R}$-space by means of the natural action

$$t:u = e^{it}u \qquad (t \in \mathbb{R}; u \in \mathbb{E}); \qquad (6)$$

and let us define $\Phi: \mathbb{E}^2 \to Z = \mathbb{E}$ to be the projection $\langle u, v \rangle \mapsto u$, which is certainly equivariant with respect to the actions (5) and (6) of $\mathbb{R}$. There is then only one $\mathbb{R}$-orbit in Z, so that every non-degenerate *-representation of $\mathscr{B}$ is associated with this one orbit. Taking 1 as a typical point of Z, and denoting its stability subgroup $\{2\pi n : n \in \mathbb{Z}\}$ by H, we learn from 3.12 that the inducing operation sets up a one-to-one correspondence between the non-degenerate *-representations T of $\mathscr{B}$ on the one hand, and on the other hand the non-degenerate *-representations S of $\mathscr{B}_H$ such that $S|B_0$ has its spectral measure carried by $\Phi^{-1}(1) = \{\langle 1, v\rangle : v \in \mathbb{E}\}$. By 2.21 these S can be identified with the non-degenerate *-representations of the transformation bundle associated with the action $\langle n, u\rangle \mapsto e^{2\pi i n\beta} u$ of $\mathbb{Z}$ on $\mathbb{E}$.

Thus we have shown that *the non-degenerate *-representations of the $\mathscr{B}$ of this number are in natural one-to-one correspondence with the non-degenerate *-representations of the transformation bundle $\mathscr{B}$ studied in 9.5–9.8 (provided we take $\alpha = 2\pi\beta$ in the latter).*

The Glimm Bundles

9.13. In VIII.17.7 we constructed a class of semidirect product bundles whose *-representations are in one-to-one correspondence with those of the Glimm algebras of VI.17.5. It is reasonable to refer to these as the Glimm bundles. The Mackey analysis applied to the Glimm bundles will thus give us information about the *-representation theory of the Glimm algebras.

9.14. Let us fix a sequence $d_1, d_2, \ldots$ of positive integers, and adopt all the notation of the construction of VIII.17.7. The transformation bundle $\mathscr{B} = A \underset{\tau}{\times} G$ will be called the $(d_1, d_2, \ldots)$ *Glimm bundle*. The information that we can obtain from the Mackey analysis about $\hat{\mathscr{B}}$ is almost identical in form with what we obtained for the $\mathscr{B}$ of 9.5. Indeed, the action of G on $\hat{A} \cong E$ is the same as multiplication by G in E; and each element of E has trivial stability subgroup. Thus, associated with the orbit θ_x of an element x of E there is one and only one irreducible *-representation

$$V^x = \underset{A\uparrow\mathscr{B}}{\text{Ind}}(\phi_x)$$

of the Glimm bundle $\mathscr{B}$. (Here $\phi_x : f \mapsto f(x)$ is the *-homomorphism of A consisting of evaluation at x.) As in (3) we have:

$$V^x \cong V^y \Leftrightarrow y = gx \qquad \text{for some } g \text{ in } G. \tag{7}$$

Now the same argument as in 9.6 and 9.8 shows that the action of G on E is not smooth, and that there are many irreducible *-representations of $\mathscr{B}$ apart from the V^x. For example, let ν stand for Haar measure on E; let S be the *-representation of A on $\mathscr{L}_2(\nu)$ sending f into multiplication by f; and let U be the unitary representation of G acting on $\mathscr{L}_2(\nu)$ by translation of E. Then the *-representation of $\mathscr{B}$ corresponding (by VIII.15.6) to the pair S, U is irreducible but is not associated with any G-orbit in E.

9.15.   Remark.   Let D stand for the $(d_1, d_2, \ldots)$ Glimm algebra. It is of some interest to describe directly, in terms of D, the irreducible *-representation of D which corresponds to V^x via VIII.17.7.

For this we need the notion of the tensor product of an infinite sequence of Hilbert spaces. Let $X_1, X_2, \ldots$ be a sequence of Hilbert spaces, and fix a unit vector ξ_n in X_n for each n. Let Y_n be the Hilbert tensor product $X_1 \otimes \cdots \otimes X_n$. Thus, for each $n = 1, 2, \ldots, L_n : \zeta \mapsto \zeta \otimes \xi_{n+1}$ is a linear isometry of Y_n into Y_{n+1}. Let us agree to regard L_n as an identification. Then the Y_n $(n = 1, 2, \ldots)$ form an increasing sequence of Hilbert spaces; and their union $Y = \bigcup_{n=1}^{\infty} Y_n$ is a pre-Hilbert space whose completion we call $\bar{Y}$. This $\bar{Y}$ is referred to as the *tensor product of the $\{X_n\}$ with respect to the $\{\xi_n\}$*, and is sometimes denoted by $\prod_n^{\otimes} X_n$. (Of course the latter symbolism is deficient, inasmuch as the construction depends essentially on the $\{\xi_n\}$.)

If $n = 1, 2, \ldots$, each q in $\mathcal{O}(Y_n)$ gives rise to an operator $\bar{q}$ in $\mathcal{O}(\bar{Y})$, namely that operator which, for each $p > n$, coincides on Y_p with

$$q \otimes 1_{n+1} \otimes \cdots \otimes 1_p$$

(1_r being the identity operator on X_r). The map $q \mapsto \bar{q}$ is clearly a *-homomorphism.

Now let us specialize X_n to be of finite dimension d_n, having an orthonormal basis $\{\eta_n^{(t)} : t \in E_n\}$ indexed by the group E_n. Fix an element x of the Cartesian product group E; take ξ_n to be the unit vector $\eta_n^{(x_n)}$ in X_n; and let $\bar{Y} = \prod_n^{\otimes} X_n$ be formed with respect to these $\{\xi_n\}$. Furthermore, let $A_n = \mathcal{O}(Y_n)$; and for $n > m$ define $F_{nm} : A_m \to A_n$ by means of the equation

$$F_{nm}(q) = q \otimes 1_{m+1} \otimes \cdots \otimes 1_n.$$

Thus $\{A_n\}, \{F_{nm}\}$ is the directed system whose direct limit is the $(d_1, d_2, \ldots)$ Glimm algebra D. The mapping $q \mapsto \bar{q}$ of $\bigcup_n A_n$ into $\mathcal{O}(\bar{Y})$ clearly extends to a *-representation W of D. Since W depends on the vectors $\{\xi_n\}$ with respect to which $\bar{Y}$ was formed, and hence on the element x of E, we shall denote it by W^x. The reader should now be able to prove the following result:

Proposition*.   *The above W^x is the irreducible *-representation of D associated with V^x by the correspondence of VIII.17.7.*

In view of (7) we have

$$W^y \cong W^x \Leftrightarrow y = gx \quad \text{for some } g \text{ in } G.$$

A "Non-Smooth" Modification of E^2

9.16. Let F be the semidirect product group $\mathbb{C} \underset{\tau}{\times} \mathbb{Z}$, the action τ of $\mathbb{Z}$ on $\mathbb{C}$ being given by

$$\tau_n(b) = e^{in}b \qquad\qquad (n \in \mathbb{Z}; b \in \mathbb{C}).$$

Thus the map $\langle b, n \rangle \mapsto \{e^{in}, b\}$ (see 8.6) is a continuous (though not bicontinuous) group-isomorphism of F onto a dense subgroup of E^2.

Let N be the closed normal subgroup $\{\langle b, 0 \rangle : b \in \mathbb{C}\}$ of F (which of course can be identified with $\mathbb{C}$). The conjugating action of the element $\langle a, n \rangle$ of F sends the character $\phi_\lambda \colon \langle b, 0 \rangle \mapsto e^{i\mathrm{Re}(\lambda b)}$ ($\lambda \in \mathbb{C}$) of N into $\phi_{\lambda \exp(-in)}$. So the orbits in $\hat{N}$ under the action of F are the sets of the form

$$\theta_\lambda = \{\phi_{\lambda \exp(in)} : n \in \mathbb{Z}\} \qquad\qquad (\lambda \in \mathbb{C}).$$

The irreducible unitary representations of F associated with the trivial orbit $\theta_0 = \{\phi_0\}$ are as usual those which are lifted from $F/N \cong \mathbb{Z}$, namely the

$$\chi_v \colon \langle b, n \rangle \mapsto v^n \qquad\qquad (v \in \mathbb{E}). \qquad (8)$$

If $\lambda \neq 0$, the stability subgroup for ϕ_λ is N; and so the only irreducible unitary representation of F associated with θ_λ is

$$V^\lambda = \underset{N \uparrow G}{\mathrm{Ind}}(\phi_\lambda). \qquad (9)$$

If λ, μ are non-zero elements of $\mathbb{C}$,

$$V^\lambda \cong V^\mu \Leftrightarrow \theta_\lambda = \theta_\mu \Leftrightarrow \mu = e^{in}\lambda \quad \text{for some } n \in \mathbb{Z}. \qquad (10)$$

The smoothness condition 5.8(13) breaks down in this example. This follows from the fact that $F/N \cong \mathbb{Z}$ acts on each circle $C_r = \{b \in \mathbb{C} : |b| = r\}$ ($r > 0$) in just the same way as $\mathbb{Z}$ acted on $\mathbb{E} \cong \hat{A}$ in 9.5(1) (if we take $\alpha = 1$ in 9.5). Indeed, this similarity leads to the following precise relationship between the unitary representation theory of F and the *-representation theory of 9.5:

Proposition*. *For every primary unitary representation T of F, there is a unique non-negative number r such that $T|N$ is concentrated on the circle C_r. Furthermore, for each $r > 0$, there is a natural correspondence (preserving equivalence, irreducibility, etc.) between the primary unitary representations T of F such that $T|N$ is concentrated on C_r and the primary *-representations of the $\mathscr{B}$ of 9.5.*

Proof. The first statement is obtained by applying Proposition 2.20 to the map $\Psi \colon b \mapsto |b|$ of $\mathbb{C}$ into $\{r \in \mathbb{R} : r \geq 0\}$. Given $r > 0$, the second statement is obtained by applying Proposition 2.21 with G, D, and I replaced by $\mathbb{Z}$, $\mathscr{C}_0(\mathbb{C})$, and $\{f \in \mathscr{C}_0(\mathbb{C}) : f(b) = 0 \text{ whenever } |b| = r\}$. ∎

In particular there are a great many irreducible unitary representations of F other than those of (8) and (9).

An Example of Non-Closed Stability Subgroups

9.17. The example of 9.16 can be modified to give an example in which the stability subgroups for the elements of the structure space of a closed normal subgroup are not closed.

Let G be the semidirect product group $\mathbb{C} \times_\tau (\mathbb{Z} \times \mathbb{E})$, where

$$\tau_{n,u}(b) = ue^{in}b \qquad\qquad (n \in \mathbb{Z}; u \in \mathbb{E}; b \in \mathbb{C}).$$

It will be convenient to identify the element $\langle b, n, 1 \rangle$ of G with the element $\langle b, n \rangle$ of the group F of 9.16. Thus F becomes a closed normal subgroup of G. The normal subgroup N of F (see 9.16) is also of course normal in G; and the character ϕ_λ of N is carried by the conjugating action of $\langle b, n, u \rangle$ into $\phi_{\lambda u^{-1} \exp(-in)}$. From this and XI.12.20 (applied to the automorphism of F generated by $\langle b, n, u \rangle$) it follows that the conjugating action of $\langle b, n, u \rangle$ sends the V^λ of (9) into $V^{\lambda u^{-1} \exp(-in)}$. Now by (10) $V^{\lambda u^{-1} \exp(-in)} \cong V^\lambda$ if and only if $u = e^{im}$ for some m in $\mathbb{Z}$. Thus, if $\lambda \neq 0$, the stability subgroup of G for the irreducible unitary representation V^λ of F is

$$\{\langle b, n, e^{im} \rangle : b \in \mathbb{C}, n, m \in \mathbb{Z}\},$$

which is dense and not closed in G.

The Conditions of 5.8 May Hold Even If 5.8(13) Fails

9.18. We shall now present an example of a nilpotent group K having a closed normal subgroup M such that the conclusion of 5.8 holds (for the unitary representation theory of K) even though 5.8(13) fails.

Let K denote the (nilpotent) multiplicative group of all 4×4 matrices of the form

$$\{r, s, t, u\} = \begin{pmatrix} 1 & r & r & u \\ 0 & 1 & 0 & s \\ 0 & 0 & 1 & t \\ 0 & 0 & 0 & 1 \end{pmatrix} \qquad (r, s, t, u \in \mathbb{R}).$$

Thus

$$\{r, s, t, u\}\{r', s', t', u'\} = \{r + r', s + s', t + t', u + u' + r(s' + t')\},$$

$$\{r, s, t, u\}^{-1} = \{-r, -s, -t, -u + r(s + t)\}.$$

The Abelian closed normal subgroup

$$N = \{\{0, s, t, u\} : s, t, u \in \mathbb{R}\}$$

of K is isomorphic with $\mathbb{R}^3$; and $\hat{N}$ is isomorphic with $\mathbb{R}^3$ under the correspondence $\langle a, b, c \rangle \mapsto \phi_{a,b,c}$ ($\langle a, b, c \rangle \in \mathbb{R}^3$), where

$$\phi_{a,b,c}(\{0, s, t, u\}) = e^{i(as + bt + cu)}.$$

Since $\{r, s, t, u\}^{-1}\{0, s', t', u'\}\{r, s, t, u\} = \{0, s', t', u' - r(s' + t')\}$, the conjugating action of $\{r, s, t, u\}$ carries $\phi_{a,b,c}$ into $\phi_{a-cr, b-cr, c}$. Thus the orbits in $\hat{N}$ are of just two kinds: (i) For each pair a, b of real numbers the one-element set $\{\phi_{a,b,0}\}$ is an orbit; (ii) if $b, c \in \mathbb{R}$ and $c \neq 0$, then

$$\theta_{b,c} = \{\phi_{d, b+d, c} : d \in \mathbb{R}\}$$

is an orbit. It is easy to see that the set of all orbits of the forms (i) and (ii) satisfies Condition (C) of 2.16. So Theorem 5.8 and 5.9 are applicable to K and N; and after the pattern of previous examples we conclude:

K has a Type I *unitary representation theory. The one-dimensional elements of $\hat{K}$ are just those lifted from the characters of the commutative quotient group $K/C \cong \mathbb{R}^3$, where C is the central subgroup $\{\{0, 0, 0, u\} : u \in \mathbb{R}\}$. The elements of $\hat{K}$ of dimension greater than one are just the*

$$T^{b,c} = \underset{N \uparrow K}{\mathrm{Ind}}(\phi_{0,b,c}),$$

where $b, c \in \mathbb{R}$ and $c \neq 0$. We have

$$T^{b',c'} \cong T^{b,c} \Leftrightarrow b' = b, c' = c.$$

Let us now try to apply the Mackey analysis with a *different* closed normal subgroup of K, namely

$$M = \{\{0, n\sqrt{2}, m, u\} : n, m \in \mathbb{Z}, u \in \mathbb{R}\}.$$

One verifies that M is indeed an Abelian closed normal subgroup of K contained in N. Now $\hat{M}$ is isomorphic with $\mathbb{E} \times \mathbb{E} \times \mathbb{R}$ under the map $\langle v, w, c \rangle \mapsto \psi_{v,w,c}$, where

$$\psi_{v,w,c}(\{0, n\sqrt{2}, m, u\}) = v^n w^m e^{icu}$$

($v, w \in \mathbb{E}$; $c, u \in \mathbb{R}$; $n, m \in \mathbb{Z}$). Notice that the restriction map $\pi : \phi \mapsto \phi|M$ of $\hat{N}$ into $\hat{M}$ sends $\phi_{a,b,c}$ into $\psi_{\exp(ia\sqrt{2}), \exp(ib), c}$. From this and the known action of K on the $\phi_{a,b,c}$ we see that the conjugating action of $\{r, s, t, u\}$ carries $\psi_{v,w,c}$ into

$$\psi_{v\exp(-icr\sqrt{2}), w\exp(-icr), c}. \tag{11}$$

But by IX.6.5(II) $\{\langle e^{ir\sqrt{2}}, e^{ir}\rangle : r \in \mathbb{R}\}$ is a dense subgroup of $\mathbb{E} \times \mathbb{E}$. Hence by VIII.19.4, for each $c \neq 0$, the action (11) of K on $\{\psi_{v,w,c} : v, w \in \mathbb{E}\}$ is ergodic with respect to Haar measure on $\mathbb{E} \times \mathbb{E}$. From this it follows by the same argument as in 9.6 that the K-orbits in $\hat{M}$ cannot satisfy Condition (C) of 2.16.

On the other hand, let T be any primary unitary representation of K. By 5.8 applied to N, the spectral measure of $T|N$ is concentrated on one of the K-orbits θ in $\hat{N}$. From this it is easily verified that the spectral measure of $T|M$ must be concentrated on $\pi(\theta)$. But $\pi(\theta)$ is a K-orbit in $\hat{M}$. Therefore T must be associated with some K-orbit in $\hat{M}$.

We have shown that, although condition 5.8(13) fails when we work with the normal subgroup M of K, nevertheless the conclusion of Theorem 5.8 holds for M. Thus, given a situation in which conditions 5.8(12) and 5.8(14) hold, the failure of 5.8(13) cannot be taken as an automatic guarantee that there exist primary *-representations not concentrated on any orbit.

10. Exercises for Chapter XII

1. Let H be a subgroup of the finite group G, S a representation of H, and T the induced representation $\mathrm{Ind}_{H \uparrow G}(S)$ of G. Set $p(\alpha)(f) = \mathrm{Ch}_\alpha f$ ($\alpha \in G/H$; $f \in X(T)$).
   (a) Show that T, p is a system of imprimitivity for G over G/H (see 1.7).
   (b) Prove that if T', p' is the system of imprimitivity induced by the unitary representation S of H, then S is equivalent to the subrepresentation of $T'|H$ acting on range $(p'(eH))$ (see 1.7 and XI.14.21).

2. Suppose that V and W are two elements of $(H/N)^{\widehat{}(\tau^{-1})}$ such that $S^{(V)} \cong S^{(W)}$ under a unitary equivalence P. Show, as stated in the proof of Proposition 1.19, that $P = 1 \otimes \Psi$, where Ψ is a unitary equivalence of V and W.

3. Show in Example 1.25 that one of the cocycles in the Mackey obstruction of J is the bicharacter σ of G/M given by

$$\sigma(\langle r, s \rangle, \langle r', s' \rangle) = \begin{cases} 1, \text{ unless } s = r' = w, \\ -1, \text{ if } s = r' = w. \end{cases}$$

4. Verify that the action of G on $\hat{N}$ in Example 1.27 is given by (33).

5. Let G be a finite group and N a normal subgroup of G. Let D be an irreducible unitary representation of N, and p the central self-adjoint idempotent element of $\mathscr{L}(N)$ corresponding to D. (That is, $p(n) = \dim(D)\overline{\mathrm{Trace}(D_n)}$ for $n \in N$; so that, if S is any unitary representation of N, range(S_p) is the D-subspace for S.) Let H be the stability subgroup of G for D (under the action of G on $\hat{N}$ by conjugation). We shall identify $\mathscr{L}(N)$ and $\mathscr{L}(H)$ with *-subalgebras of $\mathscr{L}(G)$.

(I) Defining $B = p * \mathcal{L}(G) * p$, show that

$$B \subset \mathcal{L}(H);$$

in fact show that B is a *-ideal of $\mathcal{L}(H)$.

(II) By §VII.4 there is a natural correspondence between the irreducible *-representations of B and the irreducible unitary representations of G such that $T_p \neq 0$. By (I) and §VII.4 there is a natural correspondence between the irreducible *-representations of B and the irreducible unitary representations of H which do not vanish on B. Show that the composition of these two natural correspondences is exactly the correspondence constructed in Step II of the Mackey analysis for G, N (see XII.1.28).

6. Show in the proof of Proposition 2.10 that ${}^x R$ is equivalent to the subrepresentation of ${}^x S$ acting on $\mathrm{range}((Q^S(W))^{\tilde{}})$.

7. Prove that equation (7) in the proof of Proposition 2.11 defines a unitary equivalence F of ${}^x S$ and S.

8. Show that for irreducible representations of finite groups the notion of association defined in Definition 2.13 coincides with that of 1.5.

9. Prove that any one-element subset of an almost Hausdorff space is a Borel set (see the proof of 2.17).

10. Prove Proposition 2.21.

11. Show in the proof of Proposition 4.8 that equation (9) implies the existence of a complex constant u such that equation (10) is true.

12. Verify equations (33) and (36) in the proof of Lemma 4.20.

13. Show that the Mackey obstruction as defined in 4.29 generalizes the cocycle definition 1.17 when the latter is reformulated in terms of group extensions.

14. Fill in the details of the proof of Theorem 5.6.

15. Check in the proof of 6.19 that the actions of G on θ by conjugation in $\mathscr{F}^\theta$ and in $\mathscr{B}$ are the same. Verify also that, if $D_\theta \in \theta$, the Mackey obstructions of D_θ in $\mathscr{B}$ and in $\mathscr{F}$ are the same.

16. Let G be a locally compact group, and N a closed normal Abelian subgroup of G such that G/N is compact. Let T be an irreducible unitary representation of G associated with an orbit θ in $\hat{N}$ (under G). Show that, if $\chi \in \theta$, then T is a subrepresentation of $\mathrm{Ind}_{N \uparrow G}(\chi)$.

17. Let $\mathscr{B}$ be the saturated C^*-algebraic bundle over G (with unit fiber A) constructed in Exercise 55 of Chapter XI.

    (a) Show that every D in $\hat{A}$ has trivial Mackey obstruction in $\mathscr{B}$.

    (b) Assume that G is σ-compact and satisfies 5.8(13); and suppose that for every $D \in \hat{A}$ the stability subgroup for D has a Type I unitary representation theory. Show that $\mathscr{B}$ has a Type I *-representation theory.

18. Show in the proof of Lemma 7.8 that the implication in (1) is true.

19. Verify that $W_{i(z)} = \bar{z}\mathbf{1}$ for z in $\mathbb{E}$ in the proof of Lemma 7.14.

20. Verify the statements concerning weak Frobenius Reciprocity made in 7.20.

21. Let G be the "$ax + b$" group of 8.3; and let T^+ and T^- be the two infinite-dimensional irreducible unitary representations of G (as in 8.3). Show that the

regular representation of G is unitarily equivalent to the Hilbert direct sum of countably infinitely many copies of $T^+ \oplus T^-$.

22. Give the details of the proof sketched of Proposition 8.4.

23. A finite-dimensional C^*-algebra will be called *monomial* if it is *-isomorphic with the cross-sectional C^*-algebra of some cocycle bundle.

Prove that if A is a monomial finite-dimensional C^*-algebra such that some D in $\hat{A}$ is one-dimensional, then A is *-isomorphic with the group algebra of some finite group.

24. A C^*-algebra A will be called *primitive* if there does not exist a saturated C^*-bundle structure for A over a locally compact group G having more than one element. (See VIII.17.4.)

For any positive integer n, M_n denotes as usual the $n \times n$ total matrix C^*-algebra (so that $M_1 \cong \mathbb{C}$).

   (I)   Show that $\mathbb{C} \oplus M_n$ is primitive for every $n = 2, 3, \ldots$.

   (II)   Let p be a prime number and G the cyclic group of order p. Let A be a finite-dimensional C^*-algebra. Show that the following two conditions are equivalent:

   (i)   There exists a saturated C^*-bundle structure for A over G.

   (ii) For each $m = 1, 2, \ldots, p - 1$, the number of elements of $\hat{A}$ which are of dimension m is divisible by p.

   (III) Show that a finite-dimensional C^*-algebra of dimension less than 60, such that $\hat{A}$ has one and only one element of dimension 1, is primitive.

25. Prove Proposition 8.7.

26. Show in Example 8.10 that the conjugating action of the element $\{r, s, t\}$ of H sends $\phi_{p,q}$ into $\phi_{p-qr,q}$. Furthermore, verify equation (11) of 8.10.

27. Establish Proposition 8.11.

28. Show that the conjugating action of an element $\langle u, a \rangle$ of the Poincaré group P on $\hat{N}$ (see 8.14) sends ϕ_λ into $\phi_{a\lambda}$.

29. Verify the equations in (18) of Example 8.16. Show, moreover, that the mapping $\Phi: K_\theta \cap H \to E_r^2$ defined in 8.16 is an injective homomorphism.

30. Verify the statements made in Remark 8.17.

31. Verify the details left to the reader in 8.23.

32. Prove the Proposition 9.9.

33. Prove Proposition 9.15.

34. Give the details of the proof for Proposition 9.16.

Notes and Remarks

For a brief survey of the history of the Mackey normal subgroup analysis and its generalizations, we refer the reader to the Introductions to this volume and to the present chapter.

While the terminology "normal subgroup analysis" (or "little group analysis" as it is sometimes called) is appropriate in the group context, it seems misleading in the context of Banach *-algebraic bundles, where (after an initial reduction; see XII.2.3) no normal subgroups of the base group need appear. We have therefore called the topic of this chapter "the generalized Mackey analysis".

The material of §1 of this chapter is entirely classical. That of §§2 and 3, as far as techniques of proof are concerned, is largely taken from the important article of Blattner [6].

Notice that Step II of the Mackey analysis has been "algebraized" by Rieffel [9], at least in the group context: Given a "well-behaved" orbit in $\hat{N}$ (N being a normal subgroup of G), he constructs appropriate C^*-algebras D and E and a D, E imprimitivity bimodule $\mathcal{L}$, such that the correspondence between representations asserted in Step II of the Mackey analysis is precisely that obtained from the abstract Imprimitivity Theorem (§XI.6) applied to $D, E, \mathcal{L}$.

§4 is the present authors' adaptation, to the non-separable bundle context, of the theory of the Mackey obstruction (see Mackey [8]). Notice that our exposition automatically includes the situation where the object of study is the projective representation theory of the "big" group, since projective representations of a group G are just *-representations of cocycle bundles over G.

One topic not touched upon at all in this work is the theory of virtual subgroups, designed by Mackey (in the separable group context) to deal with the situation when 5.8(13) fails (i.e., the action of the "big" group on the structure space of the normal subgroup is not "smooth"). The ideas of the theory of virtual subgroups are sketched in Mackey [13], and developed in detail by A. Ramsay [1, 2, 4, 5].

Theorem 5.6 (in the separable group context) is due to Dixmier [23]. See also Rieffel [9] and P. Green [2].

Theorem 6.19 is due to Fell. It was presented at the 19th Congress of the Scandinavian Mathematical Society at Reykjavik in 1984, and appears here in print for the first time.

§7 answers only a very few of the immense range of questions of the general form: Given a saturated C^*-algebraic bundle $\mathcal{B}$ with unit fiber algebra A over a locally compact group G, to what extent are the properties of A inherited by $C^*(\mathcal{B})$? Much further information on such questions, in the context of semidirect product bundles, will be found in Gootman [7].

A few remarks on Frobenius reciprocity are in order here. Generalizations of the classical Frobenius Reciprocity Theorem to unitary representations of non-compact groups have been long and diligently sought for, but with only limited success. Mackey [6] proved an interesting measure-theoretic version of Frobenius reciprocity (in the context of separable locally compact groups). But most of the work done in this area has centered around Weak Frobenius Reciprocity (WFR) as defined in 7.18. Henrichs [3] has obtained results (in the group context) relating WFR with conditions on the relative compactness of conjugacy classes in the group. Moscovici [2] showed that WFR holds for pairs G, H where G is a simply connected nilpotent Lie group and H is a closed connected normal subgroup of G. Nielsen [1, 4] showed that Moscovici's result becomes false if H is allowed to be non-normal.

The bundle version 7.15 of the Frobenius Reciprocity Theorem, assuming as it does the compactness of the base group and strong "discreteness" in the unit fiber, is not a radical departure from the classical Frobenius Theorem. However it does not seem to have appeared in print before.

§§8 and 9 contain only a tiny portion of all the applications found for the Mackey analysis. Many more of these will be found in the survey articles of Mackey [10, 12, 15, 18, 21]. We would like to observe especially its applications to nilpotent and solvable Lie groups.

It was pointed out in §1 that finite solvable groups are particularly well adapted to iteration of the Mackey analysis. One might expect that the same would be true for solvable Lie groups. For nilpotent Lie groups, indeed, Kirillov [4] made use of the Mackey normal subgroup analysis to give a very elegant description of their structure spaces (see also Pukanszky [1]). Such groups always have Type I unitary representation theories. But the situation is less satisfactory for solvable Lie groups. Such groups need not have a Type I unitary representation theory. The most satisfactory investigation of the structure spaces of such groups up till now is that of L. Auslander and B. Kostant [1]. They construct irreducible representations of these groups by means of "holomorphic induction" (see the Introduction to this volume).

It goes without saying that for *simple* (indeed for *semisimple*) Lie groups the Mackey normal subgroup analysis will be of little use in its classical form, because of the absence of non-trivial normal subgroups. Fortunately it turns out that such groups (at least if they have a finite center) must have a large compact subgroup (in the sense of XI.7.8); and this fact enabled Harish-Chandra and others to apply the "*compact* subgroup analysis" (mentioned at the beginning of the introduction to this chapter) with great power to such groups. Among recent achievements in the direction of the "compact subgroup analysis" is the classification by Vogan [3] of the structure space of $SL(n, \mathbb{C})$ for an arbitrary positive integer n. (The case $n = 2$, i.e., the structure space of $SL(2, \mathbb{C})$, has been well understood for more than 30 years; see for example Naimark [6].)

Though the normal subgroup analysis in its classical form is helpless for semisimple Lie groups, the generalization of the Mackey analysis presented in this work opens up new possibilities. Indeed, if G is a semisimple Lie group, one may ask the question: Do there exist useful saturated C^*-bundle structures for $C^*(G)$, constructed by methods other than forming the group extension bundle from a normal subgroup, to which the generalized Mackey analysis could then be applied? At present we do not know.

Bibliography

AARNES, J. F.

[1] On the continuity of automorphic representations of groups, *Commun. Math. Phys.* **7**(1968), 332–336. MR 36 #6542.

[2] Physical states on *C**-algebras, *Acta Math.* **122**(1969), 161–172. MR 40 #747.

[3] Full sets of states on a *C**-algebra, *Math. Scand.* **26**(1970), 141–148. MR 41 #5978.

[4] Quasi-states on *C**-algebras, *Trans. Amer. Math. Soc.* **149**(1970), 601–625. MR 43 #8311.

[5] Continuity of group representations with applications to *C**-algebras, *J. Functional Anal.* **5**(1970), 14–36. MR 41 #393.

[6] Differentiable representations. I. Induced representations and Frobenius reciprocity, *Trans. Amer. Math. Soc.* **220**(1976), 1–35. MR 54 #5392.

[7] Distributions of positive type and representations of Lie groups, *Math. Ann.* **240**(1979), 141–156. MR 80e: 22017.

AARNES, J. F., EFFROS, E. G., and NIELSEN, O. A.

[1] Locally compact spaces and two classes of *C**-algebras, *Pacific J. Math.* **34**(1970), 1–16. MR 42 #6626.

AARNES, J. F., and KADISON, R. V.

[1] Pure states and approximate identities, *Proc. Amer. Math. Soc* **21**(1969), 749–752. MR 39 #1980.

ABELLANAS, L.

[1] Slices and induced representations, *Bull. Acad. Polon. Sci. Sér. Sci. Math. Astronom. Phys.* **19**(1971), 287–290. MR 44 #4142.

ALI, S. T.

[1] Commutative systems of covariance and a generalization of Mackey's imprimitivity theorem, *Canad. Math. Bull.* **27**(1984), 390–397. MR 86i: 22007.

AKEMANN, C. A.

[1] The general Stone-Weierstrass problem, *J. Functional Anal.* **4**(1969), 277–294. MR 40 #4772.

[2] Subalgebras of C*-algebras and von Neumann algebras, *Glasgow Math. J.* **25**(1984), 19–25. MR 85c: 46059.

AKEMANN, C. A., and ANDERSON, J.

[1] The Stone-Weierstrass problem for C*-algebras, *Invariant subspaces and other topics*, pp. 15–32, *Operator Theory: Adv. Appl.*, **6**, Birkhäuser, Basel-Boston, Mass., 1982. MR 85a: 46035.

AKEMANN, C. A., and WALTER, M. E.

[1] Nonabelian Pontryagin duality, *Duke Math. J.* **39**(1972), 451–463. MR 48 #3595.

ALBERT, A. A.

[1] Structure of algebras, *Amer. Math. Soc. Colloq. Publ.*, **24**, Providence, R. I. 1939. MR 1, 99.

ALESINA, A.

[1] Equivalence of norms for coefficients of unitary group representations, *Proc. Amer. Math. Soc.* **74**(1979), 343–349. MR 80e: 22008.

ALI, S. T.

[1] Commutative systems of covariance and a generalization of Mackey's imprimitivity theorem, *Canad. Math. Bull.* **27**(1984), 390–397. MR 86i: 22007.

AMBROSE, W.

[1] Spectral resolution of groups of unitary operators, *Duke Math. J.* **11**(1944), 589–595. MR 6, 131.

[2] Structure theorems for a special class of Banach algebras, *Trans. Amer. Math. Soc.* **57**(1945), 364–386. MR 7, 126.

ANDERSON, J., and BUNCE, J.

[1] Stone-Weierstrass theorems for separable C*-algebras, *J. Operator Theory* **6**(1981), 363–374. MR 83b: 46077.

ANDLER, M.

[1] Plancherel pour les groupes algébriques complexes unimodulaires, *Acta Math.* **154**(1985), 1–104.

ANGELOPOULOS, E.

[1] On unitary irreducible representations of semidirect products with nilpotent normal subgroup, *Ann. Inst. H. Poincaré Sect. A(N.S.)* **18**(1973), 39–55. MR 49 #462.

ANKER, J. P.

[1] Applications de la p-induction en analyse harmonique, *Comment. Math. Helv.* **58**(1983), 622–645. MR 85c: 22006.

ANUSIAK, Z.

[1] On generalized Beurling's theorem and symmetry of L_1-group algebras, *Colloq. Math.* **23**(1971), 287–297. MR 49 #11147.

APOSTOL, C., FIALKOW, L., HERRERO, D., and VOICULESCU, D.

[1] *Approximation of Hilbert space operators. Volume II, Research Notes in Mathematics*, **102**, Pitman Books, London–New York, 1984. MR 85m: 47002.

APOSTOL, C., FOIAS, C., and VOICULESCU, D.

[1] Some results on non-quasitriangular operators. II, *Rev. Roumaine Math. Pures Appl.* **18**(1973); III, ibidem 309; IV, ibidem 487; VI, ibidem 1473. MR 48 #12109.

[2] On the norm-closure of nilpotents. II, *Rev. Roumaine Math. Pures Appl.* **19**(1974), 549–557. MR 54 #5876.

APOSTOL, C., HERRERO, D. A., and VOICULESCU, D.

[1] The closure of the similarity orbit of a Hilbert space operator, *Bull. New Series Amer. Math. Soc.* **6**(1982), 421–426. MR 83c: 47028.

APOSTOL, C., and VOICULESCU, D.

[1] On a problem of Halmos, *Rev. Roumaine Math. Pures Appl.* **19**(1974), 283–284. MR 49 #3574.

ARAKI, H., and ELLIOTT, G. A.

[1] On the definition of C^*-algebras, *Publ. Res. Inst. Math. Sci. Kyoto Univ.* **9**(1973), 93–112. MR 50 #8085.

ARENS, R.

[1] On a theorem of Gelfand and Naimark, *Proc. Nat. Acad. Sci. U.S.A.* **32**(1946), 237–239. MR 8, 279.

[2] Representation of *-algebras, *Duke Math. J.* **14**(1947), 269–283. MR 9, 44.

[3] A generalization of normed rings, *Pacific J. Math.* **2**(1952), 455–471. MR 14, 482.

ARMACOST, W. L.

[1] The Frobenius reciprocity theorem and essentially bounded induced representations, *Pacific J. Math.* **36**(1971), 31–42. MR 44 #6902.

ARVESON, W. B.

[1] A theorem on the action of abelian unitary groups, *Pacific J. Math.* **16**(1966), 205–212. MR 32 #6241.

[2] Operator algebras and measure-preserving automorphisms, *Acta Math.* **118**(1967), 95–109. MR 35 #1751.

[3] Subalgebras of C^*-algebras, *Acta Math.* **123**(1969), 141–224. MR 40 #6274.

[4] Subalgebras of C^*-algebras, II, *Acta Math.* **128**(1972), 271–308. MR 52 #15035.

[5] On groups of automorphisms of operator algebras, *J. Functional Anal.* **15**(1974), 217–243. MR 50 #1016.

[6] *An invitation to C*-algebras*, Graduate texts in math., **39**, Springer-Verlag, Berlin-Heidelberg-New York, 1976. MR 58 #23621.

ATIYAH, M. F.

[1] *K-Theory*, Benjamin, New York-Amsterdam, 1967. MR 36 7130.

AUPETIT, B.

[1] *Propriétés spectrales des algèbres de Banach*, Lecture Notes in Math., **735**, Springer-Verlag, Berlin-Heidelberg-New York,1979. MR 81i: 46055.

AUSLANDER, L.

[1] *Unitary representations of locally compact groups—The elementary and type I theory*, Lecture Notes, Yale University, New Haven, Conn., 1962.

AUSLANDER, L., and KOSTANT, B.

[1] Polarization and unitary representations of solvable Lie groups, *Invent. Math.* **14**(1971), 255–354. MR 45 #2092.

AUSLANDER, L., and MOORE, C. C.

[1] *Unitary representations of solvable Lie groups*, Memoirs Amer. Math. Soc., **62**, Amer. Math. Soc., Providence, R. I., 1966. MR 34 #7723.

BACKHOUSE, N. B.

[1] Projective representations of space groups, III. Symmorphic space groups, *Quart. J. Math. Oxford Ser.* (2) **22**(1971), 277–290. MR 45 #2028.

[2] On the form of infinite dimensional projective representations of an infinite abelian group, *Proc. Amer. Math. Soc.* **41**(1973), 294–298. MR 47 #6940.

[3] Tensor operators and twisted group algebras, *J. Math. Phys.* **16**(1975), 443–447. MR 52 #546.

BACKHOUSE, N. B., and BRADLEY, C. J.

[1] Projective representations of space groups, I. Translation groups, *Quart. J. Math. Oxford Ser.* (2) **21**(1970), 203–222. MR 41 #5510.

[2] Projective representations of space groups, II. Factor systems, *Quart. J. Math. Oxford. Ser.* (2) **21**(1970), 277–295. MR 43 #7517.

[3] Projective representations of abelian groups, *Proc. Amer. Math. Soc.* **36**(1972), 260–266. MR 46 #7443.

[4] Projective representations of space groups, IV. Asymmorphic space groups, *Quart. J. Math. Oxford Ser.* (2) **23**(1972), 225–238. MR 51 #5734.

BACKHOUSE, N. B., and GARD, P.

[1] On induced representations for finite groups, *J. Math. Phys.* **17**(1976), 1780–1784. MR 54 #2775.

[2] On the tensor representation for compact groups, *J. Math. Phys.* **17**(1976), 2098–2100. MR 54 #4374.

BAGCHI, S. C., MATHEW, J., and NADKARNI, M. G.

[1] On systems of imprimitivity on locally compact abelian groups with dense actions, *Acta Math.* **133**(1974), 287–304. MR 54 #7690.

BAGGETT, L.

[1] Hilbert-Schmidt representations of groups. *Proc. Amer. Math. Soc.* **21**(1969), 502–506. MR 38 #5991.

[2] A note on groups with finite dual spaces, *Pacific J. Math.* **31**(1969), 569–572. MR 41 #3658.

[3] A weak containment theorem for groups with a quotient R-group, *Trans. Amer. Math. Soc.* **128**(1967), 277–290. MR 36 #3921.

[4] A description of the topology on the dual spaces of certain locally compact groups, *Trans. Amer. Math. Soc.* **132**(1968), 175–215. MR 53 #13472.

[5] A separable group having a discrete dual is compact, *J. Functional Anal.* **10**(1972), 131–148. MR 49 #10816.

[6] Multiplier extensions other than the Mackey extension, *Proc. Amer. Math. Soc.* **56**(1976), 351–356. MR 53 #13468.

[7] Operators arising from representations of nilpotent Lie groups, *J. Functional Anal.* **24**(1977), 379–396. MR 56 #536.

[8] Representations of the Mautner group, I, *Pacific J. Math.* **77**(1978), 7–22. MR 80e: 22014.

[9] A characterization of "Heisenberg groups"; When is a particle free? *Rocky Mountain J. Math.* **8**(1978), 561–582. MR 80a: 22014.

[10] On the continuity of Mackey's extension process, *J. Functional Anal.* **56**(1984), 233–250. MR 85e: 22011.

[11] Unimodularity and atomic Plancherel measure, *Math. Ann.* **266**(1984), 513–518. MR 86a: 22004.

BAGGETT, L., and MERRILL, K.

[1] Multiplier representations of Abelian groups, *J. Functional Anal.* **14**(1973), 299–324. MR 51 #791.

BAGGETT, L., MITCHELL, W., and RAMSAY, A.

[1] Representations of the Mautner group and cocycles of an irrational rotation, *Michigan Math. J.* **33**(1986), 221–229. MR 87h: 22011.

BAGGETT, L., MITCHELL, W., and RAMSAY, A.

[1] Representations of the discrete Heisenberg group and cocycles of irrational rotations, *Michigan Math. J.* **31**(1984), 263–273. MR 86k: 22017.

BAGGETT, L., and RAMSAY, A.

[1] Some pathologies in the Mackey analysis for a certain nonseparable group, *J. Functional Anal.* **39**(1980), 375–380. MR 83b: 22007.

[2] A functional analytic proof of a selection lemma, *Canad. J. Math.* **32**(1980), 441–448. MR 83j: 54016.

BAGGETT, L., and SUND, T.

[1] The Hausdorff dual problem for connected groups, *J. Functional Anal.* **43**(1981), 60–68. MR 83a: 22006.

BAGGETT, L., and TAYLOR, K.

[1] Groups with completely reducible representation, *Proc. Amer. Math. Soc.* **72**(1978), 593–600. MR 80b: 22009.

[2] A sufficient condition for the complete reducibility of the regular representation, *J. Functional Anal.* **34**(1979), 250–265. MR 81f: 22005.

[3] On asymptotic behavior of induced representations, *Canad. J. Math.* **34**(1982), 220–232. MR 84j: 22017.

BAKER, C. W.

[1] A closed graph theorem for Banach bundles, *Rocky Mountain J. Math.* **12**(1982), 537–543. MR 84h: 46010.

BALDONI-SILVA, M. W., and KNAPP, A. W.

[1] Unitary representations induced from maximal parabolic subgroups, *J. Functional Anal.* **69**(1986), 21–120.

BARGMANN, V.

[1] Irreducible unitary representations of the Lorentz group. *Ann. Math.* (2) **48**(1947), 568–640. MR 9, 133.

[2] On unitary ray representations of continuous groups, *Ann. Math.* (2) **59**(1954), 1–46. MR 15, 397.

BARGMANN, V., and WIGNER, E. P.

[1] Group theoretical discussion of relativistic wave equations, *Proc. Nat. Acad. Sci. U.S.A.* **34**(1948), 211–223. MR 9, 553.

BARIS, K.

[1] On induced representations of *p*-adic reductive groups, *Karadeniz Univ. Math. J.* **5**(1982), 168–177. MR 85f: 22016.

BARNES, B. A.

[1] When is a *-representation of a Banach *-algebra Naimark-related to a *-representation? *Pacific J. Math.* **72**(1977), 5–25. MR 56 #16385.

[2] Representations Naimark-related to a *-representation; a correction: "When is a *-representation of a Banach *-algebra Naimark-related to a *-representation?" [*Pacific J. Math.* **72**(1977), 5–25], *Pacific J. Math.* **86**(1980), 397–402. MR 82a: 46060.

[3] The role of minimal idempotents in the representation theory of locally compact groups, *Proc. Edinburgh Math. Soc.* (2) **23**(1980), 229–238. MR 82i: 22007.

[4] A note on separating families of representations, *Proc. Amer. Math. Soc.* **87**(1983), 95–98. MR 84k: 22006.

BARUT, A. O., and RACZKA, R.

[1] *Theory of group representations and applications*, Polish Scientific Publishers, Warsaw, 1977. MR 58 #14480.

BASS, H.

[1] *Algebraic K-theory*, Benjamin, New York, 1968. MR 40 #2736.

BECKER, T.

[1] A few remarks on the Dauns-Hofmann theorems for C^*-algebras, *Arch. Math. (Basel)* **43**(1984), 265–269.

BEHNCKE, H.

[1] Automorphisms of crossed products, *Tôhoku Math. J.* **21**(1969), 580–600. MR 42 #5056.

[2] C^*-algebras with a countable dual, *Operator algebras and applications*, Part 2, Kingston, Ont., 1980, pp. 593–595, Proc. Sympos. Pure. Math., **38**, Amer. Math. Soc., Providence, R. I., 1982. MR 83j: 46004b.

BEHNCKE, H., and LEPTIN, H.

[1] Classification of C^*-algebras with a finite dual, *J. Functional Anal.* **16**(1974), 241–257. MR 49 #9638.

BEKES, ROBERT A.

[1] Algebraically irreducible representations of $L_1(G)$, *Pacific J. Math.* **60**(1975), 11–25. MR 53 #10978.

BELFI, V. A., and DORAN, R. S.

[1] Norm and spectral characterizations in Banach algebras, *L'Enseignement Math.* **26**(1980), 103–130. MR 81j: 46071.

BENNETT, J. G.

[1] Induced representations and positive linear maps of C^*-algebras, Thesis, Washington University, 1976.

[2] Induced representations of C^*-algebras and complete positivity. *Trans. Amer. Math. Soc.* **243**(1978), 1–36. MR 81h: 46068.

BERBERIAN, S. K.

[1] *Baer *-rings*, Springer-Verlag, New York-Berlin, 1972. MR 55 #2983.

[2] *Lectures on functional analysis and operator theory*, Graduate Texts in Math. **15**, Springer-Verlag, Berlin-Heidelberg-New York, 1974. MR 54 #5775.

BEREZIN, F. A.

[1] Laplace operators on semi-simple Lie groups, *Trudy Moskov. Math. Obšč.* **6**(1957), 371–463 (Russian). MR 19, 867.

BEREZIN, F. A., and GELFAND, I. M.

[1] Some remarks on the theory of spherical functions on symmetric Riemannian manifolds, *Trudy Moskov. Mat. Obšč.* **5**(1956), 311–351 (Russian). MR 19, 152.

BERKSON, E.

[1] Some characterizations of C^*-algebras, *Illinois J. Math.* **10**(1966), 1–8. MR 32 #2922.

BERNAT, P., and DIXMIER, J.

[1] Sur le dual d'un groupe de Lie, *C. R. Acad. Sci. Paris* **250**(1960), 1778–1779. MR 27 #1536.

BERNSTEIN, I. N., and ZELEVINSKY, A. V.

[1] Induced representations of reductive p-adic groups. I, *Ann. Sci. École Norm. Sup.* (4) **10**(1977), 441–472. MR 58 #28310.

BERTRAND, J., and RIDEAU, G.

[1] Non-unitary representations and Fourier transform on the Poincaré group, *Rep. Math. Phys.* **4**(1973), 47–63. MR 53 #710.

BEURLING, A.

[1] *Sur les intégrales de Fourier absolument convergentes, et leur application à une transformation fonctionnelle*, Congrès des Math. Scand., Helsingfors (1938).

BICHTELER, K.

[1] On the existence of noncontinuous representations of a locally compact group, *Invent. Math.* **6**(1968), 159–162. MR 38 #4610.

[2] A generalization to the non-separable case of Takesaki's duality theorem for *C**-algebras, *Invent. Math.* **9**(1969), 89–98. MR 40 #6275.

[3] Locally compact topologies on a group and its corresponding continuous irreducible representations, *Pacific J. Math.* **31**(1969), 583–593. MR 41 #394.

BLACKADAR, B. E.

[1] Infinite tensor products of *C**-algebras, *Pacific J. Math.* **72**(1977), 313–334. MR 58 #23622.

[2] A simple unital projectionless *C**-algebra, *J. Operator Theory* **5**(1981), 63–71. MR 82h: 46076.

[3] *K-Theory for operator algebras*, Math. Sci. Research Institute Publications, Springer-Verlag, New York-Berlin-Heidelberg, 1986.

BLATTNER, R. J.

[1] On induced representations, *Amer. J. Math.* **83**(1961), 79–98. MR 23 #A2757.

[2] On induced representations. II. Infinitesimal induction, *Amer. J. Math.* **83**(1961), 499–512. MR 26 #2885.

[3] On a theorem of G. W. Mackey, *Bull. Amer. Math. Soc.* **68**(1962), 585–587. MR 25 #5135.

[4] A theorem on induced representations, *Proc. Amer. Math. Soc.* **13**(1962), 881–884. MR 29 #3894.

[5] Positive definite measures, *Proc. Amer. Math. Soc.* **14**(1963), 423–428. MR 26 #5095.

[6] Group extension representations and the structure space, *Pacific J. Math.* **15**(1965), 1101–1113. MR 32 #5785.

BLATTNER, R. J., COHEN, M., and MONTGOMERY, S.

[1] Crossed products and inner actions of Hopf algebras, *Trans. Amer. Math. Soc.* **298**(1986), 671–711.

BOCHNER, S.

[1] *Vorlesungen über Fouriersche Integrale*, Akad. Verlag, Leipzig, 1932.

[2] Integration von Funktionen, deren Werte die Elemente eines Vectorraumes sind, *Fund. Math.* **20**(1933), 262–276.

BOE, B. D.

[1] Determination of the intertwining operators for holomorphically induced representations of SU(p,q). *Math. Ann.* **275**(1986), 401–404.

BOE, B. D., and COLLINGWOOD, D. H.

[1] A comparison theory for the structure of induced representations, *J. Algebra* **94**(1985), 511–545. MR 87b: 22026a.

[2] Intertwining operators between holomorphically induced modules, *Pacific J. Math.* **124**(1986), 73–84.

BOERNER, H.

[1] *Representations of groups*, North-Holland Publishing Co., Amsterdam London, 1970. MR 42 #7792.

BOHNENBLUST, H. F., and KARLIN, S.

[1] Geometrical properties of the unit sphere of Banach algebras, *Ann. Math.* **62**(1955), 217–229. MR 17, 177.

BOIDOL, J.

[1] Connected groups with polynomially induced dual, *J. Reine Angew. Math.* **331**(1982), 32–46. MR 83i: 22010.

[2] *-regularity of some classes of solvable groups, *Math. Ann.* **261**(1982), 477–481. MR 84g: 22016.

[3] Group algebras with unique C^*-norm, *J. Functional Anal.* **56**(1984), 220–232. MR 86c: 22006.

[4] A Galois-correspondence for general locally compact groups, *Pacific J. Math.* **120**(1985), 289–293. MR 87c: 22010.

[5] Duality between closed normal subgroups of a locally compact group G and hk-closed subduals of $\hat{G}$ (preprint).

BOIDOL, J., LEPTIN, H., SCHÜRMAN, J., and VAHLE, D.

[1] Räume primitiver Ideale von Gruppenalgebren, *Math. Ann.* **236**(1978), 1–13. MR 58 #16959.

BONIC, R. A.

[1] Symmetry in group algebras of discrete groups, *Pacific J. Math.* **11**(1961), 73–94. MR 22 #11281.

BONSALL, F. F., and DUNCAN, J.

[1] *Complete Normed Algebras*, Ergebnisse der Mathematik und ihrer Grenz gebiete. **80**. Springer-Verlag, Berlin-Heidelberg-New York, 1973. MR 54 #11013.

BORCHERS, H. J.

[1] On the implementability of automorphism groups, *Comm. Math. Phys.* **14**(1969), 305–314. MR 41 #4267.

BOREL, A.

[1] *Représentations de groupes localement compacts*, Lecture Notes in Math. **276**, Springer-Verlag, Berlin-Heidelberg-New York, 1972. MR 54 #2871.

[2] On the development of Lie group theory, *Niew Arch. Wiskunde* **27**(1979), 13–25. MR 81g: 01013.

BOREL, A., and WALLACH, N. R.

[1] *Continuous cohomology, discrete subgroups, and representations of reductive groups*, Annals of Mathematics Studies, **94**, Princeton University Press, Princeton, N. J.; University of Tokyo Press, Tokyo, 1980. MR 83c: 22018.

BOURBAKI, N.

The first twelve references are in the series *Éléments de Mathématique, Actualités Sci. et. Ind.*, Hermann et cie, Paris; each title is identified by a serial number.

[1] *Théorie des Ensembles*, Chapitres I, II, No. 1212(1966). MR 34 #7356.

[2] *Théorie des Ensembles*, Chapitre III, No. 1243(1963). MR 27 #4758.

[3] *Topologie générale*, Chapitres I, II, No. 1142(1965). MR 39 #6237.

[4] *Topologie générale*, Chapitres III, IV, No. 1143(1960). MR 25 #4021.

[5] *Algèbre*, Chapitre II, No. 1032(1967). MR 9, 406.

[6] *Algèbre*, Chapitres IV, V, No. 1102(1967). MR 30 #4576.

[7] *Algèbre*, Chapitre VIII, No. 1261(1958). MR 20, #4576.

[8] *Espaces Vectoriels Topologiques*, Chapitres I, II, No. 1189(1966). MR 34 #3277.

[9] *Espaces Vectoriels Topologiques*, Chapitres III, IV, V, No. 1229(1964). MR 17 1062.

[10] *Intégration*, Chapitres I–IV, V, VI, Nos. 1175(1965), 1244(1967), 1281(1959). MR 39 #6237.

[11] *Intégration*, Chapitres VII–VIII, No. 1306(1963). MR 31 #3539.

[12] *Théories Spectrales*, Chapitres I, II, No. 1332(1967). MR 35 #4725.

[13] *General topology*, Parts I, II, Hermann, Paris, Addison-Wesley Pub. Co., Reading, Mass., 1966. MR 34 #5044.

[14] *Theory of sets*, Hermann, Paris, Addison-Wesley Pub. Co., Reading, Mass., 1968.

[15] *Algebra*, Part I, Hermann, Paris, Addison-Wesley Pub. Co., Reading, Mass., 1974. MR 50 #6689.

[16] *Lie groups and Lie algebras*, Part I, Hermann, Paris, Addison-Wesley Pub. Co., Reading, Mass., 1975.

BOYER, R., and MARTIN, R.

[1] The regular group C^*-algebra for real-rank one groups *Proc. Amer. Math. Soc.* **59**(1976), 371–376. MR 57 #16464.

[2] The group C^*-algebra of the de Sitter group, *Proc. Amer. Math. Soc.* **65**(1977), 177–184. MR 57 #13381.

BRATTELI, O.

[1] Inductive limits of finite dimensional C^*-algebras, *Trans. Amer. Math. Soc.* **171**(1972), 195–234. MR 47 #844.

[2] Structure spaces of approximately finite-dimensional C^*-algebras, *J. Functional Anal.* **16**(1974), 192–204. MR 50 #1005.

[3] Crossed products of UHF algebras by product type actions, *Duke Math. J.* **46**(1979), 1–23. MR 82a: 46063.

[4] *Derivations, dissipations and group actions on C*-algebras*, Lecture Notes in Math. 1229, Springer-Verlag, Berlin-New York, 1986.

BRATTELI, O., and ELLIOTT, G. A.

[1] Stucture spaces of approximately finite-dimensional *C*-algebras, II., J. Functional Anal.* **30**(1978), 74–82. MR 80d: 46111.

BRATTELI, O., and ROBINSON, D. W.

[1] *Operator algebras and quantum statistical mechanics, I*, Texts and Monographs in Physics, Springer-Verlag, New York-Heidelberg, 1979. MR 81a: 46070.

[2] *Operator algebras and quantum statistical mechanics, II*, Texts and Monographs in Physics, Springer-Verlag, New York-Heidelberg, 1981. MR 82k: 82013.

BREDON, G. E.

[1] *Introduction to compact transformation groups*, Academic Press, New York-London, 1972. MR 54 #1265.

BREZIN, J., and MOORE, C. C.

[1] Flows on homogeneous spaces: a new look, *Amer. J. Math.* **103**(1981), 571–613. MR 83e: 22009.

BRITTON, O. L.

[1] Primitive ideals of twisted group algebras, *Trans. Amer. Math. Soc.* **202**(1975), 221–241. MR 51 #11011.

BRÖCKER, T., and TOM DIECK, T.

[1] *Representations of compact Lie groups*, Graduate Texts in Math. **98**, Springer-Verlag, New York-Berlin-Heidelberg-Tokyo, 1985. MR 86i: 22023.

BROWN, I. D.

[1] Representations of finitely generated nilpotent groups, *Pacific J. Math.* **45**(1973), 13–26. MR 50 #4811.

[2] Dual topology of a nilpotent Lie group, *Ann. Sci. École Norm. Sup.* (4) **6**(1973), 407–411. MR 50 #4813.

BROWN, I. D., and GUIVARC'H, Y.

[1] Espaces de Poisson des groupes de Lie, *Ann. Sci. École Norm. Sup.* (4) **7**(1974), 175–179(1975). MR 55 #570a.

BROWN, L. G.

[1] Extensions of topological groups, *Pacific J. Math.* **39**(1971), 71–78. MR 46 #6384.

[2] Locally compact abelian groups with trivial multiplier group (preprint 1968).

BROWN, L. G., GREEN, P., and RIEFFEL, M. A.

[1] Stable isomorphism and strong Morita equivalence of *C*-algebras, Pacific J. Math.* **71**(1977), 349–363. MR 57 #3866.

BRUHAT, F.

[1] Sur les représentations induites des groupes de Lie, *Bull. Soc. Math. France* **84**(1956), 97–205. MR 18, 907.

[2] Distributions sur un groupe localement compact et applications à l'étude des représentations des groupes p-adiques, *Bull. Soc. Math. France* **89**(1961), 43-75. MR 25 #4354.

[3] Sur les représentations de groupes classiques p-adiques. I, *Amer. J. Math.* **83**(1961), 321-338. MR 23 #A3184.

[4] Sur les représentations de groupes classiques p-adiques. II, *Amer. J. Math.* **83**(1961), 343-368. MR 23 #A3184.

[5] *Lectures on Lie groups and representations of locally compact groups.* Notes by S. Ramanan, Tata Institute of Fundamental Research Lectures on Math., **14**, Bombay, 1968. MR 45 #2072.

BUNCE, J. W.

[1] Representations of strongly amenable C*-algebras, *Proc. Amer. Math. Soc.* **32**(1972), 241-246. MR 45 #4159.

[2] Characterizations of amenable and strongly amenable C*-algebras, *Pacific J. Math.* **43**(1972), 563-572. MR 47 #9298.

[3] Approximating maps and a Stone-Weierstrass theorem for C*-algebras, *Proc. Amer. Math. Soc.* **79**(1980), 559-563. MR 81h: 46082.

[4] The general Stone-Weierstrass problem and extremal completely positive maps, *Manuscripta Math.* **56**(1986), 343-351.

BUNCE, J. W., and DEDDENS, J. A.

[1] C*-algebras with Hausdorff spectrum, *Trans. Amer. Math. Soc.* **212**(1975), 199-217. MR 53 #8911.

BURCKEL, R. B.

[1] *Weakly almost periodic functions on semigroups*, Gordon and Breach, New York, 1970. MR 41 #8562.

[2] *Characterizations of C(X) among its subalgebras*, Marcel-Dekker, New York, 1972. MR 56 #1068.

[3] Averaging a representation over a subgroup, *Proc. Amer. Math. Soc.* **78**(1980), 399-402. MR 80m: 22005.

BURNSIDE, W.

[1] On the condition of reducibility for any group of linear substitutions, *Proc. London Math. Soc.* **3**(1905), 430-434.

BURROW, M.

[1] *Representation theory of finite groups*, Academic Press, New York-London, 1965. MR 38 #250.

BUSBY, R. C.

[1] On structure spaces and extensions of C*-algebras, *J. Functional Anal.* **1**(1967), 370-377. MR 37 #771.

[2] Double centralizers and extensions of C*-algebras, *Trans. Amer. Math. Soc.* **132**(1968), 79-99. MR 37 #770.

[3] On a theorem of Fell, *Proc. Amer. Math. Soc.* **30**(1971), 133-140. MR 44 #814.

[4] Extensions in certain topological algebraic categories, *Trans. Amer. Math. Soc.* **159**(1971), 41-56. MR 43 #7937.

[5] On the equivalence of twisted group algebras and Banach *-algebraic bundles, *Proc. Amer. Math. Soc.* **37**(1973), 142–148. MR 47 #4018.

[6] Centralizers of twisted group algebras, *Pacific J. Math.* **47**(1973), 357–392. MR 48 #11920.

BUSBY, R. C., and SCHOCHETMAN, I.

[1] Compact induced representations, *Canad. J. Math.* **24**(1972), 5–16. MR 45 #2495.

BUSBY, R. C., SCHOCHETMAN, I., and SMITH, H. A.

[1] Integral operators and the compactness of induced representations, *Trans. Amer. Math. Soc.* **164**(1972), 461–477. MR 45 #4167.

BUSBY, R. C., and SMITH, H. A.

[1] Representations of twisted group algebras, *Trans. Amer. Math. Soc.* **149**(1970), 503–537. MR 41 #9013.

CALKIN, J. W.

[1] Two-sided ideals and congruences in the ring of bounded operators in Hilbert space, *Ann. Math.* **42**(1941), 839–873. MR 3, 208.

CAREY, A. L.

[1] Square integrable representations of non-unimodular groups, *Bull. Austral. Math. Soc.* **15**(1976), 1–12. MR 55 #3153.

[2] Induced representations, reproducing kernels and the conformal group, *Commun. Math. Phys.* **52**(1977), 77–101. MR 57 #16465.

[3] Some infinite-dimensional groups and bundles, *Publ. Res. Inst. Math. Sci.* **20**(1984), 1103–1117.

CAREY, A. L., and MORAN, W.

[1] Some groups with T_1 primitive ideal spaces, *J. Austral. Math. Soc. (Ser. A)* **38**(1985), 55–64. MR 86c: 22007.

[2] Cocycles and representations of groups of CAR type, *J. Austral. Math. Soc. (Ser. A)* **40**(1986), 20–33. MR 87d: 22010.

CARLEMAN, T.

[1] Zur Theorie der linearen Integralgleichungen, *Math. Zeit.* **9**(1921), 196–217.

CARTAN, E.

[1] Les groupes bilinéaires et les systèmes de nombres complexes, *Ann. Fac. Sc. Toulouse*, 1898, *Oeuvres complètes, pt. II*, t. 1, pp. 7–105, Gauthier-Villars, Paris, 1952.

CARTAN, H.

[1] Sur la mesure de Haar, *C. R. Acad. Sci. Paris* **211**(1940), 759–762. MR 3, 199.

[2] Sur les fondements de la théorie du potentiel, *Bull. Soc. Math. France* **69**(1941), 71–96. MR 7, 447.

CARTAN, H., and GODEMENT, R.

[1] Théorie de la dualité et analyse harmonique dans les groupes abéliens localement compacts, *Ann. Sci. École. Norm. Sup.* (3) **64**(1947), 79–99. MR 9, 326.

CASTRIGIANO, D. P. L., and HENRICHS, R. W.

[1] Systems of covariance and subrepresentations of induced representations, *Lett. Math. Phys.* **4**(1980), 169–175. MR 81j: 22010.

CATTANEO, U.

[1] Continuous unitary projective representations of Polish groups: The BMS-group, Group theoretical methods in physics, Lecture Notes in Physics, Vol. 50, pp. 450–460, Springer, Berlin, 1976. MR 58 #6051.

[2] On unitary/antiunitary projective representations of groups. *Rep. Mathematical Physics* **9**(1976), 31–53. MR 54 #10478.

[3] On locally continuous cocycles, *Rep. Mathematical Physics* **12**(1977), 125–132. MR 57 #523.

[4] Splitting and representation groups for Polish groups, *J. Mathematical Physics* **19**(1978), 452–460. MR 57 #16460.

[5] On Mackey's imprimitivity theorem, *Comment Math. Helv.* **54**(1979), 629–641. MR 81b: 22009.

CAYLEY, A.

[1] On the theory of groups, as depending on the symbolic equation $\theta^n = 1$, *Phil. Mag.* **7**(1854), 40–47. Also, Vol. II, Collected Mathematical Papers, Cambridge, 1889, pp. 123–130.

CHOJNACKI, W.

[1] Cocycles and almost periodicity, *J. London Math. Soc.* (2) **35**(1987), 98–108.

CHRISTENSEN, E.

[1] On nonselfadjoint representations of C^*-algebras, *Amer. J. Math.* **103**(1981). 817–833. MR 82k: 46085.

CIVIN, P. and YOOD, B.

[1] Involutions on Banach algebras, *Pacific J. Math.* **9**(1959), 415–436. MR 21 #4365.

CLIFFORD, A. H.

[1] Representations induced in an invariant subgroup, *Ann. of Math.* **38**(1937), 533–550.

COHEN, P. J.

[1] Factorization in group algebras, *Duke Math. J.* **26**(1959), 199–205. MR 21 #3729.

COIFMAN, R. R. and WEISS, G.

[1] Representations of compact groups and spherical harmonics, *L'Enseignement Math.* **14**(1969), 121–173. MR 41 #537.

COLEMAN, A. J.

[1] *Induced and subinduced representations, group theory and its applications*, E. M. Loebl, Ed., 57-118, Academic Press, New York, 1968.

[2] Induced representations with applications to S_n and GL(n), Queens Papers in Pure and Applied Mathematics 4 (Queens Univ., Kingston), 91 pp. 1966. MR 34 #2718.

COLOJOARĂ, I.

[1] Locally convex bundles. (Romanian). *An. Univ. Craiova Mat. Fiz.-Chim.* No. 4(1976), 11–21. MR 58 #23648.

COMBES, F.

[1] Crossed products and Morita equivalence, *Proc. London Math. Soc.* (3), **49**(1984), 289–306. MR 86c: 46081.

COMBES, F. and ZETTL, H.

[1] Order structures, traces and weights on Morita equivalent C^*-algebras, *Math. Ann.*, **265**(1983), 67–81. MR 85f: 46106.

CONNES, A.

[1] Classification of injective factors. Cases II_1, II_∞, III_λ, $\lambda \neq 1$, *Ann. Math.* (2)**104**(1976), 73–115. MR 56 #12908.

[2] On the cohomology of operator algebras, *J. Functional Anal.* **28**(1978), 248–253. MR 58 #12407.

[3] On the spatial theory of von Neumann algebras, *J. Functional Anal.* **35**(1980), 153–164. MR 81g: 46083.

CONWAY, J. B.

[1] *A course in functional analysis*, Graduate Texts in Math. no. 96, Springer-Verlag, Berlin-Heidelberg-New York, 1985. MR 86h: 46001.

CORWIN, L.

[1] Induced representations of discrete groups, *Proc. Amer. Math. Soc.* **47**(1975), 279–287. MR 52 #8329.

[2] Decomposition of representations induced from uniform subgroups and the "Mackey machine," *J. Functional Anal.* **22**(1976), 39–57. MR 54 #468.

CORWIN, L., and GREENLEAF, F. P.

[1] Intertwining operators for representations induced from uniform subgroups. *Acta Math.* **136**(1976), 275–301. MR 54 #12967.

COURANT, R. and HILBERT, D.

[1] *Methods of mathematical physics.* Vol. I, Interscience Publishers, Inc., New York, 1953. MR 16, 426.

CUNTZ, J.

[1] K-theoretic amenability for discrete groups, *J. Reine Angew. Math.* **344**(1983), 180–195. MR 86e: 46064.

CURTIS, C. W., and REINER, I.

[1] *Representation theory of finite groups and associative algebras*, Interscience Publishers, 1962. MR 26 #2519.

[2] *Methods of representation theory (with applications to finite groups and orders).* Vol. I, Wiley-Interscience (pure & applied Math.) 1981. MR 82i: 20001.

CURTO, R., MUHLY, P. S., and WILLIAMS, D. P.

[1] Crossed products of strongly Morita equivalent C^*-algebras, *Proc. Amer. Math. Soc.* **90**(1984), 528–530. MR 85i: 46083.

DADE, E. C.

[1] Compounding Clifford's Theory, *Ann. of Math.* **91**(1970), 236–290. MR 41 #6992.

DAGUE, P.

[1] Détermination de la topologie de Fell sur le dual du groupe de Poincaré, *C.R. Acad. Sci., Paris* **283**(1976), 293–296. MR 54 #12981.

DANG-NGOC, N.

[1] Produits croisés restreints et extensions de groupes, Mai 1975 (unpublished preprints).

DANIELL, P. J.

[1] A general form of integral, *Ann. of Math.* (2) **19**(1917–1918), 279–294.

VAN DANTZIG, D.

[1] Zur topologischen Algebra, II, Abstrakte b_v-adische Ringe, *Composit. Math.* **2**(1935), 201–223.

DAUNS, J.

[1] The primitive ideal space of a C^*-algebra, *Canad. J. Math.* **26**(1974), 42–49. MR 49 #1131.

DAUNS, J., and HOFMANN, K. H.

[1] *Representations of rings by sections*, Memoirs Amer. Math. Soc. **83**, Amer. Math. Soc., Providence, R. I., 1968. MR 40 #752.

[2] Spectral theory of algebras and adjunction of identity, *Math. Ann.* **179**(1969), 175–202. MR 40 #734.

DAY, M. M.

[1] *Normed linear spaces*, Ergebnisse der Math. **21**, Springer-Verlag, Berlin, Göttingen, Heidelberg, 1958. MR 20 #1187.

DEALBA, L. M. and PETERS, J.

[1] Classification of semicrossed products of finite-dimensional C^*-algebras, *Proc. Amer. Math. Soc.* **95**(1985), 557–564. MR 87e: 46088.

DELIYANNIS, P. C.

[1] Holomorphic imprimitivities, *Proc. Amer. Math. Soc.* **16**(1965), 228–233. MR 31 #625.

DERIGHETTI, A.

[1] Sur certaines propriétés des représentations unitaires des groupes localement compacts, *Comment. Math. Helv.* **48**(1973), 328–339. MR 48 #8686.

[2] Sulla nozione di contenimento debole e la proprietà di Reiter, *Rend. Sem. Mat. Fis. Milano* **44**(1974), 47–54. MR 54 #10476.

[3] Some remarks on $L^1(G)$, *Math Z.* **164**(1978), 189–194. MR 80f: 43009.

DIEUDONNÉ, J.

[1] *Treatise on Analysis. Vol. VI, Harmonic analysis* **10**, Pure and Applied Mathematics. Academic Press, New York, 1978. MR 58 #29825b.

[2] *Special functions and linear representations of Lie groups.* CBMS Regional Conf. Ser. Math. **42**, Amer. Math. Soc., Providence, R. I., 1980. MR 81b: 22002.

[3] *History of functional analysis*, North Holland Mathematics Studies, **49**, North-Holland Publishing Co., Amsterdam, 1981. MR 83d: 46001.

DIESTEL, J., and UHL, JR, J. J.

[1] *Vector measures, Math. Survey*, **15**, Amer. Math. Soc. Providence, R. I., 1977. MR 56 #12216.

DINCULEANU, N.

[1] *Vector measures*, Pergamon Press, Oxford, London, 1967. MR 34 #6011b.

[2] *Integration on locally compact spaces*, Noordhoff International Publishing Co., Leyden, 1974. MR 50 #13428.

DISNEY, S., and RAEBURN, I.

[1] Homogeneous *C**-algebras whose spectra are tori, *J. Austral. Math. Soc. Ser. A* **38**(1985), 9–39. MR 86i: 46057.

DIXMIER, J.

[1] Sur la réduction des anneaux d'opérateurs, *Ann. Sci. École Norm. Sup.* (3) **68**(1951), 185–202. MR 13, 471.

[2] Algèbres quasi-unitaires, *Comment. Math. Helv.* **26**(1952), 275–322. MR 14, 660.

[3] On unitary representations of nilpotent Lie groups, *Proc. Nat. Acad. Sci. U.S.A.* **43**(1957), 958–986. MR 20 #1927.

[4] Sur les représentations unitaires des groupes de Lie algébriques, *Ann. Inst. Fourier (Grenoble)* **7**(1957), 315–328. MR 20 #5820.

[5] Sur les représentations unitaires des groupes de Lie nilpotents. I, *Amer. J. Math.* **81**(1959), 160–170. MR 21 #2705.

[6] Sur les représentations unitaires des groupes de Lie nilpotents. II, *Bull. Soc. Math. France* **85**(1957), 325–388. MR 20 #1928.

[7] Sur les représentations unitaires des groupes de Lie nilpotents. III, *Canad. J. Math.* **10**(1958), 321–348. MR 20 #1929.

[8] Sur les représentations unitaires des groupes de Lie nilpotents. IV, *Canad. J. Math.* **11**(1959), 321–344. MR 21 #5693.

[9] Sur les représentations unitaires des groupes de Lie nilpotents. V, *Bull. Soc. Math. France* **87**(1959), 65–79. MR 22 #5900a.

[10] Sur les *C**-algèbres, *Bull. Soc. Math. France* **88**(1960), 95–112. MR 22 #12408.

[11] Sur les représentations unitaires des groupes de Lie nilpotents. VI, *Canad. J. Math.* **12**(1960), 324–352. MR 22 #5900b.

[12] Sur les structures boréliennes du spectre d'une *C**-algèbre, *Inst. Hautes Études Sci. Publ. Math.* **6**(1960), 297–303. MR 23 #A2065.

[13] Points isolés dans le dual d'un groupe localment compact, *Bull. Soc. Math. France* **85**(1961), 91–96. MR 24 #A3237.

[14] Points séparés dans le spectre d'une *C**-algèbre, *Acta Sci. Math. Szeged* **22**(1961), 115–128. MR 23 #A4030.

[15] Sur le revêtement universel d'un groupe de Lie de type I, *C. R. Acad. Sci. Paris* **252**(1961), 2805–2806. MR 24 #A3241.

[16] Représentations intégrables du groupe de DeSitter, *Bull. Soc. Math. France* **89**(1961), 9–41. MR 25 #4031.

[17] Dual et quasi dual d'une algèbre de Banach involutive, *Trans. Amer. Math. Soc.* **104**(1962), 278–283. MR 25 #3384.

[18] Traces sur les C^*-algèbres, *Ann. Inst. Fourier (Grenoble)* **13**(1963), 219–262. MR 26 #6807.

[19] Représentations induites holomorphes des groups resolubles algébriques, *Bull. Soc. Math. France* **94**(1966), 181–206. MR 34 #7724.

[20] Champs continus d'espaces hilbertiens et de C^*-algèbres. II, *J. Math. Pures Appl.* **42**(1963), 1–20. MR 27 #603.

[21] *Les C^*-algèbres et leurs représentations*, Gauthier-Villars, Paris, 1964. MR 30 #1404.

[22] *Les algèbres d'opérateurs dans l'espace hilbertien (algèbres de von Neumann)*, 2nd ed., Gauthier-Villars, Paris, 1969. *MR* 20 #1234.

[23] *Bicontinuité dans la méthode du petit groupe de Mackey, Bull. Sci. Math.* (2) **97**(1973), 233–240. MR 53 #3187.

[24] *C^*-algebras*, 15, North-Holland Publishing Co., Amsterdam, 1977. MR 56 #16388.

[25] *Von Neumann algebras*, 27, North-Holland Publishing Co., Amsterdam, 1981. MR 50 #5482.

DIXMIER, J., and DOUADY, A.

[1] Champs continus d'espaces hilbertiens et de C^*-algèbres, *Bull. Soc. Math. France* **91**(1963), 227–283. MR 29 #485.

Dô, NGOK Z'EP.

[1] The structure of the group C^*-algebra of the group of affine transformations of the line (Russian), *Funkcional Anal. Priložen.* **9**(1974), 63–64. MR 51 #793.

[2] Quelques aspects topologiques en analyse harmonique, *Acta Math. Vietnam.* **8**(1983), 35–131 (1984). MR 86j: 22005.

DOPLICHER, S., KASTLER, D., and ROBINSON, D.

[1] Covariance algebras in field theory and statistical mechanics, *Commun. Math. Phys.* **3**(1966), 1–28. MR 34 #4930.

DORAN, R. S.

[1] Construction of uniform CT-bundles, *Notices Amer. Math. Soc.* **15**(1968), 551.

[2] *Representations of C^*-algebras, by uniform CT-bundles*, Ph.D. Thesis, Univ. of Washington, Seattle, 1968.

[3] A generalization of a theorem of Civin and Yood on Banach *-algebras, *Bull. London Math. Soc.* **4**(1972), 25–26. MR 46 #2442.

DORAN, R. S., and BELFI, V. A.

[1] *Characterizations of C^*-algebras: the Gelfand-Naimark Theorems*, Pure and Applied Mathematics, 101, Marcel-Dekker Pub. Co., New York, 1986.

DORAN, R. S., and TILLER, W.

[1] Extensions of pure positive functionals on Banach *-algebras, *Proc. Amer. Math. Soc.* **82**(1981), 583–586. MR 82f: 46062.

[2] Continuity of the involution in a Banach *-algebra, *Tamkang J. Math.* **13**(1982), 87–90. MR 84d: 46086.

DORAN, R. S., and WICHMANN, J.

[1] The Gelfand-Naimark theorems for C^*-algebras, *Enseignement Math.* (2) **23**(1977), 153–180. MR 58 #12395.

[2] *Approximate identities and factorization in Banach modules.* Lecture Notes in Mathematics, **768**. Springer-Verlag, Berlin-Heidelberg-New York, 1979. MR 81e: 46044.

DORNHOFF, L.

[1] *Group representation theory, Part A, Ordinary theory*, Marcel-Dekker, New York, 1971. MR 50 #458a.

[2] *Group representation theory, Part B, Modular representation theory*, Marcel-Dekker, New York, 1972. MR 50 #458b.

DOUADY, A., and DAL SOGLIO-HÉRAULT, L.

[1] Existence de sections pour un fibré de Banach au sens de Fell, unpublished manuscript (see J. M. G. Fell [15], Appendix).

DOUGLAS, R. G.

[1] *Banach algebra techniques in operator theory*, Academic Press, New York and London, 1972. MR 50 #14335.

DUFLO, M.

[1] *Harmonic analysis and group representations*, Cortona, 1980, 129–221, Liguori, Naples 1982.

[2] Théorie de Mackey pour les groupes de Lie algébriques, *Acta Math.* **149**(1982), 153–213. MR 85h: 22022.

DUFLO, M., and MOORE, C. C.

[1] On the regular representation of a nonunimodular locally compact group, *J. Functional Anal.* **21**(1976), 209–243. MR 52 #14145.

DUNFORD, N.

[1] Resolution of the identity for commutative B^*-algebras of operators, *Acta Sci. Math. Szeged Pars B* **12**(1950), 51–56. MR 11, 600.

DUNFORD, N., and PETTIS, B. J.

[1] Linear operations on summable functions, *Trans. Amer. Math. Soc.* **47**(1940), 323–392. MR 1, 338.

DUNFORD, N., and SCHWARTZ, J. T.

[1] *Linear operators, Part I: General theory*, Interscience Publishers, New York and London, 1958. MR 22 #8302.

[2] *Linear operators, Part II: General theory*, Interscience Publishers, New York and London, 1963. MR 32 #6181.

DUNKL, C. F., and RAMIREZ, D. E.

[1] *Topics in harmonic analysis*, Appleton-Century Crofts, New York, 1971.

DUPONCHEEL, L.

[1] How to use induced representations. *Proceedings of the Conference on p-adic analysis*, pp. 72–77, Report, 7806, Math. Inst., Katolieke Univ., Niumegen, 1978. MR 80c: 22010.

[2] Non-archimedean induced representations of compact zerodimensional groups, *Composito Math.* **57**(1986), 3–13.

DUPRÉ, M. J.

[1] Classifying Hilbert bundles, *J. Functional Anal.* **15**(1974), 244–278. MR 49 #11266.

[2] Classifying Hilbert bundles. II, *J. Functional Anal.* **22**(1976), 295–322. MR 54 #3435.

[3] Duality for *C**-algebras, Proc. Conf., Loyola Univ., New Orleans, La., pp. 329–338, Academic Press, New York, 1978. MR 80a: 46034.

[4] *The classification and structure of C*-algebra bundles*, Memoirs Amer. Math. Soc. **21**, (222), 1–77, Amer. Math. Soc. Providence, R. I., 1979. MR 83c: 46069.

DUPRÉ, M. J., and GILLETTE, R. M.

[1] *Banach bundles, Banach modules and automorphisms of C*-algebras*, **92**, Pitman's Research Notes in Mathematics Series, Pitman Pub. Co., New York, 1983. MR 85j: 46127.

DURBIN, J. R.

[1] On locally compact wreath products. *Pacific J. Math.* **57**(1975), 99–107. MR 51 #13125.

DYE, H.

[1] On groups of measure preserving transformations I, *Amer. J. Math.* **81**(1959), 119–159. MR 24 #A1366.

[2] On groups of measure preserving transformations II, *Amer. J. Math.* **85**(1963), 551–576. MR 28 #1275.

EDWARDS, C. M.

[1] *C**-algebras of central group extensions. I, *Ann. Inst. H. Poincaré (A)* **10**(1969), 229–246. MR 40 #1536.

[2] The operational approach to algebraic quantum theory. I, *Commun. Math. Phys.* **16**(1970), 207–230. MR 42 #8819.

[3] The measure algebra of a central group extension, *Quart. J. Math. Oxford Ser.* (2) **22**(1971), 197–220. MR 46 #609.

EDWARDS, C. M., and LEWIS, J. T.

[1] Twisted group algebras I, II, *Commun. Math. Phys.* **13**(1969), 119–141. MR 40 #6279.

EDWARDS, C. M., and STACEY, P. J.

[1] On group algebras of central group extensions, *Pacific J. Math.* **56**(1975), 59–75. MR 54 #10480.

EDWARDS, R. E.

[1] *Functional analysis: theory and applications*, Holt, Rinehart, and Winston, New York-Toronto-London, 1965. MR 36 #4308.

[2] *Integration and harmonic analysis on compact groups*, London Mathematical Society Lecture Note Series. **8**, Cambridge University Press, London, 1972. MR 57 #17116.

EFFROS, E.

[1] A decomposition theory for representations of C^*-algebras, *Trans. Amer. Math. Soc.* **107**(1963), 83–106. MR 26 #4202.

[2] Transformation groups and C^*-algebras, *Ann. Math.* **81**(1965), 38–55. MR 30 #5175.

[3] *Dimensions and C^*-algebras*, CBMS Regional Conference Series in Mathematics, **46**, Conference Board of the Mathematical Sciences, Washington, D. C., 1981. MR 84k: 46042.

EFFROS, E. G., and HAHN, F.

[1] *Locally compact transformation groups and C^*-algebras*, Memoirs Amer. Math. Soc. **75**, Amer. Math. Soc., Providence, R. I., 1967. MR 37 #2895.

EFFROS, E. G., HANDELMAN, D. E., and SHEN, C.-L.

[1] Dimension groups and their affine representations, *Amer. J. Math.* **102**(1980), 385–407. MR 83g: 46061.

EHRENPREIS, L., and MAUTNER, F. I.

[1] Some properties of the Fourier transform on semi-simple Lie groups. I, *Ann. Math.* (2) **61**(1955), 406–439. MR 16, 1017.

[2] Some properties of the Fourier transform on semi-simple Lie groups. II, *Trans. Amer. Math. Soc.* **84**(1957), 1–55. MR 18, 745.

[3] Some properties of the Fourier transform on semi-simple Lie groups. III, *Trans. Amer. Math. Soc.* **90**(1959), 431–484. MR 21 #1541.

ELLIOTT, G. A.

[1] A characterization of compact groups, *Proc. Amer. Math. Soc.* **29**(1971), 621. MR 43 #2155.

[2] Another weak Stone-Weierstrass theorem for C^*-algebras, *Canad. Math. Bull.* **15**(1972), 355–358. MR 47 #4011.

[3] An abstract Dauns-Hofmann-Kaplansky multiplier theorem, *Canad. J. Math.* **27**(1975), 827–836. MR 53 #6334.

[4] On the classification of inductive limits of sequences of semi-simple finite dimensional algebras, *J. Algebra* **38**(1976), 29–44. MR 53 #1279.

[5] On the K-theory of the C^*-algebra generated by a projective representation of a torsion-free discrete abelian group, *Operator algebras and group representations*, 1, pp. 157–184, Pitman, London, 1984.

ELLIOTT, G. A., and OLESEN, D.

[1] A simple proof of the Dauns-Hofmann theorem, *Math. Scand.* **34**(1974), 231–234. MR 50 #8091.

EMCH, G. G.

[1] *Algebraic methods in statistical mechanics and quantum field theory*, Wiley-Interscience, John Wiley & Sons, New York, 1972.

ENFLO, P.

[1] Uniform structures and square roots in topological groups. I., *Israel J. Math.* **8**(1970), 230–252. MR 41 #8568.

[2] Uniform structures and square roots in topological groups. II., *Israel J. Math.* **8**(1970), 253–272. MR 41 #8568.

[3] A counterexample to the approximation problem in Banach spaces, *Acta Math.* **130**(1973), 309–317. MR 53 #6288.

ENOCK, M.

[1] Produit croisé d'une algèbre de von Neumann par une algèbre de Kac, *J. Functional Anal.* **26**(1977), 16–47. MR 57 #13513.

[2] Kac algebras and crossed products, *Algèbres d'opérateurs et leurs applications en physique mathematique*, Proc. Colloq., Marseille, 1977, pp. 157–166, Colloques Internat. CNRS, **274**, CNRS, Paris, 1979. MR 81e: 46051.

ENOCK, M., and SCHWARTZ, J.-M.

[1] Une dualité dans les algèbres de von Neumann, *Bull. Soc. Math. France Suppl. Mem.* **44**(1975), 1–144. MR 56 #1091.

[2] Produit croisé d'une algèbre de von Neumann par une algèbre de Kac. II, *Publ. Res. Inst. Math. Sci.* **16**(1980), 189–232. MR 81m: 46084.

[3] Algèbres de Kac moyennables, *Pacific J. Math.* **125**(1986), 363–379.

ERNEST, J.

[1] Central intertwining numbers for representations of finite groups, *Trans. Amer. Math. Soc.* **99**(1961), 499–508. MR 23 #A2467.

[2] A decomposition theory for unitary representations of locally compact groups, *Bull. Amer. Math. Soc.* **67**(1961), 385–388. MR 24 #A784.

[3] A decomposition theory for unitary representations of locally compact groups, *Trans. Amer. Math. Soc.* **104**(1962), 252–277. MR 25 #3383.

[4] A new group algebra for locally compact groups. I, *Amer. J. Math.* **86**(1964), 467–492. MR 29 #4838.

[5] Notes on the duality theorem of non-commutative, non-compact topological groups, *Tôhoku Math. J.* **16**(1964), 291–296. MR 30 #192.

[6] A new group algebra for locally compact groups. II, *Canad. J. Math.* **17**(1965), 604–615. MR 32 #159.

[7] The representation lattice of a locally compact group, *Illinois J. Math.* **10**(1966), 127–135. MR 32 #1288.

[8] Hopf-Von Neumann algebras, *Functional Analysis*, pp. 195–215, Academic Press, New York, 1967. MR 36 #6956.

[9] The enveloping algebra of a covariant system, *Commun. Math. Phys.* **17**(1970), 61–74. MR 43 #1553.

[10] A duality theorem for the automorphism group of a covariant system, *Commun. Math. Phys.* **17**(1970), 75–90. MR 42 #8298.

[11] A strong duality theorem for separable locally compact groups, *Trans. Amer. Math. Soc.* **156**(1971), 287–307. MR 43 #7555.

[12] On the topology of the spectrum of a *C**-algebra, *Math. Ann.* **216**(1975), 149–153. MR 53 #8913.

EVANS, B. D.

[1] *C*-bundles and compact transformation groups*, Memoirs Amer. Math. Soc. **269**, Amer. Math. Soc., Providence, R. I., 1982. MR 84a: 46148.

EYMARD, P.

[1] L'algèbre de Fourier d'un groupe localement compact, *Bull. Soc. Math. France.* **92**(1964), 181–236. MR 37 #4208.

[2] *Moyennes invariantes et représentations unitaires*, Lecture Notes in Math., **300**, Springer-Verlag, Berlin-New York, 1972. MR 56 #6279.

FABEC, R. C.

[1] A theorem on extending representations, *Proc. Amer. Math. Soc.* **75**(1979), 157–162. MR 80f: 22002.

[2] Cocycles, extensions of group actions and bundle representations, *J. Functional Anal.* **56**(1984), 79–98. MR 85k: 22017.

[3] Induced group actions, representations and fibered skew product extensions, *Trans. Amer. Soc.* **301**(1987), 489–513.

FACK, T., and SKANDALIS, G.

[1] Structure des idéaux de la *C**-algèbre associée à un feuilletage, *C. R. Acad. Sci. Paris Sér. A-B* **290**(1980), A1057–A1059. MR 81h: 46088.

[2] Sur les représentations et idéaux de la *C**-algèbre d'un feuilletage, *J. Operator Theory* **8**(1982), 95–129. MR 84d: 46101.

FAKLER, R. A.

[1] On Mackey's tensor product theorem, *Duke Math. J.* **40**(1973), 689–694. MR 47 #8764.

[2] Erratum to: "On Mackey's tensor product theorem," *Duke Math. J.* **41**(1974), 691. MR 49 #5224.

[3] Representations induced from conjugate subgroups, *Indiana J. Math.* **19**(1977), 167–171. MR 82e: 22012.

[4] An intertwining number theorem for induced representations, *Nanta Math.* **11**(1978), 164–173. MR 80h: 22006.

FARMER, K. B.

[1] A survey of projective representation theory of finite groups, *Nieuw Arch. Wiskunde* **26**(1978), 292–308. MR 58 #16860.

FELDMAN, J.

[1] Borel sets of states and of representations, *Michigan Math. J.* **12**(1965), 363–366. MR 32 #375.

FELDMAN, J., and FELL, J. M. G.

[1] Separable representations of rings of operators, *Ann. Math.* **65**(1957), 241–249. MR 18, 915.

FELDMAN, J., HAHN, P., and MOORE, C. C.

[1] Orbit structure and countable sections for actions of continuous groups, *Adv. Math.* **28**(1978), 186–230. MR 58 #11217.

FELIX, R.

[1] Über Integralzerlegungen von Darstellungen nilpotenter Lie-gruppen (English summary), *Manuscripta Math.* **27**(1979), 279–290. MR 81d: 22007.

[2] When is a Kirillov orbit a linear variety? *Proc. Amer. Math. Soc.* **86**(1982), 151–152. MR 83h: 22017.

FELIX, R., HENRICHS, R. W., and SKUDLAREK, H.

[1] Topological Frobenius reciprocity for projective limits of Lie groups, *Math. Z.* **165**(1979), 19–28. MR 80e: 22010.

FELL, J. M. G.

[1] C*-algebras with smooth dual, *Illinois J. Math.* **4**(1960), 221–230. MR 23 #A2064.

[2] The dual spaces of C*-algebras, *Trans. Amer. Math. Soc.* **94**(1960), 365–403. MR 26 #4201.

[3] The structure of algebras of operator fields, *Acta. Math.* **106**(1961), 233–280. MR 29 #1547.

[4] A Hausdorff topology on the closed subsets of a locally compact non-Hausdorff space, *Proc. Amer. Math. Soc.* **13**(1962), 472–476. MR 25 #2573.

[5] Weak containment and induced representations of groups, *Canad. J. Math.* **14**(1962), 237–268. MR 27 #242.

[6] A new proof that nilpotent groups are CCR, *Proc. Amer. Math. Soc.* **13**(1962), 93–99. MR 24 #A3238.

[7] Weak containment and Kronecker products of group representations, *Pacific J. Math.* **13**(1963), 503–510. MR 27 #5865.

[8] Weak containment and induced representations of groups II, *Trans. Amer. Math. Soc.* **110**(1964), 424–447. MR 28 #3114.

[9] The dual spaces of Banach algebras, *Trans. Amer. Math. Soc.* **114**(1965), 227–250. MR 30 #2357.

[10] Non-unitary dual spaces of groups, *Acta. Math.* **114**(1965), 267–310. MR 32 #4210.

[11] Algebras and fiber bundles, *Pacific J. Math.* **16**(1966), 497–503. MR 33 #2674.

[12] Conjugating representations and related results on semisimple Lie groups, *Trans. Amer. Math. Soc.* **127**(1967), 405–426. MR 35 #299.

[13] An extension of Mackey's method to algebraic bundles over finite groups, *Amer. J. Math.* **91**(1969), 203–238. MR 40 #735.

[14] *An extension of Mackey's method to Banach *-algebraic bundles*, Memoirs Amer. Math. Soc. **90**, Providence, R. I., 1969. MR 41 #4255.

[15] *Banach *-algebraic bundles and induced representations*, Actes du Congrès International des Mathématiciens, Nice, 1970, Tome 2, pp. 383–388, Gauthier-Villars, Paris 1971. MR 54 #8315.

[16] A new look at Mackey's imprimitivity theorem. *Conference on Harmonic Analysis*, Univ. Maryland, College Park, Md., 1971, pp. 43–58. Lecture Notes in Math., **266**, Springer, Berlin, 1972. MR 53 #13471.

[17] *Induced representations and Banach *-algebraic bundles*, Lecture Notes in Math., **582**, Springer-Verlag, Berlin-Heidelberg-New York, 1977. MR 56 #15825.

FELL, J. M. G., and KELLEY, J. L.

[1] An algebra of unbounded operators, *Proc. Nat. Acad. Sci. U.S.A.* **38**(1952), 592–598. MR 14, 480.

FELL, J. M. G., and THOMA, E.

[1] Einige Bemerkungen über vollsymmetrische Banachsche Algebren, *Arch. Math.* **12**(1961), 69–70. MR 23 #A2067.

FIGÀ-TALAMANCA, A., and PICARDELLO, M.

[1] *Harmonic analysis on free groups*, 87, Marcel Dekker, New York, 1983. MR 85j: 43001.

FONTENOT, R. A., and SCHOCHETMAN, I.

[1] Induced representations of groups on Banach spaces, *Rocky Mountain J. Math.* **7**(1977), 53–82.. MR 56 #15824.

FORD, J. W. M.

[1] A square root lemma for Banach *-algebras, *J. London Math. Soc.* **42**(1967), 521–522. MR 35 #5950.

FOURIER, J. B. J.

[1] La théorie analytique de la chaleur, 1822, trans. A. Freeman, Cambridge, 1878.

FOURMAN, M. P., MULVEY, C. J., and SCOTT, D. S. (eds.)

[1] *Applications of sheaves*, Lecture Notes in Math., **753**, Springer-Verlag, Berlin-Heidelberg-New York, 1979. MR 80j: 18001.

FOX, J.

[1] Frobenius reciprocity and extensions of nilpotent Lie groups, *Trans. Amer. Math. Soc.* **298**(1986), 123–144.

FRENCH, W. P., LUUKKAINEN, J., and PRICE, J. F.

[1] The Tannaka-Krein duality principle, *Adv. Math.* **43**(1982), 230–249. MR 84f: 22011.

FREUDENTHAL, H., and DE VRIES, H.

[1] *Linear Lie groups*, Academic Press, New York-London, 1969. MR 41 #5546.

FROBENIUS, G.

[1] Über lineare Substitutionen und bilineare Formen, *J. Creele*, **84**(1878), 1–63.

[2] Über Gruppencharaktere, *Berl. Sitz.* (1896), 985–1021.

[3] Über Primfaktoren der Gruppendeterminant, *Berl. Sitz.* (1896), 1343–1382.

[4] Darstellung der Gruppen durch lineare Substitutionen. *Berl. Sitz.* (1897), 994–1015.

[5] Über Relationen zwischen den Charakteren einer Gruppe und denen ihrer Untergruppen, *Sitz. Preuss. Akad. Wiss.* (1898), 501–515.

FROBENIUS, G., and SCHUR, I.

[1] Über die Äquivalenz der Gruppen linearer Substitutionen, *Sitz. Preuss. Akad. Wiss.* (1906), 209–217.

FUJIMOTO, I., and TAKAHASI, S.

[1] Equivalent conditions for the general Stone-Weierstrass problem, *Manuscripta Math.* **53**(1985), 217–224.

FUJIWARA, H.

[1] On holomorphically induced representations of exponential groups, *Proc. Japan Acad.* **52**(1976), 420–423. MR 57 #12778.

[2] On holomorphically induced representations of exponential groups, *Japan J. Math.* (n.s.) **4**(1978), 109–170. MR 80g: 22008.

FUKAMIYA, M.

[1] On B*-algebras, *Proc. Japan Acad.* **27**(1951), 321–327. MR 13, 756.

[2] On a theorem of Gelfand and Naimark and the B*-algebra, *Kumamoto J. Sci. Ser. A-1*, 1 (1952), 17–22. MR 14, 884; MR 15, 1139.

FUNAKOSI, S.

[1] Induced bornological representations of bornological algebras, *Portugal. Math.* **35**(1976), 97–109. MR 56 #12924.

[2] On representations of non-type I groups, *Tôhoku Math. J.* (2) **31**(1979), 139–150. MR 81b: 22006.

GAAL, S. A.

[1] *Linear analysis and representation theory*, Grundlehren der Math. Wiss, Band 198, Springer-Verlag, New York-Heidelberg, 1973. MR 56 #5777.

GAMELIN, T. W.

[1] *Uniform algebras*, Prentice Hall, Englewood Cliffs, N. J., 1969. MR 53 #14137.

[2] *Uniform algebras and Jensen measures*, London Mathematical Society Lecture Note Series, **32**, Cambridge University Press, Cambridge-New York, 1978. MR 81a: 46058.

GANGOLLI, R.

[1] On the symmetry of L_1-algebras of locally compact motion groups and the Wiener Tauberian theorem, *J. Functional Anal.* **25**(1977), 244–252. MR 57 #6284.

GARDNER, L. T.

[1] On the "third definition" of the topology of the spectrum of a C*-algebra, *Canad. J. Math.* **23**(1971), 445–450. MR 43 #6730.

[2] On the Mackey Borel Structure, *Canad. J. Math.* **23**(1971), 674–678. MR 44 #4532.

[3] An elementary proof of the Russo-Dye theorem, *Proc. Amer. Math. Soc.* **90**(1984), 171. MR 85f: 46017.

GELBAUM, B. R.

[1] Banach algebra bundles, *Pacific J. Math.* **28**(1969), 337–349. MR 39 #6077.

[2] *Banach algebra bundles. II, Troisième Colloque sur l'Analyse Fonctionnelle*, Liège, 1970, pp. 7–12, Vander, Louvain, 1971. MR 53 #14132.

[3] *Group algebra bundles, Problems in analysis,* Papers dedicated to Salomon Bochner, 1969, pp. 229–237. Princeton Univ. Press, Princeton, N. J., 1970. MR 50 #997.

GELBAUM, B. R., and KYRIAZIS, A.

[1] Fibre tensor product bundles, *Proc. Amer. Math. Soc.* **93**(1985), 675–680. MR 86g: 46107.

GELFAND, I. M.

[1] Sur un lemme de la théorie des espaces linéaires, *Comm. Inst. Sci. Math. Kharkoff* **13**(1936), 35–40.

[2] On normed rings, *Dokl. Akad. Nauk. SSSR* **23**(1939), 430–432.

[3] Normierte Ringe, *Mat. Sb.* **9**(1941), 3–24. MR 3, 51.

[4] Über absolut konvergente trigonometrische Reihen und Integrale, *Mat. Sb. N.S.* (51) **9**(1941), 51–66. MR 3, 51.

[5] The center of an infinitesimal group ring, *Mat. Sb. N.S.* (68) **26**(1950), 103–112(Russian). MR 11, 498.

[6] Unitary representations of the real unimodular group (principal nondegenerate series), *Izv. Akad. Nauk. SSSR* **17**(1953), 189–248 (Russian).

[7] Spherical functions in symmetric Riemann spaces, *Dokl. Akad. Nauk. SSSR* **70**(1956), 5–8 (Russian).

[8] The structure of a ring of rapidly decreasing functions on a Lie group, *Dokl. Akad. Nauk. SSSR* **124**(1959), 19–21 (Russian). MR 22 #3987.

[9] Integral geometry and its relation to the theory of representations, *Uspehi Mat. Nauk* **15**(1960), 155–164 (Russian). MR 26 #1903. [Translated in *Russian Math. Surveys* **15**(1960), 143–151.]

GELFAND, I. M., and GRAEV, M. I.

[1] Analogue of the Plancherel formula for the classical groups, *Trudy Moskov. Mat. Obšč.* **4**(1955), 375–404 (Russian). MR 17, 173. [*Amer. Math. Soc. Trans.* **9**(1958), 123–154. MR 19, 1181.]

[2] Expansion of Lorenz group representations into irreducible representations on spaces of functions defined on symmetrical spaces, *Dokl. Akad. Nauk. SSSR* **127**(1959), 250–253 (Russian). MR 23 #A1238.

[3] Geometry of homogeneous spaces, representations of groups in homogeneous spaces and related questions of integral geometry, *Trudy Moskov. Mat. Obšč.* **8**(1959), 321–390 (Russian). MR 23 #A4013.

GELFAND, I. M., GRAEV, M. I., and VERSIK, A. M.

[1] Representations of the group of smooth mappings of a manifold X into a compact Lie group, *Compositio Math.* **35**(1977), 299–334.

GELFAND, I. M., and NAIMARK, M. A.

[1] On the embedding of normed rings into the ring of operators in Hilbert space, *Mat. Sb.* **12**(1943), 197–213. MR 5, 147.

[2] Unitary representations of the group of linear transformations of the straight line, *Dokl. Akad. Nauk. SSSR* **55**(1947), 567–570 (Russian). MR 8, 563.

[3] Unitary representations of the Lorenz group, *Izv. Akad. Nauk. SSSR* **11**(1947), 411–504. (Russian). MR 9, 495.

[4] Normed rings with involution and their representations, *Izv. Akad. Nauk SSSR, Ser-math.* **12**(1948), 445–480. MR 10, 199.

[5] Unitary representations of the classical groups, *Trudy Mat. Inst. Steklov* **36**(1950), 1–288. (Russian). MR 13, 722.

GELFAND, I. M., and PYATETZKI-SHAPIRO, I.

[1] Theory of representations and theory of automorphic functions, *Uspehi Mat. Nauk* **14**(1959), 171–194 (Russian).

[2] Unitary representation in homogeneous spaces with discrete stationary subgroups, *Dokl. Akad. Nauk SSSR* **147**(1962), 17–20 (Russian).

GELFAND, I. M., and RAIKOV, D. A.

[1] Irreducible unitary representations of arbitrary locally bicompact groups, *Mat. Sb. N.S.* **13**(55) (1943), 301–316 (Russian). MR 6, 147.

GELFAND, I. M., RAIKOV, D. A., and ŠILOV, G. E.

[1] Commutative normed rings, *Uspehi Mat. Nauk* **1**: 2(12) (1946), 48–146. MR 10, 258.

[2] *Commutative normed rings* (Russian), Sovremennye Problemy Matematiki Gosudarstv. Izdat. Fiz.-Mat. Lit. Moscow, 1960. MR 23 #A1242. [Translated from the Russian, Chelsea Publishing Co., New York, 1964. MR 34 #4940.]

GELFAND, I. M., and ŠILOV, G. E.

[1] Über verschiedene Methoden der Einführung der Topologie in die Menge der maximalen Ideale eines normierten Rings, *Mat. Sb.* **51**(1941), 25–39. MR 3, 52.

GELFAND, I. M., and VILENKIN, N. JA.

[1] *Generalized functions, Vol. IV. Some applications of harmonic analysis: Rigged Hilbert spaces*, Gos. Izd., Moscow, 1961. MR 26 #4173; MR 35 #7123.

[2] *Integral geometry and connections with questions in the theory of representations*, Fitmatgiz, Moscow, 1962 (Russian).

GELFAND, I. M., MINLOS, R. A., and SHAPIRO, Z. YA.

[1] *Representations of the rotation and Lorentz groups and their applications*, Macmillan, New York, 1963. MR 22 #5694.

GHEZ, P., LIMA, R., and ROBERTS, J. E.

[1] W*-categories, *Pacific J. Math.* **120**(1985), 79–109. MR 87g: 46091.

GIERZ, G.

[1] *Bundles of topological vector spaces and their duality* (with an appendix by the author and K. Keimel). Lecture notes in Mathematics, **955**, Springer-Verlag, Berlin-New York, 1982. MR 84c: 46076.

GIL DE LAMADRID, J.

[1] Extending positive definite linear forms, *Proc. Amer. Math. Soc.* **91**(1984), 593–594. MR 85g: 46068.

GINDIKIN, S. G., and KARPELEVIC, F. I.

[1] Plancherel measure of Riemannian symmetric spaces of non-positive curvature, *Dokl. Akad. Nauk* **145**(1962), 252–255 (Russian). MR 27 #240.

GIOVANNINI, M.

[1] Induction from a normal nilpotent subgroup, *Ann. Inst. H. Poincaré Sect. A N.S.* **26**(1977), 181–192. MR 57 #535.

[2] *Induction from a normal nilpotent subgroup*, pp. 471–480. Lecture Notes in Physics, **50**, Springer, Berlin, 1976. MR 57 #12784.

GLASER, W.

[1] Symmetrie von verallgemeinerten L^1-Algebren, *Arch. Math. (Basel)* **20**(1969), 656–660. MR 41 #7448.

GLEASON, A. M.

[1] Measures on the closed subspaces of a Hilbert space, *J. Math. Mech.* **6**(1957), 885–894. MR 20 #2609.

GLICKFELD, B. W.

[1] A metric characterization of $C(X)$ and its generalization to C^*-algebras, *Illinois J. Math.* **10**(1966), 547–556. MR 34 #1865.

GLIMM, J. G.

[1] On a certain class of operator algebras, *Trans. Amer. Math. Soc.* **95**(1960), 318–340. MR 22 #2915.

[2] A Stone-Weierstress theorem for C^*-algebras, *Ann. Math.* **72**(1960), 216–244. MR 22 #7005.

[3] Type I C^*-algebras, *Ann. Math.* **73**(1961), 572–612. MR 23 #A2066.

[4] Locally compact transformation groups, *Trans. Amer. Math. Soc.* **101**(1961), 124–138. MR 25 #146.

[5] Families of induced representations, *Pacific J. Math.* **12**(1962), 855–911. MR 26 #3819.

[6] *Lectures on Harmonic analysis (non-abelian)*, New York Univ. Courant Institute of Math. Sciences, New York, 1965.

GLIMM, J. C., and KADISON, R. V.

[1] Unitary operators in C^*-algebras, *Pacific J. Math.* **10**(1960), 547–556. MR 22 #5906.

GODEMENT, R.

[1] Sur une généralization d'un théorème de Stone, *C. R. Acad. Sci Paris* **218**(1944), 901–903. MR 7, 307.

[2] Sur les relations d'orthogonalité de V. Bargmann. I. Résultats préliminaires, *C. R. Acad. Sci Paris* **225**(1947), 521–523. MR 9, 134.

[3] Sur les relations d'orthogonalité de V. Bargmann. II. Démonstration générale, *C. R. Acad. Sci. Paris* **225**(1947), 657–659. MR 9, 134.

[4] Les fonctions de type positif et la théorie des groupes, *Trans. Amer. Math. Soc.* **63**(1948), 1–84. MR 9, 327.

[5] Sur la transformation de Fourier dans les groupes discrets, *C. R. Acad. Sci. Paris* **228**(1949), 627–628. MR 10, 429.

[6] Théorie générale des sommes continues d'espaces de Banach, *C. R. Acad. Sci Paris* **228**(1949), 1321–1323. MR 10, 584.

[7] Sur la théorie des représentations unitaires, *Ann. Math.* (2) **53**(1951), 68–124. MR 12, 421.

[8] Mémoire sur la théorie des caractères dans les groupes localement compacts unimodulaires, *J. Math. Pures Appl.* **30**(1951), 1–110. MR 13, 12.

[9] A theory of spherical functions, I. *Trans. Amer. Math. Soc.* **73**(1952), 496–556. MR 14, 620.

[10] Théorie des caractères. I. Algèbres unitaires, *Ann. Math.* (2) **59**(1954), 47–62. MR 15, 441.

[11] Théorie des caractères. II. Définition et propriétés générales des caractères, *Ann. Math.* (2) **59**(1954), 63–85. MR 15, 441.

GOLDIN, G. A., and SHARP, D. H.

[1] Lie algebras of local currents and their representations, *Group Representations in Mathematics and Physics*, Battelle, Seattle, 1969, Rencontres, Lecture notes in Physics, **6**, Springer, Berlin, 1970, pp. 300–311.

GOLODETS, V. YA.

[1] Classification of representations of the anticommutation relations, *Russian Math. Surveys* **24**(1969), 1–63.

GOODEARL, K. R.

[1] *Notes on real and complex C*-algebras*, Birkhäuser, Boston, 1982. MR 85d: 46079.

GOODMAN, R.

[1] Positive-definite distributions and intertwining operators, *Pacific J. Math.* **48**(1973), 83–91. MR 48 #6319.

[2] *Nilpotent Lie groups: structure and applications to analysis*, Lecture Notes in Mathematics, **562**, Springer-Verlag, Berlin-New York 1976. MR 56 #537.

GOOTMAN, E. C.

[1] Primitive ideals of C*-algebras associated with transformation groups, *Trans. Amer. Math. Soc.* **170**(1972), 97–108. MR 46 #1961.

[2] The type of some C*- and W*-algebras associated with transformation groups, *Pacific J. Math.* **48**(1973), 93–106. MR 49 #461.

[3] Local eigenvectors for group representations, *Studia Math.* **53**(1975), 135–138. MR 52 #8327.

[4] Weak containment and weak Frobenius reciprocity, *Proc. Amer. Math. Soc.* **54**(1976), 417–422. MR 55 #8246.

[5] Induced representations and finite volume homogeneous spaces, *J. Functional Anal.* **24**(1977), 223–240. MR 56 #532.

[6] Subrepresentations of direct integrals and finite volume homogeneous spaces, *Proc. Amer. Math. Soc.* **88**(1983), 565–568. MR 84m: 22009.

[7] On certain properties of crossed products, *Proc. Symp. Pure Math.* **38**(1982), Part 1, 311–321.

[8] Abelian group actions on type I C*-algebras, *Operator algebras and their connections with topology and ergodic theory*, Busteni, 1983, pp. 152–169, Lecture Notes in Mathematics **1132**, Springer, Berlin-New York 1985.

GOOTMAN, E. C., and ROSENBERG, J.

[1] The structure of crossed product C^*-algebras,: a proof of the generalized Effros-Hahn conjecture, *Invent. Math.* **52**(1979), 283-298. MR 80h: 46091.

GRAEV, M. I.

[1] Unitary representations of real simple Lie groups, *Trudy Moskov. Mat. Obšč.* **7**(1958), 335-389 (Russian). MR 21 #3510.

GRANIRER, E. E.

[1] On group representations whose C^*-algebra is an ideal in its von Neumann algebra, *Ann. Inst. Fourier (Grenoble)* **29**(1979), 37-52. MR 81b: 22007.

[2] A strong containment property for discrete amenable groups of automorphisms on W^*-algebras, *Trans. Amer. Math. Soc.* **297**(1986), 753-761.

GRASSMANN, H.

[1] Sur les différents genres de multiplication, *J. Crelle* **49**(1855), 199-217; Leipzig, Teubner, 1904.

GREEN, P.

[1] C^*-algebras of transformation groups with smooth orbit space, *Pacific J. Math.* **72**(1977), 71-97. MR 56 #12170.

[2] The local structure of twisted covariance algebras, *Acta Math.* **140**(1978), 191-250. MR 58 #12376.

[3] Square-integrable representations and the dual topology *J. Functional Anal.* **35**(1980), 279-294. MR 82g: 22005.

[4] The structure of imprimitivity algebras, *J. Functional Anal.* **36**(1980), 88-104. MR 83d: 46080.

[5] Twisted crossed products, the "Mackey machine," and the Effros-Hahn conjecture, *Operator algebras and applications, Part* 1, Kingston, Ont., 1980, pp. 327-336, Proc. Sympos. Pure Math., **38**, Amer. Math. Soc., Providence, R. I., 1982. MR 85a: 46038.

GREENLEAF, F. P.

[1] Norm decreasing homomorphisms of group algebras, *Pacific J. Math.* **15**(1965), 1187-1219. MR 29 #2664.

[2] Amenable actions of locally compact groups, *J. Functional Anal.* **4**(1969), 295-315. MR 40 #268.

[3] *Invariant means on topological groups and their applications*, Van Nostrand-Reinhold Co., New York, 1969. MR 40 #4776.

GREENLEAF, F. P., and MOSKOWITZ, M.

[1] Cyclic vectors for representations of locally compact groups, *Math. Ann.* **190**(1971), 265-288. MR 45 #6978.

[2] Cyclic vectors for representations associated with positive definite measures: Nonseparable groups, *Pacific J. Math.* **45**(1973), 165-186. MR 50 #2389.

GREENLEAF, F. P., MOSKOWITZ, M., and ROTHSCHILD, L. P.

[1] Central idempotent measures on connected locally compact groups, *J. Functional Anal.* **15**(1974), 22-32. MR 54 #5741.

GROSS, K. I.

[1] On the evolution of noncommutative harmonic analysis, *Amer. Math. Monthly*
    85(1978), 525–548. MR 80b: 01016.

GROSSER, S., MOSAK, R., and MOSKOWITZ, M.

[1] Duality and harmonic analysis on central topological groups. I., *Indag. Math.*
    35(1973), 65–77. MR 49 #5225a.

[2] Duality and harmonic analysis on central topological groups. II. *Indag. Math.*
    35(1973), 78–91. MR 49 #5225b.

[3] Correction to "Duality and harmonic analysis on central topological groups.
    I." *Indag. Math.* **35**(1973), 375. MR 49 #5225c.

GROSSER, S., and MOSKOWITZ, M.

[1] Representation theory of central topological groups, *Trans. Amer. Math. Soc.*
    129(1967), 361–390. MR 37 #5327.

[2] Harmonic analysis on central topological groups, *Trans. Amer. Math. Soc.*
    156(1971), 419–454. MR 43 #2165.

[3] Compactness conditions in topological groups, *J. Reine Angew. Math.*
    246(1971), 1–40. MR 44 #1766.

GROTHENDIECK, A.

[1] Un résultat sur le dual d'une C^*-algèbre, *J. Math. Pures Appl.* **36**(1957),
    97–108. MR 19, 665.

GUICHARDET, A.

[1] Sur un problème posé par G. W. Mackey, *C. R. Acad. Sci. Paris* **250**(1960),
    962–963. MR 22 #910.

[2] Sur les caractères des algèbres de Banach à involution, *C. R. Acad. Sci. Paris*
    252(1961), 2800–2862..

[3] Caractères des algèbres de Banach involutives, *Ann. Inst. Fourier* (*Grenoble*)
    13(1963), 1–81. MR 26 #5437 MR 30, 1203.

[4] Caractères et représentations des produits tensoriels de C^*-algèbres, *Ann. Sci.*
    École Norm. Sup. (3) **81**(1964), 189–206. MR 30 #5176.

[5] Utilisation des sous-groupes distingués ouverts dans l'etude des représenta-
    tions unitaires des groupes localement compacts, *Compositio Math.* **17**(1965),
    1–35. MR 32 #5787.

[6] *Théorie générale des représentations unitaires.* Summer school on representa-
    tion of Lie groups, Namur, 1969, pp. 1–59, Math. Dept., Univ. of Brussels,
    Brussels, 1969. MR 58 #11214.

[7] *Représentations de G^x selon Gelfand et Delorme*, Seminaire Bourbaki 1975/76,
    no. 486, pp. 238–255. Lecture Notes in Math. **567**, Springer, Berlin, 1977. MR
    57 #9910.

[8] Extensions des représentations induites des produits semi-directs. *J. Reine*
    Angew. Math. **310**(1979), 7–32. MR 80i: 22017.

GURARIE, D.

[1] Representations of compact groups on Banach algebras, *Trans. Amer. Math.*
    Soc. **285**(1984), 1–55. MR 86h: 22007.

HAAG, R., and KASTLER, D.
[1] An algebraic approach to quantum field theory, *J. Math. Phys.* **5**(1964), 848–861. MR 29 #3144.

HAAGERUP, U.
[1] The standard form of von Neumann algebras, *Math. Scand.* **37**(1975), 271–283. MR 53 #11387.
[2] Solution of the similarity problem for cyclic representations of C^*-algebras, *Ann. Math.* (2) **118**(1983), 215–240. MR 85d: 46080.
[3] All nuclear C^*-algebras are amenable, *Invent. Math.* **74**(1983), 305–319. MR 85g: 46074.

HAAR, A.
[1] Der Messbegriff in der Theorie der kontinuerlichen Gruppen, *Ann. Math.* (2) **34**(1933), 147–169.

HADWIN, D. W.
[1] Nonseparable approximate equivalence, *Trans. Amer. Math. Soc.* **266**(1981), 203–231. MR 82e: 46078.
[2] Completely positive maps and approximate equivalence, *Indiana Univ. Math. J.*, **36**(1987), 211–228.

HAHN, P.
[1] Haar measure for measure groupoids, *Trans. Amer. Math. Soc.* **242**(1978), 1–33. MR 82a: 28012.
[2] The regular representation of measure groupoids, *Trans. Amer. Math. Soc.* **242**(1978), 34–72. MR 81f: 46075.

HAHN, P., FELDMAN, J., and MOORE, C. C.
[1] Orbit structure and countable sections for actions of continuous groups, *Adv. Math.* **28**(1978), 186–230. MR 58 #11217.

HALMOS, P. R.
[1] *Measure theory*, D. Van Nostrand Co., New York, 1950. MR 11, 504.
[2] *Introduction to Hilbert space and the theory of spectral multiplicity*, Chelsea Publ. Co., New York, 1951. MR 13, 563.
[3] *A Hilbert space problem book*, Graduate Texts in Mathematics, **19**, Springer-Verlag, Berlin-Heidelberg-New York, 1974. MR 34 #8178.

HALPERN, H.
[1] A generalized dual for a C^*-algebra, *Trans. Amer. Math. Soc.* **153**(1971), 139–156. MR 42 #5058.
[2] Integral decompositions of functionals on C^*-algebras, *Trans. Amer. Math. Soc.* **168**(1972), 371–385. MR 45 #5769.
[3] Mackey Borel structure for the quasi-dual of a separable C^*-algebra, *Canad. J. Math.* **26**(1974), 621–628. MR 52 #3973.

HALPERN, H., KAFTAL, V., and WEISS, G.
[1] The relative Dixmier property in discrete crossed products, *J. Functional Anal.* **69**(1986), 121–140.

HAMILTON, W. R.

[1] *Lectures on quaternions*, Dublin, 1853.

HANNABUSS, K.

[1] Representations of nilpotent locally compact groups, *J. Functional Anal.* **34**(1979), 146–165. MR 81c: 22016.

HARISH-CHANDRA

[1] On some applications of the universal enveloping algebra of a semi-simple Lie algebra, *Trans. Amer. Math. Soc.* **70**(1951), 28–96. MR 13, 428.

[2] Representations of semi-simple Lie groups. II, *Proc. Nat. Acad. Sci. U.S.A.* **37**(1951), 362–365. MR 13, 107.

[3] Plancherel formula for complex semi-simple Lie groups, *Proc. Nat. Acad. Sci. U.S.A.* **37**(1951), 813–818. MR 13, 533.

[4] The Plancherel formula for complex semi-simple Lie groups, *Trans. Amer. Math. Soc.* **76**(1954), 485–528. MR 16, 111.

[5] Representations of a semi-simple Lie group on a Banach space, I, *Trans. Amer. Math. Soc.* **75**(1953), 185–243. MR 15, 100.

[6] Representations of semi-simple Lie groups, II, *Trans. Amer. Math. Soc.* **76**(1954), 26–65. MR 15, 398.

[7] Representations of semi-simple Lie groups. III, *Trans. Amer. Math. Soc.* **76**(1954), 234–253. MR 16, 11.

[8] Representations of semi-simple Lie groups. IV, *Amer. J. Math.* **77**(1955), 743–777. MR 17, 282.

[9] Representations of semi-simple Lie groups. V, *Amer. J. Math.* **78**(1956), 1–41. MR 18, 490.

[10] Representations of semi-simple Lie groups. VI, *Amer. J. Math.* **78**(1956), 564–628. MR 18, 490.

[11] On the characters of a semi-simple Lie group, *Bull. Amer. Math. Soc.* **61**(1955), 389–396. MR 17, 173.

[12] The characters of semi-simple Lie groups, *Trans. Amer. Math. Soc.* **83**(1956), 98–163. MR 18, 318.

[13] On a lemma of F. Bruhat, *J. Math. Pures. Appl.* (9) **35**(1956), 203–210. MR 18, 137.

[14] A formula for semi-simple Lie groups, *Amer. J. Math.* **79**(1957), 733–760. MR 20, #2633.

[15] Differential operators on a semi-simple Lie algebra, *Amer. J. Math.* **79**(1957), 87–120. MR 18, 809.

[16] Fourier transforms on a semi-simple Lie algebra. I, *Amer. J. Math.* **79**(1957), 193–257. MR 19, 293.

[17] Fourier transforms on a semi-simple Lie algebra. II, *Amer. J. Math.* **79**(1957), 653–686. MR 20, #2396.

[18] Spherical functions on a semi-simple Lie group. I, *Amer. J. Math.* **80**(1958), 241–310. MR 20, #925.

[19] Spherical functions on a semi-simple Lie group. II, *Amer. J. Math.* **80**(1958), 553-613. MR 21, #92.

[20] Automorphic forms on a semi-simple Lie group, *Proc. Nat. Acad. Sci. U.S.A.* **45**(1959), 570-573. MR 21, #4202.

[21] Invariant eigendistributions on semi-simple Lie groups, *Bull. Amer. Math. Soc.* **69**(1963), 117-123. MR 26, #2545.

HARTMAN, N. N., HENRICHS, R. W., and LASSER, R.

[1] Duals of orbit spaces in groups with relatively compact inner automorphism groups are hypergroups, *Monatsh. Math.* **88**(1979), 229-238. MR 81b: 43006.

HAUENSCHILD, W.

[1] Der Raum Prim*C**(G) für eine abzählbare, lokal-endliche Gruppe G, *Arch. Math.* (Basel) **46**(1986), 114-117.

HAUENSCHILD, W., KANIUTH, E., and KUMAR, A.

[1] Harmonic analysis on central hypergroups and induced representations, *Pacific J. Math.* **110**(1984), 83-112. MR 85e: 43015.

HAWKINS, T.

[1] *Lebesgue's theory of integration: Its origins and development*, University of Wisconsin Press, 1970.

[2] The origins of the theory of group characters, *Arch. History Exact Sci.* **7**(1970-71), 142-170.

[3] Hypercomplex numbers, Lie groups and the creation of group representation theory, *Arch. History Exact Sci.* **8**(1971-72), 243-287.

[4] New light on Frobenius' creation of the theory of group characters, *Arch. History Exact Sci.* **12**(1974), 217-243.

VAN HEESWIJCK, L.

[1] Duality in the theory of crossed products, *Math. Scand.* **44**(1979), 313-329. MR 83d: 46082.

HELGASON, S.

[1] *Differential geometry and symmetric spaces*, Pure and Applied Mathematics, **12**, Academic Press, New York, 1962. MR 26, #2986.

[2] *Differential geometry, Lie groups, and symmetric spaces*, Pure and Applied Mathematics, **80**, Academic Press, New York-London, 1978. MR 80k: 53081.

[3] *Topics in harmonic analysis on homogeneous spaces*, Progress in Mathematics, **13**, Birkhäuser, Boston, Mass., 1981. MR 83g: 43009.

[4] *Groups and geometric analysis*, Pure and Applied Mathematics, **113**, Academic Press, New York-London, 1984. MR 86c: 22017.

HELSON, H.

[1] Analyticity on compact abelian groups, *Algebras in Analysis*, Birmingham, 1973, pp.1-62, Academic Press, London, 1975. MR 55 #989.

[2] *Harmonic analysis*, Addison Wesley Publishing Co., Reading, Mass., 1983. MR 85e: 43001.

HENNINGS, M. A.

[1] Fronsdal *-quantization and Fell inducing, *Math. Proc. Cambridge Philos. Soc.* **99**(1986), 179–188. MR 87f: 46097.

HENRARD, G.

[1] A fixed point theorem for C^*-crossed products with an abelian group, *Math. Scand.* **54**(1984), 27–39. MR 86b: 46111.

HENRICHS, R. W.

[1] Die Frobeniuseigenschaft FP für diskrete Gruppen, *Math. Z.* **147**(1976), 191–199. MR 53 #8324.

[2] On decomposition theory for unitary representations of locally compact groups, *J. Functional Anal.* **31**(1979), 101–114. MR 80g: 22002.

[3] Weak Frobenius reciprocity and compactness conditions in topological groups, *Pacific J. Math.* **82**(1979), 387–406. MR 81e: 22003.

HENSEL, K.

[1] *Zahlentheorie*, Berlin, 1913.

HERB, R.

[1] Characters of induced representations and weighted orbital integrals, *Pacific J. Math.* **114**(1984), 367–375. MR 86k: 22030.

HERB, R., and WOLF, J. A.

[1] The Plancherel theorem for general semisimple groups, *Compositio Math.* **57**(1986), 271–355. MR 87h: 22020.

HERSTEIN, I. N.

[1] Group rings as *-algebras, *Publ. Math. Debrecen* **1**(1950), 201–204. MR 21, 475.

[2] *Noncommutative rings*, Carus Monographs **15**, Mathematical Association of America, Menascha, Wis., 1968. MR 37 #2790.

HEWITT, E.

[1] The ranges of certain convolution operators, *Math. Scand.* **15**(1964), 147–155. MR 32 #4471.

HEWITT, E., and KAKUTANI, S.

[1] A class of multiplicative linear functionals on the measure algebra of a locally compact abelian group, *Illinois J. Math.* **4**(1960), 553–574. MR 23 #A527.

HEWITT, E., and ROSS, K. A.

[1] *Abstract harmonic analysis, I. Structure of topological groups, integration theory, group representations*, Grundlehren der Math. Wiss. **115**, Springer-Verlag, Berlin-Göttingen-Heidelberg, 1963. MR 28 #158.

[2] *Abstract harmonic analysis, II. Structure and analysis for compact groups, analysis on locally compact abelian groups*, Grundlehren der Math. Wiss. **152**, Springer-Verlag, New York-Berlin, 1970. MR 41 #7378.

HIGGINS, P. J.

[1] *An introduction to topological groups*, Cambridge University Press, Cambridge, 1974. MR 50 #13355.

HIGMAN, D. G.

[1] Induced and produced modules, *Canad. J. Math.* **7**(1955), 490–508. MR 19, 390.

HILBERT, D.

[1] *Grundzüge einer allgemeinen Theorie der linearen Integralgleichungen*, Leipzig and Berlin, 1912.

HILDEBRANDT, T. H.

[1] Integration in abstract spaces, *Bull. Amer. Math. Soc.* **59**(1953), 111–139. MR 14, 735.

HILL, V. E.

[1] *Groups, representations, and characters*, Hafner Press, New York, 1975. MR 54 #7596.

HILLE, E.

[1] *Functional analysis and semigroups*, Amer. Math. Soc. Coll. Publ. 31, New York, 1948. MR 9, 594.

HILLE, E., and PHILLIPS, R. S.

[1] *Functional analysis and semigroups*, Amer. Math. Soc. Coll. Publ. 31, rev. ed., Amer. Math. Soc., Providence, R. I., 1957. MR 19, 664.

HILLE, E., and TAMARKIN, J. D.

[1] On the characteristic values of linear integral equations, *Acta Math.* **57**(1931), 1–76.

HIRSCHFELD, R.

[1] Reprèsentations induites dans les espaces quasi-hilbertiens parfaits, *C. R. Acad. Sci. Paris Ser. A.-B* **272**(1971), A104–A106. MR 42 #7829.

[2] Projective limits of unitary representations, *Math. Ann.* **194**(1971), 180–196. MR 45 #6979.

[3] Duality of groups: a narrative review, *Nieuw Arch. Wisk.* (3) **20**(1972), 231–241. MR 49 #5226.

HOCHSCHILD, G.

[1] Cohomology and representations of associative algebras, *Duke. Math. J.* **14**(1947), 921–948. MR 9, 267.

HOFFMAN, K.

[1] *Banach spaces of analytic functions*, Prentice-Hall, Englewood Cliffs, N.J. 1962. MR 24 #A2844.

HOFMANN, K. H.

[1] *The duality of compact semigroups and C*-bigebras*, Lecture Notes in Mathematics, **129**, Springer-Verlag, Berlin-Heidelberg-New York, 1970. MR 55 #5786.

[2] Representations of algebras by continuous sections, *Bull. Amer. Math. Soc.* **78**(1972), 291–373. MR 50 #415.

[3] Banach bundles, Darmstadt Notes, 1974.

[4] Bundles and sheaves are equivalent in the category of Banach spaces, pp. 53–69, Lecture Notes in Mathematics, **575**, Springer-Verlag, Berlin-Heidelberg-New York, 1977.

HOFMANN, K. H., and LIUKKONEN, J. R., EDS.

[1] *Recent advances in the representation theory of rings and C*-algebras by continuous sections*, Memoirs Amer. Math. Soc. **148**, Amer. Math. Soc. Providence, R. I., 1974. MR 49 #7063.

HOLZHERR, A. K.

[1] Discrete groups whose multiplier representations are type I, *J. Austral. Math. Soc. Ser. A* **31**(1981), 486–495. MR 83g: 22003.

[2] Groups with finite dimensional irreducible multiplier representations, *Canad. J. Math.* **37**(1985), 635–643. MR 87a: 22010.

HOWE, R.

[1] On Frobenius reciprocity for unipotent algebraic groups over Q, *Amer. J. Math.* **93**(1971), 163–172. MR 43 #7556.

[2] Representation theory for division algebras over local fields (tamely ramified case), *Bull. Amer. Math. Soc.* **77**(1971), 1063–1066. MR 44 #4146.

[3] The Brauer group of a compact Hausdorff space and n-homogeneous C*-algebras, *Proc. Amer. Math. Soc.* **34**(1972), 209–214. MR 46 #4218.

[4] On representations of discrete, finitely generated, torsion-free, nilpotent groups, *Pacific J. Math.* **73**(1977), 281–305. MR 58 #16984.

[5] The Fourier transform for nilpotent locally compact groups, I. *Pacific J. Math.* **73**(1977), 307–327. MR 58 #11215.

[6] On the role of the Heisenberg group in harmonic analysis, *Bull. Amer. Math. Soc. N.S.* (2) **3**(1980), 821–843. MR 81h: 22010.

HOWE, R., and MOORE, C. C.

[1] Asymptotic properties of unitary representations, *J. Functional Anal.* **32**(1979), 72–96. MR 80g: 22017.

HULANICKI, A.

[1] Groups whose regular representation weakly contains all unitary representations, *Studia Math.* **24**(1964), 37–59. MR 33 #225.

HULANICKI, A., and PYTLIK, T.

[1] On cyclic vectors of induced representations, *Proc. Amer. Math. Soc.* **31**(1972), 633–634. MR 44 #6905.

[2] Corrigendum to: "On cyclic vectors of induced representations," *Proc. Amer. Math. Soc.* **38**(1973), 220. MR 48 #6318.

[3] On commutative approximate identities and cyclic vectors of induced representations, *Studia Math.* **48**(1973), 189–199. MR 49 #3024.

HURWITZ, A.

[1] Über die Erzeugung der Invarianten durch Integration, *Nachr. Ges. Gott. Math.-Phys. K1.* 1897, 71–90 (also Math. Werke. Vol. II, pp. 546–564. Birkhäuser, Basel, 1933).

HUSEMOLLER, D.

[1] *Fiber bundles*, Mcgraw-Hill, New York, 1966. MR 37 #4821.

HUSAIN, T.

[1] *Introduction to topological groups*, W. B. Saunders, Philadelphia-London, 1966. MR 34 #278.

IMAI, S., and TAKAI, H.

[1] On a duality for C^*-crossed products by a locally compact group, *J. Math. Soc. Japan* **30**(1978), 495–504. MR 81h: 46090.

IORIO, V. M.

[1] Hopf-C^*-algebras and locally compact groups, *Pacific J. Math.* **87**(1980), 75–96. MR 82b: 22007.

ISMAGILOV, R. C.

[1] On the unitary representations of the group of diffeomorphisms of the space R^n, $n > 2$, *J. Functional Anal.* **9**(1975), 71–72. MR 51 #14143.

JACOBSON, N.

[1] Structure theory of simple rings without finiteness assumptions, *Trans. Amer. Math. Soc.* **57**(1945), 228–245. MR 6, 200.

[2] The radical and semi-simplicity for arbitrary rings, *Amer. J. Math.* **67**(1945), 300–320. MR 7, 2.

[3] A topology for the set of primitive ideals in an arbitrary ring, *Proc. Nat. Acad. Sci. U.S.A.* **31**(1945), 333–338. MR 7, 110.

[4] *Structure of rings*, Amer. Math. Soc. Colloq. Publ. **37**, Amer. Math. Soc. Providence, R. I. 1956. MR 18, 373.

JACQUET, H., and SHALIKA, J.

[1] The Whittaker models of induced representations, *Pacific J. Math.* **109**(1983), 107–120. MR 85h: 22023.

JENKINS, J. W.

[1] An amenable group with a nonsymmetric group algebra, *Bull. Amer. Math. Soc.* **75**(1969), 359–360. MR 38 #6366.

[2] Free semigroups and unitary group representations, *Studia Math.* **43**(1972), 27–39. MR 47 #395.

JOHNSON, B. E.

[1] An introduction to the theory of centralizers. *Proc. Lond. Math. Soc.* **14**(1964), 229–320. MR 28 #2450.

[2] *Cohomology in Banach algebras*, Memoirs Amer. Math. Soc. **127**, Amer. Math. Soc. Providence, R. I. 1972. MR 51 #11130.

JOHNSON, G. P.

[1] Spaces of functions with values in a Banach algebra, *Trans. Amer. Math. Soc.* **92**(1959), 411–429. MR 21 #5910.

JORGENSEN, P. E. T.

[1] Perturbation and analytic continuation of group representations, *Bull. Amer. Math. Soc.* **82**(1976), 921–924. MR 58 #1026.

[2] Analytic continuation of local representations of Lie groups, *Pacific J. Math.* **125**(1986), 397–408.

JOY, K. I.

[1] A description of the topology on the dual space of a nilpotent Lie group, *Pacific J. Math.* **112**(1984), 135–139. MR 85e: 22013.

KAC, G. I.

[1] Generalized functions on a locally compact group and decompositions of unitary representations, *Trudy Moskov. Mat. Obšč.* **10**(1961), 3–40 (Russian). MR 27 #5863.

[2] Ring groups and the principle of duality, I. *Trudy Moskov. Mat. Obšč.* **12**(1963), 259–301 (Russian). MR 28 #164.

[3] Ring groups and the principle of duality, II. *Trudy Moskov. Mat. Obšč.* **13**(1965), 84–113 (Russian). MR 33 #226.

KADISON, R. V.

[1] *A representation theory for commutative topological algebra*, Memoirs Amer. Math. Soc., 7, Amer. Math. Soc., Providence, R. I., 1951. MR 13, 360.

[2] Isometries of operator algebras, *Ann. Math.* **54**(1951), 325–338. MR 13, 256.

[3] Order properties of bounded self-adjoint operators, *Proc. Amer. Math. Soc.* **2**(1951), 505–510. MR 13, 47.

[4] Infinite unitary groups, *Trans. Amer. Math. Soc.* **72**(1952), 386–399. MR 14, 16.

[5] Infinite general linear groups, *Trans. Amer. Math. Soc.* **76**(1954), 66–91. MR 15, 721.

[6] On the orthogonalization of operator representations, *Amer. J. Math.* **77**(1955), 600–621. MR 17, 285.

[7] Operator algebras with a faithful weakly closed representation, *Ann. Math.* **64**(1956), 175–181. MR 18, 54.

[8] Unitary invariants for representations of operator algebras, *Ann. Math.* **66**(1957), 304–379. MR 19, 665.

[9] Irreducible operator algebras, *Proc. Nat. Acad. Sci. U.S.A.* **43**(1957), 273–276. MR 19, 47.

[10] States and representations, *Trans. Amer. Math. Soc.* **103**(1962), 304–319. MR 25 #2459.

[11] Normalcy in operator algebras, *Duke. Math. J.* **29**(1962), 459–464. MR 26 #6814.

[12] Transformations of states in operator theory and dynamics, *Topology* **3**(1965), 177–198. MR 29 #6328.

[13] Strong continuity of operator functions, *Pacific J. Math.* **26**(1968), 121–129. MR 37 #6766.

[14] Theory of operators, Part II: operator algebras, *Bull. Amer. Math. Soc.* **64**(1958), 61–85.

[15] Operator algebras—the first forty years, in *Operator algebras and applications*, pp. 1–18, Proc. Symp. Pure Math., Amer. Math. Soc. Providence, R. I. 1982.

KADISON, R. V., and PEDERSEN, G. K.

[1] Means and convex combinations of unitary operators, *Math. Scand.* **57**(1985), 249–266. MR 87g: 47078.

KADISON, R. V., and RINGROSE, J. R.

[1] *Fundamentals of the theory of operator algebras*, vol. 1, *Elementary theory*, Pure and Applied Mathematics, **100**, Academic Press, New York-London, 1983. MR 85j: 46099.

[2] *Fundamentals of the theory of operator algebras*, vol. 2, *Advanced theory*, Pure and Applied Mathematics, **100**, Academic Press, New York-London, 1986.

KAJIWARA, T.

[1] Group extension and Plancherel formulas, *J. Math. Soc. Japan* **35**(1983), 93–115. MR 84d: 22011.

KAKUTANI, S., and KODAIRA, K.

[1] Über das Haarsche Mass in der lokal bikompakten Gruppe, *Proc. Imp. Acad. Tokyo* **20**(1944), 444–450. MR 7, 279.

KALLMAN, R. R.

[1] A characterization of uniformly continuous unitary representations of connected locally compact groups, *Michigan Math. J.* **16**(1969), 257–263. MR 40 #5787.

[2] Unitary groups and automorphisms of operator algebras, *Amer. J. Math.* **91**(1969), 785–806. MR 40 #7825.

[3] A generalization of a theorem of Berger and Coburn, *J. Math. Mech.* **19**(1969/70), 1005–1010. MR 41 #5982.

[4] Certain topological groups are type I., *Bull. Amer. Math. Soc.* **76**(1970), 404–406. MR 41 #385.

[5] Certain topological groups are type I. II., *Adv. Math.* **10**(1973), 221–255.

[6] A theorem on the restriction of type I representations of a group to certain of its subgroups, *Proc. Amer. Math. Soc.* **40**(1973), 291–296. MR 47 #5177.

[7] The existence of invariant measures on certain quotient spaces, *Adv. Math.* **11**(1973), 387–391. MR 48 #8682.

[8] Certain quotient spaces are countably separated, I., *Illinois J. Math.* **19**(1975), 378–388. MR 52 #650.

[9] Certain quotient spaces are countably separated, II, *J. Functional Anal.* **21**(1976), 52–62. MR 54 #5384.

[10] Certain quotient spaces are countably separated, III. *J. Functional Anal.* **22**(1976), 225–241. MR 54 #5385.

KAMPEN, E. R. VAN.

[1] Locally bicompact Abelian groups and their character groups, *Ann. Math.* (2) **36**(1935), 448–463.

KANIUTH, E.

[1] Topology in duals of SIN-groups, *Math. Z.* **134**(1973), 67–80. MR 48 #4197.

[2] On the maximal ideals in group algebras of SIN-groups, *Math. Ann.* **214**(1975), 167–175. MR 52 #6325.

[3] A note on reduced duals of certain locally compact groups, *Math. Z.* **150**(1976), 189–194. MR 54 #5386.

[4] On separation in reduced duals of groups with compact invariant neighborhoods of the identity, *Math. Ann.* **232**(1978), 177–182. MR 58 #6053.

[5] Primitive ideal spaces of groups with relatively compact conjugacy classes, *Arch. Math. (Basel)* **32**(1979), 16–24. MR 81f: 22009.

[6] Ideals in group algebras of finitely generated FC-nil potent discrete groups, *Math. Ann.* **248**(1980), 97–108. MR 81i: 43005.

[7] Weak containment and tensor products of group representations, *Math. Z.* **180**(1982), 107–117. MR 83h: 22013.

[8] On primary ideals in group algebras, *Monatsh. Math.* **93**(1982), 293–302. MR 84i: 43003.

[9] Weak containment and tensor products of group representations, II, *Math. Ann.* **270**(1985), 1–15. MR 86j: 22004.

[10] On topological Frobenius reciprocity for locally compact groups, *Arch. Math.* **48**(1987), 286–297.

KAPLANSKY, I.

[1] Topological rings, *Amer., J. Math.* **69**(1947), 153–183. MR 8, 434.

[2] Locally compact rings. I, *Amer. J. Math.* **70**(1948), 447–459. MR 9, 562.

[3] Dual rings, *Ann. of Math.* **49**(1948), 689–701. MR 10, 7.

[4] Normed algebras, *Duke Math. J.* **16**(1949), 399–418. MR 11, 115.

[5] Locally compact rings. II, *Amer. J. Math.* **73**(1951), 20–24. MR 12, 584.

[6] The structure of certain operator algebras, *Trans. Amer. Math. Soc.* **70**(1951), 219–255. MR 13, 48.

[7] Group algebras in the large, *Tohoku Math. J.* **3**(1951), 249–256. MR 14, 58.

[8] Projections in Banach algebras, *Ann. Math.* (2) **53**(1951), 235–249. MR 13, 48.

[9] A theorem on rings of operators, *Pacific J. Math.* **1**(1951), 227–232. MR 14, 291.

[10] Modules over operator algebras, *Amer. J. Math.* **75**(1953), 839–858. MR 15, 327.

KARPILOVSKY, G.

[1] *Projective representations of finite groups*, Pure and Applied Mathematics, 84, Marcel-Dekker, New York, 1985.

KASPAROV, G. G.

[1] Hilbert C^*-modules: theorems of Stinespring and Voiculescu, *J. Operator Theory*, **4**(1980), 133–150. MR 82b: 46074.

KATAYAMA, Y.

[1] Takesaki's duality for a non-degenerate co-action, *Math. Scand.* **55**(1984), 141–151. MR 86b: 46112.

KATZNELSON, Y., and WEISS, B.

[1] The construction of quasi-invariant measures, *Israel J. Math.* **12**(1972), 1–4. MR 47 #5226.

KAWAKAMI, S.

[1] Irreducible representations of nonregular semi-direct product groups, *Math. Japon.* **26**(1981), 667–693. MR 84a: 22015.

[2] Representations of the discrete Heisenberg group, *Math. Japon.* **27**(1982), 551–564. MR 84d: 22015.

KAWAKAMI, S., and KAJIWARA, T.

[1] Representations of certain non-type I *C**-crossed products, *Math. Japon.* **27**(1982), 675–699. MR 84d: 22013.

KEHLET, E. T.

[1] A proof of the Mackey-Blattner-Nielson theorem, *Math. Scand.* **43**(1978), 329–335. MR 80i: 22019.

[2] A non-separable measurable choice principle related to induced representations, *Math. Scand.* **42**(1978), 119–134. MR 58 #16954.

[3] On extensions of locally compact groups and unitary groups, *Math. Scand.* **45**(1979), 35–49. MR 81d: 22009.

[4] Cross sections for quotient maps of locally compact groups, *Math. Scand.* **55**(1984), 152–160. MR 86k: 22011.

KELLEY, J. L.

[1] *General topology*, D. Van Nostrand Co., Toronto-New York-London, 1955. MR 16, 1136.

KELLEY, J. L., NAMIOKA, I., ET AL.

[1] *Linear topological spaces*, D. Van Nostrand Co., Princeton, 1963. MR 29 #3851.

KELLEY, J. L., and SRINAVASAN, T. P.

[1] *Measures and integrals*, Springer Graduate Series, Springer, New York, in Press.

KELLEY, J. L., and VAUGHT, R. L.

[1] The positive cone in Banach algebras, *Trans. Amer. Math. Soc.* **74**(1953), 44–55. MR 14, 883.

KEOWN, R.

[1] *An introduction to group representation theory*, Academic Press, New York, 1975. MR 52 #8230.

KHOLEVO, A. S.

[1] Generalized imprimitivity systems for abelian groups. *Izv. Vyssh. Uchebn. Zaved. Mat.* (1983), 49–71 (Russian). MR 84m: 22008.

KIRCHBERG, E.

[1] Representations of coinvolutive Hopf-*W**-algebras and non-abelian duality, *Bull. Acad. Polon. Sci. Sér. Sci. Math. Astronom. Phys.* **25**(1977), 117–122. MR 56 #6415.

KIRILLOV, A. A.

[1] Induced representations of nilpotent Lie groups, *Dokl. Akad. Nauk SSSR* **128**(1959), 886–889 (Russian). MR 22 #740.

[2] On unitary representations of nilpotent Lie groups, *Dokl Akad. Nauk. SSSR* **130**(1960), 966–968 (Russian). MR 24 #A3240.

[3] Unitary representations of nilpotent Lie groups, Dokl. Akad. Nauk. SSSR **138**(1961), 283–284 (Russian). MR 23 #A3205.

[4] Unitary representations of nilpotent Lie groups, *Uspehi Mat. Nauk* **106**(1962), 57–110 (Russian). MR 25 #5396.

[5] Representations of certain infinite-dimensional Lie groups, *Vestnik Moskov . Univ. Ser. I. Mat. Meh.* **29**(1974), 75–83 (Russian). MR 52 #667.

[6] *Elements of the theory of representations*, 2nd Ed. Nauka. Moscow, 1978 (Russian). MR 80b: 22001. [English translation of first ed.: Grundlehren der math. Wiss. 220, Springer-Verlag, Berlin-Heildelberg-New York, 1976].

KISHIMOTO, A.

[1] Simple crossed products of C^*-algebras by locally compact abelian groups (preprint), University of New South Wales, Kensington, 1979.

KITCHEN, J. W., and ROBBINS, D. A.

[1] Tensor products of Banach bundles, *Pacific J. Math.* **94**(1981), 151–169. MR 83f: 46078.

[2] Sectional representations of Banach modules, *Pacific J. Math.* **109**(1983), 135–156. MR 85a: 46026.

[3] Internal functionals and bundle duals, *Int. J. Math. Sci.* **7**(1984), 689–695. MR 86i: 46077.

KLEPPNER, A.

[1] The structure of some induced representations, *Duke Math. J.* **29**(1962), 555–572. MR 25 #5132.

[2] Intertwining forms for summable induced representations, *Trans. Amer. Math. Soc.* **112**(1964), 164–183. MR 33 #4186.

[3] Multipliers on abelian groups, *Math. Ann.* **158**(1965), 11–34. MR 30 #4856.

[4] Representations induced from compact subgroups, *Amer. J. Math.* **88**(1966), 544–552. MR 36 #1577.

[5] Continuity and measurability of multiplier and projective representations, *J. Functional Anal.* **17**(1974), 214–226. MR 51 #790.

[6] Multiplier representations of discrete groups, *Proc. Amer. Math. Soc.* **88**(1983), 371–375. MR 84k: 22007.

KLEPPNER, A., and LIPSMAN, R. L.

[1] The Plancherel formula for group extensions. I. II., *Ann. Sci. École. Norm. Sup.* (4) **5**(1972), 459–516; ibid. (4) **6**(1973), 103–132. MR 49 #7387.

KNAPP, A. W.

[1] *Representation theory of semisimple groups, an overview based on examples*, Princeton Mathematical Series, **36**, Princeton University Press, Princeton, N. J., 1986.

KOBAYASHI, S.

[1] On automorphism groups of homogeneous complex manifolds, *Proc. Amer. Math. Soc.* **12**(1961), 359–361. MR 24 #A3664.

KOOSIS, P.

[1] An irreducible unitary representation of a compact group is finite dimensional, *Proc. Amer. Math. Soc.* **8**(1957), 712–715. MR 19, 430.

KOPPINEN, M., and NEUVONEN, T.

[1] An imprimitivity theorem for Hopf algebras, *Math. Scand.* **41**(1977), 193–198. MR 58 #5758.

KOORNWINDER, T. H., and VAN DER MEER, H. A.

[1] Induced representations of locally compact groups, *Representations of locally compact groups with applications*, Colloq. Math. Centre, Amsterdam, 1977/78, pp. 329–376, MC syllabus, **38**, Math. Centrum, Amsterdam, 1979. MR 81k: 22001.

KOTZMANN, E., LOSERT, V., and RINDLER, H.

[1] Dense ideals of group algebras, *Math. Ann.* **246**(1979/80), 1–14. MR 80m: 22007.

KRALJEVIĆ, H.

[1] Induced representations of locally compact groups on Banach spaces, *Glasnik Math.* **4**(1969), 183–196. MR 41 #1928.

KRALJEVIĆ, H., and MILIČIĆ, D.

[1] The C^*-algebra of the universal covering group of $SL(2, R)$, *Glasnik Mat. Ser. III* (27) **7**(1972), 35–48. MR 49 #5222.

KRIEGER, W.

[1] On quasi-invariant measures in uniquely ergodic systems, *Invent. Math.* **14**(1971), 184–196. MR 45 #2139.

KUGLER, W.

[1] Über die Einfachheit gewisser verallgemeinerter L^1-Algebren, *Arch. Math. (Basel)* **26**(1975), 82–88. MR 52 #6326.

[2] On the symmetry of generalized L^1-algebras, *Math. Z.* **168**(1979), 241–262. MR 82f: 46074.

KUMJIAN, A.

[1] On C^*-diagonals, *Can. J. Math.* **38**(1986), 969–1008.

KUNZE, R. A.

[1] L_p Fourier transforms on locally compact unimodular groups, *Trans. Amer. Math. Soc.* **89**(1958), 519–540. MR 20 #6668.

[2] On the irreducibility of certain multiplier representations, *Bull. Amer. Math. Soc.* **68**(1962), 93–94. MR 24 #A3239.

[3] *Seminar: Induced representations without quasi-invariant measures.* Summer School on Representations of Lie Groups, Namur, 1969, pp. 283–289. Math. Dept., Univ. of Brussels, Brussels, 1969. MR 57 #9904.

[4] On the Frobenius reciprocity theorem for square-integrable representations, *Pacific J. Math.* **53**(1974), 465–471. MR 51 #10530.

[5] Quotient representations, *Topics in modern harmonic analysis*, Vol. I, II, Turin/Milan, 1982, pp. 57–80, Ist. Naz. Alta Mat. Francesco Severi, Rome, 1983. MR 85j: 22009.

KUNZE, R. A., and STEIN, E. M.

[1] Uniformly bounded representations and harmonic analysis on the $n \times n$ real unimodular group, *Amer. J. Math.* **82**(1960), 1–62. MR 29 #1287.

[2] Uniformly bounded representations. II. Analytic continuation of the principal series of representations of the $n \times n$ complex unimodular group, *Amer. J. Math.* **83**(1961), 723–786. MR 29 #1288.

KURATOWSKI, K.

[1] *Introduction to set theory and topology*, Pergamon Press, Oxford-Warsaw, 1972. MR 49 #11449.

KUSUDA, M.

[1] Crossed products of C^*-dynamical systems with ground states, *Proc. Amer. Math. Soc.* **89**(1983), 273–278.

KUTZKO, P. C.

[1] Mackey's theorem for nonunitary representations, *Proc. Amer. Math. Soc.* **64**(1977), 173–175. MR 56 #533.

LANCE, E. C.

[1] Automorphisms of certain operator algebras, *Amer. J. Math.* **91**(1969), 160–174. MR 39 #3324.

[2] On nuclear C^*-algebras, *J. Functional Anal.* **12**(1973), 157–176. MR 49 #9640.

[3] Refinement of direct integral decompositions, *Bull. London Math. Soc.* **8**(1976), 49–56. MR 54 #8307.

[4] Tensor products of non-unital C^*-algebras, *J. London Math. Soc.* **12**(1976), 160–168. MR 55 #11059.

LANDSTAD, M.

[1] Duality theory for covariant systems, Thesis, University of Pennsylvania, 1974.

[2] Duality for dual covariant algebras, *Comm. Math. Physics* **52**(1977), 191–202. MR 56 #8750.

[3] Duality theory for covariant systems, *Trans. Amer. Math. Soc.* **248**(1979), 223–269. MR 80j: 46107.

[4] Duality for dual C^*-covariance algebras over compact groups, University of Trondheim (preprint), 1978.

LANDSTAD, M. B., PHILLIPS, J., RAEBURN, I., and SUTHERLAND, C. E.

[1] Representations of crossed products by coactions and principal bundles, *Trans. Amer. Math. Soc.* **299**(1987), 747–784.

LANG, S.

[1] *Algebra*, Addison-Wesley Publishing Co., Reading, Mass., 1965. MR 33 #5416.

[2] $SL_2(R)$, Addison-Wesley Publishing Co., Reading, Mass., 1975. MR 55 #3170.

LANGE, K.

[1] A reciprocity theorem for ergodic actions, *Trans. Amer. Math. Soc.* **167**(1972), 59–78. MR 45 #2085.

LANGE, K., RAMSAY, A., and ROTA, G. C.

[1] Frobenius reciprocity in ergodic theory, *Bull. Amer. Math. Soc.* **77**(1971), 713–718. MR 44 #1769.

LANGLANDS, R.

[1] The Dirac monopole and induced representations, *Pacific J. Math.* **126**(1987), 145-151.

LANGWORTHY, H. F.

[1] Imprimitivity in Lie groups, Ph.D. thesis, University of Minnesota, 1970.

LARSEN, R.

[1] *An introduction to the theory of multipliers*, Springer-Verlag New York-Heidelberg, 1971. MR 55 #8695.

LEBESGUE, H.

[1] Sur les séries trigonométriques, *Ann. École. Norm. Sup.* (3) **20**(1903), 453-485.

LEE, R. Y.

[1] On the C^*-algebras of operator fields, *Indiana Univ. Math. J.* **25**(1976), 303-314. MR 53 #14150.

[2] Full algebras of operator fields trivial except at one point, *Indiana Univ. Math. J.* **26**(1977), 351-372. MR 55 #3812.

[3] On C^*-algebras which are approachable by finite-dimensional C^*-algebras, *Bull. Inst. Math. Acad. Sinica* **5**(1977), 265-283. MR 58 #17864.

LEE, T. Y.

[1] Embedding theorems in group C^*-algebras, *Canad. Math. Bull.* **26**(1983), 157-166. MR 85a: 22012.

DE LEEUW, K., and GLICKSBERG, I.

[1] The decomposition of certain group representations, *J. d'Analyse Math.* **15**(1965), 135-192. MR 32 #4211.

LEINERT, M.

[1] Fell-Bündel und verallgemeinerte L^1-Algebren, *J. Functional Anal.* **22**(1976), 323-345. MR 54 #7694.

LEJA, F.

[1] Sur la notion du groupe abstract topologique, *Fund. Math.* **9**(1927), 37-44.

LEPTIN, H.

[1] Verallgemeinerte L^1-Algebren, *Math. Ann.* **159**(1965), 51-76. MR 34 #6545.

[2] Verallgemeinerte L^1-Algebren und projektive Darstellungen lokal kompakter Gruppen. I, *Invent. Math.* **3**(1967), 257-281. MR 37 #5328.

[3] Verallgemeinerte L^1-Algebren, und projektive Darstellungen lokal kompakter Gruppen. II, *Invent. Math.* **4**(1967), 68-86. MR 37 #5328.

[4] Darstellungen verallgemeinerter L^1-Algebren, *Invent. Math.* **5**(1968), 192-215. MR 38 #5022.

[5] *Darstellungen verallgemeinerter L^1-Algebren.* II, pp. 251-307, Lecture Notes in Mathematics **247**, Springer-Verlag, Berlin and New York, 1972. MR 51 #6433.

[6] The structure of $L^1(G)$ for locally compact groups, *Operator Algebras and Group Representations*, pp. 48-61, Pitman Publishing Co., Boston-London, 1984. MR 85d: 22015.

LIPSMAN, R. L.

[1] *Group representations*, Lecture Notes in Mathematics **388**, Springer-Verlag, Berlin and New York, 1974. MR 51 #8333.

[2] Non-Abelian Fourier analysis, *Bull. Sci. Math.* (2) **98**(1974), 209–233. MR 54 #13467.

[3] Harmonic induction on Lie groups, *J. Reine Angew. Math.* **344**(1983), 120–148. MR 85h: 22021.

[4] Generic representations are induced from square-integrable representations, *Trans. Amer. Math. Soc.* **285**(1984), 845–854. MR 86c: 22022.

[5] An orbital perspective on square-integrable representations, *Indiana Univ. Math.* **34**(1985), 393–403.

LITTLEWOOD, D. E.

[1] *Theory of group characters and matrix representations of groups*, Oxford University Press, New York, 1940. MR 2, 3.

LITVINOV, G. L.

[1] Group representations in locally convex spaces, and topological group algebras, *Trudy Sem. Vektor. Tenzor. Anal.* **16**(1972), 267–349 (Russian). MR 54 #10479.

LIUKKONEN, J. R.

[1] Dual spaces of groups with precompact conjugacy classes, *Trans. Amer. Math. Soc.* **180**(1973), 85–108. MR 47 #6937.

LIUKKONEN, J. R., and MOSAK, R.

[1] The primitive ideal space of [FC]⁻ groups, *J. Functional Anal.* **15**(1974), 279–296. MR 49 #10814.

[2] Harmonic analysis and centers of group algebras, *Trans. Amer. Math. Soc.* **195**(1974), 147–163. MR 50 #2815.

LONGO, R.

[1] Solution of the factorial Stone-Weierstrass conjecture. An application of the theory of standard split W^*-inclusions, *Invent. Math.* **76**(1984), 145–155. MR 85m: 46057a.

LONGO, R. and PELIGRAD, C.

[1] Noncommutative topological dynamics and compact actions on C^*-algebras, *J. Functional Anal.* **58**(1984), 157–174. MR 86b: 46114.

LOOMIS, L. H.

[1] Haar measure in uniform structures, *Duke Math. J.* **16**(1949), 193–208. MR 10, 600.

[2] *An introduction to abstract harmonic analysis*, D. Van Nostrand Co., Toronto-New York-London, 1953. MR 14, 883.

[3] Positive definite functions and induced representations, *Duke Math. J.* **27**(1960), 569–579. MR 26 #6303.

LOSERT, V., and RINDLER, H.

[1] Cyclic vectors for $L^p(G)$, *Pacific J. Math.* **89**(1980), 143–145. MR 82b: 43006.

LUDWIG, J.

[1] Good ideals in the group algebra of a nilpotent Lie group, *Math. Z.* **161**(1978), 195–210. MR 58 #16958.

[2] Prime ideals in the C^*-algebra of a nilpotent group, *Mh. Math.* **101**(1986), 159–165.

MACKEY, G. W.

[1] On a theorem of Stone and von Neumann, *Duke Math. J.* **16**(1949), 313–326. MR 11, 10.

[2] Imprimitivity for representations of locally compact groups, I, *Proc. Nat. Acad. Sci. U.S.A.* **35**(1949), 537–545. MR 11, 158.

[3] Functions on locally compact groups, *Bull. Amer. Math. Soc.* **56**(1950), 385–412 (survey article). MR 12, 588.

[4] On induced representations of groups, *Amer. J. Math.* **73**(1951), 576–592. MR 13, 106.

[5] Induced representations of locally compact groups. I, *Ann. Math.* **55**(1952), 101–139. MR 13, 434.

[6] Induced representations of locally compact groups. II, The Frobenius theorem, *Ann. Math.* **58**(1953), 193–220. MR 15, 101.

[7] Borel structure in groups and their duals, *Trans. Amer. Math. Soc.* **85**(1957), 134–165. MR 19, 752.

[8] Unitary representations of group extensions. I, *Acta Math.* **99**(1958), 265–311. MR 20 #4789.

[9] *Induced representations and normal subgroups*, Proc. Int. Symp. Linear Spaces, Jerusalem, pp. 319–326, Pergamon Press, Oxford, 1960. MR 25 #3118.

[10] Infinite dimensional group representations, *Bull. Amer. Math. Soc.* **69**(1963), 628–686. MR 27 #3745.

[11] Ergodic theory, group theory, and differential geometry, *Proc. Nat. Acad. Sci. U.S.A.* **50**(1963), 1184–1191. MR 29 #2325.

[12] *Group representations and non-commutative harmonic analysis with applications to analysis, number theory, and physics*, Mimeographed notes, University of California, Berkeley, 1965.

[13] Ergodic theory and virtual groups, *Math. Ann.* **166**(1966), 187–207. MR 34 #1444.

[14] *Induced representations of groups and quantum mechanics*, W. A. Benjamin, New York-Amsterdam, 1968. MR 58 #22373.

[15] *The theory of unitary group representations.* Based on notes by J. M. G. Fell and D. B. Lowdenslager of lectures given at University of Chicago, 1955, University of Chicago Press, Chicago, 1976. MR 53 #686.

[16] Products of subgroups and projective multipliers, *Colloq. Math.* **5**(1970), 401–413. MR 50 #13370.

[17] *Ergodicity in the theory of group representations*, Actes du Congrès International des Mathématiciens, Nice, 1970, Tome 2, pp. 401–405, Gauthier-Villars, Paris, 1971. MR 55 #8247.

[18] *Induced representations of locally compact groups and applications: Functional analysis and related fields*, Proc. Conf. M. Stone, University of Chicago, 1968, pp. 132–166, Springer, New York, 1970. MR 54 #12968.

[19] *On the structure of the set of conjugate classes in certain locally compact groups*, Symp. Math. **16**, Conv. Gruppi Topol. Gruppi Lie, INDAM, Rome, 1974, pp. 433–467, Academic Press, London, 1975. MR 53 #8323.

[20] Ergodic theory and its significance for statistical mechanics and probability theory, *Adv. Math.* **12**(1974), 178–268. MR 49 #10857.

[21] *Unitary group representations in physics, probability, and number theory*, Benjamin/Cummings, Reading, Mass., 1978. MR 80i: 22001.

[22] *Origins and early history of the theory of unitary group representations*, pp. 5–19, London Math. Soc., Lecture Note Series, **34**, Cambridge University Press, Cambridge, 1979. MR 81j: 22001.

[23] Harmonic analysis as the exploitation of symmetry—a historical survey, *Bull. Amer. Math. Soc.* **3**(1980), 543–698. MR 81d: 01019.

MAGYAR, Z., and SEBESTYÉN, Z.

[1] On the definition of C^*-algebras, II, *Canad. J. Math.* **37**(1985), 664–681. MR 87b: 46061.

MARÉCHAL, O.

[1] Champs mesurables d'espaces hilbertiens, *Bull. Sci. Math.* **93**(1969), 113–143. MR 41 #5948.

MARTIN, R. P.

[1] Tensor products for $SL(2, k)$ *Trans. Amer. Math. Soc.* **239**(1978), 197–211. MR 80i: 22033.

MASCHKE, H.

[1] Beweis des Satzes, dass diejenigen endlichen linearen Substitutionsgruppen, in welchen einige durchgehends verschwindende Koeffizienten auftreten, intransitiv sind, *Math. Ann.* **52**(1899), 363–368.

MATHEW, J., and NADKARNI, M. G.

[1] On systems of imprimitivity on locally compact abelian groups with dense actions, *Ann. Inst. Fourier (Grenoble)* **28**(1978), 1–23. MR 81j: 22002.

MAUCERI, G.

[1] Square integrable representations and the Fourier algebra of a unimodular group, *Pacific J. Math.* **73**(1977), 143–154. MR 58 #6054.

MAUCERI, G., and RICARDELLO, M. A.

[1] Noncompact unimodular groups with purely atomic Plancherel measures, *Proc. Amer. Math. Soc.* **78**(1980), 77–84. MR 81h: 22005.

MAURIN, K.

[1] Distributionen auf Yamabe-Gruppen, Harmonische Analyse auf einer Abelschen l.k. Gruppe, *Bull. Acad. Polon. Sci. Sér. Sci. Math. Astronom. Phys.* **9**(1961), 845–850. MR 24 #A2839.

MAURIN, K., and MAURIN, L.

[1] Duality, imprimitivity, reciprocity, *Ann. Polon. Math.* **29**(1975), 309–322. MR 52 #8330.

MAUTNER, F. I.

[1] Unitary representations of locally compact groups. I, *Ann. Math.* **51**(1950), 1–25. MR 11, 324.

[2] Unitary representations of locally compact groups. II, *Ann. Math.* **52**(1950), 528–556. MR 12, 157.

[3] The structure of the regular representation of certain discrete groups, *Duke Math. J.* **17**(1950), 437–441. MR 12, 588.

[4] The regular representation of a restricted direct product of finite groups, *Trans. Amer. Math. Soc.* **70**(1951), 531–548. MR 13, 11.

[5] Fourier analysis and symmetric spaces, *Proc. Nat. Acad. Sci. U.S.A.* **37**(1951), 529–533. MR 13, 434.

[6] A generalization of the Frobenius reciprocity theorem, *Proc. Nat. Acad. Sci. U.S.A.* **37**(1951), 431–435. MR 13, 205.

[7] Spherical functions over p-adic fields, *Amer. J. Math.* **80**(1958), 441–457. MR 20, #82.

MAYER, M. E.

[1] *Differentiable cross sections in Banach *-algebraic bundles*, pp. 369–387, Cargése Lectures, 1969, D. Kastler, ed., Gordon & Breach, New York, 1970. MR 58 #31188.

[2] Automorphism groups of C^*-algebras, Fell bundles, W^*-bigebras, and the description of internal symmetries in algebraic quantum theory, *Acta Phys. Austr. Suppl.* **8**(1971), 177–226, Springer-Verlag, 1971. MR 48 #1596.

MAYER-LINDENBERG, F.

[1] Invariante Spuren auf Gruppenalgebren, *J. Reine. Angew. Math.* **310**(1979), 204–213. MR 81b: 22008.

[2] Zur Dualitätstheorie symmetrischer Paare, *J. Reine Angew. Math.* **321**(1981), 36–52. MR 83a: 22009.

MAZUR, S.

[1] Sur les anneaux linéaires, *C. R. Acad. Sci. Paris* **207**(1938), 1025–1027.

MICHAEL, E. A.

[1] *Locally multiplicatively convex topological algebras*, Memoirs Amer. Math. Soc. **11**, Amer. Math. Soc., Providence, R. I., 1952. MR 14, 482.

MILIČIĆ, D.

[1] Topological representation of the group C^*-algebra of $SL(2, R)$, *Glasnik Mat. Ser. III* (26) **6**(1971), 231–246. MR 46 #7909.

MINGO, J. A., and PHILLIPS, W. J.

[1] Equivalent triviality theorems for Hilbert C^*-modules. *Proc. Amer. Math. Soc.* **91**(1984), 225–230. MR 85f: 46111.

MITCHELL, W. E.

[1] The σ-regular representation of $Z \times Z$, *Michigan Math. J.* **31**(1984), 259–262. MR 86h: 220C.

MONTGOMERY, D., and ZIPPIN, L.

[1] *Topological transformation groups*, Interscience, New York, 1955. MR 17, 383.

MOORE, C. C.

[1] On the Frobenius reciprocity theorem for locally compact groups, *Pacific J. Math.* **12**(1962), 359–365. MR 25 #5134.

[2] Groups with finite-dimensional irreducible representations, *Trans. Amer. Math. Soc.* **166**(1972), 401–410. MR 46 #1960.

[3] Group extensions and cohomology for locally compact groups. III, *Trans. Amer. Math. Soc.* **221**(1976), 1–33. MR 54 #2867.

[4] Group extensions and cohomology for locally compact groups. IV, *Trans. Amer. Math. Soc.* **221**(1976), 34–58. MR 54 #2868.

[5] Square integrable primary representations, *Pacific J. Math.* **70**(1977), 413–427. MR 58 #22381.

MOORE, C. C., and REPKA, J.

[1] A reciprocity theorem for tensor products of group representations, *Proc. Amer. Math. Soc.* **64**(1977), 361–364. MR 56 #8749.

MOORE, C. C., and ROSENBERG, J.

[1] Comments on a paper of I. D. Brown and Y. Guivarc'h, *Ann. Sci. École Norm. Sup.* (4) **8**(1975), 379–381. MR 55 #570b.

[2] Groups with T_1 primitive ideal spaces, *J. Functional Anal.* **22**(1976), 204–224. MR 54 #7693.

MOORE, C. C., and ZIMMER, R. J.

[1] Groups admitting ergodic actions with generalized discrete spectrum, *Invent. Math.* **51**(1979), 171–188. MR 80m: 22008.

MOORE, R. T.

[1] *Measurable, continuous and smooth vectors for semigroups and group representations*, Memoirs Amer. Math. Soc. **78**, Amer. Math. Soc. Providence, R. I., 1968. MR 37 #4669.

MORITA, K.

[1] Duality for modules and its applications to the theory of rings with minimum condition, *Tokyo Kyoiku Daigaku Sec. A* **6**(1958), 83–142. MR 20 #3183.

MORRIS, S. A.

[1] *Pontryagin duality and the structure of locally compact abelian groups*, Cambridge University Press, Cambridge, 1977. MR 56 #529.

[2] Duality and structure of locally compact abelian groups... for the layman, *Math. Chronicle* **8**(1979), 39–56. MR 81a: 22003.

MOSAK, R. D.

[1] A note on Banach representations of Moore groups, *Resultate Math.* **5**(1982), 177–183. MR 85d: 22012.

MOSCOVICI, H.

[1] Generalized induced representations, *Rev. Roumaine Math. Pures. Appl.* **14**(1969), 1539–1551. MR 41 #3671.

[2] Topological Frobenius properties for nilpotent Lie groups, *Rev. Roumaine Math. Pures Appl.* **19**(1974), 421–425. MR 50 #540.

MOSCOVICI, H., and VERONA, A.

[1] Harmonically induced representations of nilpotent Lie groups, *Invent. Math.* **48**(1978), 61–73. MR 80a: 22011.

[2] Holomorphically induced representations of solvable Lie groups, *Bull. Sci. Math.* (2) **102**(1978), 273–286. MR 80i: 22025.

MOSKALENKO, Z. I.

[1] On the question of the locally compact topologization of a group, *Ukrain. Mat. Ž.* **30**(1978), 257–260, 285 (Russian). MR 58 #11208.

MOSTERT, P.

[1] Sections in principal fibre spaces, *Duke Math. J.* **23**(1956), 57–71. MR 57, 71.

MUELLER-ROEMER, P. R.

[1] A note on Mackey's imprimitivity theorem, *Bull. Amer. Math. Soc.* **77**(1971), 1089–1090. MR 44 #6907.

[2] Kontrahierende Erweiterungen und Kontrahierbare Gruppen, *J. Reine Angew. Math.* **283/284**(1976), 238–264. MR 53 #10977.

MUHLY, P. S., and WILLIAMS, D. P.

[1] Transformation group C^*-algebras with continuous trace. II, *J. Operator Theory*, **11**(1984), 109–124. MR 85k: 46074.

MUHLY, P. S., RENAULT, J. N., and WILLIAMS, D. P.

[1] Equivalence and isomorphism for groupoid C^*-algebras, *J. Operator Theory* **17**(1987), 3–22.

MURNAGHAN, F.

[1] *The theory of group representations*, Dover, New York, 1963 (reprint).

MURRAY, F. J., and VON NEUMANN, J.

[1] On rings of operators I, *Ann. Math.* **37**(1936), 116–229.

[2] On rings of operators II, *Trans. Amer. Math. Soc.* **41**(1937), 208–248.

[3] On rings of operators IV, *Ann. Math.* **44**(1943), 716–808. MR 5, 101.

NACHBIN, L.

[1] *The Haar integral*, D. Van Nostrand, New York, 1965. MR 31 #271.

NAGUMO, M.

[1] Einige analytische Untersuchungen in linearen metrischen Ringen, *Jap. J. Math.* **13**(1936), 61–80.

NAIMARK, M. A.

[1] Positive definite operator functions on a commutative group, *Izv. Akad. Nauk SSSR* **7**(1943), 237–244 (Russian). MR 5, 272.

[2] Rings with involution, *Uspehi Mat. Nauk* **3**(1948), 52–145 (Russian). MR 10, 308.

[3] On a problem of the theory of rings with involution, *Uspehi Mat. Nauk* **6**(1951), 160–164 (Russian). MR 13, 755.

[4] On irreducible linear representations of the proper Lorentz group, *Dokl. Akad. Nauk SSSR* **97**(1954), 969–972 (Russian). MR 16, 218.

[5] *Normed rings*, Gosudarstv. Izdat. Tehn.-Teor. Lit., Moscow, 1956 (Russian). MR 19, 870. MR 22 #1824.

[6] *Linear representations of the Lorentz group*, Gosudarstv. Izdat. Fiz.-Mat. Lit., Moscow, 1958 (Russian). MR 21 #4995.

[7] Decomposition into factor representations of unitary representations of locally compact groups, *Sibirsk. Mat. Ž.* **2**(1961), 89–99 (Russian). MR 24 #A187.

[8] *Normed algebras*, Wolters-Noordhoff Publishing Co., Groningen, 1972. MR 55 #11042.

NAIMARK, M. A., and STERN, A. I.

[1] *Theory of group representations*, Springer-Verlag, New York-Heidelberg-Berlin, 1982. MR 58 #28245. MR 86k: 22001.

NAKAGAMI, Y., and TAKESAKI, M.

[1] *Duality for crossed products of von Neumann algebras*, Lecture Notes in Mathematics **731**, Springer-Verlag, Berlin-Heidelberg-New York, 1979. MR 81e: 46053.

NAKAYAMA, T.

[1] Some studies on regular representations, induced representations and modular representations, *Ann. Math.* **39**(1938), 361–369.

VON NEUMANN, J.

[1] Allgemeine Eigenwerttheorie Hermitescher Funktionaloperatoren, *Math. Ann.* **102**(1929), 49–131.

[2] Zur Algebra der Funktionaloperatoren und Theorie der normalen Operatoren, *Math. Ann.* **102**(1929), 370–427.

[3] Die Eindeutigkeit der Schrödingerschen Operatoren, *Math. Ann.* **104**(1931), 570–578.

[4] Über adjungierte Functionaloperatoren, *Ann. Math.* **33**(1932), 294–310.

[5] Zum Haarschen Mass in topologischen Gruppen, *Compositio Math.* **1**(1934), 106–114.

[6] The uniqueness of Haar's measure, *Mat. Sb. N.S.* **1**(43) (1936), 721–734.

[7] On rings of operators. III, *Ann. Math.* **41**(1940), 94–161. MR 1, 146.

[8] On rings of operators, Reduction theory, *Ann. Math.* **50**(1949), 401–485. MR 10, 548.

[9] Über einen Satz von Herrn M. H. Stone, *Ann. Math.* (2) **33**(1932), 567–573.

NIELSEN, O. A.

[1] The failure of the topological Frobenius property for nilpotent Lie groups, *Math. Scand.* **45**(1979), 305–310. MR 82a: 22009.

[2] The Mackey-Blattner theorem and Takesaki's generalized commutation relations for locally compact groups, *Duke Math. J.* **40**(1973), 105–114.

[3] *Direct integral theory*, Marcel-Dekker, New York, 1980. MR 82e: 46081.

[4] The failure of the topological Frobenius property for nilpotent Lie groups, II., *Math. Ann.* **256**(1981), 561–568. MR 83a: 22011.

OKAMOTO, K.

[1] On induced representations, *Osaka J. Math.* **4**(1967), 85–94. MR 37 #1519.

OLESEN, D.

[1] A classification of ideals in crossed products, *Math. Scand.* **45**(1979), 157–167. MR 81h: 46083.

[2] A note on free action and duality in C^*-algebra theory (preprint), 1979.

OLESEN, D., LANDSTAD, M. B., and PEDERSEN, G. K.

[1] Towards a Galois theory for crossed products of C^*-algebras, *Math. Scand.* **43**(1978), 311–321. MR 81i: 46074.

OLESEN, D., and PEDERSEN, G. K.

[1] Applications of the Connes spectrum to C^*-dynamical systems, *J. Functional Anal.* **30**(1978), 179–197. MR 81i: 46076a.

[2] Applications of the Connes spectrum to C^*-dynamical systems, II, *J. Functional Anal.* **36**(1980), 18–32. MR 81i: 46076b.

[3] On a certain C^*-crossed product inside a W^*-crossed product, *Proc. Amer. Math. Soc.* **79**(1980), 587–590. MR 81h: 46075.

[4] Partially inner C^*-dynamical systems, *J. Functional Anal.* **66**(1986), 262–281.

OLSEN, C. L., and PEDERSEN, G. K.

[1] Convex combinations of unitary operators in von Neumann algebras, *J. Functional Anal.* **66**(1986), 365–380.

O'RAIFEARTAIGH, L.

[1] Mass differences and Lie algebras of finite order, *Phys. Rev. Lett.* **14**(1965), 575–577. MR 31 #1047.

ØRSTED, B.

[1] Induced representations and a new proof of the imprimitivity theorem, *J. Functional Anal.* **31**(1979), 355–359. MR 80d: 22007.

PACKER, J. A.

[1] K-theoretic invariant for C^*-algebras associated to transformations and induced flows, *J. Functional Anal.* **67**(1986), 25–59.

PALMER, T. W.

[1] Characterizations of C^*-algebras, *Bull. Amer. Math. Soc.* **74**(1968), 538–540. MR 36 #5709.

[2] Characterizations of C^*-algebras. II, *Trans. Amer. Math. Soc.* **148**(1970), 577–588. MR 41 #7447.

[3] Classes of nonabelian, noncompact, locally compact groups, *Rocky Mountain Math. J.* **8**(1978), 683–741. MR 81j: 22003.

PANOV, A. N.

[1] The structure of the group C^*-algebra of a Euclidean motion group. *Vestnik Moskov. Univ. Ser. I Mat. Meh.* (1978), 46–49 (Russian). MR 80a: 22008.

PARRY, W., and SCHMIDT, K.

[1] A note on cocycles of unitary representations, *Proc. Amer. Math. Soc.* **55**(1976), 185–190. MR 52 #14146.

PARTHASARATHY, K. R.

[1] *Probability measures on metric spaces*, Academic Press, New York-London, 1967. MR 37 #2271.

[2] *Multipliers on locally compact groups*, Lecture Notes in Mathematics, **93**, Springer-Verlag, Berlin-New York, 1969. MR 40 #264.

PASCHKE, W. L.

[1] Inner product modules over *B**-algebras, *Trans. Amer. Math. Soc.* **182**(1973), 443–468. MR 50 #8087.

[2] The double *B*-dual of an inner product module over a *C**-algebra *B*, *Canad. J. Math.* **26**(1974), 1272–1280. MR 57 #10433.

[3] Inner product modules arising from compact automorphism groups of von Neumann algebras, *Trans. Amer. Math. Soc.* **224**(1976), 87–102. MR 54 #8308.

PASCHKE, W. L., and SALINAS, N.

[1] *C**-algebras associated with free products of groups, *Pacific J. Math.* **82**(1979), 211–221. MR 82c: 22010.

PATERSON, A. L. T.

[1] Weak containment and Clifford semigroups, *Proc. Roy. Soc. Edinburgh. Sect. A* **81**(1978), 23–30. MR 81c: 43001.

PAULSEN, V. I.

[1] *Completely bounded maps and dilations*, Pitman Research Notes in Mathematics, Vol. 146, John Wiley & Sons, New York, 1986.

PEDERSEN, G. K.

[1] Measure theory for *C**-algebras. I, *Math. Scand.* **19**(1966), 131–145. MR 35 #3453.

[2] Measure theory for *C**-algebras. II, *Math. Scand.* **22**(1968), 63–74. MR 39 #7444.

[3] Measure theory for *C**-algebras. III, *Math. Scand.* **25**(1969), 71–93. MR 41 #4263.

[4] Measure theory for *C**-algebras. IV, *Math. Scand.* **25**(1969), 121–127. MR 41 #4263.

[5] *C**-algebras and their automorphism groups, Academic Press, New York, 1979. MR 81e, 46037.

PEDERSEN, N.

[1] Duality for induced representations and induced weights, Copenhagen (preprint), 1978.

PENNEY, R. C.

[1] Abstract Plancherel theorems and a Frobenius reciprocity theorem, *J. Functional Anal.* **18**(1975), 177–190. MR 56 #3191.

[2] Rational subspaces of induced representations and nilmanifolds, *Trans. Amer. Math. Soc.* **246**(1978), 439–450. MR 80b: 22010.

[3] Harmonically induced representations on nilpotent Lie groups and auto-morphic forms on manifolds, *Trans. Amer. Math. Soc.* **260**(1980), 123–145. MR 81h: 22008.

[4] Lie cohomology of representations of nilpotent Lie groups and holomorphi-cally induced representations, *Trans. Amer. Math. Soc.* **261**(1980), 33–51. MR 81i: 22005.

[5] Holomorphically induced representations of exponential Lie groups, *J. Func-tional Anal.* **64**(1985), 1–18. MR 87b: 22018.

PERLIS, S.

[1] A characterization of the radical of an algebra, *Bull. Amer. Math. Soc.* **48**(1942), 128–132. MR 3, 264.

PETERS, F., and WEYL, H.

[1] Die Vollständigkeit der primitiven Darstellungen einer geschlossenen kontin-uerlichen Gruppe, *Math. Ann.* **97**(1927), 737–755.

PETERS, J.

[1] Groups with completely regular primitive dual space, *J. Functional Anal.* **20**(1975), 136–148. MR 52 #652.

[2] On traceable factor representations of crossed products, *J. Functional Anal.* **43**(1981), 78–96. MR 83b: 46086.

[3] Semi-crossed products of C^*-algebras, *J. Functional Anal.* **59**(1984), 498–534. MR 86e: 46063.

[4] On inductive limits of matrix algebras of holomorphic functions, *Trans. Amer. Math. Soc.* **299**(1987), 303–318.

PETERS, J., and SUND, T.

[1] Automorphisms of locally compact groups, *Pacific J. Math.* **76**(1978), 143–156. MR 58 #28263.

PETTIS, B. J.

[1] On integration in vector spaces, *Trans. Amer. Math. Soc.* **44**(1938), 277–304.

PHILLIPS, J.

[1] A note on square-integrable representations, *J. Functional Anal.* **20**(1975), 83–92. MR 52 #15026.

[2] Automorphisms of C^*-algebra bundles , *J. Functional Anal.* **51**(1983), 259–267. MR 84i: 46067.

PHILLIPS, J., and RAEBURN, I.

[1] Crossed products by locally unitary automorphism groups and principal bundles, *J. Operator Theory* **11**(1984), 215–241. MR 86m: 46058.

PIARD, A.

[1] Unitary representations of semi-direct product groups with infinite dimen-sional abelian normal subgroup, *Rep. Math. Phys.* **11**(1977), 259–278. MR 57 #12772.

PICARDELLO, M. A.

[1] Locally compact unimodular groups with atomic duals, *Rend. Sem. Mat. Fis. Milano* **48**(1978), 197–216 (1980). MR 82e: 22015.

PIER, J. P.

[1] *Amenable locally compact groups*, Pure and Applied Mathematics, John Wiley and Sons, New York, 1984. MR 86a: 43001.

PIMSNER, M. and VOICULESCU, D.

[1] K-groups of reduced crossed products by free groups, *J. Operator Theory* **8**(1982), 131–156. MR 84d: 46092.

PLANCHEREL, M.

[1] Contribution à l'étude de la représentation d'une fonction arbitraire par les intégrales définies, *Rend. Pal.* **30**(1910), p. 289.

POGUNTKE, D.

[1] Epimorphisms of compact groups are onto, *Proc. Amer. Math. Soc.* **26**(1970), 503–504. MR 41 #8577.

[2] Einige Eigenschaften des Darstellungsringes kompakter Gruppen, *Math. Z.* **130**(1973), 107–117. MR 49 #3020.

[3] Decomposition of tensor products of irreducible unitary representations, *Proc. Amer. Math. Soc.* **52**(1975), 427–432. MR 52 #5862.

[4] Der Raum der primitiven Ideale von endlichen Erweiterungen lokalkompakter Gruppen, *Arch. Math. (Basel)* **28**(1977), 133–138. MR 55 #12862.

[5] Nilpotente Liesche Gruppen haben symmetrische Gruppenalgebren, *Math. Ann.* **227**(1977), 51–59. MR 56 #6283.

[6] Symmetry and nonsymmetry for a class of exponential Lie groups, *J. Reine. Angew. Math.* **315**(1980), 127–138. MR 82b: 43013.

[7] Einfache Moduln über gewissen Banachschen Algebren: ein Imprimitivitätssatz, *Math. Ann.* **259**(1982), 245–258. MR 84i: 22008.

PONTRYAGIN, L. S.

[1] *Der allgemeine Dualitätssatz für abgeschlossene Mengen*. Verhandlungen des Internat. Math.-Kongr., Zürich, 1932. Zürich u. Leipzig: Orell Füssli, Bd. II, pp. 195–197.

[2] Sur les groupes topologiques compacts et cinquième problème de D. Hilbert, *C. R. Acad. Sci. Paris* **198**(1934), 238–240.

[3] Sur les groupes abéliens continus, *C. R. Acad. Sci. Paris* **198**(1934), 328–330.

[4] The theory of topological commutative groups, *Ann. Math.* (2) **35**(1934), 361–388.

[5] *Topological groups*, Princeton University Press, Princeton, N. J., 1939. MR 1, 44, MR 19, 867, MR 34 #1439.

[6] *Topological groups*, 2nd ed., Gordon and Breach, Science Publishers, New York, 1966.

POPA, S.

[1] Semiregular maximal abelian *-subalgebras and the solution to the factor Stone-Weierstrass problem, *Invent. Math.* **76**(1984), 157–161. MR 85m: 46057b.

POVZNER, A.

[1] Über positive Funktionen auf einer Abelschen Gruppe, *Dokl. Akad. Nauk SSSR, N.S.* **28**(1940), 294–295.

POWERS, R. T.

[1] Simplicity of the *C**-algebra associated with the free group on two generators, *Duke Math. J.* **42**(1975), 151–156. MR 51 #10534.

POZZI, G. A.

[1] Continuous unitary representations of locally compact groups. Application to Quantum Dynamics. Part I. Decomposition Theory. *Suppl. Nuovo Cimento* **4**(1966), 37–171. MR 36 #6406.

PRICE, J. F.

[1] On positive definite functions over a locally compact group, *Canad. J. Math.* **22**(1970), 892–896. MR 41 #8593.

[2] *Lie groups and compact groups*, London Mathematical Society Lecture Note Series, **25**, Cambridge University Press, Cambridge-New York-Melbourne, 1977. MR 56 #8743.

PROSSER, R. T.

[1] *On the ideal structure of operator algebras*, Memoirs Amer. Math. Soc. **45**, Amer. Math. Soc. Providence, R. I., 1963. MR 27 #1846.

PTÁK, V.

[1] Banach algebras with involution, *Manuscripta Math.* **6**(1972), 245–290. MR 45 #5764.

PUKANSZKY, L.

[1] *Leçons sur les représentations des groupes*, Monographies de la Société Mathématique de France, Dunod, Paris, 1967. MR 36 #311.

[2] Unitary representations of solvable Lie groups, *Ann. Sci. École Norm. Sup.* (4) **4**(1971), 457–608. MR 55 #12866.

[3] The primitive ideal space of solvable Lie groups, *Invent. Math.* **22**(1973), 75–118. MR 48 #11403.

[4] Characters of connected Lie groups, *Acta Math.* **133**(1974), 81–137. MR 53 #13480.

PYTLIK, T.

[1] L^1-harmonic analysis on semi-direct products of Abelian groups, *Monatsh. Math.* **93**(1982), 309–328. MR 83m: 43006.

PYTLIK, T., and SZWARC, R.

[1] An analytic family of uniformly bounded representations of free groups, *Acta Math.* **157**(1986), 287–309.

QUIGG, J. C.

[1] On the irreducibility of an induced representation, *Pacific J. Math.* **93**(1981), 163–179. MR 84j: 22007.

[2] On the irreducibility of an induced representation II, *Proc. Amer. Math. Soc.* **86**(1982), 345–348. MR 84m: 22010.

1432

[3] Approximately periodic functionals on C^*-algebras and von Neumann algebras, *Canad. J. Math.* **37**(1985), 769–784. MR 86k: 46088.

[4] Duality for reduced twisted crossed products of C^*-algebras, *Indiana Univ. Math. J.* **35**(1986), 549–572.

RAEBURN, I.

[1] On the Picard group of a continuous trace C^*-algebra, *Trans. Amer. Math. Soc.* **263**(1981), 183–205. MR 82b: 46090.

[2] On group C^*-algebras of bounded representation dimension, *Trans. Amer. Math. Soc.* **272**(1982), 629–644. MR 83k: 22018.

RAEBURN, I., and TAYLOR, J. L.

[1] Continuous trace C^*-algebras with given Dixmier-Douady class, *J. Austral. Math. Soc. Ser. A* **38**(1985), 394–407. MR 86g: 46086.

RAEBURN, I., and WILLIAMS, D. P.

[1] Pull-backs of C^*-algebras and crossed products by certain diagonal actions, *Trans. Amer. Math. Soc.* **287**(1985), 755–777. MR 86m: 46054.

RAIKOV, D. A.

[1] Positive definite functions on commutative groups with an invariant measure, *Dokl. Akad. Nauk SSSR, N.S.* **28**(1940), 296–300.

[2] The theory of normed rings with involution, *Dokl. Akad. Nauk SSSR* **54**(1946), 387–390. MR 8, 469.

RAMSAY, A.

[1] Virtual groups and group actions, *Adv. Math.* **6**(1971), 253–322. MR 43 #7590.

[2] Boolean duals of virtual groups, *J. Functional Anal.* **15**(1974), 56–101. MR 51 #10586.

[3] Nontransitive quasi-orbits in Mackey's analysis of group extensions, *Acta Math.* **137**(1976), 17–48. MR 57 #524.

[4] Subobjects of virtual groups, *Pacific J. Math.* **87**(1980), 389–454. MR 83b: 22009.

[5] Topologies on measured groupoids, *J. Functional Anal.* **47**(1982), 314–343. MR 83k: 22014.

READ, C. J.

[1] A solution to the invariant subspace problem, *Bull. London Math. Soc.* **16**(1984), 337–401. MR 86f: 47005.

RENAULT, J.

[1] *A groupoid approach to C^*-algebras*, Lecture Notes in Mathematics, **793**, Springer-Verlag, Berlin-New York, 1980. MR 82h: 46075.

REPKA, J.

[1] A Stone-Weierstrass theorem for group representations, *Int. J. Math. Sci.* **1**(1978), 235–244. MR 58 #6055.

RICKART, C. E.

[1] Banach algebras with an adjoint operation, *Ann. Math.* **47**(1946), 528–550. MR 8, 159.

[2] *General theory of Banach algebras*, D. Van Nostrand Co., Princeton, 1960. MR 22 #5903.

RIEDEL, N.

[1] Topological direct integrals of left Hilbert algebras. I., *J. Operator Theory* **5**(1981), 29–45. MR 83e: 46049.

[2] Topological direct integrals of left Hilbert algebras. II., *J. Operator Theory* **5**(1981), 213–229. MR 83e: 46050.

RIEFFEL, M. A.

[1] On extensions of locally compact groups, *Amer. J. Math.* **88**(1966), 871–880. MR 34 #2771.

[2] Induced Banach representations of Banach algebras and locally compact groups, *J. Functional Anal.* **1**(1967), 443–491. MR 36 #6544.

[3] Unitary representations induced from compact subgroups, *Studia Math.* **42**(1972), 145–175. MR 47 #398.

[4] On the uniqueness of the Heisenberg commutation relations, *Duke Math. J.* **39**(1972), 745–752. MR 54 #466.

[5] Induced representations of *C**-algebras, *Adv. Math.* **13**(1974), 176–257. MR 50 #5489.

[6] Morita equivalence for *C**-algebras and *W**-algebras, *J. Pure Appl. Algebra* **5**(1974), 51–96. MR 51 #3912.

[7] Induced representations of rings, *Canad. J. Math.* **27**(1975), 261–270. MR 55 #3004.

[8] Strong Morita equivalence of certain transformation group *C**-algebras, *Math. Ann.* **222**(1976), 7–22. MR 54 #7695.

[9] *Unitary representations of group extensions; an algebraic approach to the theory of Mackey and Blattner*. Studies in analysis, pp. 43–82. Advances in Mathematics, Suppl. Studies, **4**, Academic Press, New York 1979. MR 81h: 22004.

[10] Actions of finite groups on *C**-algebras, *Math. Scand.* **47**(1980), 157–176. MR 83c: 46062.

[11] *C**-algebras associated with irrational rotations, *Pacific J. Math.* **93**(1981), 415–429. MR 83b: 46087.

[12] *Morita equivalence for operator algebras*, Proc. Sympos. Pure Math., pp. 285–298, **38**, Part 1, Amer. Math. Soc., Providence, R. I. 1982.

[13] *Applications of strong Morita equivalence to transformation group C**-algebras, Proc. Sympos. Pure Math., pp. 299–310, 38, Part 1, Amer. Math. Soc., Prov., R.I. 1982.

[14] K-theory of crossed products of *C**-algebras by discrete groups, Group actions on rings, Proc. AMS-IMS-SIAM Summer Res. Conf., 1984, *Contemp. Math.* **43**(1985), 227–243.

RIESZ, F.

[1] Sur les opérations fonctionnelles linéaires, *C. R. Acad. Sci. Paris* **149**(1909), 974–977. (Also in Gesammelte Arbeiten, vol. I., pp. 400–402, Budapest, 1960).

[2] *Les systèmes d'equations linéaires à une infinité d'inconnus*, Paris, 1913.

[3] Über lineare Funktionalgleichungen, *Acta Math.* **41**(1918), 71–98.

[4] Sur la formule d'inversion de Fourier, *Acta Sci. Math. Szeged* **3**(1927), 235–241.

RIESZ, F., and SZ-NAGY, B.

[1] *Leçons d'analyse fonctionelle*, Académie des Sciences de Hongrie, Budapest, 1952. MR 14, 286.

[2] *Functional analysis*, Frederick Ungar Publishing Co., New York, 1955. MR 17, 175.

RIGELHOF, R.

[1] Induced representations of locally compact groups, *Acta Math.* **125**(1970), 155–187. MR 43 #7550.

RINGROSE, J. R.

[1] On subalgebras of a C^*-algebra, *Pacific J. Math.* **15**(1965), 1377–1382. MR 32 #4561.

ROBERT, A.

[1] Exemples de groupes de Fell (English summary), *C. R. Acad. Sci. Paris Ser. A-B* (8) **287**(1978), A603–A606. MR 83k: 22015.

[2] *Introduction to the representation theory of compact and locally compact groups*, London Mathematical Society Lecture Note Series, **80**, Cambridge University Press, 1983. MR 84h: 22012.

ROBINSON, G. DE B.

[1] *Representation theory of the symmetric group*, University of Toronto Press, Toronto, 1961. MR 23 #A3182.

ROELCKE, W., and DIEROLF, S.

[1] *Uniform structures on topological groups and their quotients*, McGraw-Hill, New York, 1981. MR 83a: 82005.

ROSENBERG, A.

[1] The number of irreducible representations of simple rings with no minimal ideals, *Amer. J. Math.* **75**(1953), 523–530. MR 15, 236.

ROSENBERG, J.

[1] The C^*-algebras of some real and p-adic solvable groups, *Pacific J. Math.* **65**(1976), 175–192. MR 56 #5779.

[2] Amenability of crossed products of C^*-algebras, *Comm. Math. Phys.* **57**(1977), 187–191. MR 57 #7190.

[3] A quick proof of Harish-Chandra's Plancherel theorem for spherical functions on a semisimple Lie group, *Proc. Amer. Math. Soc.* **63**(1977), 143–149. MR 58 #22391.

[4] Frobenius reciprocity for square-integrable factor representations, *Illinois J. Math.* **21**(1977), 818–825. MR 57 #12771.

[5] Square-integrable factor representations of locally compact groups *Trans. Amer. Math. Soc.* **237**(1978), 1–33. MR 58 #6056.

[6] Appendix to: "Crossed products of UHF algebras by product type actions" [*Duke Math. J.* **46**(1979), 1–23] by O. Bratteli, *Duke Math. J.* **46**(1979), 25–26. MR 82a: 46064.

ROSENBERG, J., and VERGNE, M.

[1] Hamonically induced representations of solvable Lie groups, *J. Functional Anal.* **62**(1985), 8–37. MR 87a: 22017.

ROUSSEAU, R.

[1] The left Hilbert algebra associated to a semi-direct product, *Math. Proc. Cambridge Phil. Soc.* **82**(1977), 411–418. MR 56 #16386.

[2] An alternative definition for the covariance algebra of an extended covariant system, *Bull. Soc. Math. Belg.* **30**(1978), 45–59. MR 81i: 46081.

[3] Un système d'induction pour des groupes topologiques, *Bull. Soc. Math. Belg.***31**(1979), 191–195. MR 81m: 22008.

[4] The covariance algebra of an extended covariant system, *Math. Proc. Cambridge Phil. Soc.* **85**(1979), 271–280. MR 80e: 46040.

[5] The covariance algebra of an extended covariant system. II, *Simon Stevin* **53**(1979), 281–295. MR 82g: 46105.

[6] Le commutant d'une algèbre de covariance, *Rev. Roum. Math. Pures Appl.* **25**(1980), 445–471. MR 81m: 46095.

[7] A general induction process, *Quart. J. Math. Oxford Ser.* (2) **32**(1981), 453–466. MR 83a: 22005.

[8] Tensor products and the induction process, *Arch. Math. (Basel)* **36**(1981), 541–545. MR 83h: 22014.

[9] Crossed products of von Neumann algebras and semidirect products of groups, *Nederl. Akad. Wetensch. Indag. Math.* **43**(1981), 105–116. MR 82i: 46103.

[10] Quasi-invariance and induced representations, *Quart. J. Math. Oxford* (2) **34**(1983), 491–505. MR 86i: 22008.

[11] A general induction process. II., *Rev. Roumaine Math. Pures Appl.* **30**(1985), 147–153. MR 86i: 22009.

RÜHL, W.

[1] *The Lorentz group and harmonic analysis*, W. A. Benjamin, New York, 1970. MR 43 #425.

RUDIN, W.

[1] *Fourier analysis on groups*, J. Wiley, New York, 1962. MR 27 #2808.

[2] *Functional analysis*, Mcgraw Hill, New York, 1973. MR 51 #1315.

RUSSO, R., and DYE, H. A.

[1] A note on unitary operators in C^*-algebras, *Duke Math. J.* **33**(1966), 413–416. MR 33 #1750.

SAGLE, A. A., and WALDE, R. E.

[1] *Introduction to Lie groups and Lie algebras*, Academic Press, New York-London, 1973. MR 50 #13374.

SAKAI, S.

[1] C^*-algebras and W^*-algebras, Ergebnisse der Math. **60**, Springer-Verlag, Berlin, Heidelberg, New York 1971. MR 56 #1082.

SALLY, P. J.

[1] *Harmonic analysis on locally compact groups*, Lecture Notes, University of Maryland, 1976.

[2] Harmonic analysis and group representations, *Studies in Harmonic analysis*, **13**, pp. 224–256, MAA Studies in Mathematics, Mathematical Association of America, Washington D. C., 1976.

SANKARAN, S.

[1] Representations of semi-direct products of groups, *Compositio Math.* **22**(1970), 215–225. MR 42 #722.

[2] Imprimitivity and duality theorems, *Boll. Un. Mat. Ital.* (4) **7**(1973), 241–259. MR 47 #5176.

SAUVAGEOT,, J. L.

[1] Idéaux primitifs de certains produits croisés, *Math. Ann.* **231**(1977/78), 61–76. MR 80d: 46112.

[2] Idéaux primitifs induits dans les produits croisés, *J. Functional Anal.* **32**(1979), 381–392. MR 81a: 46080.

SCHAAF, M.

[1] *The reduction of the product of two irreducible unitary representations of the proper orthochronous quantum-mechanical Poincaré group*, Lecture Notes in Physics, **5**, Springer-Verlag, Berlin-New York, 1970. MR 57 #16471.

SCHAEFFER, H. H.

[1] *Topological vector spaces*, Macmillan, New York, 1966. MR 33 #1689.

SCHEMPP, W.

[1] *Harmonic analysis on the Heisenberg nilpotent Lie group, with applications to signal theory*, Pitman Research Notes in Mathematics, 147, John Wiley & Sons, New York, 1986.

SCHLICHTING, G.

[1] Groups with representations of bounded degree, Lecture Notes in Mathematics **706**, pp. 344–348. Springer, Berlin, 1979. MR 80m: 22006.

SCHMIDT, E.

[1] Entwicklung willkürlicher Funktionen nach Systemen vorgeschriebener, *Math. Ann.* **63**(1907), 433–476.

SCHMIDT, K.

[1] *Cocycles on ergodic transformation groups*, Macmillan Lectures in Mathematics **1**, Macmillan Co. of India, Delhi 1977. MR 58 #28262.

SCHOCHETMAN, I. E.

[1] Topology and the duals of certain locally compact groups, *Trans. Amer. Math. Soc.* **150**(1970), 477–489. MR 42 #422.

[2] Dimensionality and the duals of certain locally compact groups, *Proc. Amer. Math. Soc.* **26**(1970), 514–520. MR 42 #418.

[3] Nets of subgroups and amenability, *Proc. Amer. Math. Soc.* **29**(1971), 397–403. MR 43 #7551.

[4] Kernels of representations and group extensions, *Proc. Amer. Math. Soc.* **36**(1972), 564–570. MR 47 #8761.

[5] Compact and Hilbert-Schmidt induced representations, *Duke Math. J.* **41**(1974), 89–102. MR 48 #11392.

[6] Induced representations of groups on Banach spaces, *Rocky Mountain J. Math.* **7**(1977), 53–102. MR 56 #15824.

[7] *Integral operators in the theory of induced Banach representations*, Memoirs Amer. Math. Soc. **207**, Amer. Math. Soc. Providence, R. I., 1978. MR 58 #6060.

[8] The dual topology of certain group extensions, *Adv. Math.* **35**(1980), 113–128. MR 81e: 22005.

[9] Integral operators in the theory of induced Banach representations. II. The bundle approach, *Int. J. Math. Sci.* **4**(1981), 625–640. MR 84c: 22009.

[10] Generalized group algebras and their bundles, *Int. J. Math. Sci.* **5**(1982), 209–256. MR 84f: 46094.

SCHREIER, O.

[1] Abstrakte kontinuerliche Gruppen, *Abh. Math. Sem. Univ. Hamburg* **4**(1926), 15–32.

SCHUR, I.

[1] Über die Darstellung der endlichen Gruppen durch gebrochene lineare Substitutionen, *J. Crelle* **127**(1904), 20–50.

[2] Neue Begründung der Theorie der Gruppencharaktere, *Sitz. preuss. Akad. Wiss.* (1905), 406–432.

[3] Neue Anwendungen der Integralrechnung auf Probleme der Invarianten theorie. I, II, III, *Sitz. preuss. Akad. Wiss. phys.-math. Kl.* (1924), 189–208, 297–321, 346–355.

SCHWARTZ, J.-M.

[1] Sur la structure des algèbres de Kac, I, *J. Functional Anal.* **34**(1979), 370–406. MR 83a: 46072a.

[2] Sur la structure des algèbres de Kac, II, *Proc. London Math. Soc.* **41**(1980), 465–480. MR 83a: 46072b.

SCUTARU, H.

[1] Coherent states and induced representations, *Lett. Math. Phys.* **2**(1977/78), 101–107. MR 58 #22406.

SEBESTYÉN, Z.

[1] Every C^*-seminorm is automatically submultiplicative, *Period. Math. Hungar.* **10**(1979), 1–8. MR 80c: 46065.

[2] On representability of linear functionals on *-algebras, *Period. Math. Hungar.* **15**(1984), 233–239. MR 86a: 46068.

SEDA, A. K.

[1] A continuity property of Haar systems of measures, *Ann. Soc. Sci. Bruxelles* **89**(1975), 429–433. MR 53 #6555.

[2] Haar measures for groupoids, *Proc. Roy. Irish. Sect.* **A76**(1976), 25–36. MR 55 #629.

[3] Quelques résultats dans la catégorie des groupoids d'opérateurs, *C. R. Acad. Sci. Paris* **288**(1979), 21–24. MR 80b: 22005.

[4] Banach bundles and a theorem of J. M. G. Fell, *Proc. Amer. Math. Soc.* **83**(1981), 812–816. MR 84d: 22006.

[5] Banach bundles of continous functions and an integral representation theorem, *Trans. Amer. Math. Soc.* **270**(1982), 327–332. MR 83f: 28009.

[6] Sur les espaces de fonctions et les espaces de sections, *C. R. Acad. Sci. Paris Sér. I Math.* **297**(1983), 41–44. MR 85a: 46041.

SEGAL, I. E.

[1] The group ring of a locally compact group. I, *Proc. Nat. Acad. Sci. U.S.A.* **27**(1941), 348–352. MR 3, 36.

[2] Representation of certain commutative Banach algebras, *Bull. Amer. Math. Soc.* Abst. 130, **52**(1946), 421.

[3] The group algebra of a locally compact group, *Trans. Amer. Math. Soc.* **61**(1947), 69–105. MR 8, 438.

[4] Irreducible representations of operator algebras, *Bull. Amer. Math. Soc.* **53**(1947), 73–88. MR 8, 520.

[5] Two-sided ideals in operator algebras, *Ann. Math.* **50**(1949), 856–865. MR 11, 187.

[6] The two-sided regular representation of a unimodular locally compact group, *Ann. Math.* (2) **51**(1950), 293–298. MR 12, 157.

[7] An extension of Plancherel's formula to separable unimodular groups, *Ann. Math.* (2) **52**(1950), 272–292. MR 12, 157.

[8] *Decompositions of operator algebras*, I *and* II, Memoirs Amer. Math. Soc. **9**, Amer. Math. Soc. Providence, R. I., 1951. MR 13, 472.

[9] A non-commutative extension of abstract integration, *Ann. Math.* (2) **57**(1953), 401–457. MR 14, 991. MR 15, 204.

[10] *Caractérisation mathématique des observables en théorie quantique des champs et ses conséquences pour la structure des particules libres*, Report of Lille conference on quantum fields, pp. 57–163, C.N.R.S., Paris, 1959.

[11] An extension of a theorem of L. O'Raifeartaigh, *J. Functional Anal.* **1**(1967), 1–21. MR 37 #6079.

SELBERG, A.

[1] Harmonic analysis and discontinuous groups in weakly symmetric Riemannian spaces with applications to Dirichlet series, *J. Indian Math. Soc.* **20**(1956), 47–87. MR 19, 531.

SEN, R. N.

[1] Bundle representations and their applications, pp. 151–160, Lecture Notes in Mathematics, **676**, Springer-Verlag, 1978. MR 80e: 22003.

[2] Theory of symmetry in the quantum mechanics of infinite systems. I. The state space and the group action, *Phys. A* **94**(1978), 39–54. MR 80h: 81031a.

[3] Theory of symmetry in the quantum mechanics of infinite systems. II. Isotropic representations of the Galilei group and its central extensions, *Phys. A* **94**(1978), 55–70. MR 80h: 81031b.

SERIES, C.

[1] Ergodic actions on product groups, *Pacific J. Math.* **70**(1977), 519–547. MR 58 #6062.

[2] An application of groupoid cohomology, *Pacific J. Math.* **92**(1981), 415–432. MR 84f: 22014.

SERRE, J. P.

[1] *Linear representations of finite groups*, Graduate Texts in Mathematics, **42**, Springer-Verlag, Berlin-Heidelberg-New York, 1977. MR 80f: 20001.

SHERMAN, S.

[1] Order in operator algebras, *Amer. J. Math.* **73**(1951), 227–232. MR 13, 47.

SHIN'YA, H.

[1] *Spherical functions and spherical matrix functions on locally compact groups*, Lectures in Mathematics, Department of Mathematics, Kyoto University, 7, Kinokuniya Book Store Co., Tokyo, 1974. MR 50 #10149.

[2] Irreducible Banach representations of locally compact groups of a certain type, *J. Math. Kyoto Univ.* **20**(1980), 197–212. MR 82c: 22009.

[3] On a Frobenius reciprocity theorem for locally compact groups, *J. Math. Kyoto Univ.* **24**(1984), 539–555. MR 86a: 22008.

[4] On a Frobenius reciprocity theorem for locally compact groups. II, *J. Math. Kyoto Univ.* **25**(1985), 523–547. MR 87b: 22006.

SHUCKER, D. S.

[1] Square integrable representations of unimodular groups, *Proc. Amer. Math. Soc.* **89**(1983), 169–172. MR 85d: 22013.

SHULTZ, F. W.

[1] Pure states as a dual object for C^*-algebras, *Comm. Math. Phys.* **82**(1981/82), 497–509. MR 83b: 46080.

SINCLAIR, A. M.

[1] *Automatic continuity of linear operators*, Lecture Notes Series **21**, London Math. Soc., London, 1977. MR 58 #7011.

SINGER, I. M.

[1] *Report on group representations*, National Academy of Sciences—National Research Council, Publ. **387** (Arden House), Washington, D.C. (1955), 11–26.

SKUDLAREK, H. L.

[1] On a two-sided version of Reiter's condition and weak containment, *Arch. Math. (Basel)* **31**(1978/79), 605–610. MR 80h: 43010.

SMITH, H. A.

[1] Commutative twisted group algebras, *Trans. Amer. Math. Soc.* **197**(1974), 315–326. MR 51 #792.

[2] Characteristic principal bundles, *Trans. Amer. Math. Soc.* **211**(1975), 365–375. MR 51 #13128.

[3] Central twisted group algebras, *Trans. Amer. Math. Soc.* **238**(1978), 309–320. MR 58 #7093.

SMITH, M.

[1] Regular representations of discrete groups, *J. Functional Anal.* **11**(1972), 401–406. MR 49 #9641.

SMITHIES, F.

[1] The Fredholm theory of integral equations, *Duke Math. J.* **8**(1941), 107–130. MR 3, 47.

SOLEL, B.

[1] Nonselfadjoint crossed products: invariant subspaces, cocycles and subalgebras, *Indiana Univ. Math. J.* **34**(1985), 277–298. MR 87h: 46140.

SPEISER, A.

[1] *Die Theorie der Gruppen von endlicher Ordnung*, Grundlehren der Mathematischen Wiss., **5**, Berlin, 1937.

STEEN, L. A.

[1] Highlights in the history of spectral theory, *Amer. Math. Monthly* **80**(1973), 359–381. MR 47 #5643.

STEIN, E. M.

[1] Analytic continuation of group representations, *Adv. Math.* **4**(1970), 172–207. MR 41 #8584.

ŠTERN, A. I.

[1] The connection between the topologies of a locally bicompact group and its dual space, *Funkcional. Anal. Priložen.* **5**(1971), 56–63 (Russian). MR 45 #3639.

[2] Dual objects of compact and discrete groups, *Vestnik Moskov. Univ. Ser. I Mat. Meh.* **2**(1977), 9–11 (Russian). MR 56 #8748.

STEWART, J.

[1] Positive definite functions and generalizations, an historical survey, *Rocky Mountain J. Math.* **6**(1976), 409–434. MR 55 #3679.

STINESPRING, W. F.

[1] Positive functions on *C**-algebras, *Proc. Amer. Math. Soc.* **6**(1955), 211–216. MR 16, 1033.

[2] A semi-simple matrix group is of type I, *Proc. Amer. Math. Soc.* **9**(1958), 965–967. MR 21 #3509.

STONE, M. H.

[1] Linear transformations in Hilbert space. III. Operational methods and group theory, *Proc. Nat. Acad. Sci. U.S.A.* **16**(1930), 172–175.

[2] *Linear transformations in Hilbert space and their applications to analysis*, Amer. Math. Soc. Coll. Publ. **15**, New York, 1932.

[3] Notes on integration. I, II, III, *Proc. Nat. Acad. Sci. U.S.A.* **34**(1948), 336–342, 447–455, 483–490. MR 10, 24. MR 10, 107. MR 10, 239.

[4] Applications of the theory of Boolean rings to general topology, *Trans. Amer. Math. Soc.* **41**(1937), 375–481.

[5] The generalized Weierstrass aproximation theorem, *Math. Magazine* **21**(1948), 167–184, 237–254. MR 10, 255.

[6] On one-parameter unitary groups in Hilbert space, *Ann. Math.* (2) **33**(1932), 643–648.

[7] A general theory of spectra, I, *Proc. Nat. Acad. Sci. U.S.A.*, **26**(1940), 280–283.

STØRMER, E.

[1] Large groups of automorphisms of C^*-algebras, *Commun. Math. Phys.* **5**(1967), 1–22. MR 37 #2012.

STOUT, E. L.

[1] *The theory of uniform algebras*, Bogden & Quigley, New York, 1971. MR 54 #11066.

STRASBURGER, A.

[1] Inducing spherical representations of semi-simple Lie groups, *Diss. Math. (Rozpr. Mat.)* **122**(1975), 52 pp. MR 53 #8332.

STRATILA, S.

[1] *Modular theory in operator algebras*, Abacus Press, Tunbridge Wells, 1981. MR 85g: 46072.

STRATILA, S., and ZSIDÓ, L.

[1] *Lectures on von Neumann algebras*, Abacus Press, Tunbridge Wells, 1979. MR 81j: 46089.

SUGIURA, M.

[1] *Unitary representations and harmonic analysis. An introduction*, Kodansha, Tokyo; Halstead Press, John Wiley & Sons, New York-London-Sydney, 1975. MR 58 #16977.

SUND, T.

[1] Square-integrable representations and the Mackey theory, *Trans. Amer. Math. Soc.* **194**(1974), 131–139. MR 49 #9115.

[2] Duality theory for groups with precompact conjugacy classes. I, *Trans. Amer. Math. Soc.* **211**(1975), 185–202. MR 53 #8332.

[3] Isolated points in duals of certain locally compact groups, *Math. Ann.* **224**(1976), 33–39. MR 54 #10477.

[4] A note on integrable representations, *Proc. Amer. Math. Soc.* **59**(1976), 358–360. MR 55 #3154.

[5] Duality theory for groups with precompact conjugacy classes. II, *Trans. Amer. Math. Soc.* **224**(1976), 313–321. MR 55 #12863.

[6] Multiplier representations of exponential Lie groups, *Math. Ann.* **232**(1978), 287–290. MR 57 #16467.

[7] Multiplier representations of nilpotent Lie groups (preprint) 1976.

SUTHERLAND, C.

[1] Cohomology and extensions of von Neumann algebras. I, *Publ. Res. Inst. Math. Sci.* **16**(1980), 105–133. MR 81k: 46067.

[2] Cohomology and extensions of von Neumann algebras, II, *Publ. Res. Math. Sci.* **16**(1980), 135–174. MR 81k: 46067.

[3] Induced representations for measured groupoids (preprint), 1982.

SZMIDT, J.

[1] On the Frobenius reciprocity theorem for representations induced in Hilbert module tensor products, *Bull. Acad. Polon. Sci. Sér. Sci. Math. Astronom. Phys.* **21**(1973), 35–39. MR 47 #6941.

[2] The duality theorems. Cyclic representations. Langlands conjectures, *Diss. Math. (Rozpr. Mat.)* **168**(1980), 47 pp. MR 81j: 22012.

SZ.-NAGY, B.

[1] *Spektraldarstellung linearer Transformationen des Hilbertschen Raumes,* Ergebnisse Math. V. **5**, Berlin (1942). MR 8 #276.

TAKAHASHI, A.

[1] Hilbert modules and their representation, *Rev. Colombiana Mat.* **13**(1979), 1–38. MR 81k: 46056a.

[2] A duality between Hilbert modules and fields of Hilbert spaces, *Rev. Colombiana Mat.* **13**(1979), 93–120. MR 81k: 46056b.

TAKAI, H.

[1] The quasi-orbit space of continuous C^*-dynamical systems, *Trans. Amer. Math. Soc.* **216**(1976), 105–113. MR 52 #6444.

TAKENOUCHI, O.

[1] Sur une classe de fonctions continues de type positif sur un groupe localement compact, *Math. J. Okayama Univ.*, **4**(1955), 153–173. MR 16, 997.

[2] Families of unitary operators defined on groups, *Math. J. Okayama Univ.*, **6**(1957), 171–179. MR 19, 430.

[3] Sur la facteur représentation d'un group de Lie résoluble de type (E), *Math. J. Okayama Univ.* **7**(1957), 151–161. MR 20 #3933.

TAKESAKI, M.

[1] Covariant representations of C^*-algebras and their locally compact auto morphism groups, *Acta Math.* **119**(1967), 273–303. MR 37 #774.

[2] A duality in the representation of C^*-algebras, *Ann. Math.* (2) **85**(1967), 370–382. MR 35 #755.

[3] A characterization of group algebras as a converse of Tannaka-Stinespring-Tatsuuma duality theorem, *Amer. J. Math.* **91**(1969), 529–564. MR 39 #5752.

[4] A liminal crossed product of a uniformly hyperfinite C^*-algebra by a compact Abelian group, *J. Functional Anal.* **7**(1971), 140–146. MR 43 #941.

[5] *Duality and von Neumann algebras,* Lecture Notes in Mathematics, **247** Springer, New York, 1972, 665–786. MR 53 #704.

[6] Duality for crossed products and the structure of von Neumann algebras of type III, *Acta Math.* **131**(1973), 249–308. MR 55 #11068.

[7] *Theory of operator algebras.* I, Springer-Verlag, New York-Heidelberg-Berlin, 1979. MR 81e: 46038.

TAKESAKI, M., and TATSUUMA, N.

[1] Duality and subgroups, *Ann. Math.* **93**(1971), 344–364. MR 43 #7557.

[2] Duality and subgroups. II. *J. Functional Anal.* **11**(1972), 184–190. MR 52 #5865.

TAMAGAWA, T.

[1] On Selberg's trace formula, *J. Fac. Sci. Univ. Tokyo Sect.* (1) **8**(1960), 363–386. MR 23 #A958.

TATSUUMA, N.

[1] A duality theorem for the real unimodular group of second order, *J. Math. Japan* **17**(1965), 313–332. MR 32 #1290.

[2] A duality theorem for locally compact groups, *J. Math. Kyoto Univ.* **6**(1967), 187–293. MR 36 #313.

[3] Plancherel formula for non-unimodular locally compact groups, *J. Math. Kyoto Univ.* **12**(1972), 179–261. MR 45 #8777.

[4] Duality for normal subgroups, *Algèbres d'opérateurs et leurs applications en physique mathématique*, Proc. Colloq., Marseille, 1977, pp. 373–386, Colloques Int. CNRS, 274, CNRS, Paris, 1979. MR 82b: 22009.

TAYLOR, D. C.

[1] Interpolation in algebras of operator fields, *J. Functional Anal.* **10**(1972), 159–190. MR 51 #13700.

[2] A general Hoffman-Wermer theorem for algebras of operator fields, *Proc. Amer. Math. Soc.* **52**(1975), 212–216. MR 52 #6455.

TAYLOR, K. F.

[1] The type structure of the regular representation of a locally compact group, *Math. Ann.* **222**(1976), 211–224. MR 54 #12965.

[2] Group representations which vanish at infinity, *Math. Ann.* **251**(1980), 185–190. MR 82a: 22006.

TAYLOR, M. E.

[1] *Noncommutative harmonic analysis*, Mathematical Surveys and Monographs **22**, Amer. Math. Soc., Providence, R. I., 1986.

TERRAS, A.

[1] *Harmonic analysis on symmetric spaces and applications*, Springer-Verlag, New York-Berlin, 1985. MR 87f: 22010.

THIELEKER, E.

[1] On the irreducibility of nonunitary induced representations of certain semidirect products, *Trans. Amer. Math. Soc.* **164**(1972), 353–369. MR 45 #2097.

THOMA, E.

[1] Über unitäre Darstellungen abzählbarer, diskreter Gruppen, *Math. Ann.* **153**(1964), 111–138. MR 28 #3332.

[2] Eine Charakterisierung diskreter Gruppen vom Typ I, *Invent. Math.* **6**(1968), 190–196. MR 40 #1540.

TITCHMARSH, E. C.

[1] *The theory of functions*, 2nd ed, Oxford University Press, Cambridge, 1939.

[2] *Introduction to the theory of Fourier integrals*, Oxford University Press, Cambridge, 1948.

TOMIYAMA, J.

[1] Topological representation of C^*-algebras, *Tôhoku Math. J.* **14**(1962), 187–204. MR 26 #619.

[2] *Invitation to C^*-algebras and topological dynamics*, World Scientific Advanced Series in Dynamical Systems, Vol. 3, World Scientific, Singapore, 1987.

TOMIYAMA, J., and TAKESAKI, M.

[1] Applications of fibre bundles to a certain class of C^*-algebras, *Tôhoku Math. J.* **13**(1961), 498–523. MR 25 #2465.

TURUMARU, T.

[1] Crossed product of operator algebra, *Tôhoku Math. J.* (2) **10**(1958), 355–365. MR 21 #1550.

VALLIN, J. M.

[1] C^*-algèbres de Hopf et C^*-algèbres de Kac, *Proc. London Math. Soc.* (3) **50**(1985), 131–174. MR 86f: 46072.

VAINERMAN, L. I.

[1] A characterization of objects that are dual to locally compact groups, *Funkcional. Anal. Priložen* **8**(1974), 75–76 (Russian) MR 49 #463.

VAINERMAN, L. I., and KAC, G. I.

[1] Non-unimodular ring groups and Hopf-von Neumann algebras, *Mat. Sb.* **94**(1974), 194–225. MR 50 #536.

VAN DAELE, A.

[1] *Continuous crossed products and type III von Neumann algebras*, Lecture Notes London Math. Soc. **31**, Cambridge University Press (1978). MR 80b: 46074.

VAN DIJK, G.

[1] On symmetry of group algebras of motion groups, *Math. Ann.* **179**(1969), 219–226. MR 40 #1782.

VARADARAJAN, V. S.

[1] *Geometry of quantum theory*. I, Van Nostrand, Princeton, N. J., 1968. MR 57 #11399.

[2] *Geometry of quantum theory*. II, Van Nostrand, Princeton, N. J., 1970. MR 57 #11400.

[3] *Harmonic analysis on real reductive groups*, Lecture Notes In Mathematics, **576**, Springer-Verlag, Berlin-New York, 1977. MR 57 #12789.

VARELA, J.

[1] Duality of C^*-algebras, *Memoirs Amer. Math. Soc.* **148**(1974), 97–108. MR 50 #5490.

[2] Sectional representation of Banach modules, *Math. Z.* **139**(1974), 55–61. MR 50 #5473.

[3] Existence of uniform bundles, *Rev. Colombiana Mat.* **18**(1984), 1–8. MR 86i: 46076.

VAROPOULOS, N. T.

[1] Sur les formes positives d'une algèbre de Banach, C. R. *Acad. Sci. Paris* **258**(1964), 2465–2467. MR 33 #3121.

VASIL'EV, N. B.

[1] C*-algebras with finite-dimensional irreducible representations, *Uspehi Mat. Nauk* **21**(1966), 135–154 [English translation: *Russian Math. Surveys* **21**(1966), 137–155]. MR 34 #1871.

VESTERSTRØM, J., and WILS, W.

[1] Direct integrals of Hilbert spaces. II., *Math. Scand.* **26**(1970), 89–102. MR 41 #9011.

VOGAN, D. A., JR.

[1] *Representations of real reductive Lie groups*, Progress in Mathematics, **15**, Birkhauser, Boston, Mass., 1981. MR 83c: 22022.

[2] Unitarizability of certain series representations, *Ann. Math.* (2) **120**(1984), 141–187. MR 86h: 22028.

[3] The unitary dual of GL(n) over an Archimedean field, *Invent. Math.* **83**(1986), 449–505. MR 87i: 22042.

VOICULESCU, D.

[1] Norm-limits of algebraic operators, *Rev. Roumaine Math. Pure Appl.* **19**(1974), 371–378. MR 49 #7826.

[2] Remarks on the singular extension in the C*-algebra of the Heisenberg groups, *J. Operator Theory* **5**(1981), 147–170. MR 82m: 46075.

[3] Asymptotically commuting finite rank unitary operators without commuting approximants, *Acta Sci. Math.* (*Szeged*), **45**(1983), 429–431. MR 85d: 47035.

[4] Dual algebraic structures on operator algebras related to free products, *J. Operator Theory* **17**(1987), 85–98.

VOWDEN, B. J.

[1] On the Gelfand-Naimark theorem, *J. London Math. Soc.* **42**(1967), 725–731. MR 36 #702.

WAELBROECK, L.

[1] Le calcule symbolique dans les algèbres commutatives, *J. Math. Pures Appl.* **33**(1954), 147–186. MR 17, 513.

[2] Les algèbres à inverse continu, *C. R. Acad. Sci. Paris* **238**(1954), 640–641. MR 17, 513.

[3] Structure des algèbres à inverse continu, *C. R. Acad. Sci. Paris* **238**(1954), 762–764. MR 17, 513.

VAN DER WAERDEN, B. L.

[1] *Die gruppentheoretische Methode in der Quantenmechanik*, Grundlagen der Math. Wiss., **36** Springer, Berlin, 1932.

[2] *Moderne Algebra*, Leipzig, 1937.

WALLACH, N. R.

[1] *Harmonic analysis on homogeneous spaces*, Pure and Applied Mathematics, **19**, Marcel-Dekker, New York, 1973. MR 58 #16978.

WALTER, M. E.

[1] Group duality and isomorphisms of Fourier and Fourier-Stieltjes algebras from the W^*-algebra point of view, *Bull. Amer. Math. Soc.* **76**(1970), 1321-1325. MR 44 #2047.

[2] The dual group of the Fourier-Stieltjes algebra, *Bull. Amer. Math. Soc.* **78**(1972), 824-827. MR 46 #611.

[3] W^*-algebras and nonabelian harmonic analysis, *J. Functional Anal.* **11**(1972), 17-38. MR 50 #5365.

[4] A duality between locally compact groups and certain Banach algebras, *J. Functional Anal.* **17**(1974), 131-160. MR 50 #14067.

[5] On the structure of the Fourier-Stieltjes algebra, *Pacific J. Math.* **58**(1975), 267-281. MR 54 #12966.

WANG, S. P.

[1] On isolated points in the dual spaces of locally compact groups, *Math. Ann.* **218**(1975), 19-34. MR 52 #5863.

[2] On integrable representations, *Math. Z.* **147**(1976), 201-203. MR 53 #10975.

WARD, H. N.

[1] The analysis of representations induced from a normal subgroup, *Michigan Math. J.* **15**(1968), 417-428. MR 40 #4384.

WARNER, G.

[1] *Harmonic analysis on semi-simple Lie groups.* I, Die Grundlehren der mathematischen Wissenschaften, **188**, Springer-Verlag, New York-Heidelberg, 1972. MR 58 #16979.

[2] *Harmonic analysis on semi-simple Lie groups.* II, Die Grundlehren der mathematischen Wissenschaften, **189**, Springer-Verlag, New York-Heidelberg, 1972. MR 58 #16980.

WARNER, S.

[1] Inductive limits of normed algebras, *Trans. Amer. Math. Soc.* **82**(1956), 190-216. MR 18, 52.

WAWRZYŃCZYK, A.

[1] On the Frobenius-Mautner reciprocity theorem, *Bull. Acad. Polon. Sci. Ser. Sci. Math. Astronom. Phys.* **20**(1972), 555-559.

[2] Reciprocity theorems in the theory of representations of groups and algebras, *Diss. Math. (Rozp. Mat.)* **126**(1975), 1-60. MR 56 #534.

[3] *Group representations and special functions*, C. Reidel Publishing Co., Dordrecht-Boston, Mass., PWN-Polish Scientific Publishers, Warsaw, 1984.

WEDDERBURN, J. M.

[1] On hypercomplex numbers, *Proc. London Math. Soc.* (2) **6**(1908), 77-118.

WEIERSTRASS, K.

[1] Über die analytische Darstellbarkeit sogenannter willkürlicher Funktionen reeller Argumente, *Sitz. Preuss. Akad. Wiss.* (1885), 633–640, 789–906.

WEIL, A.

[1] *L'intégration dans les groupes topologigues et ses applications*, Actualités Sci. et Ind. **869**, Hermann et Cie., Paris 1940. MR 3, 198.

[2] Sur certains groupes d'opérateurs unitaires, *Acta Math.* **111**(1964), 143–211. MR 29 #2324.

WEISS, G.

[1] Harmonic analysis on compact groups, *Studies in Harmonic Analysis*, **13**, pp. 198–223, MAA Studies in Mathematics, Mathematical Association of America, Washington, D. C., 1976. MR 57 #13383.

WENDEL, J. G.

[1] Left centralizers and isomorphisms of group algebras, *Pacific J. Math.* **2**(1952), 251–261. MR 14, 246.

WERMER, J.

[1] Banach algebras and analytic functions, *Adv. Math.* **1**(1961), 51–102. MR 26 #629.

WESTMAN, J. J.

[1] Virtual group homomorphisms with dense range, *Illinois J. Math.* **20**(1976), 41–47. MR 52 #14235.

WEYL, H.

[1] Theorie der Darstellung kontinuerlicher halb-einfacher Gruppen durch lineare Transformationen. I, II, III, *Nachtrag. Math. Z.* **23**(1925), 271–309; **24**(1926), 328–376, 377–395, 789–791.

[2] *The classical groups, their invariants and representations*, Princeton University Press, Princeton, N. J., 1939. MR 1, 42.

[3] *Theory of groups and quantum mechanics*, Dover, New York, 1964 (translated from German, 2nd ed., published 1931).

WEYL, H., and PETER, F.

[1] Die Vollständigkeit der primitiven Darstellungen einer geschlossenen kon tinuerlichen Gruppe, *Math. Ann.* **97**(1927), 737–755.

WHITNEY, H.

[1] On ideals of differentiable functions, *Amer. J. Math.* **70**(1948), 635–658. MR 10, 126.

WHITTAKER, E. T., and WATSON, G. N.

[1] *Modern analysis*, Cambridge Press, London, 1927.

WICHMANN, J.

[1] Hermitian *-algebras which are not symmetric, *J. London Math. Soc.* (2) **8**(1974), 109–112. MR 50 #8088.

[2] On the symmetry of matrix algebras, *Proc. Amer. Math. Soc.* **54**(1976), 237–240. MR 52 #8947.

[3] The symmetric radical of an algebra with involution, *Arch. Math. (Basel)*, **30**(1978), 83–88. MR 58 #2313.

WIENER, N.

[1] Tauberian theorems, *Ann. Math.* **33**(1932), 1–100, 787. (Also Selected Papers of N. Wiener, M.I.T. Press, 1964, pp. 261–360).

[2] *The Fourier integral and certain of its applications*, Cambridge University Press, Cambridge, 1933.

WIGHTMAN, A. S.

[1] *Quelques problèmes mathématiques de la théorie quantique relativiste*, Report of Lille conference on quantum fields, C.N.R.S., Paris, 1959, pp. 1–35.

WIGNER, E. P.

[1] On unitary representations of the inhomogeneous Lorentz group, *Ann. Math.* **40**(1939), 149–204.

[2] Unitary representations of the inhomogeneous Lorentz group including reflections, *Group theoretical concepts and methods in elementary particle physics* (Lectures Istanbul Summer School Theoretical Physics, 1962), pp. 37–80, Gordon and Breach, New York, 1964. MR 30 #1210.

WILCOX, T. W.

[1] A note on groups with relatively compact conjugacy classes, *Proc. Amer. Math. Soc.* **42**(1974), 326–329. MR 48 #8685.

WILLIAMS, D. P.

[1] The topology on the primitive ideal space of transformation group C^*-algebras and CCR transformation group C^*-algebras, *Trans. Amer. Math. Soc.* **226**(1981), 335–359. MR 82h: 46081.

[2] Transformation group C^*-algebras with continuous trace, *J. Functional Anal.* **41**(1981), 40–76. MR 83c: 46066.

[3] Transformation group C^*-algebras with Hausdorff spectrum, *Illinois J. Math.* **26**(1982), 317–321. MR 83g: 22004.

WILLIAMS, F. L.

[1] *Tensor products of principal series representations*, Lecture Notes in Mathematics **358**, Springer-Verlag, Berlin-New York, 1973. MR 50 #2396.

WILLIAMSON, J. H.

[1] A theorem on algebras of measures on topological groups, *Proc. Edinburgh Math. Soc.* **11**(1958/59), 195–206. MR 22 #2851.

[2] *Lebesgue integration*, Holt, Rinehart & Winston, New York, 1962. MR 28 #3135.

WILS, W.

[1] Direct integrals of Hilbert spaces. I., *Math. Scand.* **26**(1970), 73–88. MR 41 #9010.

WORONOWICZ, S.

[1] On a theorem of Mackey, Stone and von Neumann, *Studia Math.* **24**(1964-65), 101–105. MR 28 #4815, MR 30 #1205.

WULFSOHN, A.

[1] The reduced dual of a direct product of groups, *Proc. Cambridge Phil. Soc.* **62**(1966), 5–6. MR 32 #5791.

[2] The primitive spectrum of a tensor product of C^*-algebras, *Proc. Amer. Math. Soc.* **19**(1968), 1094–1096. MR 37 #6771.

[3] A compactification due to Fell, *Canad. Math. Bull.* **15**(1972), 145–146. MR 48 #3004.

YAMAGAMI, S.

[1] The type of the regular representation of certain transitive groupoids, *J. Operator Theory* **14**(1985), 249–261.

YANG, C. T.

[1] Hilbert's fifth problem and related problems on transformation groups. *Proc. Symp. Pure Math.*, 28, Part I, pp. 142–146, Amer. Math. Soc., Providence, R.I., 1976. MR 54 #13948.

YOSIDA, K.

[1] On the group embedded in the metrical complete ring, *Japan. J. Math.* **13**(1936), 7–26.

[2] *Functional analysis*, Grundlehren der Math. Wiss., **123**, Springer-Verlag, Berlin, 1965. MR 31 #5054.

ZAITSEV, A. A.

[1] Holomorphically induced representations of Lie groups with an abelian normal subgroup. *Trudy Moskov. Mat. Obšč.* **40**(1979), 47–82 (Russian). MR 80m: 22025.

[2] Equivalence of holomorphically induced representations of Lie groups with abelian normal subgroups. *Mat. Sb. N.S.* (160) **118**(1982), 173–183, 287 (Russian). MR 84j: 22019.

ZELAZKO, W.

[1] *Banach algebras*, Elsevier, Amsterdam, 1973. MR 56 #6389.

ZELEVINSKY, A. V.

[1] Induced representations of reductive p-adic groups II. On irreducible representations of GL(n), *Ann. Sci. École. Norm. Sup.* (4) **13**(1980), 165–210. MR 83g: 22012.

ZELLER-MEIER, G.

[1] Produits croisés d'une C^*-algèbre par un groupe d'automorphismes, *C. R. Acad. Sci. Paris*, **263**(1966), A20–23. MR 33 #7877.

[2] Produits croisés d'une C^*-algèbre par un groupe d'automorphismes, *J. Math. Pures Appl.* (9) **47**(1968), 101–239. MR 39 #3329.

ŽELOBENKO, D. P.

[1] A description of a certain class of Lorentz group representations, *Dokl. Akad. Nauk SSSR* **125**(1958), 586–589 (Russian). MR 21 #2920.

[2] Linear representations of the Lorentz group, *Dokl. Akad. Nauk SSSR.* **126**(1959), 935–938. MR 22 #906.

[3] *Compact Lie groups and their representations*, Translations of Mathematical Monographs, **40**, Amer. Math. Soc. Providence, R. I., 1973. MR 57 #12776b.

ZETTL, H.

[1] Ideals in Hilbert modules and invariants under strong Morita equivalence of C^*-algebras, *Arch. Math.* **39**(1982), 69–77. MR 84i: 46060.

ZIMMER, R. J.

[1] Orbit spaces of unitary representations, ergodic theory, and simple Lie groups, *Ann. Math.* (2) **106**(1977), 573–588. MR 57 #6286.

[2] Induced and amenable ergodic actions of Lie groups, *Ann. Sci. École Norm. Sup.* (4) **11**(1978), 407–428. MR 81b: 22013.

[3] Amenable actions and dense subgroups of Lie groups, *J. Functional Anal.* **72**(1987), 58–64.

ZYGMUND, A.

[1] *Trigonometric series*, I, II, Cambridge University Press, New York, 1959. MR 21 #6498.

Name Index

Subject Index

Index of Notation

$C^*(G)$	the group C^*-algebra of G, 866
$\det(A)$	the determinant of the matrix A, 44
δ_x	the function $y \mapsto \delta_{xy}$, 44, 221; the unit mass at x in G, 933
$\mathbb{E}$	the circle group, 42, 170
$\mathbb{E}^n$	the n-dimensional torus, 182
$\mathcal{E}(T)$	the linear span of $\{T_{ij}: i, j = 1, \ldots, r\}$, 939
f_x, f^x	the left and right translates of f, 933
$\hat{f}$	the Fourier transform of f, 1003
F_*	$F_* = F \setminus \{0\}$, 195
F_p	the p-adic field, 199
$\mathcal{F}$	the set of continuous functions $f : \mathbb{C} \to \mathbb{C}$, 394
$\mathcal{F}(G)$	the subspace of central functions of $\mathcal{L}(G)$, 949
$f(a), f(E)$	function of a normal element, 395, 433
$\hat{G}$	the structure space of a locally compact group G, 915; the character (or dual) group of a locally compact Abelian group G 1004
G_m	the stabilizer of m in G, 177
$GL(n, \mathbb{C})$	$n \times n$ general linear group, 170
$GL(n, \mathbb{R})$	$n \times n$ general real linear group, 170
G/H	space of left cosets, 46
$\langle G, :, \mathcal{F} \rangle$	reverse topological group, 165
$\mathrm{Hom}(T, T)$	commuting division algebra, 283
$\mathrm{Hom}(T, T')$	set of intertwining operators, 267
I_p	the p-adic integers, 199
$\mathcal{J}(G)$	the family of all (algebraic) equivalence classes of irreducible finite-dimensional representations of G, 941
$\mathrm{Ker}(f)$	kernel of a homomorphism, 46
$\mathcal{L}_1(\lambda)$	$\mathcal{L}_1$ group algebra, 229
$\langle \lambda, \mu \rangle$	multiplier of an algebra, 774, 790
$\Gamma(N, G)$	the family of isomorphism classes of central extensions of N by G, 182
$\hat{\mu}$	the Fourier transform of a measure μ, 1012
M'	commutant of M, 508
M''	double commutant, 508
$M_n(A)$	$n \times n$ matrices over an algebra A, 522
$M(n, F)$	the algebra of $n \times n$ matrices (total matrix algebra) over F, 44, 266, 293
$\eta : x \mapsto \eta_x$	the Pontragin duality isomorphism, 1020
$N \underset{i}{\to} H \underset{j}{\to} G$	group extension, 186
$O(n)$	$n \times n$ orthogonal group, 170
$\mathcal{O}(X, Y)$	continuous linear maps from X to Y, 51
$\mathcal{O}(X)$	continuous linear operators on X, 51
$\mathcal{O}_D(X_1, X_2), \mathcal{O}'_D(X_1)$	the set of D-linear maps, 46
$\mathcal{P}(G)$	the set of functions of positive type on G with $p(e) = 1$, 921
$P_{n,r}$	the family of r-dimensional subspaces of $\mathbb{C}^n$, 181
$\mathrm{Prim}(A)$	the set of primitive ideals of A, 557
$\mathrm{PSp}(a)$	the point spectrum of a, 426
R	the left regular representation, 847
$\mathbb{R}_*$	the multiplicative group of non-zero reals, 169
$\mathbb{R}_{++}$	the multiplicative group of positive reals, 169
$\mathbb{R}^n$	n-dimensional Euclidean space, 182
$\mathcal{R} = \mathcal{R}(G)$	the set of $f \in \mathcal{L}(G)$ such that the linear span of $\{f_x : x \in G\}$ is finite-dimensional, 934
$* : A \to A$	involution, 48

Direct sums

Equivalences and orderings

Induced representations, systems of imprimitivity, and Mackey analysis

Index of Notation

Inner products and norms

Linear operators

$S^\perp$	the orthogonal complement of S in $X \oplus Y$, 620	
λS	scalar multiplication, 621	
$S + T$	the sum of unbounded operators, 621	
$V \circ S$	the composition of unbounded operators, 621	
$\inf(\mathscr{P})$	the greatest lower bound of projections, 55	
$\sup(\mathscr{P})$	the least upper bound of projections, 55	
$\sum \mathscr{P}$	the sum of projections, 55	
$Q(I)$	the projection of X onto X_I, 597	
N_T	$\{\xi \in \text{domain}(T): T\xi = 0\}$, 622	
R_T	the closure of range(T), 622	
$\mathscr{O}(X, Y)$	the set of bounded linear maps from X to Y, 51	
$\mathscr{O}(X)$	the set of bounded linear operators on X, 51	
$\mathscr{O}(X, \mathbb{C})$	the adjoint or dual space of X, 51	
$\mathscr{O}_c(X, Y)$	the set of compact linear maps in $\mathscr{O}(X, Y)$, 450	
$\mathscr{O}_c(X)$	the set of compact operators in $\mathscr{O}(X)$, 450	
$\mathscr{O}_F(X, Y)$	the set of linear maps in $\mathscr{O}(X, Y)$ with finite rank, 450	
$\mathscr{O}_F(X)$	the set of operators in $\mathscr{O}(X)$ with finite rank, 450	
$\mathscr{U}(X, Y)$	the set of closed densely defined linear maps from X to Y, 620	
$\mathscr{U}(X)$	the set of closed densely defined linear operators on X, 620; the group of unitary operators on X, 833	
X^*	the adjoint or dual space of X, 51	

Linear spaces and linear topological spaces

$\mathbb{C}^n$	the space of complex n-tuples, 182
$\mathbb{R}^n$	the space of real n-tuples, 182
$H(S)$	the convex hull of S, 563
$(\;|\;)$	duality for linear spaces, 563, 1225
$[\;\cdot\;]$	the linear span, 787
LCS	locally convex space, 50
$\sigma(X_1, X_2)$	the weak topology, 1226
$s(F)$	the finite sum $\sum_{i \in F} x_i$, 50
S^π, S_π	the polar sets for S, 563
$x \perp y$	x and y are orthogonal, 53
$X^\perp$	the orthogonal complement of X, 53
$X^\#$	the set of complex-valued linear functions on X, 47
$\bar{X}$	the complex conjugate linear space, 47
$\langle X_1, X_2, (\;|\;)\rangle$	a dual system, 1225
$\sum \mathscr{W}$	the linear span of $\mathscr{W}$, 47

Measure theory

$\mathscr{B}_\mu$	the locally μ-measurable sets B such that $\sup\{|\mu|(A): A \in \mathscr{S}, A \subset B\} < \infty$, 69
$\mathscr{B}(X)$	the Borel subsets of X, 91
Δ	the modular function, 210
$dv(\alpha(x))$	the transported measure $\alpha^{-1} \cdot v$, 84
$\mathscr{E}_\mu$	a sub-δ-ring of $\mathscr{B}_\mu$, 96
f^0	$f^0(x) = \int_K f(xk)\,dvk$, 237
f^{00}	$f^{00}(xK) = f^0(x)$, 237
$f d\mu$	the measure ρ defined by $\rho(A) = \int_A f\,d\mu$, 87

Index of Notation

Positive functionals and representation theory

Products of spaces, functions, and measures

Sets and mappings

Special spaces

Tensor products

Topology

PURE AND APPLIED MATHEMATICS

* Presently out of print

Printed and bound by CPI Group (UK) Ltd, Croydon, CR0 4YY

13/10/2024

01773508-0003